Chemistry of Atmospheres

Cover illustration: Ozone concentrations measured in March 1997 with the GOME satellite instrument (see pp. 181-2). Low ozone concentrations are represented by darker colours: the dark zone at northern polar latitudes reflects the presence of an Arctic ozone hole. The authors of the picture are Kai-Uwe Eichmann, Andreas Richter and John P. Burrows, Institute of Environmental Physics and Remote Sensing (IUP/IFE) of the University of Bremen. GOME flies aboard the ESA-ERS-2 (ESA's second Earth Research Satellite), which was launched from Kourou on 20 April 1995. GOME is a small scale version of SCIAMACHY, which is a German, Dutch, and Belgian contribution to ESA ENVISAT, planned for Launch in mid 2001.

Chemistry of Atmospheres

An Introduction to the Chemistry of the Atmospheres of Earth, the Planets, and their Satellites

THIRD EDITION

Richard P. Wayne

Professor of Chemistry,
University of Oxford and
Dr Lee's Reader in Chemistry,
Christ Church, Oxford

OXFORD
UNIVERSITY PRESS

OXFORD
UNIVERSITY PRESS

Great Clarendon Street, Oxford OX2 6DP

Oxford University Press is a department of the University of Oxford
It furthers the University's aim of excellence in research, scholarship,
and education by publishing worldwide in

Oxford New York

Athens Auckland Bangkok Bogotá Buenos Aires Calcutta
Cape Town Chennai Dar es Salaam Delhi Florence Hong Kong Istanbul
Karachi Kuala Lumpur Madrid Melbourne Mexico City Mumbai
Nairobi Paris São Paulo Singapore Taipei Tokyo Toronto Warsaw
with associated companies in Berlin Ibadan

Oxford is a registered trade mark of Oxford University Press
in the UK and in certain other countries

Published in the United States
by Oxford University Press Inc., New York

Reprinted 1992, 1993, 1994, 1995, 1996

First published 1985
Second edition 1991
Third edition 2000

British Library Cataloguing in Publication Data
(data available)

ISBN 0 19 850375 X (Pbk)

Typeset by the author

Printed in Great Britain by
T.J. International Ltd,
Padstow, Cornwall

Preface to the third edition

The best part of a decade has elapsed since I wrote the second edition of this book. In that period, the study of atmospheric chemistry has attracted ever increasing numbers of researchers from diverse disciplines, and enormous quantities of new information have become available about the atmospheres of our planet and of other bodies in the solar system. One of the exciting aspects of the research is that it continues to provide new insights into the fundamental natures of the atmospheres and the ways they behave. Such insights provide the prime motivation for writing a book such as the present one, and it seemed to me that the time must therefore be ripe for the appearance of a new edition. I pointed out in my preface to the second edition that my objective in preparing it was not in itself to provide greater detail and newer quantitative information than was presented in the edition it replaced, but rather to show how the subject appeared as an even more unified and exciting one than it had before, and that remains my aim here.

Much inspiration is to be drawn from the fundamental 'pure' science of our study. There are, however, extremely important practical issues at stake, as well. The living organisms on Earth are dependent on the atmosphere in several ways, both obvious and subtle. In turn, the nature and composition of our atmosphere have been fashioned by the living organisms. It is now evident that man can modify the atmosphere substantially. Pollution of the lower atmosphere has been with us a long time, in manifestations as diverse as the smokes and fogs of industrial cities, through the generation of ozone in photochemical smogs, to the formation of acid rain and snow. The stratosphere, with its protective ozone layer, is also at risk from man. Global stratospheric ozone loss is now clearly seen to be attributed at least in part to human release of certain compounds to the atmosphere, and the dramatic polar ozone 'holes' are known to be almost entirely due to such releases. The entire climate system may, in fact, be changing in response to the changing chemical composition of the atmosphere. Control policies need to be based on sound science, and heightened public awareness of the environmental issues in which atmospheric chemistry plays a part demands well informed scientists to furnish proper explanations. Part of the aim of this book, right from the appearance of the first edition, was to provide the scientific framework for sensible discussion and interpretation of the causes and consequences of man's impact on the atmosphere. Developments over the last ten years gave me yet another reason to prepare this third edition and thus examine some of the key issues at the very beginning of the new millennium.

Although the general structure of this third edition of the book remains the same as that of its predecessors, I have taken the opportunity to revise and update all parts of it, and reorganize some of the material. The most substantial expansion is to be found in Chapter 4 (Earth's stratosphere) and Chapter 5

(Earth's troposphere), and it is in those chapters that the restructuring will be most apparent. For example, in Chapter 4, I have now more clearly separated natural perturbations of the stratosphere (solar proton events, changes in solar ultraviolet irradiance, the quasi-biennial oscillation, El Niño, and volcanoes) from man's impact on the stratosphere. Then I have discussed polar ozone holes before variation and trends in ozone concentrations, since the former may well impact on the latter. In this Preface, I shall attempt to draw attention to some of the topics that are new to the book, or that receive greater emphasis than before. I suggest, however, that the reader look first at the Prefaces to the earlier editions, reproduced here, to understand better what approach I have adopted throughout and to see how the subject has developed over 15 years.

One clear trend over the years has been our increasing awareness of the rôle of heterogeneous processes — reactions occurring on or in liquid droplets and solid particles — in the chemistry of the Earth's atmosphere, and probably in the atmospheres of other bodies as well. Curiously enough, some of the earlier discussions of atmospheric chemistry were concerned with chemical transformations occurring in clouds and droplets. But, from about 1960 onwards, many of the new researchers in atmospheric chemistry were laboratory chemists whose expertise lay in homogeneous gas-phase processes. Since there was so much to discover in homogeneous atmospheric chemistry, it is natural that many of the advances should initially lie in this area. To compound the problem, laboratory studies of the heterogeneous processes were (and still are) extremely hard to conduct properly, especially if the results are to have true relevance to the atmosphere itself. It was probably the recognition in the late 1980s that the phenomenon of the Antarctic ozone holes was intimately bound up with chemical reactions occurring on the surfaces of polar stratospheric cloud particles that forced atmospheric chemists to accept that lack of understanding of heterogeneous processes did not equate with lack of importance in the atmosphere! It is now universally accepted that heterogeneous chemical change must be considered in interpretations of most aspects of both tropospheric and stratospheric chemistry. To try to reflect this altered perception of atmospheric chemistry, I have made changes to this edition of the book in several places. In Chapter 3, which deals with the background to the physical chemistry, I have added three subsections (3.4.3—3.4.5) that develop the ideas associated with the chemical kinetics of condensed-phase, surface, and heterogeneous reactions. Then in Chapter 4 I have added Section 4.4.4 to deal explicitly with heterogeneous chemistry in the stratosphere, although the subject is alluded to throughout the chapter. Similarly, in Chapter 5, the new Section 5.8 gives a fuller account of heterogeneous chemistry in the troposphere than was hitherto provided.

One natural phenomenon that is connected with the question of heterogeneous chemistry is that of volcanic eruptions. The last part of the twentieth century seems to have been a rather active one in respect of eruptions. One of the largest of all was that of Mount Pinatubo, in the

Philippines, which erupted in June 1991. Twenty million of tonnes or so of sulphur dioxide were injected into the atmosphere and converted into sulphuric acid droplets. Vast quantities of dust also entered the atmosphere. The additional particle load in the atmosphere not only altered the balance of incoming and outgoing solar radiation, and thus global average temperatures, but also greatly increased the surface area of particles on which heterogeneous chemical processing could occur. As it happened, the eruption took place around the time of campaigns of coordinated atmospheric measurements. The eruption was initially seen as an unwanted interference that could potentially ruin the carefully planned experiments. In reality, the eruption provided a magnificent opportunity of putting to the test several aspects of the hypotheses concerning heterogeneous chemistry. The response of a system to a perturbation is often one of the very best ways of learning about the detailed behaviour of the unperturbed system. The eruption of Mount Pinatubo provided an enormous perturbation of the atmosphere, immeasurably greater than man could possibly (or ethically) achieve. Signatures of the eruption were observed in many chemical measurements, and our knowledge almost certainly advanced more rapidly and more substantially than it would otherwise have been. The eruption is thus a recurring theme in this new edition of our book.

Man might not be able to perturb the atmosphere on the scale of a volcanic eruption — nuclear explosions apart — but since the second edition of the book was published, there has been renewed and widespread interest in the effects that aircraft operations might have on the atmosphere, especially the upper troposphere and lower stratosphere. Commercial aircraft operations have increased in scale very considerably, and are projected to grow for the foreseeable future. Gases such as the oxides of nitrogen, as well as CO_2, CO, and H_2O, are released to the atmosphere in large quantities. So are solid particles and liquid aerosols such as soot, directly formed sulphate particles, and oil. The H_2O vapour may condense to form contrails, and the particles may nucleate the formation of clouds such as cirrus. Altogether, there is the potential for an impact on both homogeneous and heterogeneous chemistry, and for direct and indirect impacts on climate. Separation of the chemical effects into those on the stratosphere and those on the troposphere is rather artificial, and I have therefore chosen to bring together the discussion in Chapter 4 (Section 4.6.2). The possible influence on climate seems more appropriately treated in Chapter 9 (Section 9.6.6).

The first edition of the book was published just about the same time as the first paper drawing attention to the anomalous springtime ozone depletions in the Antarctic that have become known as ozone 'holes'. That first edition thus had no mention of the phenomenon. Six years later, by the time the second edition had appeared, the depletions had been observed in each successive year, the 'holes' being generally deeper and more widespread in the later years. The chemical explanation had also been established, and involved heterogeneous chemical processing on polar stratospheric clouds. Several expeditions had measured anomalous chemical composition in the Antarctic. The situation in the Arctic was less clear cut. Evidence was accumulating for anomalous chemistry, but it was not yet certain that any chemically induced seasonal ozone loss had been observed. In this third edition, we now report clear evidence for such depletion. In the spring periods of 1995, 1996, 1997, and 1998, ozone losses in the Arctic rivalled in extent those seen earlier in the southern hemisphere roughly a decade earlier. The years involved were characterized by unusually low polar stratospheric temperatures in the north, just those conditions likely to favour the formation of polar stratospheric clouds, amongst other important meteorological factors. It is interesting that ozone losses were relatively small in the winter of 1998/1999, when temperatures were much higher.

These new observations provide further confirmation, if such were needed, of the significance of heterogeneous reactions occurring in the special meteorological conditions of the polar stratospheres in both hemispheres. That the reactions occur at all is undoubtedly a result of man's release of halogen-containing compounds. Although the losses are less dramatic, these same compounds cause ozone depletion at mid-latitudes, and mixing of ozone-poor polar air with lower latitude air may add to the effects. These matters, and the conclusions of the Ozone Trends Panel, are set out in Section 4.8.

Bromine, as well as chlorine, compounds are involved in catalytic ozone destruction, and bromine is an even more effective catalyst than chlorine. This understanding was already emerging by the time the second edition was published, but the importance of bromine compounds has now come into even sharper focus. One matter that has been much explored recently is the relative contribution of natural and anthropogenic sources of methyl bromide. Control policy, in particular, requires this information. The present situation about legislation is set out in Section 4.6.6, and anticipated future anthropogenic ozone depletions are discussed in Section 4.6.7.

One of the measures adopted to reduce the impact of anthropogenic halogen-containing species on the stratosphere has been to replace the chlorofluorocarbons (CFCs) once widely used by hydrofluorocarbons (HFCs) and hydrochlorofluorocarbons (HCFCs). As explained in Section 4.6.6, the HFCs and HCFCs are degraded much more rapidly in the troposphere than the CFCs, and thus relatively much less of them reaches the stratosphere (the HFCs contain no catalytic chlorine, either, of course). However, a new issue arises, that of the pathways of the tropospheric removal processes and of the atmospheric impact of the degradation products. A section has been added to Chapter 5 (Section 5.10.8) to discuss the chemistry.

Halogen atoms (Cl, Br, and perhaps I) and the corresponding monoxide radicals (ClO, BrO, IO) have been recognized since 1974 as important catalytic species in *stratospheric* ozone chemistry. Much more recently, it has become apparent that these atoms and radicals may play a rôle in the *troposphere* as well. The surprising observation was made that tropospheric ozone in Arctic regions was depleted rapidly during April, at polar sunrise, over periods of hours to days. Various pieces of evidence link elevated halogen atom concentrations to the ozone losses, and direct observations have now been made of oxide radicals such as BrO and IO, as well as of Cl_2, a precursor of Cl atoms. The halogens may well have natural origins, perhaps being generated from inorganic materials such as sea-salt spray. Several potential reaction mechanisms have been identified for the tropospheric conversion of inorganic chlorides and bromides to the corresponding atoms. Most of these mechanisms involve heterogeneous steps. There remain many speculations and uncertainties about this area of atmospheric chemistry, but the presence of halogen atoms in the troposphere would have several chemical consequences besides that of ozone depletion. Section 5.7 is new to this edition, and it gives an overview of what is currently known about what promises to be an exciting field.

Significant contributions by halogen species to the initiation of chemical processing would represent a change in the oxidizing capacity of the troposphere. In the preface to the second edition, I drew attention to the way in which the concentrations of trace gases such as methane (CH_4) and carbon monoxide (CO) were increasing at rates as large as one per cent per year, and that the increased burden of such gases might either reflect or bring about an altered atmospheric oxidizing capacity. Over the past ten years, these trends appear to have changed substantially. The rate of growth of CH_4 concentrations is about one-quarter of what it was when the second edition of the book was prepared, and even seemed to be zero or slightly negative in 1992–1993. Concentrations of CO have also decreased over the last decade, perhaps by 0.5 to 0.8 per cent per year. The eruption of Mount Pinatubo, described earlier, is one explanation that has been advanced for the changed trends in the concentrations of both the gases. Immediately after the eruption, growth rates for both species were enhanced, possibly because UV penetration into the troposphere was reduced, leading to lower concentrations of the OH radical (which reacts with both CH_4 and CO). In the following year, however, stratospheric ozone concentrations were lower because of heterogeneous processing on volcanic aerosol, and tropospheric UV, and thus OH concentrations, both *increased*, with the effect of slowing or stopping the growth of CH_4 and CO. Reduced temperatures or changed circulation patterns following the eruption have also been invoked to explain the altered trends. The concentration trends, and the underlying chemistry, are documented and discussed at various points in this new edition of the book.

Methane is released to the Earth's atmosphere in quantities of around 500 million tonnes annually. About 80 per cent of this release is 'new' methane from the biosphere, and the other 20 per cent is of fossil origin (natural gas, oil, and coal). Of the biospheric methane, two-thirds can be attributed to man's activities, such as agriculture and biomass burning. Nevertheless, this attribution still leaves some 160 million tonnes of CH_4 released annually by 'natural' biogenic processes. In addition to the CH_4, around one thousand other volatile organic compounds (VOCs) enter the atmosphere as a result of biogenic processes (mostly from terrestrial plants). These VOCs include ethene, isoprene, terpenes, alcohols, aldehydes, ketones, acids, and esters. In the case of isoprene and the terpenes, hundreds of millions of tonnes are released annually, so that the source strengths of these biogenic compounds equal or exceed those of CH_4 itself. Many of the compounds are highly reactive in the atmosphere, and there has been a growing awareness of the contribution that the species make to tropospheric chemistry. Section 5.4 of this edition has accordingly been newly written to provide an introduction to the nature of the compounds involved, to the chemistry in which they participate, and to the impact that they have on the atmosphere. The opportunity is also taken to present some ideas about how and why plants might emit such a variety of compounds, several of them rather exotic.

Aromatic hydrocarbons (Section 5.5) such as benzene, toluene, and the isomers of xylene, represent another class of compound treated in this edition for the first time. They represent a major class of organic compound associated with the urban environment. Model calculations suggest that the aromatic species might contribute more than 30 per cent to the formation of photochemical oxidant in urban areas. The chemical behaviour of these compounds in the atmosphere shows some distinct differences from that of the aliphatic species considered so far, and it therefore deserves explicit consideration. Motor vehicles are the dominant contributors to the emissions of aromatic VOCs, at least in the USA and densely populated parts of Europe. The release of the aromatic species is thus a form of *air pollution*. Other aspects of air pollution treated for the first time in the present edition are pollution by polycyclic aromatic hydrocarbons (Section 5.10.9), and by the degradation of HFCs and HCFCs described earlier (Section 5.10.8). Biomass burning has become increasingly recognized as a major anthropogenic source of trace species, and as a serious pollution threat, and I have accordingly put some of the most salient facts together in the new Section 5.10.10.

Tropospheric oxidation of the minor oxidizable gases present is started largely by the attack of the hydroxyl (OH) radical during the day complemented by reactions of the nitrate (NO_3) radical at night. Both these radicals indirectly owe their existence to ozone in the troposphere. However, unsaturated VOCs such as alkenes and dienes may be attacked in the atmosphere not only by OH and NO_3, but by O_3 itself. Organic radicals, and even OH, may be formed as products, and a chain oxidation follow the initial step. In circumstances when solar intensity is low (resulting in low [OH]) or NO_x levels are small (resulting in low [NO_3]), the direct reactions with O_3 can make a significant contribution to the loss of VOC. The chemistry involved has become much more clearly understood over the past few years, and I have included a description of it (Section 5.3.7) in this new edition.

The comments made in the last paragraph and elsewhere indicate that the OH radical is an absolutely central part of tropospheric oxidation chemistry, and one of the long-term goals of field measurement studies has been to determine concentrations of the radical by direct and explicit means. This goal has been achieved with resounding success since the last edition was published; both laser-induced fluorescence and long-path absorption spectroscopy have been used. I describe the experimental techniques in Section 5.9.2, and present some results of the measurements in Section 5.9.3.

In a more general way, I have thought it desirable to give a slightly more extended account than previously of the methods used for making measurements of closed-shell species, as well as atoms and radicals, in both the stratosphere (Section 4.4.6) and the troposphere (Section 5.9.2). The identification of the species, and the determination of their concentrations, is an essential component of the study of atmospheric chemistry, and often the most expensive and difficult aspect. Ultimately, the results of the measurements are compared with the predictions of numerical models. I have now expanded my treatment of modelling, with the principles being set out in Sections 3.6 and 5.9.1, and comparisons between measurements and model predictions being presented in Sections 4.4.6 (stratosphere) and 5.9.3 (troposphere). Some other alterations to this edition of the book are, in part, bound up with the question of models. Vertical transport (Section 2.3.3) and winds (Section 2.4) redistribute chemical species, and an understanding of them is obviously central to sophisticated modelling. Stratosphere–troposphere exchange is one aspect of vertical transport, and brings about the transfer of chemical constituents in both directions across the tropopause. These areas have all seen significant developments in recent years, so that the possibility of updating the text has been opportune. One source of information about stratosphere–troposphere exchange has been the measurement of the water content of air in the tropical stratosphere, which retains the signature of the tropopause temperature for months or even years on layers of slowly ascending gas. The system has been entertainingly likened to an 'atmospheric tape recorder'. Chapter 5 itself sees some further discussion of relevant aspects of physics and physical chemistry with the new Sections 5.2.1 (dry and wet deposition), 5.2.2 (the boundary layer), and 5.2.3 (transport in the troposphere).

Atmospheric chemical composition affects several aspects of the absorption, scattering, and re-radiation of solar energy, and thus can be expected to have a direct influence on climate. Evidence about climates of the past can provide valuable clues about the composition of our atmosphere in earlier times. Perhaps of more immediate practical importance, because man may influence the outcome, changed composition in the future may be one factor that leads to alterations in global climate. Climatic change is a subject that excites great interest and concern, and previous editions of this book have sought to explain the possible connections between atmospheric chemistry and climate. However, there have been several developments since the publication of the second edition that have made it essential to update and expand Section 9.5 (climates of the past) and, especially, Section 9.6 (climates of the future).

Carbon dioxide, and certain other 'greenhouse' gases, are increasing in concentration in the atmosphere at an unprecedented rate, and the effect is largely of man's making. An obvious question to ask is if we have yet experienced a global warming attributable to the increases in concentrations of the gases known to have occurred since the industrial revolution. Theoretically, warmings of 0.5 – 1.0 °C's should have arisen over the course of this century. Some of the data certainly seem to show a signal of temperature increase emerging from the noise of year-to-year variability. Whether it can be attributed to net positive radiative forcing by greenhouse gases is another matter. The evidence has been hotly debated, of course. In response to the perceived impacts of global increases in temperature, a concerted international effort has been made to assess the current state of knowledge by the establishment of an Intergovernmental Panel on Climate Change (IPCC). Even in 1990, the IPCC was disinclined to make an unequivocal statement linking climate change to anthropogenic releases of greenhouse gases. In their 1995 report, however, the IPCC cautiously concluded that "the balance of evidence suggests a discernible influence on a global climate." To attempt to present the currently accepted view in context, as well as to explain why there are dissenting opinions, I have included four new sub-sections (Sections 9.6.1–9.6.4). Hypothetical projections about changes in atmospheric concentrations and climate are discussed in Section 9.6.5, while the effects of aircraft appear as a separate topic in Section 9.6.6. Finally, I have described briefly the impacts of climate change (Section 9.6.7) and some legislation and policy issues relating to the control of greenhouse-gas release (Section 9.6.8).

It would be remiss of me to dwell so long on the advances made in our understanding of our own atmosphere that I could not mention briefly the continuing excitement of improving knowledge about the atmospheres possessed by other planets and certain of their satellites. Chapter 8 looks at the fascinating topic of the chemistry of these varied atmospheres. Some of the new material concerns more sophisticated interpretation and modelling of results from earlier planetary missions as well as terrestrial observations. But there has been further planetary exploration, and the fully operational Hubble Space Telescope has also played its part in gleaning information about the distant atmospheres. Following on from the Voyager encounters with Jupiter and the outer solar system, the Galileo mission to Jupiter and the satellites Europa, Io, and Callisto has made a further huge contribution to knowledge. Galileo was launched from the space shuttle Atlantis in 1989, and on the voyage to Jupiter took the first close-up images of an asteroid (Gaspra), and discovered the first known moon of an asteroid, Dactyl at Ida. Galileo also had the unique opportunity of being able to obtain images of the crash of comet Shoemaker–Levy 9 into Jupiter's atmosphere. This visually dramatic event also provided an opportunity to observe how the atmosphere reacted to a massive perturbation. An atmospheric probe was released in July 1995, and reached the planet in December. During the probe's survival period of 59 minutes, data were returned on atmospheric temperature, pressure, composition, winds, and lightning. Meanwhile the orbiter started a tour of the Jovian system that was extended until the end of 1999 as the Galileo-Europa mission, with focused objectives on 'ice, water, and fire': the ice on Europa, the water in Jupiter's atmosphere, and the fire of Io's volcanoes, along with investigations of Io's torus and of the properties of Callisto. Several surprises about the satellites are presented in Section 8.5.2.

The turn of the new millennium is an exciting time, too, for studies of Saturn (Section 8.4) and its satellite Titan (Section 8.5.1). The Cassini mission is, perhaps, the most ambitious effort in planetary space exploration ever mounted, and will study the Saturnian system over a four-year period. Cassini was launched in 1997, and will reach Saturn in 2004. The twelve scientific instruments will conduct studies of the planet, its moons, rings, and magnetic environment. Cassini carries the probe Huygens that will be released in November 2004 from the main spacecraft to parachute for more than two hours through the atmosphere to the surface of Saturn's moon Titan.

If all goes according to plan, there will be several further space missions during the next decade or so that address questions closely related to atmospheric composition and chemistry. They include visits to Mars, and probes to comets that will land to study composition, isotopic abundances, and complex organic molecules, and perhaps even return samples of cometary material to Earth. These are grand and ambitious plans indeed, but they will bring great rewards.

As in the earlier editions, I have given literature citations within the text only in exceptional circumstances. In the place of references, I have provided bibliographies for each chapter that are sorted by topic and by section number. I believed that this approach would both make the text more readable and render the citations more useful, and many comments that I have received have encouraged me to continue in this way. One of the difficulties that faced me in tackling this third edition was the enormous quantity of published literature that now exists on virtually every aspect of atmospheric chemistry. Original research papers appear in large quantities every week, and many of them make worthwhile contributions to our subject. I decided that, to reduce the bibliographies to manageable proportions, the most useful solution to the problem would be to make a selection of the publications that have appeared in recent years, while retaining a good number of the citations given in the earlier editions. The recent references provide an entry point to older work for the reader who wishes to pursue a particular topic. Those retained from previous editions are often seminal articles about research that spawned lines of enquiry, or are specially good overviews of topics that have undergone only modest change in the intervening years. I hope that I have judged correctly a satisfactory balance between old and new, and between acceptable length and exhaustive coverage.

It is one of the very pleasant privileges for an author writing a preface to be able to thank those who have provided help and encouragement. Many of those whom I acknowledged in the prefaces to the first two editions of this book have continued to feed me with comments, information, and constructive criticism, and I thank them for it. I am particularly grateful, on this occasion, for the detailed advice that I received from my good friends and colleagues Peter Borrell, Peter Brimblecombe, John Burrows, Daniel Gautier, Nick Hewitt, Jim Lovelock, Howard Roscoe, Tom Slanger and Adrian Tuck. They told me things that I did not know, they explained things that I did not understand, and they showed me how to express my ideas better. Three former research associates of mine in Oxford provided much help. John Burrows (Universität Bremen) gave me privileged access to results from the GOME satellite instrument, Neil Harris (Ozone Coordinating Unit, Cambridge) answered interminable questions about the stratosphere, and Dudley Shallcross (University of Exeter) ran some entirely new model calculations for the purposes of this book while he was working with John Pyle in Cambridge. Yet others provided me with new diagrams and data, and I trust that I have acknowledged their contributions at the appropriate places correctly if not adequately. I give thanks to all of them. My Oxford research group has, as usual, had to accept that time I have spent on the book has necessarily meant time that I could not spend with them. How far that might be a loss to them is another matter. Those most involved during the preparation of this third edition were Carlos Canosa-Mas, Eimear Cotter, Mark Flugge, Martin King, Dina Shah, Katherine Thompson, and Alison Vipond. They were the best group that I could have wanted, helpful, conscientious, and cooperative. I chose to produce myself camera-ready copy of the text. Inevitably, I had to call on Pete Biggs to explain how WordPerfect might be made to generate the copy with the appearance that I wanted. Pete gave me the necessary advice and instructions with his customary expertise and unfailing good humor. My good friend Ann Kiddle provided me with several valuable suggestions for the improvement of the book, and has kindly proof-read parts of it.

I wrote in my first edition that, in its preparation, my children had "tolerated a father sometimes a little absent in spirit even if only too present in the flesh." Fifteen years have gone past since I wrote that, and the 'children' have made positive contributions to this third edition. Andrew, a physicist, keeps me on track with the latest developments in solar-system exploration, while Carol, a chemist, has proof-read parts of the book with an expert eye. Finally, and with the deepest sense of gratitude, I thank my wife, Brenda. First, she has given freely of her professional skills as an editor to help me in preparing this and the two earlier editions of the book. Much more than that, however, she has been a constant source of strength and support during the quite difficult times we have both been through since I started this revision of the book. All these people have made the task of preparing the third edition exciting, enjoyable, and worthwhile.

Oxford
October 1999

R.P.W.

Preface to the second edition

Research activity in atmospheric chemistry has continued to accelerate in the few years since the first edition of this book was published. An enormous quantity of detailed factual information has become available in areas where previously our knowledge was less well defined, and new measurements have given some of the quantitative information a more secure foundation. For a book like the present one, the greater detail and the modifications to numerical values would not, in themselves, justify the publication of a new edition. The purpose of the book is to lay down the principles of atmospheric chemistry and to provide the necessary background for study of the subject in greater depth: the older examples might be just as good as the newer in illustrating these principles. Some of the recent advances in our subject are, however, of the kind that affords new insights and interpretations, and it is to record and place in proper context these exciting, dramatic, and sometimes worrying, developments that this revised edition of the book has been prepared. In the paragraphs that follow, I have tried to highlight the most important of these developments, each of which is discussed in appropriate parts of the text.

Probably the most startling discovery about our own atmosphere was that of the 'Antarctic ozone hole' by scientists of the British Antarctic Survey. Each Antarctic spring, there now occurs substantial depletion of ozone in the stratosphere. Since we depend on the stratospheric ozone layer for protection from damaging ultraviolet solar radiation, there is obviously more than academic interest in the chemistry that leads to its thinning over the Antarctic. Although the science turns out to be fascinating, more important are the interpretations that have followed the experimental observations; the depletion could not have been predicted by our understanding of atmospheric chemistry before this discovery.

One part of the explanation for the behaviour of Antarctic ozone is associated with chemical changes occurring on the surface of the particles in polar stratospheric clouds. Chemical reactions on the surfaces of solid or liquid aerosols, or within the droplets of clouds, have long been known to play a role in a variety of atmospheric processes, although in many cases our understanding has lacked quantitative foundations. The unravelling of the mystery of the ozone hole has served to focus much attention on chemical transformations in and on particles. By chance, the period has overlapped with one in which there has been continued analysis of the atmospheric effects of large quantities of dust being injected into the atmosphere by a large volcanic eruption (El Chichón, in the spring of 1982). The heightened awareness of surface phenomena in the Earth's atmosphere has encouraged scientists to consider their influence in other atmospheres, especially those of notably dusty (Mars) or cloudy (Venus) planets. Photochemical hazes, attributable to formation of light-scattering particles, are familiar in our atmosphere, but they also seem to be present in the atmospheres of even the outermost planets of the solar system and their satellites.

There has been a growing emphasis on the way in which the composition of Earth's atmosphere is changing. The concentrations of trace gases are now known to be altering (generally increasing), not on geological time-scales, but at rates as large as one per cent per year in some cases. Noteworthy in this respect are the gases methane (CH_4), carbon monoxide (CO), and nitrous oxide (N_2O). The chemical effect of the changed burden of trace gases may extend to altering the oxidizing capacity of the atmosphere, and thus to reducing the ability of the atmosphere to attack chemically other substances released to it. The time-scales of the changes, and the period when they appear to have begun, point to man's activities as a source of the substances. Other evidence points in the same direction: for example, there is a marked geographical asymmetry in the increase of CO, which is confined largely to the highly populated and industrialized northern hemisphere. It is not only industry, however, that leads to the increases. Changes in land use and agricultural practices are also culprits, with micro-organisms acting as intermediaries in the release of many trace gases. Biomass burning has become firmly established over the last few years as another significant contributor to the total input of many trace gases.

Some compounds released to the atmosphere have no natural source, and are entirely a consequence of manufacture by man. The chlorofluorocarbons (CFCs) are typical of these species, and they have important impacts on the atmosphere. They are, for example, active in destroying stratospheric ozone; it is now fairly certain that there has been a small trend towards lower ozone concentrations, and the depletions are likely to be associated with the release of CFCs. So far as the Antarctic ozone anomaly is concerned, the causal relation between the increased atmospheric load of CFCs and the deepening and widening of the ozone hole seems definitely established.

The CFCs are also what are known as 'greenhouse gases'. That is, they trap solar infrared radiation in the lower atmosphere and so increase temperatures. Many of the other trace gases of the atmosphere whose concentrations are increasing are also greenhouse gases. Most important of all in this context is carbon dioxide; the concentration of this gas in the atmosphere has increased by about 25 per cent since the industrial revolution, mainly because of the burning of fossil fuels. Carbon dioxide and the other trace gases act synergistically in trapping infrared radiation, and there is the very real risk that the global climate may be altered significantly, with adverse socio-economic consequences. Argument continues about whether atmospheric temperatures yet show any response to the increased load of greenhouse gases, with the balance of opinion perhaps saying that they do not.

Atmospheres of bodies in the solar system other than our own planet have received their share of attention over the last five years. Voyager 2 completed its proposed odyssey through the solar system by visiting Uranus in early 1986, and then Neptune in August 1989, thus bringing resounding success to a mission that had started more than twelve years earlier. Only part of the research was concerned with atmospheric science, of course, but our understanding of the atmospheres of the outermost planets and their satellites, especially Triton, has improved immeasurably as a result of the encounters. The year 1986 also saw the 'Halley Armada' approach and investigate the best known of the comets. Some of the experiments were designed to study the gases present in the coma and the tail of the comet, and the results provide new insights into the composition and chemistry of primitive bodies in the solar system.

Alongside the new discoveries and interpretations in atmospheric science, the last five years have seen a heightened public awareness of environmental problems in which atmospheric chemistry plays a large part. Greenhouse warming and climate change, deforestation and 'slash-and-burn' agriculture, stratospheric ozone depletion, acid rain, and photochemical smog are becoming familiar terms, even if the non-scientific public feels somewhat hazy about the details. As I said in the preface to the first edition of this book, it is necessary for scientists to be well informed about these environmental issues so that they can judge both the reality of the threats and the viability of hypothetical control strategies.

In some cases, control measures have already been adopted; use of catalytic converters on automobiles has certainly had a marked effect on photochemical smog in California. But scientific advances have a way of outpacing the legislation. For example, the Montreal protocol that seeks to limit the release of CFCs to the atmosphere was based on old evidence for ozone depletion, and was concluded just as the most dramatic ozone depletions ever were being observed in the Antarctic. It seems now that a much more stringent reduction in CFC release may be needed to protect the ozone layer. Perhaps even more important than the question of ozone depletion is that of climatic change. Concentrations of trace gases, including carbon dioxide, have indubitably increased, but current evidence for a related increase in global temperatures is weak. Yet it is now that action needs to be taken if the untoward possible consequences of climate change are to be alleviated or averted. This book does not attempt to suggest answers to the problems that we now face, but it does try to lay the foundations for the study of that atmospheric chemistry on which rational decisions will have to be based.

It is once again a real pleasure to express my thanks to all those who have helped with the preparation of this book. My wife, Brenda, has continued to support me unwaveringly and to encourage me at all times. Many scientific colleagues have helped in one way or another. Some whose help was recorded in the preface to the first edition have also helped with this one. I particularly value the advice given by Karl Becker, Tony Cox, Dieter Kley, Joel Levine, Susan Solomon, Richard Stolarski, Georg Witt, and Don Wuebbles, but I hope that those not mentioned by name here will also accept my gratitude. I am also most appreciative of the support and interest of my research group in Oxford, especially for the suggestions and constructive criticism of Anne Brown and Carlos Canosa-Mas.

Oxford
April 1990

R.P.W.

Preface to the first edition

Our knowledge of the atmospheres of Earth and the planets has improved so dramatically over the last two decades that what was, in the early sixties, a rather esoteric branch of learning, has now become a 'hard' science that is a major area of interest to physicists and chemists. Two factors have brought the new-found information. Exploration of atmospheres, first our own and then those of other bodies, has been made possible by the imaginative rocket and satellite programmes of the last decades. At the same time, man has become increasingly aware that the results of many of his activities—ranging from driving motor cars to fertilizing the soil—could have damaging effects on the chemical balance of the atmosphere, and perhaps even lead to an irreversible global climatic change. Considerable effort and ingenuity have gone into separating reality or probability from speculation. The labours of laboratory kineticists and photochemists, physicists, meteorologists, mathematicians, geologists, and biologists have gone into this effort. Many uncertainties remain, and more problems have been unearthed, but we now possess a body of fact that permits of rational discussion. It seems to me that an educated scientist of the present day must have an understanding of how physics and chemistry control atmospheric behaviour. The scientist must be able to judge how a new piece of information on atmospheric behaviour fits into the procession of ideas and discoveries in the world of science. Certain of the interpretations that are now being offered are sufficiently creative and innovative to be considered as real advances in understanding our relation to the rest of the Universe. At the same time, the scientist must be concerned with the less elevated matter of man's comfort and even survival. He must certainly use this knowledge and judgement to assess possible threats to the environment. Some threats are no more than threats: others may be real, but the cure may have socio-economic consequences worse than the environmental damage; others yet again may be so serious that they must be averted at all costs. Who is to judge? Better the well-informed scientist than someone who comes to the problem with some prejudice—for *or* against action—but little knowledge. I have written this book to try to provide the connecting links between atmospheric chemistry and the traditional formal education of physics, chemistry, or biology, in the hope that the reader shall be better able to place in context new advances and new problems in atmospheric behaviour.

Meteorology has a long history; its study has been of practical importance for centuries because of the need to forecast the weather. The subject is now a fully-fledged science, and weather prediction remarkably accurate. Bit by bit the descriptive and synoptic meteorologist has become less distinguishable from his colleague, the atmospheric physicist. Most University courses in physics now include atmospheric physics as at least an option. My approach is that of a physical chemist: the idea of teaching atmospheric chemistry to chemistry undergraduates is new because our knowledge of the subject has been so recently gained (at least on a University time-scale!). For five years now, I have given a short course to third-year Oxford undergraduates on the 'Chemistry of Planetary Atmospheres'. Indeed, the content of that course, considerably expanded and rearranged, forms the basis of this book. From the point of view of a chemist, the subject actually offers an excellent vehicle for the teaching of physico-chemical concepts, quite apart from the specific interest attaching to atmospheric studies. Planetary atmospheres are giant photochemical reactors, and the atoms and 'small' molecules involved fit in well with precise descriptions of spectroscopy, reaction kinetics, photochemistry, excited states, and thermodynamics. Many of the ideas of atmospheric physics should be familiar to chemists, and teaching them can serve a useful educational function by showing a practical application of fundamental physical chemistry. What is frequently not familiar is the manner of presentation. Physicists and chemists often treat the same material in different ways, and the chemist can find unnecessary difficulties in physics, and vice versa. I have tried to write this book to be intelligible to readers approaching atmospheric chemistry from any scientific discipline.

Atmospheres in the solar system show great variety, ranging from the hydrogen and helium of the giant planets, Jupiter and Saturn, to the carbon dioxide of Venus and Mars. Along the way are the tenuous sulphur dioxide and atomic sodium atmosphere of Jupiter's satellite Io, and the dense nitrogen, argon, and methane atmosphere of Saturn's satellite Titan. Earth's atmosphere stands unique in the solar system, being oxidizing, but not completely oxidized. Life is responsible for our unusual atmosphere, which in turn is essential for the maintenance of life. Chapter 1 of this book discusses the variety of atmospheres, and the special nature of Earth's gaseous mantle. The relationship between the biosphere and atmosphere are explored, and consideration is given to the biological control of composition. Topics in atmospheric physics and meteorology that are essential to a quantitative understanding of atmospheric chemistry are presented in Chapter 2. They include the interpretation of pressure and temperature structure in the atmosphere, circulation patterns, and the formation and optical properties of clouds and aerosols. This chapter explains the conventional division of atmospheres into regions such as troposphere, stratosphere, mesosphere, thermosphere, and exosphere. Chapter 3 similarly describes topics in chemistry that are of particular importance in atmospheric processes. We deal in this chapter with the elements of photochemistry and reaction kinetics that will be used repeatedly in later parts of the book. The chapter ends with an explanation of how chemical rate processes and physical transport are incorporated into mathematical models of atmospheres that are used for diagnostic and prognostic applications.

Chapters 4 to 7 are concerned with different aspects of the chemistry of Earth's atmosphere. By and large, the energetics of chemical processes that can occur decrease with decreasing altitude, because the ultraviolet radiation that initiates photochemical change is filtered out by the atmosphere. Shorter wavelengths (higher photon energies) penetrate less deeply, so that their chemical effects are confined to high altitudes. The photons reaching the stratosphere (about 15–55 km) are energetic enough to split molecular oxygen to atoms; the atoms can then add to molecules to make ozone. This ozone is concentrated in a layer, and because it removes ultraviolet that would be hazardous for living organisms, it is an exceedingly important part of our atmosphere. Chapter 4 examines the details of stratospheric chemistry, and discusses how some of man's activities might compromise the ozone layer. Most of the mass of the atmosphere is lower down, in the troposphere (Chapter 5). In this region, the chemistry is dominated by radical reactions initiated by the hydroxyl radical. Trace gases released from the surface, mainly as a result of biological activity, undergo oxidation and conversion. Man supplements the natural emissions of hydrocarbons and the oxides of carbon, nitrogen, and sulphur, most particularly by the combustion of fossil fuels. Because of the intensity of release compared with the natural background, the delicate chemical balance of the troposphere can be upset, and 'air pollution' results. The causes of, and remedies for, common forms of air pollution are included in the discussion of Chapter 5. Ions exist throughout the atmosphere, but their concentration is small below the mesosphere. Above about 70 km altitude, ion and electron concentrations become high enough to exert an effect on radio-waves that has been recognized for sixty years or more. The region where ions are abundant is called the ionosphere, and its chemistry is presented in Chapter 6. Electronically and vibrationally excited species can emit radiation, and so contribute to the 'air-glow' that exists by day and by night. Chapter 7 reviews some of the most significant emission features of the airglow of Earth, Venus, and Mars, and examines the mechanisms by which the radiating species become excited.

Planetary exploration has led to a blossoming of interest in and knowledge about atmospheres other than our own. Chapter 8 describes the picture as it appears at the time of writing, although progress is so rapid that changes in detail are inevitable. The emphasis throughout the chapter is on explaining why a particular type of chemistry operates in a certain atmosphere. Why an atmosphere should have a particular composition, and where that atmosphere came from, are the concern of the first parts of Chapter 9, which looks at the evolution of atmospheres. The present-day composition may even hold clues about the origin of the planet or satellite to which the atmosphere belongs. Geological and geochemical evidence about the past of our own planet is sufficiently rich to enable us to construct detailed 'scenarios' for the evolution of the atmosphere, climate, and even life itself. Projections into the future are also possible. One generally accepted view, described towards the end of Chapter 9, is that man may be warming his planet up by converting back to carbon dioxide in a few centuries the fossil fuel deposits that photosynthesis created from carbon dioxide over hundreds of millions of years.

In deciding on the order of presentation just described, I tried to make the progression of topics orderly and logical, and to avoid, as far as possible, the need to 'look ahead' to later portions of the book from earlier ones. Sometimes I have chosen to delay part of a discussion, as for example the question of increasing carbon dioxide concentrations and concomitant climatic change (Chapter 9), which is an extension of the air pollution topics of Chapter 5. I have been quite liberal with cross-references—both back and forward—in order to keep the reader aware of where he has come across a subject before, or will see it again. Generally, these cross-references are to sections of a chapter, but occasionally a specific page is more appropriate. With regard to literature citations, I have been guided by my undergraduate and postgraduate students. They were unanimous in their condemnation of any form of referencing within the text, on the grounds that numbers, names, or whatever, break the flow and distract the reader's attention. At first, I was slightly uneasy about not providing chapter and verse for my statements, especially where I have chosen to present one of several disputed views. On the other hand, the book has been written as an introductory textbook rather than as a research monograph. Accordingly, there are only very few references in the text, and these are mostly to historically important work. Instead, I have provided a fairly generous set of references at the end of each chapter. I have arranged the citations more or less in the order in which topics to which they refer appear in the chapter. The references are collected in small groups preceded by a short explanation of the content of the papers. It is thus easy for the reader to find the justification for every point of substance if he has a mind to, but not if he would prefer to read on uninterrupted. Referencing in this way has the advantage that I can introduce mention of additional important topics that do not fit in to the development of the main text. Of course, I have provided attributions wherever appropriate for the tables and figures. Most references are to easily-found books or journals, although some sources (e.g. *The Stratosphere 1982*, published by the World Meteorological Organization, and some of the NASA reports) are so important that they have been included, even though they may need to be hunted down in the library. The papers and books cited themselves carry extensive literature references, so that a student can easily build up a list of original works if he has a mind to explore a subject in greater depth than is provided in this book. Where possible, I have included references to reviews or commentary-style papers.

Atmospheric chemistry is a lively and active field in which development is rapid and in which quantitative data are being modified continuously. This book presents a snapshot picture of the discipline as it appeared in 1984. Most of the underlying concepts are well established, but I am well aware of the dangers inherent in quoting 'up-to-date' numbers to illustrate the various topics treated, since refinements and new discoveries will inevitably follow soon. Speculative ideas abound in atmospheric chemistry, and I have, as far as possible, distinguished reasonably well-established fact from speculation.

A brief word about units and nomenclature may be in order. I have tried, where it is sensible, to use SI units. However, since one of the aims of this book is to enable the reader to appreciate published works in the atmospheric sciences, some exceptions are inevitable. For example, the bar ($10^5\,N\,m^{-2}$) and mbar are used almost universally as units of pressure in meteorological and atmospheric studies. Similarly, systematic chemical names are not commonly used in the literature, and I should be failing my reader if I insisted on calling, for instance, HCHO 'methanal' rather than 'formaldehyde'.

Now I come to the very pleasant task of expressing my gratitude to the many people who have helped me to make this book a reality. First, and most important, I thank my wife, Brenda, who not only provided the support and encouragement I needed to sustain me while writing, but also gave me very real practical help with her editorial skills. Then I thank my children, Carol and Andrew, who tolerated a father sometimes a little absent in spirit even if only too present in the flesh; they took a lively and informed interest in what I was writing, and made the task more enjoyable. Next, I wish to thank my research group: Tim Wallington, Martin Fowles, Peter Biggs, and Mary Davies. They accepted that, while preparing the book, their research supervisor would be more distracted than usual from their work. They also provided valuable ideas about the content and presentation, and were sympathetic listeners when I conceived an enthusiasm for, or, indeed, a passionate dislike of, some topic I was pursuing. I am grateful to the many colleagues and friends who gave me advice, information, and their time. Perhaps I might mention a few by name: Rudi Burke, Tony Cox, Bill DeMore, Hugh Ellsaesser, Garry Hunt, Don Hunten, Joel Levine, Mike McElroy, Darrel Strobel, and Georg Witt. I hope the more numerous body of un-named friends will not take this selection amiss, and will know how much I appreciate their efforts. Finally, I should like to tell my typist, Mrs Margot Long, how pleased I am that she accepted the tiresome job of typing my words. As always, she was efficient and cheerful even when faced with a much-amended manuscript that might reduce others to tears.

Oxford
July 1984

R.P.W.

Contents

1 Chemical composition: a preliminary survey

1.1 Earth's atmosphere in perspective

Planet Earth possesses an atmosphere that, for hundreds of millions of years, does not seem to have been obeying the laws of physics and chemistry! Since it is the purpose of this book to interpret the composition and properties of atmospheres in terms of physico-chemical principles, an explanation of Earth's peculiar behaviour must be sought at the outset. Apparently alone amongst the planets of the solar system, Earth supports life. The composition of our atmosphere is determined by biological processes acting in concert with physical and chemical change. At the same time, our unique atmosphere seems essential for the support of life in the forms we know it. In later parts of the book we shall speculate about the origins and evolution of atmospheres, especially our own. We shall, however, start our discussion by considering the atmospheres as we find them today.

Atmospheres (Greek 'atmos' = 'vapour' + 'sphaira' = 'ball'), the gaseous envelopes around heavenly bodies, show an amazing diversity in the solar system. Venus, Earth, Mars, Jupiter, Saturn, Uranus, and Neptune all have substantial atmospheres, as does Saturn's largest satellite, Titan. Table 1.1 shows the relative abundances of various species in these atmospheres, together with information for the Sun. Data have, for the most part, been obtained directly in planetary missions (e.g. Viking, Pioneer Venus, Venera, Voyager 1, 2, and Galileo), although in some cases the suggested compositions depend on rational interpretation of observations made from afar. The concentrations are given as *mixing ratios* by volume; the unit is the one commonly used by atmospheric scientists, and is identical to the chemists' *mole fraction*. In the table, the mixing ratios are given in parts per million, p.p.m. (p.p.m.v. is another frequently used abbreviation). The mixing ratios quoted are really averages for the lower atmosphere of each planet; as we shall see, the relative importance of different species can vary with altitude. However, since pressure and density decrease more or less exponentially with increasing altitude (Chapter 2), most of the atmosphere is near the surface of a planet, and the lower atmosphere composition is almost the same as the average throughout the atmosphere. For the Earth, half the mass of the atmosphere lies below about 5.5 km altitude, and 99 per cent below 30 km. The total mass of our atmosphere, by the way, is about 5×10^{18} kg. A more formal division of the regions of the atmosphere, based on temperature structure, is explained in Chapter 2. It is convenient to introduce here the names of the two lowest regions, the *troposphere* and *stratosphere*, so that we can distinguish between these parts of the atmosphere. The terms reflect an important aspect of physical behaviour since 'tropos' is Greek for 'turning' and 'stratus' is Latin for 'layered'. Rather strong vertical mixing characterizes the troposphere;

Table 1.1 Solar system bodies with substantial atmospheres

Body	Surface temperature (K)	Surface pressure (Earth atm)	H_2	He	H_2O	CH_4	NH_3	Ne	H_2S	CO_2	N_2	O_2	CO	SO_2	Ar	N_2O
Sun	-	-	890000	110000	1000	600	150	140	25	-	-	-	-	-	-	-
Venus	735	93	10^a	12	60	0.6^a	-	7	2^a	965000	35000	$< 0.3^a$	30	0.1	70	-
Earth	288	1	0.53	5.2	0–40000	1.7	<0.01	18	10^{-4}	350	780840	209460	0.04–0.2	10^{-4}	9340	0.3
Mars	223	0.006	-	-	300	-	-	2.5	-	953200	27000	1300	700	-	16000	-
Jupiter	$165^{b,c}$	-	898000	102000	5	3000	2600	-	?	-	-	-	0.002	-	-	-
Saturn	$134^{b,c}$	-	963000	32500	5	4500	200	-	<0.4	-	-	-	-	-	-	-
Uranus	76^b	-	825000	152000	-	23000^d	-	-	-	-	-	-	-	-	-	-
Neptune	72^b	-	800000	190000	-	15000^d	-	-	-	-	-	-	1.2	-	-	-
Titan	95	1.6	2000	-	< 0.001	20000	-	-	-	0.0015	> 970000	-	60	-	< 10000	-

Notes: Values are given in parts per million (p.p.m.) by volume.
[a]Disputed identification.
[b]Values given for altitude where pressure is that at Earth's surface.
[c]No true surface.
[d]Troposphere.

Data from: Encrenaz, T. and Combes, M. *Icarus* **52**, 54 (1982); Holland, H.D. *The chemistry of the atmosphere and oceans*, John Wiley, Chichester, 1978; Hudson, R. (ed.-in-chief) *The stratosphere 1981*, WMO, Geneva, 1981; Hunt, G.E. *Annu. Rev. Earth & Planet. Sci.* **11**, 415 (1983); Moroz, V.I. *Space Sci. Rev.* **29**, 3 (1981); Owen, T. *Planet Space Sci.* **30**, 833 (1982); Owen, T., Biemann K., Rushnek, D.R., Biller, J.E., Howarth, D.W., and Lafleur, A.L. *J. geophys. Res.* **82**, 4635 (1977); Jakosky, B.M. Chapter 13 in Beatty, J.K., Petersen, C.C. and Chaikin, A. (eds.) *The new solar system*, 4th edn, Cambridge University Press, 1999; Strobel, D.F. *Int Rev. phys. Chem.* **3**, 145 (1983); Trafton, L. *Rev. Geophys. & Space Phys.* **19**, 43 (1981); Stone, E.C. and Miner, E.D., *Science* **233**, 147 (1989); Fegley, B., Jr. Properties and composition of the terrestrial oceans and of the atmospheres of the Earth and other planets pp. 320–345 in Ahrens, T. (ed.) *AGU handbook of physical constants*, AGU, Washington, 1995; Yung, Y.L and DeMore, W.B. *Photochemistry of planetary atmospheres*, OUP, Oxford, 1999.

individual molecules can traverse the entire depth in periods between a few days in clear air and a few minutes in the updraughts of large thunderstorms. Conversely, the stratosphere is characterized by very small vertical mixing, the time-scale for transport being of the order of years. The troposphere extends through roughly the first 10–17 km of the atmosphere, and the stratosphere through the next 30–40 km. However, the altitude of the *tropopause* (the hypothetical boundary between troposphere and stratosphere) is rather variable, and depends on season and latitude. Abrupt changes in the concentrations or mixing ratios of some of the minor constituents of the atmosphere occur at the tropopause, and it is important, therefore, to specify the atmospheric region when discussing the chemistry. Although the tropospheric composition is dominant in determining total mixing ratios, chemical (and physical) processes occurring at higher altitudes can affect the atmosphere as a whole.

So far as Jupiter and Saturn are concerned, there is probably no surface marking a transition from gaseous to solid material. Instead, the pressure increases with depth, and the gases of the atmosphere become more dense (perhaps liquid or even metallic). We have given quantities in Table 1.1 for these planets appropriate to that altitude in the atmosphere where the pressure is the same as that at Earth's surface. The planets missing from our table are Mercury and Pluto. Mercury has almost no atmosphere (about 10^{15} times less dense than Earth's; what there is consists of 98 per cent He and 2 per cent H). Pluto is surrounded by methane at a pressure of around 10^{-4} of an Earth atmosphere, together with some other 'heavy' gas such as neon. The volcanism of Jupiter's satellite Io indicates that gases are being released, and there is a tenuous atmosphere containing, amongst various components, neutral and ionized atomic sodium. Some other satellites of the outer planets may possess similarly tenuous atmospheres, perhaps not firmly attached to the satellite, but strung out as a torus around the parent planet. Our main concern, however, is with the massive atmospheres listed in the table.

A cursory inspection of Table 1.1 shows that the outer planets (Jupiter, Saturn, Uranus, Neptune) have atmospheres very different from those of the inner planets (Venus, Earth, Mars). Hydrogen and helium are present on the outer planets in abundances not too far removed from those on the Sun, and the atmospheres are certainly reducing. In contrast, the atmospheres of the inner planets are either oxidized or oxidizing.

At this point, the unexpected nature of the Earth's atmosphere becomes apparent. Since Earth lies in the solar system between Venus and Mars, Earth's atmosphere might have been expected to consist primarily of the *oxidized* compound, carbon dioxide. But CO_2 is only a minor (although very important) constituent of our atmosphere. It is the presence of elemental oxygen as a major constituent that poses the most serious problems. There is too much oxygen in the presence of too many gases that react with oxygen, and our atmosphere appears to be a combustible mixture. Oxygen reacts with hydrogen to form water, with nitrogen to form nitrate, with methane to form carbon dioxide and

Table 1.2 Actual and equilibrium concentrations of constituents of the Earth's atmosphere

Species	Expected fractional equilibrium concentration[a]	Present fractional concentration	Output[b] kg yr^{-1}	Residence time[c]
N_2	$<10^{-10}$	0.78	2.5×10^{11}	1.6×10^7 yr
CH_4	$<10^{-35}$	1.7×10^{-6}	5.4×10^{11}	9 yr
N_2O	$<10^{-20}$	3.1×10^{-7}	2.5×10^{10}	94 yr
NH_3	$<10^{-35}$	10^{-9}	5.5×10^{10}	20 days
H_2	$<10^{-35}$	5.3×10^{-7}	5.0×10^{10}	4 yr

Notes:

[a]Equilibrium concentrations are from Margulis, L. and Lovelock, J.E. in Billingham, J. (ed.) *Life in the universe,* MIT Press, Cambridge, Mass., 1981. The values are based on current oxygen concentrations and production rates.

[b]Annual outputs are estimates based on the reasoning developed towards the end of this chapter. The values depend heavily on the assumptions made in their derivation.

[c]Residence time is calculated here in a simplified way as the atmospheric mass of the compound divided by the rate of supply, and assuming a steady state for the atmospheric concentration. For some compounds, such as CH_4 and N_2O, there is a slight imbalance between production and loss, so that concentrations in the atmosphere are slowly increasing. See also Section 3.6.1.

Representative concentrations and outputs given here are taken from Seinfeld, J.H. and Pandis, S.N. *Atmospheric chemistry and physics*, John Wiley, Chichester, 1998, and Bolin, B. and Cook, R.B. (eds.) *The major biogeochemical cycles and their interactions*, SCOPE **21**, John Wiley, Chichester, 1983.

water, and so on. The extent to which the atmosphere departs from equilibrium is shown in Table 1.2, in which expected equilibrium concentrations, calculated for the actual oxygen content of the atmosphere, are compared with the measured values. Rates of processes are sometimes slow enough to prevent achievement of thermodynamic stability, as in the conversion of diamond to graphite. Such arguments cannot be advanced to explain the discrepancies in atmospheric composition. By identifying sources of the various chemical species, as discussed in the following sections of the chapter, the annual input to the atmosphere can be estimated. Since concentrations are roughly constant over a period of a few years, there must be corresponding loss processes, and there is a residence time (or lifetime) for the species, calculable in the simplest way as the total atmospheric mass ('burden') of that species divided by the rate of its supply, also in mass units.[a] Some rough values of residence times are

[a]The burdens and rates can also equally well be expressed in concentration units. A more detailed explanation of the concept of lifetime is given in Section 3.6.1.

given in Table 1.2, and show clearly that there is no kinetic limitation on the establishment of chemical equilibrium. Only for the longest-lived species, nitrogen, is the lifetime comparable with geological time-scales of typically 10^7 to 10^9 years. Residence times for the other gases are much shorter, and in some cases are measured in days. The Earth's atmosphere thus maintains a steady-state disequilibrium composition with the fuel being continuously replenished.

Biological processes are dominant in the production of the oxidizable components of our atmosphere. That is, they bring about the thermodynamic disequilibrium, and effectively reduce the entropy of the atmosphere. Energy is needed to drive the entropy reduction, and it is supplied virtually entirely by the radiation from the Sun that reaches the Earth and its atmosphere. Not only does biology provide the fuel, but also the oxidant, oxygen. In the absence of life, the photochemical sources of oxygen limit the surface concentration to a value perhaps 10^{13} times smaller than the amount that the atmosphere actually contains. Consideration of the biological release rate of oxygen to the atmosphere suggests that present-day concentrations are maintained with a residence time between 5000 and 10 000 years.

1.2 Land, sea, and air

Table 1.1 shows that, of all the planets, Earth has the largest relative abundance of water vapour. Although a typical concentration is 1 per cent, common experience tells us that water vapour is one of the most variable components of the atmosphere. In the tropics, water may account for up to 4 per cent (by volume) of the atmosphere, while in polar regions or in desert air the abundance may be a fraction of one per cent. Water is of peculiar interest with regard to the Earth, not only because of its importance to life, but also because it is the only atmospheric component that can exist in all three phases—solid, liquid, gas—at the temperatures found near the Earth's surface. Most of Earth's water is liquid; the term *hydrosphere* ('hydro' is Greek for water) is used to describe the condensed water environment. Only about 0.001 per cent of the total water budget is in the atmosphere at any one time, the rest being largely divided between oceans (97 per cent) and fresh water lakes or rivers (0.6 per cent). Snow and polar ice sheets account for 2.4 per cent of the water. The condensed water behaves as an enormous reservoir for atmospheric water vapour. Water fulfils two important non-biological functions. First, phase transformations in water are accompanied by large latent heat changes. Evaporation, subsequent transport by wind, and recondensation of water thus provide for heat transport in the atmosphere; the released heat in turn may ultimately drive the wind system. Currents in the oceans themselves provide an additional means of heat redistribution. Secondly, the oceans can hold gases in solution and provide a supply of atmospheric gases and act as a buffer against atmospheric change. Carbon dioxide is the prime example of a gas buffered by the oceans, because it undergoes reversible chemical change to bicarbonate ions on solution

$$CO_2 + H_2O \rightleftharpoons HCO_3^- + H^+ \qquad (1.1)$$

and is therefore abnormally soluble. Indeed, the oceans contain around fifty times more CO_2 than the atmosphere.

Comparison of the behaviour of water vapour on Earth with that on the neighbouring planets, Mars and Venus, proves interesting. Mars is further from the Sun than is Earth, and its surface temperature is lower (typical values for Mars and Earth are 223 K and 288 K). Liquid water does not, therefore, exist on Mars, although various channel and valley forms observed on the planet's surface suggest it may have done in the past. Rather, a water-ice deposit in the north polar region of Mars is the chief water reservoir. Indeed, it is so cold in the winter polar regions of Mars that seasonal polar caps of solid carbon dioxide form there. Since CO_2 is the main constituent of the Martian atmosphere, condensation or evaporation of the solid reservoir can lead to changes in total pressure, at least locally. Venus, by contrast with Mars, has a much higher surface temperature (~735 K) than Earth. There are thus no oceanic reservoirs for water on Venus, and the low water-vapour content of the atmosphere (Table 1.1) represents a low planetary abundance of water. How the Venusian atmosphere came to be so hot and dry we shall explore further in Chapter 2.

The land mass—the *lithosphere* ('lithos' is Greek for stone')—itself exchanges gases with the atmosphere either directly or indirectly via the hydrosphere. Let us use carbon dioxide as our example again. On Earth, carbonate rocks such as limestone ($CaCO_3$) account for 100 000 times as much CO_2 as there is in the atmosphere. Thus the lithosphere is a potentially greater reservoir of CO_2 than are the oceans, although the rates of release may be much smaller. One indirect release process on Earth involves weathering of limestone by CO_2 dissolved in water; the process returns bicarbonate ions to the oceans

$$CaCO_3 + H_2O + CO_2 \rightleftharpoons Ca^{2+} + 2HCO_3^-. \qquad (1.2)$$

A most important source of the carbonate minerals is secretion by animals and plants, and the deposition of calcite ($CaCO_3$) shells. Inorganic weathering reactions can also convert silicate rocks such as diopside ($CaMgSi_2O_6$) to carbonate

$$CaMgSi_2O_6 + CO_2 \rightleftharpoons MgSiO_3 + CaCO_3 + SiO_2. \qquad (1.3)$$

We note that on Venus the biological source of carbonate rocks is missing, that the high surface temperature favours the reactant side of equilibrium (1.3), and that there are no oceans to store CO_2. Since the total mass of CO_2 in the

Venusian atmosphere (~ 3×10^{20} kg) is very close to that stored on Earth as carbonate rocks, it is certainly very tempting to speculate that the total inventory of CO_2 is the same for the two planets.

1.3 Particles, aerosols, and clouds

Suspensions of particles in a gas are called *aerosols*. In principle, if the particles are liquid we refer to the aerosol as a *cloud* or *mist*, while if they are solid, the aerosol is *smoke* or *dust*. Usage of the words is, however, rather loose. More important, from our point of view, is what we mean by 'suspension' and 'particles'. Particles have a tendency to fall under the influence of gravity; this sedimentation is opposed by the kinetic energy of the surrounding molecules. For very small particles, the rate of sedimentation is so slow that macro-scale fluid motions completely dominate, and keep the particles thoroughly mixed (cf. pp. 48–49). Larger droplets reach a terminal sedimentation velocity determined by viscous drag. Some idea of the numbers of aggregated molecules, particle radii, and terminal velocities for water molecules is given in Table 1.3.

Even a typical cloud drop of 10 μm diameter would take a day to fall through a cloud 1 km thick, and is, for practical purposes, permanently suspended. On the other hand, a typical raindrop falls through the cloud in under three minutes, and is not in suspension. What is it that distinguishes a 'particle' from a gas phase molecule or aggregate of molecules? The distinction is conventionally made on the basis of the light scattering properties of a body (Section 2.6). If a particle has a size comparable with, or greater than, the wavelength of light, then scattering can occur from different parts of the same particle. The consequent interference effects lead to an angular distribution of scattered intensity markedly different from that obtained with

Table 1.3 Size and terminal velocities of water droplets in the atmosphere.

Radius (μm)	Name of droplet	Approximate number of H_2O molecules per drop	Terminal fall velocity (m s^{-1})
0.1	Condensation nucleus	10^8	10^{-6}
1	Cloud	10^{11}	10^{-4}
10	Cloud	10^{14}	10^{-2}
50	Large cloud	10^{16}	0.27
100	Drizzle	10^{17}	0.70
1000	Rain	10^{18}	6.50

Data from: Wallace, J.M. and Hobbs, P.V. *Atmospheric science*, Academic Press, New York, 1977.

scattering from a point. Changes in scattering behaviour can be detected for particle sizes about one order of magnitude smaller than the wavelength of light, so that, for visible radiation ($\lambda \sim 500$ nm $\equiv$ 0.5 μm) the corresponding detectable particle size is ~0.05 μm. We must emphasize that 'particle-like' properties in respect to scattering obviously depend on the wavelength of radiation used. At the lower size end of the scale, therefore, there is an imperceptible merging between what is an aerosol and what is a large molecule or cluster of molecules.

Aerosols are found in most of the atmospheres we shall discuss, and they play an important part in the physics and chemistry. Water clouds cover about 50 per cent of the Earth's surface at any one time, while the surface of Venus is almost completely obscured by several cloud layers that are apparently composed of sulphuric acid droplets. Cloud cover on Mars is much more patchy, although water-ice clouds are formed in the winter polar regions. Thin carbon dioxide clouds are found at higher altitudes in many places, and thicker CO_2 clouds form lower down near the winter poles. Dust storms on Mars can grow to global proportions, and particles of a few μm in size are placed in suspension in the atmosphere. Compounds such as CH_4, NH_3, NH_4SH, and H_2O in the atmospheres of the outer planets condense out in cloud layers where the temperature and pressure are appropriate. Colours seen in the outer cloud layers of these planets are likely to be associated with solid aerosols of sulphur, phosphorus, or organic compounds. Titan has methane clouds, but in addition there is a haze brought about by aerosol particles with sizes in the range 0.1 to 0.5 μm.

Clouds obviously play a central part on Earth in the precipitation of water as rain, hail, snow, and so on. The fascinating subject of cloud physics and precipitation lies largely outside the scope of this book. We shall, however, touch upon the formation of clouds in Chapter 2. A conclusion we shall reach is that condensation of water to droplets usually starts by nucleation on foreign solid aerosol particles, known as *cloud condensation nuclei* (CCN). Hygroscopic or soluble nuclei are particularly effective, so that the presence of such aerosols in the Earth's atmosphere must be regarded as an important component of the evaporation–condensation–precipitation cycle of the Earth's water.

There is no doubt that human activities greatly increase the concentrations of solid aerosols, especially the smallest ones (radii less than about 0.5 μm, known as *Aitken nuclei*). Typical Aitken nucleus counts near the Earth's surface are 10^5 particles cm^{-3} or more over cities, 10^4 particles cm^{-3} over rural land, and 10^3 particles cm^{-3} over the sea. These particles originate primarily from deliberate combustion processes, and will be discussed further in relation to air pollution (Chapter 5). Appreciable concentrations of particles in continental and marine air indicate the existence of natural sources. So far as the global sources of all solid particles are concerned, man at present probably contributes around 20 per cent. Natural sources include evaporation of sea-spray from the ocean, wind-blown dust from surface erosion, forest fires, meteoric debris, and volcanic emissions. In addition, gas-to-particle conversion can occur in (photo)chemical processes. One of the most important examples of such a reaction is the oxidation of sulphur dioxide (natural, or released as a pollutant) to SO_3 and ultimately sulphate-containing aerosols. The colouring particles in the clouds of the outer planets, and the haze in Titan's atmosphere, must certainly come from photochemical gas-to-particle conversion.

Clouds and other aerosols may greatly modify the atmospheric balance of incoming and outgoing radiation; atmospheric and surface temperatures may be altered. Both the reflectivity and the absorptivity of the atmosphere are involved. Massive volcanic eruptions have long been thought to have a potential influence on climate. The spectacular sunsets over the entire Earth that follow a large eruption are well documented, and they are, of course, a result of increased scattering of sunlight by suspended particles. Recent large eruptions have provided an interesting comparison of climatic effects. The eruption of Mount Pinatubo in the Philippines in June 1991 was the largest so far in the 20th century. About 20 million tonnes of SO_2 were injected into the atmosphere, and were converted to H_2SO_4 droplets. Vast quantities of dust also entered the atmosphere. The injection penetrated the tropopause, and the droplets and particles were in the stratosphere from which loss is very slow. Average mid-latitude aerosol concentrations were up to 40 times the 'natural' level in the following March. Solar radiation reaching the lower atmosphere was reduced by about two per cent, and global average temperatures may have been reduced by as much as 0.5 °C during the year following the eruption, with a reduction of 0.25 °C persisting during 1992 and 1993. The El Chichón eruption (Spring 1982) in SE Mexico similarly spewed vast amounts of SO_2 into the atmosphere that may have led to a hemispheric cooling of the Earth of up to 0.5 °C between the end of 1983 and 1985. The effects of the Pinatubo and El Chichón eruptions are on a par with the largest climatic perturbations recorded over the last 150 years (including the eruption of Krakatoa in 1883). By way of contrast, the Mount St Helens explosion (May 1980) is not thought to have had a major effect on temperatures, even though huge amounts of ash were injected high into the atmosphere. The difference is that the Mount St Helens explosion lacked the gases needed for the production of lasting aerosol.

As chemists, we should be alerted to the possibility of aerosols altering the course or rates of chemical change. Heterogeneous and catalytic processes in the atmosphere are not well identified or understood, but that does not make them unimportant! A few examples are known. For example, the rate of oxidation of SO_2 increases several-fold as the air becomes more nearly saturated with water. Sulphates can be formed by the reaction of sulphur dioxide and ammonia in cloud droplets: when the water evaporates, ammonium sulphate aerosol is left in suspension. The combined presence of soot particles

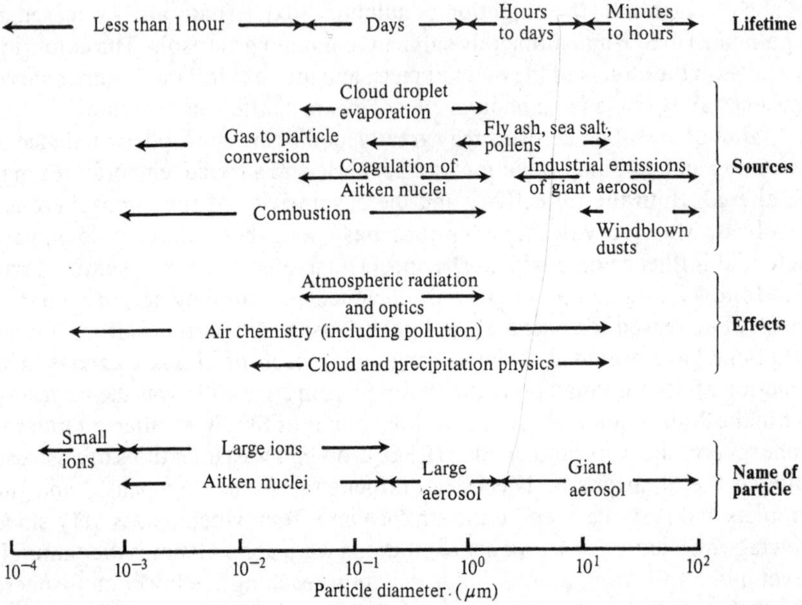

Fig.1.1. Names of atmospheric particles, together with effects, sources, and lifetimes. The lifetime of very small particles is short because they coagulate rapidly to form larger particles. Giant aerosol particles are short-lived because they precipitate out of the atmosphere. Figure drawn from data of Wallace, J. M. and Hobbs, P. V. *Atmospheric science*, Academic Press, New York, 1977.

and sulphur dioxide leads to enhanced oxidation rates as well as a greatly increased health hazard (see Chapter 5). The following section will describe how dependent we are on the presence of ozone in the stratosphere. An 'ozone hole' has been evident during early spring in the Antarctic for more than a decade (see Section 4.7). It owes its origin to anomalous chemistry involving heterogeneous reactions occurring on the surfaces of polar stratospheric clouds. Water-ice particles make up these clouds, but there are indications that sulphate aerosols may also provide sites for surface reactions in the stratosphere. In that case, massive volcanic eruptions may also result in depletion of stratospheric ozone (Sections 4.5.5 and 4.7.7), as certainly appears to have occurred following the eruption of Mount Pinatubo. The involvement of particles in tropospheric chemistry is explored further in Section 5.8. Figure 1.1 summarizes some information about the sources, lifetimes, and effects of aerosols of different sizes.

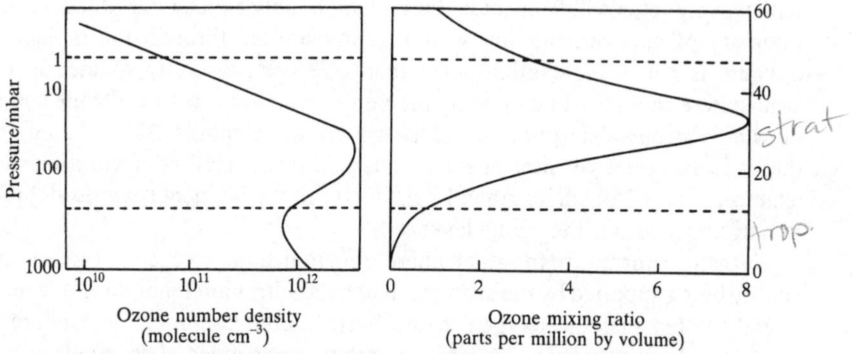

Fig. 1.2. Variation of atmospheric ozone concentration with altitude, expressed as an absolute number density and as a mixing ratio. From *Stratospheric Ozone 1988*, UK Stratospheric Ozone Research Group, HMSO, London, 1988.

1.4 Ozone

Ozone (trioxygen, O_3) plays a peculiarly significant part in the chemistry of the Earth's atmosphere, even though it is a 'minor' species in terms of abundance. Concentrations are rather variable, but the mixing ratio with respect to the entire atmosphere is a few tenths of a part per million. If the ozone in a column of the atmosphere were collected and compressed to 1 atm pressure, it would occupy a column about 3 mm tall. In fact, these statements about concentration conceal a very interesting feature about atmospheric ozone. Instead of being found in a constant fractional abundance, the ozone concentrations are very sharply dependent on altitude. Figure 1.2 shows a typical ozone altitude profile in two ways: as an absolute number density (concentration), and as the mixing ratio (fractional composition). The mixing ratio peaks sharply, and the ozone is pictured as being contained in an *ozone layer* about 20 km thick, and centred on an altitude of about 25–30 km. The dashed lines represent the approximate positions of the tropopause and stratopause, thus showing that the ozone layer lies within the stratosphere: indeed, as we shall see later, it is the presence of ozone that is responsible for the existence of the stratosphere. Peak fractional abundances in the ozone layer can approach 10^{-5} (10 p.p.m.), and peak concentrations reach more than 10^{12} molecule cm^{-3}.

Several factors contribute to ozone's importance, and they will be a recurrent theme in our later discussions. Perhaps the outstanding feature is the relationship between the absorption spectrum of ozone and the protection of living systems from the full intensity of solar ultraviolet radiation. The macromolecules, such as proteins and nucleic acids, that are characteristic of living cells, are damaged by radiation of wavelength shorter than about 290 nm. Major components of the atmosphere, especially O_2, filter out solar ultraviolet with wavelengths < 230 nm; at that wavelength, only about 1 part in 10^{16} of the intensity of an overhead sun would be transmitted through the molecular oxygen. But at wavelengths longer than ~230 nm, the only species in the atmosphere capable of attenuating the Sun's radiation is ozone. Ozone has an unusually strong absorption just at the critical wavelengths (230–290 nm), so that it is an effective filter in spite of its relatively small concentration. For example, at $\lambda = 250$ nm less than 1 part in 10^{30} of the incident (overhead) solar radiation penetrates the ozone layer.

Ozone is formed in the atmosphere from molecular oxygen, the necessary energy being supplied by the absorption of solar ultraviolet radiation. We have already noted that the oxygen in the Earth's contemporary atmosphere is largely biological in origin. Now we see that ozone, needed as an ultraviolet filter to protect life, is itself dependent on the atmospheric oxygen. These links further emphasize the special nature of Earth's atmosphere. Actually, the interactions are even more subtle than we have suggested. Absolute concentrations, and, indeed, the height distribution of ozone depend on a competition between production and loss. Loss of ozone is, as we shall see in Chapter 4, regulated by chemistry involving some of the other trace gases of the atmosphere, such as the oxides of nitrogen, which are themselves at least partly of biological origin. Biological processes thus influence both the generation and destruction of ozone. We shall certainly have to examine if human activities could interfere with the delicate balance (Section 4.6).

Energy absorbed by ozone from the solar ultraviolet radiation is ultimately degraded to heat; so, indeed, is the solar energy originally used in the formation of the ozone. The net result is a heating of the Earth's atmosphere in the region of the ozone layer that has a profound influence on atmospheric temperature structure and vertical stability (Section 2.3). Examination of the chemical routes that lead to the release of thermal energy is instructive. Ozone is an endothermic substance (positive enthalpy of formation from the standard state of the element)

$$3/2\ O_2 \rightarrow O_3;\ \Delta H^{\ominus}_{298} = 143\ \text{kJ mol}^{-1}. \qquad (1.4)$$

Reactions involving ozone are thus often exothermic. Further, rates of these reactions tend to be relatively high, so that ozone is an important chemical reaction partner. In the particular case where ozone itself absorbs a quantum of light ($\lambda \leq 310$ nm), the energy of the system is yet further elevated. A manifestation of this energization is that the ozone *dissociates* to form an atomic and a molecular fragment (i.e. O and O_2), *both of which are electronically excited* (cf. p. 105). The oxygen atom has an excitation energy of 190 kJ mol^{-1}, and this excess energy makes the atom highly reactive. Processes such as

$$O^* + H_2O \rightarrow OH + OH; \quad \Delta H^{\ominus}_{298} = -119\,\mathrm{kJ\,mol^{-1}} \tag{1.5}$$

$$O^* + CH_4 \rightarrow OH + CH_3; \quad \Delta H^{\ominus}_{298} = -178\,\mathrm{kJ\,mol^{-1}} \tag{1.6}$$

$$O^* + N_2O \rightarrow NO + NO; \quad \Delta H^{\ominus}_{298} = -340\,\mathrm{kJ\,mol^{-1}} \tag{1.7}$$

(where O^* represents an electronically excited oxygen atom), are all exothermic; we note that reactions (1.5) and (1.6) are endothermic with ground-state oxygen atoms for which enthalpies of reaction are 190 kJ mol^{-1} less. All the reactions are fast because they have low activation energies. Water vapour, CH_4, and N_2O are important minor atmospheric constituents, while the radicals OH, CH_3, and NO are themselves highly reactive and are involved in atmospheric chemical changes of paramount significance. In each case, a driving force for radical production can be the absorption of solar radiation by ozone.

1.5 Cyclic processes

Our discussion of the photochemical production of O_3 from O_2 in the last section serves to introduce the concept of *cyclic processes* in atmospheric chemistry. Destruction of O_3 in the atmosphere ultimately yields O_2, so that the number of elemental O atoms is conserved in the O_2–O_3 system by a cyclic process occurring solely within the atmosphere (largely within the ozone layer). Several oxygen-containing species are involved in the cycles: O, O_2, O_3 and, indeed, OH. O_2 is also involved in another key cycle, with CO_2 being converted to O_2 by photosynthesis, and oxidation of organic species regenerating CO_2 (see Sections 1.5.1 and 1.5.2).

Production and loss of atmospheric constituents have to be balanced if concentrations are not to vary. Over the lifetime of the Earth, the atmospheric composition has almost certainly undergone considerable modification (Chapter 9), and the balance is not perfect. On short time-scales, however, a steady state obviously more or less holds for components such as nitrogen or oxygen. To a first approximation, the land, sea, air—lithosphere, hydrosphere, atmosphere—system is a 'closed' one: that is, material substance neither enters nor leaves it. In that case, the total quantity of each chemical element is fixed, although the distribution between the elemental and combined forms can alter. This conservation of elemental quantity means that if a species appears in the atmosphere (at a rate equalled by its disappearance), then the elements involved must be passing through a series of *cyclic* chemical and/or physical transformations. Evaporation, transport, condensation, and precipitation of water are steps in a physical cycle linking the atmosphere and hydrosphere. Lithosphere, hydrosphere, and atmosphere are all involved in a cycle of which the weathering of rocks by CO_2 (described on p. 6) forms a part.

This chapter continues with a discussion of some of the major cycles occurring in the Earth's troposphere. Our concern will be to see what reservoirs there are for the species, and what determines the rate of physical and chemical interconversions. In other words, we wish to study the *budgets* involved in the cycles. Quite apart from the fundamental issue of how the Earth's atmospheric disequilibrium is sustained, an understanding of the budgets will also allow us to assess whether *anthropogenic* (man-made) sources of various species could be comparable with or larger than natural ones, and thus pose a threat of local or even global pollution (Sections 4.6 and 5.10).

Before going on to a consideration of individual cycles, it is worth emphasizing again the role played by life in determining the composition of our atmosphere. With the exception of the noble gases and water vapour, *all* the gases listed in Table 1.1 for Earth have a biological or microbiological source, and for species such as oxygen, methane, and ammonia, such sources may be the *only* significant ones in the contemporary atmosphere. Even apparently abiological changes such as the weathering of silicate rocks by CO_2 can be modulated by biological influences. Partial pressures of CO_2 in the soil where weathering occurs are 10–40 times higher than the atmospheric pressure, and these high partial pressures are maintained by soil bacteria. The cycles are thus *biogeochemical* in nature, and the term *biosphere* is used (by analogy with lithosphere, etc.) to represent the biological component of the Earth's surroundings. The source regions of the biosphere are divided into oxic (containing free oxygen) and anoxic. Upper layers of oceans and soils are oxic, and produce fully oxidized species (e.g. CO_2) as well as partially reduced species (e.g. N_2O from bacterial processes and NH_3 from decay of animal excreta). Reduced species (e.g. H_2S, CH_4) are produced in anoxic environments such as lower soil regions or the interiors of animals. Transport and change in the biosphere itself may influence the nature and amount of gas reaching the atmosphere. Thus bacterial reduction of continental shelf sediments probably produces about $10^{12}\,kg\,yr^{-1}$ of H_2S. Less than one per cent reaches the atmosphere, the remainder being taken up by bacterial oxidation. Similarly, the decay of organic materials under anaerobic conditions gives rise to H_2 as the major primary product. Several groups of micro-organisms generate the intermediate (e.g. H_2S) or released (e.g. CH_4, N_2O, N_2) product.

Biospheric sinks exist for many of the atmospheric gases at the land/sea interface with the atmosphere. Respiration and other oxidative processes remove O_2, photosynthetic organisms (Section 1.5.1) remove CO_2, and certain species of plant—micro-organism systems fix nitrogen, for example. Within the troposphere itself, the chemical fates of almost all the trace gases are governed by reactions with hydroxyl radicals, which lead ultimately to oxidation. Hydroxyl radicals are themselves formed largely in reaction (1.5) between excited oxygen and water vapour; some (but not all) of the photochemical precursor of O*, ozone, is transported down from the stratospheric ozone layer. Thus one aspect of the importance of ozone photochemistry and of OH radical production begins to emerge. Methane reacts first to form methyl radicals

$$CH_4 + OH \rightarrow CH_3 + H_2O \tag{1.8}$$

which are oxidized by O_2 in a complex set of reactions, involving formaldehyde (methanal) formation and photolysis, to carbon monoxide. Carbon monoxide from this source, as well as that released directly to the atmosphere, is also oxidized by OH in a cyclic set of reactions starting with the process

$$CO + OH \rightarrow CO_2 + H, \tag{1.9}$$

to yield CO_2 as the end product. Hydroxyl radicals are further implicated in the conversion of SO_2 to H_2SO_4, and of NO and NO_2 to HNO_3. Rain can dissolve the acids (which may perhaps serve as condensation nuclei) and rainout returns the sulphur- and nitrogen-containing compounds to the ground. Ammonia is very important as one of the few trace gas constituents that are bases. Although it can be dissolved and rained out of the atmosphere, it can also react with the acids to form solid aerosols of NH_4NO_3 and $(NH_4)_2SO_4$, touched on earlier (Section 1.3).

Of the gases in our atmosphere, only the 'inert' or 'noble' gases have neither a biological nor an atmospheric source. Chemical cycles are precluded because of the inertness of the noble gases. Budgets of the inert gases may therefore allow us to probe how far the lithosphere—hydrosphere—atmosphere system is truly closed. Argon is a surprisingly abundant element in the atmosphere (Table 1.1) at least so far as its isotope of atomic weight 40 is concerned. Interestingly, ^{40}Ar is *not* the natural isotope of argon (^{36}Ar with some ^{38}Ar are the natural isotopes). Radioactive decay of potassium, ^{40}K, however, does yield ^{40}Ar, and it is this radiogenic source that has provided the vast bulk of our atmospheric argon. The potassium in the Earth's crust, and deeper in the mantle, has decayed over the lifetime of the Earth to form ^{40}Ar, which has degassed from the solid. According to different estimates of the terrestrial potassium abundance, the present argon load (6.8×10^{16} kg) in the atmosphere corresponds to something between one-half and all of the ^{40}Ar generated within the Earth. No chemical sinks exist, and the oceans can dissolve only one per cent of the atmospheric argon. The mean annual input of ^{40}Ar to the atmosphere has thus been the present load divided by the lifetime of the Earth ($\sim 4.5 \times 10^9$ yr), and is 1.5×10^7 kg yr^{-1} globally. We shall use this result shortly. Helium is much less abundant than argon in the atmosphere: the ^4He isotope has a mixing ratio of 5.24 p.p.m., and ^3He is nearly a million times yet less abundant. Radioactive decay, this time of the ^{238}U, ^{235}U, and ^{232}Th series, is again the source of the noble gas. The relative atomic ratio of ^4He to ^{40}Ar in many samples of natural gas lies between 0.2 and 5, so that the order of magnitude of $^4He/^{40}Ar$ in gases entering the atmosphere is unity. Why, then, is present-day ^4He so much less abundant—by a factor of 1782—than ^{40}Ar? The answer must be that some helium has escaped from the Earth altogether: from the top of the atmosphere to interplanetary space. Possible mechanisms for such escape will be discussed in Section 2.3.2; other things being equal, a relatively light atom such as He will find it much easier to escape than a heavier one. Some idea of the time-scale of escape can be obtained by making an estimate of the atmospheric residence time. Assuming, as discussed above, that the volume (or atomic) rates of release of ^4He and ^{40}Ar are identical, then from the figures calculated earlier, the rate of release of ^4He must be $1.5 \times 10^7 \times (4/40)\,kg\,yr^{-1}$, where the (4/40) term reflects the differing atomic masses. The mass of helium in the atmosphere is $(6.8 \times 10^{16}/1782) \times (4/40)$ kg. For our estimate, we equate residence time with the atmospheric load divided by the rate of release, and obtain 2.6×10^6 yr. This lifetime is three orders of magnitude less than the Earth's age, so that most of the helium ever released from the crust and mantle has escaped. With a different emphasis, it can also be said that 99.996 per cent of the particular helium atoms in the atmosphere when a man is born are still there when he dies!

The examination of noble gas behaviour has highlighted two new features of the fluxes of elements into and out of atmospheres. First, material can be released, for example by degassing, from the solid body of the planet. Gases can be formed, as in the cases of ^4He and ^{40}Ar, radiogenically, or they may have been trapped as the planet was created. Secondly, the extreme limit of the atmosphere forms an interface for the exchange of matter with 'space'. Our example was of escape, but material can also enter the Earth system at this interface. The Sun emits a continual stream of electrically charged particles, the *solar wind*, which impinges on the outer layers of a planetary atmosphere, or on the surface of a body without an atmosphere. Then meteorites, asteroids, and even comets occasionally enter our gravitational field and burn up (*ablate*) or evaporate in the atmosphere.

Figure 1.3 summarizes the complex interactions that go to make up the biogeochemical cycles. The figure shows the many ways in which volatile materials can be delivered to the atmosphere or be removed from it. Solid–gas cycling occurs on a time-scale of hundreds of millions of years. Volcanic gases, primarily CO_2 and H_2O, are released to the atmosphere from the Earth's crust. Carbon dioxide participates in weathering reactions that result in the deposition of carbonate sediments. Sea-floor *plates* are finally *subducted* into the mantle, volatiles trapped are released at high temperature and pressure, and recycled into the atmosphere through volcanoes along the plate interface. On a shorter time-scale, the hydrologic cycle exchanges water between the atmosphere and the condensed phases on the surface. Water vapour precipitates into the oceans, and on average the CO_2 and H_2O cycles are balanced so that hydrosphere and atmosphere are maintained at roughly constant volume. Interactions with the biosphere determine the detailed composition of the atmosphere, both with regard to major components such as

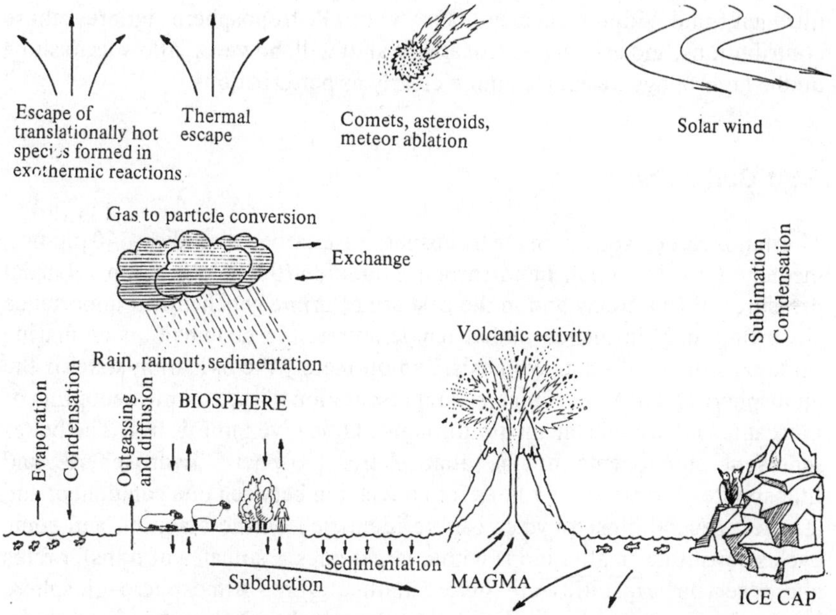

Fig. 1.3. Representation of the cyclic processes of biogeochemistry that exchange constituents between air, land, and sea. At the top of the picture, solar particles and extraplanetary objects bring matter into the atmosphere, while a certain amount escapes. Most material, however, is recycled. Even solids deposited on land and on the ocean beds can eventually be subducted to become molten, and components returned to the atmosphere through volcanic activity. Drawn by Sophie McLaughlin, with biospheric species from Carol Wayne and David Koslow.

O_2, and in respect of almost all the trace gases. The figure also shows that the cycles are not completely closed. Some material, *juvenile* in the sense that it has not hitherto been cycled, is released from the Earth's interior; the solar wind and debris left by stray bodies entering the atmosphere also contribute to the inward flux. Escape from the Earth's gravitational field to the interplanetary medium constitutes an outward flux.

Important biogeochemical cycles of the natural troposphere are considered in the sections that follow. The approach is concerned with budgets and lifetimes, but it should be emphasized that the actual numbers are representative only, and many are very uncertain. Some difficulty also arises over the concept of the 'natural' troposphere. It is certain that interactions with the biosphere largely determine atmospheric composition. Human life can potentially alter the composition *out of proportion* to human life's biological importance. We wish to reserve a fuller presentation of the topic for a discussion of tropospheric pollution (Section 5.10). Man's activities undoubtedly contribute to present-day budgets and reservoirs in carbon, nitrogen, and sulphur cycles. Our 'natural' troposphere ignores these contributions, and is to that extent artificial. It will, however, allow us to show up the anthropogenic sources more clearly as perturbations.

1.5.1 Carbon cycle

The major carbon species in the troposphere are carbon dioxide (~340 p.p.m.), methane (1.6–1.7 p.p.m.) and carbon monoxide (0.04–0.20 p.p.m.). Partial pressures of CO_2 today and in the past are of prime geochemical importance since they may influence global temperatures, the composition of marine sediments, rates of photosynthesis, and ultimately the oxidation state of the atmosphere and oceans. A pictorial representation of the complex coupling of inorganic and organic chemistry through CO_2 is given in Fig. 1.4. The boxes represent components of the atmosphere, biosphere, hydrosphere, and lithosphere. Arrows show routes for conversion between one component and another; several closed cycles can be identified on the diagram, and some cycles constitute smaller loops within larger ones. Estimates of transfer rates and reservoir capacities are given in the figure. Atmosphere–biosphere interactions completely dominate over the geochemical parts of the cycle. About 150×10^{12} kg of carbon are transferred each way each year, so that about 20 per cent of the CO_2 content of the atmosphere is converted annually. Photosynthesis by plants and micro-organisms is responsible for the intake of CO_2, while respiration and decay account for the reverse process. From the point of view of generation of organic material, the overall photosynthetic process consists of the formation of carbohydrates by the reduction of carbon dioxide,

$$n\mathrm{CO_2} + n\mathrm{H_2O} \xrightarrow{\text{light}} (\mathrm{CH_2O})_n + n\mathrm{O_2}, \qquad (1.10)$$

where $(CH_2O)_n$ is a shorthand for any carbohydrate. The essence of the process is the use of photochemical energy to split water and, hence, to reduce CO_2. Molecular oxygen is liberated in the reaction, although it appears at an earlier stage in the sequence of steps than the reduction of CO_2. Light absorption is achieved by pigment systems involving various chlorophyll-cell structures. Chlorophylls are peculiarly suited to this purpose since their optical absorption is in the visible region, just where the photochemically active part of the solar radiation is most intense at ground level (see Section 3.1). Furthermore, the structures of the chlorophylls make them particularly efficient photosensitizers. The biochemical reactions and cycles involved in the photosynthetic process are fascinating, but not directly relevant to our present interest. A multistep mechanism, with at least two absorption events, is needed

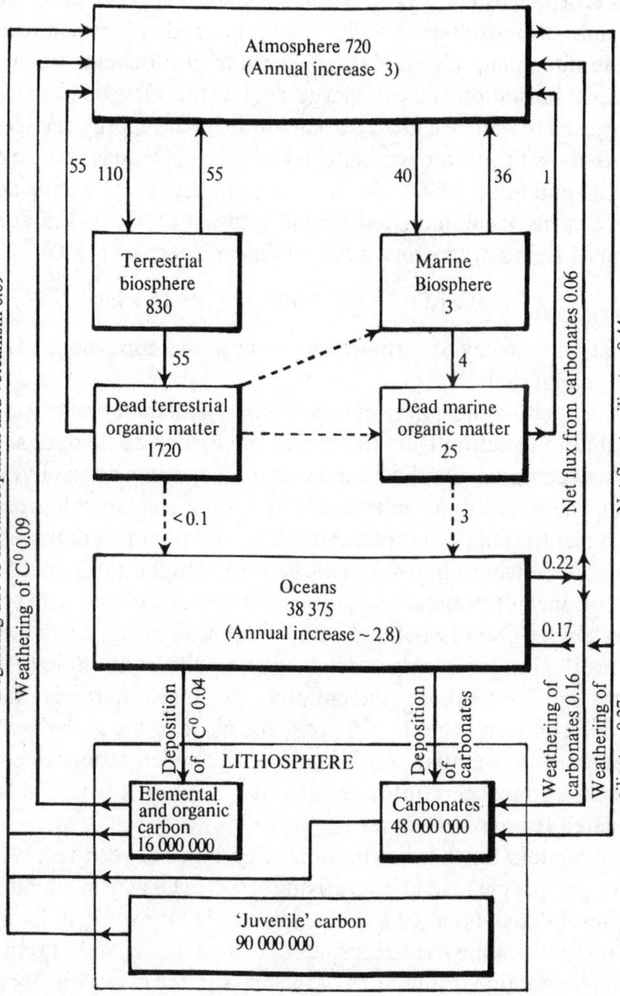

Fig. 1.4. The carbon cycle. Arrows show the transfer of elemental carbon (C^0) between the various carbon reservoirs (boxes), and indicate the closed loops that make up the cycle. Transfer rates are given in units of 10^{12} kg of C per year, and reservoir contents in 10^{12} kg of C. Burning of fossil fuels releases an additional 5×10^{12} kg yr^{-1} to yield the net increase in atmospheric burden. The figure is due originally to Holland, H.D. *The chemistry of the atmospheres and oceans*, John Wiley, Chichester, 1978, and is updated with information reviewed by Clark, W.C. (ed.) *Carbon dioxide review: 1982*, Oxford University Press, Oxford, 1982; and by Bolin, B. and Cook, R.B. (eds.) *The major biogeochemical cycles and their interactions*, SCOPE 21, John Wiley, Chichester, 1983.

to allow the relatively low energy photons (~200–300 kJ mol^{-1}) to split the O–CO (531 kJ mol^{-1}) or H–OH (498 kJ mol^{-1}) bonds. The organic chemical cycle itself is driven by energy-rich triphosphates or reduced phosphates, such as adenosine triphosphate (ATP) and reduced nicotinamide adenine dinucleotide phosphate (NADPH), so familiar in biochemistry. It is in the photochemical formation of these energy carriers that O_2 is liberated from H_2O; hydrogen is used to reduce CO_2 to carbohydrate by the energy carriers. Isotope experiments show that photosynthetically produced O_2 does come exclusively from the H_2O and not the CO_2. So far as the atmospheric cycle is concerned, this result can be accommodated by the return of CO_2 to the atmosphere. Respiration of living organisms is the reverse of reaction (1.10),

$$(CH_2O)_n + nO_2 \rightarrow nCO_2 + nH_2O + \text{heat}, \qquad (1.11)$$

while oxidation or decay of carbohydrates likewise returns both CO_2 and H_2O out of the biosphere.

The biosphere–atmosphere cycles in Fig. 1.4 are very nearly closed, and can be treated in isolation from the rest of the carbon cycle over short times. Balances between photosynthesis and respiration/decay rates obviously alter diurnally and seasonally. As an example of diurnal change, it has been shown that CO_2 concentrations in a forest can rise to 400 p.p.m. at night, and drop to 305 p.p.m. at noon when photosynthetic activity is highest. Figure 1.5(a) shows how the CO_2 concentrations at a site in Hawaii have oscillated with season over many years: high CO_2 is associated with winter and spring, when there is least photosynthesis. Similar observations made near the South Pole (Fig. 1.5(b)) show very much smaller fluctuations, because there is little local photosynthetic activity, and the CO_2 remains close to the global average. We note in the figure an overall upward trend in CO_2 concentrations over the years. This increase is almost certainly a result of the burning of fossil fuels, and will be investigated further in Chapter 9.

Simple models for the distribution of carbon between reservoirs in the closed biological cycle lead to a surprising result. The way in which the rate of photosynthesis depends on CO_2 concentration is known; it is likely that the rates of transfer from the living biosphere to dead matter and of return back to the atmosphere are linear functions of the mass in the reservoir. It can then be shown that the carbon content of each reservoir is determined only by the total carbon content of all of them. That is, the carbon content of the atmosphere is determined, and the rate of photosynthesis has no direct influence. In reality, of course, there are slow leaks to the hydrosphere and lithosphere: some dead matter (mainly marine, with a smaller terrestrial contribution) does not return to the atmosphere, and about 0.1×10^{12} kg yr^{-1} becomes buried. Losses from the rapidly cycling biosphere–atmosphere system are what determine atmospheric CO_2 pressure and the carbon content of the biosphere.

Chemical weathering both adds and removes CO_2 from the atmosphere. Oxidation of elemental carbon (C^0) and organic compounds in rocks adds CO_2;

decomposition of calcium and magnesium silicates [reaction (1.3)] and solution of carbonate minerals [reaction (1.2)] removes it. The weathering processes for C° and carbonates are included in Fig. 1.4. Weathering of silicates consumes an additional 0.27×10^{12} kg yr^{-1} of C from CO_2, while the reverse decomposition yields 0.16×10^{12} kg yr^{-1}. Thus the total budget, in 10^{12} kg yr^{-1} to the atmosphere, is made up of: C°, (0.09–0.12) = - 0.03; carbonates, (0.22–0.16) = + 0.06; silicates, (0.16–0.27) = -0.11. Carbon leaks from the atmosphere–hydrosphere–upper lithosphere system at a rate of 0.08×10^{12} kg yr^{-1}, or roughly 0.1 per cent of the turnover by photosynthesis. This rate of removal, if continued, would deplete the atmosphere in (720/0.08) ~ 9000 yr, and the oceans in (38 375/0.08) ~ 480 000 yr. No geological evidence exists for *large* fluctuations of CO_2 over such periods (cf. Fig. 9.5), and it is likely that the shortfall of 0.08×10^{12} kg yr^{-1} is made up by degassing and volcanic release from the interior of the Earth of juvenile carbon or subducted carbonates. Whatever the explanation, it is clear from the calculations that the balance of CO_2 in atmospheres and oceans is a delicate one, and that a serious alteration in the leak rates could change CO_2 concentrations quite rapidly.

The lifetime of CO_2 in the atmosphere appears to be determined by the biospheric interaction, since that is the largest contributor to the annual turnover shown in Fig. 1.4. According to the numerical values used in this discussion, the residence time would be (720/110) ~ 6.5 yr if land photosynthesis alone were involved, or (720/150) ~ 4.8 yr if the marine component is included. Direct evidence suggests that the actual lifetime is shorter. Nuclear weapon testing in the 1960s doubled the atmospheric level of ^{14}C, which is also produced naturally by cosmic radiation. Subsequent return to the earlier concentrations after the tests ended was much more rapid than can be explained by the rates of photosynthetic turnover and the sizes of the biospheric reservoirs. Figure 1.4 does not, however, show the exchange of CO_2 between atmosphere and ocean resulting from reversible solution, since the dissolving and evolution processes are in dynamic equilibrium. About 120×10^{12} kg yr^{-1} of C are transferred each way, so that with this term the true atmospheric lifetime of any CO_2 molecule becomes (720/270) ~ 2.7 yr. The oceanic reservoir of dissolved CO_2 (and HCO_3^- ions) is about 60 times bigger than the atmospheric reservoir, with the result that the partitioning of artificial ^{14}C can virtually remove it from the atmosphere in a few years.

Methane and carbon monoxide participate in cycles that are linked with each other and the main carbon cycle. Concentrations of methane are at present about 1.72 p.p.m. They are essentially independent of altitude within the troposphere but show a slight latitudinal variation: mean concentrations are about six per cent higher in the Northern Hemisphere than in the Southern Hemisphere, largely because the sources are situated on land. Primary 'natural' CH_4 production mechanisms are enteric fermentation in animals, mostly cattle, and microbiological anaerobic decomposition of organic matter in wetlands, swamps, and paddy fields. In radiocarbon (^{14}C) dating, it is assumed that living

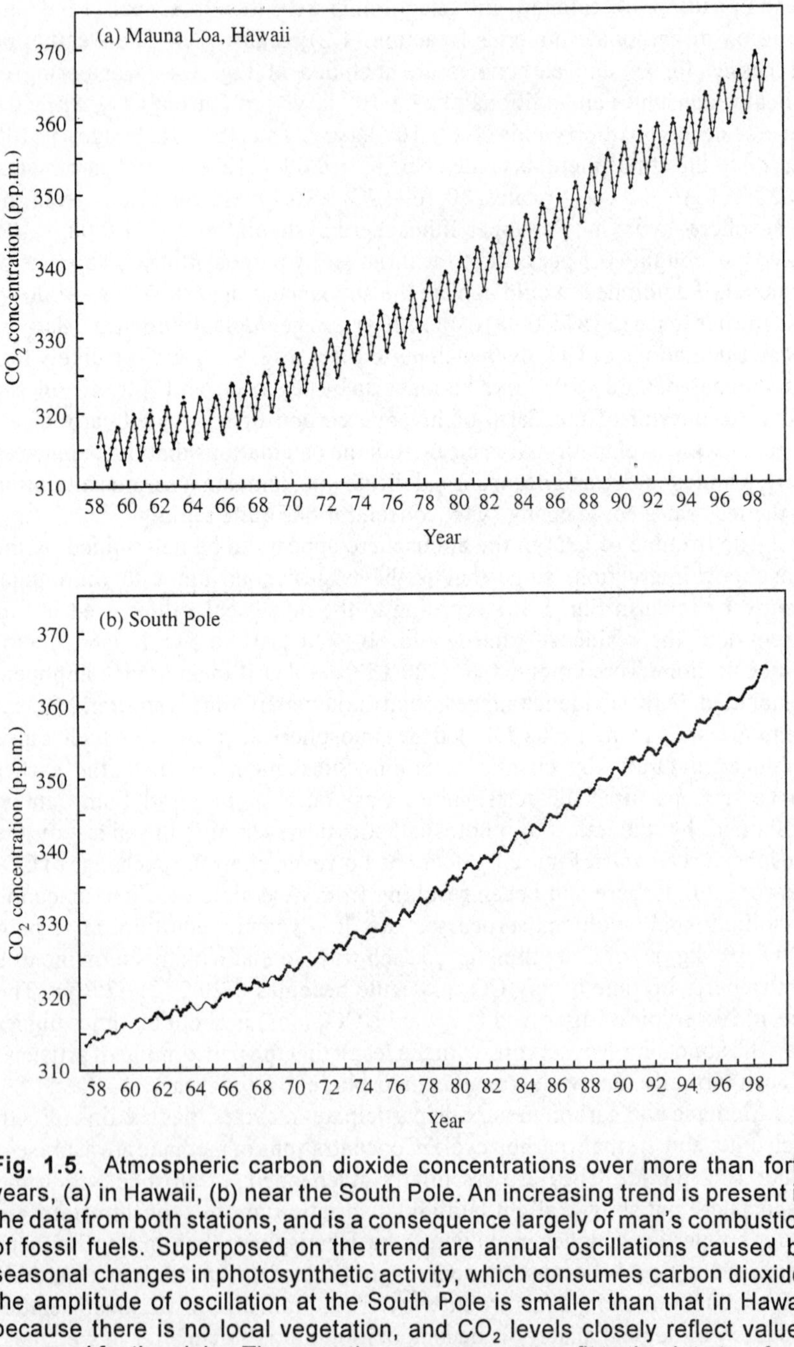

Fig. 1.5. Atmospheric carbon dioxide concentrations over more than forty years, (a) in Hawaii, (b) near the South Pole. An increasing trend is present in the data from both stations, and is a consequence largely of man's combustion of fossil fuels. Superposed on the trend are annual oscillations caused by seasonal changes in photosynthetic activity, which consumes carbon dioxide: the amplitude of oscillation at the South Pole is smaller than that in Hawaii because there is no local vegetation, and CO_2 levels closely reflect values averaged for the globe. The smooth curve represents a fit to the data to a four-harmonic annual cycle, which increases linearly with time, and a spline fit of the

organisms possess the same $^{14}C/^{12}C$ content as the atmosphere, in which the ^{14}C is continually replenished by cosmic ray bombardment of ^{14}N. Dead or fossilized material no longer incorporates new ^{14}C, so that radioactive decay reduces the ^{14}C content. The half-life of ^{14}C is only about 5600 years, so that the radiocarbon has almost completely disappeared in fossil gases, oils, and solids. The ^{14}C content of atmospheric methane is about 80 per cent that of modern wood, so that less than 20 per cent of the total CH_4 supply comes from natural gas leakage. Current estimates put the total CH_4 source strength at 0.4 to $0.7 \times 10^{12}\,kg\,yr^{-1}$. These figures imply that roughly one per cent of all photosynthetically produced organic matter decays to produce atmospheric methane. Given the atmospheric load of 4.8×10^{12} kg, the methane could be supplied in 7 to 12 years. In fact, there is some evidence that production and loss are not quite balanced, and that the concentration of CH_4 has been increasing by about 0.01 p.p.m. yr^{-1}, starting about 150 years ago. Long-term changes could alter both the chemistry and the climate of the atmosphere, and will be discussed in Chapter 9. Loss processes for CH_4 include bacterial consumption in the soil, but by far the most important loss mechanism is the inorganic oxidation chain (p. 15 and pp. 331–2) initiated by hydroxyl radicals (reaction 1.8). One net destruction cycle can be written in the simplified form

$$CH_4 + O_2 \xrightarrow[\text{light}]{\text{OH}} H_2O + CO + H_2 \qquad (1.12)$$

to show that CO may be a product. A few per cent of the methane molecules produced ($\sim 0.02 \times 10^{12}\,kg\,yr^{-1}$) cross the tropopause to enter the stratosphere. Above the tropopause, CH_4 concentrations drop quite rapidly by a factor of about four between 15 and 40 km, suggesting that the CH_4 is rapidly destroyed. Stratospheric loss is not, in itself, important for tropospheric methane budgets, but the process does represent a chemical transport of hydrogen-containing species from troposphere to stratosphere.

Oxidation of methane is a major natural source of carbon monoxide. Non-methane organic compounds such as terpenes may also be oxidized, and contribute up to 50 per cent of the atmospheric carbon monoxide. Smaller amounts of CO are emitted by plants and micro-organisms. It used to be thought that most of the atmospheric CO was anthropogenic in origin, and that the difference in concentrations for Northern (0.07–0.20 p.p.m.) and Southern

---

interannual component of the variation. The dots indicate monthly average concentrations. Source: Keeling, C.D., Bacastow, R.B., Carter, A.F., Piper, S.C., Whorf, T.P., Heimann, M., Mook, W.G., and Roeloffzen, H., Appendix A in *Aspects of climate variability in the Pacific and western Americas*, Geophysical Monograph, American Geophysical Union, **55**, 1989 (Nov), and Keeling, C.D., Whorf, T.P., Wahlen, M., and van der Plicht, J., *Nature, London*, **375**, 666 (1995). Updated with information provided by C.D. Keeling and T.P. Whorf at Scripps Institute of Oceanography (SIO) in April 1999. The results were obtained in a cooperative programme of the US National Oceanic and Atmospheric Administration (NOAA) and SIO.

(0.04–0.06 p.p.m.) Hemispheres reflected the differences in location of industrial sources. However, the asymmetry in the natural oxidation sources could also match the concentration asymmetry. According to recent estimates, the global source of CO is about 2.2×10^{12} kg yr^{-1}, of which 60 per cent is natural. For steady-state CO concentrations globally averaged at 0.12 p.p.m., the corresponding residence time is about two months; without the natural sources, the residence time would have to be up to years. Strong hemispheric asymmetries are obviously more easily maintained if lifetimes are short. Loss of CO is, as with methane, largely (~90 per cent) dependent on reaction with OH in the very fast process (1.9),

$$CO + OH \rightarrow CO_2 + H. \quad (1.9)$$

The hydrogen atom can re-form OH radicals in a number of ways (Chapter 5), so that the oxidation of CO to CO_2 may be regarded as catalysed by hydroxyl radicals. A microbiological sink for CO at the surface of the soil accounts for about 10 per cent of the total loss.

1.5.2 Oxygen cycle

Short-term oxygen fluxes are dominated by the photosynthetic cycle. Carbon dioxide transfers between atmosphere and biosphere will correspond stoicheiometrically to oxygen transfers in the opposite direction: thus photosynthesis releases, while respiration and decay consume, atmospheric oxygen. Photosynthesis accounts for the annual consumption of 150×10^{12} kg of atmospheric carbon (Fig. 1.4) and the release of $150 \times (32/12) \times 10^{12} = 400 \times 10^{12}$ kg of oxygen. The atmosphere contains 1.2×10^{18} kg of O_2, so that oxygen cycles through the biosphere in ($1.2 \times 1018/400 \times 1012$) ~ 3000 yr, a period much longer than the equivalent one of a few years for carbon because of the much greater atmospheric oxygen reservoir. Seasonal fluctuations in concentration, seen for CO_2 (pp. 20–22), are therefore damped out in the case of oxygen.

As we discussed in Section 1.5.1, there is a small leak of carbon out of the atmosphere–biosphere system which has a strong influence on atmospheric CO_2 concentrations. This same leak is also important in the geochemistry of oxygen. Marine organic sediment deposition buries ~0.12×10^{12} kg yr^{-1} of C without decay, thus releasing ~0.32×10^{12} kg yr^{-1} of O_2. That is, the leak could cause atmospheric O_2 to double in concentration in ($1.2 \times 1018/0.32 \times 10^{12}$) ~$4 \times 10^6$ yr. Because O_2 is less soluble in water than CO_2, the oceans provide far less of a stabilizing reservoir for O_2. Marked variations in the atmospheric O_2 level may have occurred over geological time. Nevertheless, in the contemporary atmosphere oxygen is consumed in oxidation of rocks. Weathering rates of elemental carbon to carbon dioxide, sulphide rocks to sulphate, and iron (II) rocks to iron (III) roughly match the leak rate. Oxidation of reduced volcanic gases (e.g. H_2 or CO) is a smaller, but not negligible, additional balancing process.

1.5.3 Nitrogen cycle

Molecular nitrogen is chemically rather inert, partly because the large N–N bond energy of 945 kJ mol^{-1} makes most reactions endothermic, or at least kinetically limited because of a large activation energy. Natural processes can *fix* nitrogen (bring it into combination). Lightning within the atmosphere can produce the higher oxides of nitrogen (NO, NO_2, etc.) which are converted to acids (HNO_2, HNO_3) and rained out. Biological fixation is, however, of even greater importance, at least over land. Independent micro-organisms can fix nitrogen into soils, but, for the planet as a whole, the greatest sources of naturally fixed nitrogen are symbiotic organisms found in the root nodules of the pulses or leguminous plants (peas, beans, etc.). Assimilation of nitrogen is catalysed by the enzyme complex *nitrogenase* that brings about the reduction of N_2 to NH_4^+ ions. As in the photosynthetic conversion of CO_2 to carbohydrate, the energy-rich phosphate ATP drives the process, several molecules probably being required. The first step can be represented by the reaction

$$N_2 + 8H^+ + 6e^- \xrightarrow{\text{ATP}} 2NH_4^+. \tag{1.13}$$

In the symbiotic micro-organisms, the ATP derives from the host plant, which receives up to 90 per cent of the fixed nitrogen in return. Other links in the soil microbiological chain involve *nitrifying* bacteria that oxidize NH_4^+ to NO_3^-, *denitrifying* bacteria that reduce NO_3^- to N_2, and *ammonifying* bacteria in which wastes and remains of animals and dead plants are reconverted to ammonia. The microbiological chains maintain the enormous disequilibrium between atmospheric N_2 and O_2 concentrations and those of NO_3^- ions in sea-water, while the greater solubility of common nitrate minerals compared with common carbonates is a partial cause of the dominance of N_2 over CO_2 in the Earth's atmosphere.

Estimates of global budgets and reservoirs are even more imprecise for N_2 than those for carbon, but they still indicate the major aspects of the nitrogen cycle. Weight ratios for C/N are about 7.9 for terrestrial plants and 5.7 for marine plants, so that the biospheric reservoirs and transfer rates in Fig. 1.4 may be converted to provide rough values for the nitrogen cycle. Figure 1.6 shows the main aspects of the cycle. In both terrestrial and marine organic systems, most of the nitrogen is recycled within the system, although a smaller fraction is transferred between biosphere and atmosphere. For the hydrosphere, some 130×10^9 kg yr^{-1} of nitrogen are transferred from the atmosphere, as against 4.0×10^{12} kg yr^{-1} cycling in the marine biosphere. Without this

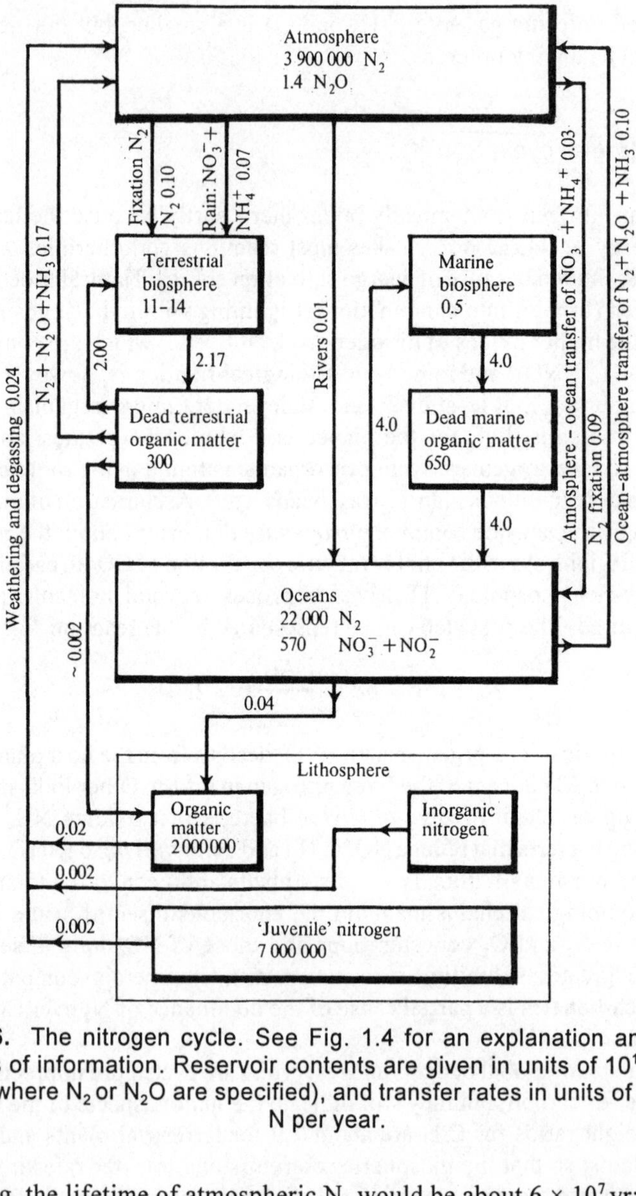

Fig. 1.6. The nitrogen cycle. See Fig. 1.4 for an explanation and for the sources of information. Reservoir contents are given in units of 10^{12} kg (of N except where N_2 or N_2O are specified), and transfer rates in units of 10^{12} kg of N per year.

recycling, the lifetime of atmospheric N_2 would be about 6×10^7 yr; with the numbers used in the figure for atmospheric transfer the calculated lifetime is nearer 1.6×10^6 yr. Again according to the figure, about 40×10^9 kg yr^{-1} of N_2 are buried as dead organic matter, mostly from the marine biosphere. Exposure, weathering, and conversion of organic and inorganic deposits, and release of juvenile nitrogen, can balance the burial rate. However, even without the restoring sources, it would take $(7.80 \times 106/0.040) \sim 2 \times 10^8$ yr to consume all atmospheric nitrogen. That is, the imbalance is small even on geologically long time-scales because of the large nitrogen content of the atmosphere.

Nitrous oxide (dinitrogen oxide, N_2O) is liberated from soils as a result of incomplete microbiological nitrification or denitrification. Indications are that the biological source is dominated by nitrification. The yield of N_2O depends in a non-linear way on oxygen concentrations, ranging from 0.03 per cent (referred to overall NH_4^+ oxidation) at high oxygen levels to more than 10 per cent at low oxygen levels. Global mixing ratios seem to have increased from 0.292 p.p.m. in 1961 to 0.311 p.p.m. by 1992. Nitrous oxide is chemically significant in the atmosphere, largely as a precursor of the higher oxides of nitrogen. Reaction (1.7) with excited oxygen from ozone photolysis produces nitrogen monoxide (nitric oxide, NO) which may be further oxidized to nitrogen dioxide (NO_2). Both NO and NO_2 are involved in much of the chemistry of the troposphere and stratosphere (Chapters 5 and 4). Transport across the tropopause provides a tropospheric supply of N_2O, and hence of NO and NO_2 in the stratosphere.

1.5.4 Sulphur cycle

The abundances, sources, sinks, budgets, and photochemistry of atmospheric sulphur compounds are poorly understood compared to those of the carbon, oxygen, and nitrogen species considered so far. Sulphur in the atmosphere can be converted to SO_2, SO_3, and H_2SO_4. It therefore acts as an aerosol precursor, may have an effect on cloud production, and may be of climatological significance. The H_2SO_4 may also lower the pH of rain-water with deleterious consequences. In the case of the atmospheric sulphur budget, it is clear that anthropogenic sources are at least comparable with the natural ones. Volcanic activity probably produces 7×10^9 kg yr^{-1} of sulphur in the form of SO_2 (but see p. 28). Decay of biospheric organic matter probably yields 58×10^9 kg yr^{-1} of sulphur over land, and 48×10^9 kg yr^{-1} over the ocean, in the reduced forms of H_2S, $(CH_3)_2S$ (dimethyl sulphide), and $(CH_3)_2S_2$ (dimethyl disulphide). Sea-spray might contribute another 44×10^9 kg yr^{-1} of sulphur compounds. Atmospheric mixing ratios for the sulphur-containing compounds are rather small. Sulphur dioxide is present at a globally averaged mixing ratio of 167 p.p.t. (parts per trillion or parts per million million) corresponding to a total atmospheric load of 0.9×10^9 kg S. Hydrogen sulphide is very variable, typical global mixing ratios lying between zero and 100 p.p.t. ($\equiv$ 0 to 0.6×10^9 kg S). Lifetimes for H_2S must be only a few days even for the highest concentrations observed, and are probably more usually a few hours. Similar remarks apply to SO_2. The mechanisms for the rapid destruction of these species are uncertain, but may involve reaction with OH, or heterogeneous oxidation steps. Two more sulphur-containing gases, carbonyl sulphide (COS) and carbon disulphide (CS_2), are present in the troposphere, and it has been suggested that these are the main precursors of SO_2 and sulphate aerosol. Carbonyl sulphide is thought to be the most abundant (500 p.p.t.) sulphur gas in the troposphere. Its distribution with latitude is uniform, consistent with a rather long lifetime of more than one year. While carbon disulphide is much less abundant, and variable (2–120 p.p.t.) because its atmospheric lifetime is only a few weeks, tropospheric oxidation appears to produce COS, so that CS_2 contributes to the total COS available. Most sulphur-containing gases emitted into the troposphere from natural or artificial sources are too reactive or too soluble to reach the stratosphere, but COS is an important exception. Apart from volcanic injection, tropospheric COS is the main source of stratospheric sulphur. The stratospheric sulphate layer, which influences temperature structure in the lower stratosphere and may have some effect on ozone concentrations (see Sections 1.3, 4.5.5 and 4.7.7), thus depends on this source of sulphur. Oceans are a global source of COS contributing about $0.15 \times 10^9\,\mathrm{kg\,yr^{-1}}$. Ash collected from the eruption of Mount St Helens (18 May 1980) gave off large amounts of COS and CS_2 at room temperature, although these gases were much less concentrated than H_2S in the gaseous part of the plume. Interestingly, SO_2 became the dominant gas only after the eruption of 15 June. Such results may be suggestive of sources, but the origins of COS and CS_2 are really unknown at present, even to the extent of whether the gases are natural or anthropogenic.

Dimethyl sulphide (DMS) is an important sulphur-carrying gas that has natural sources. Large quantities are produced by algae in the oceans. For example, in spring and summer, algae along the coasts of the North Sea produce enough DMS during April and May to make a contribution after oxidation of up to 25 per cent of the H_2SO_4 burden carried in the troposphere over some parts of Europe. The mechanism of the oxidation of DMS is discussed in Section 5.6.

1.6 Linking biosphere and atmosphere

Figure 1.3 illustrated the cyclic processes that exchange constituents between air, land, and sea. We now present in Fig.1.7 a diagram, devised by Professor Ø. Hov, that shows in more detail some major chemical systems within the atmosphere. Some of these systems have already been mentioned briefly, and they will be the main subjects of Chapters 4 and 5 of this book. The diagram indicates clearly that not only are the systems coupled, but that there are feedback loops between them. Some of the loops are direct in the sense that they concern the chemical species present and their concentrations. Others are indirect. Carbon dioxide and several other trace gases in the atmosphere 'trap' radiation in the troposphere to warm it. This is the so-called 'greenhouse effect', and we shall explore it in greater detail in Sections 2.2 and 9.6. In the stratosphere, the same gases can lead to a cooling effect. The modification of

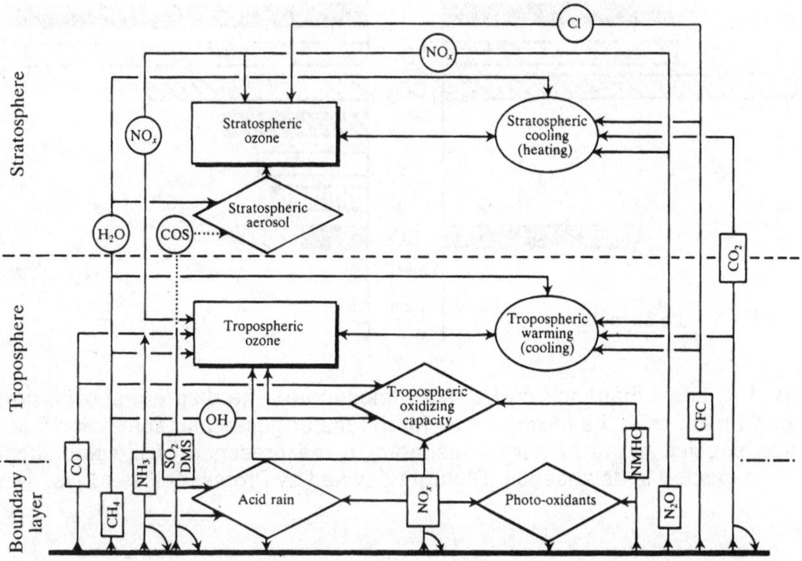

Fig. 1.7. Interactions and feedbacks between important chemical systems in the troposphere and stratosphere. The figure particularly emphasizes the factors that influence ozone concentrations in both regions as well as showing the formation of important pollutants in the boundary layer (roughly the first km of highly turbulent air next to the surface). Many of the chemical species modify atmospheric temperatures, which in turn determine the rates and relative importance of the different processes. Not shown on the diagram, but also an important feedback mechanism, is the control that the chemical compounds may exert on cloud formation and thus indirectly on the biogenic source gases. Source species are shown in the small oblong boxes, while radicals and other secondary intermediate species are shown in circles: DMS = dimethylsulphide CFC = chlorofluorocarbons, NMHC = non-methane hydrocarbons. Based on a diagram presented by Professor Ø. Hov, University of Bergen, in Varese, September 1989.

temperature in both regions can affect the rates of chemical reactions and thus the relative importance of different chemical steps. Again, the formation of aerosols can alter the amount of photochemically active light in the stratosphere and troposphere as well as influencing the thermal radiation budget and thus temperatures. If such aerosols act as cloud condensation nuclei, then the effects can be further exaggerated. Not only may the atmospheric chemical transformations be altered, but the biological sources of the trace gases themselves are likely to respond to changes in temperature and light intensity. Many other loops of this kind can be identified.

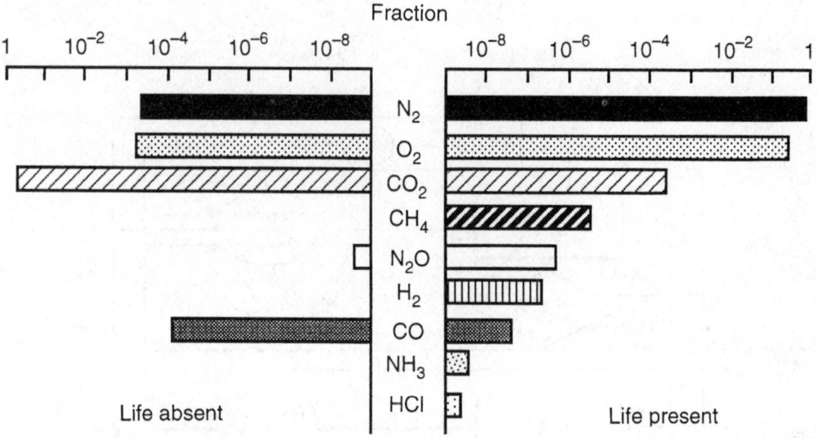

Fig. 1.8. Life's influence on Earth's atmosphere. The diagram shows, on a logarithmic scale, the mixing ratios for the major gases and some significant trace species found in our atmosphere in the presence of life and those expected in its absence. Diagram devised by Professor Peter Liss.

We have emphasized, throughout this introductory chapter, the influence that life has on the composition of the Earth's atmosphere. It is highly instructive to examine what the probable composition of the atmosphere would be if life were not present. Figure 1.8 shows the mixing ratios actually in our atmosphere, and those that might be expected if life were absent. Bearing in mind that the scale in this diagram is logarithmic, the enormous influence of the biota is immediately evident.

To end this chapter, we discuss an idea originally presented by Lovelock and Margulis in 1974 (see Bibliography), more to provoke thought than as an accepted theory. We have shown that the composition of Earth's atmosphere is displaced from equilibrium because of interactions with the biosphere. Lovelock and Margulis go further, and see the interaction between life and the atmosphere as so intense that the atmosphere can be regarded as an extension of the biosphere. The atmosphere is not living, but is a construction maintained by the biota. From this concept comes the *Gaia hypothesis* ('Gaia' = Earth Mother), which postulates that the climate and the chemical composition of the Earth's surface and atmosphere are kept at an optimum by and for the biosphere. The relationship between composition and the biosphere is seen as analogous to that between the circulatory blood system and the animal to which it belongs. In the case of the atmosphere, if highly improbable arrangements (equivalent to low entropy) extend beyond the boundaries of living entities so as to include also their planetary environment, then the environment and life taken together can be considered to constitute a single larger entity. It is the 'operating system' of life and its environment that is called Gaia. Chemistry, pressures, and temperatures are all regulated. Indeed, following the line of thought further, the expected efficiency of evolution would mean that every trace gas in the atmosphere had a purpose as a chemical information carrier. Control systems developed by humans have required feedback mechanisms and amplifiers. Some biogeochemical amplifiers can be identified. Chemical weathering of silicate rocks takes place where high partial pressures of CO_2 are maintained by biological oxidation. Rates of oxidation double for every 10 °C temperature rise. The biota act as temperature sensors and amplifiers to control the rate of CO_2 pumping from the air. Dimethyl sulphide from oceanic sources leads to substantial and amplified changes in cloud cover (see p. 29) so that biological activity in the surface waters of the oceans may be an important link in an ecosystem regulation of temperature and illumination over the Earth. Again, we have seen the non-linear and amplified response of N_2O production to changed oxygen concentrations in the microbiological nitrification process.

Naturally, the Gaia hypothesis does not receive universal acclaim! A more cautious view (Holland, 1978: see Bibliography) does not see the close links between atmosphere, oceans, and biosphere as implying the existence of an adaptive control system. Rather, the ocean–atmosphere system has adjusted to biological activity such as photosynthesis, and the biosphere has responded by optimizing the use of available free energy. The Gaian counter-argument asks why living organisms, which generally exhibit an economy of function, should produce, for example, trace gases such as N_2O or CH_4, unless there is some evolutionary advantage to the organism. The potential control of atmospheric composition and climate by these trace gases is then seen, in the Gaian context, as their regulatory role. In the engineering control-system analogy, the difference between Gaian and non-Gaian views is concerned with whether or not there are closed feedback loops. Although resolution of the different views is not, at present, possible, the concept of closed-loop control is of more than passing interest in predicting the atmosphere's response to natural or anthropogenic disturbance.

Bibliography

Books and articles concerned with atmospheric chemistry in general

Atmospheric chemistry and physics. Seinfeld, J.H. and Pandis, S.N. (John Wiley, Chichester, 1998).

Introduction to atmospheric chemistry. Hobbs, P.V. (Cambridge University Press, Cambridge, 1999).

Photochemistry of planetary atmospheres. Yung, Y.K. and DeMore, W.B. (Oxford University Press, Oxford, 1999).

Atmosphere, climate and change. Graedel, T.E. and Crutzen, P.J. (Scientific American Library, New York, 1997).

Atmospheric change: an Earth system perspective. Graedel, T.E. and Crutzen, P.J. (W.H. Freeman, New York, 1993).

Composition, chemistry, and climate of the atmosphere. Singh, H.B. (ed.) (Van Nostrand, New York, 1995).

Principles of atmospheric physics and chemistry. Goody, R.M. (Oxford University Press, Oxford, 1995).

Progress and Problems in Atmospheric Chemistry. Barker, J.R. (ed.) (World Scientific Publishing, Singapore, 1995).

Earth under siege: from air pollution to global change. Turco, R.P. (Oxford University Press, Oxford, 1997).

Properties and composition of the terrestrial oceans and of the atmospheres of Earth and other planets. Fegley, B., Jr. in *AGU handbook of physical constants*, Ahrens, T. (ed.) (AGU, Washington, DC, 1995).

Low temperature chemistry of the atmosphere. Moortgat, G.K., Barnes, A.J., Le Bras, G., and Sodeau, J.R. (eds.) (Springer-Verlag, Berlin, 1994).

Atmospheric chemistry. Wayne, R.P. *Sci. Prog.* **74**, 379 (1990).

Chemistry of the natural atmosphere. Warneck, P. (Academic Press, San Diego, 1988).

Chemistry of the atmosphere. McEwan, M. J. and Phillips, L. F. (Edward Arnold, London, 1975).

The chemistry of the atmosphere and oceans. Holland, H. D. (John Wiley, Chichester, 1978).

Chemistry of the lower atmosphere. Rasool, S. I. (ed) (Plenum Press, New York, 1973).

Atmospheric chemistry. Heicklen, J. (Academic Press, New York, 1976).

Energy and the atmosphere: a physical-chemical approach. Campbell, I. M., 2nd edn. (John Wiley, London, 1986).

The photochemistry of atmospheres. Levine, J. S. (ed.) (Academic Press, New York, 1984).

The planets and their atmospheres. Lewis, J. S. and Prinn, R. G. (Academic Press, Orlando, 1984.

Air: composition and chemistry. Brimblecombe, P., 2nd edn. (Cambridge University Press, Cambridge, 1996).

Chemical compounds in the atmosphere. Graedel, T. E. (Academic Press, New York, 1978).

These two books are specifically concerned with air pollution, but also contain much about atmospheric chemistry in general, especially in the earlier chapters.

Atmospheric chemistry. Finlayson-Pitts, B. J. and Pitts, J. N. Jr., (John Wiley, Chichester, 1986).

Atmospheric chemistry and physics of air pollution. Seinfeld, J. L. (John Wiley, Chichester, 1986).

The next book also gives a good general introduction to atmospheric chemistry, but is more specifically directed to the question of whether the biosphere and atmosphere control each other's behaviour and composition in a more than casual way (see Section 1.6).

Gaia: a new look at life on earth. Lovelock, J. E. (Oxford University Press, Oxford, 1979).

This issue of Scientific American *is devoted to a discussion of the lithosphere, hydrosphere, atmosphere, and biosphere, and the interactions between them.*

The dynamic earth. *Scient. Am.* **249**, No. 3, pp. 30–144 (Sept. 1983).

Section 1.1

The mass of the Earth's atmosphere

The total mass of the atmosphere. Trenberth, K.E. and Guillemot, C.J. *J. geophys. Res.* **99**, 23079 (1994).

Overviews of the solar system

Physics and chemistry of the solar system. Lewis, J.S., revised edn. (Academic Press, San Diego, 1997).

The solar system. Jones, B. W. (Pergamon Press, Oxford, 1984).

The chemistry of the solar system. Lewis, J. S. *Scient. Am.* **230**, 50 (March 1974).

Atmospheres and evolution. Margulis, L. and Lovelock, J. E., in *Life in the universe* Billingham, J. (ed.) pp. 79–100. (M.I.T. Press, Cambridge, Mass., 1981).

Atmospheres of the terrestrial planets. Jakosky, B.M., in *The new solar system.* Beatty, J. K., Petersen, C.C., and Chaikin, A. (eds.) 4th edn. (Cambridge University Press, Cambridge, 1999).

The atmospheres of Venus, Earth, and Mars: a critical comparison. Prinn, R. G. and Fegley, B., Jr. *Ann. Rev. Planet. Space Sci.* **15**, 171 (1987).

The atmospheres of the outer planets and satellites. Trafton, L. *Rev. Geophys. & Space Phys.* **19**, 43 (1981).

The atmospheres of the planets. Mason, B. J. *Observatory* **97**, 217 (1977).

Atmospheric composition: influence of biology. McElroy, M. B. *Planet. Space Sci.* **31**, 1065 (1983).

Section 1.2

The atmosphere and ocean: a physical introduction. Wells, N. (Taylor and Francis, London, 1986).

Air-sea exchange of gases and particles. Liss, P. S. and Slinn, W. G. N. (eds.) (D. Reidel Co., Dordrecht, 1983).

The role of the ocean in the global atmospheric cycle. Nguyen, B. C., Bonsang, B., and Gaudry, A., *J. geophys. Res.* **88**, 10903 (1983).

Role of oceans in atmospheric chemistry (Conference, Hamburg, August 1982). *J. geophys. Res.* **87**, 8769 (1982).

The dynamics of the coupled atmosphere and ocean. Charnock, H. and Philander, S. G. H. (eds.) (The Royal Society, London, 1990).

The biogeochemistry of the air-sea interface. Lion, L. W. and Leckie, J. O. *Annu. Rev Earth & Planet. Sci.* **9**, 449 (1981).

The carbonate–silicate geochemical cycle and its effect on atmospheric carbon dioxide over the past 100 million years. Berner, R. A., Lasaga, A. C., and Garrels, R. M. *Am. J. Sci.* **283**, 641 (1983).

Feedbacks between weathering and atmospheric CO_2 over the last 100 million years. Volks, T. *Am. J. Sci.* **287**, 763 (1987).

Coordination chemistry of weathering: kinetics of surface-controlled dissolution of oxide minerals. Stumm, W. and Wollast, R. *Rev. Geophys.* **28**, 53 (1990).

Section 1.3

Clouds, rain and aerosols. Warneck, P. in *Low temperature chemistry of the atmosphere.* Moortgat, G.K., Barnes, A.J., Le Bras, G., and Sodeau, J.R. (eds.) (Springer-Verlag, Berlin, 1994).

Atmospheric aerosols: biogeochemical sources and role in atmospheric chemistry. Andreae, M.O. and Crutzen, P.J. *Science* **221**, 744 (1997).

A two-dimensional model of sulfur species and aerosols. Weisenstein, D.K., Yue, G.K., Ko, M.K.W., Sze, N.-D., Rodriguez, J.M., and Scott, C.J. *J. geophys. Res.* **102**, 13019 (1997).

The atmospheric aerosol system: an overview. Prospero, J. M., Charlson, R. J., Mohnen, V., Jaenicke, R., Delany, A. C., Moyers, J., Zoller, W., and Rahn, K. *Rev Geophys. & Space Phys.* **21**, 1607 (1983).

Aerosols and atmospheric chemistry. Hidy, G. M. (ed.) (Academic Press, New York, 1972).

Elemental constituents of atmospheric particulates and particle density. Sugimae, A. *Nature, Lond.* **307**, 145 (1984).

Natural organic atmospheric aerosols of terrestrial origin. Zenchelsky, S. and Youssefi, M. *Rev. Geophys. & Space Phys.* **17**, 459 (1979).

Heterogeneous interactions of the C, N, and S cycles in the atmosphere: the role of aerosols and clouds. Taylor, G. S., Baker, M. B., and Charlson, R. J., SCOPE **21**, 115 (1983).

Kinetic studies of raindrop chemistry. 1, Inorganic and organic processes. Graedel, T. E. and Goldberg, K. I. *J. geophys. Res.* **88**, 10865 (1983).

Chemistry with aqueous atmospheric aerosols and raindrops. Graedel. T. E. and Weschler, C. J. *Rev. Geophys. & Space Phys.* **19**, 505 (1981).

Review: Atmospheric deposition and plant assimilation of gases and particles. Husker, R. P., Jr., and Lindberg, S. E. *Atmos. Environ.* **16**, 889 (1982).

Volcanoes are one source of atmospheric gases and aerosols. A fascinating finding of the Voyager mission to Jupiter was the existence of intense volcanic activity on the moon Io. Major volcanic activity on Venus is strongly suspected. Recent major eruptions on Earth have greatly increased the aerosol burden of the atmosphere, at least temporarily, apparently with meteorological and chemical consequences.

The Mount Pinatubo eruption: effects on the atmosphere and climate. Fiocco, G., Fu, D. and Visconti, G. (Springer-Verlag, Berlin 1996).

Global evolution of the Mt Pinatubo volcanic aerosols observed by the infrared limb-sounding instruments CLAES and ISAMS on the Upper Atmosphere Research Satellite. Lambert, A., Grainger, R.G., Rodgers, C.D., Taylor, F.W., Mergenthaler, J.L., Kumer, J.B., and Massie, S.T. *J. geophys. Res.* **102**, 1495 (1997).

Comparison of Mount Pinatubo and El Chichón volcanic events: Lidar observations at 10.6 and 0.69 μm. Post, M.J., Grund, C.J., Weickmann, A.M., Healy, K.R., and Willis, R.J. *J. geophys. Res.* **101**, 3929 (1996).

Sulfur emissions to the stratosphere from explosive volcanic eruptions. Pyle, D.M., Beattie, P.D., and Bluth, G.J.S. *Bull. Volcan.* **57**, 663 (1996).

Atmospheric effects of the Mt. Pinatubo eruption. McCormick, M.P., Thomason, L.W., and Trepte, C.R. *Nature* **373** 399 (1995).

Radiatively forced dispersion of the Mt. Pinatubo volcanic cloud and induced temperature perturbations in the stratosphere during the first few months following the eruption. Young, R.E., Howben, H., and Toon, O.B. *Geophys. Res. Letts.* **21**, 369 (1994).

Comparisons of stratospheric warming following Agung, El Chichón and Pinatubo volcanic eruptions. Angell, J.K. *Geophys Res. Lett.* **20**, 715 (1993).

The El Chichón volcanic cloud: An introduction. Pollack, J.B., Toon, O.B., and Danielsen, E.F. *Geophys. Res. Letts.* **19**, 805 (1983).

A comparison of volcanic eruption processes on Earth, Moon, Mars, Io and Venus. Wilson, L. and Head, J. W., III. *Nature, Lond.* **302**, 663 (1983).

Measurements of the evolution of the Mt. Pinatubo aerosol cloud by ISAMS. Lambert, A., Grainger, R.G., Remedios, J.J., Rodgers, C.D., Corney, M. and Taylor, F.W. *Geophys. Res. Lett.* **20**, 1287 (1993).

The impact of Mount Pinatubo on world-wide temperatures. Parker, D.E., Wilson, H., Jones, P.D., Christy, J.R., and Folland, C.K. *Int. J. Climatol.* **16**, 487 (1996).

Changes in CH_4 and CO growth rates after the eruption of Mt. Pinatubo and their link with changes in tropical tropospheric UV flux. Dlugokencky, E.J., Dutton, E.G., Novelli, P.C., Tans, P.P., Masarie, K.A., Lantz, K.O., and Madronich, S. *Geophys. Res. Lett.* **23**, 2761 (1996).

Sulfur emissions to the stratosphere from explosive volcanic eruptions. Pyle, D.M., Beattie, P.D., and Bluth, G.J.S. *Bull. Volcan.* **57**, 663 (1996).

The contribution of explosive volcanism to global atmospheric sulfur dioxide concentrations. Bluth, G.J.S., Schnetzler, C.C., Krueger, A.J., and Walter, L.S. *Nature, Lond.* **366**, 327 (1993).

Stratospheric loading of sulfur from explosive volcanic eruptions. Bluth, G.J.S., Rose, W.I., Sprod, I.E., and Krueger, A.J. *J. Geol.* **105**, 671 (1997).

Infrared heating rates in the stratosphere due to volcanic sulfur dioxide. Zhong, W.Y., Haigh, J.D., Toumi, R. and Bekki, S. *Q. J. Roy. Meteor. Soc.* **122**, 1459 (1996).

Ionospheric disturbances observed during the period of Mount Pinatubo eruptions in June 1991. Chen, K. and Huang, Y.N. *J. geophys. Res.* **97**, 16995 (1992).

Changes in the character of polar stratospheric clouds over Antarctica in 1992 due to the Pinatubo volcanic aerosol. Deshler, T., Johnson, B.J., and Rozier, W.R. *Geophys. Res. Lett.* **21**, 273 (1994).

Mount Pinatubo aerosols, chlorofluorocarbons, and ozone depletion. Brasseur, G. and Granier, C. *Science* **257**, 1239 (1992).

An analysis of the effects of Mount Pinatubo aerosols on atmospheric radiances. Nair, P.R. and Moorthy, K.K. *Int. J. Remote Sensing* **19**, 697 (1998).

Observed episodic warming at 86 and 100 km between 1990 and 1997: effects of Mount Pinatubo eruption. She, C.Y., Thiel, S.W., and Krueger, D.A. *Geophys. Res. Lett.* **25**, 497 (1998).

Response of summertime odd nitrogen and ozone at 17 mbar to Mount Pinatubo aerosol over the southern midlatitudes: observations from the halogen occultation experiment. Mickley, L.J., Abbatt, J.P.D., Frederick, J.E., and Russell, J.M. *J. geophys. Res.* **102**, 23573 (1997).

Estimated impact of Agung, El Chichón and Pinatubo volcanic eruptions on global and regional total ozone after adjustment for the Quasi-Biennial Oscillation (QBO). Angell, J.K. *Geophys. Res. Lett.* **24**, 647 (1997).

Ozone and temperature-changes in the stratosphere following the eruption of Mount Pinatubo. Randel, W.J., Wu, F., Russell, J.M., Waters, J.W., Froidevaux, L. *J. geophys. Res.* **100**, (16753).

The atmospheric effects of El Chichón. Rampino, M. R. and Self, S. *Scient. Am.* **250**, 34 (Jan. 1984).

El Chichón: composition of plume gases and particles. Kotra, J. P., Finnegan, D. L., Zoller, W. H., Hart, M. A., and Moyers, J. L. *Science* **222**, 1018 (1983).
El Chichón volcanic aerosols: impact of radiative, thermal, and chemical perturbations. Michelangeli, D. V., Allen, M., and Yung, Y. L. *J. geophys. Res.* **94**, 18429 (1989).
An assessment of the impact of volcanic eruptions on the Northern Hemisphere's aerosol burden during the last decade. Michalsky, J. J., Pearson, E. W., and LeBaron, B. A. *J. geophys. Res.* **95**, 5677 (1990).
Volcanic winters. Rampino, M. R., Self, S., and Stothers, R. B. *Ann. Rev. Earth planet. Sci.* **16**, 73 (1988).
Climatic effects of the eruption of El Chichón. (Many papers collected in part of a special issue.) *Geophys. Res. Lett.* **10**, 989–1060 (1983).
Increases in the stratospheric background sulfuric acid aerosol mass in the past ten years. Hofmann, D.J. *Science* **248**, 996 (1990).
COS in the stratosphere: El Chichón observations. Leifer, R., Juzdau, Z. R., and Larsen, R. *Geophys. Res. Lett.* **11**, 549 (1984).

Section 1.4

Ozone in the stratosphere is the subject of Chapter 4, and tropospheric ozone is discussed in detail in Chapter 5. References at the end of those chapters supplement the introductory articles listed here.

The chemistry of the stratosphere. Thrush, B. A. *Rev. Prog. Phys.* **51**, 1341 (1988).
Stratospheric ozone: an introduction to its study. Nicolet, M. Rev. *Geophys. & Space Phys.* **13**, 593 (1975).
Photochemical reactions initiated by and influencing ozone in unpolluted tropospheric air. Crutzen, P. J. *Tellus* **26**, 47 (1974).

Section 1.5

Cyclic processes are of paramount importance in the exchange of chemical constituents between the troposphere and the planetary surface. Several of the citations for this section are to SCOPE, which is an acronym for Scientific Committee on Problems of the Environment. An example is

Evolution of the global biogeochemical sulphur cycle. Brimblecombe, P. and Lein, A. Yu. SCOPE 39 (John Wiley, Chichester, 1989).

Earlier SCOPE *publications directly relevant to this section are:*

Nitrogen, phosphorus, and sulphur: global cycles, SCOPE 7 (1975).
The global carbon cycle, SCOPE 13 (1979).
Carbon cycle modelling, SCOPE 16 (1981).
Some perspectives of the major biogeochemical cycles, SCOPE 17 (1981).
The global biogeochemical sulphur cycle, SCOPE 19 (1983).

The major biogeochemical cycles and their interactions, SCOPE 21 (1983).

The next seven references provide a general introduction to biogeochemical cycles, and they are followed by two references that show the way in which atmospheric residence times of gases are estimated.

Global Environment: water, air, and geochemical cycles. Berner, E.K. and Berner, R.A. (Prentice-Hall, Upper Saddle River, NJ, 1996).

Global biogeochemical cycles. Butcher, S.S., Charlson, R.J., Orians, G.H., and Wolfe, G.V. (eds.) (Academic Press, San Diego, 1992).

The natural and polluted troposphere. Stewart, R. W., Hameed, S., and Pinto, J., in *Man's impact on the troposphere* Levine, J. S. and Schryver, D. R. (eds.) (NASA Reference Publication, No. 1022, 1978).

Interactions of biogeochemical cycles. Bolin, B., Crutzen, P. J., Vitousek, P. M., Woodmansee, R. G., Goldberg, E. G., and Cook, R. B. SCOPE **21**, 1 (1983).

Atmospheric interactions—Homogeneous gas reactions of C, N, and S containing compounds. Crutzen, P. J. SCOPE **21**, 67 (1983).

The global troposphere: biogeochemical cycles, chemistry and remote sensing. Levine, J. S. and Allario, F. *Environ. Monitg. & Assessm.* **1**, 263 (1982).

Influence of the biosphere on the atmosphere. (Symposium covering most gases and cycles). Dutsch, H. U. (ed.) *Pure & appl. Geophys.* **116**. 452 (1978).

Residence time and variability of tropospheric trace gases. Junge, C. E. *Tellus* **26**, 477 (1974).

Residence time and spatial variability for gases in the atmosphere. Hamrud, M. *Tellus* **35B**, 295 (1983).

There is considerable evidence that concentrations of some trace gases may be increasing. There follow references to books and articles that address this issue specifically and that also contain much information about atmospheric concentrations and lifetimes of the compounds.

Global atmospheric chemical change. Hewitt, C.N and Sturges, W.T. (eds.) (Chapman and Hall, London, 1994).

The changing atmosphere. Rowland, F. S. and Isaksen, I.S.A. (eds.) (John Wiley, Chichester, 1988).

Distribution and trends of carbon monoxide in the lower troposphere. Novelli, P.C., Maserie, K.A., and Lang, P.M. *J. geophys. Res.* **103**, 19015 (1998).

Possible causes for the 1990-1993 decrease in the global tropospheric CO abundances: a three-dimensional sensitivity study. Granier, C., Muller, J.F., Madronich, S., and Brasseur, G.P. *Atmos. Env.* **30**, 1673 (1996).

Increase in the atmospheric nitrous oxide concentration during the last 250 years. Machida, T., Nakazawa, T., Fujii, Y., Aoki, S., and Watanabe O. *Geophys. Res. Lett.* **22**, 2921 (1995).

Recent changes in atmospheric carbon monoxide. Novelli, P.C., Masarie, K.A., Tans, P.P., and Lang, P.M. *Science* **263**, 1587 (1994).
The growth rate and distribution of atmospheric CH_4. Dlugokencky, E.J., Steele, L.P., Lang, P.M. and Mesarie, K.A. *J. geophys. Res.* **99**, 17021 (1994).
Trace gas trends are their possible role in climate change. Ramanathan, V., Cicerone, R. J., Singh, H. B., and Kiehl, J. T. *J. geophys. Res.* **90**, 5547 (1985).
The changing atmosphere. Graedel, T. J. and Crutzen, P. J. *Scient. Am.* **261**, 28 (Sept 1989).

References follow to detailed discussions of individual cycles or important aspects of them.

Carbon

The global carbon cycle. Martin, H. (ed.) (Springer-Verlag, Berlin, 1993).
The carbon cycle. Bolin, B., SCOPE **21**, 41 (1983).
The carbon cycle. Bolin, B. *Scient. Am.* **223**, 124 (Sept. 1970).
Changes of land biota and their importance for the carbon cycle. Bolin, B. *Science* **196**, 613 (1977).
Increased activity of northern vegetation inferred from atmospheric CO_2 measurements. Keeling, C.D., Chin, J.F.S., and Whorf, T.P. *Nature, Lond.* **382**, 146 (1996).
Global and hemispheric CO_2 sinks deduced from changes in atmospheric O_2 concentration. Keeling, R.F., Piper, S.C., and Heimann, M. *Nature, Lond.* **381**, 218 (1996).
Natural sources of atmospheric CO. McConnell, J. C., McElroy, M. B., and Wofsy, S. C. *Nature, Lond.* **233**, 187 (1971).
The cycle of atmospheric CO. Seiler, W. *Tellus* **26**, 116 (1974).
The global methane cycle. Wahlen, M. *Ann. Rev. Earth Planet. Sci.* **21**, 407 (1993).
Atmospheric methane: sources, sinks, and role in global change. Khalil, M.A.K. (ed.) (Springer-Verlag, 1993).
An inverse modeling approach to investigate the global atmospheric methane cycle. Hein, R., Crutzen, P.J., and Heimann, M. *Global Biogeochem. Cycles* **11**, 43 (1997).
Three-dimensional model synthesis of the global methane cycle. Fung, I., John, J., Lerner, J., Matthews, E., Prather, M., Steele, L.P., and Frazer, P.J. *J. geophys. Res.* **99**, 13033 (1991).
Atmospheric cycle of methane. Ehhalt, D. H. *Tellus* **26**, 58 (1974).
Elemental carbon in the atmosphere: cycle and lifetime. Ogren, J. A. and Charlson. R. J. *Tellus* **35B**, 241 (1983).
Continuing worldwide increase in tropospheric methane, 1978 to 1987. Blake, D. R. and Rowland, F. S. *Science* **239**, 1129 (1988).
Volatile organic compounds in the atmosphere. Hester, R.E. and Harrison, R.M. (eds.) (Royal Society of Chemistry, Cambridge, 1995).

Microbiology of atmospheric trace gases: sources, sinks and global change processes. Murrell, J.C. and Kelly, D.P. (eds.) (Springer-Verlag, Berlin, 1996).

Trace gas emissions by plants. Sharkey, T.J., Holland, E.A., and Mooney, H.A. (eds.) (Academic Press, San Diego, 1991).

Measurements of atmospheric hydrocarbons and biogenic emission fluxes in the Amazon boundary layer. Zimmerman, P. R., Greenberg, J. P., and Westberg, C. E. *J. geophys. Res.* **93**, 1407 (1988).

Oxygen

The first article explains the elements of the photosynthetic process.

How plants make oxygen. Govindjee and Coleman, W. J. *Sci. Amer.* **262** (2), 42 (Feb 1990).

The atmospheric oxygen cycle — the oxygen isotopes of atmospheric CO_2 and O_2 and the O_2/N_2 ratio. Keeling, R.F., *Rev. Geophys.* **33**, 1253 (1995).

The oxygen cycle. Walker, J. C. G. *The handbook of environmental chemistry*, **1A**, 87 (Springer-Verlag, Berlin, 1980).

Nitrogen

A three-dimensional model of the global ammonia cycle. Dentener, F. J. and Crutzen, P. J. *J. atmos. Chem.* 19, 331(1994).

Global gridded inventories of anthropogenic emissions of sulfur and nitrogen. Benkovitz, C.M., Scholtz, T.M., Pacyna, J., Rarrason, L., Dignon, J., Voldner, E.C., Spiro, P.A., Logan, J.A., and Graedel, T.E. *J. geophys. Res.* **101**, 29239 (1996).

Evidence for an additional source of atmospheric N_2O. McElroy, M.B. and Jones, D.B.A. *Global Biochem. Cycles* **10**, 651 (1996).

The nitrogen cycle. Rosswall, T., SCOPE **21**, 46 (1983).

The nitrogen cycle. Delwiche, C. C. *Scient. Am.* **223**, 136 (Sept. 1970).

Uncertainties in the global source distribution of nitrous oxide. Bouwman, A.F., van der Hoeck, K.W. and Oliver, J.G.J. *J. geophys. Res.* **100**, 2785 (1995).

Increase in the atmospheric nitrous oxide concentration during the last 250 years. *Geophys. Res. Lett.* **22**, 2921 (1995).

Analysis of sources and sinks of atmospheric nitrous oxide (N_2O). Cicerone, R. J. *J. geophys. Res.* **94**, 18265 (1989).

Production of nitrogen oxides by lightning discharges. Tuck, A. F. *Q. J. R. Meteorol. Soc.* **102**, 749 (1976).

Nitrogen fixation by lightning. Dawson, G. A. *J. atmos. Sci.* **37**, 174 (1980).

Nitrous oxide production by lightning. Hill, R. D., Rinker, R. G., and Coucouvinos, A. *J. geophys. Res.* **89**, 1411 (1984).

Sulphur

Sulfur in the atmosphere. Berresheim, H., Wine, P.H., and Davis, D.D., in *Composition, chemistry, and climate of the atmosphere* Singh, H.B. (ed.) (Van Nostrand, New York, 1995).

A geophysiologist's thoughts on the natural sulphur cycle. Lovelock, J. *Phil. Trans. Roy. Soc. Lond.* **B352**, 143 (1997).

The sulfur cycle. Charlson, R.J., Anderson, T.L., and McDuff, R.E., in *Global biogeochemical cycles.* Butcher, S.S., Charlson, R.J., Orians, G.H. and Wolfe, G.V. (eds.) (Academic Press, San Diego, 1992).

The atmospheric sulfur cycle in ECHAM-4 and its impact on the shortwave radiation. Feichter, J., Lohmann, U., and Schult, I. *Climate Dynamics* **13**, 235 (1997).

Global gridded inventories of anthropogenic emissions of sulfur and nitrogen. Benkovitz, C.M., Scholtz, T.M., Pacyna, J., Rarrason, L., Dignon, J., Voldner, E.C., Spiro, P.A., Logan, J.A., and Graedel, T.E. *J. geophys. Res.* **101**, 29239 (1996).

Global sources and sinks of OCS and CS_2 and their distributions. Chin, M. and Davis, D.D. *Global Biogeochem. Cycles* **7**, 321 (1993).

A global three-dimensional model of the tropospheric sulfur cycle. Langner, J. and Rodhe, H. *J. atmos. Chem.* **13**, 225-263 (1991).

Global sources, lifetimes and mass balances of carbonyl sulfide (OCS) and carbon disulfide (CS_2) in the Earth's atmosphere. Khalil, M.A.K. and Rasmussen, R.A. *Atmos. Env.* **18**, 1805 (1984).

Ocean–atmosphere interactions in the global biogeochemical sulfur cycle. Andreae, M.O. *Marine Chem.* **30**, 1 (1990).

The sulphur cycle. Freney, J. R., Ivanov, M. V., and Rohde, H. SCOPE **21**, 56 (1983).

The sulfur cycle. Kellogg, W. W., Cadle, R. D., Allen, E. R., Lazrus, A. L., and Martell, E. A. *Science* **175**, 587 (1972).

The global sulfur cycle. Friend, J. P., in *Chemistry of the lower atmosphere* Rasool, S. I. (ed.) (Plenum Press, New York, 1973).

The homogeneous chemistry of atmospheric sulfur. Graedel, T. E. *Rev. Geophys. & Space Phys.* **15**, 421 (1977).

Human influence on the sulphur cycle. Brimblecombe, P., Hammer, C., Rohde, H., Ryaboshapko, A., and Boutron, C. F. SCOPE **39**, 77 (1989).

Photochemistry of COS, CS_2, CH_3, SCH_3, and H_2S: implications for the atmospheric sulfur cycle. Sze, N. D. and Ko, M. K. W. *Atmos. Environ.* **14**, 1223 (1980).

Marine biological controls on climate via the carbon and sulphur geochemical cycles. Watson, A.J. and Liss, P.S. *Phil. Trans. Roy. Soc. Lond,* **B353**, 1365 (1998).

Marine sulphur emissions. Liss, P.S., Hatton, A.D., Malin, G., Nightingale, P.D., and Turner, S.M. *Phil. Trans. Royal Soc.* **B352**, 159 (1997).

Dimethyl sulfide in the surface ocean and the marine atmosphere: a global view. Andreae, M. O. and Raemdonck, H. *Science* **211**, 744 (1983).

Are global cloud albedo and climate controlled by marine phytoplankton? Schwarz, S. E. *Nature* **336**, 441 (1988).

Atmospheric dimethyl sulphide and the natural sulphur cycle. Lovelock, J. E., Maggs, R. J., and Rasmussen, R. A. *Nature* **237**, 452 (1972).
Oceanic phytoplankton, atmospheric sulfur, cloud albedo, and climate. Charlson, R. J., Lovelock, J. E., Andreae M. O., and Warren, S. J. *Nature* **326**, 655 (1987).

Phosphorus

The phosphorus cycle. Richey, J. E., SCOPE **21**, 51 (1983).

Section 1.6

On the possibility that the influence on atmospheric composition of the biota may be part of a 'deliberate' control mechanism (see also book by Lovelock above).

Atmospheric homeostasis by and for the biosphere: the Gaia hypothesis. Lovelock, J. E. and Margulis, L. *Tellus* **26**, 2 (1974).
Biogeochemistry: an analysis of global change. Schlesinger, W.H., 2nd edn. (Academic Press, San Diego, 1997).
The molecular biology of Gaia. Williams, G.R. (Columbia University Press, New York, 1996).
Direct effects of CO_2 and temperature on silicate weathering — possible implications for climate control. Brady, P.V. and Carroll, S.A. *Geochim. Cosmochim. Acta* **58,** 1853 (1994).
Biogeochemistry — its origins and development. Gorham, E. *Biogeochem.* **13,** 199 (1991).
Oceanic phytoplankton, atmospheric sulfur, cloud albedo and climate. Charlson, R.J., Lovelock, J.E., Andreae, M.O., and Warren, S.G. *Nature, Lond.* **326,** 655 (1987).
Biocontrolled thermostasis involving the sulfur cycle. Rodhe, H. *Climatic Change* **8,** 91 (1986).
Bio-controlled thermostasis involving the sulfur cycle. Shaw, G.E. *Climatic Change* **5,** 297 (1983).
Biological homeostasis of the global environment: the parable of Daisyworld. Watson, A. J. and Lovelock, J. E. *Tellus* **35B**, 284 (1983).
Chaos in daisyworld. Zeng, X., Pielke, R.A., and Eykholt, R. *Tellus* **42B**, 309 (1990).
The world as a living organism. Lovelock, J. E. *New Scientist* **112**, 25 (1986).
Geophysiology: the science of Gaia. Lovelock, J. E. *Rev. Geophys.* **27**, 215 (1989).

Lovelock's newer book counters some of the arguments that have been advanced against the concept of Gaia, and explores some new ideas. The article by Kirchner is representative of those that claim the Gaia hypothesis to be ill-defined, untestable, and potentially misleading. Other entries are to books and papers discussing one aspect or another of Gaia.

The ages of Gaia: A biography of our living Earth. Lovelock, J. E. (Oxford University Press, Oxford, 1988).

The Gaia hypothesis: can it be tested? Kirchner, J. W. *Rev. Geophys.* **27**, 223 (1989).

Scientists on Gaia. Schneider, S.H. and Boston, P.J. (eds.) (MIT Press, Cambridge, MA, 1992).

Gaia's Body: toward a physiology of Earth. Volk, T. (Springer-Verlag, New York, 1998).

Gaia in action: science of the living Earth. Bunyard, P. (ed.) (Floris, Edinburgh, 1996).

Gaia and natural selection. Lenton, T.M. *Nature, Lond.* **394,** 439 (1998).

Self-organization of the Earth's biosphere — geochemical or geophysiological. Schwartzman, D.W., Shore, S.N., Volk, T., and McMenamin, M. *Origins Life Evol. Biosphere* **24,** 435 (1994).

Open systems living in a closed biosphere — a new paradox for the Gaia debate. Barlow, C. and Volk, T. *Biosystems* **23,** 371 (1990).

2 Atmospheric behaviour as interpreted by physics

Chemical and physical processes in atmospheres are closely interdependent. Temperatures, for example, may depend on chemical composition, while the chemical processes occurring, and their rates, may depend on temperature. It is obvious that the distinction between atmospheric chemistry and atmospheric physics is artificial. Historically, however, the atmospheric sciences have emphasized the disciplines of meteorology and climatology, the first dealing with the time-dependent behaviour of atmospheric phenomena, the second with properties averaged over the long term. Both disciplines have been treated as extensions of applied physics. Chemical change is considered within this framework mainly for the way in which it influences physical behaviour. For example, the heating of the atmosphere (Section 2.2) is strongly dependent on the presence of trace polyatomic molecules such as O_3 (Section 1.4) or CO_2 (Section 1.5.1); the abundances of such species is determined almost wholly (O_3) or partially (CO_2) by chemistry. In turn, this heating drives the meteorology and physics of the atmosphere. The approach in this book is generally somewhat different, since the primary emphasis is on the chemistry of atmospheres. We ask how the physical structure of the atmosphere affects chemical processes and composition, at the same time as recognizing that the chemistry itself plays a part in determining weather, climate, and physical behaviour in general. The aim of this chapter is to explain those aspects of the physics of atmospheres that have a bearing on the chemistry. Familiar concepts of physics and physical chemistry are applied to interpret pressures, absorption of radiation, temperatures, mixing, and so on. Winds, the circulation, and cloud formation may redistribute matter and energy and they are treated briefly.

2.1 Pressures

The mean atmospheric pressure at a planet's surface (p_0) is the total atmospheric force (mass, M_A, multiplied by the acceleration due to gravity at the surface, g_0) divided by the surface area, or

$$p_0 = \frac{M_A g_0}{4\pi R_p^2} \tag{2.1}$$

for a smooth planet of radius R_p. Table 2.1 gives some physical data for those bodies with atmospheres. Pressures have now been expressed in the SI-related units of Pa (Pascal: $1\,\mathrm{Pa} = 1\,\mathrm{N\,m^{-2}}$) rather than in multiples of Earth Atmospheres (Table 1.1). One Standard Atmosphere on Earth is defined as 101 325 Pa. We may as well admit here that atmospheric scientists in general, and meteorologists in particular, have not yet adopted the units of Pa or $\mathrm{N\,m^{-2}}$.

Table 2.1 Physical data for bodies with atmospheres

Body	Planetary radius R_p (km)	Surface pressure $10^{-2}p_0$ (Pa) $\equiv p_0$ (hPa) $\equiv p_0$ (mbar)[b]	Albedo	Effective temperature T_e(K)	Surface temperature T_s (K)	Surface acceleration due to gravity g_0(m s^{-2})	Escape velocity (km s^{-1})
Venus	6052	92100	0.77	227	735	8.60	10.3
Earth	6378	1013.25	0.31	256	288	9.78	11.2
Mars	3393	6.3	0.15	217	223	3.72	5.0
Jupiter	71490[a]	[1000[a]]	0.33	110	165[a]	22.88	59.5
Saturn	60270[a]	[1000[a]]	0.36	80	134[a]	9.05	35.6
Uranus	25500	[1000[a]]	~0.4	56	76[a]	7.77	21.2
Neptune	24765	[1000[a]]	~0.4	44	72[a]	11.00	23.6
Titan	2560	1500	0.2	85	95	1.25	2.1

[a]At the 1000 mbar level.

[b]100 Pa (Pascal) ≡ 1 hPa (hectoPascal) ≡ 1 mbar.

Compiled from Beatty, J.K., Petersen, C.S., and Chaikin, A. (eds.) *The new solar system*, 4th edn, Cambridge University Press, Cambridge, 1999; Kondratyev, K.Y. and Hunt, G.E. *Weather and climate on planets*, Pergamon Press, Oxford, 1982; Stone, E.C. and Miner, E.D. *Science* **212**, 159 (1981); *Science* **215**, 499 (1982); Trafton, L. *Rev. Geophys. & Space Phys.* **19**, 43 (1981); Henbest, N. *The planets*, Penguin Books, London, 1992, revised 1994.

Instead, the millibar (mbar) is in almost universal use. The bar, and hence millibar, are exactly defined in terms of the Pascal (1 mbar = 10^2 Pa), but they are not part of the SI. One millibar is thus equal to one hectoPascal (hPa), a unit that is finding increasing favour. For conformity with existing practice, we shall use the millibar throughout most of this book. We note from the definition that 1 atm is very roughly equivalent to 1 bar, or more exactly to 1013.25 mbar.

Equation (2.1) can be rearranged in order to calculate M_A from the planetary pressure and radius. For Earth, the figures in Table 2.1 give $M_A = 5.3 \times 10^{18}$ kg; the calculation does not, of course, allow for variability of pressure, of g, or of surface elevation.

Gaseous components in the atmosphere are compressible. Unlike liquids, they therefore do not settle down on the planetary surface under the influence of gravitational attraction because the translational kinetic energy of the particles competes with the sedimentation forces. As a result of the competition, the density of gas falls with increasing altitude in the atmosphere. The vertical pressure profile may readily be predicted by considering the change in overhead atmospheric force, dF, for a change in atmospheric altitude, dz

$$dF = -g\rho A\,dz, \tag{2.2}$$

in a column of gas whose density is ρ and area A. Hence

$$dp = -g\rho dz. \tag{2.3}$$

For an ideal gas at temperature T

$$\rho = Mp/RT = mp/\mathbf{k}T, \tag{2.4}$$

where M, m are the relative molar mass and molecular mass, respectively, and R, $\mathbf{k}$ are the gas and Boltzmann's constants. Substitution of (2.4) in (2.3) yields

$$\frac{dp}{p} = -\frac{dz}{(\mathbf{k}T/mg)} \tag{2.5}$$

Integration then gives the result

$$p = p_0 \exp\left\{-\int_0^z \frac{dz}{(\mathbf{k}T/mg)}\right\} \tag{2.6}$$

if p_0 is the surface pressure. Equation (2.6) is known as the *hydrostatic equation*. For a planetary atmosphere, the acceleration due to gravity, g, is nearly constant at $\sim g_0$, since the atmospheric thickness is much less than the planetary radius. In a hypothetical atmosphere of constant temperature, eqn (2.6) then reduces to

$$p = p_0 \exp(-mgz/\mathbf{k}T). \tag{2.7}$$

Numbers of particles are proportional to pressures for a fixed temperature, so

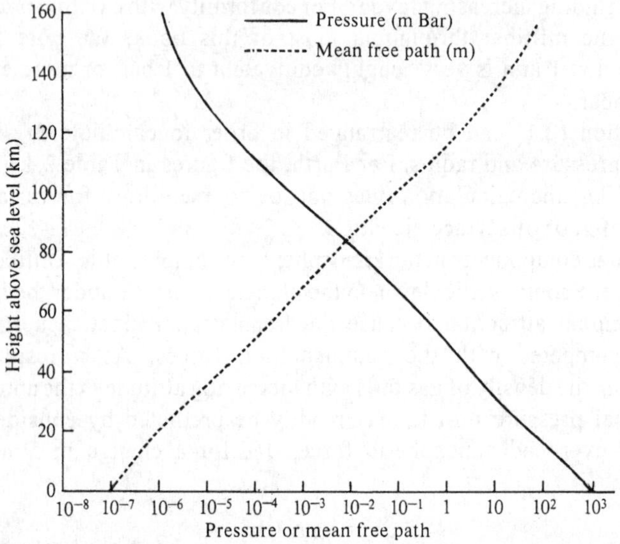

Fig. 2.1. Pressure and mean free path as a function of altitude in the Earth's atmosphere. Redrawn from Wallace, J.M. and Hobbs, P.V. *Atmospheric science*, Academic Press, New York, 1977.

that this form is equivalent to the Boltzmann distribution, with mgz corresponding to (geopotential) energy. The quantity ($\mathbf{k}T/mg$) has the units of length, and represents a characteristic distance over which the pressure drops by a factor l/e. It is given the symbol H_s, and is called the *scale height*.

Temperature is not independent of altitude in a real atmosphere, so that the scale height is not constant. For the Earth's lower atmosphere, the scale height, H_s, varies between 6 km at T ~210 K to 8.5 km at T ~290 K. The corresponding values near the surface of other planets are Venus: 14.9 km; Mars: 10.6 km; Jupiter (cloud tops): 25.3 km. Figure 2.1 shows pressure as a function of altitude for a 'standard' Earth atmosphere (one which represents the horizontal and time-averaged structure of the atmosphere). Since the pressure axis is logarithmic, the simplified expression (2.7) would predict a straight line, and the observed deviations reflect temperature variations in the real atmosphere.

According to the development so far, the scale height depends on the molecular or molar mass. At first sight, then, it would appear that each component of a planetary atmosphere would have its own scale height, and the pressure distribution would be specific to that species. In that case, mixing ratios even of unreactive gases would be a function of altitude. Yet, at least in the Earth's lower atmosphere, observation shows the gas composition, in the absence of sources and sinks, to be constant (water is an exception, because it can condense). In fact, the lower atmosphere behaves as though it is composed of a single species of relative molar mass ~$0.2 \times 32 + 0.8 \times 28 = 28.8$. Similarly, in the lower atmospheres of the other planets, all components behave as though they had a single relative molar mass determined by the relative compositional abundances: 44 for Venus and Mars, 2.2 for Jupiter. The homogeneity of lower atmospheres is a consequence of mixing due to fluid motions. Mixing on a macro-scale, by convection, turbulence, or small eddies, does not discriminate according to molecular mass. It thus redistributes chemical species that gravitational attraction is trying to separate on a molecular scale by diffusion. The relative importance of the molecular and bulk motions depends on the relative distances moved between transport events in each case. For molecular motions, that distance is clearly the *mean free path*, λ_m, the average distance a particle travels between collisions. The equivalent quantity for bulk fluid motions is the *mixing length*. A simple expression for mean free path in terms of pressure, p, and molecular diameter, d, can be derived from the kinetic theory of gases

$$\lambda_m = (1/\sqrt{2}\pi d^2)(\mathbf{k}T/p) \tag{2.8}$$

The dotted line in Fig. 2.1 shows mean free paths calculated for the Earth's atmosphere, and emphasizes the inverse pressure dependence suggested by eqn (2.8). Fluid mixing lengths tend to decrease with increasing altitude, so that there is an altitude for which molecular diffusion and bulk mixing are of comparable importance. In the Earth's atmosphere, the mean free path and mixing lengths are both of the order of 0.1–1 m at z = 100–120 km. At higher altitudes, where the mean free path becomes larger than the mixing length, we may expect to see mass discrimination. Observations agree with these predictions. Above 100 km, molecular nitrogen begins to exceed its ground level mixing ratio (see p. 1). Atomic fragments are favoured gravitationally (as well as chemically—see later) at high altitudes. Finally, at the highest altitudes, only the lightest species (H, H_2, He) are present.

The region of transition in an atmosphere between turbulent mixing and molecular diffusion is known as the *turbopause* (or, sometimes, *homopause*). Luminous rocket trails often reveal the turbopause as a region below which the trail is violently disturbed, but above which it is not. Atmospheric structure is sometimes described in terms of composition, the well-mixed region below the turbopause being called the *homosphere*, and the gravitationally separated region being called the *heterosphere*. More often, temperature structure is used to define atmospheric regions, as discussed briefly in Chapter 1, and we turn in the next sections to an examination of atmospheric temperature profiles.

2.2 Radiative heating

2.2.1 Solar and planetary radiation

Almost all the energy balance for the inner planets (Venus, Earth, Mars) is determined by solar heating; Jupiter and Saturn have an internal heat source as well. Radiation that reaches the planets as heat, visible light, and near ultraviolet is emitted from the Sun's *photosphere*, which behaves nearly as a black body of temperature ~5785 K in these spectral regions (although there is much greater emission in the X-ray, far ultraviolet, and radio spectral regions than a black body would allow). The total amount of energy of all wavelengths intercepted in unit time by unit surface area at the top of the Earth's atmosphere, corrected to the Earth's mean distance from the Sun, is known as the *solar constant*. The degree of variability of this 'constant', which can now be investigated by satellites in orbit outside the atmosphere, is of great potential importance in assessing climatic changes. At present, the solar flux through a surface normal to the beam is approximately 1368 $\mathrm{W\,m^{-2}}$ near the Earth. Some of the radiation is reflected by the surface and by the atmosphere; the overall reflectivity, A, of a planet is called the *albedo*, and is shown in Table 2.1. The fraction absorbed is thus $(1-A)$. A planet also radiates thermal energy itself, and if the overall temperature of the planet is to remain constant, then the inward and outward fluxes must be the same. Let us make the simplifying assumption that the atmosphere does not itself absorb any radiation. An estimate of the effective or equilibrium temperature, T_e, of the planet can then be obtained by assuming that the planet is a black-body radiator which obeys the Stefan–Boltzmann 'fourth power' law

$$E = sT_e^4, \tag{2.9}$$

where E is the radiated flux (per unit area) and s is the Stefan–Boltzmann constant ($5.67 \times 10^{-8}\ \mathrm{W\,m^{-2}K^{-4}}$). For a planet of radius R_p, the total emitting area is $4\pi R_p^2$, but the absorbing disc presented normal to the solar beam has an area πR_p^2, so that energy balance demands that

$$4\pi R_p^2 s T_e^4 = \pi R_p^2 (1 - A) F_s, \tag{2.10}$$

where F_s is the solar flux at the edge of the planet's atmosphere. That is,

$$T_e = \left\{ \frac{(1 - A)F_s}{4s} \right\}^{1/4}. \tag{2.11}$$

The albedo of the Earth is ~0.3; using the values of F_s and s already quoted leads to a value of $T_e = 256$ K. Values of F_s for other planets can be estimated from that for Earth by using the inverse-square law in conjunction with the planetary distances from the Sun. Effective temperatures calculated in this way from eqn (2.11) are listed in Table 2.1. At first sight, the agreement between

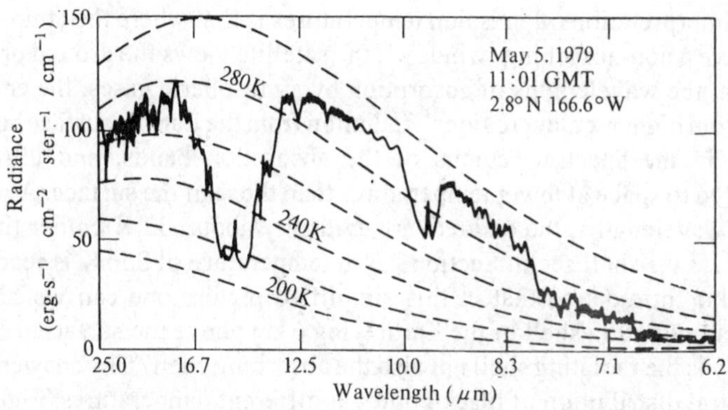

Fig. 2.2. Spectrum of infrared emission escaping to space, as observed from outside the Earth's atmosphere by the Nimbus 4 satellite. Dashed lines represent the spectrum expected from a black body at different temperatures. Atmospheric absorbers cause the escaping radiation to come from different altitudes at different wavelengths, so that the effective temperature of the Earth's spectrum is wavelength dependent. Figure reproduced from Dickinson, R.E., in Clark, W.C. (ed.) *Carbon dioxide review: 1982*, Oxford University Press, Oxford, 1982.

the calculated effective temperatures, T_e, and the surface temperatures, T_s, might seem reasonable for Earth and Mars. However, it should be remembered that a change of 32 K at the Earth's surface, up or down, would make the planet largely uninhabitable. In any case, it is obvious that the calculation is grossly in error for Venus, where surface temperatures are 500 K higher than the effective temperatures.

2.2.2 Radiation trapping: the 'greenhouse effect'

Our calculation of T_e deliberately excluded any absorption of solar or planetary radiation by atmospheric gases, and it seems reasonable to look first at atmospheric absorption as a way of reconciling effective and surface temperatures. This idea is supported by the measured average infrared emission temperatures of the planets as seen from *outside* their atmospheres (230, 250, 220 K for Venus, Earth, and Mars), which are quite close to the calculated values of T_e. A more detailed look at the Earth's planetary emission gives a further indication of what is happening. Figure 2.2 shows a low-resolution spectrum obtained from a satellite in a cloud-free field of view, the dashed lines indicating the expected black-body radiance at different temperatures. Over some of the spectral region, the temperature corresponds to near-surface temperatures. However, there is a huge emission-temperature dip between 12 and 17 μm and smaller dips at 9.6μm and at less than 8 μm. The three spectral regions correspond to infrared active bands of CO_2, O_3, and H_2O, respectively, and the interpretation of emission temperatures is that where the atmospheric gases have a non-absorbing 'window', the satellite views the ground or layers near it; at the wavelengths of absorption by atmospheric gases, the emission comes from higher, colder regions. Radiation from the Earth's surface has been *trapped* in the spectral regions of the absorption bands, and ultimately re-radiated to space at lower temperatures than those of the surface. Averaged over all wavelengths, the emitters are evidently about 32 K colder than the surface. As we shall see in Section 2.3, a temperature of 256 K is reached at about 6 km altitude, so that in this simplified picture one can visualize an equivalent radiating shell of the Earth lying 6 km above the surface.

Why is the radiating shell not also the absorbing shell? The answer lies in the spectral distribution of black bodies at different temperatures. Figure 2.3 shows the Planck black-body curves for 5780 K (Sun's photospheric temperature) and 256 K (T_e for Earth). The bulk of the Sun's radiation lies at wavelengths where the atmospheric gases absorb only weakly. In contrast, throughout most of the wavelength region emitted by the low temperature Earth, the atmosphere is opaque. The idea is that the atmosphere lets shorter wavelength radiation in, but does not let the longer wavelength radiation out. By supposed analogy with the behaviour of panes of glass, the effect is often called the *greenhouse effect*. (In reality, greenhouses are effective almost entirely because they inhibit convection rather than because they trap radiation.) Absorption and emission for a particular transition occur at the same wavelengths, of course, so that one molecule can absorb radiation emitted by another molecule of the same chemical compound. Radiation is therefore passed back and forth between the infrared active molecules, and escapes to space only from layers in the atmosphere high enough for the absorption to

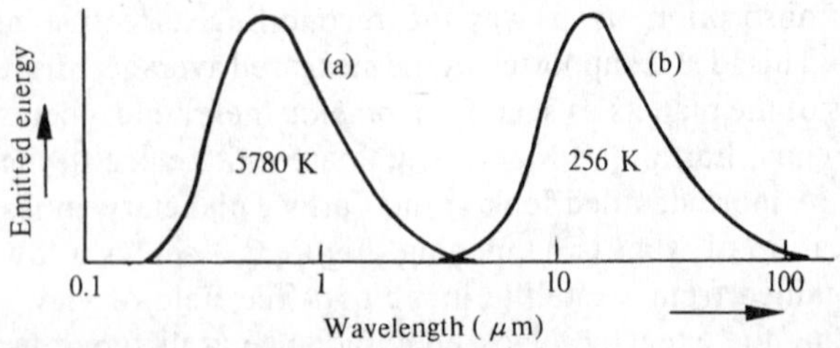

Fig. 2.3. Curves of black-body emission plotted as a function of wavelength for (a) a temperature approximating to that of the Sun; and (b) the mean temperature of the Earth's atmosphere. The scales are relative, the absolute emission energy being much larger for curve (a) than for curve (b).

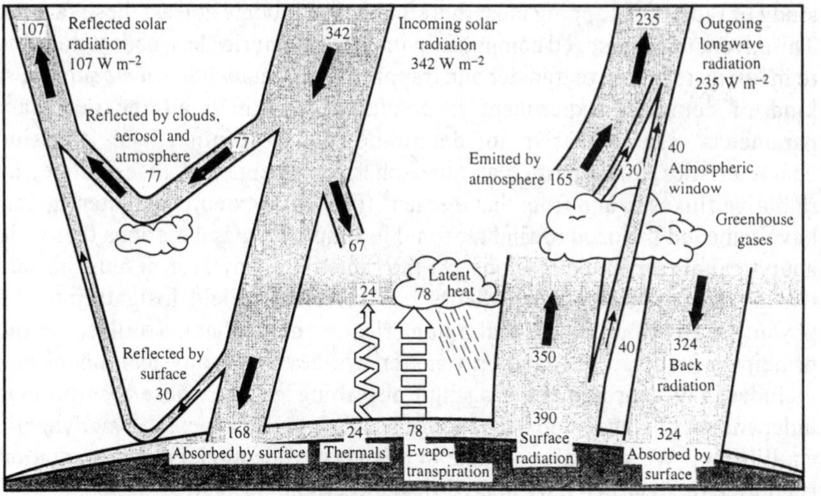

Fig. 2.4. Heat balance of the Earth. Incoming and outgoing radiation (globally averaged) is shown for each major contributor as power per unit area (W m^{-2}). From Houghton, J.T. *et al.* (eds.) *Climate change 1995*, Cambridge University Press, Cambridge, 1996.

have become weak. The upper layers of the atmosphere suffer a net loss of energy that is relatively more than that for the lower ones, and they are therefore cooler.

The total heat budget of a planet with an atmosphere that absorbs and emits is shown pictorially in Fig. 2.4. The numerical values apply to Earth. Of 342 units (W m^{-2}) of radiation reaching Earth, 30 units are reflected by the surface, and a further 77 units are reflected to space by the cloud cover and the atmosphere. The average albedo of the Earth is thus a result of the (30 + 77) = 107 units reflected by clouds, atmosphere, and surface, and is numerically equal to 107/342 = 0.31 as reported in Table 2.1. Non-radiative heat losses from surface to atmosphere are due to convection of warm air (*sensible heat*) and condensation of moist air (*latent heat*).

2.2.3 Models of radiation trapping and transfer

The discussion of the last section has indicated that several gases can lead to radiation trapping in a planetary atmosphere. Each of these species is likely to be present with a different mixing ratio and possibly different altitude distribution. Each will absorb or emit with an intensity, and with spectral features (vibrational, rotational, isotopic, etc.), corresponding to its identity. Spectral lines have widths determined by collisional (pressure-dependent) and

Doppler (temperature-dependent) effects. Light scattering by molecules, and especially by particles, alters the optical behaviour of the atmosphere. A proper study of radiation trapping must therefore accommodate at least these factors. The advent of high-speed computers with large memories has made it possible to investigate radiative transfer and trapping by direct *numerical modelling*. A kind of computer experiment is conducted in which all the necessary parameters can be used to determine the absorption and emission characteristics for a particular atmospheric element, and to calculate the radiative fluxes to and from that element. To a large extent, computer models have removed the need to find reasonable simplifying assumptions that yield analytical algebraic expressions. At the same time, however, it must be said that an elegantly simplified theory can sometimes yield insights into the physical basis of a process that a numerical model can not. To illustrate the principles involved in radiation transfer, we develop a physical model that includes in its approximations a single absorbing species whose absorption is independent of wavelength, absence of scattering, and a local thermodynamic equilibrium. We shall also assume in this very crude model that all radiation is emitted or absorbed only in a vertical direction.

Absorption of radiation is governed by the interaction between photons and matter, so that if radiation of intensity I traverses thin slabs of absorber of unit area and thickness dz, the decrease in intensity is given by

$$-\,\mathrm{d}I = I n \sigma_{\mathrm{a}} \mathrm{d}z, \tag{2.12}$$

where n is the number density of absorbers, and σ_{a} a constant for the absorbing species and the wavelength of radiation. The constant σ_{a} has the units of area, and is called the *absorption cross-section*. Integration of eqn (2.12) for incident intensity I_0, and intensity, I_{t}, transmitted through a slab of thickness z yields

$$I_{\mathrm{t}} = I_0 \exp\left\{-\int_0^z n\sigma_{\mathrm{a}}\,\mathrm{d}z\right\}. \tag{2.13}$$

In the case where n is independent of z, then eqn (2.13) simplifies to

$$I_{\mathrm{t}} = I_0 \exp(-n\sigma_{\mathrm{a}} z), \tag{2.14}$$

the familiar *Beer–Lambert* expression.[a]

Any species that absorbs radiation also emits, and *Kirchhoff's Law* states that the emissivity and absorptivity will be identical. For short-wavelength (e.g. visible or ultraviolet) spectroscopy carried out in the laboratory, the emission intensity is almost always dominated by the absorption. In reality, however, eqn (2.12) should be modified to read

[a] In practice, the Beer–Lambert law does not apply for real measurements on 'saturated' lines in the infrared region except at the very highest resolutions. This shortcoming emphasizes one aspect of the crudeness of the present approach to obtaining realistic estimates of the extent of trapping.

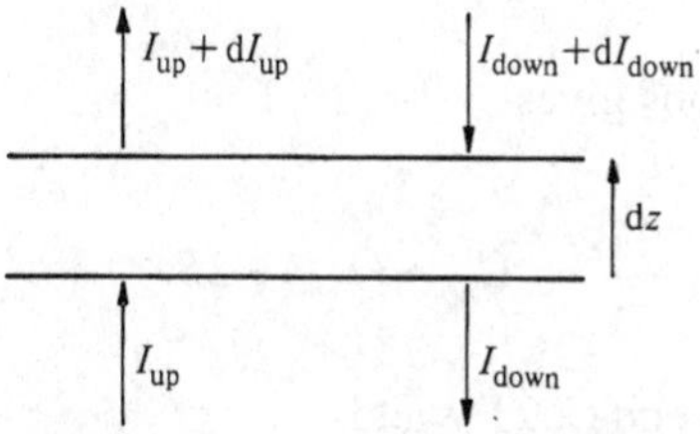

Fig. 2.5. Diagram to illustrate the upward and downward radiation fluxes discussed in the treatment of radiation transfer.

$$-\,dI = In\sigma_a dz - Bn\sigma_a dz, \tag{2.15}$$

where B is the black-body emission (per unit solid angle per unit surface area) and is a function of temperature. The modified equation can be used straightforwardly in one simplified picture of atmospheric radiative transfer. Let us first write the product $n\sigma_a z = \chi$, the *optical depth*. Reference to eqn (2.14) shows that, in a homogeneous medium, an optical depth of unity corresponds to an intensity attenuation by a factor 1/e. With χ measured from the top of the atmosphere downwards, eqn (2.15) can therefore be expressed in the form

$$\frac{dI}{d\chi} = I - B. \tag{2.16}$$

Figure 2.5 represents a small thickness of the atmosphere through which the upward and downward radiation fluxes are represented by I_{up} and I_{down}. The net flux upwards through the layer, ϕ_r, is given by

$$\phi_r = I_{up} - I_{down}. \tag{2.17}$$

There is no net change in energy, since the layer does not heat up or cool down with time, so that $dI_{up}/dz = dI_{down}/dz$. Thus, $(dI_{up} - dI_{down})/dz = d\phi_r/dz = 0$, and ϕ_r is a constant independent of z (or χ). Equation (2.16) can be written explicitly for the upward and downward fluxes

$$\frac{dI_{up}}{d\chi} = I_{up} - B, \tag{2.18}$$

and

$$-\frac{dI_{down}}{d\chi} = I_{down} - B. \tag{2.19}$$

Subtracting eqn (2.19) from (2.18) gives

$$\frac{d(I_{up} + I_{down})}{d\chi} = I_{up} - I_{down} = \phi_r, \tag{2.20}$$

and adding the equations gives

$$\frac{d\phi_r}{d\chi} = (I_{up} + I_{down}) - 2B. \tag{2.21}$$

Since ϕ_r is a constant, eqn (2.21) yields

$$I_{up} + I_{down} = 2B \tag{2.22}$$

Substitution into eqn (2.20) and integration then gives

$$B = \tfrac{1}{2}\phi_r\chi + \text{constant}. \tag{2.23}$$

At the 'top' of the atmosphere, where $\chi = 0$, $I_{down} = 0$ and equations (2.17) and (2.22) combine to yield $I_{up} = 2B = \phi_r$ for this situation. Substitution of $\chi = 0$ and $B = \tfrac{1}{2}\phi_r$ into equation (2.23) shows that the constant is also $\tfrac{1}{2}\phi_r$ so that

$$B = \tfrac{1}{2}\phi_r(\chi + 1). \tag{2.24}$$

The result allows us to calculate the black-body function B for any optical depth in the atmosphere, and, since B is proportional to the black-body emission rate in eqn (2.9), the temperature can also be calculated as a function of χ (and, from a knowledge of n and σ_a, of z). In particular, if the temperature of the emitting edge of the atmosphere is T_e, that of a near-surface layer is T_0, and the total optical thickness of the atmosphere is χ_0, then

$$\frac{sT_0^4}{sT_e^4} = \frac{\tfrac{1}{2}\phi_r(\chi_0 + 1)}{\tfrac{1}{2}\phi_r} \tag{2.25}$$

or

$$T_0^4 = T_e^4(\chi_0 + 1). \tag{2.26}$$

Figure 2.6 shows how eqns (2.24) and (2.26) might work in the Earth's atmosphere. Since $T_0 \sim 288$ K and $T_e \sim 256$ K, χ_0 is ~ 0.6 on our model. Assuming in this very crude model that $\chi = \chi_0 \exp(-z/H_s)$, we can then generate the altitude profile of temperature predicted by the radiative equilibrium model. In comparison with the real atmosphere, the simple model predicts too steep a fall in temperature in the lowest few kilometres of the atmosphere, although it subsequently matches quite well for the next few kilometres. Convection in the lowest region reduces the *lapse rate* (rate of decrease of temperature with altitude), as we shall see in Section 2.3.

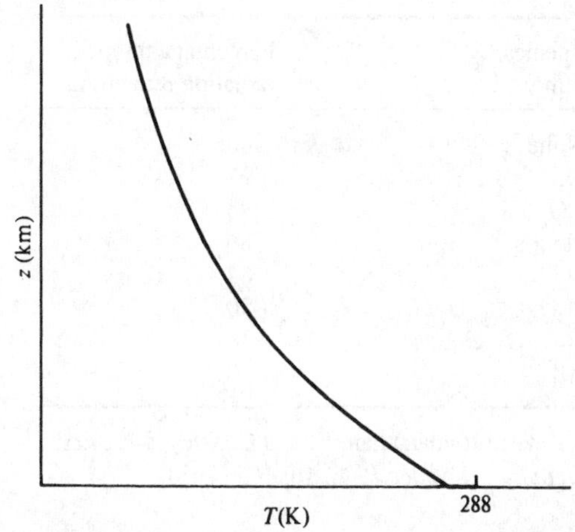

Fig. 2.6. Radiative equilibrium temperature *T* as a function of altitude, z. There is a temperature discontinuity at the lower boundary. The altitude scale is arbitrary at this stage: units will appear in Fig. 2.10 (p. 65).

2.2.4 Trapping in real atmospheres

A real atmosphere may contain several species with infrared active modes, and a full model will include their contributions to radiation trapping, together with the trapping produced by clouds and aerosols. Partly because the infrared bands of the various components overlap, the contributions of the individual absorbers do not add linearly. Table 2.2 shows the percentage of trapping that would remain if particular absorbers were removed from the atmosphere. We see that the clouds only contribute 14 per cent to the trapping with all other species present, but would trap 50 per cent if the other absorbers were removed. Carbon dioxide adds 12 per cent to the trapping of the present atmosphere: that is, it is a less important trapping agent than water vapour or clouds. On the other hand, on its own CO_2 would trap three times as much as it actually does in the Earth's atmosphere. The point is of importance in seeing how far increases in the greenhouse effect could provoke climatic response to changed carbon dioxide concentrations. In this context, it is interesting to note that, since the upper layers of the atmosphere leak relatively more radiation to space than they trap, additional carbon dioxide leads to atmospheric cooling rather than warming for atmospheric layers above about 20 km. Many of the trace atmospheric gases, such as CH_4, N_2O, and O_3, have infrared modes active in the trapping region. These gases may therefore have a direct effect on global temperatures quite distinct from that exercised through their possible modification of concentration of major absorbers.

Table 2.2. Contribution of absorbers to atmospheric thermal trapping.

Species removed	Percentage trapped radiation remaining
None	100
O_3	97
CO_2	88
clouds	86
H_2O	64
H_2O, CO_2, O_3	50
H_2O, O_3, clouds	36
All	0

Data of Ramanathan, V. and Coakley, J.A., *Rev. Geophys. & Space Phys.* **16**, 465 (1978).

The influence of minor constituents is particularly marked if they absorb where there is otherwise an atmospheric window. This 'stopping up' of windows shows itself in the Venusian atmosphere. Over several years, doubt existed about whether a greenhouse effect on Venus could plausibly explain the high surface temperatures. The problem was in part to know how much solar radiation penetrated below the cloud tops, and in part to find infrared active molecules that possessed an optical depth $\chi = (732/227)^4 - 1 = 107$. Carbon dioxide alone cannot provide the necessary depth over the emitting spectral region. However, the Pioneer Venus and Venera 11/12 probes of 1978 showed that not only does enough sunlight reach the surface to fuel the effect, but that also the small H_2O and SO_2 concentrations are sufficient, together with pressure-induced transitions in CO_2, to close the spectral windows. For Mars, which also has an atmosphere predominantly composed of CO_2, the constraints are less severe, since $T_s \sim 223$ K is not so very much larger than $T_e \sim 217$ K. The requirement is for $\chi = (223/217)^4 - 1 \sim 0.12$. Radiative transfer can, in fact, account for the gross features of temperatures throughout most of the Martian atmosphere.

2.2.5 Unstable greenhouses: Venus, Earth, and Mars compared

Water vapour makes a sizeable contribution to atmospheric heating on Earth (Table 2.2). Since vapour pressures rapidly increase with increasing temperature, thus further increasing trapping, there exists a mechanism for positive feedback in the greenhouse effect. Evaporation from a planetary surface will proceed either until the atmosphere is saturated with water or until

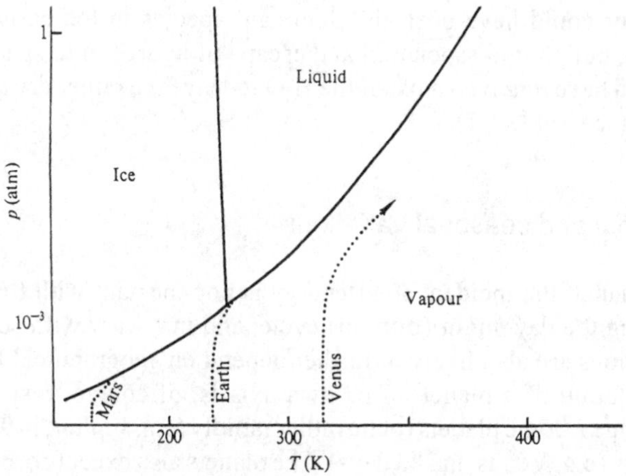

Fig. 2.7. The 'Runaway Greenhouse Effect'. The figure is a phase diagram for water onto which has been superposed as a dotted curve the 'greenhouse' temperatures produced as a function of water vapour pressure on Mars, Earth, and Venus. Starting temperatures are 'effective temperatures' calculated for the three planets using eqn (2.11) with identical values of albedo of 0.15 (that of Mars today: Table 2.1). When water vapour becomes saturated with respect to ice or liquid, the increase in greenhouse effect is halted. On Venus this saturation never occurs, and the temperature rises until all water is vaporized.

all the available water has evaporated. What happens on any particular planet will depend on the starting temperature in the absence of radiation trapping, since that will decide whether the vapour ever becomes saturated at the temperatures reached. That is, we need to compare the temperature–pressure relation with the phase diagram for the water system. Starting temperatures can be estimated in the way described for T_e in Section 2.2.1, but using throughout the albedo for cloudless Mars (0.15), instead of the real values for cloudy Venus and Earth. To illustrate the effect of water vapour, consider an initially dry atmosphere to which more and more water is added. Figure 2.7 shows the phase diagram for water, together with the planetary temperatures expected for differing water vapour pressures. On Mars and Earth, the additional heating due to liberation of water vapour is not sufficient to prevent the vapour reaching saturation as ice or liquid. However, on Venus there comes a critical vapour pressure (~10 mbar) when the rate of heating begins to increase dramatically: that vapour pressure is never reached at the lower temperatures on Earth or Mars. As a result, the P–T curve for the atmospheric water vapour increases more slowly than the vapour–liquid equilibrium curve. Condensation never occurs on Venus, and additional burdens of H_2O serve to increase the temperature even further. The effect is sometimes called the *Runaway Greenhouse Effect*. Certainly this positive feedback mechanism would explain why there is no surface water on Venus at the present day. Large amounts of water vapour could have been the dominant species in the early Venusian atmosphere, but photodissociation and escape of hydrogen to space (Section 2.3.2) would have removed most of the H_2O to leave the rather dry atmosphere now found (see Table 1.1).

2.2.6 Diurnal and seasonal variations

On a real planet, the incident solar energy is not constant with time, since it varies during the day–night (diurnal) cycle, and may vary with season: both these variations are also likely to further depend on geographical location.

The rotation of a planet on its own axis is, of course, responsible for diurnal changes. Some planets rotate rather rapidly (e.g. Jupiter, in 9.8 h), some rather slowly (e.g. Venus, in 243 days). The planets also execute orbital motion around the Sun. If the equator of the planet is not in the same plane as the orbit, or if the orbit is elliptic, then the planet will experience seasonal variations of average incident radiation intensity, as illustrated in Fig. 2.8. Venus has an almost circular orbit, and very small equatorial *inclination*. It does not, therefore, experience seasons. Curiously, its slow rotation is in the sense opposite to that of most planets: that is, the rotation is *retrograde*, in the direction opposite to the orbital motion, and the Sun rises in the west. The contrarotation of the axial period of 243 Earth days with the orbital period of 224.7 days gives a mean Venusian day of about 117 Earth days, and illumination for half that period each cycle. Earth and Mars possess similar equatorial inclinations (23–24°), and both therefore experience seasons. In

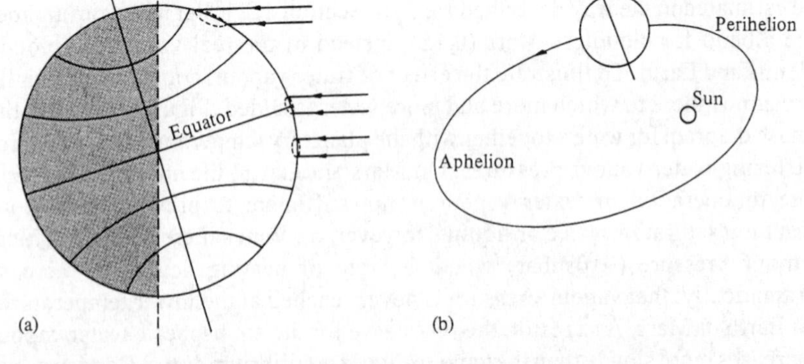

Fig. 2.8. Effects of (a) inclined, and (b) elliptic orbits on solar intensity. With an inclined orbit, the highest intensity (photons per unit area) is not necessarily at the equator, but where incident rays are normal to the surface; there are seasonal variations in the noontime angle of incidence and the length of day. An elliptic orbit causes variations as a result of changing Sun–planet distances.

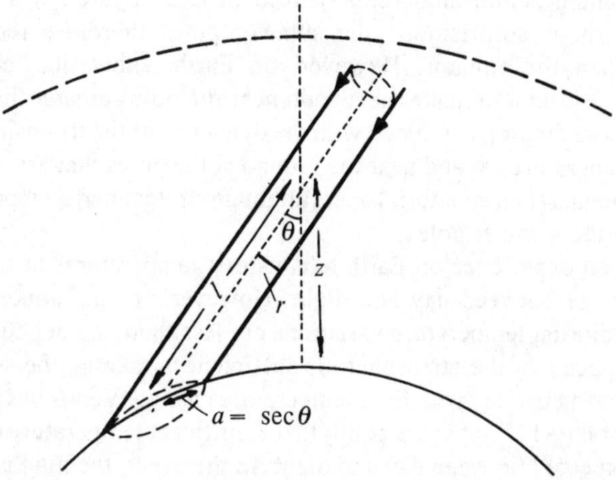

Fig. 2.9. Intensity at the surface of a planet depends on the area illuminated. For a zenith angle θ, the incident intensity is diminished by a factor sec θ compared with overhead illumination. The path, l, traversed through the atmosphere by the rays is equal to $z\sec\theta$ from any vertical altitude z.

addition, the orbits are slightly eccentric (elliptic). For example, Earth is slightly closer to the Sun during northern winter than during summer, so that the incident solar radiation is more intense in northern than in southern winter; seasonal changes due to inclination are thus reduced in the north. The opposite is true in the Southern Hemisphere, where seasonal changes due to eccentricity and inclination augment each other: other things being equal, seasonal changes are thus more extreme in the Southern than in the Northern Hemisphere. Similar effects on Mars allow the polar cap in the north to persist throughout the year, while that in the south disappears during the southern summer.

Inclination has two effects on the incident solar intensity averaged over a day. Only at the equinoxes are days and nights 12 h long everywhere. At other times of year, the day period depends on date and latitude, which also determine the maximum elevation that the Sun reaches. The angle that the Sun's rays make to a normal drawn from the surface is usually referred to as the *solar zenith angle*, θ (Fig. 2.9). For any particular place, date, and time during day, θ is defined, the angle being minimum at any season near local noon, and largest at any time of day during local winter. Figure 2.9 shows the effect of solar zenith angle on incident intensity. A cylinder of solar radiation of area πr^2 falls on an area $\pi r(r/\cos\theta) = \pi r^2 \sec\theta$ of the planet or atmosphere. Intensity corresponds to power per unit area, so that the intensity is reduced by the factor sec θ. For $\theta \geq 90°$, we have night, and there is no radiation. A day's averaged intensity depends on a value of sec θ averaged over the period for which $\theta < 90°$. The longer day (24 h) at the summer poles more than offsets the larger minimum zenith angle experienced there compared with the (12 h) equator. Perhaps surprisingly, the summer poles therefore receive more radiation than the equator. However, on Earth, snow, ice, and general cloudiness also tend to make the albedo near the poles greater than near the equator. These factors, combined with the dynamics of the troposphere, result in polar temperatures at and near the ground at the poles that are not actually higher in summer than equatorial ones (although stratospheric temperatures *are* higher near the summer poles).

Common experience on Earth tells us that temperatures at and near the surface change between day and night. However, for the atmosphere as a whole, the diurnal temperature variations are less than one per cent, because the heat capacity of the atmosphere is sufficient to take up the solar energy absorbed during the day with little temperature change. Venus has such a long diurnal period (~ 117 days: see p. 60) that significant temperature differences might be expected between day and night. In the event, the high atmospheric pressures (~90 atm—see Table 2.1), and hence mass, buffer the atmosphere against temperature changes. Entry probes from the Pioneer-Venus mission showed very small day–night variation of temperature in the lower atmosphere: at nominal zero altitude the atmospheric temperatures were 732.4 K from the day probe and 733.8 K from the night probe. There are, however, marked altitude fluctuations of day–night temperature differences. It may be that wind, eddy, and wave transport modify the effects of high thermal inertia. The Martian day, at 24.6 h, is comparable with the Earth's, but several factors encourage relatively large diurnal atmospheric temperature changes. First, the atmospheric pressure is very low—less than one per cent of Earth's surface pressure (Table 2.1)—so that the atmospheric thermal inertia is much smaller than is ours. Secondly, the surface is arid and dry, with small thermal conductivity or capacity. Like in Earth's desert regions, ground surface temperatures show large day–night variations, but, unlike Earth, there are no huge oceanic areas where temperatures stay nearly constant. Finally, the major component of the Martian atmosphere is the 'greenhouse' gas CO_2, so that the atmosphere is sensitive to changing infrared flux from the variable temperature surface. The two Viking landers descended through the Martian atmosphere at different local times of day (as well as at somewhat different latitude and season). Differences in lower atmospheric temperatures measured by the two landers seem consistent with the picture of diurnal as well as seasonal temperature variability in the Martian atmosphere.

Solar zenith angle has been viewed in our treatment as affecting the intensity of the solar radiation only through the influence on illuminated area. There is, however, a second influence of zenith angle in an atmosphere that absorbs or scatters solar radiation. Absorption of light depends on the total number of absorbing atoms or molecules that are encountered by a light beam, the form of relationship being shown in eqn (2.13). For any depth of atmosphere, d_a, that an overhead beam traverses, an oblique ray passes through a longer layer of path $d_a \sec\theta$. In general, then, the larger the zenith angle, the more will be the absorption in upper layers of the atmosphere, and the less radiation will penetrate to the lower levels. The translation of these ideas into quantitative terms requires a knowledge of altitude–concentration profiles and absorption cross-sections of all species capable of absorbing or scattering at the wavelength of interest. Qualitatively, experience tells us that more sunlight is scattered at sunrise or sunset, and the intensity of near ultraviolet radiation at the surface (at, say, $\lambda > 310$ nm) falls with increasing zenith angle much faster than does visible radiation because the ultraviolet passes obliquely through the ozone layer. To illustrate the magnitude of the effect, consider an unreal ozone layer on a flat Earth that is 20 km thick containing 3×10^{12} molecule cm^{-3} of ozone homogeneously mixed throughout so that the simplified eqn (2.14) can be used. At $\lambda = 310$ nm, σ_a is $\sim 10^{-19}$ cm^2 for ozone, so that the transmitted intensity is a fraction $\exp(-3 \times 10^{12} \times 10^{-19} \times 2 \times 10^6)$ ~ 0.55 of the incident intensity for overhead illumination ($\theta = 0$). Now if the zenith angle is say 75.5° ($\sec\theta = 4$), the optical path becomes 80 km rather than 20 km, and the transmitted light is a fraction 0.09 of the incident radiation. The dramatic fall is partly a result of the exponential in eqn (2.13) or (2.14), but $\sec\theta$ itself becomes very sensitive to θ at large angles, becoming 5.75 at $\theta = 80°$ and 11.5 at $\theta = 85°$. Thus for a zenith angle of 85° (near sunset), our hypothetical transmitted intensity at $\lambda = 310$ nm becomes 10^{-3} of the incident intensity. Remember, however, that this illustration assumes a flat Earth and a flat ozone layer. In the real world, and with a curved ozone layer, the optical path increases by substantially less than the factor $\sec\theta$. A final point we should note about radiation absorption in atmospheres is that upper layers may be illuminated long after the ground and lower atmospheric layers are in shadow.

2.3 Temperature profiles

2.3.1 Troposphere, stratosphere, and mesosphere

Radiative transfer in the atmosphere tends to produce the highest temperatures at the lowest altitudes, as illustrated by the model calculation represented in Fig. 2.6. At first sight, it might seem inevitable that convection would arise in this situation, since the hot, lighter, air initially lies under the cold, heavier air. In an atmosphere, the behaviour is somewhat more complex because gases are compressible, and pressures decrease with increasing altitude. A rising air parcel therefore expands, does work against the surrounding atmosphere, and is somewhat cooled. It can be that the temperature drop resulting from expansion would exceed the decrease in temperature of the surrounding atmosphere: in that case convection will not occur.

Simple thermodynamic arguments allow us to calculate the temperature decrease with altitude that is expected from the work of expansion. We imagine a packet of dry[a] gas that is in pressure equilibrium with its surroundings, but thermally isolated from them. This packet of gas is imagined to be moveable up or down in the atmosphere, and the rate of change of temperature with altitude (lapse rate) can be calculated. Since no heat flows (adiabatic conditions) and the gas is dry, the temperature profile calculated is the *dry adiabatic lapse rate*. The First Law of Thermodynamics can be expressed as

$$dU = dq + dw, \tag{2.27}$$

where dU is the change in internal energy, dq the heat supplied to the system, and dw the work done on it. From the definition of enthalpy, H, for a system of volume V at pressure p

$$dH = dU + p\,dV + V\,dp. \tag{2.28}$$

For our system, $dq = 0$ (adiabatic) and $dw = -p\,dV$. In our calculation, therefore,

$$dH = V\,dp. \tag{2.29}$$

The heat capacity of the gas at constant pressure, C_p, is defined as $(dH/dT)_p$, so that

$$C_p\,dT = V\,dp. \tag{2.30}$$

We already have an expression for dp in the atmosphere from the differential form of the hydrostatic equation (2.3), and on substitution into (2.30) we obtain

$$C_p\,dT = -V\rho g\,dz. \tag{2.31}$$

For unit mass of gas, for which $V = 1/\rho$, we replace C_p by the value appropriate to unit mass (c_p), with the result

$$-\frac{dT}{dz} = \frac{g}{c_p} = \Gamma_d. \tag{2.32}$$

The dry adiabatic lapse rate, Γ_d, thus depends only on the acceleration due to gravity on a planet, and the average heat capacity per unit mass of the atmospheric gases. For Venus, Earth, Mars, and Jupiter, the calculated values of Γ_d are 10.7, 9.8, 4.5, and 20.2 K km^{-1}. This information now allows us to determine whether any particular dry atmosphere is stable or unstable with respect to convection. If the actual temperature gradient in the atmosphere,

[a] 'Dry' here means containing no condensable material, so that the discussion can be applicable to planets other than Earth.

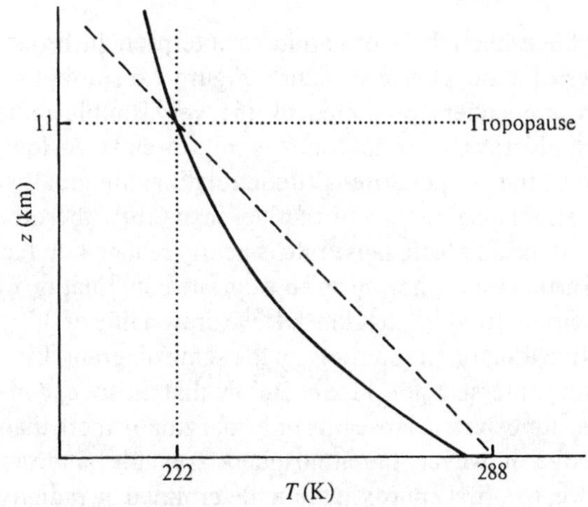

Fig. 2.10. A dry adiabatic lapse line (~6 K km^{-1}) has been added to the radiative transfer model results of Fig. 2.6. The atmosphere is convectively unstable below ~11 km.

$-(dT/dz)_{atm}$, is less than Γ_d, then any attempt of an air packet to rise is counteracted by cooling due to expansion that makes it colder and more dense than its surroundings. The atmosphere is stable. Conversely, if there were a tendency for $-(dT/dz)_{atm}$ to be greater than Γ_d, convection would be set up. Such convection will restore the temperature gradient until the atmosphere is stable again, so that actual atmospheric lapse rates rarely exceed Γ_d by more than a very small amount.

Presence of condensable vapours in the atmospheric gases complicates matters. Condensation to liquid or solid releases latent heat to our hypothetical air parcel. For a saturated vapour, every decrease in temperature is accompanied by additional condensation. Qualitatively, it is obvious that the *saturated adiabatic lapse rate*, Γ_s, must be smaller than Γ_d. The derivation of an expression, similar to (2.32), for Γ_s proceeds essentially as before, but with an additional term for the latent heat of condensation as the vapour load changes with temperature. Γ_s is not constant, but depends on temperature and pressure. In the Earth's atmosphere, where the important condensable vapour is water, Γ_s varies from about 4 K km^{-1} near the ground at 25 °C to 6–7 K km^{-1} at around 6 km and −5 °C. The stability conditions in saturated atmospheres are the same as in unsaturated ones, but with Γ_s replacing Γ_d. Partial saturation is treated in two stages, with Γ_d being applicable until temperatures are low enough to cause condensation, and Γ_s thereafter. In these circumstances, a situation known as *conditional instability* can arise. If the atmospheric lapse rate is less than Γ_d but more than Γ_s, the atmosphere is stable unless some kind of forced lifting (e.g. by winds) raises gas to an altitude where condensation occurs. From this point on, the atmosphere is unstable.

We now have enough information to interpret, in broad terms, some further features of atmospheric structure. Figure 2.6 shows the temperatures in the Earth's atmosphere predicted by the very simple radiative transfer model. More sophisticated models show similar trends. At low altitudes, the negative slope of the temperature–altitude relationship greatly exceeds even the dry adiabatic lapse rate. Convective instability therefore mixes the atmosphere until the adiabatic lapse rate is nearly reached. In fact, the average lapse rate on Earth (resulting from an appropriate combination of Γ_d and Γ_s, as well as global circulation) is 6.5 K km^{-1}. If we draw a line of this slope together with the radiative transfer predictions on the same diagram (Fig. 2.10), we see that the two lines intersect at ~ 11 km. Below that altitude, convection keeps even the largest atmospheric lapse rate only marginally more than the adiabatic lapse rate. Above, however, the atmosphere is stable, and the atmospheric temperatures are, to a first approximation, determined by radiative transfer for that part of the atmosphere where trapping rather than escape of radiation occurs. The critical level corresponds to the tropopause, while below is the turning, turbulent troposphere and above the layered, stable stratosphere, as described without explanation in Chapter 1. In reality, temperatures begin to rise again in the stratosphere as a consequence of the presence of ozone (Section 1.4), thus conferring additional stability. We shall return to this issue at the end of the present section.

A Martian altitude–temperature profile, obtained in the Viking 1 experiments, is shown in Fig. 2.11. The mean lapse rate up to 40 km is about 1.6 K km^{-1}, which is markedly subadiabatic, as indicated by the adiabatic line.

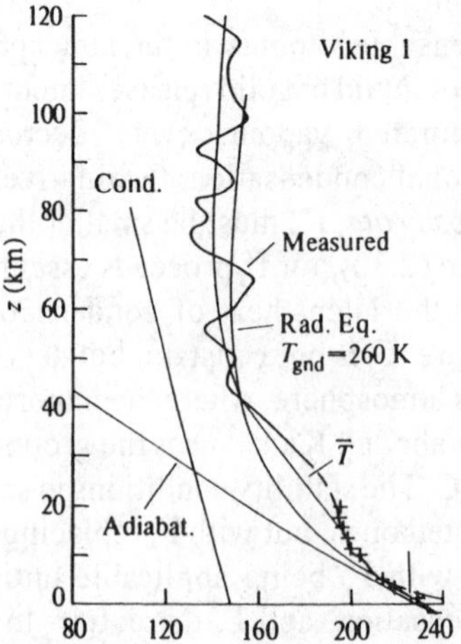

Fig. 2.11. Temperature structure of the Martian atmosphere as measured by instruments on board the Viking 1 lander. From Seiff, A. and Kirk, D.B. *J. geophys. Res.* **82**, 4364 (1977).

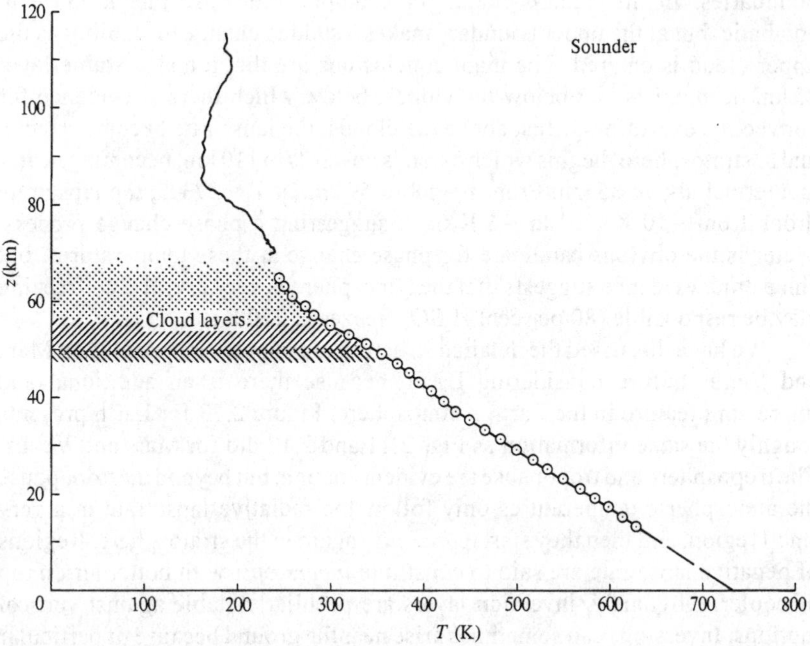

Fig. 2.12. Temperature structure of the Venusian atmosphere as observed by the Pioneer-Venus 'sounder' probe. Data below ~ 65 km, shown by the circles, were obtained by direct sensing of temperature and pressure. Data at higher attitudes were derived from measurements of atmospheric densities defined by probe deceleration during high-speed entry into the atmosphere. From Seiff, A., Kirk, D.B., Young, R.E., Blanchard, R.C., Findlay, J.T., Kelly, G.M., and Sommer, S.C. *J. geophys. Res.* **85**, 7903 (1980).

It is also clear that the CO_2 condensation boundary lies well below the temperatures measured in these low latitudes (22.3 °N) at mid-afternoon (landed 4:13 p.m. local time). It follows that clouds and hazes seen in the Mars atmosphere at similar places and seasons are probably water ice rather than CO_2. Contrary to some earlier predictions, there is no sharply defined tropopause, and radiative transfer can explain the temperature profiles at least down to 1.5 km from the surface. However, there is a clear change in slope of the profile a few kilometres above the surface, which suggests a turbulent region, produced by winds rather than by natural convection. The subadiabatic profile itself may be 'stabilized' by direct absorption of sunlight by the widely distributed atmospheric dust on Mars. Thermal 'tides', caused by the diurnal fluctuations discussed earlier, are thought to be responsible for the oscillatory excursions of the temperature profile above ~ 40 km. Venusian atmospheric temperatures determined by the Pioneer-Venus sounder are displayed in Fig. 2.12. Pronounced changes in lapse rate below 60 km are evident at higher resolution, and several changes can be correlated with the clouds and cloud boundaries. In the middle cloud, for example, the lapse rate is close to adiabatic, but at the upper boundary makes a sudden change to stability as the upper cloud is entered. The main conclusions are that a major stable layer 25 km deep exists just below the clouds, below which there is evidence for convective overturning. Just above the clouds, the lapse rate becomes stable, and a 'stratosphere' begins which extends upwards to 110 km, becoming nearly isothermal above 85 km. From $z = 58$ to 59 km, at $T = 273$ K, the lapse rate drops from ~ 10 $\mathrm{K\,km^{-1}}$ to ~4 $\mathrm{K\,km^{-1}}$, suggesting a phase change process. Water is the obvious candidate for phase change at these temperatures, but since other evidence suggests that the atmosphere is very dry, H_2SO_4 freezing may be responsible (80 per cent H_2SO_4 freezes at 270 K).

We have discussed the detailed structure of temperature profiles on Mars and Venus before considering Earth, because there is an additional and interesting feature in the Earth's atmosphere. Figure 2.13 for Earth presents roughly the same information as Figs 2.11 and 2.12 did for Mars and Venus. The troposphere and tropopause are evident enough, but beyond the tropopause the atmospheric temperatures only follow the radiative lapse rate in a very small region, and then they start to *increase* again in the stratosphere. Regions of negative lapse rate are said to constitute *inversions*, with hotter air on top of cooler. Obviously, inversion layers are peculiarly stable against vertical motions. Inversions can sometimes arise near the ground because of particular meteorological and geographical conditions. Such inversions can trap pollutants and prevent their dissipation, as we shall see in Chapters 4 and 5. Inversions have also been detected in the atmospheres of Jupiter and Saturn (at pressures and temperatures of about 135 mbar, 110 K; 70 mbar, 82 K, respectively). On Earth, the stratospheric inversion is a result of heating by absorption of solar ultraviolet and visible radiation in the ozone layer. Ozone is formed photochemically from O_2 and the 'layer' structure owes its existence to a peak in absorption and in reaction rates (Section 4.3.2). Too low in the atmosphere, there are insufficient short-wavelength photons left to dissociate much O_2, while too high there are insufficient O_2 molecules to absorb much light and to associate with O atoms to make O_3. A series of chemical reactions concerned in the formation and destruction of ozone ultimately releases the chemical energy of O_2 dissociation, while the solar radiation absorbed by ozone itself is also liberated as heat. As a consequence, the heating in the stratosphere is related to the ozone concentration profile (although modified according to the exact mechanism of conversion of absorbed radiation to heat energy). On planets such as Mars and Venus, with little oxygen, there is virtually no ozone layer and no stratospheric heating. It is noteworthy that the ozone layer absorbs wavelengths from the Sun that do not reach the Earth's surface, and that the heating is achieved *in situ* (except during the long polar nights) rather than by absorption of re-radiated infrared. As well as its absorption bands in the ultraviolet region, ozone also absorbs in the visible region: although the absorption is weaker, the solar radiation is more intense, and the heating is comparable to that produced by the ultraviolet absorption. Figure 2.13 shows that at altitudes above about 50 km, the heating effect is too weak to compete with the cooling processes. and temperatures decrease again. Conventional nomenclature ascribes the name *mesosphere* to this next atmospheric region, and *stratopause* to the upper boundary of the stratosphere. The lapse rate is subadiabatic in the mesosphere (about 3.75 K km^{-1} in the mid-profile of our figure), so that the mesosphere is stable with respect to convective motions. The lowest temperatures in the entire atmosphere of the Earth are found in the mesosphere, and are lowest over the summer polar regions. Finally, the temperature stops falling at the *mesopause*.

2.3.2 Thermosphere, exosphere, and escape

All atmospheres eventually become so thin that collisions between gaseous species are very infrequent. One consequence may be that energy is not equilibrated between the available degrees of freedom. In particular,

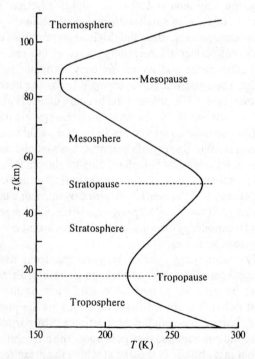

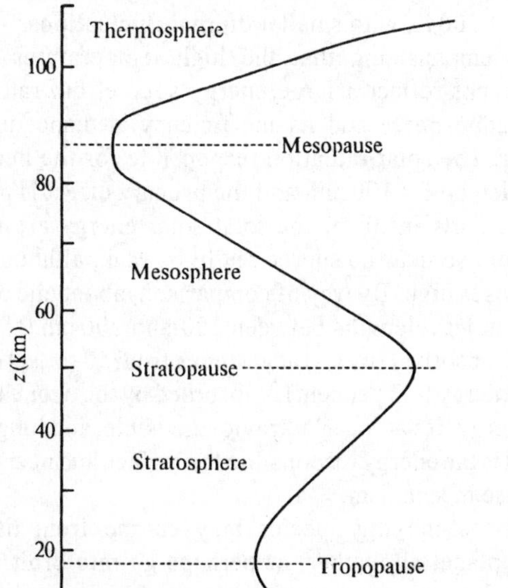

Fig. 2.13. Temperature–altitude profile for the Earth. The curve shown represents the mean structure for latitude 40°N during June. Data from Houghton, J.T. *The physics of atmospheres*, Cambridge University Press, 1977.

translational temperatures may be much higher than vibrational or rotational temperatures, and atomic species with high kinetic energy may not be sharing that energy with molecular species. Since the loss of energy to space from atmospheric constituents has been interpreted as radiation from infrared-active molecular vibrations, it follows that inefficient re-equilibration between energy modes will lead to a bottle-neck for energy loss and excess translational temperatures. The region of a planetary atmosphere that shows, from these causes, increasing temperature with altitude is called the *thermosphere*. For Earth, the residual cooling mechanisms are probably infrared emission from transitions between the angular momentum levels of atomic oxygen, together with some emission from nitric oxide. The cooling efficiency is low, and thermospheric temperatures are correspondingly high. Thermal inertia is very small, so that there is a big diurnal temperature variation. At $z = 250$ km, translational temperatures are typically 850 K at night and 1100 K during the day; but at these altitudes the energetic particles that constitute the solar wind interact with the atmosphere, so that temperatures also depend on whether the Sun is 'quiet' or 'active'. In contrast, the thermospheric temperatures of both Mars and Venus are much lower than Earth's because of the preponderance of CO_2 that can itself radiate. Venusian thermospheric temperatures are only ~300 K during the day and ~100 K at night; Martian thermospheric temperatures are ~300 K, with smaller diurnal fluctuations.

It is worth emphasizing that the high temperatures in the Earth's thermosphere do not reflect a large energy source, but rather the extreme thinness of the atmosphere and its inefficiency at disposing of energy by radiative transfer. The solar radiation responsible for the heating lies in the extreme ultraviolet, at $\lambda \leq 100$ nm, and the primary effect is photoionization. Only about three parts in 10^6 of the total solar energy are radiated in this wavelength region, so that the source really is weak, although the energy of individual photons is high. By way of comparison, about one part in 10^4 of the solar energy lies at wavelengths between 100 and 200 nm ('far' or 'vacuum' ultraviolet) and is absorbed (by O_2) at altitudes from 50 to 110 km. Much more ultraviolet solar energy (~2 per cent) is absorbed by the ozone layer, while ~98 per cent of the energy lies at 'near' ultraviolet, visible, and longer wavelengths ($\lambda > 310$ nm) and is the energy responsible for surface and near-surface heating *via* the greenhouse mechanism.

Sufficiently fast-moving species may escape from the gravitational influence of the planet altogether, or perhaps go into orbit around it. Two conditions must be met for escape to occur. First of all, the escaping constituent must not suffer any collisions as it leaves the planet. A *critical level*, z_c, can be defined for which a proportion e^{-1} of particles capable of escape experience no collisions as they pass out of the atmosphere. Intuitively, it would seem that this altitude is the one at which the mean free path, λ_m [cf. eqn (2.8)] becomes equal to the scale height for the particle in question, and, indeed, this condition can easily be shown to fit the definition of z_c. For the

Earth's atmosphere, z_c is 400–500 km, and beyond that altitude the region of the atmosphere is the *exosphere*. The second condition to be fulfilled for escape is that the particle has sufficient translational energy (i.e. starting velocity) to escape completely the gravitational attraction of the parent planet. An analogy can be seen with the ionization potential of an atom, where we are interested in removing an electron from the coulombic attraction of the nucleus. In the gravitational case, the potential energy of a mass m at a distance a from the centre of a planet, mass M_p, is GmM_p/a, where G is the gravitational constant. If the few hundred kilometres contribution made by z_c to a are ignored, a can be set at the planetary radius R_p, and $GM_p = gR_p^2$. For a velocity v of the particle, the kinetic energy is $\frac{1}{2}mv^2$, so that the condition for escape is

$$\tfrac{1}{2}mv^2 > mgR_p, \tag{2.33}$$

or

$$v > (2gR_p)^{1/2}. \tag{2.34}$$

The minimum velocities meeting this condition are the escape velocities given in Table 2.1. Although it is only a critical (mass-independent) velocity that is needed, the energy is still mass-dependent. For Earth and Venus, the velocities are 10–11 km s^{-1}, but, put in energy terms, that corresponds to ~0.6 eV or ~58 kJ mol^{-1} per atomic mass unit. From Mars, the escape energies are one-fifth of those from Earth, and from Titan only one twenty-eighth.

We now have to consider escape mechanisms. Sir James Jeans formulated the details of the process that bears his name. Escape is seen to involve the high-velocity particles in the tail of the Maxwell velocity distribution. Atomic hydrogen has a most probable velocity of ~3 km s^{-1} at 600 K, and the fraction of atoms with $v > 11.2$ km s^{-1} is just greater than 10^{-6}. Atomic oxygen, on the other hand, has a most probable velocity of ~0.8 km s^{-1}, and the fraction with a velocity greater than the escape velocity is 10^{-84}! Escape of hydrogen is reasonably probable, but little oxygen can have escaped from the Earth over its lifetime by the thermal mechanism. Several non-thermal mechanisms have been proposed for escape. Many of these employ the energy of a chemical (particularly photochemical or ionic) reaction to impart high translational energy to product fragments. A typical example is *dissociative recombination* between ions and electrons (see Chapter 6). The process is essentially the reverse of ionization, and the ionization energy has to be carried off as translation of fragments, as in the process

$$O_2^+ + e \rightarrow O^t + O^t, \tag{2.35}$$

where the superscript represents translational excitation. Even though some energy also goes into electronic excitation of atomic oxygen, 2.5 eV are released as translational energy to each O atom. The equivalent velocity is ~5.5 km s^{-1}, sufficient to escape from Mars, for example. Ions are particularly important both because of their excess energy over the neutral parents, and also because they can be accelerated in planetary electric fields. A few tens of volts acceleration can dominate over any available photon or chemical energies. In fact, the *charge-exchange* process

$$H^{+t} + H \rightarrow H^+ + H^t, \tag{2.36}$$

with accelerated H^+ ions, is thought to be a more important source of 'hot' H^t atom escape from the Earth's atmosphere than the Jeans mechanism.

There are three major stages in the escape process: transport of the constituent through the atmosphere, conversion to the escaping form (usually atom or ion), and the actual escape. Normally one of these will be the slowest, and rate-determining. On Earth, for example, hydrogen loss is limited by the upward diffusion flux and not by any process involved in the conversion to escaping H atoms. With this knowledge, the flux of escaping hydrogen can be estimated from the diffusion rate through the stratosphere (and hence through the mesosphere and to the exosphere). Mixing ratios for H_2O, H_2, and CH_4 yield an escape flux of $\sim 2.7 \times 10^8$ H atoms $cm^{-2} s^{-1}$ from the atmosphere, *regardless* of the escape mechanism.

Preferential escape of a lighter isotope is possible, since for a given available energy the lighter isotope will have a larger velocity. Isotopic fractionation can then occur, and give valuable clues about the history of an atmosphere. The changes in isotopic composition depend, for example, on whether all the original inventory of an element was in the early atmosphere, or whether much was in surface or interior reservoirs. We shall meet several examples in later chapters, and defer further discussion until then.

2.3.3 Vertical transport

Vertical mixing and redistribution of atmospheric constituents is assured in the troposphere because of the turbulence that characterizes the region. Above the tropopause, however, where lapse rates are subadiabatic (or even negative, as in some regions of the Earth's atmosphere), how, and how fast, are chemical species transported in the vertical plane?

In the last section, we suggested that the escape of hydrogen from our atmosphere is kinetically limited by the rate of diffusive motion through the upper atmosphere. Diffusion in one form or another does indeed provide a mechanism for transport through a non-turbulent atmosphere. The type of diffusion most familiar to chemists is *molecular diffusion*, in which molecules (or atoms) move in response to a concentration gradient, and in such a direction as to try to remove the gradient. At the level of individual particles, random molecular motion is destroying a special arrangement (higher concentration, and thus more molecules, in one volume element than in another). Thermodynamically, the system is behaving so as to maximize the entropy in accordance with the Second Law. Similar remarks apply to the velocities of particles as to concentrations. A system in which molecules with high velocity are separated from those with low velocity possesses a temperature gradient, and the statistical re-randomization corresponds to a heat flow, or thermal conduction, in response to that gradient. If the 'hot' molecules are also *chemically* different from the surroundings, as they might well be in an atmosphere, then thermal conduction also corresponds to an identifiable redistribution of matter, and the process is called *thermal diffusion*.

Calculation of the rates of transport in an atmosphere by diffusion is only slightly different from the procedure familiar in laboratory chemical problems. We shall illustrate the problem for diffusion in one dimension. The starting point for molecular diffusion is *Fick's First Law*, which for one dimension is

$$\Phi = -D(\partial n/\partial z). \tag{2.37}$$

The law thus states that the flux, Φ, of molecules across unit area in unit time is proportional to the concentration gradient in z direction, $\partial n/\partial z$, that drives the process. The constant of proportionality, D, is the *diffusion coefficient*; for gases, elementary kinetic theory gives

$$D = \tfrac{1}{3}\,\bar{c}\,\lambda_{\mathrm{m}}, \tag{2.38}$$

where $\bar{c}$ is the mean molecular velocity and λ_{m} the mean free path. Equation (2.8) for λ_{m} thus shows that D is inversely proportional to pressure. The flux of molecules leads to redistribution, and the rate of change in an element ∂z is the difference between the flux entering and the flux leaving. Hence

$$\frac{\partial n}{\partial t} = -\frac{\partial \Phi}{\partial z} = -\frac{\partial}{\partial z}\left(-D\frac{\partial n}{\partial z}\right). \tag{2.39}$$

If D is independent of z, then the second equality on the right of eqn (2.39) can be written $D(\partial^2 n)/\partial z^2$, and we have *Fick's Second Law* of diffusion.

In an atmosphere consisting of a compressible fluid, the modification to the simple diffusion equations in the vertical direction starts from a recognition that $\partial n/\partial z$ is not zero in a completely mixed atmosphere because pressure and number density of all components decrease with z according to the hydrostatic equation (2.5) or (2.6). That is, the driving force for diffusive separation of a component i is not the actual concentration gradient $\partial n_i/\partial z$ (where the subscript refers to the particular component), but rather the difference between the actual concentration and the equilibrium (perfectly mixed) gradient $(\partial n_i/\partial z)_e$. Equation (2.37) is then rewritten

$$\Phi_i = -D\left\{\frac{\partial n_i}{\partial z} - \left(\frac{\partial n_i}{\partial z}\right)_{\mathrm{e}}\right\}. \tag{2.40}$$

Let us write, for convenience, $n_i = f_i N$, where f_i is the mixing ratio of the species and N the total atmospheric number density. Remembering that $(\partial f_i/\partial z)_e = 0$ for perfect mixing, we can rewrite eqn (2.40) in the form

$$\Phi_i = -D\left(N\frac{\partial f_i}{\partial z}\right), \tag{2.41}$$

and eqn (2.39) for the rate of change of concentration becomes, with D independent of z,

$$\frac{\partial n_i}{\partial t} = N\frac{\partial f_i}{\partial t} = -\frac{\partial \Phi_i}{\partial z} = D\left(\frac{\partial N}{\partial z}.\frac{\partial f_i}{\partial z} + N\frac{\partial^2 f_i}{\partial z^2}\right). \tag{2.42}$$

The differential $\partial N/\partial z$ is equivalent to the differential form of the hydrostatic equation (2.5) in terms of number density and scale height $H_s = \mathbf{k}T/mg$,

$$\frac{\partial N}{\partial z} = -\frac{N}{H_s} - \frac{N}{T}.\frac{\partial T}{\partial z} \tag{2.43}$$

and the term in $\partial T/\partial z$ coming from the temperature-dependent relationship between number density and pressure. Substitution of eqn (2.43) into (2.42) leads, after cancelling N on both sides, to

$$\frac{\partial f_i}{\partial t} = D\left(\frac{\partial^2 f_i}{\partial z^2} - \frac{1}{H_s}.\frac{\partial f_i}{\partial z} - \frac{1}{T}\frac{\partial T}{\partial z}.\frac{\partial f_i}{\partial z}\right). \tag{2.44}$$

The differential equation thus shows how an atmospheric mixing ratio will evolve with time under the influence of diffusion, and is the quantitative expression of the rate of diffusive transport in the vertical direction.

Molecular diffusion, and where appropriate thermal diffusion, are able to account for the rates of vertical transport in the upper mesosphere and thermosphere of the Earth's atmosphere. A number of indications suggest, however, that vertical mixing in the stratosphere is much more rapid than can be explained by simple diffusion mechanisms, even though the rates are also much less than those operating in the troposphere. Methods for investigating vertical transport in the stratosphere include measurements from satellites or balloons of the vertical concentration profiles of various minor constituents (e.g. CH_4 or N_2O). Observations of tracer species injected by volcanic eruptions or by nuclear weapon testing have also been used.

Full three-dimensional models (see Section 3.6) of atmospheric behaviour include advection of chemical species within the atmosphere. Before such models were available, the concept of *eddy diffusion* was introduced. Although rather discredited as an artificial method of treating the mixing, it is instructive to examine the idea in the context of ordinary diffusion. Instead of individual molecules moving independently, small packets of gas are envisaged as executing eddy motions that can still be treated by the methods of the kinetic theory of gases. An eddy diffusion equation can then be derived, identical in all respects to eqn (2.44), with the exception that the molecular diffusion coefficient, D, is replaced by an eddy diffusion coefficient, K_z. An idea of the relative magnitudes of the rates of diffusion by molecular and eddy mechanisms is given by comparing the values of D and K_z for a mid-stratospheric altitude. At say 35 km, $D \sim 2 \times 10^{-3}$ m^{2}s^{-1}, while values of K_z typically adopted are about 10^2 m^{2}s^{-1}. A given 'tagged' molecule might then be expected to move 1 km vertically in a period measured in hours for the eddy mechanism, but in years for the molecular mechanism.

Transfer of chemical constituents in both directions across the tropopause is clearly of great importance. Many of the minor reactants in the stratosphere are initially released, naturally or anthropogenically, to the troposphere. Downward transport may be an important stratospheric loss mechanism for those species or their products, as well as a source of ozone to drive tropospheric chemistry. For about fifty years, it has been supposed that the stratosphere contains very little water vapour (a few parts per million) because gases entered the stratosphere through a region cold enough to trap out the water. The requisite cold trap temperature is in the region of 190 K, and the proposal is that troposphere to stratosphere exchange takes place in the tropics, which is where the lowest tropopause temperatures are experienced. These tropical regions are, as we shall see in Section 2.4, associated with rapidly rising moist tropospheric air masses. Mixing ratios of H_2O in the tropical stratosphere vary seasonally as the tropopause temperature passes through its annual cycle. Layers of air that have risen from the troposphere possess a water-vapour content determined by the tropopause temperature. Horizontal transport out of the tropical stratosphere is rather slow, so that this signature is retained for months or even years. Alternating layers of high and low water content reflect the annually varying saturated vapour pressure at the tropical tropopause, and these layers slowly move upwards within the stratosphere at velocities ranging from 0.2 to 0.4 mm s^{-1}. The system has been likened to an 'atmospheric tape recorder'; the air is 'marked' as it enters the stratosphere like a signal recorded on an upwards moving magnetic tape.

Exactly how the transfer takes place remains unclear. Trace constituents, and water vapour in particular, could be transferred either by a steady rising motion or in violent tropical cumulonimbus cloud convection. There is also known to be an exchange of stratospheric and tropospheric air at middle and high latitudes as a result of *tropopause folding*. Thin (~ 1 km) laminar intrusions of stratospheric air enter the troposphere for perhaps 1000 km parallel to the tropospheric jet stream (Section 2.4), and then become mixed with the turbulent tropospheric air. Perhaps more generally important than these mechanisms is *wave-driven pumping* (see Section 2.4) within the stratosphere at mid-latitudes. This extratropical pumping leads to steady large-scale ascent of air in the tropics, and large-scale descent near the poles. The memory effect referred to in the previous paragraph may explain observations that the lowest water-vapour concentrations are found at altitudes several kilometres above the tropopause, since minima from the seasonal H_2O cycle could be propagated vertically. Another explanation that has been advanced is that cumulonimbus convection sometimes overshoots the tropopause by several kilometres.

The recent observations, including those of hemispheric asymmetries in water-vapour mixing ratios, suggest that the stratosphere can be divided into three regions when discussing transport. First, there is the 'overworld', in which transport is controlled by non-local dynamical processes, such as those involving the pumping mechanism. Secondly, there is a tropically controlled transition region, which consists of relatively young air that has passed through, and been dehydrated by, the cold tropical tropopause. Finally, there is the lowermost stratosphere, in which stratosphere–troposphere exchange occurs by processes such as tropopause folding.

2.4 Winds

Horizontal motions of the atmosphere are central to the interests of meteorology. From the viewpoint of atmospheric chemistry, winds are responsible for the transport and mixing of chemical constituents. We shall examine in this section the barest outline of wind systems.

At and near ground level, winds are very variable and turbulent, gusting and changing direction frequently. Away from surface features and friction, the motions are more regular, and averaged over many days or weeks a clear and reproducible pattern emerges. The term *general circulation* is used to describe the global average winds that we need to discuss. Regular wind regimes have been recognized since the early days of sailing ships, with the tropical 'trade' winds in the Atlantic and Pacific (north-east in the Northern Hemisphere, south-east in the Southern) being very important. Even today, aircraft flying the transatlantic routes at mid-latitudes take longer in the east–west direction than on the return journey because of the prevailing high-speed westerly (from the west) *jet stream* winds at aircraft altitudes near the top of the troposphere.

Winds may be regarded as the air flow that is a response to pressure differentials between different locations on a planet. Temperature differences may be the cause of the pressure variations, as we shall see shortly. Thus the winds, by transporting heat (in both sensible and latent forms), tend to remove those temperature differences, in accordance with thermodynamic expectations. Equator-to-pole temperature differences on Earth are certainly smaller than would exist in the absence of atmospheric motions, although the oceans also play a large role. The Martian atmosphere is so thin that heat transport is ineffective in diminishing latitudinal temperature variations, and the winter polar regions become so cold (~ 150 K) that CO_2 can condense out. In turn, this seasonal condensation leads to a horizontal transport phenomenon unique to Mars, that of condensation flow, which is still a response to temperature differentials. Venus has an atmosphere so massive that it can maintain surface temperatures identical, within a few degrees, between poles and equator.

For Earth, we first consider the basic observations of circulation patterns, and turn later to possible explanations of them. On a global scale, hot air rises near the equator and sinks at higher latitudes. Each hemisphere has its own circulation cell—named, after its discoverer, a *Hadley cell*—and they converge near the equator in the *Intertropical Convergence Zone* (ITCZ). Until recently, it was thought that the tropical trade winds were separated by a region of calm winds called the 'doldrums', but it now seems that the transition occurs a few degrees away from the equator. The ITCZ migrates with the Sun, being north of the equator in the northern summer, and south in the southern summer, its exact position being modified by the distribution of sea and land masses. The very narrow ITCZ belt is characterized by very strong upward motion and heavy rainfall. The returning surface air becomes saturated with water as it passes over the oceans. As it rises in the ITCZ, water condenses and heavy precipitation occurs. Release of latent heat by condensation increases the convective instability, reduces the lapse rate, and hence increases the driving pressure differential. It can readily be seen now why the ITCZ is a prime candidate region for the convective overshoot from troposphere to stratosphere suggested in Section 2.3.3.

Hadley circulation might well consist of two cells each encompassing a hemisphere from equator to pole, and with the flows directly north–south, were it not for the rotation of the planet. For Earth, the real Hadley circulation only extends within the tropics (up to the 'horse latitudes'), and it has a strong westerly component aloft, and a corresponding easterly component on the return surface flow. Both the westerly directional component and the limited span of the Hadley cell are a consequence of planetary rotation. Atmospheric gases possess mass, and if they rotate more or less with the planet they therefore possess angular momentum. North–south motions imply a change in radius of rotation, decreasing to near zero at the poles. Yet angular momentum must be conserved, and the atmosphere achieves this conservation by developing zonal motion (that is, in the direction of rotation of the Earth). For example, an air parcel at rest with respect to the equator must develop a zonal (west to east) velocity by the time it reaches 30°N of 134 m s^{-1}. The hypothetical force producing this motion perpendicular to the initial direction of transport is called the *Coriolis force*. The horizontal component of the Coriolis force is directed perpendicular to the horizontal velocity vector: to the right in the Northern Hemisphere and to the left in the Southern Hemisphere. It can readily be shown that for an air parcel of horizontal velocity V_h at latitude ϕ on a planet with angular velocity Ω, the horizontal component of the Coriolis force, F_c, has a magnitude

$$F_c = 2\Omega V_h \sin\phi \tag{2.45}$$

That is, the force is minimum at the equator and maximum at the poles. Trade winds on Earth clearly fit the requirements of angular momentum conservation. The winds aloft, from equator to poles, gain a component from west to east in both hemispheres; the returning surface trade winds are northeasterly (from the north-east) and south-easterly, north and south of the equator. At mid and high latitudes, the Coriolis force increases [cf. the $\sin\phi$ term in eqn (2.45)] and winds flow mostly in an east–west direction.

We must now ask what drives the lifting of air near the equator and its subsequent sinking at higher latitudes. The conventional explanation has been in terms of pressure differentials. Our experience on Earth tells us that surface pressures do not vary dramatically from one place to another. The same is not true aloft in the atmosphere, and the various forms of the hydrostatic equations (2.5)–(2.7) all indicate that pressures fall off more rapidly with altitude when the temperatures are low. If one location experiences a higher surface temperature than another, then, for similar lapse rates, the temperatures higher in the troposphere will bear the same relationship, and the pressure aloft over the warmer area will be higher. The consequent force will accelerate the air mass until kinetic energy losses (e.g. by friction) are matched. The concept then is that hot air rises near the equator and is forced towards the cooler poles

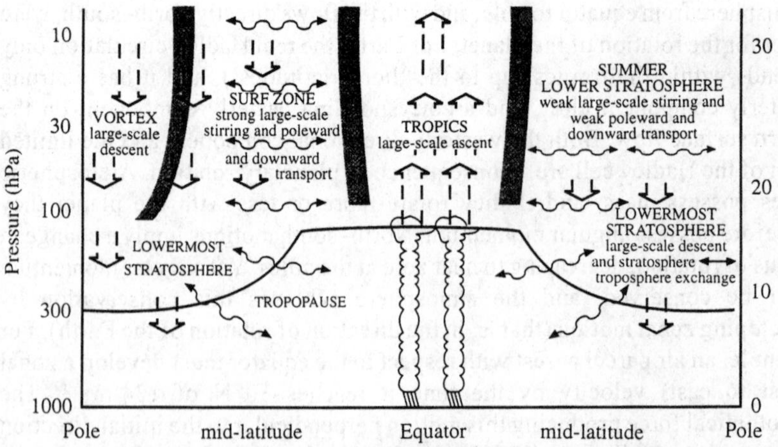

Fig. 2.14. Schematic representation of the principal regions of the lower stratosphere, illustrating the several different transport characteristics. The diabatic circulation is represented by broad arrows, while the wavy arrows are indicative of a motion akin to stirring (along surfaces of constant entropy). The thick solid line is the tropopause. From *Scientific assessment of stratospheric ozone: 1998,* World Meteorological Organization, Geneva, 1999.

by the higher pressures in the higher temperature regions. However, there is increasing evidence for the flows being generated by a 'pump' driven by atmospheric waves. Several kinds of *atmospheric oscillations* or wave motions have been identified. *Rossby waves*, or *planetary waves*, owe their existence to the variation of the Coriolis force with latitude. Such waves can cause air parcels to oscillate back and forth about their equilibrium latitude, and a pumping mechanism can be envisaged that will lead to mass transport of the kind seen in the Hadley cells.

Fluid-dynamic instabilities produce a different wave-like progression of low- and high-pressure regions (*cyclones* and *anticyclones*) around middle latitude areas of the Earth. Disturbances of this kind can be both reproduced in the laboratory in experiments with heated rotating fluids, and predicted by numerical models. The instabilities arise when latitudinal temperature differentials become too large, and they are called *baroclinic instabilities*. Eddy motions (*baroclinic waves*), resulting from the instability, transport heat both poleward and vertically upward (thus reducing the tropospheric lapse rate). They also transport momentum to the upper troposphere and thus maintain the high-velocity jet stream. Four to six pairs of waves typically encircle the Earth at any one time.

One feature of stratospheric winds will turn out to be of particular importance with the discussion of polar ozone 'holes' (Section 4.7). This feature is the development of a strong cyclonic vortex over the winter pole. In the Antarctic, the vortex consists of a westerly wind motion, developing speeds of up to 100 m s^{-1} by the end of winter. The vortices arise because there is no internal heating of the stratosphere during the polar night (it will be recalled that absorption of solar radiation by ozone normally heats the stratosphere). The air thus cools and descends and, again subject to the Coriolis force, picks up an appropriate rotational component.

Figure 2.14 is an attempt to represent one view of transport characteristics in the lower stratosphere. Above an altitude of about 16 km, the inner vortex region in both hemispheres is isolated from mid-latitudes. There is greater exchange between the *edge* of the vortex region and mid-latitudes, and also at lower altitudes. In the 'surf zone', air is drawn out of the polar vortex or the tropics, and is stirred by the large-scale flow. This behaviour leads to the formation of sloping sheet-like structures that are observed in measurements of the distributions of chemical tracers. Horizontal cross sections through such sheets show filamentary structures, while vertical profiles demonstrate the presence of laminae. The filaments of polar or tropical air appear to survive for 20–25 days before they become mixed with their surroundings.

Large-amplitude planetary-scale waves in the stratosphere can play an important role in the transport and variability of inert and photochemically active species. A dynamical phenomenon of considerable interest in the Earth's upper atmosphere is the *Sudden Stratospheric Warming* (SSW) that occurs about once every one to three years in the Northern Hemisphere winter. A large growth in wave amplitude occurs over a two-week period, and the normal cyclonic polar vortex then becomes highly distorted. There are strong poleward fluxes of heat, and temperatures at 40 km altitude can increase by as much as 80 K at high latitudes. This heating may be sufficient to reverse the usual latitudinal temperature gradients and winds, the normal westerlies temporarily changing to easterlies. Sudden stratospheric warmings play an important role in the budgets of trace species, as well as of heat, momentum, and energy in the stratosphere.

Having established that the east–west component of the Earth's winds is intimately bound up with the effects of planetary rotation, it is interesting to compare the circulation patterns with those of other planets. Rapid rotations of Jupiter and Saturn (periods 9.8 h, 10.23 h) seem to be responsible for the alternating cloud bands of different colours, seen so beautifully in the Voyager pictures. The cloud bands are separated by jet streams, and the Great Red Spot and other such features seem to have a fluid-dynamic origin. There is little pole-to-equator energy transfer at the level of the clouds. Venus rotates very slowly (period 243 days), so that Coriolis forces and associated instabilities are weak. Nevertheless, the winds in its lower and middle atmosphere blow primarily in an east–west direction in the sense of the planet's rotation. Close to the cloud tops, the wind velocities are a phenomenal 100 $m\,s^{-1}$ over the entire planet (but only about 1 $m\,s^{-1}$ near the surface). Transport of momentum by a combination of a hemispheric Hadley circulation and eddies is thought to generate the high east–west velocities. The centrifugal force due to the winds themselves may be responsible for the east–west directionality; the north–south component is characterized by much more modest wind speeds (5–10 $m\,s^{-1}$) at the cloud tops.

Winds on the planets seem, then, to be explicable largely in terms of a thermally driven circulation modified by Coriolis forces and baroclinic instabilities. Additional features, such as stationary eddies (resulting from topographical or temperature contrast) and thermal tides (especially important on Mars—see Section 2.2.6) also influence transport of heat and momentum (e.g. to the stratosphere) to produce the observed general circulation at all altitudes.

2.5 Condensation and nucleation

Cloud formation is intimately coupled, *via* optical and latent heat effects, to the radiation balance, atmospheric temperature structure, and circulation patterns that we have been discussing in the preceding sections. Just in terms of albedo, clouds have a remarkable influence: a cloudless Earth, with the albedo of Mars (0.15), would have an average ground-level temperature of ~304 K, some 16 degrees higher than the present-day value, with consequent melting of polar ice caps.

In this section, we wish to examine rather more closely how condensation can occur. The principles apply to the formation of liquid droplets or solid particles, and so are relevant to cloud or aerosol formation in any atmosphere. We shall, however, illustrate the problem initially with respect to water clouds. Section 1.3 contains a hint that condensation requires the presence of small particles to act as condensation nuclei. The problem concerns the vapour pressure produced over a curved surface. As we shall show shortly, the equilibrium vapour pressure at any temperature is larger for a drop than the saturated vapour pressure over a plane surface, and the excess vapour pressure increases as the radius of the drop becomes smaller. The atmosphere must therefore be supersaturated with respect to the thermodynamic (plane surface) vapour pressure if droplets are to condense. A water droplet of radius 0.01 μm, for example, requires a surrounding atmosphere supersaturated[a] by about 12.5 per cent if it is to survive. However, a droplet this size contains about 10^5 water molecules, a number exceedingly unlikely to come together at the same time by chance. A smaller initial droplet would contain fewer molecules, but require an even larger supersaturation. Atmospheres are sometimes supersaturated, but 12.5 per cent is about the maximum ever observed. One way out of the apparent dilemma is provided by condensation on nuclei. It is certainly suggestive that rainfall is greater over industrial areas where particle counts are highest, and that more rain falls on weekdays than at weekends. Before exploring the action of condensation nuclei, however, we should see why vapour pressure is influenced by curvature.

All liquids (and solids) possess an energy at the interface with their vapour that we call the *surface energy*. The energy is a function of surface area, σ, and can conveniently be expressed as $\mathrm{d}G = \gamma \mathrm{d}\sigma$ where $\mathrm{d}G$ is the surface contribution to Gibbs free energy and the constant γ is called the *surface tension*. It is in order to minimize the surface energy that liquids try to form spherical drops (the shape with minimum surface to volume ratio). The surface energy augments the chemical potential of the liquid phase, so that the equilibrium pressure of the vapour must be correspondingly larger. Some quantitative manipulation may be carried out to determine the free energy change associated with condensation, ΔG_{cond} in producing a droplet of radius r. For a plane surface, we would write

$$G_{\mathrm{liq}} = G^{\ominus}_{\mathrm{liq}} \tag{2.46}$$

$$G_{\mathrm{vap}} = G^{\ominus}_{\mathrm{vap}} + nRT \ln p, \tag{2.47}$$

where n is the number of moles in the system. Thus,

$$\Delta G_{\mathrm{cond}} = \Delta G^{\ominus}_{\mathrm{cond}} - nRT \ln p. \tag{2.48}$$

[a] If the saturated vapour pressure over a plane surface is p_{sat} and the vapour pressure actually present is p, then a supersaturation of 12.5 per cent means $p/p_{\mathrm{sat}} = 1.125$. The term *relative humidity* is given to the ratio expressed as a percentage ($100p/p_{\mathrm{sat}}$), so that percentage supersaturation is equal to the relative humidity minus 100.

At equilibrium $\Delta G_{\text{cond}} = 0$ and $p = p_{\text{sat}}$, so that

$$\Delta G^{\ominus}_{\text{cond}} = -nRT\ln p_{\text{sat}}. \tag{2.49}$$

For the curved-surface droplet, of surface area σ, eqn (2.46) must be modified to

$$G_{\text{liq}} = G^{\ominus}_{\text{liq}} + \gamma\sigma, \tag{2.50}$$

and

$$\Delta G_{\text{cond}} = \Delta G^{\ominus}_{\text{cond}} + \gamma\sigma - nRT\ln p \tag{2.51}$$

$$= \gamma\sigma - nRT\ln(p/p_{\text{sat}}). \tag{2.52}$$

For a droplet of radius r, $\sigma = 4\pi r^2$, and the number of moles, n, is $(4/3)\pi r^3\rho/M$ where ρ is the liquid density and M the relative molar mass. Hence

$$\Delta G_{\text{cond}} = 4\pi r^2\gamma - \tfrac{4}{3}\pi r^3\frac{\rho}{M}RT\ln\left(\frac{p}{p_{\text{sat}}}\right). \tag{2.53}$$

Figure 2.15 shows ΔG_{cond} calculated from this expression for a droplet in an undersaturated ($p < p_{\text{sat}}$) and a supersaturated ($p > p_{\text{sat}}$) vapour. In the first case, ΔG_{cond} becomes more positive with increasing drop radius, so there is no tendency for a drop to form. On the other hand, for the supersaturated vapour, ΔG_{cond} passes through a maximum at a droplet radius r_c, and then becomes smaller for increasing droplet radius. The maximum thus corresponds to a position of *unstable equilibrium*. If an embryonic droplet has radius $r < r_c$ it tends to evaporate, but if $r > r_c$ it tends to grow spontaneously at the expense of the vapour. Differentiation of eqn (2.53) with respect to r to obtain the position of maximum ΔG_{cond} allows calculation of the critical radius for any degree of supersaturation. The result is

$$r_c = \frac{2\gamma}{(\rho/M)RT\ln(p/p_{\text{sat}})}, \tag{2.54}$$

and is known as the *Kelvin equation*. It can be rewritten in terms of the vapour pressure, p_{drop}, needed to sustain any drop of radius r in (unstable) equilibrium

$$p_{\text{drop}} = p_{\text{sat}}\exp(2\gamma M/r\rho\mathbf{k}T) \tag{2.55}$$

The positive exponential in eqn (2.55) now shows explicitly that $p_{\text{drop}} > p_{\text{sat}}$, and that p_{drop} increases with decreasing r. Figure 2.16 shows supersaturation as a function of droplet radius calculated from eqn (2.55).

We are now in a position to see why condensation nuclei can promote condensation. Suppose water is condensed on a completely wettable nucleus of diameter 0.2 μm. The water film would be in unstable equilibrium with air

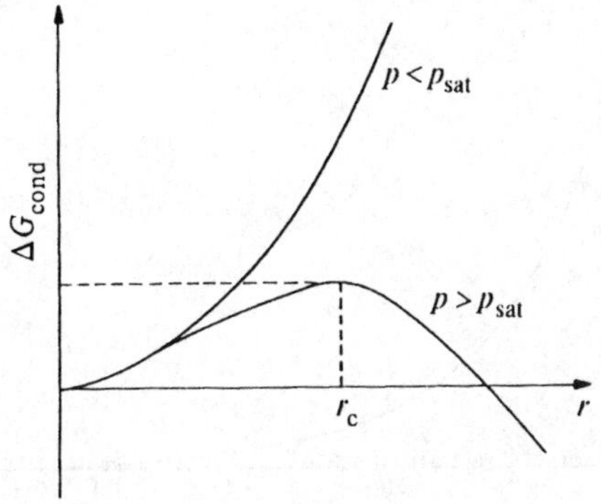

Fig. 2.15. Change in Gibbs free energy, ΔG_{cond} for condensation as a function of droplet size. If the vapour pressure is greater than the saturation vapour pressure with respect to a plane surface, droplet formation can be stable once a critical radius, r_c, has been exceeded.

supersaturated by as little as 0.6 per cent, a very reasonable atmospheric value. Slight increases in drop size would now lead to additional growth of the droplet as the free energy decreases yet further.

A most interesting situation arises if a condensation nucleus is soluble in water. Saturated vapour pressures are lower over solutions than over pure solvent, as a result of the lowered chemical potential of the solution. Solution droplets can therefore exist in equilibrium with their surroundings even when the relative humidity is less than 100 per cent (unsaturated air). The thermodynamic advantage gained by dissolving explains why salts can *deliquesce* in moist, but not necessarily saturated, air. Growth of the droplet may be limited, because at some stage the solution becomes sufficiently dilute that the droplet vapour pressure equals that of the surroundings. Additional condensation would tend to increase the droplet vapour pressure beyond that of the surroundings, and evaporation would follow. Conversely, evaporation from the equilibrium droplet would make the solution more concentrated and favour more condensation. Droplet size is therefore stabilized, and the droplet is in stable equilibrium with its surroundings. *Haze* droplets are often of this stabilized kind. Only if the stable equilibrium diameter is also near the critical value, r_c, for the unstable equilibrium of surface-tension-controlled growth can the particle go on condensing yet more material.

Heterogeneous nucleation on pre-existing particles, although one of the most important mechanisms, is not the only way in which clouds and aerosols

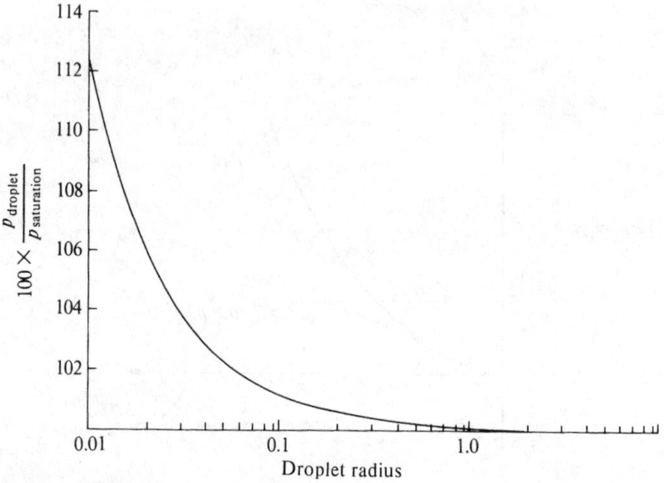

Fig. 2.16. Supersaturation needed to sustain drops in (unstable) equilibrium. Data are calculated from eqn (2.55) for pure water at 5°C and expressed as a percentage relative humidity with respect to equilibrium vapour pressure over a plane surface. From Wallace, J.M. and Hobbs, P.V. *Atmospheric science*, Academic Press, New York, 1977.

can be formed. Ion nucleation may be an important source of aerosol, especially in the stratosphere. Because of charge-induced dipole coulombic forces, ions have a greater tendency to attract molecules around them than do neutral species, between which the intermolecular forces are much weaker. Large 'cluster' ions are well known in the laboratory, and will be discussed in Chapter 6. In the stratosphere, balloon-borne mass spectrometers find the most abundant ions to be of the type $HSO_4^-(H_2SO_4)_n(H_2O)_m$ with n and m up to 4 or 5. For $n > 3$, these ions already develop properties similar to H_2SO_4–H_2O solution droplets. Ultimately, the ions can become neutralized to yield sulphuric acid aerosol, although there are undoubtedly species other than H_2SO_4 and H_2O present. The essential feature of such ionic processes is the reduction of free energy due to the electrostatic term; this reduction allows particle growth to start at lower supersaturations or smaller radii than would otherwise be possible.

Chemical composition measurements have also been made of single tropospheric aerosol particles, using laser-ionization mass spectrometry. Potassium, iron, and organic species are among the most common positive ions, while SO_4^-, NO_3^-, O^-, OH^-, and again organic species are the most common peaks in the negative-ion spectra.

Once a liquid or solid particle has passed its critical size for condensation, it may grow by several mechanisms. Continued condensation is one possibility, larger drops (lower saturated vapour pressure) growing at the expense of smaller ones (higher pressure). Vapour pressure differences are even more important in the growth of ice and snow crystals in water supercooled with respect to freezing. At say −10 °C, saturated vapour pressures over water are about 10 per cent larger than over ice, so that ice grows at the expense of liquid. *Agglomeration* or *coalescence* represents another very important mechanism for growth. A large particle moving through a field of smaller ones can sweep them up. Raindrops coalesce in clouds in this way, and if updraughts compete with gravitational settling to keep drops in the cloud, there may be time for considerable growth. A large raindrop (~ 2.5 mm radius at the beginning of a shower) will need to have collected up the equivalent of $(2500/2.5)^3 = 10^9$ droplets of initial size 2.5 μm! At least it is clear why heavy rain is associated with clouds of considerable thickness (e.g. cumulus) and drizzle with thinner ones (e.g. stratus).

2.6 Light scattering

Electromagnetic radiation may be scattered by a medium even when it is not absorbed, and, as suggested in Section 1.3, the scattering behaviour depends on the size of the scattering body relative to the wavelength of radiation. Electrons in any object are driven into oscillation by incident light, and in due course those electrons re-emit radiation. For a perfectly homogeneous non-absorbing medium, all secondary re-emitted waves interfere destructively except in the original direction of the incident beam. To an observer, therefore, the medium will appear to have had no effect. If, however, there are inhomogeneities in the medium, scattered radiation in other directions does not cancel exactly, so that light can be observed away from the original direction of propagation. At the same time, the intensity of the forward beam is decreased by an amount equivalent to the light scattered. Analogous equations to those for absorption can be written, (2.12) being replaced by

$$-\mathrm{d}I = In\sigma_s \mathrm{d}z, \tag{2.56}$$

where σ_s, the *scattering cross-section*, replaces σ_a, the absorption cross-section. Integrated forms are exactly equivalent to (2.13) and (2.14). A quantity referred to as the *scattering area coefficient*, κ, compares (as a ratio) the scattering cross-section with the geometrical cross-sectional area of the scattering particles.

Particles which are small compared with the wavelength of the incident light have their entire electronic distribution distorted by the electromagnetic radiation. Scattering in these circumstances is called *Rayleigh scattering*. Conversely, several different waves can be scattered from large particles, and interference of the various waves produces a different scattering behaviour, both in magnitude and angular distribution. The comparison between wavelength λ and particle radius r can be expressed most conveniently in terms

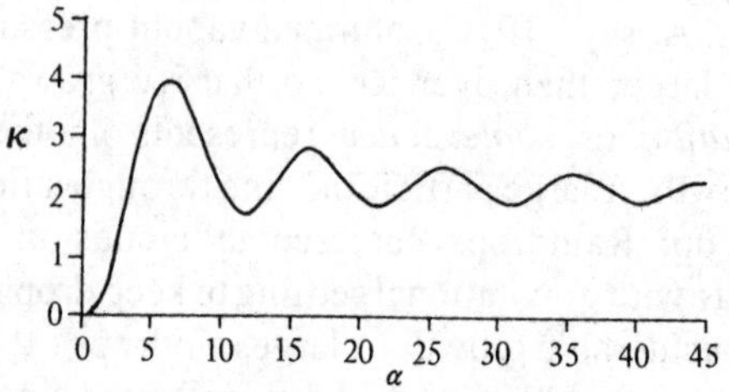

Fig. 2.17. Scattering area coefficient, κ, as a function of size parameter, α, for a refractive index of 1.33. From Wallace J.M. and Hobbs, P.V. *Atmospheric science*, Academic Press, New York, 1977.

of the dimensionless size parameter $\alpha = 2\pi r/\lambda$. For $\alpha \ll 1$, we are in the Rayleigh scattering regime. Molecular diameters for gases are typically one thousand times smaller than visible and near ultraviolet wavelengths, so that non-aggregated species in the atmosphere cause Rayleigh scattering at these wavelengths. Similarly, weather radar operates at 'microwave' wavelengths of a few centimetres, so that scattering by raindrops ($r < 2.5$ mm) is also Rayleigh in nature.

The scattering parameter, κ, depends not only on α, but on the refractive index of the particles. For the Rayleigh scattering case of $\alpha \ll 1$, it can be shown that κ is proportional to α^4: the scattering intensity is thus proportional to $1/\lambda^4$ for a fixed-radius particle. That result explains why light scattered from an aerosol-free sky is blue, while that transmitted through long pathlengths at sunrise and sunset is red. Fourth-power proportionality emphasizes the wavelength dependences. Red light ($\lambda \sim 650$ nm) and blue light ($\lambda \sim 450$ nm) are scattered in the relative ratio $1{:}(650/450)^4 = 1{:}4.4$. Rayleigh scattering theory predicts that the scattered intensity at any angle θ to the incident beam is proportional to $(1 + \cos^2\theta)$ for unpolarized radiation. Maximum intensity therefore occurs in the forward ($\theta = 0$) and backward ($\theta = 180°$) directions. As the inequality $\alpha \ll 1$ ceases to hold, so the angular dependence of scattering ceases to follow the simple Rayleigh formula, and κ also deviates from its α^4 dependence. For $\alpha \gtrsim 50$, κ becomes ~ 2, and the angular distribution of scattered radiation can be described by conventional reflection, refraction, and interference. Visible radiation ($\lambda \sim 500$ nm) thus undergoes scattering of this kind by particles of radius greater than about 4 μm. The size range includes 'typical cloud drops' (~ 10 μm) and larger particles. Distinctive optical phenomena such as glories, halos, and rainbows are produced in this regime.

We have so far avoided discussion of the region for which α is neither much less than nor much greater than unity. For visible radiation, the range of sizes involved is from a few nm to a few μm: that is, it spans exactly that range of sizes conventionally associated with Aitken particles, condensation nuclei, and aerosols in general (see Section 1.3). A theory of scattering applicable to all values of α was worked out by Gustav Mie as long ago as 1908; the term *Mie scattering* is used more restrictively for α typically in the range 0.05 to 50. The angular distribution of scattered radiation is very complicated, and varies rapidly with α in this range. Scattering area coefficients (κ) show an oscillatory dependence on α, as illustrated in Fig. 2.17 for refractive index 1.33 (water). Note that for $\alpha \sim 5$, corresponding to $r \sim 0.4$ μm in visible light, the scattering is particularly intense. Colours associated with Mie scattering of sunlight by haze, smoke, dust, etc., depend first on whether the particle sizes cover several oscillations of the κ–α relationship: if they do, the scattering is fairly neutral. For uniform particles, however, the colour cast will depend on whether κ increases or decreases with α (and, thus, in the reverse sense, with λ).

Bibliography

Books on general atmospheric physics and meteorology

The atmosphere and ocean: a physical introduction. Wells, N. (Wiley, Chichester, 1997).

Clouds, chemistry and climate. Crutzen, P.J. (ed.) (Springer-Verlag, Berlin, 1996).

Fundamentals of atmospheric physics. Salby, M.L. (Academic Press, San Diego, 1996).

Principles of atmospheric physics and chemistry. Goody, R.M. (Oxford University Press, Oxford, 1995).

Global physical climatology. Hartmann, D.L. (Academic Press, San Diego, 1994).

An introduction to dynamic meteorology. Holton, J.R., 3rd edn. (Academic Press, San Diego, 1992).

Atmosphere, weather and climate. Barry, R.G. and Chorley, R.J., 5th edn. (Routledge, London, 1987).

Middle Atmosphere Dynamics. Andrews, D.G., Holton, J.R., and Leovy, C.B. (Academic Press, New York, 1987).

The solar-terrestrial environment. Hargreaves, J.K. (Cambridge University Press, Cambridge, 1992).

The atmosphere. Lutgens, F. K. and Tarbuck, E. J. (Prentice-Hall, Englewood Cliffs, New Jersey, 1982).

Atmospheres. Goody, R. M. and Walker, J. C. G. (Prentice-Hall, Englewood Cliffs, New Jersey, 1972).

Atmospheric science. Wallace, J. M. and Hobbs, P. V. (Academic Press, New York, 1977).

The physics of atmospheres. Houghton, J. T. 2nd edn. (Cambridge University Press, Cambridge, 1986).

Aeronomy (Parts A & B). Banks, P. M. and Kockarts, G. (Academic Press, New York, 1973).

Theory of planetary atmospheres. Chamberlain, J. W. and Hunten, D. M. 2nd edn. (Academic Press, New York, 1987).

The two books listed below describes the major techniques available for determining temperatures, densities, and composition in the atmosphere. The special issue of Phil. Trans. discusses in situ methods as well as remote sounding.

Remote sensing of the lower atmosphere. Stephens, G.L. (Oxford University Press, Oxford, 1994).

Remote sounding of atmospheres. Houghton, J. T., Taylor, F. W., and Rodgers, C. D. (Cambridge University Press, Cambridge, 1984).

The middle atmosphere as observed from balloons, rockets and satellites. *Phil. Trans. R. Soc.* **A296**, 1–268 (1980).

Section 2.2

Radiation and cloud processes in the atmosphere. Liou, K.-N. (Oxford University Press, Oxford, 1995).

Atmospheric Radiative Transfer. Lenoble, J. (A. Deepak Publishing, Hampton, VA, 1993).

The Sun in time. Sonnett, C.P., Giampapa, M.S., and Mattews, M.S. (eds.) (University of Arizona Press, Tucson, AZ, 1991).

Atmospheric Radiation. Goody, R. M. and Yung, Y. L. (Oxford University Press, Oxford, 1989).

Water vapour and greenhouse trapping: the role of far infrared radiation. Sinha, A. and Harries, J.E. *Geophys. Res. Lett.* **22**, 2147 (1995).

Models for infrared atmospheric radiation. Tiwasi, S. N. *Adv. Geophys.* **20**, 1 (1978).

A comparison of the contribution of various gases to the greenhouse effect. Rohde, H. *Science* **248**, 1217 (1990).

A search for human influences on the thermal structure of the atmosphere. Santer, B.D., Taylor, K.E., Wigley, T.M.L., Johns, T.C., Jones, P.D., Karoly, D.J., Mitchell, J.F.B., Oort, A.H., Penner, J.E., Ramaswamy, V., Schwarzkopf, M.D., Stouffer, R.J., and Tett, S. *Nature* **382**, 39 (1996).

The effect of clouds on the Earth's solar and infrared radiation budget. Herman, G. F., Man-Li, Wu. C., and Johnson, W. T. *J. atmos. Sci.* **37**, 1251 (1980).

The runaway greenhouse: a history of water on Venus. Ingersoll, A. P. *J. atmos. Sci.* **26**, 1191 (1969).

Response of Earth's atmosphere to increases in solar flux and implications for loss of water from Venus. Kasting, J. F., Pollack, J. B., and Ackerman, T. P. *Icarus* **57**, 335 (1984).

Section 2.3.1

The atmospheric boundary layer. Garratt, J.R. (Cambridge University Press, Cambridge, 1992).

Retrieval of upper stratospheric and mesospheric temperature profiles from Millimeter-Wave Atmospheric Sounder data. von Engeln, A., Langen, J., Wehr, T., Buhler, S., and Kunzi, K. *J. Geophys. Res.* **103**, 31735 (1998).

The Brewer-Dobson circulation: Dynamics of the tropical upwelling. Plumb, R.A. and Eluszkiewicz, J. *J. atmos. Sci.* **56**, 868 (1999).

Dynamics of the tropical middle atmosphere: A tutorial review. Hamilton, K. *Atmos. Ocean*, **36**, 319 (1998).

Laminae in the tropical middle stratosphere: Origin and age estimation. Jost, H., Loewenstein, M., Pfister, L., Margitan, J.J., Chang, A.Y., Salawitch, R.J., and Michelsen, H.A. *Geophys. Res. Letts.* **25**, 4337 (1998).

Section 2.3.2

Escape of gases from the atmosphere: see also Bibliographies for Chapters 8 and 9

Nonthermal escape of the atmospheres of Venus, Earth, and Mars. Shizgal, B.D., and Arkos, G.G. *Rev. Geophys.* **34,** 483 (1996).

Upper atmospheric waves, turbulence, and winds — importance for mesospheric and thermospheric studies. Killeen, T.L. and Johnson, R.M. *Rev. Geophys.* **33**, 737 (1995).

Escape of atmospheres and loss of water. Hunten, D. M., Donahue, T. M., Walker, J. C. G., and Kasting, J. F. in *Origin and evolution of planetary and satellite atmospheres*. Atreya, S. K., Pollack, J. B., and Matthews, M. S. (eds.) (University of Arizona Press, Tucson, 1989).

Hydrogen and deuterium loss from the terrestrial atmosphere: a quantitative assessment of non-thermal escape fluxes. Yung, Y. L., Wen. J-S., Moses, J.I., Landry, B. M., Allen, M., and Hsu, K.-J. *J. geophys. Res.* **94**, 14971 (1989).

Thermal and non-thermal escape mechanisms for terrestrial bodies. Hunten, D. M. *Planet. Space Sci.* **30**, 773 (1982).

Escape of atmospheres, ancient and modern. Hunten, D.M., *Icarus* **85**, 1 (1990).

Section 2.3.3

Vertical transport mechanisms

The first paper describes mathematical models for vertical transport (and is thus also relevant to the models discussed in Section 3.6). Exchange of minor components between troposphere and stratosphere is discussed in the next papers, and the possibility that water vapour is removed at a 'cold trap' near the tropopause is considered.

Dynamic effects on atomic and molecular oxygen distributions in the upper atmosphere: a numerical solution to equations of motion and continuity. Shimazaki, T. *J. atmos. terr. Phys.* **29**, 723 (1967).

A new perspective on the dynamical link between the stratosphere and troposphere. Hartley, D.E., Villarin, J.T., Black, R.X., and Davis, C.A. *Nature, Lond.* **391**, 471 (1998).

An atmospheric tape-recorder—the imprint of tropical tropopause temperatures on stratospheric water-vapor. Mote, P.W., Rosenlof, K.H., McIntyre, M.E., Carr, E.S., Gille, J.C., Holton, J.R., Kinnersley, J.S., Pumphrey, H.C., Russell, J.M. and Waters, J.W. *J. geophys. Res.* **101**, 3989 (1996).

A globally balanced two-dimensional middle atmosphere model: Dynamical studies of mesopause meridional circulation and stratosphere–mesosphere exchange. Zhu, X., Swaminathan, P.K., Yee, J.H., Strobel, D.F., and Anderson, D. *J. geophys. Res.* **102**, 13095 (1997).

Troposphere to stratosphere transport at low latitudes as studied using HALOE observations of water vapour 1992-1997. Jackson, D.R., Driscoll, S.J., Highwood, E.J., Harries, J.E., and Russell, J.M. *Q.J. Royal Meteor. Soc.* **124**, 169)1998).

High-resolution lidar measurements of stratosphere-troposphere exchange. Eisele, H., Scheel, H.E., Sladkovic, R., and Trickl, T. *J. atmos. Sci.* **56**, 319 (1999).

Diagnosing extratropical synoptic-scale stratosphere-troposphere exchange: A case study. Wirth, V. and Egger, J. *Quart. J. Royal Meteor. Soc.* **125,** 635 (1999).

The Brewer-Dobson circulation: dynamics of the tropical upwelling. Plumb, R.A. and Eluszkiewicz, J. *J. atmos. Sci.* **56**, 868 (1999).

Use of on-line tracers as a diagnostic tool in general circulation model development 2. Transport between the troposphere and stratosphere. Rind, D., Lerner, J., Shah, K., and Suozzo, R. *J. geophys. Res.* **104,** 9151 (1999).

Phosphorus 32, phosphorus 37, beryllium 7, and lead 210: Atmospheric fluxes and utility in tracing stratosphere troposphere exchange. Benitez-Nelson, C.R. and Buesseler, K.Q. *J. geophys. Res.* **104,** 11745 (1999).

Transport of tropospheric air into the lower midlatitude stratosphere. Vaughan, G. and Timmis, C. *Quart. J. Royal Meteor. Soc.* **124**, 1559 (1998).

Troposphere-to-stratosphere transport in the lowermost stratosphere from measurements of H_2O, CO_2, N_2O and O_3. Hintsa, E.J., Boering, K.A., Weinstock, E.M., Anderson, J.G., Gary, B.L., Pfister, L., Daube, B.C., Wofsy, S.C., Loewenstein, M., Podolske, J.R., Margitan, J.J., and Bui, T.P. *Geophys. Res. Letts.* **25**, 2655 (1998).

Vertical velocity, vertical diffusion, and dilution by midlatitude air in the tropical lower stratosphere. Mote, P.W., Dunkerton, T.J., McIntyre, M.E., Ray, E.A., Haynes, P.H., and Russell, J.M., III. *J. geophys. Res.* **103**, 8651 (1998).

Stratosphere-troposphere exchange of ozone associated with the equatorial Kelvin wave as observed with ozonesondes and rawinsondes. Fujiwara, M., Kita, K., and Ogawa, T. *J. Geophys. Res.* **103**, 19173 (1998).

Tracer lamination in the stratosphere: a global climatology. Appenzeller, C. and Holton, J.R. *J. geophys. Res.* **102**, 13555 (1997).

Tracer transport in the tropical stratosphere due to vertical diffusion and horizontal mixing. Hall, T.M. and Waugh, D.W. *geophys. Res. Letts.* **24**, 1383 (1997).

In situ evidence of rapid, vertical, irreversible transport of lower tropospheric air into the lower troposphere by convective cloud turrets and by larger scale upwelling in tropical cyclones. Danielsen, E.F. *J. geophys. Res.* **98**, 8665 (1993).

Radon measurements in the lower tropical stratosphere: evidence for rapid vertical transport and dehydration of tropospheric air. Kritz, M. A., Rosner, S.W., Kelly, K.K., Loewenstein, M., and Chan, K.R. *J. geophys. Res.* **99**, 8725 (1993).

Stratosphere–troposphere exchange. Holton, J.R., Haynes, P.H., McIntyre, M.E., Douglass, A.R., Rood, R.B., and Pfister, L. *Rev. Geophys.* **33**, 403 (1995).

Hemispheric asymmetries in water vapor and inferences about transport in the lower stratosphere. Rosenlof, K.H., Tuck, A.F., Kelly, K.K., Russell, J.M., III, and McCormick, M.P. *J. geophys. Res.* **102**, 13213 (1997).

Measurements of stratospheric carbon dioxide and water vapor at northern midlatitudes: implications for troposphere-to-stratosphere transport. Boering, K.A., Hintsa, E.J., Wofsy, S.C., Anderson, J.G., Daube, J.C., Jr., Dessler, A.E., Loewenstein, M., McCormick, M.P., Podolske, J.R., Weinstock, E.M. and Yue, G.K. *Geophys. Res. Lett.* **22**, 2737 (1995).

Interannual variability in the stratospheric–tropospheric circulation in a coupled ocean–atmosphere GCM. Kitoh, A., Koide, H., Kodera, K., Yukimoto, S., and Noda, A. *Geophys. Res. Lett.* **23**, 543 (1996).

Seasonal-variation of isentropic transport out of the tropical stratosphere. Waugh, D.W. *J. geophys. Res.* **101**, 4007 (1996).

The transport of minor atmospheric constituents between troposphere and stratosphere. Robinson, G. D. *Q. J. R. Meteorol. Soc.* **106**, 227 (1980).

Stratospheric-tropospheric exchange processes. Reiter, E. R. *Rev. Geophys. & Space Phys.* **13**, 459 (1975).

Why is the stratosphere so dry? Newell, R. E. *Nature, Lond.* **300**, 686 (1982).

The distribution of water vapour in the stratosphere. Harries, J. E. *Rev. Geophys. & Space Phys.* **14**, 565 (1976).

Stratospheric H_2O. Ellsaesser, H. W., Harries, J. E., Kley, D., and Penndorf, R. *Planet. Space Sci.* **28**, 827 (1980).

Mesospheric water vapor. Deguchi, S. *J. geophys. Res.* **87**, 1343 (1982).

Section 2.4

Winds, circulations, and models for them

Dynamics of atmospheric motion. Dutton, J.A. (Dover, New York, 1996).

Dynamics of meteorology and climate. Scorer, R.S. (Wiley, Chichester, 1997).

The kinematics of mixing. Ottino, J.M.. (Cambridge University Press, Cambridge, 1990).

The fluid mechanics of the atmosphere. Brown, R.A. (Academic Press, San Diego, 1991).

Atmospheric convection. Emanuel, K.A. (Oxford University Press, Oxford, 1994).

Symmetric circulations of planetary atmospheres. Koschmieder, E. L. *Adv. Geophys.* **20**, 131 (1978).

The range and unity of planetary circulations. Williams, G. P. and Holloway, J. L., Jr. *Nature, Lond.* **297**, 295 (1982).

On the sensitivity of a general circulation model to changes in cloud structure and radiative properties. Hunt, G. E. *Tellus* **34**, 29 (1982).

Changing lower stratospheric circulation: The role of ozone and greenhouse gases. Graf, H.F., Kirchner, I., and Perlwitz, J. *J. Geophys. Res.* **103**, 11251 (1998).

Timescales for the stratospheric circulation derived from tracers. Hall, T.M. and Waugh, D.W. *J. geophys. Res.* **102**, 8991 (1997).

Quantification of lower stratospheric mixing processes using aircraft data. Balluch, M.G. and Haynes, P.H. *J. geophys. Res.* **102**, 23487 (1997).

Quantifying transport between the tropical and mid-latitude lower stratosphere. Volk, C.M., Elkins, J.W., Fahey, D.W., Salawitch, R.J., Dutton, G.S., Gilligan, J.M., Proffitt, M.H., Loewenstein, M., Podolske, J.R., Minschwaner, K., Margitan, J.J., and Chan, K.R. *Science* **272**, 1763 (1996).

Sulfur hexafluoride–a powerful new atmospheric tracer. Maiss, M., Steele, L.P., Francey, R.J., Fraser, P.J., Langenfels, R.L., Trivett, N.B.A., and Levin, I. *Atmos. Environ.* **30**, 1621 (1996).
Isentropic mass exchange between the tropics and extratropics in the stratosphere. Chen, P., Holton, J.R., O'Neill, A., and Swinbank, R. *J. atmos. Sci.* **51**, 3006 (1994).
Model study of atmospheric transport using carbon 14 and strontium 90 as inert tracers. Kinnison, D.E., Johnston, H.S., and Wuebbles, D.J. *J. geophys. Res.* **99**, 20647 (1994).

Global circulation and Global Circulation Models (GCMs)

A very high resolution general circulation model simulation of the global circulation in Austral winter. Jones, P.W., Hamilton, K., and Wilson, R.J. *J. atmos. Sci* **54**, 1107 (1997).
The forcing and maintenance of global monsoonal circulations: an isentropic analysis. Johnson, D.R. *Advan. Geophys.* **31**, 43 (1989).
Climatology of the SKYHI troposphere–stratosphere–mesosphere general-circulation model. Hamilton, K., Wilson, R.J., Mahlman, J.D., and Umscheid, L.J. *J. Atmos. Sci.* **52**, 5 (1995).
Evaluation of the SKYHI general-circulation model using aircraft N_2O measurements. 1. Polar winter stratospheric meteorology and tracer morphology. Strahan, S.E. and Mahlman, J.D. *J. geophys. Res.* **99** (1994).
A further analysis of internal variability in a perpetual January integration of a troposphere-stratosphere-mesosphere GCM. Yoden, S., Naito, Y., and Pawson, S. *J. Meteor. Soc. Japan.* **74**, 175 (1996).
Simulations of stratospheric sudden warmings in the Berlin troposphere–stratosphere–mesosphere GCM. Erlebach, P., Langematz, U., and Pawson, S. *Ann. Geophys.* **14**, 443 (1996).
Dynamics of transport processes in the upper troposphere. Mahlman, J.D. *Science* **276**, 1079 (1997).
The Berlin troposphere–stratosphere–mesosphere GCM: Climatology and forcing mechanisms. Langematz, U. and Pawson, S. *Q. J. Roy. Meteorol. Soc.* **123**, 1075 (1997).

Stratospheric behaviour and circulation

Physics of the upper polar atmosphere. Brekke, A. (Wiley, Chichester, 1997).
The Brewer-Dobson circulation in the light of high altitude *in situ* aircraft observations. Tuck, A.F., Baumgardner, D., Chan, K.R., Dye, J.E., Elkins, J.W., Hovde, S.J., Kelley, K.K., Loewenstein, M., Margitan, J.J., May, R.D., Podolske, J.R., Proffitt, M.H., Rosenlof, K.H., Smith, W.L., Webster, C.R., and Wilson, J.C. *Q. J. Roy. Meteorol. Soc.* **123**, 1 (1997).
The climatology of the stratospheric thin air model. Kinnersley, J.S. *Q. J. Roy. Meteorol. Soc.* **122**, 219 (1996).
Variability in daily, zonal mean lower-stratospheric temperatures. Christy, J.R. and Drouilhet, S.J. *J. Climate* **7**, 106 (1994).

Interhemispheric differences in stratospheric water vapour during late winter, in version 4 MLS measurements. Morrey, M.W. and Hanwood, R.S. *Geophys. Res. Lett.* **25**, 147 (1998).

Characteristics of stratospheric winds and temperatures produced by data assimilation. Coy, L. and Swinbank, R. *J. geophys. Res.* **102**, 25763 (1997).

Tracer lamination in the stratosphere: A global climatology. Appenzeller, C. and Holton, J.R. *J. geophys. Res.* **102**, 13555 (1997).

Stratospheric thermal damping times. Newman, P.A. and Rosenfield, J.E. *Geophys. Res. Lett.* **24**, 433 (1997).

Long-lived tracer transport in the Antarctic stratosphere. Strahan, S.E., Nielsen, J.E., and Cerniglia, M.C. *J. geophys. Res.* **101**, 26615 (1996).

Comparison of analyzed stratospheric temperatures and calculated trajectories with long-duration balloon data. Knudsen, B.M., Rosen, J.M., Kjome, N.T., and Whitten, A.T. *J. geophys. Res.* **101**, 19137 (1996).

The influences of zonal flow on wave breaking and tropical–extratropical interaction in the lower stratosphere. Chen, P. *J. Atmos. Sci.* **53**, 2379 (1996).

Space-borne H_2O observations in the Arctic stratosphere and mesosphere in the spring of 1992. Aellig, C.P., Bacmeister, J., Bevilacqua, R.M., Daehler, M., Kriebel, D., Pauls, T., Siskind, D., Kampfer, N., Langen, J., Hartmann, G., Berg, A., Park, J.H., and Russell, J.M. *Geophys. Res. Lett.* **23**, 2325 (1996).

Stratospheric climate and variability from a general-circulation model and observations. Manzini, E. and Bengtsson, L. *Climate Dynamics* **12**, 615 (1996).

Physics and chemistry of the stratosphere. Pommereau, J.P. *Compt. Rend. Ser II* **A322**, 811 (1996).

Rossby-wave phase speeds and mixing barriers in the stratosphere. 1. Observations. Bowman, K.P. *J. Atmos. Sci.* **53**, 905 (1996).

Dynamics of wintertime stratospheric transport in the Geophysical Fluid-Dynamics Laboratory SKYHI general-circulation model. Eluszkiewicz, J., Plumb, R.A., and Nakamura, N. *J. geophys. Res.* **100**, 20883 (1995).

Millimeter-wave spectroscopic measurements over the south pole. 1. A study of stratospheric dynamics using N_2O observations. Crewell, S., Cheng, D.J., DeZafra, R.L., and Trimble, C. *J. geophys. Res.* **100**, 20839 (1995).

Reconsideration of the cause of dry air in the southern middle latitude stratosphere. Mote, P.J. *Geophys. Res. Lett.* **22**, 2025 (1995).

Subsidence of the Arctic stratosphere determined from thermal emission of hydrogen fluoride. Traub, W.A., Jucks, K.W., Johnson, D.G., and Chance, K.V. *J. geophys. Res.* **100**, 11261 (1995).

Sudden stratospheric warming

Rayleigh lidar observations of thermal structure and gravity-wave activity in the high Arctic during a stratospheric warming. Whiteway, J.A. and Carswell, A.I. *J. Atmos. Sci.* **51**, 3122 (1994).

Simulations of the February 1979 stratospheric sudden warming—model comparisons and 3-dimensional evolution. Manney, G.L., Farrara, J.D., and Mechoso, C.R. *Month. Weather Rev.* **122**, 1115 (1994).

Wave mean flow interaction and stratospheric sudden warming in an isentropic model. Tao, X. *J. Atmos. Sci.* **51**, 134 (1994).
The dynamics of sudden stratospheric warmings. Holton, J.R. *Ann. Rev. Earth & planet. Sci.* **8**, 169 (1980).
The dynamics of stratospheric warmings. O'Neill, A. *Q. J. Roy. Meteorol. Soc.* **106**, 659 (1980).
Tropospheric circulation changes associated with stratospheric sudden warmings—a case-study. Kodera, K. and Chiba, M. *J. geophys. Res.* **100**, 11055 (1995).
Downward transport in the upper-stratosphere during the minor warming in February 1979. Kouker, W., Beck, A., Fischer, H., and Petzoldt, K. *J. geophys. Res.* **100**, 11069 (1995).

The polar vortex

See Bibliography for Section 4.7.2

Sections 2.5 and 2.6

The formation of aerosols and clouds, and the optical properties of molecules and particles in the atmosphere. See also the references for Section 1.3.

Clouds and Storms. Ludlam, F.H. (Pennsylvania State University Press, University Park, 1980).
Clouds, chemistry and climate. Crutzen, P.J. (ed.) (Springer-Verlag, Berlin, 1996).
Radiation and cloud processes in the atmosphere. Liou, K.-N. (Oxford University Press, Oxford, 1995).
Light scattering by small particles. Van der Hulst, H.C. (Dover, New York, 1981).
Rainbows, halos and glories. Greenler, R. (Cambridge University Press, Cambridge, 1980).
Tropospheric clouds and lower stratospheric heating rates: Results from late winter in the Southern Hemisphere. Hicke, J., and Tuck, A. *J. geophys. Res.* **104**, 9309 (1999).
An interactive cirrus cloud radiative parameterization for global climate models. Joseph, E., and Wang, W.C. *J. geophys. Res.* **104**, 9501 (1999).
Nucleation: measurements, theory, and atmospheric applications. Laaksonen, A., Talanquer, V., and Oxtoby, D.W. *Annu. Rev. Phys. Chem.* **46**, 489 (1995).
Nucleation processes in the atmosphere. Mirabel, P. and Miloshev, N. in *Low temperature chemistry of the atmosphere.* Moortgat, G.K., Barnes, A.J., Le Bras, G., and Sodeau, J.R. (eds.) (Springer-Verlag, Berlin, 1994).

Aerosol and particle composition and behaviour

Chemical composition of single aerosol particles as Idaho Hill. Murphy, D.M. and Thomson, D.S. *J. geophys. Res.* **102**, 6341, 6353 (1997).
Evolution of the stratospheric aerosol in the northern hemisphere following the June 1991 volcanic eruption of Mount Pinatubo—role of tropospheric–stratospheric exchange and transport. Jonsson, H.H., Wilson, J.C., Brock, C.A., Dye, J.E., Ferry, G.V., and Chan, K.R. *J. geophys. Res.* **101**, 1553 (1996).

3-Dimensional transport simulations of the dispersal of volcanic aerosol from Mount Pinatubo. Fairlie, T.D.A. *Q. J. Roy. Meteorol. Soc.* **121**, 1943 (1995).

Particulate matter in the atmosphere: primary and secondary particles. Chapter 12 in *Atmospheric chemistry*. Finlayson-Pitts, B. J. and Pitts, J. N., Jr. (John Wiley, Chichester, 1986).

Growth laws for the formation of secondary ambient aerosols: implications for chemical conversion mechanisms. McMurry, P. H. and Wilson, J. C. *Atmos. Environ. 16, 121 (1982).*

Stratospheric aerosols: observation and theory. Turco, R. P., Whitten, R. C., and Toon, O. B. *Rev. Geophys. & Space Phys.* **20**, 233 (1982).

Stratospheric aerosol particles and their optical properties. Cadle, R. D. and Grams, G. W. *Rev. Geophys. & Space Phys.* **13**, 475 (1975).

Light scattering in planetary atmospheres. Hansen, J. E. and Travis, L. D. *Space Sci. Rev.* **16**, 527 (1974).

The scattering of light and other electromagnetic radiation. Kerker, M. (Academic Press, New York, 1969).

Optical properties of aerosols—comparison of measurements with model calculations. Sloane, C. S. *Atmos. Environ.* **17**, 409 (1983).

Rayleigh scattering by air. Bates, D. R. *Planet. Space Sci.* **32**, 785 (1984).

3 Photochemistry and kinetics applied to atmospheres

Solar radiation not only heats planetary atmospheres, in the ways described in Chapter 2, but it also drives much of the disequilibrium chemistry of those atmospheres through photochemically initiated processes. The consequences of primary photochemical change are often decided by the relative rates of competing chemical reactions: that is to say, kinetically. It is the purpose of this chapter to summarize the principles of photochemistry and kinetics as they are applied to atmospheric chemistry. Attention is paid to features that would be thought of as esoteric detail in laboratory chemistry, but that are of major significance in atmospheres.

3.1 Photochemical change

An atmosphere is a giant photochemical reactor, in which the light source is the Sun. Radiation, generally in the visible and ultraviolet regions, either fragments the atmospheric constituents to produce atoms, radicals, and ions, or excites the constituents, without chemical change, to alter their reactivity.

To understand why visible and ultraviolet radiation are so important, we must first examine the magnitude of one quantum of electromagnetic radiation, the photon. According to *Planck's Law*, the energy of one photon of frequency is $h\nu$, where h is Planck's constant. Photochemists usually write '$h\nu$' in chemical equations as a shorthand for the photon that is a reactant, e.g.

$$O_3 + h\nu \rightarrow O_2 + O. \tag{3.1}$$

Planck's constant is known, so that the energy per photon can be calculated immediately. Chemists often find it convenient to consider energies (and other thermodynamic quantities) for one mole (i.e. ~6.022×10^{23} molecules or atoms) of material. Scaled in this way, the energy per 'mole' of photons, E, is given by

$$E = Lh\nu, \tag{3.2}$$

where L is the Avogadro constant; in terms of wavelength of radiation, λ, and velocity of light, c, eqn (3.2) becomes

$$E = Lhc/\lambda. \tag{3.3}$$

Substituting values for L, h, and c, we find that with E and λ expressed in $kJ\,mol^{-1}$ and nanometres (nm), respectively,

$$E = 119625/\lambda. \tag{3.4}$$

This equation can be written in terms of the *electron volt* (eV), a unit of energy commonly used in physics and chemical physics, especially where charged species are involved. It is the work done in moving *L* electrons through a potential difference of 1 volt, and is numerically equivalent to 96.485 kJ mol^{-1}. Using this conversion factor, the energy is then 1239.8/λ eV. Thus the red extreme of the visible spectrum (~ 800 nm) corresponds to about 150 kJ mol^{-1}, and the violet extreme (~ 400 nm) to twice that energy. At shorter wavelengths lies the ultraviolet, conventionally subdivided into 'near' (λ ~ 400–200nm), 'vacuum' (VUV, λ ~ 200–100 nm) and 'extreme' (EUV, λ ~ 100–10 nm) regions, with successively higher energies. X-rays, gamma rays, and galactic cosmic radiation (GCR) constitute the shortest (λ < 10 nm) wavelengths of the electromagnetic spectrum. The very high photon energies are associated with an ability to *penetrate* as well as to ionize atmospheric gases. We shall have to consider such radiation (especially GCR) as a source of ionization (Chapter 6), but the interactions are not usually thought of as photochemical. The photon energy of red light is comparable with the bond energies of rather loosely bound chemical species. Of common gaseous inorganic species, ozone is, in fact, the only compound with such a small bond energy (the O–O_2 energy is ~ 105 kJ mol^{-1}); nitrogen dioxide, with an O–NO bond energy of ~ 300 kJ mol^{-1} (≅ 399 nm) is more typical. Ionization becomes possible at the shorter wavelengths (e.g. ionization of NO at ~ 135 nm). The point is that the visible region contains the lowest energy photons of the entire electromagnetic spectrum that are capable of promoting chemical change in single quantum events. Many *more* photons arrive each second at a planet at longer wavelengths, but they can only heat the atmosphere up.

It so happens that the wavelengths at which chemical change becomes possible also correspond roughly to the energies at which *electronic transitions* are excited in atoms and molecules. Longer wavelengths tend to excite molecular vibrations or rotations. Although high vibrational levels are involved in some photochemical processes, electronic excitation is the spectroscopic step most frequently associated with photochemical change. Much of the discussion of the following sections will therefore centre around the fates of electronically excited species. We should note here that electronic excitation not only carries with it the energy absorbed, but also a new electronic structure of the chemical species. A profound influence on the chemistry and reactivity can result. Electronic transitions are governed by the ordinary rules of spectroscopy. We suggest here, as a reminder, the types of consideration that apply. First of all, absorption of radiation only occurs if an upper energy level of the species exists which is separated from the lower level by an energy equal to that of the incident photon ('*resonance condition*'). Intensities of transitions are determined in part by the *electronic transition moment*, which itself depends on the wavefunctions of lower and upper states. *Selection rules* aim to avoid the calculation of the transition moment by stating conditions which need to be satisfied if the transition moment is not to vanish. These rules are often formulated for *electric dipole* interactions, which are the most intense. Familiar examples of the rules are that $\mathbf{\Delta S} = 0$, or that, in atoms, $\mathbf{\Delta L} = \pm 1$.

Forbidden transitions can be observed either because the quantum number for which the rule is stated does not rigorously describe the system (e.g. **L** and **S** are not 'good' quantum numbers in heavy atoms because of spin–orbit coupling), or because the transition occurs by an interaction (e.g. magnetic dipole) other than that envisaged. On top of the electronic contribution to intensity, the vibrational and rotational components must be considered for molecular species. Overall intensities of absorption depend on the populations of the lower (and upper) states of the species. Even very weak, highly forbidden, transitions can contribute to atmospheric absorption because of the large optical paths involved.

Small, light chemical species generally show intense electronic absorption at shorter wavelengths than more complex compounds. Molecular oxygen absorbs strongly for $\lambda < 200$ nm, H_2O for $\lambda < 180$ nm, CO_2 for $\lambda < 165$ nm, while N_2 and H_2 absorb significantly only for $\lambda < 100$ nm. It is the limitation on laboratory experiments in air, resulting from the O_2 absorption, that has led to the term 'vacuum' ultraviolet for $\lambda < 200$ nm. Atmospheres tend to act as filters cutting out short-wavelength radiation, since the absorptions of their major constituents are generally strong at wavelengths shorter than the threshold value. As a result, photochemically active radiation that penetrates deeper into an atmosphere is of longer wavelength, and the chemistry characterized by lower energies, than that absorbed higher up. The principle is well exemplified by the chemistry of Earth's atmosphere. Tropospheric photochemistry is dominated by species such as O_3, NO_2, SO_2, and HCHO which absorb in the near-ultraviolet. At progressively higher altitudes, photodissociation of O_2 and photoionization phenomena become the most important processes. At ground level, only radiation with $\lambda > 300$ nm ($\cong 400$ kJ mol^{-1}) remains, and the peak intensity is at $\lambda \sim 500$ nm ($\cong 24$ kJ mol^{-1}). The particular significance of energy storage in the chlorophyll photosynthetic system, alluded to in Section 1.5.1, now becomes clear, since direct, single photon, decomposition of H_2O or CO_2 is energetically impossible near the Earth's surface.

3.2 Photochemical primary processes

Absorption of a photon of photochemically active radiation leads to electronic excitation, a process that we will represent symbolically as

$$AB + h\nu \rightarrow AB^*. \qquad (3.5)$$

Many fates of the excited AB* molecule are recognized, and several of them occur in atmospheres. Figure 3.1 summarizes the processes most frequently encountered. Routes (i) and (ii) lead to *fragmentation* of one kind or another. Process (iii) is the re-emission of radiation: if the optical transition is allowed

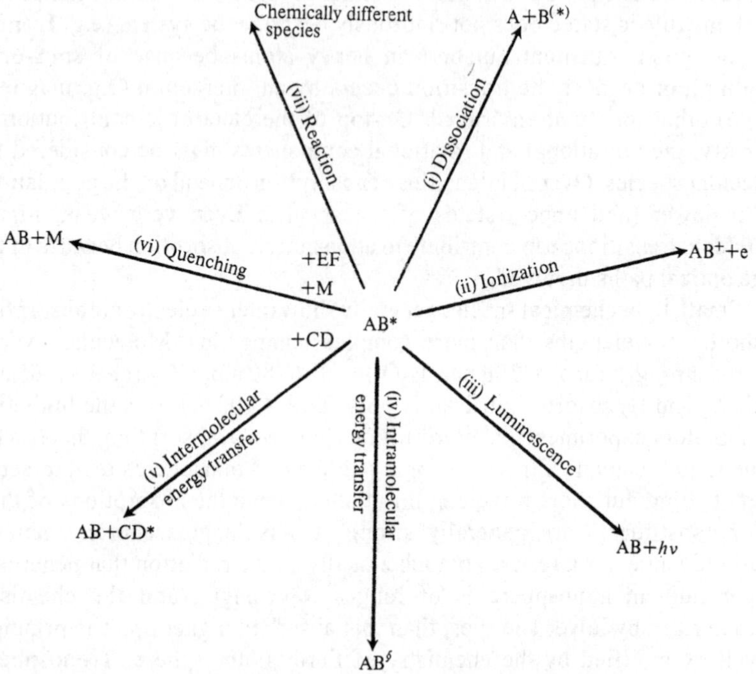

Fig. 3.1. Pathways for loss of electronic excitation that are of importance in atmospheric chemistry. The use of the symbols * and § illustrates the presence of electronic excitation: the products of any of the processes may be excited. With the exception of pathways (i) and (iv), excited atoms can participate as well as excited molecules.

the luminescence is called *fluorescence*, and otherwise *phosphorescence*. Routes (iv) and (v) involve population of excited species other than those first produced by excitation. *Intramolecular energy transfer* (iv) generates a new electronic state of the same molecule by a *radiationless transition*, while *intermolecular transfer* (v) excites a different molecule, often chemically distinct from the absorbing species. *Quenching* (vi) is a special case of intermolecular energy transfer, where electronic excitation is degraded to vibrational, rotational, and translational modes. Pathway (vii), *chemical reaction*, includes all processes where reaction is possible for, or rates are enhanced with, electronically excited reactants.

The *quantum yield* of a photochemical reaction is defined in general as the number of reactant molecules decomposed for each quantum of radiation absorbed, although it is often more helpful to define the quantum yield for a particular reaction pathway. One reactant species is electronically excited for each quantum of radiation absorbed. If the *primary* photochemical step is seen as the excitation step (3.5) taken together with the subsequent immediate fate of the excited species, then the sum of primary quantum yields for all the steps (i) to (vii) in Fig. 3.1 is obviously unity. The primary quantum yield for loss of AB has a maximum value of unity, since, for example, quenching or fluorescence, which restore AB, may compete with dissociation.

We turn now to a more detailed examination of the processes leading to chemical change.

3.2.1 Photodissociation and photoionization

Fragmentation of a chemical species following absorption of light is one of the most important photochemical processes in atmospheric chemistry. *Photodissociation* may come about when the energy of the absorbed photon exceeds the binding energy of the chemical bond under consideration. That is, the species AB* excited initially in the absorption event (3.5) can lie energetically above the dissociation threshold in the molecule, and the bond can then rupture in some way. Two main mechanisms are recognized for this photochemical rupture, and they are referred to as *optical dissociation* and *pre-dissociation*. We shall now consider the essential features of the two mechanisms, and illustrate the principles with reference to the molecules O_2 and O_3 that are of such central significance in the chemistry of Earth's atmosphere. Potential energy curves for some electronic states of O_2 are shown in Fig. 3.2. The labelling of states X, A, B, etc. is explained in the caption.

Optical dissociation is characterized by dissociation from the electronic state to which absorption first occurred. The spectrum of the absorption leading to dissociation is a continuum, since the fragments may bear translational energy away from the dissociating molecule, and such energy is essentially continuous. At some longer wavelength, the spectrum may be banded as a result of absorption to the discrete vibrational levels in a spectral region where dissociation does not follow absorption. The bands get closer together as the restoring force for the vibration gets weaker. Ultimately, a continuum is reached: the energy corresponding to the onset of the continuum ('*convergence limit*') is the dissociation energy to the products formed. Absorption from the 'X' to the 'B' state in O_2 (the so-called Schumann–Runge system) is of this kind, and Fig. 3.3 shows how the sharp banded spectrum that starts at $\lambda \sim 200$ nm converges to a limit at $\lambda \sim 175$ nm.

Examination of the potential energy curve of Fig. 3.2 shows that one of the two atomic fragments produced by extending the O–O bond length in the B state is, in fact, itself excited. The observed convergence limit at $\lambda \sim 175$ nm thus corresponds to the formation of one ground-state (^3P) and one excited (^1D) atom. When two fragments lie on a smooth extension of the parent's potential energy curve (or surface), the fragments and parents are said to *correlate* with each other (see Section 3.3). For allowed optical absorption in diatomic

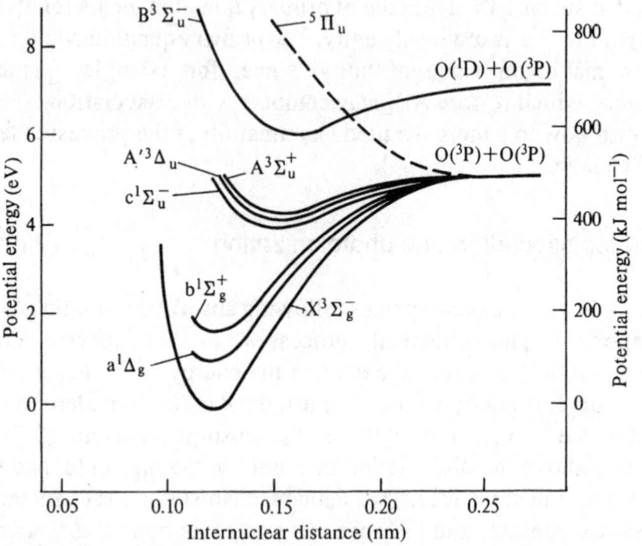

Fig. 3.2. Potential energy curves for some of the states of molecular oxygen. The ground state is labelled X, and successive higher levels of the same spin multiplicity are identified by the letters A, B, C, etc. Levels of different spin multiplicity are identified by the lower case letters a, b, c, etc. Spectroscopic term symbols for atoms and simple molecules provide more specific information about the electronic states. The symbols are constructed according to the form ^{2S+1}L (atoms) or $^{2S+1}\Lambda$ (linear molecules) where **2S** + 1 is the spin multiplicity and the angular momentum quantum number **L** (or Λ) is replaced by S, P, D, F . . . (or Σ, Π, Δ, Φ . . .) according to its value 0, 1, 2, 3, . . . Additional elements of the molecular term symbol may be provided to describe symmetry properties of the wavefunction. In general, the actual term symbols are of little consequence for our purposes, but they still provide the most convenient way of identifying particular electronic states.

The dashed curve is representative of several repulsive states (e.g. $^5\Pi_u$, $^1\Pi_u$, $^3\Pi_u$) that correlate with O(^{3}P) + O(^{3}P) and that pre-dissociate the $B^3\Sigma_u^-$ state. After Krupenie, P.H. *J. Phys. Chem. Ref. Data* **1**, 423 (1972).

molecules, the upper molecular state usually correlates with atoms of which at least one is excited, as in the O_2 case just discussed. Weak, 'forbidden', transitions can populate a state that correlates with ground-state fragments. For example, the A state in O_2 correlates with two O(^{3}P) atoms (Fig. 3.2). The A ← X absorption ('Herzberg I' system) is weak because it is forbidden (for an electric dipole interaction). It is, however, important in the atmosphere, since optical dissociation can occur for photon energies corresponding to the formation of two ground state oxygen atoms at wavelengths (~ 185 nm to 242 nm) where solar intensities are relatively higher than at the wavelengths needed for Schumann–Runge dissociation.

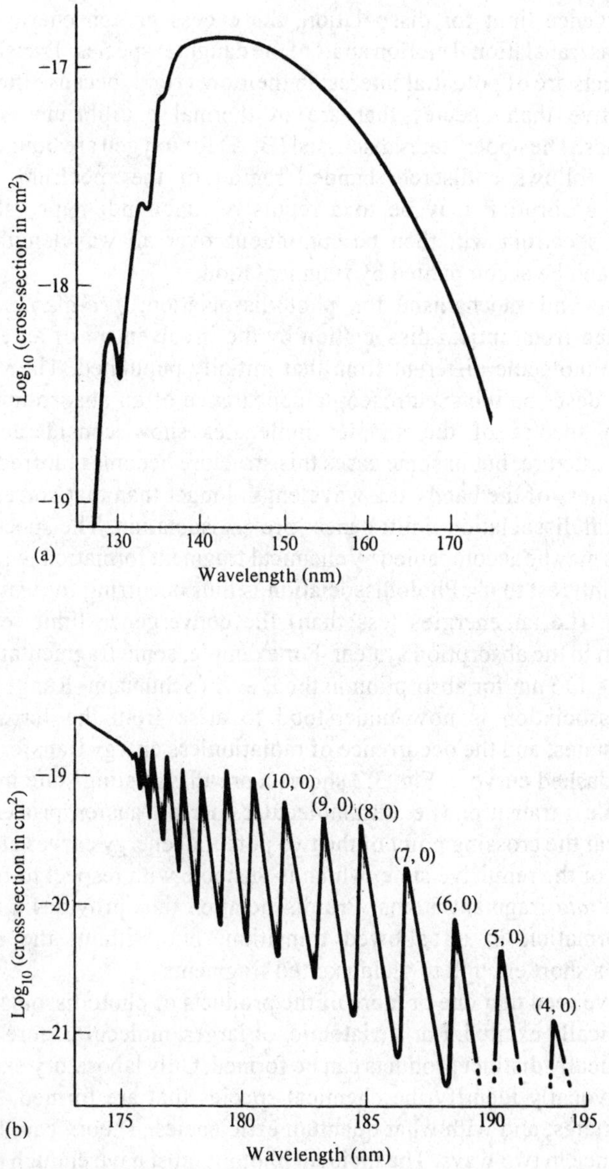

Fig. 3.3. Absorption spectrum of molecular oxygen in the $B^3\Sigma_u^- \leftarrow X^3\Sigma_g^-$ Schumann-Runge system: (a) the continuum, and (b) the discrete banded region. From McEwan, M.J. and Phillips, L.F. *Chemistry of the atmosphere*. Edward Arnold, London, 1975.

We explained the continuum in the absorption spectrum seen in optical dissociation as a consequence of the nearly continuous—properly speaking, very closely spaced—translational energy levels. For wavelengths shorter than the convergence limit for dissociation, the excess photon energy must be dissipated as translational motion apart of the daughter species. Translationally 'hot' products are of potential interest in their own right, because they may be more reactive than species that are in thermal equilibrium with their surroundings. The upper states discussed (B, A) for oxygen are bound, and the continuum follows a discrete banded region of the spectrum. In other molecules, absorption may be to a repulsive, unbound, upper state. The absorption spectrum will then be continuous over all wavelengths of the transition, and be accompanied by fragmentation.

Our second mechanism for photodissociation, *pre-dissociation,* is distinguished from optical dissociation by the involvement of an electronic state of the molecule different from that initially populated. The word was adopted to describe the spectroscopic appearance of an absorption system. Absorption spectra of the simpler molecules show considerable sharp rotational structure, but in some cases this structure becomes blurred, leading to a diffuseness of the bands at a wavelength longer than that corresponding to the optical dissociation limit: hence 'pre'-dissociation. The spectroscopic diffuseness may be accompanied by chemical fragment formation, which is the process of interest to us. Photodissociation is thus occurring for wavelengths longer than (i.e. at energies less than) the convergence limit for optical dissociation in the absorption system. For example, some fragmentation of O_2 arises at $\lambda > 175$ nm for absorption in the B ← X (Schumann–Runge) system.

Pre-dissociation is now understood to arise from the 'crossing' of electronic states, and the occurrence of radiationless energy transfer between them. The dashed curve in Fig. 3.2 shows a pre-dissociating state in oxygen. A radiationless transition (i.e. intramolecular energy transfer, process (iv) of Fig. 3.1) near the crossing point of the two potential energy curves can lead to population of the repulsive state, which is unstable with respect to formation of *ground-state* fragment atoms. Pre-dissociation thus provides a route for product formation in an allowed transition, but without the need for wavelengths short enough to yield excited fragments.

We have seen that one or more of the products of photodissociation may be electronically excited. For a triatomic, or larger, molecule, more than one set of chemically distinct products can be formed. Only laboratory experiment can unequivocally identify the chemical species that are formed, in which electronic states, and with what quantum efficiencies. Theory can, however, offer guidance in two ways. The incident photons must have enough energy to bring about any proposed change, so that thermochemical information shows if a particular photodissociative channel is energetically *possible*. Then, considerations such as the need to conserve quantum mechanical spin and orbital angular momentum indicate if the channel is *probable*.

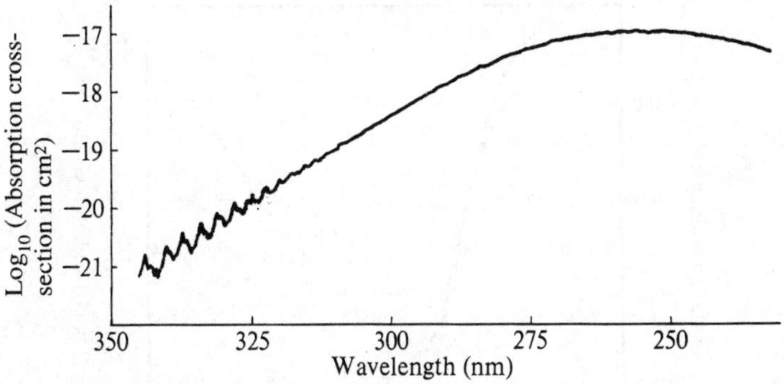

Fig. 3.4. Absorption spectrum of ozone in the ultraviolet region at 300 K. Data of Bass, A.M. and Paur, R.J., presented at the *International Workshop on Atmospheric Spectroscopy*, Rutherford Appleton Laboratory, July 1983.

We shall discuss this latter point more fully in Section 3.3, but, for the time being, we illustrate it in terms of the photolysis of ozone, reaction (3.1)

$$O_3 + h\nu \rightarrow O_2 + O. \qquad (3.1)$$

A knowledge of the efficiency of $O(^1D)$ production is of particular importance in atmospheric chemistry, because $O(^1D)$ can initiate many other reactions (see Section 1.4 for a preliminary discussion). Ozone absorbs strongly in the ultraviolet region, as shown in Fig. 3.4. A much weaker absorption exists in the red region, with a peak at $\lambda \sim 600$ nm. Depending on the photon energy, the atomic and molecular oxygen fragments of reaction (3.1) can themselves be electronically excited. Table 3.1 shows the wavelength thresholds required thermodynamically for the production of various combinations of states of O and O_2. The table does not indicate whether or not the particular fragments really are formed. However, the spin conservation theory suggests that the products of photolysis must *both* be triplets or *both* be singlets. The lowest energy singlet pair is $O(^1D) + O_2(^1\Delta_g)$ and hence, from Table 3.1, the expected

Table 3.1 Wavelength thresholds (in nm) for O_3 photodissociation channels

Atom \ Molecule	$O_2(^3\Sigma_g^-)$	$O_2(^1\Delta_g)$	$O_2(^1\Sigma_g^+)$	$O_2(^3\Sigma_u^+)$	$O_2(^3\Sigma_u^-)$
$O(^3P)$	1180	612	463	230	173
$O(^1D)$	411	310	267	168	136
$O(^1S)$	237	199	181	129	109

Data derived from: Baulch, D.L., Cox, R.A., Hampson, R.F., Jr., Kerr, J.A., Troe, J., and Watson, R.T. *J. phys. Chem. Ref. Data* **9**, 295 (1980); and Moore, C.E. *Atomic energy levels*, Vol. 1, NSRDS-NBS 35, Washington, DC, 1971.

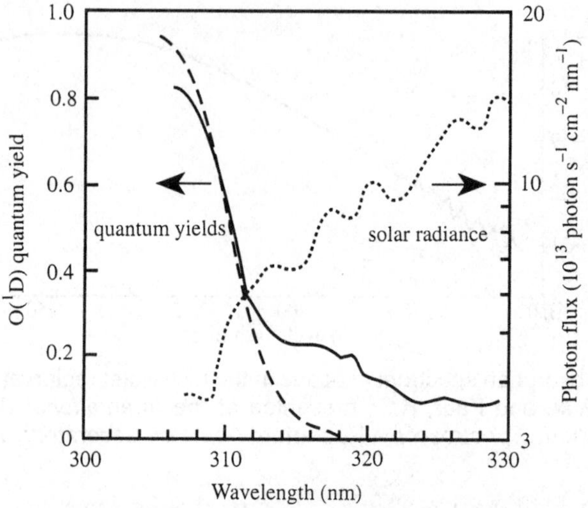

Fig. 3.5. Primary quantum yield for O(¹D) formation in the photolysis of ozone displayed as a function of photolysis wavelength. The solid line is the result of a recent experimental determination for $T = 250\,K$, while the dashed curve is that recommended previously by data evaluation panels. The dotted curve shows how solar radiance increases sharply in the same wavelength region and emphasizes how important the details of quantum yield are at wavelengths close to 310 nm. From Ravishankara, A.R., Hancock, G., Kawasaki, M., and Matsumi, Y. *Science* **280**, 60 (1998).

threshold wavelength for spin-allowed $O(^1D)$ formation is ~310 nm. Figure 3.5 shows experimentally measured quantum yields for $O(^1D)$ production as a function of wavelength. Direct determinations also show that $O_2(^1\Delta_g)$ is actually formed with very nearly the same efficiency as $O(^1D)$ at $\lambda < 300$ nm; exact correspondence would be predicted by the spin-conservation arguments.

The quantum yield–wavelength curve of Fig. 3.5 leads us to two further topics. It might be thought that, since the energy for a particular dissociative channel is well defined, the onset of dissociation ought to be abrupt. That is, diagrams such as Fig. 3.5 ought to be step functions, with the step rising at the wavelength corresponding to the critical energy. Although reasonably steep, the curve for $O(^1D)$ formation from O_3 is certainly not a step function. In this particular photodissociation process, one of the most important in all of atmospheric chemistry, there appear to be two different factors at work. The basic shape of the curve can be explained if, in polyatomic molecules, energy contained in internal vibrations and rotations can assist the photon energy in causing dissociation. Not all modes are equally 'active' in this way, and the efficiency of energy utilization may be dependent on the internal energy. Nevertheless, photon energies somewhat smaller than the bond dissociation threshold can be effective in rupturing the bond. Internal energy depends on temperature, so that the exact shape of the quantum yield–wavelength curve will also be temperature dependent. Experimental determinations show that yields of both $O(^1D)$ and $O_2(^1\Delta_g)$ exhibit a marked temperature dependence, and the numerical values for wavelengths just greater than the thermodynamic limit of 310 nm confirm that photolysis of internally excited ozone provides the major source of products.

The utilization of vibrational excitation is of potential importance for atmospheric chemistry in a number of ways beyond that of influencing the absorption–wavelength and photodissociation yield–wavelength dependences. For example, reaction (3.1) is believed to be capable of generating some ground-state $O_2(^3\Sigma_g^-)$ (together with $O(^3P)$) in vibrational levels as high as $v = 23$. Such highly *vibrationally excited* molecules possess a considerable excess energy over the ground-state species, and photodissociation can thus occur at wavelengths much longer than the conventional thermodynamic limit for 'cold' molecules. In the case of excited O_2, the process

$$O_2 + h\nu \rightarrow O + O \qquad (3.6)$$

might then be a significant source of atomic oxygen at wavelengths longer than the expected threshold for ground-state O_2, and would thus be a contributor lower in the atmosphere, where only the longer wavelengths penetrate. Removal (*relaxation*, *deactivation*, or *quenching*) of the vibrational energy by collisions with the major atmospheric constituents may be rapid, so that the quantitative evaluation of the contribution requires a knowledge of the rate parameters for relaxation.

Since the energies of individual vibrational quanta are rather small in comparison with chemical bond energies, significant chemical effects are expected only with excitation of molecules to high vibrational quantum numbers, as in the case just discussed. In this example, the vibration was released in an earlier photodissociation step, that of the photolysis of ozone in reaction (3.1). Another route to high vibrational quantum levels is excitation of *vibrational overtones*. Although the selection rule $\Delta v = \pm 1$ applies to strictly harmonic vibrational oscillators, real *anharmonic oscillators* exhibit weak overtone absorption with $\Delta v > 1$. For some atmospherically interesting molecules, these overtones lie in the visible region of the spectrum. For example, predictions have been made that significant enhancements to the photodissociation rates of HNO_3 ($HONO_2$), HNO_4 (HO_2NO_2), and H_2O_2 could result from visible-wavelength excitation of OH overtone vibrations containing sufficient energy to cleave the O–O and N–O bonds.

One further feature of Fig. 3.5 demands comment. Although we have stressed the importance of spin conservation in the ozone photolysis process (3.1), and will return to a more detailed examination of conservation in Section 3.3, there appear to be *small* contributions to the photolysis at $\lambda > 310$ nm from spin-forbidden channels yielding the product pairs $O(^3P) + O_2(^1\Delta_g)$ and $O(^1D)$

$+O_2(^3\Sigma_g^-)$. While the point will not be laboured here, it is presented to illustrate the complexity of the issue of photodissociation products and yields.

Considerable emphasis has been placed on the exact magnitude of quantum yields for primary dissociative processes and on the wavelength and temperature dependences. The reason is that calculated photolytic production rates of species such as $O(^1D)$ in atmospheres are particularly sensitive to these factors. A change in the nature of products, or, indeed, the onset of photolysis, is very often associated with a rapid change in absorption cross-section, for good spectroscopic reasons. Comparison of Figs 3.4 and 3.5 for ozone will illustrate the effect clearly: for $\lambda = 304$ to 319 nm, the absorption cross-section drops by a factor of 10, while the quantum yield for $O(^1D)$ formation falls from 0.9 to ~0. The rate of $O(^1D)$ formation depends on the product of quantum efficiency, absorption cross-section, and incident flux. The product of the first two terms changes extremely rapidly with wavelength, so that the rate of $O(^1D)$ production will be critically dependent on how the incident intensity varies with wavelength. As an example, near the ground the incident intensity increases by a factor of more than ten between $\lambda = 300$ and $\lambda = 320$ nm. A small error in the wavelength scale for Fig. 3.5 could thus have an enormous effect on the predicted rates of $O(^1D)$ formation in the troposphere.

Photoionization may be regarded as a special case of photodissociation, but one in which the products are a positively charged ion and an electron. The processes

$$O + h\nu \rightarrow O^+ + e \tag{3.7}$$

$$O_2 + h\nu \rightarrow O_2^+ + e \tag{3.8}$$

are typical of photoionization. In general, the energies (ionization potentials) required to remove an electron from an atom or molecule are larger than those needed to split a molecule into chemical fragments. Some typical ionization potentials are shown in Table 3.2, together with the corresponding wavelengths. Even the molecules with the lowest ionization potentials (NO, NO_2) require vacuum ultraviolet radiation for ionization. Photoionization is thus only a significant source of ions in the upper atmosphere of Earth. Electronically excited atoms or molecules can often use their excitation energy as a contribution towards the ionization potential. Thus the photoionization wavelength threshold for $O_2(^1\Delta_g)$, which has 0.98 eV excitation energy, is raised to 111.8 nm. As with dissociative processes, the effect of small wavelength shifts can be large in atmospheric chemistry, because of the rapid variation of incident intensity with wavelength. In the ionizing ultraviolet region, the most abundant gases such as O_2, N_2, and CO_2 absorb strongly, so that their absorption is imposed on top of the spectral distribution of the solar radiation, even at high atmospheric altitudes. Since the molecular absorptions generally show some structure, there are spectral 'windows' through which

Table 3.2 Ionization potentials of some atmospherically important species, together with equivalent threshold wavelengths

Species	Ionization energy		Equivalent wavelength
	eV	kJ mol^{-1}	(nm)
Na	5.1	496	241.2
Mg	7.7	738	162.1
NO	9.3	886	135.0
NO_2	9.8	940	127.2
O_2	12.1	1165	102.7
H_2O	12.6	1214	98.7
O_3	12.8	1233	97.0
N_2O	12.9	1246	96.0
SO_2	13.1	1264	94.6
H	13.6	1312	91.2
O	13.6	1314	91.1
CO_2	13.8	1331	89.9
CO	14.0	1352	88.5
N	14.5	1403	85.2
H_2	15.4	1488	80.4
N_2	15.6	1503	79.9

Data obtained from: (atoms), Moore, C.E. *Ionization potentials and ionization limits derived from the analysis of optical spectra*, NSRDS-NBS **34**, Washington, DC (1970); (diatomic species), Huber, K.P. and Herzberg, G. Constants of diatomic molecules, Van Nostrand-Reinhold, New York, 1979; (triatomic species), Herzberg, G. *Electronic spectra and structure of polyatomic molecules,* Van Nostrand-Reinhold, New York, 1966.

radiation is transmitted to greater depth in the atmosphere, and these windows may be the major contributors to ionization. One of the most curious coincidences in Earth's atmosphere is the almost perfect spectral match between a sharp dip in the absorption of O_2 and the Lyman-α (resonance) line of atomic hydrogen at $\lambda = 121.59$ nm. At wavelengths below ~ 190 nm, the solar emission is almost entirely in the form of atomic lines, and the strongest individual feature is the Lyman-α line. Because of the chance existence of the O_2 window, the strongest solar feature is also least attenuated in Earth's atmosphere. Table 3.2 shows that NO (as well as NO_2 and some metal atoms) can be ionized by Lyman-α radiation.

Mechanisms of photoionization are analogous to those for dissociation, both direct ionization and *pre-ionization* (*auto-ionization*) being recognized. Excited electronic (and, in molecules, vibrational and rotational) states of the ions may be generated. Any excess of energy between the ionizing photon and the resultant ion is carried off as translation by the electron, a feature utilized in *photoelectron spectroscopy*.

3.2.2 Reactions of electronically excited species

Electronically excited atoms and molecules can be formed photochemically either directly by optical pumping (absorption) or as products of a photofragmentation process (see preceding section). Electron impact can also lead to electronic excitation, and is often important in planetary atmospheres. A typical example is the process

$$N_2 + e \rightarrow N_2^* + e, \tag{3.9}$$

where N_2^* can be $N_2(^3\Sigma_u^+)$. However, our immediate concern is the reactivity of excited states, regardless of the excitation mechanism.

There are two ways in which the reactivity of a species can be influenced by its electronic state. First, the energetics of a reaction are altered, with thermodynamic and kinetic consequences. A reaction favours products only if $\Delta G^\ominus$ for the reaction is negative. If a process is endothermic ($\Delta H^\ominus$ positive) then the condition for $\Delta G^\ominus$ will be met only if the enthalpy term is dominated by a large increase in entropy (i.e. $T\Delta S^\ominus > \Delta H^\ominus$). In general, this thermodynamic condition is likely to be met only if the endothermicity is small or if the reactions take place at high temperatures. However, if a reactant is excited, and if the excitation energy can contribute to the reaction energetics, then $\Delta H^\ominus$ may become negative for the excited state reaction where it was positive for the ground state. Kinetic limitations may be even more important than thermodynamic ones. As we shall see in Section 3.4.1, rates of reaction are often proportional to a multiplying factor of $\exp(-E_a/RT)$, where E_a is the activation energy for the reaction. For an endothermic reaction, it can be shown that $E_a \geq \Delta H^\ominus$ and the exponential term can make the rate of reaction insignificant. Activation energies are usually positive even for exothermic reactions. In most cases, then, excess energy in the reactants will increase the rate of reaction so long as that energy can be used in overcoming the activation barrier.

Let us now consider some examples. We have already mentioned several times the production of hydroxyl radicals by excited oxygen atoms. The process

$$O(^3P) + H_2O \rightarrow OH + OH, \tag{3.10}$$

is about 70 kJ mol^{-1} endothermic, while the reaction with $O(^1D)$,

$$O(^1D) + H_2O \rightarrow OH + OH, \tag{3.11}$$

is exothermic by 120 kJ mol^{-1} because of the 190 kJ mol^{-1} excitation energy of $O(^1D)$. Similarly, the reaction of $O(^1D)$ with CH_4 is exothermic, but would be endothermic by 11 kJ mol^{-1} for ground state $O(^3P)$. Both reactions

$$O(^3P) + O_3 \rightarrow 2O_2 \tag{3.12a}$$

$$O(^1D) + O_3 \rightarrow 2O_2 \text{ and } 2O + O_2 \tag{3.12b}$$

are exothermic. However, at a temperature of 220 K (typical of the lower stratosphere), the second reaction proceeds more than 3×10^5 times as fast as the first for equal reactant concentrations. Most of the difference (but not all: see later in this section) results from the activation energies of the reactions: 17.1 kJ mol^{-1} in (3.12a), ~ 0 in (3.12b). Obviously, the 17.1 kJ mol^{-1} can be supplied by the 190 kJ mol^{-1} excitation energy of $O(^1D)$. Another interesting example is the reaction of excited N_2 in the $^3\Sigma_u^+$ state (produced by optical pumping, or by electron impact in reaction (3.9)) with O_2

$$N_2^* + O_2 \rightarrow N_2O + O. \tag{3.13}$$

It has been suggested that the reaction may produce N_2O about once in every two collisions between the reactants. For ground-state N_2, the reaction is endothermic and negligibly slow. Process (3.13) is a possible source of N_2O at altitudes above 20 km, and could appreciably augment the biogenic sources from the Earth's surface.

We turn now from the influence of excess energy to that of electronic structure on the reactivity of electronically excited states. Chemists are certainly inclined to believe that the properties and reactions of a substance are influenced by the number and arrangement of electrons, especially in the outer shell. Since electronic excitation implies some alteration in structure, species might be expected to show a particular reactivity for a specific electronic state. Nothing at this stage suggests that an excited state will be more (or, indeed, less) reactive than the ground state. It is, of course, necessary to disentangle the two influences of energy and of structure, since any excited state is bound to be energy rich. The easiest way to determine the influence of structure itself is to extrapolate measured reaction rates to infinite temperature, which removes the effect of activation energy since $\exp(-E_a/RT)$ is now unity. Carrying out the extrapolation with rate data for reactions (3.12a) and (3.12b) leads to the conclusion that $O(^1D)$ is intrinsically about twenty times more reactive than $O(^3P)$ in the reaction with ozone. Conversely, the intrinsic reactivity of nitrogen atoms with ground-state oxygen

$$N(^4S) + O_2(^3\Sigma_g^-) \rightarrow NO + O, \tag{3.14}$$

is several orders of magnitude greater than the intrinsic reactivity with the first excited $^1\Delta_g$ state of O_2. Very often the effects of electronic arrangement on reactivity are a reflection of the ability to pass smoothly from reactants to products without a change of, for example, spin angular momentum. We have already encountered another facet of the same phenomenon when discussing the spin states of the products of ozone photolysis. It is appropriate to consider now the underlying principles of momentum-conserved reactions.

3.3 Adiabatic processes and the correlation rules

Potential energy curves, such as those shown in Fig. 3.2, represent the molecular potential energy as a function of internuclear distance between the constituent atoms of a diatomic molecule. We have seen in Section 3.2.1 that a molecular state is said to *correlate* with particular atomic fragments if a continuous curve connects atoms and the diatomic molecule. Thus, in the oxygen case to which Fig. 3.2 applies, the B state correlates, for example, with $O(^1D) + O(^3P)$, while the ground X state correlates with two ground-state, ^3P, oxygen atoms. Bringing together a ^1D and a ^3P atom cannot, therefore, generate O_2 in the ground state in a process involving a continuous potential energy curve. This argument can be extended to the reaction

$$O(^1D) + O(^3P) \rightarrow O(^3P) + O(^3P) \qquad (3.15)$$

which is clearly exothermic. To get from reactants to products, it is necessary to start on the B curve, and then cross to the pre-dissociating repulsive state. In the context of reaction mechanisms, the term *adiabatic* has the special meaning of 'remaining on the same curve', so that reaction (3.15) can only occur non-adiabatically. Non-adiabatic reactions, which involve curve crossing, are nearly always less efficient than adiabatic ones.

At the end of this section, we shall discuss how the potential energy of a polyatomic system can be described by a *potential energy surface* rather than by the potential energy *curve* used in connection with diatomic systems. For the time being, we need note only that, with reactions involving more than two atoms in total, an adiabatic reaction will be one that is conducted on a single potential energy surface. That is, reactants and products are connected by a continuous surface. Where such connection does not exist, non-adiabatic surface crossing must occur, which may be very inefficient.

Why is it that crossing from one curve or surface to another is sometimes so inefficient? The answer lies in the probabilities of radiationless transitions, for which selection rules may be formulated just as for optical (radiative) transitions. In the example with which we started this section, reaction (3.15) cannot occur adiabatically because of the selection rule that spin must remain unchanged ($\Delta \mathbf{S} = 0$) in a radiationless transition. Reference again to Fig. 3.2 shows that the dashed curve is of different spin multiplicity (quintet: i.e. $\mathbf{S} = 2$) from the B state (triplet: i.e. $\mathbf{S} = 1$), so that the radiationless transition is formally forbidden and acts as a bottle-neck in the non-adiabatic process. In this very simple case, the electronic states and term symbols for all the atomic and molecular species can be identified. For polyatomic reaction systems, such detailed and specific information may not exist. On the other hand, it may still be possible to say that a reaction *cannot* proceed adiabatically on the basis of the information available.

Correlation or conservation rules are formulated in terms of the angular momentum or symmetry properties that do exist for reactants or products.

Often, only spin angular momentum, **S**, can be specified and our discussion will be illustrated in terms of this quantum number. However, it must always be borne in mind that orbital momentum, parity, and so on, must all be conserved where they have a meaning.

Consider a hypothetical reaction between two species A and BC in which, at some stage, A, B, and C are all interacting as a transient reaction complex that we can represent as [ABC]

$$A + BC \rightarrow [ABC] \rightarrow AB + C. \tag{3.16}$$

The basis of the conservation rules is that, for an adiabatic reaction, [ABC] can be reached without change of potential energy surface, either from the reactants or from the products. In that case, the transient must have its total spin, $\mathbf{S}_{ABC}$, produced either by a combination of the reactant spins, $\mathbf{S}_A$ and $\mathbf{S}_{BC}$, or by a combination of the product spins, $\mathbf{S}_{AB}$ and $\mathbf{S}_C$. The only possible difficulty concerns the addition of spins. Angular momenta sum vectorially, but with quantized momenta the resultant must also be quantized. Spins $\mathbf{S}_A$ and $\mathbf{S}_{BC}$ therefore produce resultants $|\mathbf{S}_A + \mathbf{S}_{BC}|$, $|\mathbf{S}_A + \mathbf{S}_{BC} - 1|$, . . . $|\mathbf{S}_A - \mathbf{S}_{BC}|$, while $\mathbf{S}_{AB}$ and $\mathbf{S}_C$ produce $|\mathbf{S}_{AB} + \mathbf{S}_C|$, $|\mathbf{S}_{AB} + \mathbf{S}_C - 1|$, ... $|\mathbf{S}_{AB} - \mathbf{S}_C|$. The question now is whether or not the first list and the second have one or more values in common. If not, then adiabatic reaction cannot occur. If there are common values then adiabatic reaction *may* be possible. Let us consider a concrete example. What combinations of the singlet and triplet spin states can the O_2 products of the reactions

$$O(^3P, {}^1D) + O_3(^1A) \rightarrow O_2 + O_2 \tag{3.17}$$

possess? For $O(^3P)$, the spins of the reactants are 1 and 0, which combine to give a total spin of 1. Two triplet products ($\mathbf{S}_{O_2} = 1$, $\mathbf{S}_{O_2} = 1$) can give a total spin of 2, 1, or 0, so that the reaction to two triplets could occur adiabatically on a triplet surface ($\mathbf{S} = 1$). One triplet and one singlet ($\mathbf{S}_{O_2} = 1$, $\mathbf{S}_{O_2} = 0$) give uniquely a total spin of 1, so that again adiabatic reaction is possible on a triplet surface. Two singlet products ($\mathbf{S}_{O_2} = 0$, $\mathbf{S}_{O_2} = 0$) can only combine to $\mathbf{S} = 0$, so that adiabatic reaction is not possible. With $O(^1D)$, the surface correlating with the reactants is a singlet, so that the product O_2 molecules in this case must be both singlets or both triplets for adiabatic reaction to occur (on the singlet surface). All other reaction possibilities involve spin-forbidden crossings between singlet and triplet reaction surfaces, and these non-adiabatic processes are likely to be inefficient.

Let us now see how the spin conservation rule could be applied to a *photochemical* process. In particular, we shall examine what are the 'spin-allowed' fragments of ozone photolysis in reaction (3.1), since the outcome has already been presented without proof in Section 3.2.1. The argument goes in the following way. Ground-state ozone is a singlet ($\mathbf{S} = 0$), and the strong ultraviolet absorption is likely to be spin allowed ($\Delta\mathbf{S} = 0$), so that the transient excited state of ozone yielding $O + O_2$ is also a singlet. Products lying on the same potential energy curve (or surface) as a precursor have to have spins that can add to give the spin of the precursor. A value of **S** = 0 can be obtained from products both possessing **S** = 0 or both possessing **S** = 1, but not from one triplet and one singlet. In the ozone case, then, $O(^1D)$ production, if spin allowed, must be accompanied by a singlet state O_2 product. Alternatively, both atom and molecule can be triplets.

A certain amount of caution is needed in applying the spin and other conservation rules. First, they are valid only so far as the quantum number, such as **S**, to which they pertain is a good description of the system. Where elements of large atomic number are concerned, **S** is a notoriously poor quantum number. In a reactive interaction ('collision': see later) **S** may be a poor representation even where only 'light' elements are concerned. The other point that should be made is that the rules show *excluded* possibilities rather than permitted ones. For example, that a triplet surface can connect reactants with products does not necessarily mean that reaction occurs over the triplet surface. Let us revert to our original two-atom reaction (3.15). According to the spin rules developed in connection with reaction (3.17), singlet + triplet reactants can give two triplet products (on a triplet surface). We know, however, from the potential curves (Fig. 3.2) that reactants and products are *not*, in fact, connected by a continuous curve, triplet or otherwise. The reaction as written could occur non-adiabatically, *via* the pre-dissociating (dashed) curve. The curve actually drawn is a quintet state, so that the curve crossing is forbidden and likely to be slow.

This section ends with a short digression about potential energy surfaces for polyatomic systems, since we have already alluded to such surfaces and will need to refer to the subject in more detail in the following section. It is easy to represent, on a two-dimensional curve, interactions and reactions between atoms, since we need to display the energy as a function of only one interatomic distance. Potential energies for polyatomic molecules are not so easily represented. Even with a triatomic system, say ABC [Fig. 3.6(a)], we need either three distances (r_{AB}, r_{BC}, r_{AC}) or two distances and an angle (e.g. r_{AB}, r_{BC}, $\angle ABC$) to define the relative positions of A, B, and C. Four dimensions are then required to show potential energy as a function of atomic coordinate. An improvement for the triatomic molecule can be achieved if one variable, such as the bond angle, is fixed. In that case, the potential energy can be described by a three-dimensional *potential energy surface*, using one coordinate for energy, and the other two for the two interatomic distances. The potential energy surface can be displayed in two dimensions by a contour map. Without the restriction on bond angle, or with more than three atoms in the molecule, then the three-dimensional surface becomes instead a multi-dimensional potential energy *hypersurface*. There is no real difficulty in concept here, since we wish only to know what the potential energy is for the atoms in given positions. The difficulty is in making drawings in two- or three-dimensional space!

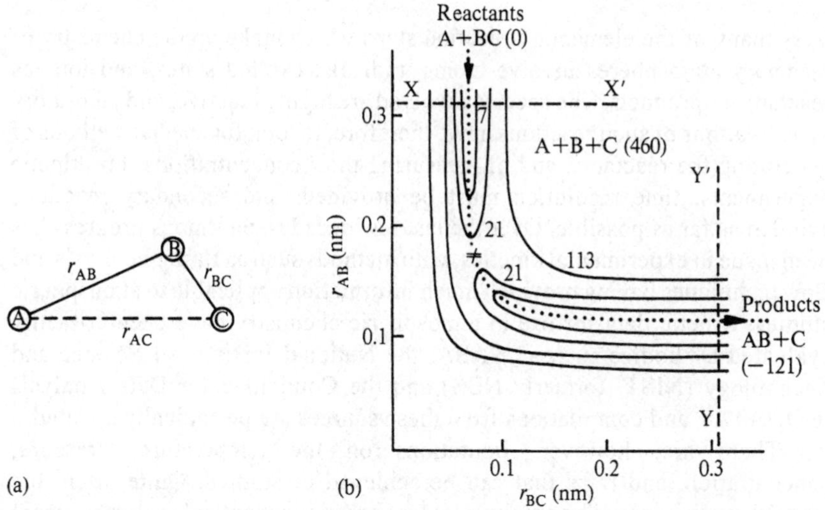

Fig. 3.6. (a) Distances in the A ... B ... C system. (b) Typical potential energy surface (contour map) for a particular ABC angle. The lowest energy path from reactants (A + BC) to products (AB + C) is shown by the dotted line, which represents the 'reaction coordinate'. On this path, the highest energy is reached at the point marked '≠', which is the 'transition state' or 'activated complex'. The potential energy surface is actually one computed *ab initio* for the reaction $F + H_2 \rightarrow H + HF$, and numerical values on the contour lines are energies in kJ mol^{-1} referred to the reactants. After Muckerman, J.T. *J. Chem. Phys.* **54**, 115 (1971).

Our remarks about potential energy surfaces apply both to surfaces for bound polyatomic molecules and to polyatomic systems created transiently during the course of a chemical reaction. For example, in the reaction represented by eqn (3.16), a transient triatomic molecule ABC is formed as the reactant A approaches BC. In this case, however, the energy of the ABC system increases as the AB distance decreases, rather than reaching the minimum characteristic of the stable configuration of a bound triatomic molecule. The contour map of Fig. 3.6(b) is of this repulsive (reactive) type. At large r_{AB} (reactants far apart) a cross-section through the surface along the line XX' would be the potential energy curve for the free diatomic reactant molecule BC. When r_{BC} is large, the products have separated, and the cross-section along YY' is the potential energy curve for the product AB.

The rate of passage over the potential energy surface from reactants to products is the subject of reaction kinetics, and it is to this topic that we now turn.

3.4 Chemical kinetics

Very many of the elementary reaction steps which make up the chemistry of planetary atmospheres involve atoms, radicals, excited states, and ions as reactants or products. The species involved are highly reactive, and laboratory investigations of such reactions have, therefore, to look for special methods of generating the reactants, and of measuring their concentrations. For kinetic experiments, time resolution must be provided, and secondary reactions avoided as far as possible. Over the last few decades, enormous progress has been made in experimental kinetics, with methods such as flash photolysis and flow techniques having provided much information applicable to atmospheric studies. Kinetic data for use in atmospheric chemistry have been critically evaluated by bodies such as NASA, the National Institute of Science and Technology (NIST, formerly NBS) and the Committee for Data Analysis (CODATA), and compilations from these sources are periodically updated.

There are, however, limitations on the temperature, pressure, concentration, and rates that can be achieved or studied. Quite often, the laboratory data cannot be obtained under conditions identical to those present in an atmosphere. Extrapolation of the laboratory results is therefore necessary in order that the kinetic data may be used in interpreting atmospheric chemistry. Dangers abound in long-range extrapolation of data obtained over a limited span of experimental conditions. At least the extrapolation should be based on a believable theory for the process concerned. To highlight the areas of importance, this section reviews some aspects of chemical kinetics.

A reaction,

$$\mathrm{A + B + \ldots \rightarrow products}, \tag{3.18}$$

proceeds with a rate proportional to the concentrations[a] raised to some power

$$Rate = -\frac{\mathrm{d[A]}}{\mathrm{d}t} = -\frac{\mathrm{d[B]}}{\mathrm{d}t} = k\mathrm{[A]}^{\alpha}\mathrm{[B]}^{\beta} \tag{3.19}$$

The powers α and β are the *order* of reaction with respect to reactants A and B, and $\alpha + \beta$ is the overall order; the constant of proportionality, k, is the rate coefficient (rate 'constant'). The *molecularity* of a reaction is the number of reactant molecules written in the stoicheiometric equation. Order is thus an experimental quantity, molecularity an arbitrary theoretical one. An *elementary*

[a]The concentrations are written here in normal chemical nomenclature, [A] and [B]. The units are molecules or moles per unit volume; atmospheric chemists dealing with gas-phase species normally use the molecular units, while solution-phase chemists employ the molar quantities. The volumes are usually given in cm^3 in the first case, and dm^3 ($1\ \mathrm{dm}^3 \cong 1$ litre) in the second. Physicists frequently write concentrations in the form n_A, n_B, again representing molecules per cm^3, to emphasize the *number density* aspect of the quantities.

reaction step is conceived as one that cannot be split into any chemically simpler processes. For truly elementary steps, order and molecularity are in general identical. Thus, if reaction (3.18) is elementary, and the only reactants are A and B, it is both bimolecular and overall second-order: first-order in each of the components A and B. However, a special case often arises in atmospheric chemistry. If the second reactant, B, is in great excess over A, then its concentration is effectively constant throughout the reaction. We can then combine the concentration with the rate coefficient, k, and write the rate of reaction as $k'[A]$, where $k' = k[B]$. Such a process is termed a *pseudo-first order* reaction, and k' is the *pseudo-first order rate coefficient*.

We now review briefly some theories of elementary reaction steps in order to explain certain of the ways that data are presented in studies of atmospheric chemistry. Bimolecular processes are discussed first; unimolecular and termolecular reactions are then considered later.

3.4.1 Bimolecular reactions

As two reactant molecules approach each other closely enough, the energy of the reaction system rises, as indicated by the potential energy surface (Fig. 3.6(b)) for the A + BC reaction (3.16). The contours of the surface show that there is a 'valley' that provides the lowest energy approach of the reactants, and the dotted line in the figure is that path. There comes a point, marked '≠', beyond which the energy starts to decrease again, and so product formation is now energetically favourable. In our 'mountain' analogy, the point ≠ is a 'col' or 'pass'. Figure 3.7 shows the energy of the ABC system as a function of distance travelled along the low path for an exothermic reaction. The abscissa labelled 'reaction coordinate' is essentially the dotted line of Fig 3.6(b) pulled out straight. The diagram shows the relationship between the critical energies (and presumably the measured activation energies: see p. 110 and p. 119) for forwards (E_c^f) and reverse (E_c^r) directions of a reaction. For an exothermic reaction as shown (left to right), ΔH is negative, and $E_c^r = E_c^f - \Delta H_{reaction}$. Furthermore, an endothermic reaction (right to left) must be accompanied by an activation energy (E_c^r) at least as large as the endothermicity, as stated in Section 3.2.2.

The reaction pathway shown is only one of an infinite number of possibilities. In principle, if the potential energy surface is known, it is possible to calculate, using the laws of mechanics, the path followed for any initial 'starting' distance and direction of approach of the reactants A and BC. For any given speed, and internal excitation (vibration, rotation) of the reactants, the fraction of 'trials' leading to product formation can then be assessed, and is related to the probability of reaction. The ordinary macroscopic rate coefficient, k, of eqn (3.19) can then be determined from a sum over the distributions of translational velocity, and vibrational and

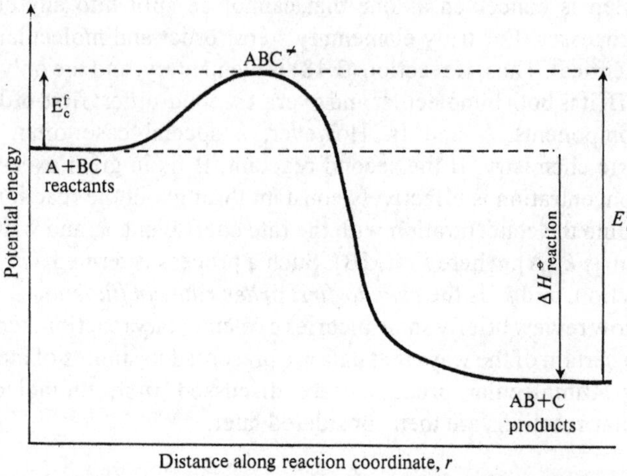

Fig. 3.7. The relationship between barrier energies for the forward (E_c^f) and reverse (E_c^r) reactions and the enthalpy of reaction ($\Delta H^{\ominus}_{reaction}$). The diagram is essentially a cross-section of a potential energy surface (e.g. Fig. 3.6 (b)) along the reaction coordinate. If the potential energies are expressed as enthalpies, then

$$\Delta H^{\ominus}_{reaction} = H_{products} - H_{reactants}$$
$$E_c^f = H_{ABC\neq} - H_{reactants}$$
$$E_c^r = H_{ABC\neq} - H_{products}$$

so that

$$E_c^f - E_c^r = \Delta H^{\ominus}_{reaction}$$

rotational excitation, appropriate to any temperature T: for thermal equilibrium, the *Maxwell–Boltzmann distribution* will be used. Such methods of *molecular dynamics* would be ideal for predicting rate coefficients if only it were possible to calculate from first principles the potential energy surfaces for all reactions. Unfortunately, it is not yet feasible to perform these '*a priori*' calculations of surfaces, except for the very simplest reactions. Modern experimental kinetics can show the probability of passing from one set of reactant states (translational, vibrational, rotational, etc.) to one set of product states: so-called '*state-to-state*' *kinetics*. The results of such experiments can then be used to test hypothetical potential energy surfaces. In this case, however, the molecular dynamic calculations have lost their predictive value.

Two simplifications are commonly adopted to overcome the lack of knowledge of complete potential energy surfaces for reaction. The first of these is the *collision theory* (CT). Reactant molecules are assumed to be hard spheres (radii r_A and r_{BC} in our example), and reaction is taken to be possible only if two conditions are met: a collision must occur, and the energy of collision along the line of centres must equal or exceed the energy required, E_C, to reach a critical configuration ($ABC^{\neq}$ in Fig. 3.7). The rate of reaction according to this theory is readily shown to be given by

$$-\frac{dn_A}{dt} = -\frac{dn_{BC}}{dt} = n_A n_{BC} \sigma_c \bar{c} \exp(-E_c/RT), \tag{3.20}$$

where σ_c is the cross-sectional area for collision (collision cross-section) given by

$$\sigma_c = \pi(r_A + r_{BC})^2, \tag{3.21}$$

and $\bar{c}$ is the mean relative velocity of molecules for temperature T

$$\bar{c} = \left(\frac{8kT}{\pi\mu}\right)^{1/2}; \quad \mu = \frac{m_A m_{BC}}{m_A + m_{BC}}. \tag{3.22}$$

The quantities n_A and n_{BC} are the concentrations of A and BC in molecular units ('number densities': see footnote on p. 116 for an explanation). Equation (3.20) certainly has the correct concentration dependence for an elementary bimolecular reaction, so that the rate coefficient can be written

$$k = \sigma_c \bar{c} \exp(-E_a/RT). \tag{3.23}$$

Experimentally, many second-order rate constants are found to follow a temperature law embodied in the Arrhenius expression

$$k = A \exp(-E_a/RT), \tag{3.24}$$

where E_a is an experimental activation energy and A is the 'pre-exponential' factor. At this stage in the development, it is usual to identify E_c with E_a, and thus to ask if $\sigma_c \bar{c}$ is to be compared with A. However, it should not be forgotten that $\bar{c}$ is dependent on $T^{1/2}$ [cf. eqn (3.22)], while A, in the simplest formulation, is not temperature dependent. Over limited temperature ranges, any $T^{1/2}$ dependence of experimentally determined A factors may be undetectable. That does not mean, though, that eqn (3.24) will be suitable for extrapolation over extended temperature ranges. A more telling difficulty concerns the absolute magnitudes of A and $\sigma_c \bar{c}$. For typical atmospheric reactants, with collision radii ~400 pm and relative molecular masses ~30, $\sigma_c \bar{c}$ is ~3×10^{-10} cm^3 $molecule^{-1}$ s^{-1} at 300 K. The product $\sigma_c \bar{c}$ is called the *collision frequency factor*. Except for the very simplest of reactants, experimental A factors are usually less than, and often much less than, the collision frequency factor. An explanation for the lack of agreement is sought in terms of molecular complexity, with the existence of special geometric arrangements that are needed during the collision to bring reactive parts of the molecules together (*steric* requirements) and of special needs for the distribution of internal energy. That explanation takes us well away from the idea of hard-sphere reactants.

The alternative simplification adopted in the interpretation of bimolecular reactions is that of the *Transition State Theory* (TST) or *Activated Complex Theory* (ACT), a theory sometimes rather hopefully also called 'absolute rate theory'. Quasi-equilibrium is assumed between reactants and the critically configured ABC molecule, in order to calculate concentrations of $ABC^{\neq}$ (the *transition state*). Rates of reaction can then be obtained from the rate at which $ABC^{\neq}$ passes to products (as a result of translational or vibrational motions along the reaction coordinate). Equilibrium constants can be expressed in statistical thermodynamic terms, and if the formulation is also valid for the quasi-equilibrium, where the system is at a (free) energy maximum rather than minimum, then the resultant rate coefficient, k, is given by

$$k = \frac{\mathbf{k}T}{\mathrm{h}} \frac{q''_{ABC^{\neq}}}{q'_A \, q'_{BC}} \exp(-E_c/RT) . \tag{3.25}$$

Total (volume-independent) partition functions are written as q'_A, q'_{BC}, for reactants and transition state. The double prime on $q''_{ABC^{\neq}}$ indicates that the motion along the reaction coordinate has been factorized out (and a numerical constant introduced). In TST, then, the internal motions neglected in CT are expressly taken into account through the use of the partition functions. For the special case where both reactants are monatomic, and thus hard spheres, both CT and TST give identical algebraic expressions even though the underlying concepts are quite different. TST concentrates only on that region of the potential energy surface around the transition state (for the calculation of the partition function $q''_{ABC^{\neq}}$) while CT is interested only in the height of the energy barrier at the transition state. It is the calculation of $q''_{ABC^{\neq}}$ that offers most difficulty in the practical implementation of TST. Spectroscopic parameters for the reactant molecules are usually available, so that q'_A, q'_{BC} are readily estimated. However, knowledge of the shape of, and the forces acting at, the transition state implies that the potential energy surface is itself known, at least in the region of $ABC^{\neq}$. Usual practice is to make an 'informed guess' at the magnitude of $q''_{ABC^{\neq}}$ based on a hypothetical interaction mechanism and a corresponding model for the transition state. Considerable differences in predicted pre-exponential factors naturally follow from models of the transition state that are, for example, linear, bent, or cyclic. In a more limited way, TST can suggest a sensible order of magnitude for the pre-exponential factor. The three total partition functions in eqn (3.25) are each the product of translational, rotational, and vibrational partition functions. The translational parts can all be calculated, and orders of magnitude for rotational and vibrational parts employed in accordance with the number of each of these modes that exist in A, BC, and $ABC^{\neq}$. More important, from our point of view, is that the temperature dependence for every partition function can be evaluated as a power law. That is, eqn (3.25) can be rewritten as

$$k = A'T^n \exp(-E_c/RT) \tag{3.26}$$

where A' is the temperature-independent part of the pre-exponential function, and n some exponent chosen from the nature of the reactants (monatomic, diatomic, etc.) and a model of the transition state. For the hard sphere (CT) case, $n = 0.5$, from eqns (3.20) and (3.22). In the more general case, n can be positive or negative. The most sensible procedure in temperature extrapolation thus seems to be first to predict n from a model of the reaction, and then to fit the experimental data to equation (3.26) with that value of n.

Regardless of whether the CT or TST simplifications are made, the rate of reaction is, in part, controlled by the energy of a critical, transition state, configuration, an energy which has as its counterpart the activation energy of experimental kinetics. The energy barrier arises because the reactant molecules are forced close together (closer than the sum of their radii in the hard sphere collision approximation), and reactant bonds have to be broken while product bonds are made. Certainly, the energy required is less than that required first to break reactant bonds and then to form product molecules in separate steps. There is, however, no suggestion that the energy of the reactant system decreases at any stage in the passage from separated reactants to the transition state. Such a decrease in energy would correspond to long-range attractive forces, and might lead to an increased collision frequency, and to an A factor that exceeded $\sigma_c\bar{c}$. Many examples of this type of behaviour are in fact known, even with neutral reactants. The effects are, however, strongest and most common with charged reactants. In *ion–molecule* reactions, such as

$$O^+ + CO_2 \rightarrow O_2^+ + CO, \tag{3.27}$$

the ion can induce a dipole in the neutral reactant, and the resultant attractive force can both balance the ordinary chemical activation barrier as well as make the real encounter rate greater than the gas kinetic collision frequency factor for neutral molecules. Near-zero activation energies are thus often found in this type of reaction, and the pre-exponential factors are typically $\sim 10^{-9}$ $cm^3\,molecule^{-1}\,s^{-1}$ (i.e. ~ 3–4 times larger than the values for neutral reactants). Because the long-range attractive forces dominate the potential energy, high velocities of approach are counterproductive in promoting reaction, and some *negative temperature coefficient* of rate constant may be observed. The stronger (or longer range) the interaction, the larger the rate coefficient. For ion reactions with neutral molecules possessing permanent (rather than induced) dipoles, pre-exponential factors are increased by another two or three times. Thus charge transfer from O^+ to the dipolar molecule H_2O,

$$O^+ + H_2O \rightarrow H_2O^+ + O, \qquad (3.28)$$

has a rate coefficient of 2.3×10^{-9} cm^3 molecule^{-1} s^{-1} at 298K, and no activation energy. The long-range interactions are yet larger, of course, for two reactants both of which are charged. Positive ion–negative ion, or positive ion–electron reactions are characterized by rate coefficients 3–4 orders of magnitude larger than typical gas kinetic collision frequency factors. For example, the rate coefficient (298 K) for neutralization of NO^+ by an electron

$$NO^+ + e \rightarrow N + O \qquad (3.29)$$

is 4.5×10^{-7} cm^3 molecule^{-1} s^{-1}.

The possibility was raised in the last paragraph that high approach velocities, associated with high temperatures or energies, might reduce reactivity rather than increase it, particularly if the activation barrier is rather small. Both experimental measurements, and reaction dynamic calculations using suitable potential energy surfaces, have shown that this effect can also occur with neutral reactants. Furthermore, vibrational or rotational energy in the reactants can affect reactivity in unexpected ways. The relative importance of translational and internal energy in overcoming an activation barrier depends on the form of the potential energy surface for reaction. We seem to be returning here to the 'state-to-state' approach to kinetics, or at least to kinetics from particular reactant states to all product states. In atmospheric chemistry the question is important, since pressures and thus non-reactive collision rates may not be sufficient to keep reactants in thermal equilibrium. Much interest surrounds the possibility of species being created energy-rich and showing an enhanced reactivity. Our treatment of electronically excited species (Section 3.2.2) had as one of its premises that excess energy might be capable of overcoming an activation barrier. Translational and vibrational reactant excitation have both been proposed as possible ways in which reactivity could be enhanced. For example, 'hot' $O(^3P)$ could be created in the quenching of $O(^1D)$ by atmospheric N_2, or directly in ozone photolysis at wavelengths longer than the $O(^1D)$ production limit (310 nm: see Section 3.2.1). Translational excitation might then help overcome the endothermicity and the barrier to reaction of a process such as reaction (3.10) of $O(^3P)$ with H_2O to form OH. Similarly, absorption of solar radiation by H_2O in overtone and combination bands has been considered as a mechanism for enhancing the rate of reaction (3.10), some energy being carried in this case by the internal vibrational excitation of the water molecule. Another typical example of this reasoning concerns the production of NO in the Earth's atmosphere. Reaction of ground-state $N(^4S)$ with oxygen

$$N(^4S) + O_2(^3\Sigma_g^-) \rightarrow NO + O \qquad (3.14)$$

has a substantial activation barrier. Nitrogen monoxide (nitric oxide, NO) is a key compound in both neutral and ionic atmospheric chemistry, and suggestions have abounded about possible enhancement of the rate of (3.14) at various levels in the atmosphere as a result of electronic excitation in N or in O_2. It now appears that photolysis of N_2

$$N_2 + h\nu \rightarrow N(^4S) + N(^2D), \tag{3.30}$$

gives not only an electronically excited $N(^2D)$ atom, but also a hot $N(^4S)$ atom with sufficient translational excitation to produce additional NO via reaction (3.14) in the Earth's lower thermosphere. Assessment of the reactivity of vibrationally or translationally excited reactants almost always assumes that all the excitation energy can be used to reduce an activation energy. The more complicated behaviour revealed by experimental and theoretical state-to-state kinetics for energies beyond the critical threshold suggests that the atmospheric calculations of reactivity should be approached with some diffidence.

3.4.2 Unimolecular and termolecular reactions

If chemical reaction requires collision, or at least close proximity, between the reactants, then it might seem that all thermal processes ought to be kinetically of second order. Unimolecular, first-order, elementary processes appear to lack the necessary approach of reactants, and termolecular, third-order, steps suffer from the impossibility of a simultaneous collision between three hard-sphere reactants. The explanation for first- and third-order thermal kinetics shares common ground, and a simple introduction is provided here.

No obstacle exists to understanding how single-step unimolecular and first-order decomposition occurs in a molecule AB that *already* has more than enough energy in it to break one of its bonds. We have seen examples in photodissociation of diatomic and triatomic molecules, where optical dissociation or predissociation populates vibrational levels of AB sufficient to cause fragmentation. The rate of fragmentation may depend on the rate at which energy can accumulate in the bond to be broken, but the reaction will be kinetically of first order. *Chemical activation* offers another route to high vibrational excitation. For example, the reaction of ClO with NO can produce a highly excited $ClONO^{\dagger}$ molecule (the dagger representing vibrational excitation)

$$ClO + NO \rightarrow ClONO^{\dagger}. \tag{3.31}$$

This excited $ClONO^{\dagger}$ can then either split up to the reactants again, or form Cl and NO_2

$$ClONO^{\dagger} \rightarrow Cl + NO_2. \tag{3.32}$$

Reaction of the excited $ClONO^{\dagger}$ is a unimolecular, first-order, elementary reaction.

It is in interpreting *thermal* unimolecular reactions that some difficulty arises, since the formation of an excited $AB^{\dagger}$ molecule involves collisions between the AB species, and might therefore be expected to show *second*-order kinetics. A basic understanding was provided by Lindemann, who suggested that thermal first-order reactions were not true elementary steps, but rather involved at least three elementary processes

$$AB + AB \xrightarrow{k_a} AB^{\dagger} + AB \quad \text{collisional activation} \tag{3.33}$$

$$AB^{\dagger} + AB \xrightarrow{k_d} AB + AB \quad \text{deactivation} \tag{3.34}$$

$$AB^{\dagger} \xrightarrow{k_r} A + B \quad \text{reaction.} \tag{3.35}$$

If reaction (3.34) dominates as a loss process for $AB^{\dagger}$ over (3.35) then the concentration of $AB^{\dagger}$ is almost at its thermal equilibrium value, while the rate-determining step for reaction is the first-order process (3.35). Overall first-order kinetics follow. It is obvious, however, that at sufficiently low concentrations of AB, there becomes a point at which reaction (3.33) is rate limiting, and the kinetic behaviour will be second order. Transition from first- to second-order behaviour is, indeed, seen at low enough pressures in this kind of thermal unimolecular reaction. Quantitative expression of these ideas can be obtained by a steady-state treatment for the concentration of $AB^{\dagger}$ (see Section 3.5). The result for the rate of loss of AB is

$$-\frac{d[AB]}{dt} = k_I[AB] = \frac{k_a k_r [AB]^2}{k_d[AB] + k_r}, \tag{3.36}$$

where k_I is the experimentally defined pseudo-first-order rate coefficient. So long as $k_d[AB] \gg k_r$, the reaction is first order, but if [AB] is reduced to the point at which the reverse inequality holds, then the reaction becomes second order. At high concentration, the limiting value of k_I (referred to as k_I^{∞}) is equal to $(k_a k_r/k_d)$ and is thus truly first order, being independent of [AB]. The low pressure limit, k_I^0, is equal to $k_a[AB]$ and is itself first order in pressure, or second order overall.

Considerations about high- and low-pressure extrapolations of rate data are most frequently met in atmospheric chemistry in connection with termolecular reactions. As with unimolecular reactions, termolecular processes have orders variable with pressure, being third order at 'low' pressure and second order at 'high' pressure. Such reactions are extremely important in combination processes, and we can see why by first looking at the reaction of two atoms to form a diatomic molecule. A typical case is the combination of two $O(^3P)$ atoms, for which some potential curves were given in Fig. 3.2. Even if the combining atoms have no relative translational energy, the newly formed O_2 molecule has the O + O combination energy stored in it: that energy is the O–O bond energy, and the O_2 is chemically activated $O_2^{\dagger}$ at its dissociation limit. *Unless* some energy is removed within one vibrational period, the molecule will fall apart again as the internuclear distance increases on the first oscillation. Energy can be removed in collisions, and we usually represent the species that dissipates energy by the symbol M. In atmospheric chemistry, the most important M may be the most abundant 'bath' gas (e.g. N_2 or O_2 on Earth). The overall reaction is now written

$$O + O + M \rightarrow O_2 + M, \tag{3.37}$$

which is a termolecular step. The redissociation that has been prevented is the unimolecular dissociation of $O_2^\dagger$ equivalent to step (3.35), and the process deactivating $O_2^\dagger$ is the equivalent of (3.34). We shall see shortly that internal energy seems to flow fairly freely between different vibrational modes of a polyatomic molecule. If the newly formed molecule is larger than diatomic, there are such modes into which the bond combination energy can flow. The lifetime of the newly formed molecule can thus correspond to many vibrational periods before the energy flows back to the critical bond. With a large enough polyatomic molecule, the lifetime can be so great that collisional removal of excess energy (*stabilization*) is no longer rate determining, and combination then exhibits second-order kinetics. Because reactions (3.34) and (3.35) are common to both unimolecular and termolecular reactions, the same general considerations about flow of energy apply to both types of process.

The analogue of expression (3.36) can be derived from the single excitation level kinetic scheme

$$A + B \xrightarrow{k_c} AB^\dagger \quad \text{combination} \tag{3.38}$$

$$AB^\dagger + M \xrightarrow{k_s} AB + M \quad \text{stabilization} \tag{3.39}$$

$$AB^\dagger \xrightarrow{k_r} A + B \quad \text{reaction.} \tag{3.35}$$

The result is

$$\frac{d[AB]}{dt} = k_{II}[A][B] = \frac{k_c k_s [A][B][M]}{k_s[M] + k_r}, \tag{3.40}$$

where k_{II} is the experimentally defined pseudo-second-order rate coefficient corresponding to k_I in eqn (3.36). We see straightaway that if $k_r \gg k_s[M]$, the reaction is third order, with $k_{II}^{\circ} = (k_c k_s / k_r)[M]$. If, however, $k_r \ll k_s[M]$, then the reaction is second order, with $k_{II}^{\infty} = k_c$. Increased complexity in the molecule AB reduces the value of k_r, because the combination energy is distributed amongst more vibrational modes. The concentration, or pressure, of third-body M at which third-order behaviour turns over to second-order kinetics is thus lower the more complex the molecule produced. 'Complex' is only a relative term here: combination of two hydrogen atoms to form H_2 is third order up to 104 atm, while combination of two CH_3 radicals to form C_2H_6 is second order at all but the lowest pressures. However, it so happens that the reactants in several combination reactions of great atmospheric importance, such as

$$OH + NO_2 + M \rightarrow HNO_3 + M, \tag{3.41}$$

or

$$O + O_2 + M \rightarrow O_3 + M, \tag{3.42}$$

are of just that molecular size that complex intermediate-order kinetics are displayed at some point in the atmospheric pressure range.

Expressions (3.36) and (3.40) represent the variations of experimentally determined rates of reaction with pressure. The pseudo-first- or second-order rate coefficients k_I or k_{II} can be conveniently expressed in terms of the high- and low-pressure limiting values k_I^∞, k_I^0, or k_{II}^∞, k_{II}^0. For example, k_{II} in eqn (3.40) can be expressed as

$$k_{II} = \frac{k_{II}^0 k_{II}^\infty}{k_{II}^0 + k_{II}^0} \tag{3.43}$$

Remembering that k_{II}^0 is itself first order in pressure, it can be seen that eqn (3.43) represents in outline the variation of k_{II} with pressure that is found experimentally. Unfortunately, however, the equation does not match experimental data in detail, so that it cannot be applied directly to the calculation of rates at intermediate pressures. The reasons for the failure are known. The reactions and the rate coefficients k_a or k_c, k_d or k_s, and k_r should have been defined for each individual quantized vibrational level of $AB^\dagger$, and the individual rates summed to give the total rate. It is, perhaps, easy to see that the more energy available (beyond the critical amount needed to break a particular bond), the more rapid will be the fragmentation (i.e. the larger will be k_r). Related to this point is the implication that energy stored in any vibrational mode can be made available to the critical bond. Experimental evidence largely favours the flow of energy between modes as being fairly free, and the distribution as being near statistical. An additional complication involves the interconversions of vibrations and rotations in the fragmenting molecule. The theory has been extended, modified, and manipulated over the years by Rice, Ramsperger, Kassel, and Marcus, and the familiar initials RRKM are used to designate their formulation. With sufficient sophistication of the input information, very good agreement can be obtained between theory and experiment. Correspondingly, one could have confidence in the extrapolation of data obtained in an intermediate concentration regime to either high-pressure (first-order) or low-pressure (second-order) limits. However, application of RRKM theory to real processes of atmospheric importance is in practice rather difficult, and an alternative, much simpler, approach is now almost universally adopted. This approach has its origins in work by Troe (see Bibliography) on the theoretical prediction of unimolecular reaction rate parameters. However, with k_{II}^∞, k_{II}^0 known, Troe has shown that a simplification of his theory allows the right-hand side of eqn (3.43) to be multiplied by a broadening factor, F, that is a function of $(k_{II}^{0}/k_{II}^{\infty})$. For many atmospherically important termolecular reactions F may be calculated from a simple mathematical expression.

Third-order reactions often show decreasing rate with increasing temperature: they have a negative temperature coefficient. The reason is that the larger the thermal kinetic energy possessed by the reactants A and B in process (3.38), the more internal vibrational energy will be stored in the $AB^{\dagger}$ molecule produced. As pointed out earlier, the chance of the critical bond energy finding its way back to a breakable bond is thus increased, and k_r is larger. Since k_c and k_s are only slightly affected by temperature, it follows from eqn (3.40) that the rate of reaction will decrease with increasing temperature. Thermal energy in effect assists the newly formed molecule to split up again, thus slowing the rate of combination. In the third-order limit, k_{II}^{0} is inversely proportional to k_r (see above). Theory suggests that the temperature variations of k_r should be better expressed in terms of a power, T^n, rather than as a conventional activation energy. Hence experimental measurements of k_{II}^{0} as a function of temperature should be fitted against a T^{-n} law to allow rational interpolation or extrapolation to atmospheric temperatures. Typical measured values of n are 2.5 to 3.1 for reaction (3.41) and 1.7 for reaction (3.42). Models of the transition state for bond association reactions also suggest that, at the high-pressure limit, k_{II} should possess a negative exponent of temperature.

Two other combination mechanisms of atmospheric interest can be included in this section, although they are not termolecular. Both mechanisms highlight the need to dissipate the energy of the combination process. Energy can be carried off as photons, so that if the combination can populate, directly or indirectly, an excited state (usually electronically excited) that can radiate, then the nascent molecule can be stabilized. An example of this process is the two-body combination of O and NO

$$O + NO \rightarrow NO_2 + h\nu. \tag{3.44}$$

The process is probably responsible for some continuum emission in the atmosphere. However, three-body combination of O and NO is much more efficient than two-body radiative combination at all but very low pressures. In ion-neutralization processes such as reaction (3.29), the initial step is the formation of a neutral molecule which is highly excited by an amount equivalent to the ionization potential. Unless the energy can be dissipated, the diatomic neutral molecule will re-ionize. We are not interested here in chemical stabilization, but rather in stabilizing the removal of charged species. Energy can be conserved in the neutralization by fragmentation of the excited NO, with translational motion carrying off the excess of ionization energy (9.2 eV) over bond-dissociation energy (6.4 eV). Ionized *atoms* obviously cannot combine with electrons in this way. Radiative stabilization thus becomes the only (non-collisional) neutralization mechanism. However, the radiative transitions needed are usually forbidden for the neutral atoms formed, so that the rates of neutralization are much slower, by about five orders of magnitude, for atomic than for molecular ions.

3.4.3 Condensed-phase, surface, and heterogeneous reactions

In the next two sections, we shall examine some features of interest in the kinetics of reactions occurring within liquid-phase systems, such as cloud droplets, and at the interfaces between solid or liquid particles and their gaseous surroundings.

Reactions within the liquid droplets of clouds and fogs are important in several aspects of tropospheric chemistry, such as the oxidation of sulphur dioxide (see Section 5.10.4). More recently, it has become evident that liquid or solid particles can play a critical role in the chemistry of the stratosphere (see Sections 4.4.4, 4.5.5 and 4.7) under certain conditions. Particles such as those of sulphate aerosol or clouds formed from water-ice and hydrates of nitric acid (*polar stratospheric clouds, PSCs)* are implicated in such processes. This chemistry may involve surface reactions, or reactions within the bulk material, but the interface between gas and condensed phases is involved in some way, and the reactions are thus known as *heterogeneous reactions*.

In many respects, the kinetics of liquid, surface, and heterogeneous reactions are governed by the same principles that we have established for gas-phase processes. There are, however, some key differences. Here, we present a necessarily brief outline of some of the most interesting features, with the objective of providing some familiarity with the vocabulary employed in dealing with these areas of reaction kinetics.

3.4.4 Liquid-phase reactions

The solvent obviously has the potential to exert a considerable influence on the course of chemistry in the liquid phase. In air at one atmosphere pressure, and at ambient temperature, the molecules themselves occupy only roughly 0.2 per cent of the total volume; in liquids, the molecules can make up half the volume. At pressures of one atmosphere and below, we have been able to assume that the reactant molecules undergo essentially unhindered motion, and that assumption lies behind the various formulations of kinetics that we have discussed in previous sections. In distinction, in liquids, the reactive molecules must squeeze past the solvent molecules (or each other, if one species is also the bulk liquid) if they are to reach each other and undergo reaction. Reactants, activated complexes or intermediates, and products can also all interact with the solvent. One manifestation of the interaction with intermediates is that energy removal in association reactions, such as the combination processes

(3.41) or (3.42), is virtually instantaneous, and the systems always display pure second-order kinetics in the liquid phase, in contrast to the behaviour described in the last section for gas-phase reactants. Interactions of the reactants and the solvent (especially water) may make the formation of ions energetically more favourable than in the gas phase. New reaction channels may thus become accessible, and the kinetics of the processes can be influenced by the attractive or repulsive electrostatic interactions between the reactants, amongst many other factors.

Simple treatments of liquid-phase kinetics often start from the concept of the *encounter pair* of reactants that find themselves together within a *solvent cage*. Two extreme cases can be envisaged. In the first, the two species are very highly reactive towards each other, and undergo chemical transformation within a very few 'collisions' within the cage. The rate-determining process is then the diffusion of the reactants through the solvent to form the encounter pair, and the process is a *diffusion-controlled reaction*. At the other extreme, the activation energy for reaction may require the partners to pick up appreciable amounts of energy as they shake against each other within the cage, so that the kinetics are controlled by the rate of reaction within the cage, rather than by the rate at which they reach it. *Activation controlled reaction* kinetics then result.

For many of the liquid-phase reactions of interest in atmospheric chemistry, the intrinsic reactivity of the partners is, indeed, very high, leading to diffusion-controlled kinetic behaviour. A very elementary treatment of the diffusion-controlled rate constant, k_d, leads to the equation

$$k_d = 4\pi r_{AB} D_{AB}, \qquad (3.45)$$

where r_{AB} is a hypothetical *encounter distance* at which two partners A and B will react, and D_{AB} is the *diffusion coefficient* for the reactants.

The encounter distance *may* be roughly the sum of the gas-kinetic radii of the partners for neutral reactants, while the appropriate diffusion coefficient *may* be similar to a mean bulk diffusion coefficient of the reactants in the solvent. Making the assumptions that these values may be taken, and with typical values of $r_{AB} = 0.5$ nm and $D_{AB} = 1.3 \times 10^{-9}\ m^2 s^{-1}$ (for Na^+ in H_2O), k_D is calculated as ca. $8 \times 10^{-18}\ m^3 molecule^{-1} s^{-1}$ or, in the units that we have been using for rate coefficients so far, $8 \times 10^{-12}\ cm^3 molecule^{-1} s^{-1}$. In liquid-phase kinetics, it is more conventional to use molar units for concentrations, so that the equivalent figure is $(6 \times 10^{23} \times 10^3 \times 8 \times 10^{-18}) \approx 5 \times 10^9\ dm^3 mol^{-1} s^{-1}$. In whatever units this rate coefficient is expressed, it is evidently about 40 times slower that the maximum gas-kinetic rate coefficient (p. 119). In general, a rate coefficient of $> 10^9\ dm^3 mol^{-1} s^{-1}$ for an aqueous-phase reaction is taken to be indicative of a diffusion-controlled mechanism.

One of the largest known rate coefficients for a condensed-phase process is that for the very important reaction

$$H^+ + OH^- \rightarrow H_2O \tag{3.46}$$

(1.4×10^{10} $dm^3 mol^{-1} s^{-1}$ at 298 K). The magnitude mainly reflects the large diffusion coefficients in water of OH^- and, especially, of H^+; the rapid diffusion is itself a consequence of the special mechanisms by which these ions migrate in liquid H_2O.

Although the diffusion coefficient is most important in making reaction (3.46) so fast, there is another factor operating that may be dominant in other reactions. The positive and negative ions attract each other, so that the effective encounter distance can be much greater than the gas-kinetic collision distance; that is, r_{AB} has to be replaced by r_{eff} in equation (3.45). For $r_{AB} = 0.5$ nm, straightforward electrostatic calculations indicate that r_{eff} ought to be about 0.2 nm for oppositely-charged ions (and 0.7 nm for like charges) in water with a relative permittivity of 78. However, it seems that this bulk permittivity is inappropriate to the highly ordered solvent molecules in the immediate vicinity of the ions, and that more realistic values of r_{eff} ought to be 10 nm and 10^{-9} nm for unlike and like charges. These values mean, of course, that oppositely charged ions will react twenty times faster than their neutral analogues, under similar conditions, while similarly charged ions can be assumed not to react at all.

While our discussion has so far centred on the behaviour of the atmospherically dominant class of diffusion-controlled reactions, some processes of interest are activation controlled. One characteristic of such reactions is that the activation energy may be smaller than for the equivalent gas-phase reaction, because the reactant pair undergoes many individual 'collisions' at each encounter, whereas, in the gas phase, the collision and the encounter are the same thing. A particularly interesting property shown by activation-controlled ionic reactions is that of the *kinetic salt effect*. Rate coefficients are affected by the presence of other ionic species present in the solution that do not themselves participate in the reaction. Interactions between oppositely charged partners are slowed down by the presence of such salts. In the atmosphere such effects may be of significance, since water droplets may contain substantial amounts of sea-salt or other similar species.

3.4.5 Heterogeneous reactions

Atmospheric chemistry involving clouds, fogs, rain droplets, ice particles, and other solid and liquid aerosol particles is necessarily *heterogeneous* in the sense that it involves transfer of gas-phase molecules to the condensed-phase system across the particle interface. We have just examined some of the factors that determine the rates of reaction *within* the condensed-phase system, and now focus our attention more specifically on the influence of the interface between condensed and gaseous phases.

In fact, reactions occurring *inside* particles are really confined to the liquid phase, since diffusion coefficients within solids are too small to allow significant reaction rates. On the other hand, reactions *on* solid surfaces are thought to be of very considerable atmospheric significance. The PSCs involved in stratospheric chemistry consist, for example, in part, of solid water-ice and solid nitric acid trihydrate. A convenient starting point in the present discussion will thus be an examination of surface reactions themselves. An added degree of complexity arises when the particle is liquid, as is the case for droplets in the troposphere, and possibly for stratospheric sulphate aerosol, which may be in the form of supercooled liquid sulphuric acid.

The first parameter that we consider in the discussion of surface processes is the *uptake coefficient*, γ, which is the ratio of molecules lost to a surface to the number of gas–surface collisions that occur. If the rate of collision of a molecule X with an area A of the surface is ω, then the rate of loss of X per unit volume, $-\mathrm{d}[\mathrm{X}]/\mathrm{d}t$, is equal to $\gamma.\omega/V$, where V is the volume of the system. The kinetic theory of gases shows that

$$\omega = \bar{c}A[\mathrm{X}]/4, \tag{3.47}$$

so that

$$-\mathrm{d}[\mathrm{X}]/\mathrm{d}t = \gamma\bar{c}A[\mathrm{X}]/4V. \tag{3.48}$$

Now the loss of X may also be described in terms of phenomenological rate equations of the type

$$-\mathrm{d}[\mathrm{X}]/\mathrm{d}t = k_{\mathrm{S}}\,\{\mathrm{S}\}[\mathrm{X}] \equiv k'_{\mathrm{S}}\,[\mathrm{X}], \tag{3.49}$$

where $\{\mathrm{S}\}$ represents the number of active surface sites per unit area and k_{S} and k'_{S} are second-order and the corresponding pseudo-first-order rate coefficients (see p. 117) for the surface loss process. It follows, from a comparison of eqns (3.48) and (3.49) that

$$k'_{\mathrm{S}} = \tfrac{1}{4}\gamma\bar{c}(A/V). \tag{3.50}$$

Uptake coefficients may be determined by a variety of experimental methods. Regardless of whether the molecule is removed by reaction on or within the particle, or by dissolving in it, eqn (3.50) provides the link between the kinetics of the uptake process and the uptake coefficient. It will be evident that the reactive uptake coefficient is equivalent to the *reaction probability*.

A complication obviously arises if a molecule does not react irreversibly, but can desorb again from a surface, or come out of solution to re-enter the gas phase. In such cases, γ can apparently be time dependent, and the measurement of the variation of γ with time provides one way of examining these reversible processes. In the case of the atmosphere, the most important aspect concerns the partitioning of molecules between gaseous and liquid phases. Solubilities of gases at low solute concentrations obey *Henry's law*

Henry's Law;

$$[X(s)] = H_X p_X \tag{3.51}$$

where $[X(s)]$ is the concentration of X in solution, p_X is its pressure in the gas phase, and H_A is the Henry's law coefficient (which is a function of temperature). Henry's law expresses an equilibrium situation, in which the fluxes of molecules into and out of the liquid are equal. However, it is straightforward to calculate the forward and reverse fluxes, and thus the net flux into the liquid, under non-equilibrium conditions. The first term comes immediately from eqn (3.50), while the second requires use of the diffusion equation for transport of the molecules from the bulk liquid to the interface. If the coefficient of diffusion for this latter process is D, then it may be shown that

$$\frac{1}{\gamma_t} = \frac{1}{\gamma_0} + \frac{\pi^{1/2}\bar{c}}{4H_A RTD^{1/2}}\, t^{1/2} \tag{3.52}$$

where γ_0 and γ_t are the uptake coefficients at time 0 and time t. The equation shows how H_A can be calculated from measurements of uptake coefficient as a function of time, or, conversely, how the variation of uptake coefficient with time may be estimated from a knowledge of the solubility of the gas. Note that at 'infinite' time, γ_t becomes zero: the system has reached equilibrium.

The material that we have developed so far is applicable to both physical processes—adsorption, absorption, or solution—or to chemical change. In the particular case of chemical change, we can envisage two possibilities. Either the reaction may involve an interaction of the gas-phase reactant with the surface or the bulk constituent of the particle, or it may involve reaction with some second species already adsorbed on, or dissolved in, the particle. The concepts set out earlier remain applicable in the second, 'bimolecular', situation, but the value of {S} at the surface (eqn (3.49)), or the concentration of the partner reactant Y, in solution, will be determined by factors similar to those already determining the adsorption or solubility of Y.

Let us now consider more explicitly the steps involved in a heterogenous reaction of a gas-phase species with either the bulk constituent of a droplet or with another species that is already dissolved in it. Figure 3.8 illustrates the separate steps that together lead to reactive or non-reactive uptake. Chemical change corresponds to loss of the gas-phase molecule; uptake that is non-reactive can arise from physical dissolution or from reversible chemistry. The steps indicated are (i) gas-phase transport of the reactant to the surface of the droplet; (ii) accommodation at the surface; (iii) diffusion into the liquid; (iv) chemical reaction; (v) diffusion of unreacted molecules and products to the surface; and (vi) desorption of species from the interface. Characterization of each of these individual steps is obviously a formidable task, although one that may be simplified—as often happens in kinetics—by one of the steps being rate determining.

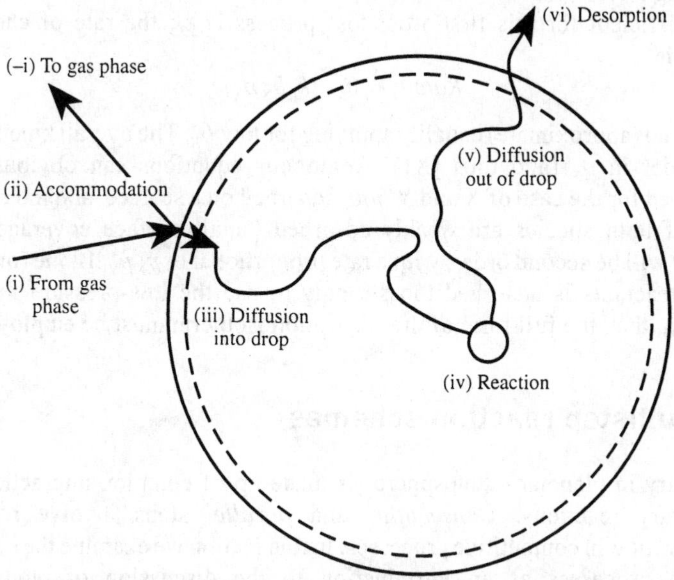

Fig. 3.8. Processes involved in the uptake and chemical reaction of a gas-phase molecule by a liquid droplet, and chemical reaction in it. From Ravishankara, A.R. and Hanson, D.R. in *Low-temperature chemistry of the atmosphere*, Moortgat, G.K., Barnes, A.J., Le Bras, G. and Sodeau, J.R. (eds.) NATO ASI Series **21**, Springer-Verlag, Berlin, 1994.

Finally, as a contribution to this section, we consider reactions *on the surface* of atmospheric particles. Formal kinetic equations for surface reactions often start from the *Langmuir adsorption isotherm*, which is the simplest of equations that expresses the partitioning of gas between surface and gas phases. The isotherm makes several assumptions, including one that states that all surface sites are equivalent and that there are no interactions between molecules adsorbed on them. With these assumptions, it may be shown that the *surface coverage*, θ_X, is given by the equation

$$\theta_X = bp_X/(1 + bp_X), \tag{3.53}$$

where b is a constant equal to the ratio of rate coefficients for adsorption onto the surface and desorption from it. For the low partial pressures of adsorbates present as minor constituents in the atmosphere, θ_X is likely to be very nearly a linear function of p_X, although the full equation might be needed for very strongly adsorbed reactants or for high partial pressures in the atmosphere. An extended treatment for the situation where *two different* species, X and Y, are adsorbed is straightforward, with the surface coverages θ_X and θ_Y both entering into the equation.

The kinetics of reaction for a single reactant are then developed by including a loss (probably decomposition) of the adsorbed molecules. If the rate coefficient for this first-order loss process is k_L, the rate of chemical change is

$$Rate = k_L\,\theta_X \simeq k_L\,b_X p_X, \tag{3.54}$$

the second (approximate) equality applying for low θ_X. The overall kinetics are first order in p_X (and thus [X]). Analogous equations can obviously be developed for the case of X and Y *both* adsorbed on a surface, and interacting there. If both species are weakly adsorbed (small surface coverage), the kinetics will be second order, with a rate proportional to $p_X p_Y$. If one (or both) of the reactants is adsorbed too strongly to use the low-pressure limiting equation, then the full form of the adsorption isotherm must be employed.

3.5 Multistep reaction schemes

Chemistry in planetary atmospheres is made up of complex interactions of elementary reactions. *Consecutive* and *parallel* steps involve reactive *intermediates* in competitive processes. In this section we examine the kinetics of such processes as an introduction to the discussion of models of atmospheric chemistry in Section 3.6. Reaction intermediates of particular interest include atoms, radicals, ions, and excited species. Most of these intermediates are highly reactive, and, with one or two exceptions, cannot be 'stored' in a laboratory for long periods because they are lost on the walls of the containing vessel, or react with each other. Such intermediates are not necessarily *unstable*, and chemical lifetimes of isolated atoms or radicals in the absence of surfaces can be virtually infinite. Many excited-state species are unstable, since they may possess enough internal energy to fragment, and they may also be able to lose their energy by emission of radiation. An excited species that cannot undergo loss by an allowed radiative transition is said to be *metastable*.

Multistep reaction schemes are interpreted kinetically by writing down the differential equations, such as eqn (3.19), for all the species of interest, including the intermediates. Solution of these equations then allows prediction of the concentration–time variation of each of the species. Unfortunately, analytical solution of the many simultaneous differential equations is rarely possible. Numerical solution has become a widely used alternative since the advent of high-speed computers and the development of good techniques for dealing with differential equations. Such methods do not, however, afford much insight into the underlying chemistry of the system. For some highly reactive intermediates, the *Stationary State Hypothesis* (SSH) provides a simplification that will permit algebraic solutions to the kinetic equations. Consider an intermediate X that is created in a process whose rate is constant, and whose loss rate increases with increased [X]. After the reaction is started, [X] will increase until the rate of loss is equal to the rate of formation. A steady state for [X] has been reached, and $d[X]/dt = 0$. To illustrate the steady-state method let us calculate the concentration of $O(^1D)$, assumed to be formed at some altitude in the Earth's atmosphere solely by oxygen photolysis and lost by quenching with nitrogen and oxygen

$$O_2 + h\nu \longrightarrow O(^1D) + O(^3P); \; \text{Rate} = I_{abs} \tag{3.55}$$

$$O(^1D) + M \xrightarrow{k_q} O(^3P) + M \tag{3.56}$$

$$M = N_2 \text{ or } O_2.$$

The kinetic equation that describes these processes is

$$\frac{d[O(^1D)]}{dt} = I_{abs} - k_q[O(^1D)][M] \tag{3.57}$$

If $O(^1D)$ is in a stationary state, then we set the differential equal to zero, and

$$[O(^1D)]_{ss} = I_{abs}/k_q[M], \tag{3.58}$$

where the subscript indicates a steady-state concentration. Such a value might now be used to estimate the rate of a minor process involving $O(^1D)$, such as reaction (3.11) which generates OH. The problem is to know if the concentration of $O(^1D)$ calculated using the SSH bears any relationship to actual concentrations. Our two-reaction example has been chosen because it can also be solved analytically. So long as I_{abs} and M are independent of time, eqn (3.57) can be integrated to yield

$$[O(^1D)] = (I_{abs}/k_q[M])(1 - \exp(-k_q[M]t)), \tag{3.59}$$

where t is the time for which the system has been illuminated. This expression for $[O(^1D)]$ approaches the steady state expression so long as $k_q[M]t \gg 1$, the error in applying the SSH being less than one per cent for $k_q[M]t > 4.6$. At an atmospheric altitude (~ 80 km) where $[M] \sim 3 \times 10^{14}$ molecule cm^{-3}, and the composite k_q for N_2 and O_2 is $\sim 3 \times 10^{-11}$ cm^3 molecule^{-1} s^{-1}, the non-steady and steady-state concentrations of $O(^1D)$ are therefore nearly identical for illumination times exceeding $4.6/(3 \times 10^{14} \times 3 \times 10^{-11}) \sim 5 \times 10^{-4}$ s. The SSH can thus be applied so long as I_{abs} and [M] remain constant over, say, 10^{-3} s. In our example, [M] (and $[O_2]$, which is the $O(^1D)$ precursor) are both time invariant, and solar intensities (I_0) change diurnally over periods of hours rather than milliseconds, so that $[O(^1D)]_{ss}$ is a good approximation for $[O(^1D)]$. As solar intensity varies during the day, so the steady-state concentration of $O(^1D)$ adjusts (on the time-scale of milliseconds) to a new value. Now let us consider what happens with an intermediate that is much less reactive than $O(^1D)$. A good example is the ground-state $O(^3P)$ formed photolytically in reaction (3.55); for simplicity, we ignore the $O(^3P)$ atom from reaction (3.56). Three-body combination of O with O_2,

$$O + O_2 + M \rightarrow O_3 + M, \tag{3.42}$$

is the major loss process for $O(^3P)$, so that eqns (3.58) and (3.59) for $O(^1D)$ are replaced by

$$[O(^3P)]_{ss} = I_{abs}/k_t[O_2]\,[M] \tag{3.60}$$

and

$$[O(^3P)] = (I_{abs}/k_t[O_2]\,[M])(1 - \exp(-k_t[O_2]\,[M]\,t)), \tag{3.61}$$

where k_t is the third-order rate coefficient for reaction (3.42). For temperatures ~ 200 K, appropriate to our hypothetical reaction altitude (80 km), $k_t \sim 1.4 \times 10^{-33}\ cm^6\,molecule^{-2}\,s^{-1}$. The exponential in eqn (3.61) is thus <0.01 for $t > 4.6/[1.4 \times 10^{-33} \times 0.2 \times (3 \times 10^{14})^2] \sim 1.8 \times 10^5\ s \sim 50\ h$. Steady-state conditions take much longer to achieve than the time-scale over which solar intensities vary; a steady state for $O(^3P)$ is not set up, and the SSH is inappropriate. The difference in behaviour of $O(^1D)$ and $O(^3P)$ arises because of the relative values of the terms $k_q[M]$ and $k_t[O_2]\,[M]$, which are the *pseudo-first-order* loss rates for the two intermediates, respectively. For the highly reactive excited atom, the loss rate is large, and a steady state is achieved rapidly; for the less reactive ground-state atom, the steady state is *not* reached on the time-scale over which reaction conditions remain constant.

3.6 Models of atmospheric chemistry

3.6.1 Lifetimes and transport

Interpretations of atmospheric behaviour can be tested by comparing the predictions of a *model* of the atmosphere with results from measurements on the atmosphere itself. Models of atmospheric physics have been implicit in much of our earlier discussion (e.g. of radiative heating, Section 2.2). We now indicate how chemical changes can be accommodated in the models. A convenient starting point for the discussion is to examine the concepts of *lifetime* or *residence time* of a chemical species in the atmosphere. These ideas have been used several times already in this book, but without much critical examination. It should become apparent during the present development that considerable subtlety attaches to what appear to be simple ideas, and that what will appear as models have to be used to calculate lifetimes properly.

In Section 1.1 it was stated that the residence time or lifetime for a species could be calculated in the simplest possible way by dividing the total atmospheric burden or loading of the species by the rate of its supply. Since, as we have emphasized frequently before, the concentrations of most species remain essentially constant, at least over periods small compared with these lifetimes, it follows that the rate of supply is very approximately the same as the rate of loss, so that the definition of lifetime could be extended to the atmospheric concentration divided by the rate of loss.

The rationale behind such a definition is straightforward enough. Consider a simple situation in which a liquid flows through a pipe without becoming mixed. The time that any volume element entering the tube 'lives' in it is the time it takes for it to pass from one end to the other of the tube. The total amount of material in the tube is thus the rate at which the material is supplied multiplied by this lifetime; rearranged, this statement reads that the lifetime is equal to the amount of material present divided by the rate of supply. It is convenient to express both the burden and the rates in concentration units (molecules or moles per unit volume). We could then write these definitions in the form

$$\tau_u = [A]/S = [A]/L, \tag{3.62}$$

where τ_u is the lifetime of a species A in the unmixed system, [A] its atmospheric concentration, and S and L the rates of supply and loss in terms of concentration per unit time. In the atmospheric context, the supply is the source of material and the loss is made up of the sinks, such as chemical loss, and physical removal by transport and deposition.

In reality, it seems very *un*reasonable to assume that the atmosphere is not mixed at all. If the atmosphere were *completely* mixed, any group of molecules introduced would slowly decrease in number as some were lost as a result of the chemical and physical removal processes. Although there is then some kind of characteristic mean time for which the molecules are in the atmosphere, certain of the molecules are lost at much earlier times, and some at much later ones. How then shall the lifetime be defined?

To reach a solution, we turn back to the ideas of reaction kinetics, which deal with the reaction and loss of large numbers of molecules. Let us assume that a trace constituent A decays by a (pseudo-) first-order process. The differential and integrated forms of the rate equation are

$$-\mathrm{d}[A]/\mathrm{d}t = k'[A]; \quad [A]_t = [A]_0 \exp(-k't), \tag{3.63}$$

where $[A]_t$ is the concentration of A at time t and $[A]_0$ the initial concentration. In a time $\tau_m = 1/k'$, the concentration of A will have decreased to a fraction 1/e (0.37) of its initial value (and in a time $\ln 2/k'$ to one-half of the initial value: that time is the *half-life*) in this fully mixed system. But

$$\tau_m = \frac{1}{k'} = \frac{[A]}{-\mathrm{d}[A]/\mathrm{d}t} \tag{3.64}$$

and $-\mathrm{d}[A]/\mathrm{d}t$ is the same as the loss rate L of equation (3.62). Comparison of equations (3.62) and (3.64) then shows that τ_u in the unmixed system is numerically identical to τ_m in the mixed one, although the definitions of the two lifetimes are quite different. We recall that τ_u is the time required for *all* the material to be removed from the system, while τ_m is the time taken for $(1 - 0.37) \times 100 = 63$ per cent of A to be removed.

The difference between unmixed and fully mixed systems has been emphasized in this treatment so that the situation with regard to the atmosphere may be properly understood. In a real atmosphere, a particular trace species may, of course, exhibit behaviour lying anywhere between the two extremes. An atmospheric lifetime can always be expressed as the concentration of some species divided by the rate of its loss, but what that lifetime really means depends on the mixing behaviour within the atmosphere. We shall return to this aspect shortly when we consider the relation between lifetime and transport.

Whatever the particular situation, the mean lifetime apparently provides a clear measure of how long it would take a small perturbation to atmospheric concentrations to disappear. The only question to be resolved is whether 'disappear' means 100 per cent of the perturbation, 63 per cent, or something in between. In this sense, the lifetimes we have described are *adjustment times*. One interesting complication that can arise in the real atmosphere is seen if addition (or removal) of one trace gas has a profound effect on those parts of atmospheric chemistry that themselves dominate the removal steps. An illustration of this phenomenon can be seen in the lifetimes of CH_4. In section 5.3.2, we shall see that CH_4 is lost in the troposphere largely by reaction with the OH radical. Furthermore, one of the oxidation products is CO, which itself reacts rapidly with OH. Thus, addition of CH_4 depletes OH, and the rate of loss of CH_4 is consequently reduced. The result is that the lifetime of CH_4 is lengthened by the very act of increasing its atmospheric concentration. For this particular example, the lifetime of CH_4 estimated from the expected chemical and photochemical loss rates for the unperturbed situation would be about 10 years. But global models including the feedback indicate that the true adjustment time is nearer to 14 years.

An extension to the concepts is needed if there is more than one loss process for a species. For example, the species may be lost by physical processes, such as transport and wet and dry deposition, as well as by several different chemical and photochemical reactions. The method is entirely straightforward. An effective pseudo first-order rate coefficient is written for each of the individual loss processes, and an overall value of k' obtained as the sum of them all.

It is now appropriate to examine more carefully the interconnection between atmospheric transport and mean lifetimes. The particular point of interest is how far a given air parcel might move within the mean lifetime of trace-gas species within it. Vertical motions within the troposphere, within the stratosphere, and exchange between the two regions, as well as E–W and N–S and hemisphere–hemisphere motions all occur on very different time-scales. If the lifetime of the species of interest is very short compared with the rate at which transport occurs between one part of the atmosphere and another, then the chemical kinetic behaviour of that species can be assessed as though it was

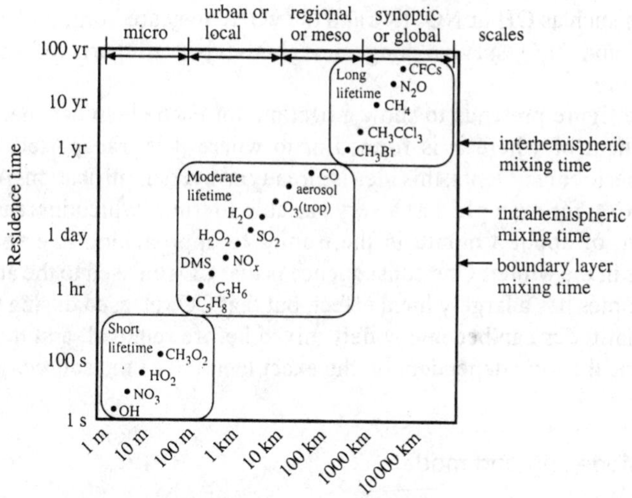

Fig 3.9. Comparison of residence times and spatial scales for some key atmospheric species. Derived from *The atmospheric sciences entering the twenty-first century*, National Academy Press, Washington, DC,1998.

confined to a well-defined volume element, just as in a laboratory system in which the reactant gases are held within a reaction vessel. At the other extreme, a trace species introduced to, or created within, the atmosphere at some location that is defined in altitude, latitude, and longitude, may become fully mixed throughout the atmosphere if its lifetime is sufficient. Reaction kinetics and atmospheric motions may thus be decoupled at the extremes of very short and very long mean lifetimes. Examples are easily found. The reactions of $O(^1D)$ described by equations (3.55) and (3.56) are rapid ones. The data presented on p. 135 for the rate coefficient of reaction (3.56) suggest a k' of roughly $10^4\ s^{-1}$ at an altitude of 80 km, and very much less at lower altitudes: according to eqn (3.64), the lifetime of $O(^1D)$ is then 10^{-4} s. The chance is thus negligible that atmospheric motions would transport an atom of $O(^1D)$ a significant distance in the period between the creation and the destruction of the excited atom. At the other extreme, a molecule such as N_2O may have a lifetime of nearly 100 years in the troposphere, so that it must become completely mixed and redistributed in terms of geographical location before an 'average' molecule is lost. It is the situations in between that offer the difficulties.

Figure 3.9 provides a simple way of visualizing the comparison between the mean residence times of a variety of species and the time-scales of different atmospheric transport processes. The feature to note especially is the distance or geographical scale over which a trace species can redistribute itself during its lifetime, because that will give an indication of the horizontal and vertical distances over which averaging of concentrations might be acceptable. Radicals such as OH or NO_3 live and die where they are born; molecules such as CH_4 and N_2O survive long enough to visit all parts of the (lower) atmosphere.

The figure pretends to show a lifetime for each chemical species that is independent of where it is formed or to where it is transported. For some atmospheric constituents, this idea is already an oversimplification. A molecule such as CO, for example, has a very variable lifetime, which increases from a minimum of about a month in the tropics to approaching one year at high latitudes in the winter. One consequence is that CO released to the atmosphere in the tropics has a largely local effect, but that CO released during the winter to high latitudes can become widely mixed before removal, and the apparent lifetime is then *not* dependent on the exact location of the source.

3.6.2 Modelling and models

One of the reasons for emphasizing the complexity of the atmospheric systems with regard to lifetimes and transport is to provide an explanation for the key role that atmospheric models play in the way that we currently think about atmospheric chemistry. The chemical reactions on their own generate a multitude of coupled differential equations, not necessarily linear, for the chemical aspects of the processes alone. This chemistry now has to be fitted into the motions of air parcels moving up and down, north and south, and east and west. The chemical reaction partners, their concentrations, the temperature, and the intensity of photolysing ultraviolet radiation, and so on, also all depend on the horizontal and vertical position of a volume element within the atmosphere. Then, we have seen that there are feedbacks between most, if not all, of the components of this complex system: changes in chemistry alter greenhouse warming and thus temperatures, as well as altering both solar irradiance and the concentration fields of other trace constituents. The chemistry alone is not susceptible to straightforward algebraic treatment, except in the very simplest circumstances. Add the other factors, and the situation is obviously far too complex to be treated by analytical mathematical methods. Recourse must therefore be had to numerical models which can be solved by computer methods (although the most complete treatments offer almost insurmountable obstacles even to the highest capacity of present-day computing power).

3.6.3 Numerical models

Numerical models are currently used very widely in the sciences in order to describe complex, interacting, and non-linear systems such as those encountered in atmospheric chemistry. In one sense, such models attempt to replicate by computer the behaviour of the natural system. The results of the computer simulation can then illuminate various aspects of the real-life atmosphere, such as the causes of certain observations or the probable effects of projected or actual perturbations.

The starting point in the simulations is often to construct *diagnostic models*. These models are used to assess hypotheses about the physics and chemistry of atmospheres from a knowledge of present-day physical and chemical structure. Information appropriate to the type of model is fed into the calculations, which provide as output further information about the other parameters in the system. These parameters can then be compared with measurements in the atmosphere. For example, such a model might have as input the sources and sink strengths, and perhaps the concentrations, of a range of trace gases, and the predictions might be of the concentrations of other trace and intermediate chemical species. Agreement between prediction and experiment lends confidence in the chemistry and physics that were originally fed as input to the model, while disagreement shows that something is wrong, quite possibly in our understanding of the underlying science. Diagnostic tests of atmospheric chemistry have frequently been conducted, for example, by comparing concentrations of reaction intermediates such as atoms or radicals with experimental determinations in the atmosphere. These concentrations are often highly sensitive to the nature of the chemical reactions incorporated into the model, as well as to the kinetic parameters chosen for them. Sometimes, modellers even extract proposed rate coefficients by comparing observation and prediction, but such a procedure rather negates the real objective of modelling, which should *start* with kinetic data obtained in independent laboratory experiments. One problem with working the other way round is that complete trust has to be placed in the hypothetical reaction scheme; another is that very great demands are placed on the accuracy of the atmospheric measurements. The question of the reaction scheme itself is highlighted by the discovery of the Antarctic ozone 'hole', which is discussed in detail in Section 4.7. Models current at the time of the discovery showed no hint of the atmospheric behaviour actually observed. The reason was that their chemical reaction schemes were incomplete, because they omitted the reactions on ice and other particles that have subsequently been recognized as playing a key role (see Section 3.4.5). In this instance, then, a lack of compatibility between the observations and model predictions required a revision of our interpretation of the fundamental physical chemistry. The new understanding has been employed to further refine models of atmospheric behaviour so that they can incorporate heterogeneous chemistry in their schemes. We note also that, since the chemical reaction schemes in the models perforce calculate concentrations of all the reaction intermediates, the predictions are often used to estimate atmospheric concentrations for species for which present-day experimental techniques cannot provide the true measurements.

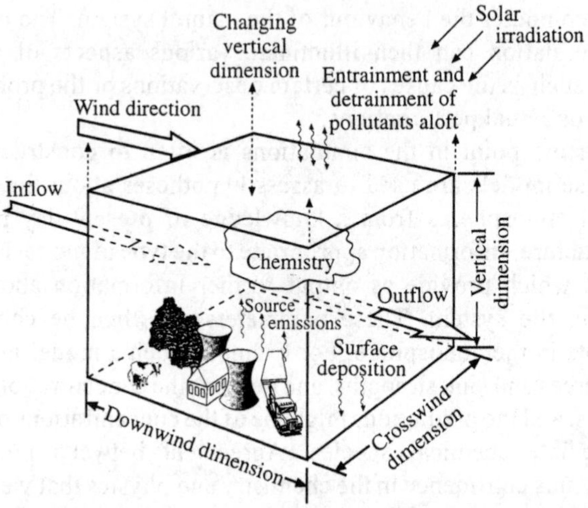

Fig 3.10. Components of a zero-dimensional box model. Adapted from Graedel, T.E. and Crutzen, P.J. *Atmospheric change*, W.H. Freeman, New York, 1993.

Once a model has proved itself diagnostically, and confidence has been reached about its validity, it may be employed to predict atmospheric response to situations that have not yet arisen. *Prognostic models* are of interest in evaluating the future behaviour and evolution of an atmosphere subject to changes in inputs of trace species produced, for example, by the biosphere, by volcanic eruption, or by man, and to changes in natural parameters such as solar intensity. The predictions of reliable prognostic models can (or at least should) be used to formulate a variety of policy decisions. The ability to perform *numerical experiments* also allows models to be used to determine whether an assumed set of 'starting' conditions could lead to observed present-day atmospheres (see Section 9.4).

It is now sensible to examine how typical models are constructed. One counsel of perfection would be to consider a volume element in an atmosphere small enough to be uniform with respect to all variables such as temperature, density, composition, and so on. If the rate of flux to and from that element of heat, radiation, matter, etc., from all other atmospheric elements is calculated, then the rate of change of the various physical parameters can be established. Chemical change within the volume element requires only slight modification for the local alterations in composition, and possibly in physical conditions. The *continuity equations* involved in the model are typified by the one for matter, which states that the net flow of mass into unit volume per unit time is equal to the local rate of change of density. Solution of the continuity equations for every physical and chemical parameter of interest, and for every volume element in the atmosphere, should then lead to a self-consistent model of atmospheric behaviour that mimics in all respects the temporal and spatial changes in the real atmosphere. According to this view, with sufficient input information, and a fine enough grid size for the volume elements, all meteorological as well as chemical phenomena could be simulated by the model. Such *three-dimensional models* (3-D) would be ideal for studying atmospheric chemistry. Computer-numerical solutions are naturally used in models, but the 3-D models are extremely demanding of computer time and memory, and the chemical content may have to be limited. Simpler models may thus provide more useful and detailed chemical information than the 3-D models. For this reason, we base the present brief overview on models employing the greatest approximations, and build up to models of greater complexity.

The simplest of all possible models is the *box model.* The box model is a *zero-dimensional* model, and assumes uniform mixing of individual constituents of an atmosphere. Figure 3.10 shows the components of such a model. Chemical species are brought into the box horizontally and vertically. In the horizontal dimension, the flow is along the wind direction. Vertically, gases may enter or leave by the top and bottom of the box. If the box is situated at ground level, as is the case for the box shown in our illustration, then there may be surface sources and sinks of the various molecules.

In the calculation, each chemical species is handled separately. An equation is written for the rate of change of concentration as determined by physical input and output in horizontal and vertical dimensions and by chemical and photochemical change within the box.

The chemistry is treated exactly as in a well-mixed laboratory system. The time evolution of each species is controlled by the chemical interactions alone. Rates of reaction are calculated as described in the previous section, with the steady-state hypothesis applied if necessary and appropriate. For long-lived species [e.g. the $O(^3P)$ atom for the conditions specified in the example of Section 3.5], diurnal (but probably not seasonal) averaging may be reasonable. Such models are most useful in the analysis of global budgets of long-lived species, and of local budgets of short-lived ones.

Despite the gross oversimplifications inherent in box models, they do allow a first appraisal of many atmospheric problems, and the relatively small demand on computing power makes them very attractive for preliminary investigations. A recent development has been the construction of a box model for tropospheric chemistry that is accessible on the World Wide Web. The intention is that researchers from different laboratories and disciplines can examine the atmospheric consequences of their latest results using *the same* model so as to ensure direct comparability. For readers wishing to try the model out, it is mounted on a computer at the author's laboratory, and the URL is *http://physchem.ox.ac.uk/sbox.html.*

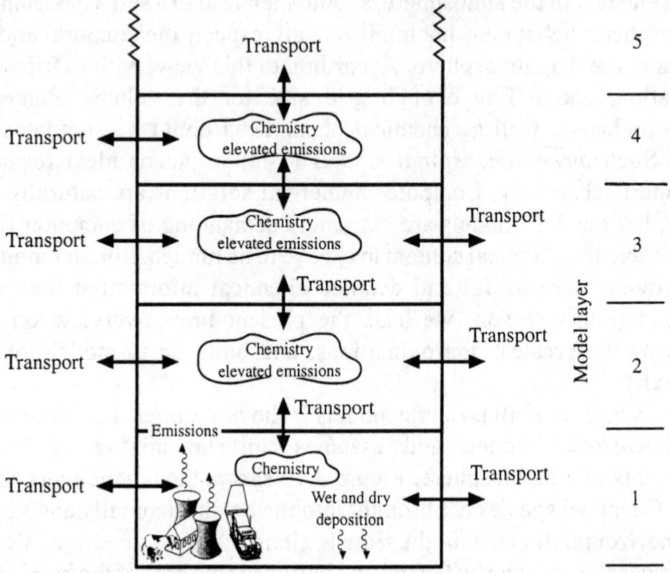

Fig 3.11. A one-dimensional box model: several of the boxes of Fig. 3.10 are stacked on top of each other. Adapted from Graedel, T.E. and Crutzen, P.J. *Atmospheric change*, W.H. Freeman, New York, 1993.

One further feature of atmospheric models should be introduced at this stage. Two distinct types of model have been developed for use in atmospheric chemistry, which will be illustrated here for the box model so far developed, although they can be extended to models of higher dimensionality. In the first approach, the box is located at a fixed geographical location, as implied by the description already given. Such a formulation is a *Eulerian model*. If atmospheric motions and the transport of chemical species are to be treated more or less realistically, that is achieved in Eulerian models of higher dimensionality by providing realistic terms for transfer in and out of the box from some or all of its faces. In a *Lagrangian model*, on the other hand, a completely different approach is taken. The hypothetical box containing the chemical components is now allowed to follow the atmospheric motions themselves, and the chemistry in this air parcel is allowed to occur as it moves around. A knowledge of the meteorology, obtained by observation or by forecasting, provides the information about the route taken by the box. Each type of model has its virtues and its disadvantages of course. One great advantage of the Lagrangian approach is that there are now no terms for input or output of chemical species from the faces of the box. On the other hand, factors such as the source strengths, temperatures, pressures, and solar irradiances may all vary as the box moves from place to place. Another problem is that it may be difficult to make real time-dependent atmospheric measurements, because the instruments would have to follow the air parcel around.

One-dimensional (1-D) *models* are the next step up in the model hierarchy. They are designed to simulate the vertical distribution of atmospheric species, but not any horizontal variations. The models thus represent horizontal averages, but do include atmospheric transport, attenuation and scattering of solar radiation, and detailed chemistry. Where vertical winds are weak or absent, as in the Earth's stratosphere, vertical transport is assumed to occur *via* the eddy diffusion mechanism (see Section 2.3.3). Equation (2.44), which is appropriate to molecular diffusion of a particular species, i, in the absence of chemical reaction, has to be modified. Allowance is made for the rate of production, P_i, and of loss, L_i, of the species in all possible chemical reactions, and the molecular diffusion coefficient is replaced by the (vertical) eddy diffusion coefficient, K_z

$$\frac{\partial f_i}{\partial t} = K_z\left(\frac{\partial^2 f_i}{\partial z^2} - \frac{1}{H_s}\frac{\partial f_i}{\partial z} - \frac{1}{T}\frac{\partial T}{\partial z}\cdot\frac{\partial f_i}{\partial z}\right) + (P_i - L_i)/N. \tag{3.65}$$

As pointed out in Section 2.3.3, K_z is a phenomenologically constructed parameter based on observations of a minor constituent whose chemistry is presumably well known. Chemical tracers (e.g. CH_4, N_2O) and radionucleides from past atmospheric nuclear tests (^{14}C, ^{90}Sr, ^{95}Zr, ^{185}W) have been used to provide source data. Typical K_z values are 10–500 $m^2\,s^{-1}$ between 50 and 80 km in the Earth's atmosphere, and 10 $m^2\,s^{-1}$ in the dynamically active troposphere. The stability of the lower stratosphere leads to values of 0.1 to 1 $m^2 s^{-1}$.

Averaging of the 1-D model results must be performed to achieve a link with physical reality. A sensible altitude spacing must be adopted; frequently, a greater number of thinner layers are set up at the lower altitudes, where changes of concentration with altitude are most rapid, and where the chemical reactions are fastest. The results must also be calculated over averaged time steps. An extreme case puts the chemistry in a stationary-state competition with vertical transport (i.e. $\partial f_i/\partial t = 0$ in eqn (3.65) for one or more constituents). Diurnal averaging procedures, on the other hand, permit considerable savings of computer time when solution is required of time-dependent problems over long durations. Fully time-dependent solutions are more useful diagnostically and prognostically, and must be used for many short-lived species. Such solutions are, however, complicated, and expensive computationally.

Two-dimensional (2-D) *models* recognize that atmospheric conditions are a strong function of latitude as well as altitude. They therefore provide spatial resolution in vertical and meridional directions (i.e. latitudinal, but not longitudinal, resolution). Mean horizontal motions must now be incorporated into the two-dimensional analogue of eqn (3.65), and the eddy diffusion coefficient has to be treated as a tensor (2 × 2 matrix) quantity. Averaged meridional velocities can be obtained, from direct observation, for the Earth's troposphere and stratosphere. Some problems exist about interpretations of eddy motions, which include, in two dimensions, large-scale internal waves. Nevertheless, 2-D models are far more realistic than 1-D models. The motivation for constructing 2-D models is that meteorological quantities on a rotating planet are not expected to show variations with longitude if averaged over several days. In practice, latitudinal gradients in the Earth's atmosphere are generally larger than those in the longitudinal direction. However, at low altitudes, where the sources and sinks of trace gases may be highly localized, ignoring the longitudinal variations may make the models unrealistic. Two-dimensional models are thus possibly most suited to the study of stratospheric problems. The use of zonal means is even then a limitation, since processes such as those involving heterogeneous chemistry can also be highly localized. Some attempts have been made to overcome this difficulty by using parametrizations that permit the incorporation of some effects of zonal asymmetries. Coupling of transport and heating effects is an important part of the models. Feedback between chemistry and other aspects of the model should ideally be accommodated because of the potential interactions. For example, the reactions determining the concentration of a 'greenhouse' gas (Section 2.2.2) such as ozone may be highly temperature sensitive, and the concentration of the gas itself will influence atmospheric temperatures, radiation balance, and dynamics (cf. Sections 4.6.1 and 4.6.9).

Because atmospheres are three-dimensional fluids, it is clear that complete simulation of the radiative–chemical–dynamical behaviour of an atmosphere requires a *three-dimensional* (3-D) *model*. One of the problems with the one- and two-dimensional models explained so far is that vertical motions and mixing are described by the artificial concept of eddy diffusion. Yet it is not generally possible to *measure* the vertical motions. A 3-D model, such as those now used in weather prediction, may yield the vertical components of the wind fields as one of the explicit or implicit results. *Regional*, *mesoscale*, and even global *general circulation models* (GCMs) have been developed to simulate weather and climate. Addition of chemistry thus becomes the next goal but, as pointed out earlier, full 3-D models with chemistry place enormous demands on computer power. A set of kinetic equations must be solved for each of up to hundreds of chemical species in the model, and at each of possibly thousands of three-dimensional grid points for which the physical equations must also be solved. However, the rewards of performing the calculations are considerable. 3-D models can, in principle, include feedback mechanisms, such as those from changes in composition, in the transport process, and they are indispensable in examining the coupling effects (e.g. radiative—transport) that are incompletely treated in simpler models. They are also of great value in assessing the accuracy of spatial- and time-averaging techniques needed in the 1-D and 2-D models. Progress in 3-D modelling is currently rapid both because of increasing programming efficiency and the ever-increasing power of computers.

Early attempts at 3-D chemical modelling involved adding simplified chemistry schemes that utilized just a small subset of the important reactions. Another way of dealing with the complexities inherent in 3-D chemical models allows the chemistry to be run *'off line'*, in *decoupled chemical transport models* (CTMs). In this approach, the model is first run without chemistry, in order to obtain the meteorological parameters such as winds and temperatures. This information is then fed into another 3-D *chemical* model, which then advects the trace species and updates the chemistry. Yet another approach uses analyses of meteorological observations to provide the wind and temperature fields. Large reductions in the complexity of the calculations can be obtained in this way, but the decoupling necessarily leads to a drift away from realism. In particular, the feedbacks between chemistry and meteorology are lost. Since the two-way impact of chemistry, radiation, and dynamics on each other are evidently one of the proclaimed virtues of the full 3-D models, the simplification is bought at considerable cost.

As computer technology advances, it seems certain that future fully coupled 3-D models will be susceptible to calculations even when full chemical schemes are run 'on line'. Such chemical models are already being run, and they can typically be integrated for simulations covering periods of a few months up to years. The task of reducing approximations, and thus increasing realism, looks like one that can be tackled with continuing hope.

3.6.4 Families

Even for 1-D and 2-D approximations, the chemistry may impose an excessive computational load. The chemistry of the stratosphere is 'simple', yet involves about fifty chemical species in nearly 200 reactions. There are thus fifty (simultaneous) continuity equations to be solved at each grid point. Many more reactants and reactions may be required to describe the troposphere completely. Direct numerical solution of the differential equations may require inordinate computer memory and run times. The problem is obviously particularly pressing when a model is used to test prognostically the effects on an atmosphere of altering concentrations of trace species, because the calculation must be repeated many times. A particular difficulty in the computations is that the differential equations for the kinetics are often *'stiff'*. That means that the concentration of one species may change very rapidly, while the concentration of another may vary much more slowly. Although there are numerical methods designed to tackle stiff differential equations, there is always a price to be paid in choosing, for example, sensible time steps over which to perform the integrations. One approach used to reduce the computational needs consists of identifying closely coupled chemical species that can be summed into *families*. The idea is that the members of the family can be interconverted on a time-scale short compared with that over which other concentrations vary. For

Earth's atmosphere, 'odd oxygen' (i.e. O, O_3 but not O_2: 'odd' means 'not even' number of oxygen atoms) is one such family. O_3 is rapidly converted photolytically to O (reactions (3.1) + (3.56)), while in the lower stratosphere and below, reaction (3.42) converts O back to O_3. Other important families, as we shall see, include NO_x ($x = 1, 2$), HO_x ($x = 0, 1, 2$) and ClO_x ($x = 0, 1, 2$). A reduced set of equations is solved for the families, and then each family is partitioned into its components. This latter stage requires some (inspired) guesswork, but 1-D models using 'family' and direct solutions compare well.

Bibliography

Sections 3.1 and 3.2

Books on general photochemistry

Photochemistry. Calvert, J. G. and Pitts, J. N., Jr. (John Wiley, New York, 1966).
Photochemistry of small molecules. Okabe, H. (John Wiley, New York, 1978).
Photochemistry. Wayne, C.E. and Wayne, R.P. (Oxford University Press, Oxford, 1996).
Principles and applications of photochemistry. Wayne, R. P. (Oxford University Press, Oxford, 1988).
Photochemistry and spectroscopy. Simons, J. P. (John Wiley-Interscience, London, 1971).

Many of the quantitative data in this classic and seminal work by Leighton are, of course, superseded. However, the book still provides a sound exposition of photochemical change in the atmosphere and of the principles determining solar irradiances and ultraviolet penetration.

Photochemistry of air pollution. Leighton, P. A. (Academic Press, New York, 1961).

Articles on photochemical change in the atmosphere

The role of solar radiation in atmospheric chemistry. Madronich, S. and Flocke, S. in *Environmental photochemistry*. Boule, P. (ed.) (Springer-Verlag, Berlin, 1999).
Atmospheric chemistry and spectroscopy. Weaver, A. and Ravishankara, A.R. in *Low temperature chemistry of the atmosphere*. Moortgat, G.K., Barnes, A.J., Le Bras, G., and Sodeau, J.R. (eds.) (Springer-Verlag, Berlin, 1994).
The atmosphere as a photochemical system. Chapter 5 in McEwan, M. J. and Phillips, L. F. *Chemistry of the atmosphere*. (Edward Arnold, London, 1975).
Photodissociation effects of solar uv radiation. Simon, P. C. and Brasseur, G. *Planet. Space Sci.* **31**, 987 (1983).
Kinetic data imprecisions in photochemical rate calculations: Means, medians and temperature. Stewart, R.W. and Thompson A.M. *J. geophys. Res.* **101,** 20953 (1996).

An example of the use of satellites to monitor the intensity and spectral distribution of ultraviolet radiation incident on the atmosphere

18 months of UV irradiance observations from the Solar Mesosphere Explorer. London, J., Bjarnason, G. G., and Rottman, G. J. *Geophys. Res. Lett.* **11**, 54 (1984).

Section 3.2.1

These articles discuss in detail the photochemistry of ozone, a molecule central to much of the chemistry of the Earth's atmosphere.

Laboratory studies of the photochemistry of ozone. Wayne, R.P. *Current Science* **63**, 711 (1992).
The photochemistry of ozone. Wayne, R. P. in *The handbook of environmental chemistry; volume 2, part E* Hutzinger, O. (ed.) (Springer-Verlag, Berlin, 1989).
The photochemistry of ozone. Wayne, R. P. *Atmos. Env.* **21**, 1683 (1987).
Photochemistry of ozone: surprises and recent lessons. Ravishankara, A.R., Hancock, G., Kawasaki, M., and Matsumi, Y. *Science* **280**, 60 (1998).
Wavelength and temperature dependence of the absolute $O(^1D)$ production yield from the 305–329 nm photodissociation of ozone. Takahasi, K., Taniguchi, N., Matsumi, Y., Kawasaki, M., and Ashfold, M.N.R. *J. chem. Phys.* **108**, 7161 (1998).
Translational energy and angular distributions of $O(^1D)$ and $O(^3P_j)$ fragments in the UV photodissociation of ozone. Takahasi, K., Taniguchi, N., Matsumi, Y., and Kawasaki, M. *Chem. Phys.* **231**, 171 (1998).
A direct measurement of the $O(^1D)$ quantum yields from the photodissociation of ozone between 300 and 328 nm. Ball, S.M., Hancock, G., Martin, S.E., and de Moira, J.C.P. *Chem. Phys. Letts* **264**, 531-538 (1997).
Production of $O(^1D)$ from photolysis of O_3. Michelsen, H. A., Salawitch, R. J., Wennberg, P. O., and Anderson, J. G. *Geophys. Res. Letts.* 21, 2227 (1994).

On the application of detailed information about the dynamics of photodissociation to atmospheric systems

Photodissociation dynamics and atmospheric chemistry. Wayne, R.P. *J. geophys. Res.* **98**, 13119 (1993).

Section 3.2.2

Consideration of enhanced reactivity generally emphasizes the effects of electronic excitation, as described in the first review. In the next papers, the possibility is explored that vibrational and translational excitation might contribute to increased reaction rates.

The excited state in atmospheric chemistry. Marston, G. *Chem. Soc. Revs.* **25**, 33 (1996).
Atmospheric radical production by excitation of vibrational overtones *via* absorption of visible light. Donaldson, D.J., Frost, G.J., Rosenlof, K.H., Tuck, A.F., and Vaida, V. *Geophys. Res. Letts.* **24**, 2651 (1997).

The possible effects of translationally excited nitrogen atoms on lower thermospheric odd nitrogen. Solomon, S. *Planet. Space Sci.* **31**, 135 (1983).

Section 3.3

Although the title of this chapter in a book seems rather restricted, the chapter provides an excellent introduction to the subject of the conservation rules and correlation diagrams. Further explanations are to be found in the texts given as references in connection with Section 3.4.

On the rapidity of ion–molecule reactions. Talrose, V. L., Vinogradov, P. S., and Larin, I. K., in *Gas phase ion chemistry* Bowers, M. T. (ed.) Vol. 1, pp. 305–47. (Academic Press, New York, 1979).

Section 3.4

Books on the theories of chemical kinetics. The first book contains chapters covering many topics of central importance to atmospheric chemistry, including the influence of temperature on bimolecular, unimolecular, and termolecular reactions, the reactions of electronically and vibrationally excited species, and on ion–molecule reactions and the influence of translational and internal energy on them.

Reactions of small transient species. Fontijn, A. and Clyne, M. A. A. (eds.) (Academic Press, New York, 1984).

Gas-phase and heterogeneous kinetics of the troposphere and stratosphere. Molina, M.J., Molina, L.T., and Kolb, C.E. *Annu. Rev. Phys. Chem.* **47**, 327 (1996).

Gas-phase reactions and energy transfer at very low temperatures. Sims, I.R. and Smith, I.W.M. *Annu. Rev. Phys. Chem.* **46**, 109 (1995).

Radical–radical reactions. Pilling, M.J. *Annu. Rev. phys. Chem.* **47**, 81 (1996).

Temperature averages and rates of stratospheric reactions. Murphy, D.M. and Ravishankara, A.R. *Geophys. Res. Letts.* **21**, 2471 (1994).

Gas phase homogeneous kinetics. Golden, D.M. in *Low temperature chemistry of the atmosphere.* Moortgat, G.K., Barnes, A.J., Le Bras, G., and Sodeau, J.R. (eds.) (Springer-Verlag, Berlin, 1994).

Thermochemical kinetics Benson, S. W., 2nd edn. (Wiley, New York, 1976).

The theory of the kinetics of elementary gas phase reactions. Wayne, R. P. in *The theory of kinetics* Bamford, C. H. and Tipper, C. F. H. (eds.) *Comprehensive chemical kinetics*, Vol. 2. (Elsevier, Amsterdam, 1969).

Kinetics and dynamics of elementary gas reactions. Smith, I. W. M. (Butterworth, London, 1980).

Molecular reaction dynamics and chemical reactivity. Levine, R. D. and Bernstein, R. B. (Oxford University Press, Oxford, 1987).

Theoretical and observational aspects of the reactions of ionic species in the gas phase. Descriptions of some experimental techniques are to be found in references given at the end of Chapter 6.

Classical ion–molecule collision theory. Su, T. and Bowers, M. T., in *Gas phase ion chemistry* Bowers, M. T. (ed.) Vol. 1, p. 84. (Academic Press, New York, 1979).

Temperature and pressure effects in the kinetics of ion molecule reactions. Meot–Ner, M., in *Gas phase ion chemistry* Bowers, M. T. (ed.) Vol. 1, p. 198. (Academic Press, New York, 1979).

Chemical reactions of anions in the gas phase. DePuy, C. H., Grabowski, J. J., and Bierbaum, V. M. *Science* **218**, 955 (1982).

Isotope exchange in ion–molecule reactions. Smith, D. and Adams, N. G., in *Ionic processes in the gas phase* Almoster Ferreira, M. A. (ed.) (D. Reidel, Dordrecht, 1984).

Elementary plasma reactions of environmental interest. Smith, D. and Adams, N. G. *Topics curr. Chem.* **89**, 1 (1980).

The prediction of rate coefficients from calculated (or measured) molecular parameters, such as ionization potentials, may be a valuable tool.

Correlations between rate parameters and molecular properties. Marston, G., Monks, P.S., and Wayne, R.P. in *General aspects of the chemistry of radicals*, Alfassi, Z.B. (ed.) (Wiley, Chichester, 1999).

Section 3.4.2

The last three references for this section describe the development of Troe's simplified treatment for the prediction of unimolecular reaction rates; the final one provides an accessible synthesis of the ideas.

Practical methods for calculating rates of unimolecular reactions. Thompson, D.L. *Int. Rev. phys. Chem.* **17**, 547 (1998).

Unimolecular reactions. Robinson, P.J. and Holbrook, K.A. (Wiley, New York, 1972).

Current aspects of unimolecular reactions. Holbrook, K. A. *Chem. Soc. Rev.* **12**, 163 (1983).

Current aspects of unimolecular reactions. Quack, M. and Troe, J. *Int. Rev. phys. Chem.* **1**, 97 (1981).

Predictive possibilities of unimolecular rate theory. Troe. J. *J. phys. Chem.* **83**, 114 (1979).

Specific rate constants k(E,J) for unimolecular bond fissions. Troe, J. *J. chem. Phys.* **79**, 6017 (1983).

The first of these three references is a review that discusses the care needed in applying laboratory kinetic data to atmospheric systems and that presents critically the types of data available. The second two references are to typical data compilations that are used by atmospheric modellers.

Kinetics of thermal gas reactions with application to stratospheric chemistry. Kaufman, F. *Annu. Rev. phys. Chem.* **30**, 411 (1979).

Chemical kinetic and photochemical data for use in stratospheric modeling. DeMore, W.B., Sander, S.P., Golden, D.M., Hampson, R.F., Kurylo, M.J., Howard, C.J., Ravishankara, A.R., Kolb, C.E., and Molina, M.J. Evaluation No. 12, JPL Publication 97-4 (Jet Propulsion Lab., Pasadena, CA, 1997).

Evaluated kinetic and photochemical data for atmospheric chemistry, Supplement V, IUPAC subcommittee on gas kinetic data evaluation for atmospheric chemistry. Atkinson, R., Baulch, D.L., Cox, R.A., Hampson, R.F., Kerr, J.A., Rossi, M.J, and Troe, J. *J. phys. Chem. Ref. Data* **26,** 521 (1997).

These reviews discuss the reactions of some specific radicals and atoms of special significance to atmospheric chemistry.

Halogen oxides: radicals, sources and reservoirs in the laboratory and in the atmosphere. Wayne, R.P. (ed.) (European Commission, Brussels, 1995); *Atmos. Environ.* **29**, 2675 (1995).

Organic peroxy radicals: kinetics, spectroscopy and tropospheric chemistry. Lightfoot, P. D., Cox, R. A., Crowley, J. N., Destriau, M., Hayman, G. D., Jenkin, M. E., Moortgat, G. K., and Zabel, F. *Atmos. Environ.* **26A**, 1805 (1992).

The nitrate radical: physics, chemistry and the atmosphere. Wayne, R. P. (ed.) (European Commission, Brussels, 1991); *Atmos. Environ.* **25A**, 1 (1991).

Kinetics of gaseous hydroperoxyl radical reactions. Kaufman, M. and Sherwell, T. *Prog. Reaction Kin.* **12**, 1 (1983).

Gas phase reactions of hydroxyl radicals. Baulch, D. L. and Campbell, I. M. *Gas kinetics and energy transfer* Specialist Periodical Reports Chem. Soc., **4**, 137 (1981).

Kinetics and mechanisms of the reactions of the hydroxyl radical with organic compounds in the gas phase. Atkinson, R., Darnall, K. R., Lloyd, A. C., Winer, A. M., and Pitts, J. N., Jr. *Adv. Photochem.* **11**, 375 (1979).

Elementary reactions of atoms and small radicals of interest in pyrolysis and combustion. Baggott, J. E. and Pilling, M. J. *Annu. Rep. Pro. Chem.* **C79**, 199 (1983).

Atom and radical recombination reactions. Troe, J. *Annu. Rev. phys. Chem.* **29**, 223 (1978).

The kinetics of radical–radical processes in the gas phase. Howard, M. J. and Smith, I. W. M. *Prog. Reaction Kin.* **12**, 55 (1983).

Sections 3.4.3–3.4.5

Aquatic Chemistry Stumm, W. and Morgan, J. J., 3rd edn. (Wiley, New York, 1996).

Aqueous solution chemistry. Warneck, P. in *Low temperature chemistry of the atmosphere.* Moortgat, G.K., Barnes, A.J., Le Bras, G., and Sodeau, J.R. (eds.) (Springer-Verlag, Berlin, 1994).

Laboratory Studies of Atmospheric Heterogeneous Chemistry. Kolb, C.E., Worsnop, D.R., Zahniser, M.S., Davidovits, P., Keyser, L.F., Leu, M.-T., Molina, M.J., Hanson, D.R., Ravishankara, A.R., Williams, L.R., and Tolbert, M.A. in *Progress and Problems in Atmospheric Chemistry*, Barker, J.R. (ed.) (World Scientific Publishing, Singapore, 1995).

Laboratory studies of heterogeneous reactions. Tolbert, M.A. in *Low temperature chemistry of the atmosphere.* Moortgat, G.K., Barnes, A.J., Le Bras, G., and Sodeau, J.R. (eds.) (Springer-Verlag, Berlin, 1994).

Heterogeneous chemistry and kinetics. Golden, D.M. and Williams, L.R. in *Low temperature chemistry of the atmosphere.* Moortgat, G.K., Barnes, A.J., Le Bras, G., and Sodeau, J.R. (eds.) (Springer-Verlag, Berlin, 1994).

Heterogeneous reaction in sulfuric acid: HOCl + HCl as a model system. Hanson, D.R. and Lovejoy, E.R. *J. phys. Chem.* **100**, 6397 (1996).

Section 3.6

The first reference provides a straightforward introduction to the practice of atmospheric modelling, and the others discuss particular aspects of constructing models. Many further references to models will be found in the Bibliographies for later chapters, especially Chapter 4 and Chapter 5.

Modelling chemical processes in the stratosphere. Chang, J. S. and Duewer, W. H. *Annu. Rev. phys. Chem.* **30**, 443 (1979).
A comparison of one-, two- and three-dimensional representations of stratospheric gases. Tuck, A. F. *Phil. Trans. R. Soc.* **A290**, 477 (1979).
A modified diabatic circulation model for stratospheric tracer transport. Pyle, J. A. and Roger, C. F. *Nature, Lond.* **287**, 711 (1980).

4 Ozone in Earth's stratosphere

4.1 Introduction

Ozone plays several extremely important roles in the Earth's atmosphere, as we have already seen in Chapters 1 and 2. First, ozone absorbs virtually all solar ultraviolet radiation between wavelengths of about 240 and 290 nm which would otherwise be transmitted to the Earth's surface. Such radiation is lethal to simple unicellular organisms, and to the surface cells of higher plants and animals. Ultraviolet radiation in the wavelength range 290 to 320 nm (so-called UV-B) is also biologically active; for example, it may destroy or modify DNA. Prolonged exposure to it may cause skin cancer in susceptible individuals (see Section 4.6.1). Secondly, upper atmospheric meteorology is greatly influenced by the heating that follows absorption by ozone of UV and visible radiation.[a] Stratospheric air is statically stable because of the increase in temperature with altitude that results from this heating. Although the importance of atmospheric ozone has been recognized for more than seventy years, research has intensified dramatically over the last thirty. Interest has been stimulated by concern that a variety of human influences might lead to detectable changes in the abundance of stratospheric ozone. An increasing awareness of the part played by minor atmospheric constituents in the determination of natural ozone concentrations has coincided with an understanding that man could inject into the stratosphere substances that, because of the stratospheric stability, could build up to levels where significant depletion of ozone might occur. The truly global problem posed by anthropogenic pollution of the stratosphere captured the imagination of the scientific community and led to the identification, much more clearly than before, of the factors that control ozone concentrations in the unpolluted atmosphere. It is now apparent that stratospheric ozone concentrations have been declining over at least the past three, and possibly the last four, decades (Section 4.8), and dramatic depletions of ozone over the Antarctic (Section 4.7) each year—the so-called 'Antarctic ozone hole'—can only be explained in terms of chemistry perturbed by man's release of halogen-containing compounds. In this chapter, we discuss the 'natural' stratosphere first, and then consider how the ozone layer might be modified by man. We must, however, remember that the present understanding of the natural stratosphere has grown out of the active programme of research motivated by the perceived threat of man's activities. Facing that programme was the formidable task of bringing together the several different aspects of the

[a]The influence of ozone on atmospheric temperatures is complex. While the heating effects of UV and visible radiation produce the temperature inversion that characterizes the stratosphere, thermal IR (9.6 μm) emission from ozone may leads to heat loss in the upper stratosphere. Ozone densities are low enough at these altitudes for the system to be 'optically thin' (see p. 57) and radiation escapes from the atmosphere rather than being retained in it.

problem: transport and dynamics, the complex chemistry and photochemistry of the stratosphere, sources and sinks of trace constituents, and the intensity and variability of solar radiation, to name a few. An elaborate series of field measurements has been carried out to verify and consolidate the understanding gained. One clear result is that it is not enough to examine the stratosphere in isolation, since its chemistry and its dynamics are linked with those of the troposphere.

4.2 Observations

The aeronomy and meteorology of atmospheric ozone have been studied for nearly a century and a half, since the suggestion[a] by Schönbein in 1840 of an atmospheric constituent having a peculiar odour (Greek for 'to smell' is *ozein*). Ozone's existence in the troposphere was established in 1858 by chemical means.[b] Subsequent spectroscopic studies in the visible and ultraviolet regions showed, as early as 1881,[c] that ozone is present at a higher mixing ratio in the 'upper' atmosphere than near the ground. In the early part of this century, quantitative analysis, particularly by Fabry and by Dobson, had shown the existence of the ozone layer (although its estimated altitude was misplaced).

We now know that ozone is found in trace amounts throughout the atmosphere, with the largest concentrations in a well-defined layer at altitudes between about 15 and 30 km (cf. Sections 1.4, 4.3.2, and Fig. 1.2 (p. 11)). The ozone layer is, however, a highly variable phenomenon. Total ozone densities in an atmospheric column near the pole can increase by 50 per cent in ten days in spring, and daily variations of up to 25 per cent have been recorded. Concentrations at particular altitudes show even greater variability, eight-fold increases in a few days being not unusual in the lower stratosphere (say 13–19 km altitude).

Average column densities of ozone vary with season and latitude. Figure 4.1(a) shows some typical column density measurements[d] in contour map

[a]Schönbein, C.F., Recherches sur la nature de l'odeur qui se manifeste dans certaines actions chimiques. *C. R. Acad. Sci. Paris* **10**, 706 (1840).

[b]Houzeau, A., Preuve de la présence dans l'atmosphère d'un nouveau principe gazeux, l'oxygène naissant. *C. R. Acad. Sci. Paris* **46**, 89 (1858).

[c]Hartley, W.N., On the absorption of solar rays by atmospheric ozone. *J. Chem. Soc.* **39**, 111 (1881).

[d]The geophysical literature almost always uses 'Dobson units' (DU) to describe atmospheric column ozone abundances in order to honour Dobson's pioneering work on stratospheric ozone. The definition of 1 DU is that thickness in units of hundredths of a millimetre that the ozone column would occupy at standard temperature and pressure (STP: 273 K and one atmosphere). As stated at the beginning of Section 1.4, the ozone column compressed in this way would typically be 3 mm thick, so that the atmospheric abundance is then 300 DU. In this book, we keep to molecular units more familiar to chemists and physicists for concentrations, but use DU for column abundances. The conversion factor is $1\ \text{DU} = 2.69 \times 10^{16}$ molecule cm^{-2}.

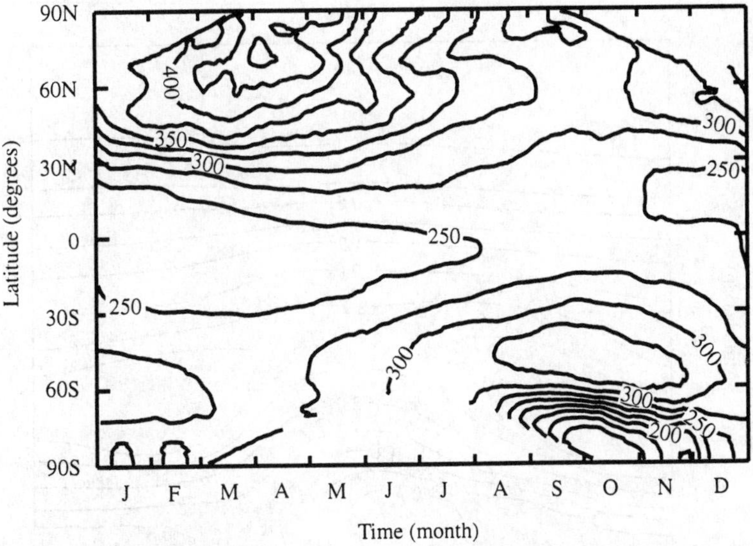

Fig. 4.1(a). Observed atmospheric ozone abundance as a function of latitude and time of year. The contour lines are total overhead column abundances in Dobson units (see footnote *a*, p. 156). The data are averages for 1998 obtained by the Global Ozone Monitoring Experiment (GOME) satellite device (see p.181); the plot was calculated using the GDP Level 2 Version 2.3 dataset from DLR-DFD and ESA. Plot kindly provided by J.P. Burrows and K.-U. Eichmann, Institute for Remote Sensing, University of Bremen, June 1999.

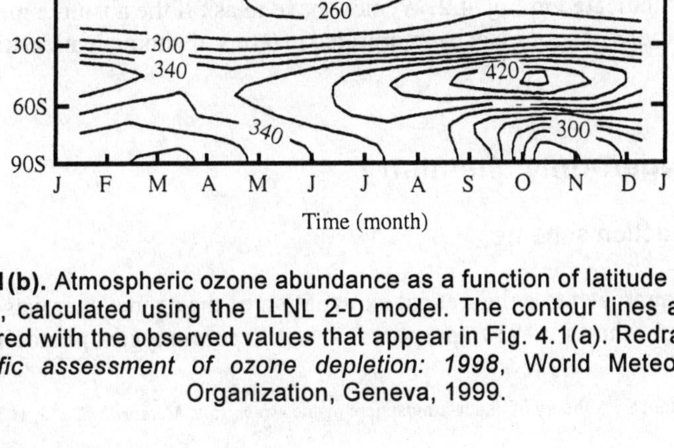

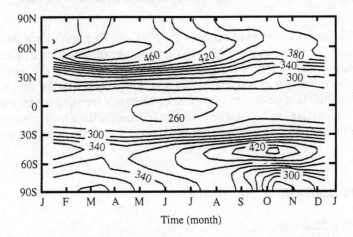

Fig. 4.1(b). Atmospheric ozone abundance as a function of latitude and time of year, calculated using the LLNL 2-D model. The contour lines are to be compared with the observed values that appear in Fig. 4.1(a). Redrawn from *Scientific assessment of ozone depletion: 1998*, World Meteorological Organization, Geneva, 1999.

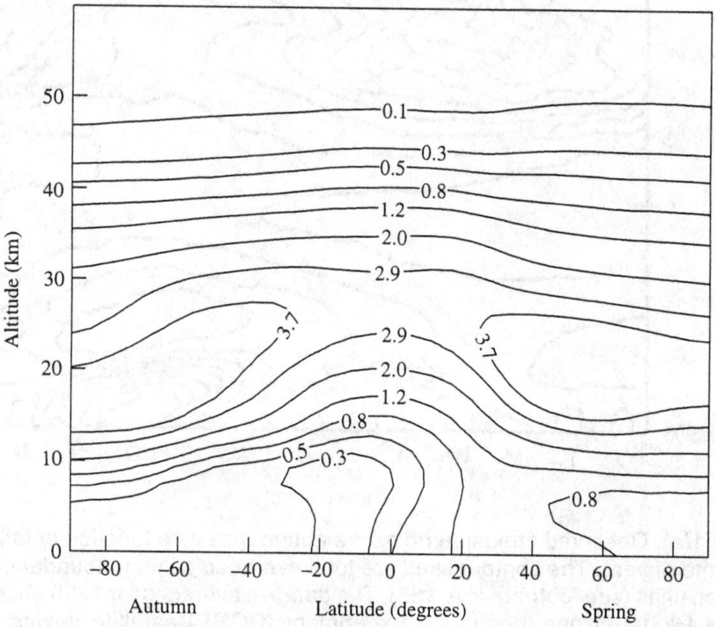

Fig. 4.2. Ozone concentrations (zonally averaged) as a function of altitude for spring (22 March). Units are 10^{12} molecule cm^{-3}. Drawn from data calculated by D.E. Shallcross using the Cambridge 2-D model, May 1999.

form. The largest amounts of ozone in the Northern Hemisphere are found at polar latitudes in spring. Figure 4.1 (b) shows some calculated abundances for comparison, and it will be discussed in Section 4.4.6. In the Southern Hemisphere, the spring maximum occurs at mid-latitudes rather than at the pole. An altitude distribution of ozone for Northern Hemisphere spring (22 March) is indicated in Fig. 4.2. We now have to ask if the absolute measured concentrations of ozone can be interpreted in terms of photochemical kinetics.

4.3 Oxygen-only chemistry

4.3.1 Reaction scheme

The first approach to a theoretical explanation of the ozone layer was that of Chapman,[a] who, in 1930, proposed a static pure-oxygen photochemical

[a]Chapman, S., A theory of upper-atmosphere ozone. *Mem. Roy. Meteorol. Soc.* **3**, 103 (1930).

steady-state model. The reactions in Chapman's scheme were

Reaction	Change in odd oxygen	
$O_2 + h\nu \rightarrow O + O$	+2	(4.1)
$O + O_2 + M \rightarrow O_3 + M$	0	(4.2)
$O_3 + h\nu \rightarrow O + O_2$	0	(4.3)
$O + O_3 \rightarrow 2O_2$	−2	(4.4)
[$O + O + M \rightarrow O_2 + M$	−2.	(4.5)]

Reaction (4.5) is now known to be too slow for it to play a part in stratospheric chemistry. Both photolytic reactions (4.1) and (4.3) can yield excited fragments, but collisional deactivation to $O(^3P)$ is the (almost) exclusive fate of any $O(^1D)$ formed. For the purposes of discussing oxygen-only chemistry, no account need be taken of excitation.

Reactions (4.2) and (4.3) interconvert O and O_3. Even at the top of the stratosphere, where pressures are lowest, reaction (4.2) has a half-life of as

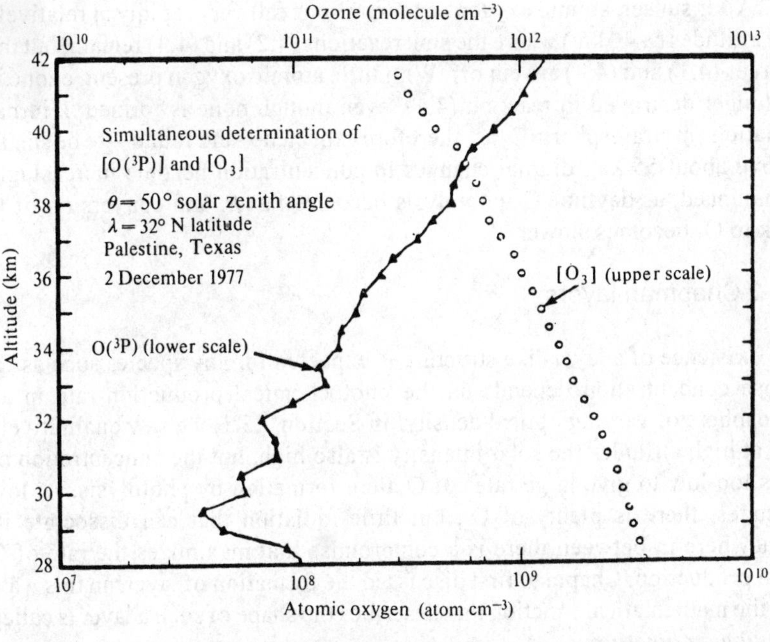

Fig. 4.3. Concentrations of $O(^3P)$ and of O_3 measured simultaneously within the same element of the stratosphere. From Hudson, R. (ed.-in-chief) *The stratosphere 1981*, World Meteorological Organization, Geneva, 1981.

little as ~ 100 s. Ozone likewise has a very short photolytic lifetime in reaction (4.3) during the day. It is this rapid interconversion of O and O_3 that provides the rationale for the concept of 'odd' oxygen (O and O_3, see p. 148). If the odd oxygen concentration is defined as [O] + [O_3], then odd oxygen is produced only in reaction (4.1) and lost only in reactions (4.4) (and (4.5)). Reactions (4.2) and (4.3) 'do nothing' so far as odd oxygen is concerned, but merely determine the ratio [O]/[O_3]. Changes in the number of odd-oxygen species are shown alongside the chemical reactions in eqns (4.1) to (4.5). Reaction (4.2) becomes slower with increasing altitude, while reaction (4.3) becomes faster. Atomic oxygen is thus favoured at high altitudes and ozone at lower ones. Simultaneous measurements of [O(^{3}P)] and [O_3], plotted in Fig. 4.3, show that these expectations are borne out experimentally. Ozone is the dominant form of odd oxygen below ~ 60 km, and in the stratosphere (< 50 km) constitutes more than 99 per cent of odd oxygen. The rate of production of ozone in the stratosphere may thus be equated with the rate of the O atom formation, which is twice the rate of O_2 photolysis in reaction (4.1). Steady-state analysis for odd oxygen shows that the concentration is equal to $(2P/k_1)^{1/2}$, where P is the rate of photolysis of O_2 in reaction (4.1) and k_1 is the rate coefficient for loss of odd oxygen in reaction (4.4). Given that most odd oxygen is in the form of ozone, it follows that stratospheric ozone concentrations are proportional to the square root of the oxygen photolysis rate, according to the steady-state model.

After sunset, atomic oxygen concentrations fall very rapidly at relatively low altitudes ($\lesssim$ 40 km) where the sink reactions (4.2) and (4.4) remain, but the sources (4.1) and (4.3) are cut off. With little atomic oxygen present, ozone is no longer destroyed in reaction (4.4), even though none is formed. Diurnal variations in stratospheric O_3 are therefore expected—and found—to be small. Above about 55 km, diurnal changes in concentration become increasingly pronounced, as daytime O_3 photolysis becomes faster, and conversion of O back to O_3 becomes slower.

Night FX

4.3.2 Chapman layers

The existence of a layer-like structure is expected for any species such as O_3 whose concentration depends on the photochemical production rate in an atmosphere of varying optical density. In Section 2.3.1, we saw qualitatively that at high altitudes the solar intensity is also high, but the concentration of O_2 is too low to give large rates of O atom formation by photolysis. At low altitudes, there is plenty of O_2, but little radiation that can dissociate it. Somewhere in between there is a compromise that maximizes the rate of O atom production. Chapman first discussed the formation of layers in this way, and the mathematical function that describes the shape of such a layer is called a *Chapman function*.

Let us show a simplified version of the derivation of the Chapman function. Atmospheric number density, n, can be obtained for this purpose,

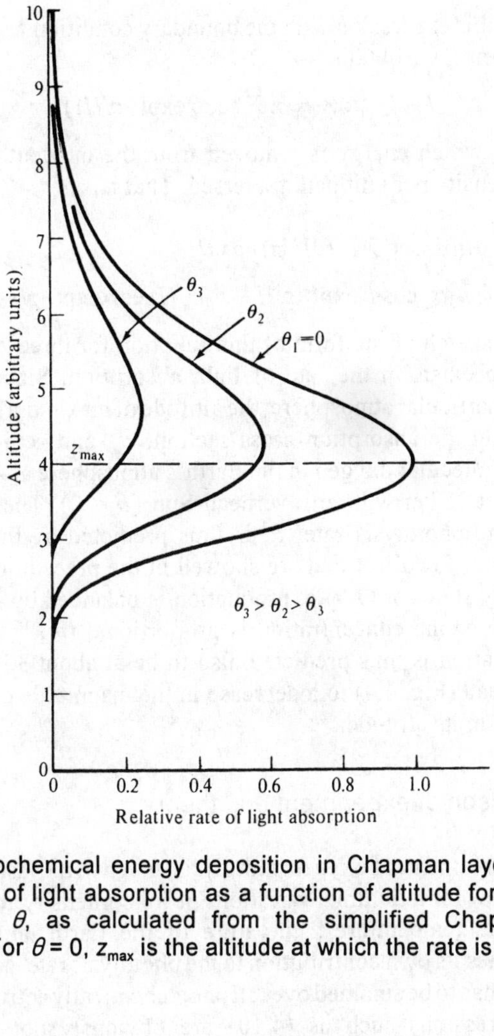

Fig. 4.4. Photochemical energy deposition in Chapman layers. The figure shows the rate of light absorption as a function of altitude for three different zenith angles, θ, as calculated from the simplified Chapman function, eqn (4.10). For $\theta = 0$, z_{max} is the altitude at which the rate is a maximum.

from eqn (2.7), p. 47, for constant scale height $H = \mathbf{k}T/mg$, and with n substituted for p

$$n = n_0 \exp(-z/H). \tag{4.6}$$

For an *increase* in altitude dz, the atmospheric path that the Sun's rays have to traverse is *decreased* by dz secθ for a zenith angle θ (Fig. 2.9, p. 61), so that eqn (2.12) for optical absorption becomes

$$\mathrm{d}I = In\sigma_a(\mathrm{d}z \sec\theta). \tag{4.7}$$

Combining eqns (4.6) and (4.7) yields

$$dI/I = d(\ln I) = n_0\sigma_a \sec\theta \exp(-z/H)\, dz. \tag{4.8}$$

On integrating this expression with the boundary condition $I = I_\infty$ at $z = \infty$ (limit of the atmosphere), we obtain

$$I = I_\infty \exp(-n_0\sigma_a H \sec\theta \exp(-z/H)). \tag{4.9}$$

The rate, P, at which energy is removed from the incident radiation is the decrease in intensity per unit path traversed. That is,

$$P = dI/(dz \sec\theta) = (dI/dz) \cos\theta$$

$$= I_\infty n_0\sigma_a \cos\theta \exp(-z/H - n_0\sigma_a H \sec\theta \exp(-z/H)). \tag{4.10}$$

Figure 4.4 is a sketch of the form of this function for three zenith angles. A clear maximum exists in the rate of light absorption. Since H and n_0 are properties of a particular atmosphere, the altitude of maximum light absorption depends only on the absorption cross-section, σ_a, and zenith angle, θ. For absorption by molecular oxygen in the Earth's atmosphere at $\lambda \sim 220$ nm, that altitude is about 35 km with an overhead Sun ($\theta = 0$). The altitude of the maximum in the photolysis rate, P, is thus predicted by this theory to be centred on 35 km. Recollect that we showed in the preceding section that if there is a steady state for O_3 (i.e. production is balanced by loss in reaction (4.4)), then the ozone concentration is proportional to $P^{1/2}$. The maximum ozone concentration is thus predicted also to be at about 35 km. Increasing zenith angles lead (Fig. 4.4) to a decrease in the magnitude of the maximum and an increase in its altitude.

4.3.3 Comparison of experiment and theory

The simplified treatment of oxygen photolysis, given in the last section, can be improved to allow for latitudinal variations of solar intensity and zenith angle, real atmospheric temperatures, curvature of the Earth, and so on. Each wavelength makes its own contribution to the photolysis rate, so that the ozone production rate has to be summed over all photochemically active wavelengths. Analytical expressions such as (4.10) are obviously not available, and numerical methods are used instead. Figure 4.5 gives rates of *formation* of ozone *via* reactions (4.1) and (4.2) calculated in this way. The contour lines are zonal averages (over longitude at a given latitude) for the Northern Hemisphere spring equinox (22 March). The production rates are thus relevant to the conditions for which ozone *concentrations* are shown in Fig. 4.2.

Comparison of Figs 4.2 and 4.5 shows straight away that the highest oxygen photolysis rates do not correspond with the highest ozone concentrations, even though the balance between production and loss of odd oxygen suggests that concentrations and production rates ought to be directly linked through the $P^{1/2}$ relation presented earlier (p. 160). Photolysis rates are

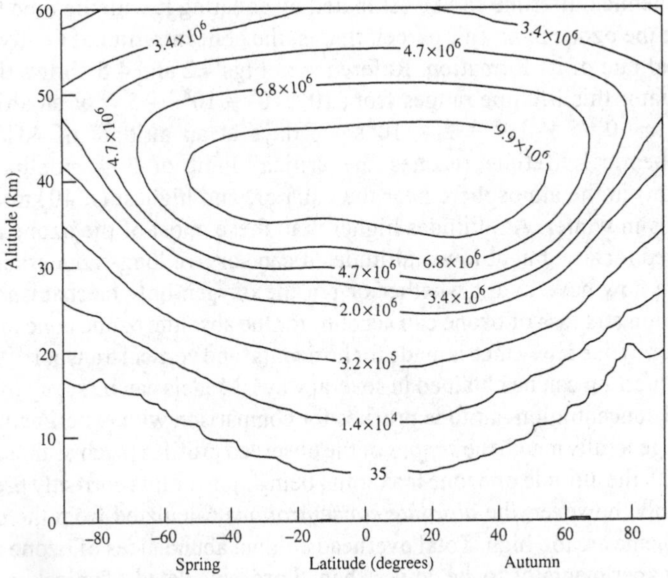

Fig. 4.5. Rate (zonally averaged) of ozone formation from the photolysis of O_2 in units of molecule $cm^{-3} s^{-1}$. Data calculated by D.E. Shallcross, using the Cambridge 2-D model, May 1999.

highest near the equator, while ozone concentrations are at a maximum in northern regions. At the equator, the ozone layer is centred on about 25 km where the production rate is insignificant, while the production of atomic oxygen reaches a maximum at ~ 40 km. Near the poles, the maximum production rate is displaced to higher altitudes, but the largest ozone concentrations are found at lower altitudes, and there is a marked north–south asymmetry. The steady state oxygen-only photochemical model predicts that ozone concentrations should be proportional to $P^{1/2}$, so that the lack of correspondence of $[O_3]$ and P indicates an inadequacy of the model. Vertical and horizontal transport are clearly of great importance in redistributing stratospheric ozone. Indeed, the comparison of Figs. 4.2 and 4.5 can be said to reveal the occurrence and direction of air motions in the stratosphere. Observations confirm this view. For example, the rapid build-up of ozone at the spring pole (Section 4.2) is strongly correlated with northward and downward air transport that could bring ozone in from high-altitude equatorial regions where the production rate is highest.

If a large amount of horizontal north–south transport is to be responsible for redistribution of ozone, then the (photo)chemical lifetime of the ozone must clearly be greater than the time taken for the transport to occur. Since seasonal changes, and north–south reversals, are seen in the ozone concentrations, it follows that the transport time-scale of interest must be 3–4 months. The photochemical lifetime can be estimated by equating it with the time taken to replace the ozone in any air parcel: that is, the concentration of O_3 divided by the local rate of its formation. Reference to Figs 4.2 and 4.5 shows that near the equator this lifetime ranges from $10^{12}/10^4 = 10^8$ s ~ 3 yr at an altitude of 15 km to $10^{12}/5 \times 10^6 = 2 \times 10^5$ s ~ 2 days at an altitude of 40 km. The photochemical lifetime reaches the critical limit of 3–4 months lowest (~ 20 km) in the atmosphere near the equator, and highest (≳ 40 km) at high latitudes in winter. At altitudes higher than these, most of the ozone must be produced locally, but at lower altitudes it can survive long-range transport.

We now have to ask whether or not the oxygen-only mechanism for the production and loss of ozone can account for the absolute ozone concentrations observed, if due allowance is made for horizontal and vertical transport. Answers to this question can be obtained in several ways. Models can be set up to predict vertical concentration–altitude profiles for comparison with experiments. Such models generally match the *shapes* of the observed profiles (such as those in Fig. 1.2), with the altitude of ozone maximum being more or less correctly predicted. Invariably, however, the *absolute concentrations* calculated from the oxygen-only scheme are too high. Total overhead column abundances of ozone are also found experimentally to be lower than those calculated. Typical measured column abundances are 8.8×10^{18} molecule cm^{-2}, while 1-D models including vertical transport give about twice this value with oxygen-only chemistry. Because the rate of reaction (4.4) is approximately proportional to $[O_3]^2$, the result suggests that the reaction accounts for only a quarter of the actual ozone loss. Similar results are obtained by examining the predicted global rates of production and destruction of ozone in order to remove errors arising from horizontal transport. For the spring equinox data, the integrated rate of ozone production is 4.86×10^{31} molecule s^{-1} of which 0.06×10^{31} molecule s^{-1} are transported to the troposphere. The calculated rate of loss of odd oxygen in reaction (4.4) is 0.89×10^{31} molecule s^{-1}, leaving 3.91×10^{31} molecule s^{-1} unaccounted for. If this figure really represented an unbalanced rate of ozone formation, the present global quantity of ozone would double in two weeks! Ozone is being produced five times faster than it is being destroyed by the Chapman mechanism. Since ozone concentrations are not rapidly increasing, we conclude that something other than the Chapman reactions is very important in destroying ozone in the natural stratosphere.

4.4 Influence of trace constituents

4.4.1 Catalytic cycles

Until about 1964, it was thought that the Chapman, oxygen-only, reactions could explain atmospheric ozone abundances. However, improved laboratory measurements of rate coefficients, especially for reaction (4.4), revealed the discrepancy discussed in the last section. Reaction (4.4) is too slow to destroy ozone at the rate it is produced globally. An additional, or faster, loss process for odd oxygen is needed.

At first sight, no trace component in the stratosphere could be responsible for loss of odd oxygen, since the species involved would be rapidly consumed. This objection disappears if the trace constituent participates in a catalytic process that removes O or O_3. The idea of catalytic atmospheric loss processes had originated with a suggestion of Bates and Nicolet[a] in 1950 about the influence of H and OH from water photolysis on ozone concentrations in the upper mesosphere. Serious consideration of the influence of such processes on ozone concentrations followed the discovery that the laboratory rate measurements could also be affected by trace impurities reacting in a cyclic manner.

The essence of catalytic schemes for loss of odd oxygen is the provision of a more efficient route than the direct one for reactions (4.4) and (4.5). A chain mechanism that achieves the same result as reaction (4.4) can be represented simplistically by the pair of reactions

$$X + O_3 \rightarrow XO + O_2 \tag{4.11}$$

$$XO + O \rightarrow X + O_2 \tag{4.12}$$

$$\text{Net} \qquad O + O_3 \rightarrow O_2 + O_2.$$

The reactive species X is regenerated in the second reaction of the pair, so that its participation in odd-oxygen removal does not lead to a change in its abundance. Since the overall reaction consists of two separate steps, each one must be exothermic to be efficient (see Section 3.4.1). That requirement places a constraint on the X–O bond energy: $107\,\text{kJ}\,\text{mol}^{-1} < D(\text{X–O}) < 498\,\text{kJ}\,\text{mol}^{-1}$. Rates of the catalytic destruction of ozone may then exceed those of the elementary step (4.4) either because [X] exceeds [O], or because the rate coefficient for reaction (4.11) exceeds that for (4.4). The rather large activation energy for reaction (4.4) of $17.1\,\text{kJ}\,\text{mol}^{-1}$ does, as we shall see shortly, often favour the indirect route at low stratospheric temperatures.

Several species have been suggested for the catalytic 'X' in the atmosphere. The most important of these for the natural stratosphere are X = H and OH, X = NO, X = Cl, X = Br, and possibly X = I. We shall see later that man has made significant contributions to the atmospheric burdens of these species, especially those involving chlorine and bromine. Reactions (4.11) and (4.12) are then replaced by specific reactions of these X and XO species. For the catalysts just mentioned, the reactions become

[a]Bates, D.R. and Nicolet, M. The photochemistry of atmospheric water vapour. *J. geophys. Res.* **55**, 301 (1950).

Cycle 1:

$$H + O_3 \rightarrow OH + O_2 \qquad (4.13)$$

$$OH + O \rightarrow H + O_2 \qquad (4.14)$$

Net $$O + O_3 \rightarrow O_2 + O_2.$$

Cycle 2:

$$OH + O_3 \rightarrow HO_2 + O_2 \qquad (4.15)$$

$$HO_2 + O \rightarrow OH + O_2 \qquad (4.16)$$

Net $$O + O_3 \rightarrow O_2 + O_2$$

Cycle 3:

$$NO + O_3 \rightarrow NO_2 + O_2 \qquad (4.17)$$

$$NO_2 + O \rightarrow NO + O_2 \qquad (4.18)$$

Net $$O + O_3 \rightarrow O_2 + O_2$$

Cycle 4:

$$Cl + O_3 \rightarrow ClO + O_2 \qquad (4.19)$$

$$ClO + O \rightarrow Cl + O_2 \qquad (4.20)$$

Net $$O + O_3 \rightarrow O_2 + O_2.$$

The catalytic cycles are then said to involve HO_x, NO_x, and ClO_x species.[a] A bromine ('BrO_x') cycle can be written that is exactly analogous to the pair of reactions (4.19) and (4.20). An IO_x cycle has also been suggested, but its efficiency is limited because IO, the chain carrier, is rapidly photolysed in the atmosphere. There is some experimental evidence from both satellite and ground-based spectroscopy that demonstrates the presence of IO itself in the stratosphere (and also in the troposphere: see p. 378). The measurements from the ground suggest a stratospheric abundance of only 5 p.p.t. (mixing ratio of 5×10^{-12}), values that, taken in conjunction with measured rate coefficients,

[a]The use of the representation NO_x refers to NO and NO_2, and if a concentration of NO_x is specified it means the sum of the concentrations of these two oxides. Another representation, NO_y, is often also found in the literature; it means 'odd nitrogen', and is defined as the sum of NO_x and all oxidized nitrogen species that represent sources or sinks of NO_x through processes that occur on relatively short time-scales. Important compounds, besides NO and NO_2, that contribute to NO_y include N_2O_5, HNO_3, HNO_4, $ClONO_2$, and the compound PAN that will be discussed in Section 5.3.4. Although N_2O is an NO_x source in the upper atmosphere, it is *not* normally considered a component of the NO_y compounds.

Table 4.1 Rate parameters for some reactions in catalytic cycles

Reaction	$X + O_3$			$XO + O$		
X	A	E_a	k_{220}	A	E_a	k_{220}
			$O_3 + O$:	8.0×10^{-12}	17.1	6.9×10^{-16}
H	1.4×10^{-10}	3.9	1.7×10^{-11}	2.2×10^{-11}	−1.0	3.8×10^{-11}
OH	1.6×10^{-12}	7.8	2.2×10^{-14}	3.0×10^{-11}	−1.7	7.5×10^{-11}
HO_2	1.1×10^{-14}	4.2	1.1×10^{-15}			
NO	1.8×10^{-12}	11.4	3.5×10^{-15}	6.5×10^{-12}	−1.0	1.1×10^{-12}
Cl	2.8×10^{-11}	2.1	8.7×10^{-12}	3.0×10^{-11}	−0.6	4.1×10^{-11}
Br	1.7×10^{-11}	6.6	4.5×10^{-13}	1.9×10^{-11}	−1.9	5.4×10^{-11}

A and E_a are the Arrhenius parameters, and k_{220} is the rate coefficient calculated for a temperature of 220 K. Units for A and k_{220} are $cm^3\,molecule^{-1}\,s^{-1}$, and those for E_a are $kJ\,mol^{-1}$. The apparent small negative activation energies for the XO + O reactions essentially mean that E_a is near zero for these atom–radical processes, but the values represent current recommendations; it might be more realistic to set $k_{220} = A$, at least as a lower limit. Rate parameters are those recommended by W.B. DeMore *et al. Chemical kinetics and photochemical data for use in atmospheric modeling, Evaluation no. 12*, JPL, Pasadena, 1997.

indicate at most a very small contribution to O_3 removal from the IO chains. Rate parameters for the key reactions are given in Table 4.1, and show how the catalytic cycles can be faster than the direct reaction (4.4). The entry for the reaction '$O_3 + O$' shows that the activation energy for reaction (4.4) with ozone is far higher than that for any of the other reactions of atomic oxygen. At a temperature of 220 K, typical of the stratosphere, the rate coefficient for the direct reaction is thus orders of magnitude smaller than the corresponding value for the catalytic reactions. Whether or not the catalytic reactions will be actually *faster* then depends on the relative concentrations of XO and of O_3.

Figure 4.6 shows the relative contribution to ozone destruction of several of the catalytic cycles as well as of the oxygen-only 'Chapman' reactions for a particular location (mid-latitude) and time of year (equinox). The model on which the figure is based contained a total of 438 reactions, including 43 heterogeneous ones. It is apparent that the catalytic cycles all make a major contribution to destruction of odd oxygen, but that different cycles dominate at different altitudes. The calculations demonstrate clearly that the catalytic cycles *can* explain atmospheric ozone concentrations if the trace constituents are present at levels consistent with atmospheric observations.

The set of catalytic cycles based on the $X \rightarrow XO \rightarrow X$ processes (4.11) and (4.12) gives a first indication of how the natural ozone balance is maintained in the atmosphere, although the simple ideas have to be modified before a fully quantitative result emerges. First of all, even within a 'family' (e.g. HO_x), catalytic cycles can be identified which do not fit into the category so far discussed. For example, the sequence of

Cycle 5:

$$OH + O \rightarrow H + O_2 \qquad (4.14)$$

$$H + O_2 + M \rightarrow HO_2 + M \qquad (4.21)$$

$$HO_2 + O \rightarrow OH + O_2 \qquad (4.16)$$

$$\text{Net} \qquad O + O + M \rightarrow O_2 + M$$

has the same overall effect as reaction (4.5), which destroys two odd-oxygen species. Direct reaction was disregarded as being of no consequence in the atmosphere, but the catalytic cycle 5 is of major importance above 40 km. Similarly, at lower altitudes (< 30 km) the cycle

Cycle 6:

$$OH + O_3 \rightarrow HO_2 + O_2 \qquad (4.15)$$

$$HO_2 + O_3 \rightarrow OH + O_2 + O_2 \qquad (4.22)$$

$$\text{Net} \qquad O_3 + O_3 \rightarrow O_2 + O_2 + O_2$$

becomes important because it does not utilize atomic oxygen, whose concentration is very low at those altitudes. Partitioning of the overall odd-hydrogen budget between the reactive species H, OH, and HO_2 thus determines which sets of reactions really lead to destruction of odd oxygen.

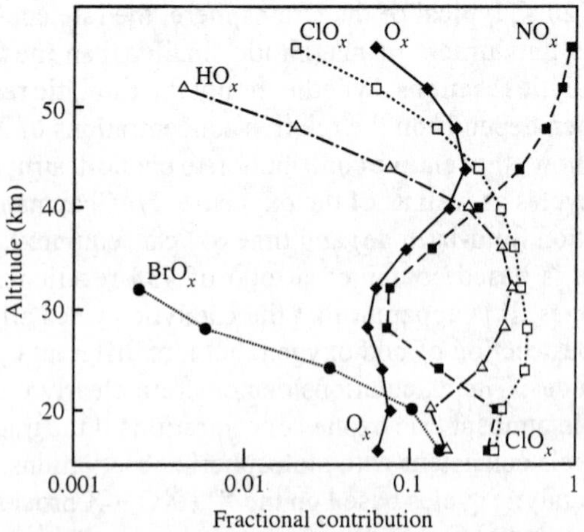

Fig. 4.6. Fraction of the odd-oxygen loss rate due to the 'Chapman', HO_x, NO_x, ClO_x, and BrO_x mechanisms. Figure devised from modelled data presented by D.J. Lary. *J. geophys. Res.* **102**, 21515 (1997).

The identification of particular cycles and of where they are atmospherically important is somewhat artificial but does give an insight into the chemistry. One important conclusion is that the effects of the various catalytic families are definitely not additive. Indeed, adding NO_x to the calculation leads to *less* destruction of O_3 than is obtained with ClO_x, or, especially, $HO_x + ClO_x$ alone. The explanation is that the various families are not isolated; rather, members of one family can react with members of another. Two reactions have shown themselves to be of particular importance in this context, and, as it turns out, to the overall balance of odd oxygen in the stratosphere. They are

$$HO_2 + NO \rightarrow OH + NO_2, \qquad (4.23)$$

and

$$ClO + NO \rightarrow Cl + NO_2. \qquad (4.24)$$

Reaction (4.23) is so central to stratospheric chemistry that each change in rate coefficient reported from different laboratory experiments has necessitated drastic revision of stratospheric models. The reactions do not appear to have any unusual characteristics, and we shall have to examine the chemistry of the catalytic cycles in more detail to discover why the two processes are so important.

4.4.2 Null cycles, holding cycles, and reservoirs

In competition with every catalytic cycle that destroys odd oxygen, cycles can be written that interconvert the species X and XO without odd-oxygen removal. Such processes are null cycles and can be illustrated by writing catalytic and null cycles alongside each other for NO_x

Cycle 3 (catalytic)		Cycle 7 (null)	
$NO + O_3 \rightarrow NO_2 + O_2$	(4.17)	$NO + O_3 \rightarrow NO_2 + O_2$	(4.17)
$NO_2 + O \rightarrow NO + O_2$	(4.18)	$NO_2 + h\nu \rightarrow NO + O$	(4.25)
Net $O + O_3 \rightarrow O_2 + O_2$		$O_3 + h\nu \rightarrow O_2 + O$	

Reaction (4.17) converts NO to NO_2 in both cases, but photolysis of NO_2 in cycle 7 leads to atomic oxygen formation. Cycle 7 has the overall effect of ozone photolysis, reaction (4.3), formally photosensitized by NO_2: it is a 'do-nothing' cycle so far as odd oxygen is concerned. However, that fraction of the NO_x that is tied up in the null cycle is ineffective as a catalyst. During the day, when reaction (4.25) can occur, a given atmospheric NO_x concentration then destroys less odd oxygen than originally anticipated because of the null cycle.

Competitive processes abound in stratospheric chemistry. Reactions (4.18) and (4.25) are not, for example, the only reactions open to NO_2. The nitrate radical, NO_3, can be formed by reaction of NO_2 with ozone

$$NO_2 + O_3 \rightarrow NO_3 + O_2. \tag{4.26}$$

Most atmospheric NO_3 is removed by photolysis during daytime. The products of photodecomposition are $O + NO_2$ (in which case there is no net loss of odd oxygen), and O_2 + NO (in which case odd oxygen is consumed): the competition between these processes depends on the quantum yields for the different channels at the wavelengths available for photolysis. In addition, some NO_3 reacts in the three-body process

$$NO_3 + NO_2 + M \rightarrow N_2O_5 + M. \tag{4.27}$$

In the gas phase, N_2O_5 is rather unreactive in the stratosphere. Ultimately, it decomposes back to NO_2 + NO_3 thermally or photochemically, so that its formation does not constitute a permanent loss of odd nitrogen. On the other hand, it does behave as an unreactive reservoir of NO_x, containing typically 5 to 10 per cent of the total NO_x budget. The cycle involving its formation and destruction is then a holding cycle. However, in the presence of stratospheric aerosol, the situation is altered dramatically, since the process

$$N_2O_5 + H_2O \rightarrow 2HNO_3 \tag{4.28}$$

becomes efficient on surfaces (see Sections 3.4.5 and 4.4.4), and the nitrogen is diverted into a new reservoir, nitric acid.

Holding cycles involving members of two families also turn out to be important routes for conversion to HNO_3. Three-body formation of nitric acid (HNO_3) is an important storage step in the HO_x–NO_x system

$$OH + NO_2 + M \rightarrow HNO_3 + M. \tag{4.29}$$

Nitric acid is photolysed to regenerate $OH + NO_2$, but the process is relatively slow, and about half the stratospheric load of NO_x is stored in the nitric acid reservoir. For the ClO_x family, hydrochloric acid (HCl) is the main reservoir, reaction of Cl with stratospheric methane being the major source. The holding cycle for Cl involving HCl as reservoir can then be written

Cycle 8:

$$Cl + CH_4 \rightarrow CH_3 + HCl \tag{4.30}$$

$$OH + HCl \rightarrow H_2O + Cl \tag{4.31}$$

Net $$CH_4 + OH \rightarrow CH_3 + H_2O$$

About 70 per cent of stratospheric ClO_x is thought to be present as HCl (although in the lower stratosphere, the partitioning is strongly dependent on heterogeneous processes, as described in Section 4.7.5).

As the understanding of stratospheric ozone chemistry has become more complete, so yet more exotic molecules have been called into service as reservoir species. The list includes HOCl (hypochlorous acid), HNO_4 (pernitric acid), and $ClONO_2$ (chlorine nitrate): the 'trivial' names of the compounds are almost always used in the published literature, and are given here for identification. Suggested routes to the compounds mentioned once again emphasize the interaction between families

$$ClO + HO_2 \rightarrow HOCl + O_2 \tag{4.32}$$

$$HO_2 + NO_2 + M \rightarrow HO_2NO_2 + M \tag{4.33}$$

$$ClO + NO_2 + M \rightarrow ClONO_2 + M. \tag{4.34}$$

Of these three reservoir species, $ClONO_2$ is the best characterized in the stratosphere, measured mixing ratios reaching about 1×10^{-9} at an altitude of 25–35 km. Chlorine nitrate plays several important roles in stratospheric chemistry. Formation of $ClONO_2$ strongly couples the chlorine and nitrogen cycles, and at high concentrations of stratospheric chlorine could lead to non-linear responses of ozone to chlorine perturbations (see Section 4.7). The observed diurnal variations in [ClO] (Section 4.4.6) appear to be driven by $ClONO_2$ chemistry. HOCl has also been identified positively, and provides further information for testing current interpretations of stratospheric chlorine chemistry.

We defer most of our discussion of bromine chemistry in the stratosphere until Sections 4.6.4 and 4.7.5, but observe here that bromine reservoirs are worthy of note, in part because they are so inefficient! In the gas phase, conversion of Br to HBr in the analogue of reaction (4.30)

$$Br + CH_4 \rightarrow CH_3 + HBr \tag{4.35}$$

is extremely slow, because the reaction is endothermic for Br (while reaction (4.30) itself is exothermic). Furthermore, although the bromine analogue of reaction (4.34) can generate bromine nitrate, $BrONO_2$, this compound photolyses

$$BrONO_2 + h\nu \rightarrow BrO + NO_2 \tag{4.36a}$$

at longer wavelengths than $ClONO_2$, in the visible region where there is higher intensity. Reversal of the formation of $BrONO_2$ is thus favoured, and neither of the reservoirs HBr or $BrONO_2$ are anything like as important as the chlorine analogues. Furthermore, another channel for photolysis

$$BrONO_2 + h\nu \rightarrow Br + NO_3 \tag{4.36b}$$

is thermodynamically accessible. If this reaction occurs to a significant extent, then a very efficient new cycle can be envisaged.

Cycle 9:

$$BrO + NO_2 + M \rightarrow BrONO_2 + M \qquad (4.37)$$

$$BrONO_2 + h\nu \rightarrow Br + NO_3 \qquad (4.36b)$$

$$NO_3 + h\nu \rightarrow NO + O_2 \qquad (4.38)$$

$$NO + O_3 \rightarrow NO_2 + O_2 \qquad (4.17)$$

$$Br + O_3 \rightarrow BrO + O_2 \qquad (4.39)$$

$$\text{Net} \qquad 2O_3 \rightarrow 3O_2$$

Yet another feature of the bromine reservoirs is that *heterogeneous* hydrolysis of $BrONO_2$

$$BrONO_2 + H_2O \rightarrow HOBr + HNO_3 \qquad (4.40)$$

is apparently efficient on the surface of stratospheric sulphate aerosol or of polar stratospheric clouds (see Sections 4.7.4 and 4.7.5). HOBr is another photochemically labile species

$$HOBr + h\nu \rightarrow OH + Br, \qquad (4.41)$$

and this reaction therefore enhances the OH (and, indirectly, the HO_2) concentration. In turn, the enhanced HO_x concentration reduces the lifetime of HCl, and thus its efficiency as a reservoir. One consequence of each of these considerations is that the BrO_x chains are even more efficient than the ClO_x ones in catalysing the destruction of stratospheric ozone.

The reservoir compounds are evidently of very great importance in stratospheric chemistry. They act to divert potentially catalytic species from active to inactive forms, but the compounds remain available to liberate active catalysts. The assumed rates of production and destruction have a large influence on the predictions of stratospheric models. It has recently become increasingly apparent that heterogeneous chemistry occurring on aerosol particles present in the stratosphere plays a very important part in the partitioning between active and inactive species. We shall see in Section 4.7 that unexpected release of active catalytic species from reservoir compounds is probably responsible for the phenomenon of the Antarctic ozone hole, and that the surface reactions are central to the explanation.

We now return to the importance of reactions (4.23) and (4.24) in stratospheric chemistry. In chemistry involving O_x and NO_x reactions alone, there are null cycles (e.g. cycle 7) and holding cycles. For ClO_x on its own, there are no corresponding null cycles, and those for HO_x, which involve H_2O_2, are rather inefficient. Reactions (4.23) and (4.24) are thus critical in providing effective null cycles for the HO_x and ClO_x families. Cycles 10 and 11 show these null cycles

Cycle 10:

$$OH + O_3 \rightarrow HO_2 + O_2 \quad (4.15)$$

$$HO_2 + NO \rightarrow OH + NO_2 \quad (4.23)$$

$$NO_2 + h\nu \rightarrow NO + O \quad (4.25)$$

Net $$O_3 + h\nu \rightarrow O_2 + O$$

Cycle 11:

$$Cl + O_3 \rightarrow ClO + O_2 \quad (4.19)$$

$$ClO + NO \rightarrow Cl + NO_2 \quad (4.24)$$

$$NO_2 + h\nu \rightarrow NO + O \quad (4.25)$$

Net $$O_3 + h\nu \rightarrow O_2 + O$$

Not only are these cycles effective null paths for HO_x or ClO_x, but they also provide an additional null cycle for NO_x. It is the photolytic step (4.25) that completes the null cycle in each case. In the real atmosphere, photolysis of NO_2 is, of course, in competition with reaction (4.18) which would lead to an overall loss of two odd oxygens.

Reaction (4.23) can also participate in rather unusual cycles that generate odd oxygen. Oxidation of CO to CO_2 usually destroys odd oxygen, but the sequence

Cycle 12:

$$OH + CO \rightarrow H + CO_2 \quad (4.42)$$

$$H + O_2 + M \rightarrow HO_2 + M \quad (4.21)$$

$$HO_2 + NO \rightarrow OH + NO_2 \quad (4.23)$$

$$NO_2 + h\nu \rightarrow NO + O \quad (4.25)$$

Net $$CO + O_2 + h\nu \rightarrow CO_2 + O$$

creates an oxygen atom. Similarly, the combination of OH and HO_2 as a direct step is a loss process for HO_x

$$OH + HO_2 \rightarrow H_2O + O_2. \quad (4.43)$$

However, catalysed by NO_x, and with the participation of reaction (4.23), a cyclic process with the effect

$$OH + HO_2 + 2h\nu \rightarrow H_2O + O + O \quad (4.44)$$

can be written. The importance of a knowledge of the rate coefficient for reaction (4.23) can now be appreciated. Sequences of steps that would destroy odd oxygen if the rate coefficient were small become diverted into null or even generating cycles for larger values. We shall see in Sections 4.6.2 and 4.6.4 how predicted effects on ozone of anthropogenic pollution have depended on the rate coefficients used in the models.

4.4.3 Natural sources and sinks of catalytic species

The catalytic families HO_x, NO_x, ClO_x, and BrO_x appear to be present in the 'natural' atmosphere unpolluted by man's activities. In the contemporary atmosphere, however, the background concentrations, especially of ClO_x and BrO_x, have already been supplemented by anthropogenic sources. We shall consider possible perturbations of the stratosphere in Section 4.6; measurements on, and models of, the atmosphere are concerned with actual total loadings of trace species, regardless of their origin.

Slow changes, over periods of tens of years or more, may be affecting concentrations of trace atmospheric constituents, but since the lifetimes (chemical or physical) of *most* of the species in the stratosphere are shorter, sources and sinks must be more or less in balance. Precursors of the atoms and radicals that participate in the catalytic cycles themselves have no source above the Earth's surface, and therefore have to be transported through the troposphere to the stratosphere. The probable exception to this generalization is a small contribution to NO_x, since N_2 and O_2 are the major atmospheric components.

Most of the stratospheric NO_x originates with tropospheric N_2O, whose source is largely biological (see Chapter 1). Excited atomic oxygen, $O(^1D)$, derived largely from ozone photolysis, reaction (4.3), and to a lesser extent from oxygen photolysis (reaction 4.1) at higher altitudes, can react with N_2O to yield NO,

$$O(^1D) + N_2O \rightarrow NO + NO, \qquad (4.45)$$

and thus initiate the NO_x chemistry. The photolytic production of $O(^1D)$ and the reasons for its reactivity were reviewed in Chapter 3, and will not be discussed further here. A great deal of ingenuity has gone into finding upper atmospheric sources of N_2O. Suggestions have included the reaction of electronically excited N_2 with ground-state O_2 (p. 111), or of excited O_2 with ground-state N_2. Another possibility is that *vibrationally* excited O_3, produced in the O + O_2 association reaction (4.2), reacts with N_2

$$O_3^* + N_2 \rightarrow N_2O + O_2. \qquad (4.46)$$

Some recent experimental evidence suggests that reaction (4.46) occurs, although it remains to be shown *directly* that such processes are of stratospheric significance.

Several other sources of NO_x exist, but they are of lesser importance than the N_2O source. Cosmic rays penetrate the atmosphere and contribute to NO production in the altitude range 10–30 km. Ionization reactions involving N_2 lead, after several steps, to NO. The source is probably the major one during the polar night, although it is not important on a global scale. Short-wavelength ultraviolet, and solar particle, ionization in the mesosphere and thermosphere also lead to NO production. However, rapid photodissociation at levels above the stratopause seems to prevent this high-altitude source from reaching the stratosphere. Tropospheric NO or NO_2 (e.g. that produced by lightning) generally does not survive long enough to be transported in significant quantities to the stratosphere, although some lightning-derived NO_x from the ITCZ (see p. 77) does seem to reach the lower stratosphere.

Reactions of $O(^1D)$ with water vapour and methane, *main source of OH radicals*

$O_3 + h\nu \rightarrow O(^1D) + O_2$

$$O(^1D) + H_2O \rightarrow OH + OH \qquad (4.47)$$

$$O(^1D) + CH_4 \rightarrow OH + CH_3, \qquad (4.48)$$

are the main sources of OH radicals. As discussed in Chapter 2, the stratosphere is very dry, probably because tropospheric water vapour has to pass through a 'cold trap' at the tropopause. Much stratospheric H_2O is in fact a product of CH_4 oxidation. Nevertheless, CH_4 constitutes more than a third of the total hydrogen ($4 \times [CH_4] + 2 \times [H_2O]$) in the lower stratosphere, so that reaction (4.48) is important. In fact, the reaction contributes more than one HO_x species, since subsequent oxidation of CH_3 to CO yields two or three more odd-hydrogen species. We shall examine this oxidation in more detail in connection with tropospheric chemistry (Chapter 5).

By far the most abundant *natural* precursor of ClO_x is methyl chloride (CH_3Cl). Tropospheric mixing ratios show little difference between Northern and Southern Hemispheres, suggesting a non-industrial source. The main contributor appears to be the world's oceans. Burning of vegetation also produces CH_3Cl, and forest fires must be regarded as natural events. Primitive agriculture employs 'slash and burn' techniques that lead to enhanced methyl chloride levels near the practising communities, but, properly speaking, that is an anthropogenic source. Volcanic emissions can also show high CH_3Cl levels. Tropospheric lifetimes for methyl chloride are estimated as ~ 1 year, and some CH_3Cl can be transported across the tropopause. Stratospheric concentrations show a decrease with altitude, as expected if CH_3Cl is reactive, but extrapolate well at lower altitudes to the tropospheric mixing ratios. Once in the stratosphere, CH_3Cl is primarily removed by reaction with OH,

$$CH_3Cl + OH \rightarrow CH_2Cl + H_2O, \qquad (4.49)$$

although above ~ 30 km about one-third is photolysed. In either case, atomic chlorine ultimately becomes available to enter the ClO_x cycle.

Alternative natural sources of ClO_x have been investigated. These include volcanic release of HCl, either slowly emitted to the troposphere followed by transport to the stratosphere, or, in major volcanic eruptions, directly to the stratosphere. Acidification (e.g. by H_2SO_4) of marine aerosols containing NaCl is another potential source of HCl. None of these processes seems capable, however, of providing enough ClO_x to have a marked influence on atmospheric ozone concentrations.

Natural bromine enters the stratosphere mainly as CH_3Br, which is produced by algae in the oceans, together with smaller amounts of species such as $CHBr_3$, CH_2Br_2, CH_2BrCl, and $CHBrCl_2$. Substantial additional amounts of bromine are attributable to man's activities (cf. Section 4.6.4). There is still considerable uncertainty about the relative contributions, but one current estimate puts the natural mixing ratio of CH_3Br at about 10×10^{-12}, with man responsible for perhaps a further 5×10^{-12} to the mixing ratio.

Sink processes for the catalytic families largely involve slow transport of the reservoir species across the tropopause. In the troposphere, the species then dissolve in water, and are subsequently 'rained out'. Reservoir compounds such as $HONO_2$, HOCl, HCl, HBr and probably $ClONO_2$, $BrONO_2$, and HO_2NO_2, are all highly soluble, and are thus specially good candidates for rainout. Radical–radical reactions such as process (4.43) are important in determining the total concentrations of active reaction intermediates; in this case, of course, the products O_2 and H_2O are hardly conventional reservoir compounds that must be removed from the stratosphere. A role for metals has also been suggested as a possibility in chlorine chemistry. Metals such as sodium and calcium are deposited by ablating meteors (p. 500 and p. 543) near the mesopause. A steady downward flux of metal atoms must be present throughout the mesosphere and stratosphere. Compounds such as NaO, NaOH, CaO, and CaOH might be formed in the stratosphere. Reaction with HCl, Cl, and ClO might then lead to the sequestering of ClO_x as NaCl. However, if NaCl were itself photolysed to Na + Cl, the overall effect would be conversion of HCl to Cl, so that, far from diminishing the impact of ClO_x on stratospheric ozone, the meteoric metals would augment it! Recent experiments have shown that particles of NaCl can react, at least at 298 K, with the reservoir compounds N_2O_5 and $ClONO_2$ to release chlorine in photochemically active forms (as $ClNO_2$ and Cl_2 for the two reactions, respectively). The processes are analogous to particle reactions that are important in producing the Antarctic ozone hole (cf. Section 4.7.5), but demonstrate here the existence of routes for converting metal halides into catalytically active gaseous species.

4.4.4 Heterogeneous chemistry

At several points in the preceding explanation, it has been emphasized that heterogeneous processes (see Section 3.4.5 for a discussion of the term) play a most important part in stratospheric chemistry. The awareness of this importance has come relatively slowly, and was prompted initially by the need to find new mechanisms that would account for massive depletions of ozone in the Antarctic (Section 4.7). It is now quite evident that heterogeneous reactions occur in regions of the stratosphere distant from the polar regions, and that particles of ice, soot, nitric acid hydrates, and sulphuric acid and sulphates can all be active. Volcanic eruptions that inject particles into the stratosphere are thus expected to have an effect on chemistry, and clear signatures are now recognized that demonstrate the influence of eruptions such as that of Mount Pinatubo.

Hints have appeared earlier that heterogeneous processing is particularly significant in relation to the reservoir compounds. Major alterations are brought about in the partitioning of species in the catalyst families between active chain carriers and temporarily inactive reservoirs, and in the partitioning between reservoirs of differing reactivity. Examples that we have already met include the hydrolysis reactions

$$N_2O_5 + H_2O \rightarrow 2HNO_3 \tag{4.28}$$

$$BrONO_2 + H_2O \rightarrow HOBr + HNO_3. \tag{4.40}$$

Reaction (4.34) has an analogue for $ClONO_2$

$$ClONO_2 + H_2O \rightarrow HOCl + HNO_3, \tag{4.50}$$

and a critical conversion of $ClONO_2$ and HCl

$$ClONO_2 + HCl \rightarrow Cl_2 + HNO_3 \tag{4.51}$$

occurs on particles in the polar stratosphere (Section 4.7.5). Note that the reaction influences two reservoirs at the same time. Other surface reactions proposed include

$$HOCl + HCl \rightarrow H_2O + Cl_2 \tag{4.52}$$

$$HOBr + HCl \rightarrow H_2O + BrCl \tag{4.53}$$

$$N_2O_5 + HCl \rightarrow ClNO_2 + HNO_3 \tag{4.54}$$

$$HNO_4 \rightarrow HONO + O_2. \tag{4.55}$$

It is now useful to summarize the gas-phase chemistry that has been introduced earlier, but it must always be interpreted with the proviso that heterogeneous reactions are likely to intervene as well. It will become apparent in Section 4.4.6 that these surface reactions are essential to a reconciliation between theory and experimental observation.

4.4.5 Summary of homogeneous chemistry

The information presented in the previous sections can most conveniently be summarized in flow charts showing source, radical, and sink species, and their

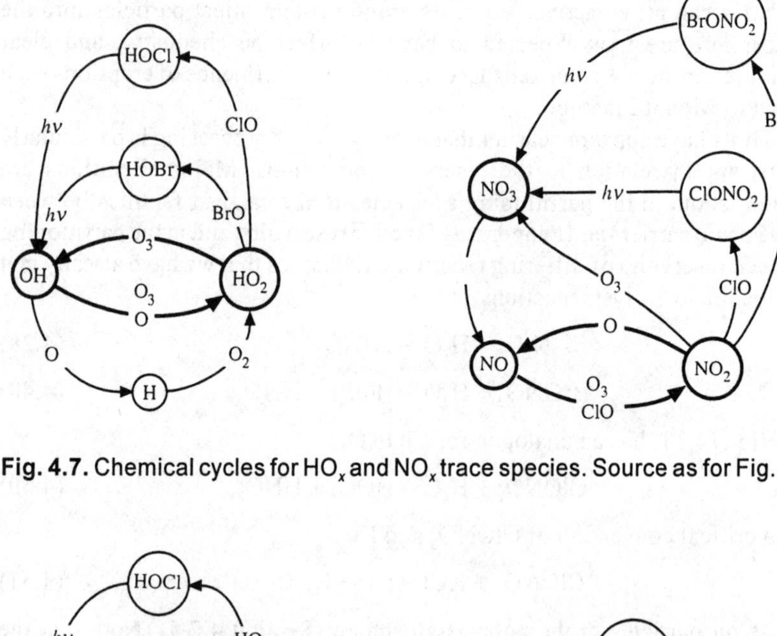

Fig. 4.7. Chemical cycles for HO_x and NO_x trace species. Source as for Fig. 4.6.

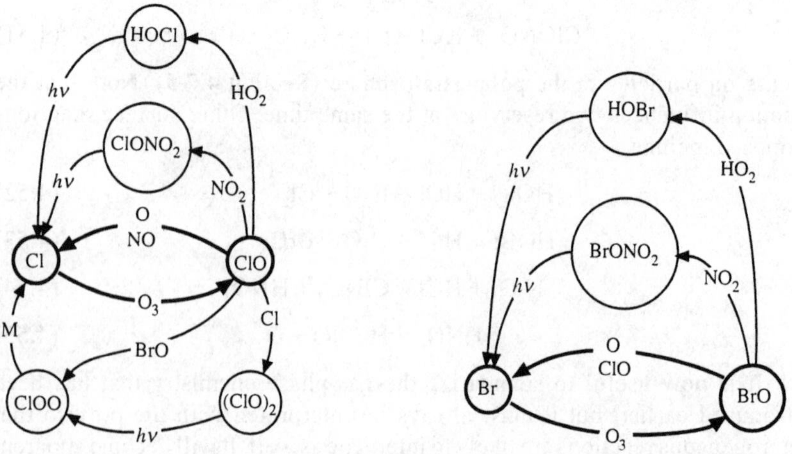

Fig. 4.8. Chemical cycles for ClO_x and BrO_x species. Source as for Fig. 4.6.

Table 4.2 Percentage relative contributions of the major rate-limiting chemical reactions to the removal of ozone.

Altitude (km)	Reaction					
	$O + O_3$	$O + NO_2$	$O + ClO$	$O + HO_2$	$HO_2 + O_3$	$O + BrO$
16	6.4	16.5	2.3	0.6	28.9	0.4
20	7.6	13.1	11.0	1.9	31.3	0.5
24	6.8	26.5	14.8	2.8	11.9	0.4
28	5.5	36.0	12.7	3.4	3.4	0.2
32	6.2	40.1	14.4	1.3	1.0	0.2
36	9.3	35.1	22.9	7.3	0.3	0.3
40	17.2	22.2	29.1	5.2	0.1	0.4
44	21.3	7.5	24.1	2.4		0.3
48	18.5	2.1	14.3	1.1		0.2
52	12.4	0.4	6.2	0.5		
56	6.0	0.1	2.1	0.2		

Data derived from models presented by D.J. Lary *J. geophys. Res.* **102**, 21515 (1997).

reactive interconnections: such diagrams are sometimes referred to as *Nicolet diagrams* to honour a pioneer in atmospheric chemistry. Figures 4.7 and 4.8 are simplified flow charts for the HO_x, NO_x, ClO_x and BrO_x species respectively. The diagrams presented here are deliberately restricted to the homogeneous processes described so far. The reaction partner that effects a transformation is given within the arrows, photochemical change being represented by $h\nu$. Cycles are readily identifiable as closed loops of arrows, and the reservoirs are generally the non-radical species in the centre of the diagrams that are formed and destroyed by radical and photochemical processes. The coupling between cycles is particularly clearly seen for the case of ClO_x (left side of Fig. 4.8), where HO_x, NO_x, and BrO_x all play a role in the interconversions.

Table 4.2 summarizes contributions of the major catalytic reactions to ozone loss at different altitudes as estimated by a particular model. The shortfall of the totals from 100 per cent reflects the occurrence of other cycles not listed (especially those involving two catalytic families). An interesting observation is that, although there are many catalytic cycles, their relative efficiency at different altitudes depends on the availability of O and O_3. Thus, in the lower stratosphere, where $[O]/[O_3]$ is low (Fig. 4.3), cycles having the net effect $2O_3 \rightarrow 3O_2$ are dominant, while in the upper stratosphere, where $[O]/[O_3]$ is high, the cycles producing the conversion $2O \rightarrow O_2$ are the most efficient. In the mid-stratosphere, the most effective cycles are those where the net reaction is $O + O_3 \rightarrow 2O_2$.

4.4.6 Comparison of experiment and theory

Over recent decades, a great deal of effort has been expended on making measurements of the trace constituents of the stratosphere, using a wide range of *in-situ* and remote-sensing techniques. Such measurements provide, of course, the basic information on atmospheric composition. In this sense, they can be employed as input data for models of atmospheric chemistry. But they also demonstrate the effect of chemical change on species that enter the stratosphere. As explained earlier, the link between experiment and theory is provided by a model that incorporates trace constituents and their chemical reactions. The observed concentrations of the trace species can thus also be used for comparison with the predictions of such models, with the aim first of testing the models themselves, and ultimately of probing the completeness of our understanding of the chemical behaviour of the atmosphere. The output of models includes information about the concentrations of the trace constituents and reactive intermediates in the chemical scheme. Confidence in the completeness of a chemical mechanism, and in the correctness of the rate parameters used, is obviously increased as the number of observable variables that can be matched with prediction becomes larger. Conversely, good agreement between experiment and theory tends to lend credibility to the modelling procedures used, and to the approximations inevitably adopted. Depending on the detail available from the model and in the observations, comparisons may be made at various levels of refinement, ranging from globally averaged column abundances to values for a particular geographical location, altitude, and time.

Measurements of trace species in the atmosphere can be categorized as *in-situ* determinations of local concentrations, and *remote sounding* determinations of non-local abundances. Table 4.3 gives some of the important characteristics of each type of measurement (see also Section 5.9.2).

Catalytic cycles for stratospheric odd-oxygen chemistry are thought to explain a loss of odd oxygen beyond that provided by the Chapman, oxygen only, reactions. Since these cycles are invoked to reconcile production and destruction rates of ozone (Section 4.4.1), a first test of the extended chemistry must be to see if experimental ozone concentrations are correctly predicted. A more sophisticated test of the concepts of stratospheric ozone chemistry is provided by measurements of the concentrations of the minor trace species that play a part in the theories as sources, sinks, or, especially, reservoirs and intermediates. Critical tests of photochemical theories require the simultaneous measurement of several coupled species within the same air mass. Observations of concentration–altitude profiles, and of the seasonal and even diurnal variations of abundance, provide excellent material for comparison with the predictions of models. Above all, the degree of match between experiment and prediction for the atomic and radical intermediates of the postulated reactions indicates how complete and realistic the theory is.

Table 4.3 Methods of measuring concentrations in the atmosphere.

	In situ	Remote
Platforms	aircraft balloons (ground)	ground aircraft; balloons satellites; space shuttle
Main methods	grab sampling/later analysis spectroscopy: absorption or emission: μwave to UV resonance fluorescence chemiluminescence electrochemical	spectroscopy: absorption or emission: μwave to UV lidar
Advantages	high spatial and temporal coverage (especially with satellites)	measurements can be made of regions some distance from instrument high spatial and temporal resolution
Disadvantages	measurements obtained only in vicinity of instrument: limited coverage in altitude, latitude, longitude and time	poor spatial resolution

An important advance of the 1980s and 1990s has been the harnessing of satellite and space-shuttle platforms to provide long-term and global measurements of trace gases in the atmosphere. An important shuttle-borne instrument is ATMOS (a high-resolution infrared spectrometer).

Several satellites have carried or currently carry the Total Ozone Mapping Spectrometer (TOMS) that provides crucially important daily global maps of stratospheric ozone concentration. These craft sense ozone concentration by looking down at backscattered ultraviolet light. Additionally, the Microwave Limb Sounder onboard NASA's Upper Atmosphere Research Satellite (UARS) measures the concentration of stratospheric ozone and chlorine monoxide by looking through the atmosphere as a thin coating at the edge (or limb) of the Earth's full-disc.

Another satellite, the European ERS-2 was launched in April 1995, carrying additional equipment called Global Ozone Monitoring Experiment (GOME). GOME is a nadir-looking instrument which measures solar radiation transmitted through or scattered by the Earth's atmosphere. In the year 2001, the European Space Agency (ESA) should be able to launch ENVISAT-1 as one

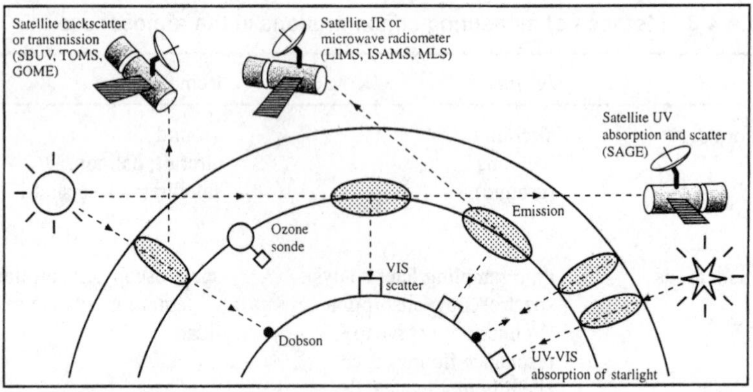

Fig. 4.9. Some of the techniques that can be used for measuring stratospheric ozone concentrations from the ground, from airborne platforms, and from satellites.

of their major satellite missions. This large satellite will carry many different instruments, of which those of major importance in studies of atmospheric chemistry are GOMOS (Global Ozone Monitoring by Occultation of Stars), MIPAS (Michelson Interferometer for Passive Atmospheric Sounding), and SCIAMACHY (Scanning Imaging Absorption Spectrometer for Atmospheric Cartography).

The glamour (and expense) of these satellites should not obscure the value of more mundane methods of making atmospheric measurements, which have, at least until now, provided most of the information that we have on chemical composition. The techniques range from ground-based measurements to experiments carried aloft by balloons, aircraft, and rockets.

A wide variety of experimental techniques now exists for the measurement of atmospheric trace species, as indicated in Table 4.3. They include absorption and emission spectroscopy in spectral regions ranging from the microwave to the ultraviolet. In addition, various grab-sampling and matrix-isolation experiments have been described. Figure 4.9 illustrates the deployment of some of the methods for the measurement of stratospheric ozone concentrations. Ground-based and satellite-platform techniques have different advantages and drawbacks. Ground-based instruments can be calibrated periodically, and the series of measurements dates back to 1957 in some cases. However, the distribution of stations is not adequate to give a truly global average. Satellite observations can overcome the averaging problem, but so far have been available only for relatively short periods. The backscatter UV (BUV) instrument on the Nimbus 4 satellite was in operation for seven years from April 1970, but suffered from an apparent drift in sensitivity. A solar backscatter UV (SBUV) and total ozone mapping spectrometer (TOMS) on the

Nimbus 7 satellite started returning ozone measurements in 1978, and subsequent satellite instruments (see earlier) subsequently took over the task.

One technique for the *in-situ* detection of atomic and radical intermediates which has proved particularly valuable in atmospheric studies is that of *resonance fluorescence*. The basis of the technique is that electronic excitation of a species by absorption of light may be followed by fluorescent emission [pathway (iii), Fig. 3.1] at the same wavelength: that is, the fluorescence is 'resonant' in energy with the exciting radiation. The wavelength(s) of absorption and emission are sharply defined for atoms and small radicals or molecules. For example, the ground-state hydrogen atom, $H(^2S)$, can absorb radiation at $\lambda = 121.6$ nm (the Lyman-α line of atomic hydrogen) to reach the first excited, ^2P, electronic state. One fate of $H(^2P)$ is radiation of the resonant Lyman-α line

$$H(^2S) + h\nu \rightarrow H^*(^2P) \qquad \text{absorption} \qquad (4.56)$$

$$H^*(^2P) \rightarrow H(^2S) + h\nu \qquad \text{resonance fluorescence.} \qquad (4.57)$$

Since the radiation is isotropic, it can be detected off-axis from the exciting beam. Observation of Lyman-α fluorescence from a system illuminated by $\lambda = 121.6$ nm radiation (obtained from, say, a hydrogen discharge lamp) thus demonstrates the presence of H atoms. The intensity of fluorescence is proportional to the concentration of hydrogen under suitable conditions. In the atmospheric experiments, a lamp and a shaped tube that is to act as a fluorescence cell are borne aloft on a balloon (or rocket) and parachuted down. Air passes through the tube, and the intensity of resonance fluorescence provides a measure of the concentration of the species for which the lamp provides specific excitation. Many of the atmospheric experiments have been carried out by Dr Jim Anderson, of Harvard University, in whose skilled hands atoms and radicals can be detected at concentrations of 10^6 particles cm^{-3} (1.7×10^{-15} M, for chemists used to thinking in molarities!). A clever trick enables detection of radicals such as HO_2 and ClO, which do not fluoresce. By injection of NO, carried with the payload, into the test cell, OH or Cl can be produced stoicheiometrically. Both OH and Cl are fluorescent. The conversion involves the two reactions that are also so important in the chemistry of the atmosphere itself

$$HO_2 + NO \rightarrow OH + NO_2 \qquad (4.23)$$

high NOx envit

$$ClO + NO \rightarrow Cl + NO_2. \qquad (4.24)$$

A modification[a] to the balloon experiments allows the fluorescence probe to be lowered, like a Yo-Yo, more than ten kilometres from the balloon by a

[a]See, for example, Stratospheric ClO *in situ* detection with a new approach. Brune, W.H., Weinstock, E.M., Schwab, J.J., Stimpfle, R.M., and Anderson, J.G. *Geophys. Res. Lett.* **12**, 441 (1985).

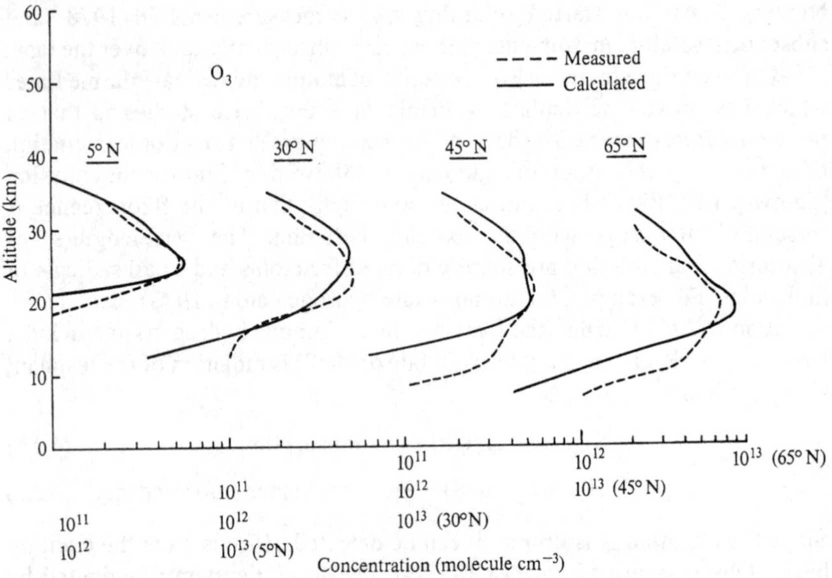

Fig. 4.10. Observed and calculated vertical concentration profiles of ozone as a function of latitude in the Northern Hemisphere for mid-spring. From Miller, C., Filkin, D.L., Owens, A.J., Steed, J.M., and Jesson, J.P. *J. geophys. Res.* **86**, 12039 (1981).

winch. The tethered instrument can thus make repetitive measurements of altitude profiles by reeling down or up, and the technique obviously represents a substantial advance over a conventional 'one-off' balloon experiment.

Aircraft campaigns have assumed an increasingly significant place in conducting stratospheric measurements. One aircraft that has been used is the ER-2 (it was formerly designated U-2, in an earlier incarnation when it was a spy plane). This aircraft can fly at such great heights that it can enter the stratosphere, and it has been equipped with a variety of instruments, including those that determine key free radical species such as OH, HO_2, ClO, and BrO, as well as the oxides of nitrogen and ozone itself. A more recent stratospheric aircraft is the Russian Geophysica (Myasishchev M55), which can operate for up to five hours at an altitude of 21 km, well into the stratosphere. The level of instrumentation possible is illustrated by a campaign based on Ushuaia, Tierra del Fuego, in September–October 1999. The Geophysica carried three experiments designed for remote sensing of chemistry, and six for *in-situ* chemistry, as well as another five experiments for remote-sensing and *in-situ* physics.

We now look at a few specific examples to give an indication of types of information available from both modelling and atmospheric observations. For ozone, the observational database is sufficiently large that sophisticated tests of two-dimensional models can be made. The largest and most well-established variation of ozone occurs on a seasonal time-scale. An important test of 2-D models is therefore their ability to simulate seasonal variations of ozone. Historically, the most extensive part of the ozone database is the record of columnar ozone obtained from a network of ground stations observing ultraviolet absorption. Satellite measurements of backscattered ultraviolet over a few years suggest that the principal features of these long-term records are realistic. Figure. 4.1(a) has already presented seasonal and latitudinal variation of column ozone abundances obtained from satellite observations. Results from a 2-D model are given in Fig. 4.1(b). The model employed an advanced chemical reaction scheme that included heterogeneous processes. All the general features of Fig. 4.1(a) are reproduced, and the model exhibits excellent qualitative agreement with measurements. Spring–autumn asymmetries show up well in the model results.

Altitude profiles are also quite well predicted by models, at least for northern latitudes, as Fig. 4.10 indicates for some mid-spring comparisons. Although this particular illustration dates from a time when the model did not include heterogeneous chemistry, the measured and calculated profiles match quite well for this season, except perhaps at the highest latitudes. Such small discrepancies as there are between model and observation are usually ascribed to errors in transport contributions.

A critical test of models and the chemistry they incorporate is the ratio of the atomic oxygen concentration to the ozone concentration. Fortunately, there exists a simultaneous measurement of $[O(^3P)]$ and $[O_3]$, the results of which have been given in Fig. 4.3. The ratio $[O(^3P)]/[O_3]$ is plotted out in Fig. 4.11. Since the experiments refer to a particular place and time, a l-D model can be used to generate the theoretical ratio. The solid line in the figure shows how good the agreement is between the calculated and measured ratios.

Hydroxyl radicals are not only central to the HO_x catalytic chain, but they are also responsible for coupling between the NO_x and ClO_x families. Thus OH is the most important member of the HO_x family, and a key species in atmospheric chemistry, although the hydroperoxyl (HO_2) radical is another species of central significance. Several techniques now exist for determining stratospheric concentrations of OH, including the *in-situ* resonance fluorescence method, and its variants, and far-infrared spectroscopy. Figure 4.12 shows measurements of HO_x obtained by a balloon-borne far-infrared Fourier transform spectrometer (FIRS-2). The solid line is the altitude profile generated by a model in which O_3, H_2O and CH_4 were constrained by simultaneous measurements. These model calculations demonstrate that, by making certain adjustments to the kinetic parameters, existing chemical reaction schemes can reproduce the observations quite well. One particular conclusion from these results is that HO_x observations in the upper stratosphere appear to require a reduction in the rate coefficient for the reaction

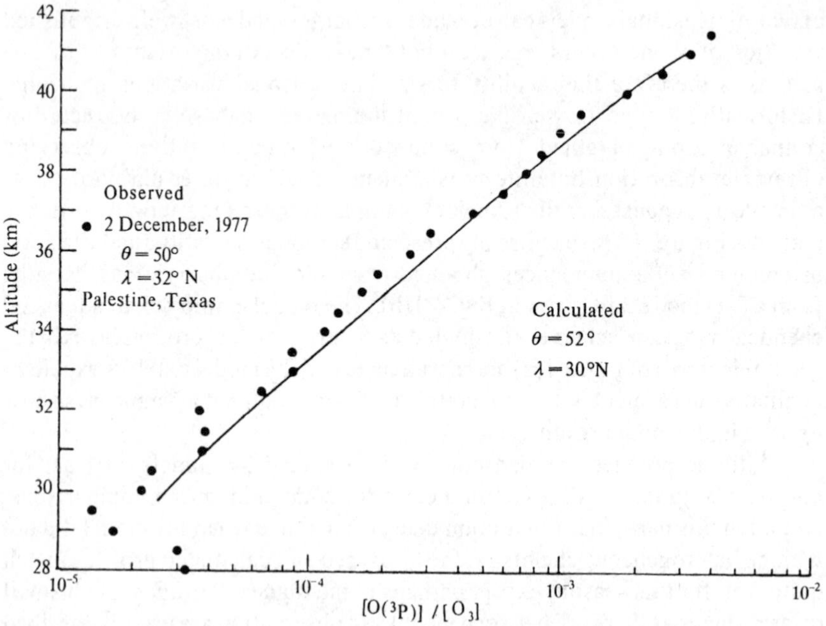

Fig. 4.11. Ratio $[O(^3P)]/[O_3]$ derived from the measurements of Fig. 4.3, and as calculated by Logan, J.A., Prather, M., Wofsy, S.C., and McElroy, M.B. *Phil. Trans. R. Soc.* **A290**, 187 (1978).

$$HO_2 + O \rightarrow OH + O_2 \tag{4.16}$$

by about 25 per cent from the currently recommended value.

Despite the apparently good agreement between observation and theory that the results of Fig 4.12 imply, there remain uncertainties about how well HO_x concentrations are understood. Although balloon-borne observations of OH generally agree with modelled concentrations to within about 10 per cent, shuttle-borne observations are 30 to 40 per cent *lower* than the predicted values, while observations of total column concentrations are up to 30 per cent *higher* than the modelled ones. Different techniques, platforms, and instruments sometimes produce conflicting data about HO_x concentrations. At present, it seems that the most sensible interpretation is that the results of these extremely difficult experiments confirm expectations well enough to lend confidence to the reaction schemes presented so far. Nevertheless, there is doubtless a place for further refinements in both observation and theory.

Odd-nitrogen species are represented in the observations by species such as NO, NO_2, and HNO_3, for which latitudinal, altitudinal, seasonal, and diurnal variations have been reported. Simultaneous measurements of N_2O and NO_y provide excellent diagnostics for stratospheric behaviour. An essentially linear anticorrelation is expected between the two sets of measurements since the

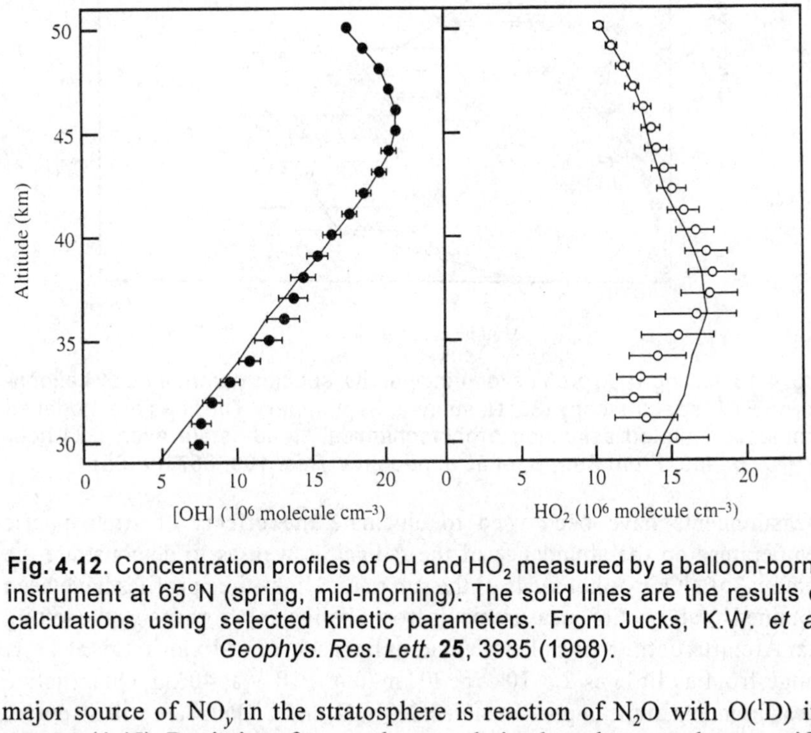

Fig. 4.12. Concentration profiles of OH and HO_2 measured by a balloon-borne instrument at 65°N (spring, mid-morning). The solid lines are the results of calculations using selected kinetic parameters. From Jucks, K.W. *et al.* *Geophys. Res. Lett.* **25**, 3935 (1998).

major source of NO_y in the stratosphere is reaction of N_2O with $O(^1D)$ in process (4.45). Deviations from such a correlation have been used to quantify permanent removal of NO_y by heterogeneous processing in the lower stratosphere.

In the altitude region where loss of O_3 is dominated by the NO_x cycle (say 25–40 km), there is good agreement between observation and theory. Figure 4.13 shows mixing-ratio profiles for NO, NO_2, HNO_3, and $ClONO_2$ obtained from a balloon-borne FTIR instrument (35°N, dawn, September), and compares them with model calculations. The only potentially serious discrepancy is that modelled concentrations of NO_2 are somewhat lower than the observed ones, a point noted in many other studies. One possible explanation is an instrumental one concerning the spectral linewidths of the NO_2 absorptions used for the FTIR measurements.

The total column abundance of the nitrate radical, NO_3, shows a maximum in spring (40°N), and decreases by a factor of three at other seasons. Polar

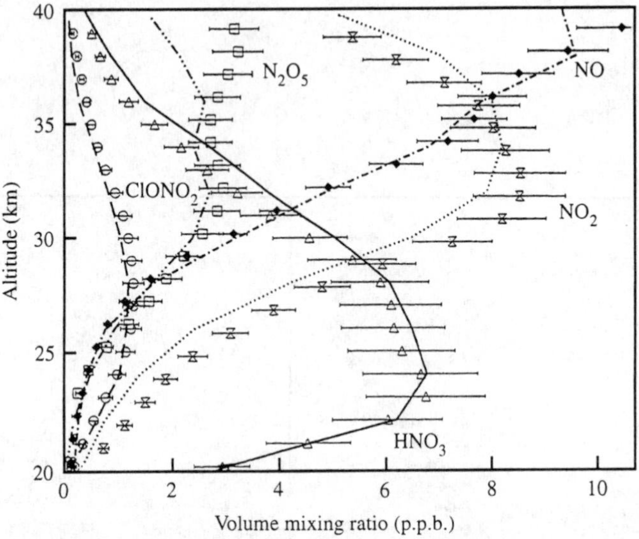

Fig. 4.13. Mixing-ratio profiles for principal NO_y species determined by balloon-borne FTIR spectroscopy (35°N, sunrise, September). The lines are modelled profiles calculated assuming a photochemical steady state over a 24-hour period. From Sen, B. *et al. J. geophys. Res.* **103**, 3571(1998).

measurements have been used to elucidate the effects of stratospheric temperature on the abundance of the radical as well as to demonstrate the removal of NO_3 by photolysis as the sun rises. Night-time profiles have been obtained by absorption measurements from a balloon platform using the setting star Arcturus or the rising planet Venus as light sources. Mixing ratios at 44°N range from as little as 2×10^{-12} at 20 km to 4×10^{-10} at 40 km. Quantitative measurements of the important molecules N_2O_5 and $ClONO_2$ are still disputed, but for the latter molecule measured altitude profiles (mixing ratios of 0.5 to just over 1×10^{-9} in the altitude range 19–30 km) seem to fit model predictions quite well.

Chlorine-containing species measured in the stratosphere include halocarbons, HCl, HOCl, and $ClONO_2$, as well as the important intermediates Cl and ClO. Halocarbon (p. 216–218) and $ClONO_2$ (above) concentrations are discussed elsewhere in this chapter. Hydrogen chloride has been determined as a function of altitude, and in a more limited way as a function of latitude and season. Figure 4.14 shows profiles obtained from instruments (ATMOS and MAS) carried on the space shuttle. Once again, models can be constrained by actual measurements of species such as total chlorine, NO_y, O_3, H_2O, CH_4 and aerosol surface area, and yet provide excellent matches with the observations if a certain amount of flexibility is allowed with regard to the kinetic parameters determined in the laboratory.

Organic molecules bear chlorine to the stratosphere, as will become abundantly evident in later sections. Once there, they are converted to the active and reservoir species, mainly by photolysis. The concentration thus decreases sharply once in the stratosphere, dropping from a mixing ratio of around 3×10^{-9} at 17 km to 1×10^{-10} and less at about 32 km. The total chlorine, inorganic and organic, is present with a mixing ratio that is very roughly constant throughout the altitude range 17–50 km; the ratio is currently just below 4×10^{-9}.

Mixing ratios of HCl at mid-latitudes range from 1×10^{-9} at 20 km to nearly 4×10^{-9} at 40–50 km: at these higher altitudes, almost all the atmospheric chlorine is thus in the form of HCl. Model calculations predict similar concentrations at most altitudes. Balloon-borne FTIR measurements of HOCl give mixing ratios from 3.5×10^{-11} (25 km) to 1.6×10^{-10} (35 km); below 31 km, the mixing ratios are lower at night. There is excellent agreement between these experimental values of [HOCl] and the predictions of models. A few measurements of atomic Cl concentrations have been made, although data obtained by direct methods (resonance fluorescence) are sparse because of the extreme difficulty of the experiments. However, the concentrations match the predictions of models within a factor of two.

The model curves shown in Figs 4.12–4.14 are all obtained using reaction schemes that include the heterogeneous chemistry discussed in Section 4.4.4.

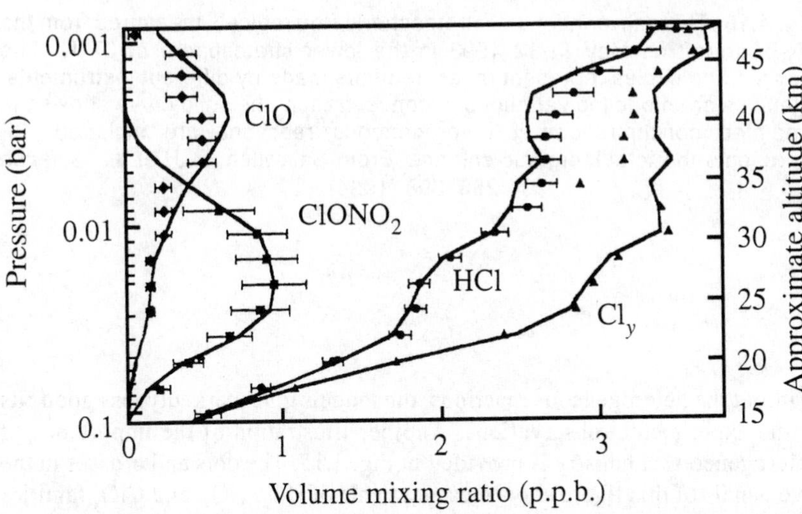

Fig. 4.14. Stratospheric inventory of some chlorine-containing species obtained by shuttle-borne instruments. The solid lines are model values constrained by measurements of other key species, and including heterogeneous reactions with measured surface areas. From Michelsen, H.A. *et al. Geophys. Res. Lett*. **23**, 2361 (1996).

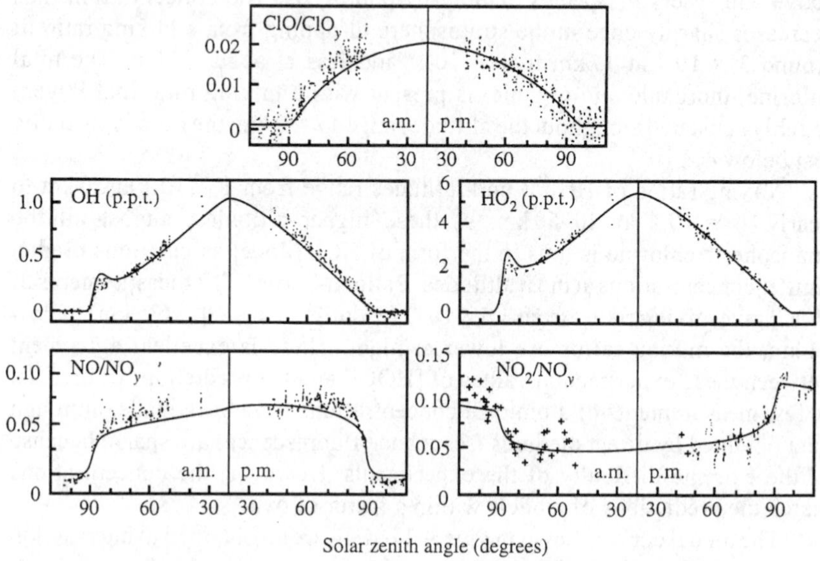

Fig. 4.15. Diurnal variations of stratospheric free radicals measured from the ER-2 aircraft on May 11/12 1993 in the lower stratosphere at 37°N. The crosses and circles represent measurements made by different instruments. The lines are modelled variations of concentration: the solid curve shows the predicted concentrations if heterogeneous reactions are included. Cl_y represents the total inorganic chlorine. From Salawitch, R.J. *et al. Science* **266**, 398 (1994).

Without the heterogeneous reactions, the models give markedly less good fits to the experimental observations. Another illustration of the importance of heterogeneous chemistry is provided in Fig. 4.15. The dots and crosses in the five panels of this figure show measurements of HO_x, NO_x, and ClO_x families made during two flights of the ER-2 aircraft. The determinations show first of all how the concentrations of these species vary during the course of a day. It is abundantly evident that the radicals that have a photochemical source (OH, HO_2, NO_2, and ClO) show a rapid increase in concentration from near zero as near zero as the Sun rises, and fall again after sunset. On the other hand, NO_2, which is itself photolysed to yield NO, decreases in concentration after sunrise. Models can predict concentration–time profiles that match the experimental observations remarkably well. The solid lines for each of the species follow the observed variations closely, even in some of the points of detail. What is notable is the failure of predictions using reaction schemes that do *not* include heterogeneous chemistry to provide anything like the same degree of quantitative agreement.

4.5 Perturbations of the stratosphere

The last section was concerned with the validation of our view of stratospheric chemistry using measurements of concentrations of the chemical species that participate in the expected reactions. An alternative, and additional, source of information can come from a study of the response of the atmosphere, and especially its ozone content, to perturbations. Man cannot ethically, of course, embark deliberately on potentially damaging experimental perturbation of the atmosphere, and, in any case, the desired perturbations may not be feasible because of the scale of the exercise. However, atmospheric nuclear tests in the past might have inadvertently perturbed the stratosphere. Since ozone data records exist for comparison with the models, the influence of such tests has been studied in depth, although the conclusions reached are conflicting. A variety of natural phenomena can also affect the stratosphere. Included in these are impulsive phenomena such as solar particle storms and volcanic eruptions, as well as more gradual variations of influences such as solar UV output or galactic cosmic ray intensity, and various oscillations (e.g. annual, semi-annual, SAO, quasi-biennial, QBO, and the El Niño–Southern Oscillation, ENSO). In terms of total ozone, the amplitude of the annual cycle ranges from about six per cent in the tropics to 30 per cent at 60 °N or S. Changes in solar activity over the solar cycle (Section 4.5.2) can bring about changes in total ozone between one and two per cent. The amplitude of the SAO is about four per cent in the northern hemisphere and two per cent in the southern, while that of the QBO ranges from four to seven or eight per cent (Section 4.5.3). In those regions of the globe where ENSO affects ozone, the variations can reach five per cent (Section 4.5.4). We shall now consider some of these perturbations and compare the expected with the observed effect on stratospheric ozone.

4.5.1 Solar proton events

Charged particles from the Sun are continually entering the Earth's atmosphere, generally at high latitudes because the moving charges are guided by the Earth's magnetic field lines (see also Section 6.1.1). Bombardment of the atmosphere by the particles is a source of odd hydrogen and nitrogen. Every few years, a burst of solar activity leads to great enhancements of the particle flux known as a *solar proton event* (SPE). During an SPE, NO_x and HO_x are produced very rapidly, and only above about 60° latitude. Any resulting changes in ozone concentration are relatively easy to identify because of this SPE 'signature', and SPEs thus act as natural 'experiments' to check elements of the stratospheric chemistry problem. Between 19 and 27 October 1989, a series of SPEs of unusually large magnitude occurred, producing large amounts of HO_x and NO_x in the stratosphere at high latitudes. A rocket-borne instrument observed enhancements of NO at altitudes between 50 and 90 km during the same period of 2.6×10^{15} molecule cm^{-2} at southern polar latitudes, while calculations based on measured solar proton fluxes provide a similar column increase, 3.0×10^{15} molecule cm^{-2}, on the assumption that every ion pair generated by the solar protons yields 1.25 N atoms. Peak production of NO_x in this event reached 2.3×10^{9} molecule cm^{-3} at an altitude of 30 km. Each ion pair is thought to produce two HO_x species as well as the NO_x. Taken together, the enhanced NO_x and HO_x are predicted by both 2-D and 3-D models to produce more than 20 per cent ozone depletions in the upper stratosphere at polar latitudes during and immediately after the SPE. Satellite measurements confirm such ozone depletions. Although the HO_x lasts for only a few hours after the SPE has abated, the NO_x persists, and ozone decreases of more than two per cent may have lasted for up to one and a half years after the SPEs of 1989.

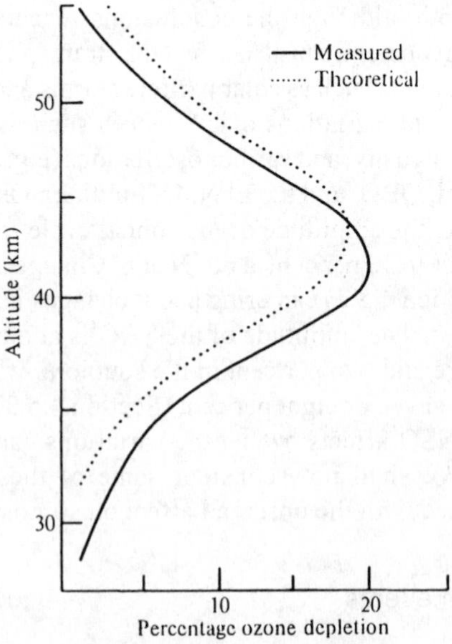

Fig. 4.16. Measured and theoretical depletion of ozone two weeks after the Solar Proton Event (SPE) of August 1972. The data are corrected for the seasonal variation. From McPeters, R.D., Jackman, C.H., and Stassinopoulos, E. G. *J. geophys. Res.* **86**, 12071 (1981).

Another comparison is provided by the SPEs of 4–9 August 1972. Although substantially less intense than the 1989 events, large depletions of ozone were recorded by the backscatter ultraviolet instrument on the Nimbus 4 satellite. At an altitude of 42 km, ozone concentrations were reduced by 15 per cent, and the depletion persisted for almost thirty days. The SPE thus demonstrated qualitatively the validity of the theory of catalytic destruction by NO_x of ozone. Quantitative comparisons of observed and predicted ozone depletions are also in good agreement. Figure 4.16 compares the ozone depletions measured two weeks after the August 1972 SPE with some model predictions. The two curves are displaced by only one kilometre in the vertical, giving considerable confidence in the modelling assumptions. Chlorine and HO_x chemistry have to be included in the reaction scheme, and the models showed the importance of including feedback mechanisms. Depletion of ozone will lead to a decrease in stratospheric temperature, since solar heating by ozone is a major contributor to the radiative balance. According to the model, cooling of 6–10 K in the upper stratosphere would follow the 1972 SPE ozone depletions. Rocketsonde temperature measurements indeed show a cooling of ~6 K after the SPE. Lower temperatures reduce the rate of odd-oxygen destruction, and there is thus a tendency for ozone depletion to be damped out. That is, there is a negative feedback mechanism which opposes changes in ozone concentrations.

4.5.2 Solar ultraviolet irradiance

The solar ultraviolet irradiance determines the overall level of stratospheric ozone, principally through the rate of odd-oxygen production in photodissociation processes, but also by affecting atmospheric temperature and dynamics. Some time variations exist in the solar UV irradiance as a result of solar rotation, changes in the Sun–Earth distance between summer and winter, and variations in solar activity (coupled to the sunspot cycle). Natural changes in ozone concentrations due to solar UV variability must be taken into account in any attempt to isolate long-term anthropogenic effects on the ozone layer. Analysis of the infrared and ultraviolet data from the Nimbus satellites shows sizeable annual, semi-annual, and quasi-biennial variations (see p. 191) in ozone abundances. These variations can be filtered from the data to seek a possible relationship between global ozone and solar activity (see Section 4.8).

Most solar UV variations occur with periods of about 11 years for the solar cycle and of 27 days for solar rotation. Over the 11-year cycle, the intensity of Lyman-α ($\lambda = 121.6$ nm) radiation varies by a factor of about two. In the mid-ultraviolet ($\lambda = 200$–300 nm), measured intensities increased by about 9 per cent from the solar minimum of 1986 to the maximum of 1990. There is a close correlation between observed ozone levels and solar ultraviolet intensity. Models indicate an ozone response that depends on latitude, with 1 per cent increases in the tropics between minimum and maximum of the cycle, and 15 per cent increases at high latitudes. Globally, ozone column amounts are expected to vary by 1.2 per cent over the cycle, and this prediction is in good agreement with the observations. Despite this excellent accord for the total ozone columns, however, prediction and measurement of vertical profiles of ozone fail to match, especially in the upper stratosphere where the largest changes occur. There are thus apparently still some inadequacies in current models in this respect, and the results from them must be interpreted with caution.

4.5.3 Quasi-biennial oscillation (QBO)

Tropical winds in the lower stratosphere switch from being easterly to westerly in a rather irregular cycle that has an average period of 27 months. This *quasi-biennial oscillation* (QBO) dominates variability in the equatorial lower stratosphere (below an altitude of about 35 km). Not only are meteorological parameters such as winds and thermal structure affected, but the distributions of ozone and other minor constituents are altered at all latitudes. A QBO modulation is observed in the concentrations of NO_2 and O_3 in the middle stratosphere. The NO_2 signal is a consequence of variations in NO_y concentrations linked to the altered meteorology, and the O_3 signal is an indirect chemical response to the modulation in NO_2, at least at altitudes above 30 km.

Yet another periodic phenomenon, the *semi-annual oscillation* (SAO), dominates the seasonal variation of winds in the tropics at higher altitudes (above about 35 km). Evidence of the influence of the SAO has been adduced from satellite measurements of stratospheric O_3.

4.5.4 El Niño

The *El Niño Southern Oscillation* (*ENSO*) is a phenomenon in which there is a significant shift in the location of the warmest water from its 'average' location in the western equatorial Pacific across to the central and eastern area. A particularly strong El Niño was observed in 1997–98, with an extensive warm anomaly east of the date-line (a classic El Niño signature). Temperatures up to 5 °C above average were observed. Such unusually warm water produces very heavy rain-showers and flooding locally (in Peru, for example) where such conditions do not normally occur. Incidentally, the western Pacific was average or cooler than average during this period. El Niño influences not only the equatorial Pacific region. Because the jet-streams in the atmosphere are strengthened during these warm phases, there are characteristic warm–cool and wet–dry anomalies outside the Pacific that are forced by El Niño. Amongst the consequences are changes in stratospheric ozone concentrations.

The ENSO phenomenon appears to be aperiodic, and its effect on total ozone does not show zonal symmetry. Rather, a wave-train propagation is followed that is known over the northern hemisphere as the *Pacific–North American teleconnection pattern.* Major ENSO warm events have been linked to observed decreases in ozone columns, while cold ENSO events have their teleconnection pattern in a global map of ozone departures that mirrors the pattern produced by the warm events. The changes in ozone column are closely associated with anomalies in the circulation of upper air known as *centres of action*, and can be traced back to a wave-train propagating into the winter hemisphere. At the centres of action of ENSO, total deficiencies of ozone of as much as four per cent can be expected. These expectations seem to matched by observations of satellite measurements for the 1997–98 event. Components from seasonal, QBO, and solar influences must be filtered out, illustrating the difficulty in pin-pointing any one influence on ozone amounts. The vertical structure of the ENSO effect on ozone has also been determined. For the 1997–98 event, the amplitude of the ENSO signal in ozone was as much as 18–19 per cent at altitudes between 13 and 20 km.

Regionally, then, the ENSO-induced variations in ozone concentrations can be large, but they are located longitudinally in the teleconnection patterns. Zonal averages of ozone changes may thus remain small during ENSO events.

4.5.5 Volcanoes

Stratospheric aerosols of natural and artificial origin can affect the chemistry and climatology of the Earth (see Bibliography). One source of the aerosols is the volcanic eruptions of significant size that occur every few years. The explosive eruptions of El Chichón (1982), and especially Mount Pinatubo (1991), are examples of recent events that have injected large quantities of material into the stratosphere (see p. 9). Appreciable effects on stratospheric ozone might be expected, but quantitative modelling is difficult because of the many potential ways in which O_3 might be influenced. Water and hydrochloric acid injected above the local tropopause add to the HO_x and Cl_x inventories. Sulphur compounds (SO_2, OCS, CS_2) are likely to increase the loading of stratospheric aerosol and thus alter, by scattering and absorption, the ultraviolet flux available for odd-oxygen production as well as changing stratospheric temperature and dynamics. More important, it is now apparent that aerosol can have a direct chemical effect on stratospheric chemistry (Sections 4.4.4 and 4.7).

The eruption of Mount Pinatubo produced what was probably the largest natural perturbation of the stratosphere during this century. It is estimated to have injected up to 20 Mtonne of SO_2 into the lower stratosphere, and the SO_2 cloud encircled the Earth in tropical regions in 22 days. Within two months, virtually all the SO_2 had been converted into liquid H_2SO_4 aerosol, and the resultant aerosol surface area reached 100 times that of the background aerosol present before the eruption. The enhanced aerosol loading led to dynamical, radiative, and chemical perturbations. A major European field measurement campaign (EASOE: European Arctic Stratospheric Ozone Experiment) had already been planned to start in the autumn of 1991, and thus was ideally timed to observe the effects of the eruption of Mount Pinatubo. In terms of chemistry, it soon became evident that the eruption rapidly led to a strong modification of the partitioning between nitrogen compounds in the stratosphere. In both the Arctic and at middle latitudes, the total NO_2 column was reduced by up to 35 per cent, and full recovery was not seen until 1995. Figure 4.17 illustrates the results of profile measurements: reductions of NO_2 concentrations by as much as 60 per cent were found at altitudes of 22 km one year after the eruption, and recovery took two to three years. The anticorrelation between NO_2 reduction and aerosol surface area is most striking. Entirely similar behaviour was observed for NO, and the stratosphere was said to have been 'denoxified' as a consequence of the eruption. Concentrations of some other compounds *increased* after the eruption, with both local concentrations and total column amounts of HNO_3 showing an enhancement. The obvious interpretation of the observations taken together is that the heterogeneous reaction

$$N_2O_5 + H_2O \rightarrow 2HNO_3 \qquad (4.28)$$

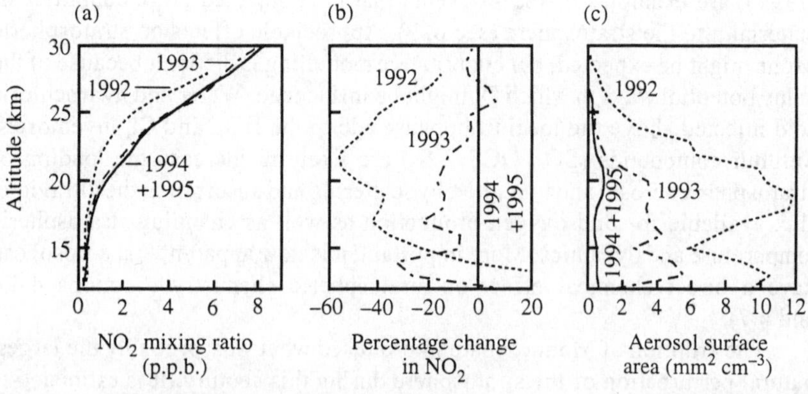

Fig. 4.17. NO_2 and volcanic aerosol: observations made over southern France in June of 1992–95. (a) Sonde profiles of NO_2; (b) percentage reductions in NO_2; (c) aerosol surface area. From Mégie, G., *et al. European Research in the Stratosphere*, EC Publications, Luxembourg, 1997.

was acting as a route for the conversion of NO_x to HNO_3. Reactions (4.26) and (4.27) convert the NO_x into N_2O_5, and that any NO_x remained in the stratosphere at all is a result of the rather slow and temperature-dependent nature of these two processes.

It will be recalled that the reservoir $ClONO_2$ is formed from ClO and NO_2 in reaction (4.34). Strong denoxification as a consequence of the volcanic aerosol loading thus also has a major impact in the partitioning between chlorine-containing species, because the latter are denied access to one of their most important reservoirs. Furthermore, there is substantial evidence that $ClONO_2$ is hydrolysed on volcanic aerosol surfaces in reaction (4.50). Observations of the amount of active chlorine (ratio of ClO to Cl_y) showed the expected enhancement after the eruption, thus confirming the influence of the heterogeneous reactions. The oxide OClO was also observed in 1992 in larger quantities than in previous years, again suggesting the activation of chlorine on sulphate aerosols.

Since reservoirs for HO_x also depend in part on the availability of NO_2 (see, for example, reactions (4.29) and (4.33)) they, too, are influenced by the reductions of NO_x following the eruption, with the HO_x radical species showing a general increase. The general picture presented by Fig. 4.6 is that of reactions involving HO_x being responsible for over half the photochemical destruction of ozone in the lower stratosphere, and halogen chemistry accounting for a further third. Thus, although NO_x catalytic cycles themselves are relatively unimportant, NO_x species are critical in regulating the other cycles. The effect of increased aerosol loading is thus to enhance the halogen and HO_x cycles at the expense of the NO_x cycles. A net increase of about 20 per cent in the ozone loss rate is anticipated for the peak aerosol loading.

Anomalous chemistry evidently followed the eruption of Mount Pinatubo, and that chemistry is predicted to lead to enhanced ozone loss. The key question is thus whether or not changes in ozone in the stratosphere were actually observed. Low ozone amounts in the tropics were detected in the first few months after the eruption, and they were roughly coincident with the region of greatest aerosol loading. On a longer time-scale, ozone reductions were observed at mid-latitudes by a variety of different instruments. Not all the ozone losses can be attributed to heterogeneous chemical processing. For example, the rapid tropical decline is probably associated both with dynamical effects as a result of increases in infrared radiative heating in the presence of aerosol, and with a reduction in the rate of UV photolysis of O_2 as a consequence of absorption of solar UV by SO_2. The longer-term ozone decreases are probably brought about by enhanced heterogeneous chemistry, accompanied by changes in the radiation field and in stratospheric dynamics. Two-dimensional model calculations of ozone concentrations reproduce well the global TOMS observations once other influences such as that of the QBO have been filtered out from the latter. Since the model needs to include heterogeneous chemistry to achieve this good match, it appears that much of the additional ozone loss seen in the years following the eruption of Mount Pinatubo was, indeed, chemical.

One of the curious anomalies highlighted by the eruption of Mount Pinatubo is that ozone losses were apparently smaller in the Southern Hemisphere than in the Northern, despite the more rapid movement of the aerosol cloud to the south. A possible explanation centres on the different altitudes to which the cloud penetrated in the two hemispheres. The peak concentrations of aerosol in the south were found at or above 22 km, where NO_x catalytic cycles are important in ozone loss (see above, and Fig. 4.6). Reductions in NO_x would thus be expected to *increase* ozone concentrations so far as these cycles are concerned, and the overall effect for the higher altitude injections might be a compensation between enhanced halogen and HO_x losses and decreased NO_x losses. In the north, however, the injections were lower, where NO_x cycles are relatively much less important. The compensatory mechanism would not apply; substantial ozone depletions would be expected and are, indeed, observed.

The strong evidence linking heterogeneous chemistry with ozone losses seen after the eruption of Mount Pinatubo suggests that smaller, but similar, depletions seen after the eruption of El Chichón should be attributed to the same cause. The loading and nature of the stratospheric injection from El Chichón were somewhat different from those of Mount Pinatubo. So far as homogeneous chemistry is concerned, the HCl injected might also be expected to provide enough chlorine to add to the background depletion of ozone in the ClO_x cycle. In the six months following the El Chichón eruptions, column HCl under the cloud of volcanic debris was measured to be enhanced by 40 per cent. Ozone depletions of as much as 10 per cent were found at altitudes between 18 and 25 km that could be attributed, through their timing and properties, to the El Chichón event. However, the period involved in 1982–83 was also a period of tropospheric climatic anomalies, making it difficult to ascribe the stratospheric changes to the volcanic injections alone.

Volcanic eruptions in the future may offer scope for investigating the response of the stratosphere to sudden disturbances that will inevitably occur, and those of the past as well as the future may complicate predictions of how the stratosphere might respond to the decrease (or increase) of man's own contribution to the chemical inventory of the stratosphere.

4.6 Man's impact on the stratosphere

Human impact on the global environment was stressed by the Study of Critical Environmental Problems (SCEP) in 1970. SCEP noted the vulnerability of the stratosphere, where the air is both thin and stable against vertical mixing. Pollutants introduced into the stratosphere by man would have a lifetime for physical removal by transport of several years, and might therefore build up to globally damaging levels. The situation is quite different from that in the troposphere or the boundary layer. Lifetimes of many pollutant species that have an impact there are small and effects are thus often localized (see p.404). SCEP considered particularly the role of supersonic stratospheric transport (SST) aircraft that could emit, into the stratosphere, H_2O, CO, and NO_x. The special feature of these aircraft is not that they are supersonic, but that for efficiency they fly in the stratosphere. Most of the immediate interest centred on the injection of water vapour into the stratosphere, and SCEP felt that the quantity of artificial NO_x was so much less than that of stratospheric O_3 that only very small reductions of O_3 could result. However, in the same year, Paul Crutzen, then working in Oxford. had drawn attention[a] to the catalytic role of NO_x in the natural stratosphere, and Harold Johnston in California pointed out[b] that artificial injection of NO_x could bring about a disproportionately large reduction in ozone that would not be calculable from a simple reaction stoicheiometry. So started the 'SST controversy'! The US set up a Climatic Impact Assessment Program (CIAP) of interdisciplinary research, and the UK and France followed with COMESA (Committee on the Meteorological Effects of Stratospheric Aircraft) and COVOS (Consequences des Vols Stratosphériques). An enormous, and extraordinarily valuable, scientific effort was put in motion, and it is hard to overestimate the advance in understanding of our atmosphere that has followed. At the same time, a curious 'political' current underlay many of the early discussions, perhaps because the US Congress had refused to fund the Boeing SST prototype programme in March 1971, while a decision had to be reached about landing rights in the US for the Anglo–French Concorde. Environmental opposition to the SSTs had focused on noise pollution in the vicinity of airports, and on sonic booms. The ozone depletion aspect of the SST debate appears to have been an 'eleventh hour' consideration. Even at the end of the CIAP programme, there was controversy concerning the timing of the release, as well as the wording, of the executive summary.

One consequence of the SST studies was the identification of a series of potential modifiers of the ozone layer, some of which are anthropogenic. The agents considered include a BrO_x cycle, as well as changes in the NO_x, HO_x, and ClO_x catalytic cycles that we have already discussed for the natural stratosphere. Other influences on ozone result from the release of infra-red active gases that can modify stratospheric temperatures, and of species such as CO that can indirectly modify (*via* reaction with OH) stratospheric composition. Certain of the agents can be injected directly into the stratosphere, while others may be of tropospheric origin, but of sufficient tropospheric stability to be transported to the stratosphere. In the subsequent parts of Section 4.6, we look at some of the individual mechanisms by which stratospheric

[a] Crutzen, P.J. The influence of nitrogen oxides on the atmospheric ozone content *Q. J. Roy. Meteorol. Soc.* **96**, 320 (1970).

[b] Johnston, H.S. Reduction of stratospheric ozone by nitrogen oxide catalysts from supersonic transport exhaust *Science* **173**, 517 (1971).

ozone can be influenced, and we also attempt to identify the effects of coupled perturbations by several agents. Section 4.8 will examine the evidence that stratospheric ozone is already showing a response to increased loads of catalytically active species. The most spectacular response of all, the development of ozone 'holes', warrants a section of its own (Section 4.7). First, however, we must see why possible changes to atmospheric ozone have aroused so much concern.

4.6.1 Consequences of ozone perturbation

A reduction in stratospheric ozone could have biological consequences because of increased intensities of UV-B (280–315 nm) that would reach the ground (see Section 4.1). Changes in ozone concentration could also have climatological effects because of altered stratospheric heating.

Environmental worries about increased UV-B radiation are centred on deleterious effects both on human beings and on all other plant or animal species. Fear of cancer has made human skin cancer the most publicized potential effect of ozone depletion. It should, however, be pointed out that some of the kinds of cancer that can definitely be attributed to sunlight (basal cell and squamous cell cancers) are not terribly dangerous, since, if caught in time, they may be successfully treated. There are much more dangerous cancers of the skin, the *melanomas*. The melanomas used to be relatively rare, and often found on parts of the body that are not exposed to sunlight. However, the incidence of melanomas seems to be increasing as humans deliberately spend more time in the sun, and a causal link with exposure to ultraviolet radiation is emerging. In particular, there appears to be an increase in the incidence of malignant melanoma in adults who suffered one or two episodes of sunburn when they were less than 11 years old.

Biological damage, other than the skin cancer response, is quite clearly produced by radiation in the UV-B region. The so-called *action spectrum* for biological response generally falls off by four or five orders of magnitude from $\lambda = 280$ nm to $\lambda = 315$ nm. This range is, of course, exactly the one over which ozone absorption also falls off, and solar irradiance at the ground thus increases. Above all, the observations demonstrate how important the ozone layer is to survival of the biota on the surface of the planet. However, the details of the behaviour depend on the particular species or response being investigated. We shall return to this point shortly.

Many investigators have used a rule of thumb that a one per cent depletion in ozone leads to a two per cent increase in UV-B. The ratio of percentage change of UV intensity incident on the surface to the change of ozone is called the *radiation amplification factor* (RAF). Two-dimensional radiative transfer models show that the assumption of an RAF of two is a great oversimplification. Figure 4.18 shows some calculated values of RAF. It is

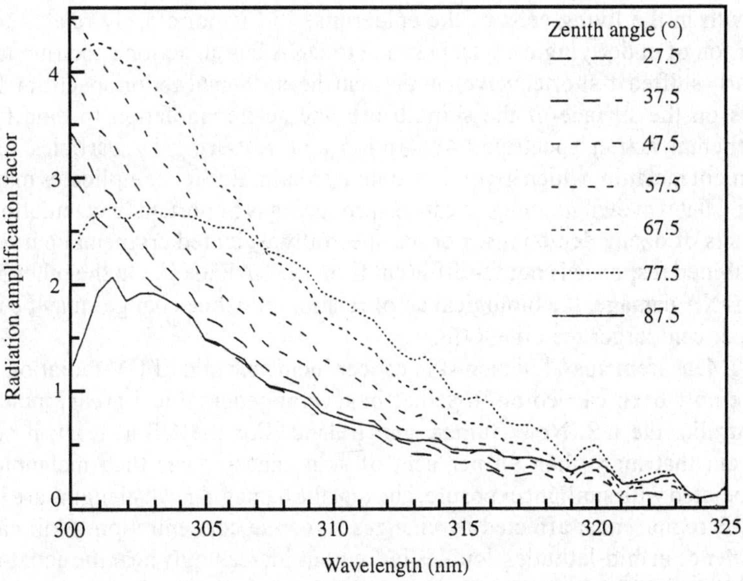

Fig 4.18. Radiation amplification factor, RAF, for a one per cent decrease in ozone shown as a function of wavelength for different solar zenith angles. From *Scientific assessment of ozone depletion: 1998*, World Meteorological Organization, Geneva, 1999.

abundantly obvious that the values are dependent not only on wavelength, but also on solar zenith angle. Note that, although the RAF generally increases with increasing zenith angle (corresponding to longer pathlengths that the sun's radiation has to traverse the atmosphere), it drops off again at the highest zenith angles.

The importance of the action spectra of the response of the living organism should now be apparent. The wavelength-dependent RAF must be convoluted with the wavelength-dependent biological action spectrum. To exemplify the differences in the action spectra, we present just a few cases. Damage to plants decreases rather suddenly by six orders of magnitude between wavelengths of 310 and 315 nm. Phytoplankton, marine organisms critical in the food chain, show a much more slowly decreasing sensitivity over the entire range, with a factor of 10^2 greater response at $\lambda = 280$ nm than at 315 nm. These organisms continue to exhibit continued damage at wavelengths as long as 360 nm. A generally adopted, but not necessarily justifiable, measure of the action of ultraviolet radiation on human skin is the spectral efficiency curve for producing *erythema*. Erythema is a reddening resulting from dilation of small blood vessels just below the epidermis. The erythemal response falls off (by roughly three orders of magnitude) between wavelengths of about 300 nm and 320 nm, but the reddening effect can still be detected at much longer wavelengths. Skin cancer, on the other hand, is a manifestation of abnormal growth in the living cells of the epidermis, and is not closely related to the dilation of underlying vessels. Damage to DNA has an action spectrum that is clearly shifted to shorter wavelengths than the erythemal action spectrum. Dead cells on the surface of the skin absorb and scatter radiation to modify the erythemal action spectrum. Absorption and scattering by particles of the pigment melanin, which gives dark colour to skin, further complicates matters. Just what wavelength range is cancer-producing is important in calculating the effects of ozone depletions. For the spectrally weighted erythemal curve, the combined response is not far different from the RAF itself. On the other hand, for DNA damage, the biological amplification of ozone changes may be up to 50 per cent larger than the RAF.

Measurements of human skin cancer incidence and of UV radiation dose have now been carried out together in several geographical areas, including Australia, the US, New Guinea, and Ireland. For the US at least, it seems certain that more than 90 per cent of skin cancer other than melanoma is associated with sunlight exposure, and that the damaging wavelengths are in the UV-B region most affected by changes in ozone concentration. Skin cancer incidence in mid-latitudes doubles for regions increasingly near the equator for each 8° to 11° of latitude, or about 1000 km. Incidence of the serious skin cancer melanoma also increases with decreasing latitude, suggesting that UV radiation is a contributing factor, but this contribution is compounded by occupational exposure and other factors. Figure 4.19 shows data on melanoma mortality rates as a function of latitude: there appears to be a statistical correlation, but not necessarily a causal one. In the US, a two per cent increase in UV-B is epidemiologically connected with a two to five per cent increase in basal skin cancer, and a four to ten per cent increase in (the more serious) squamous cell cancer. Estimates such as these are highly dependent on location, sex, skin-type, life-style, and so on. During 1990, there were in the US about 500 000 cases of basal cell cancer, 100 000 cases of squamous cell cancer, and 27 600 cases of melanoma.

The epidemiological data are usually used to show how many additional cases of skin cancer would be caused by a certain reduction of stratospheric ozone. According to the figures given in the last paragraph, a depletion of 7 per cent ozone, for example, would be expected to lead to an additional 20 000 new cases of skin cancer a year in the US. This number seems intolerably large. However, there is another way of thinking about the data. The statistics themselves come from cases of cancer found at different latitudes. A 1 per cent decrease in ozone corresponds to about a 12-mile (~ 19-km) displacement equatorward, or a 300-ft (100-m) increase in elevation. That is, from a purely statistical point of view, the chances of contracting skin cancer are increased identically by a perturbation leading to 7 per cent ozone reduction or by moving 84 miles south! Of course, one does not move populations around to compensate for controllable pollution, but the ideas might help to put the cancer

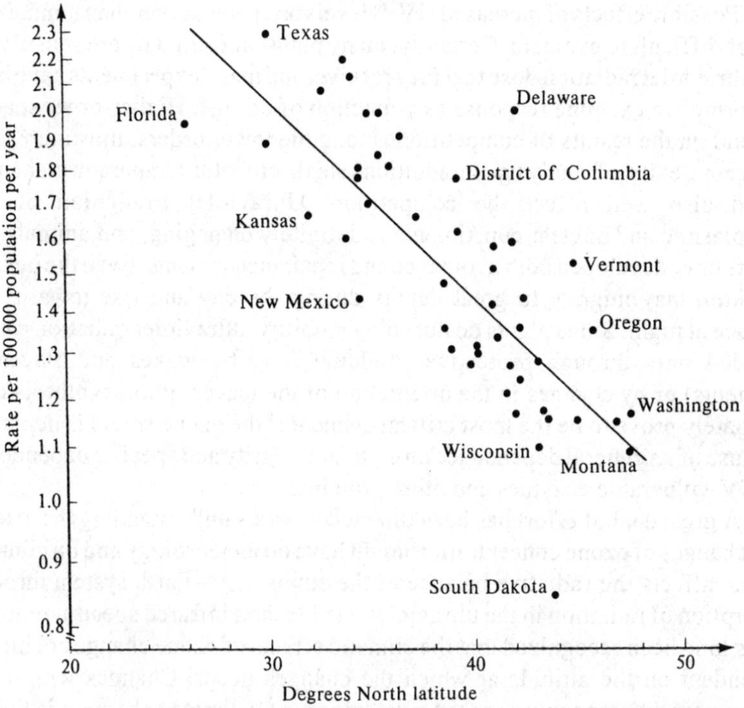

Fig. 4.19. Variation with latitude of human death due to skin melanoma among white males in the United States excluding Alaska and Hawaii. From Rowland, F.S., Chapter 4.6 in Coyle, J.D., Hill R.R., and Roberts, D.R. (eds.) *Light, chemical change and life*, Open University Press, 1982.

scare in perspective. In any case, human skin cancer would be fairly controllable if some natural disaster should deplete the ozone layer by several per cent. Only for a few hours during the middle of the day need one avoid exposure. Window glass gives sufficient protection even at midday. Actually, acute sunburn may be more of a hazard than skin cancer, and it is responsible for the loss of many man-hours of work (although, as mentioned earlier, it may also predispose to contraction of melanoma later in life).

A potentially hazardous response to increased ultraviolet exposure is that of the immune system of both animals and humans. Even a mild sunburn has been found to decrease the lymphocyte viability and function in humans for as long as 24 hours. Exposure of animals to otherwise tolerable doses of ultraviolet radiation can produce changes in allergic reaction, skin-graft rejection rates, and so on; uninhibited growth of normally rejected skin tumours also occurs. It may be this suppression of the immune system that makes possible the subsequent development of cancers. The notion that 'sunlight is good for you' and that a well-tanned skin is a sign of good health has certainly suffered a set-back!

Possible effects of increased UV-B levels on species other than humans are rather difficult to evaluate. Certainly, many plants and animals are sensitive to the ultraviolet radiation dose that they receive, and many experiments have been performed to examine response as a function of dosage. Higher-order species depend on the results of competition among the lower orders, thus increasing the complexity of analysis. In addition, small climatic temperature changes could also well affect the competition. Ultraviolet irradiation of key cytoplasmic and nuclear constituents is definitely damaging, and animals and plants have developed both avoidance and repair mechanisms. Even the humble plankton may migrate to great depths during the day and rise towards the surface at night. Since plants do not enjoy mobility, ultraviolet radiation can be avoided only through protective shielding (e.g. by waxes and flavenoid pigments) or by changes in the orientation of the leaves. Photosynthesis may ultimately prove to be the most critical element if the ozone shield is depleted, because of its general dependence on cellular integrity and specific dependence on UV-vulnerable enzymes and other proteins.

A great deal of effort has been directed towards understanding the effects that changes of ozone concentration might have on meteorology and on climate. Ozone affects the radiative balance of the atmosphere–Earth system through absorption of radiation in the ultraviolet, visible, and infrared spectral regions. It has long been recognized that the climatic effects of ozone change are highly dependent on the altitude at which the changes occur. Changes within the region near the tropopause are the most important. Climate behaviour is linked to ozone amounts in a highly complex manner. A loss of ozone in the lower stratosphere leads to an increase in visible and ultraviolet radiation reaching the troposphere, which tends to warm the climate. On the other hand, ozone also absorbs and emits infrared radiation, and is thus an active 'greenhouse' gas (see Section 2.2); loss of ozone will tend to *cool* the climate through the loss of greenhouse heating. Cooling of the lower stratosphere means that there is reduced emission of infrared radiation from stratosphere to troposphere. In simple models, at least, the net effect of the ozone loss is to cool the climate. As we shall see in subsequent sections of this chapter, there is considerable evidence that ozone concentrations in the stratosphere have been decreasing over recent decades, almost certainly as a result of man's activities. Several studies have now shown that there has been a negative trend in temperatures in the lower stratosphere of about 0.4 °C per decade over the same period.

One point of particular interest is that many of the species known to be implicated in stratospheric ozone depletion (CH_4, N_2O, the halocarbons: see the rest of Section 4.6) are also important contributors to the greenhouse effect. Global warming as a result of increased releases of such compounds is a very real concern, as explained in Sections 9.5 and 9.6. But it seems that at least a part—perhaps 20 to 25 per cent—of the increased *direct* greenhouse heating by these gases is offset by the reduced heating by ozone that these gases also bring about *indirectly*.

The feedbacks and interactions described in the preceding two paragraphs may already appear complicated enough, but there are yet other factors that must be taken into account! Increased penetration of UV radiation to the troposphere that follows a decreased stratospheric ozone shield will lead to changes in *tropospheric* chemistry. Chapter 5 of this book is concerned with the details of chemistry in the troposphere, but we have already seen in Section 1.5 that hydroxyl radicals, OH, play a key role in initiating oxidation there. In several respects, OH can be said to 'cleanse' the lower atmosphere, since it attacks a wide variety of the trace constituents present. The source of OH is a pair of reactions that we have encountered already in the present chapter

$$O_3 + h\nu \rightarrow O(^1D) + O_2 \quad (4.3a)$$

$$O(^1D) + H_2O \rightarrow OH + OH \quad (4.47)$$

that starts with the photolysis of O_3 to generate excited atomic oxygen in one branch of reaction (4.3) (see also Sections 1.4, 3.2.2, and 5.3.1). An increase in UV radiation reaching the troposphere increases the rate of photolysis in the first of these reactions, and thus the rate of production of OH in the second.

Amongst the trace species that OH attacks in the troposphere are radiatively active greenhouse gases such as CH_4. Ozone itself is also radiatively active and is, of course, removed in the photolysis step (4.3a). Recent 2-D model calculations indicate that both CH_4 and O_3 will be reduced in the troposphere as a consequence of the *observed* depletion of ozone in the stratosphere. Other factors may tend to lead to increases in each of the species, but the changes in stratospheric ozone will nevertheless mean a significant reduction in the greenhouse heating that would otherwise be experienced. Current calculations suggest that this tropospheric chemical influence may add a further 30 to 50 per cent to the reduction in greenhouse warming that the stratospheric temperature changes themselves produce.

Yet another way that stratospheric ozone changes may influence the global radiation balance through chemistry involves cloud condensation nuclei. Oxidation of dimethyl sulphide (Sections 1.5, 1.6 and 5.6) and of SO_2 leads ultimately to the formation of liquid H_2SO_4 particles, which act as very efficient nuclei. Cloud formation may thus be enhanced, and heating reduced because of the increased cloud reflectivity (albedo: Section 2.2.1). Since the oxidation of both the sulphate precursor compounds depends critically on the presence of OH, changes in stratospheric ozone levels may then influence the global cloud cover.

Quite apart from the effect that alterations in tropospheric chemistry and composition may have on climate, the chemical effects may themselves be of major significance. We have drawn attention to the key parts that O_3 and OH play in initiating almost all chemistry in the lower atmosphere. The whole question of the oxidizing capacity of the atmosphere is bound up with the concentrations of these species. Lifetimes and concentrations of natural and anthropogenic trace compounds are determined by this oxidizing capacity: in essence, they are controlled by the amount of solar radiation that the stratospheric ozone layer allows to penetrate to lower altitudes. As we pointed out earlier, there is potentially a most important feedback present, as well, because several of the trace species are precursors of the radicals that catalyse ozone destruction in the stratosphere.

So far in this section, we have examined the deleterious consequences of perturbing stratospheric ozone levels. Instinct tells us that it would be foolish not to avoid any activity that is likely to alter our ozone shield. Curiosity, on the other hand, forces us to ask whether the present-day concentrations of ozone are the 'best' that can be envisaged. Ozone levels have certainly been much smaller than they are now, and probably they have been larger, during the time that life has existed on Earth. Even today, the average intensity of UV-B reaching the surface varies by a factor of 50 in going from poles to the equator. Any effects of reduction in ozone are uncovered by presuming a relationship between response and exposure to ultraviolet radiation. Some 'bad' effects do show a correlation: could there be 'good' effects? One widely discussed 'good' effect of sunlight is the production of vitamin D in the skin. There is no doubt that vitamin-D deficiency (rickets) is frequently seen in town-dwellers at relatively high latitudes, but the disease is also associated with an inadequate diet. More convenient sources exist for the treatment and prevention of rickets than exposure to sunlight. The real argument for avoiding what amounts to deliberate change in ozone levels is that anthropogenic perturbations, such as those to be discussed in the next sections, can lead over a few years to changes comparable with the variations occurring naturally over evolutionary periods (10^5 to 10^7 years). Avoidance responses in living organisms have been learnt over the course of evolution of the species. Most important, those responses are generally triggered by particular combinations of visible light intensity and temperature. So long as UV-B and visible light bear a constant intensity relationship, the longer wavelength light is a measure of the ultraviolet. Depletion of the ozone layer changes the balance. Of all species, only humans could perhaps directly measure the UV-B intensities and adjust their life-style rapidly. Without making the measurement, enough humans at present get sunburnt by misjudging their ultraviolet exposure from the visible intensity (or temperature) on misty beaches and snowy mountains! Most flora and fauna have adapted to existence at particular latitudes, so that the 50-fold greater UV dose at the equator is of little interest to polar species: a factor of two increase might be intolerably large.

The conclusion that has to be drawn from this section is that rather little is yet known about the physical effects or the ecological consequences of stratospheric ozone changes in either downward or upward direction. It may well be that negative feedback mechanisms make the atmosphere 'self-healing' and it may be that life can cope with changed ultraviolet intensities. However, the identification of areas where damage might be done, coupled with our lack of knowledge about them, demand that we do not make unwarranted perturbations of the ozone layer. In the sections that follow, we look at some of the ways in which man is now presumed capable of modifying his ultraviolet screen. We start by considering a series of individual perturbations. These perturbations are hypothetical in the sense that they do not represent any change to the real atmosphere, in which several parameters vary simultaneously. They do, however, provide a basis for realistic 'scenarios' of simultaneous multiple perturbations.

4.6.2 Aircraft

This section began with an explanation about how concerns about the environmental impact of SST aircraft stimulated intense research on stratospheric processes that continues to this day. High-temperature combustion in air leads to formation of NO by 'fixing' of atmospheric N_2 and O_2, as well as the ordinary combustion products CO_2, CO, and H_2O. Injection of these species, especially the catalytically active NO, into the stratosphere could destroy ozone. Because of the vertical stability of the stratosphere, physical lifetimes against removal of injected species are considerable, ranging from 1–2 years at 17 km to 2–4 years at 25 km altitude. Stratospheric aircraft envisaged in 1970 included the Concorde type flying at 17 km, the US SST flying at ~ 20 km, and a conceptual hypersonic transport with a flight level of about 25 km.

For the 100-passenger Concorde, about 19 100 kg h^{-1} of fuel is consumed during cruise, and as much as 23 g of NO_x may be emitted from its Olympus engines for every kg of fuel burned. Thus up to 440 kg h^{-1} of NO_x are emitted, or about 1.1×10^5 molecule cm^{-2} s^{-1} averaged over the globe. Early estimates of growth of aircraft anticipated several hundred Concorde and US SSTs in use by 1990. Early models, based on 500 US SSTs emitting a total of 1.2×10^9 kg of NO per year at 20 km altitude predicted a reduction in the global ozone column of roughly 12 per cent, with a worst-case reduction of 25 per cent near the flight corridor. Subsequent models, based on improved chemistry, predicted much smaller depletions of ozone, as will appear later in this section. Equally important, the projected numbers of aircraft now seem like a fantasy. The proposed fleet was perhaps an unreasonable projection, since it would have been capable of transporting the entire population of London to New York in a week or so! In reality, only 16 production Concordes were built, of which 13 remained in commercial operation in 1998.

Despite the removal of the immediate concern about the impact of Concorde-like aircraft, more recently there has been renewed and intensified examination of the impact that aircraft operations may have on the chemistry of the atmosphere and on climate. In part, this research has been stimulated by discussion of a proposed second-generation high-speed civil transport (HSCT) aircraft. A fleet size of at least 500 is likely to be required for economic viability, a number that once again raises questions about the demand for air

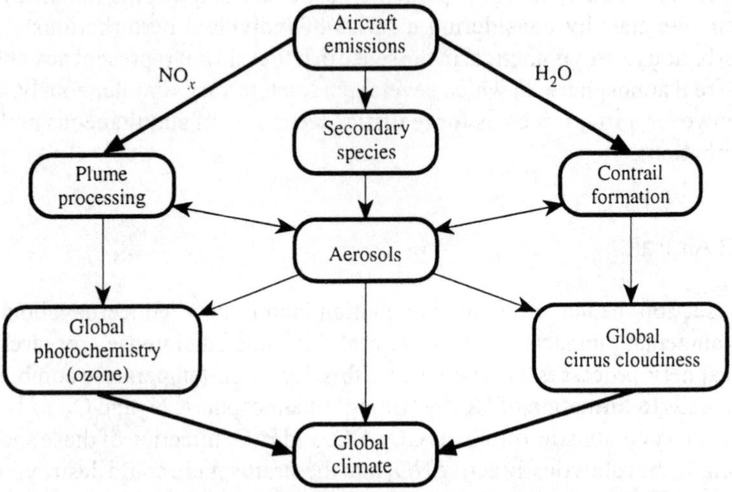

Fig. 4.20. Some potential atmospheric impacts of aircraft emissions. From *European assessment of the atmospheric effects of aircraft emissions* Brasseur, G., Amanatidis, G.T., and Angeletti, G. (eds.) *Atmos. Environ.* **32**, 2327, 1998.

transport on this scale. However, it has also become apparent that the cruising altitude of *current* commercial (subsonic) aircraft is, for 30 to 40 per cent of the time, above the tropopause. Emissions of combustion products into the stratosphere should therefore already be considered. The contributions of military aircraft, often flying even higher, may make an additional contribution, although sensitivity about security makes it difficult to obtain quantitative information.

Table 4.4 Typical emission indices for some products of combustion in aircraft engines

Species		Emission index (g kg^{-1} fuel)		
		Idle	Take-off	Cruise
NO_x (as NO_2)	short haul	3–6	20–65	7.9–11.9
	long haul	3–6	10–53	11.1–15.4
CO		10–65	< 1	1–3.5
Hydrocarbons		0–12	< 0.5	0.2–1.3
H_2O		1230	1230	1230
CO_2		3160	3160	3160
SO_2		1.0	1.0	1.0

From *European assessment of the atmospheric effects of aircraft emissions* Brasseur, G., Amanatidis, G.T., and Angeletti, G. (eds.) *Atmos. Environ.* **32**, 2327 (1998).

Table 4.5 Typical fuel consumption in kg on short- and long-haul flights

Phase of flight	Short-haul (555 km) London–Glasgow Boeing 757	Long-haul (8750 km) Los Angeles–Tokyo DC 10-30
Taxi–take off–climbout	1000	1680
Climb–cruise–descent	2400	82060
Approach–landing–taxi	430	660
Total flight	3830	84400
LTO/total flight (%)[a]	37	3

[a] LTO is the 'landing and take off' cycle.

From *European assessment of the atmospheric effects of aircraft emissions* Brasseur, G., Amanatidis, G.T., and Angeletti, G. (eds.) *Atmos. Environ.* **32**, 2327 (1998).

Alongside the revival of interest in the stratospheric effects of aircraft, there has been increased concern about what effect the release of aircraft effluents might have on the troposphere. Chapter 5 of this book has the troposphere as its subject, but in the context of the impact of aircraft on the atmosphere, it seems rather fruitless and artificial to divide the discussion into that on stratospheric and tropospheric impacts. The largest amounts of aircraft exhaust gases are released at altitudes between 10 and 12 km, a region that might encompass the upper troposphere and lower stratosphere (UTLS). As it happens, the UTLS is rather poorly understood in several ways, so that assessment of the effects of aircraft is made all the more difficult. Several research initiatives have recently sought to address the problems: for example, the AERONOX project in Europe and the US Subsonic Assessment (SASS) have specifically addressed the aircraft issues, and have spawned further research programmes. One thing seems certain, regardless of the future of SSTs, and that is that the fleet of conventional aircraft will grow in the foreseeable future. Current projections show growths ranging from 255 per cent (Boeing) to 270 per cent (Airbus) over the 20-year period 1996–2016; by 2016, it is expected that the enormous number of 7×10^{12} km per year will be flown by revenue-paying passengers.

Figure 4.20 indicates some of the major impacts that aircraft emissions might have on the atmosphere. As we have come to expect in atmospheric chemistry, there are interactions and feedbacks between the various components of this diagram; aerosols and other particles are seen to be central to several of the phenomena. Certain chemical species emitted by aircraft engines are listed in Table 4.4. The *emission index* for each substance is listed for different flight conditions, and for typical modern engines. This index is the number of grams of a material emitted for each kilogram of fuel consumed. The figure of 23 g NO_x per kg of fuel burnt, quoted for the Concorde engines earlier, is just such an emission index. Table 4.5 provides the additional numbers required to

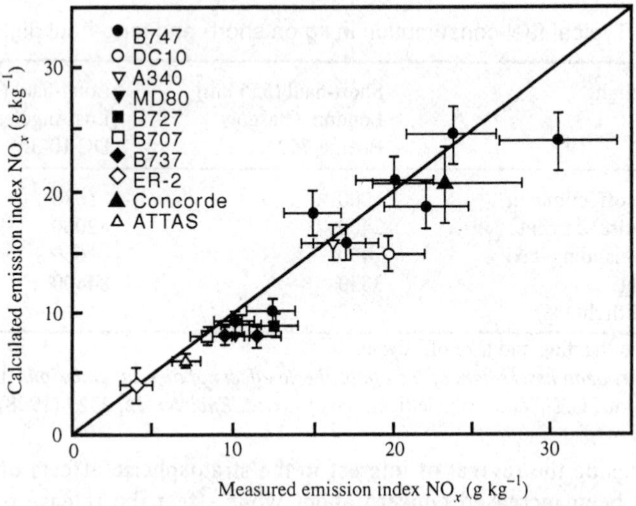

Fig. 4.21. Comparison of cruise-level NO_x emission indices calculated and measured using a chase aircraft. From *European assessment of the atmospheric effects of aircraft emissions*, Brasseur, G., Amanatidis, G.T., and Angeletti, G. (eds.) *Atmos. Environ.* **32**, 2327 (1998)

convert the emission indices into actual emissions in kg of the different species for certain types of flight. Notable for their absence from Table 4.4 are data about the emissions of solid particles and liquid aerosols such as soot, directly formed sulphate particles, and oil. Although, as we shall see soon, such particles may be of great importance, there are considerable uncertainties about the emissions: one estimate puts a typical value of 0.12 on the emission index.

The significance of the entries for the 'LTO' (landing and take off) cycle in Table 4.5 is that it is only for these phases of flight that regulatory standards exist at present, and there are no regulations covering flight phases such as cruise. Furthermore, the vast majority of emission data derive from test-bed experiments conducted at ground level, and are thus not directly relevant to the much longer high-altitude cruise conditions. Engine manufacturers have developed methods for the calculation of in-flight emissions from the ground-level certification data, and recent experiments confirm the essential validity of these calculations. A very few, extremely expensive, test facilities exist for altitude simulations. In-flight measurements have been made by remote optical examination of exhaust gases, and by sampling from the engine exhausts by a chase aircraft. Figure 4.21 shows the results of several chase-aircraft experiments for NO_x emission indices. There seems to be good agreement between the calculations and the direct experimental determinations. Several other recent investigations have confirmed the essentially reliable predictions provided by modern sophisticated computer simulations of engine behaviour.

Major improvements have been made in jet-engine technology since the 1960s. A gain in fuel conversion efficiency of roughly a factor of two has been achieved as various types of turbofan have replaced turbojet types. These fan engines by-pass part of the gas flow around the main combustion chamber; the GE90 engine of 1995 has a by-pass ratio of more than 9. The outcome of the improved efficiency is, of course, reflected in a reduction of emission of CO_2 (and H_2O vapour). However, there is a trade-off with NO_x emissions, because the gains in efficiency have generally depended on higher temperatures and pressures in the advanced engine cycles, which results in higher conversions of N_2 and O_2 to NO_x. As a result, there has more recently been an effort to achieve simultaneous reductions in NO_x emissions by using improvements in combustor technology. Target values for the emission index for NO_x of 5 are quoted, an index that should be compared with 23 for the Concorde engines, or 7.9–15.4 (Table 4.4) for present-day engines. The main conclusion relevant to our own examination of atmospheric chemistry is that, although commercial airline operations may double in terms of passenger-kilometres by 2016, the emissions may well not increase by anything like the same amount.

The importance of the NO_x that is released by aircraft has provided the focus so far of our discussion of impacts. The oxides of nitrogen have the potential of modifying the abundance and distribution of ozone in the UTLS and middle stratosphere, with the consequences, including perturbation of climate, that were explained in Section 4.6.1. NO_x is one of the most important catalysts in ozone removal in the middle stratosphere (Section 4.4). In the UTLS, on the other hand, NO_x increases may lead to an *increase* in ozone concentrations. In the first instance, NO_x in the troposphere acts as a photochemical source of atomic oxygen, and thence of ozone, *via* the normal addition process (reaction (4.2)); this aspect is examined in much more detail in Section 5.3.3. and associated parts of Chapter 5. Even in the lower stratosphere, some O_3 generation may occur, depending on the concentrations of other radical species. A further reason for enhancement of ozone concentrations in the lower stratosphere is that ozone losses are dominated there by chlorine-radical catalysis (Section 4.4). NO_x is essential to the formation of one of the most important reservoir species, $ClONO_2$ (Section 4.4.2). Increased NO_x thus means reduced efficiency of chlorine losses. Models show that heterogeneous chemistry occurring on aerosol particles (Section 4.4.4) may also considerably diminish the impact of NO_x injected by aircraft.

Uncertainties thus abound in establishing the likely effects of injections of NO_x. The altitude of injection is evidently an important factor. Within the troposphere, the relative strengths of NO_x from aircraft engines and from lightning remains poorly quantified. In spite of its importance to the ozone budget, the distribution of NO_x from sources other than aircraft is not well understood. The future effects of stratospheric injections will also depend on future stratospheric burdens of chlorine-containing species (Sections 4.6.4–4.6.7), for the reasons explained in the preceding paragraph.

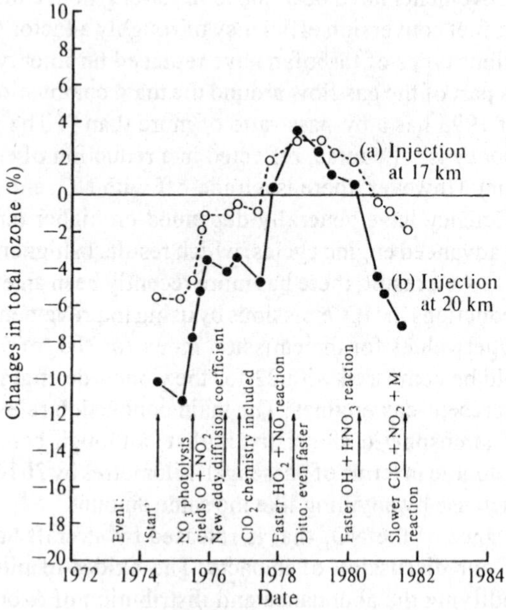

Fig. 4.22. Time history of calculations of the expected change in total ozone concentration from NO_x injection at (a) 17 km; (b) 20 km. All calculations are from the Lawrence Livermore National Laboratory one-dimensional model. Major changes in model parameters are labelled as 'events'. Unlabelled alterations generally result from minor changes in the chemistry or kinetic parameters. From Hudson, R. (ed.-in-chief) *The stratosphere 1981*, World Meteorological Organization, Geneva, 1981.

An entertaining illustration of how the estimates of stratospheric ozone depletion evolve as new information comes to light is provided by Fig. 4.22. It is a rather unusual diagram that shows the depletion of ozone predicted by a single model (the Lawrence Livermore National Laboratory 1-D model) for a fixed NO_x injection (2×10^8 molecule $cm^{-2} s^{-1}$). The unusual feature is that the depletions are plotted as a function of year of research from 1974 to 1982. Two curves are shown for NO_x injection, at the 'standard' altitudes of 17 km and 20 km. As will be seen, by mid-1978, NO_x apparently had the effect not of destroying ozone but of making it! The reason, as indicated on the diagram, was mainly the discovery of a high rate coefficient for reaction (4.23), between HO_2 and NO, a process discussed many times before in this chapter. It would be a brave reader indeed who would have dared to extrapolate this graph to the present day! Heterogeneous chemistry was not yet regarded as significant, and chlorine burdens in the atmosphere were quite substantially lower in 1982 than they are today.

So what do present-day observations and calculations tell us about the effects of NO_x from aircraft on the atmosphere? So far, there is no unambiguous observational evidence for changes in the ozone abundance that can be attributed to aircraft operations. State-of-the-art atmospheric models suggest that the impact of aviation on stratospheric ozone will be small. When allowance is made for the effects of sulphate aerosols present in the stratosphere, the models indicate that the response of O_3 to NO_x injected by aircraft will be negative (O_3 will decrease) in the middle stratosphere, but be positive below. However, the perturbations in local ozone concentrations would probably not exceed a few per cent even for the projected fleet of 500 HSCTs (see above). Operations in the troposphere may be more important. The models suggest that NO_x abundances in the upper troposphere may have increased by 20 to 50 per cent as a result of current activity. Corresponding increases in the ozone concentration of 4 to 8 per cent might be expected during summer, when photochemical activity is highest.

As indicated by Fig. 4.20, increases in NO_x represent only one aspect of the possible effects of aircraft emissions, although the consequences are of particular relevance to our enquiry about stratospheric ozone because of the importance of this species in ozone chemistry. It is appropriate, therefore, to examine briefly the possible effects of the other engine effluents.

By far the largest output is of CO_2; because of the connection between this gas and greenhouse heating, aircraft emissions could conceivably be of significance. However, it turns out that the current worldwide aircraft fleet produces only two to three per cent of the total CO_2 released to the atmosphere as a result of the burning of fossil fuel, so that the aircraft contribution is rather small. Even future projected increases in air traffic will leave the relative amount small, especially if the improvements in combustion efficiency are taken into account. We return to this topic in Chapter 9 (Section 9.6.6).

After CO_2, the next most abundant exhaust gas is H_2O vapour. This vapour condenses at high altitudes as contrails, which cover some 0.4 per cent of the sky over Central Europe, and as much as one per cent over the north Atlantic, and begin to represent a small but noticeable addition to the 20 per cent cover provided by cirrus clouds. The radiative effect of the contrails might be significant, with thin contrails expected to exert a net warming effect on the Earth's climate, and thick contrails a cooling influence. Perhaps even more important, the release of water by high-flying aircraft might increase the frequency with which polar stratospheric clouds (Sections 4.7.2 and 4.7.4) occur. These clouds are an important component of the special and perturbed chemistry (Sections 4.7.4 and 4.7.5) that gives rise to the formation of the Antarctic (and probably Arctic) ozone holes. Sulphur dioxide in engine exhausts can be converted to sulphuric acid aerosol and, together with the sulphates, soot, and other solid and liquid particles, can have an effect in altering the radiative balance of the atmosphere as well as playing a role in cloud formation. In addition, the particles may enhance substantially the extent of heterogeneous chemical processing (Section 4.4.4), whose place in the chemistry of the atmosphere has become well established.

The present section has demonstrated that much research in the past has been aimed at quantifying the atmospheric perturbations associated with commercial subsonic and supersonic aircraft operations. At the same time, it will have become evident that there remain a large number of uncertainties that remain to be addressed in the future. One major concern is the potential for redistribution of ozone in the upper troposphere and lower stratosphere, and the consequential climatic changes. Another is related to climate, this time the impact of increasing abundances and changing size distributions of particles released by aircraft that may alter cloudiness and influence the radiative balance. The problems are probably well established now, and research in the future should at least indicate the effects of what is an almost inevitable increase in aircraft traffic; with good fortune, it will also show how to alleviate the most hostile influences.

4.6.3 Rockets and the space shuttle

Space-shuttle and other rocket launch and re-entry operations generate NO_x, although the amounts are probably too small to produce a noticeable change in stratospheric ozone. Ammonium perchlorate solid-fuel boosters on the shuttle launch vehicles release HCl vapour to the stratosphere as an exhaust constituent, and could therefore influence the ClO_x catalytic cycle for the destruction of ozone. Nearly 70 tonnes of inorganic chlorine, calculated as Cl, is deposited at altitudes between 15 and 50 km for each shuttle launch, and about 38 tonnes for each launch of a Titan IV rocket. Particles are also emitted or produced, and there is thus the potential for enhancements in heterogeneous chemical processing (Section 4.4.4).

An actual response of the stratosphere to rocket launches was first reported in 1997. Comprehensive measurements followed the evolution of stratospheric ozone in the wake of two separate Titan IV rocket launches. In each case, ozone concentrations dropped nearly to zero in the plume wake, and across regions 4 to 8 km wide. Ozone loss developed within 30 minutes, and persisted for a further 30 minutes, after which the concentrations returned to normal ambient amounts. The data indicate that more than one ozone molecule is lost for each chlorine molecule deposited by the rockets, and thus suggest the occurrence of a catalytic cycle of ozone destruction that is based on chlorine-containing compounds, but that might be unique to solid rocket motor plumes. Although the local ozone losses were very large, the limited spatial and temporal extent of these experiments implies a negligible *global* impact of solid-fuel rockets on stratospheric chemistry.

Models incorporating reasonable scenarios for launch patterns also predict rather small global changes of total ozone. For example, in one calculation a launch scenario of nine shuttles and three Titans a year (close to historical launch rates for the 1990s) gave rise to a maximum enhancement of inorganic chlorine of 12 p.p.t. (or roughly 0.4 per cent on a 3 p.p.b. background level). In turn, this enhanced chlorine loading was found to produce a maximum local decrease in ozone of 0.14 per cent (middle to upper stratosphere; northern middle to high latitudes), and a maximum column decrease of 0.05 per cent (northern high latitudes; spring). Simulations including heterogeneous reactions on stratospheric sulphate aerosol (SSA) and on polar stratospheric clouds (PSC) showed an annually averaged global decrease of total ozone of only 0.014 per cent. An obvious expectation was that incorporation of the heterogeneous reactions would lead to larger predicted ozone depletions because of the effects on odd-chlorine and odd-nitrogen constituents, as described in Section 4.4.4. These effects are, indeed, apparent in the more recent calculations: ozone depletions of 0.0056 per cent are predicted for gas-phase chemistry alone, 0.010 per cent if reactions on SSA are taken into account, and the figure of 0.014 per cent already given when reactions on SSA and PSCs are both included.

Solid rocket motors inject aluminium oxide particles directly into the stratosphere, as well as HCl and the other gaseous constituents. Launches of the space shuttle and of a Titan IV rocket place 112 tonnes and 69 tonnes of Al_2O_3 in the stratosphere, or almost twice as much by mass as the HCl emitted. The known importance of heterogeneous processes in other aspects of stratospheric chemistry clearly suggests that the effects of the Al_2O_3 are worthy of consideration. Aluminium oxide is known to promote the chlorine activation reaction

$$ClONO_2 + HCl \rightarrow Cl_2 + HNO_3, \qquad (4.51)$$

which assumes such a central role when it occurs on PSCs in the perturbed chemistry of polar regions during winter (Sections 4.4.4 and 4.7.5). The reaction probability is about 0.02 on the particle surfaces; this probability is higher, and the activation reaction thus faster, than on SSA at most stratospheric temperatures. The Al_2O_3 particles are present at all latitudes in all seasons, unlike the PSCs. At one time, it was thought that the particles would become coated by H_2SO_4, and thus make only a very minor addition to the background sulphate particle burden. However, it now seems as though the Al_2O_3 particles will remain uncoated throughout most of their stratospheric residence time. Models show that the Al_2O_3 is then expected to make a small but significant contribution to the overall effects of rocket activities on stratospheric ozone. For example, taking the historical launch rates for 1997, the calculated decrease in annually averaged global total ozone is about 0.008 per cent as a result of the Al_2O_3 injection, and 0.017 per cent from the HCl; the depletions are roughly additive.

One interesting result from the model simulations is that the fractional contribution from Al_2O_3 is calculated to have increased from about one-quarter to about one-third over the period 1975 to 1997. The explanation for the increase is that the background amounts of inorganic chlorine in the upper stratosphere have increased from about 1.5 to 3.5 p.p.b. over the same period. This additional chlorine almost all derives from man-made industrial halocarbons: in 1990, for example, halocarbons contributed about 300 ktonne of chlorine to the stratosphere, while the rocket motors added only 0.725 ktonne. It is thus to the halocarbons and their impact on the stratosphere that we should turn next.

4.6.4 Halocarbons: basic chemistry

Methyl chloride is the dominant naturally occurring chlorine compound to be found in the atmosphere. Global average mixing ratios are at present about 0.55 p.p.b. (0.55×10^9), but, of that amount, only roughly 10 per cent comes from natural oceanic sources, while 80 per cent originates in biomass burning and another 10 per cent or more are a result of industrial emissions. Thus, perhaps 0.06 p.p.b. does not owe its presence to man. Meanwhile, total chlorine mixing ratios in the atmosphere are hovering around 3 p.p.b., largely contributed by man-made fluorinated chlorocarbons and some other chlorinated solvents (CH_3CCl_3 and CCl_4, for example).

The fluorinated hydrocarbons were developed in 1930 by the General Motors Research Laboratories in a search for a non-toxic, non-flammable refrigerant to replace the sulphur dioxide and ammonia that were then in use. Dichlorodifluoromethane, CF_2Cl_2, is a typical member of the class of compounds. 'Freon' (Du Pont, US) and 'Arcton' (ICI, UK) are trade names for the CFCs. Chemical inertness has made the CFCs valuable as aerosol propellants, as blowing agents for plastic foam production, and as solvents, in addition to their use as refrigerants. CFC production was until quite recently a world-wide industry with an estimated 3.6×10^8 kg of CFC-11 and 4.5×10^8 kg of CFC-12 being manufactured in 1988.[a] From that time on, production decreased, and the industry estimates of reporting companies were 1.2×10^8 kg of CFC-11 and 2.0×10^8 kg of CFC-12 in 1995. In1995–96, regulatory action (see Section 4.6.6) required that production should cease in the developed countries. The uses of the CFCs all lead ultimately to atmospheric release, since even 'hermetically sealed' refrigerators and closed-cell foams finally leak to the air. 90 per cent or more of all the CFC-11 and CFC-12 produced up to 1996 is believed to have been released.

[a]In the usual nomenclature, CFC (chlorofluorocarbon) is followed by a coded two- or three-digit number. The hundreds digit is the number of carbon atoms in the molecules less one, the tens is the number of hydrogen atoms plus one, the units is the number of fluorine atoms, and the residue is assumed to be chlorine. If the first digit is zero it is dropped. Thus CFC-11 is $CFCl_3$, CFC-12 is CF_2Cl_2, and CFC-115 is CF_3CF_2Cl.

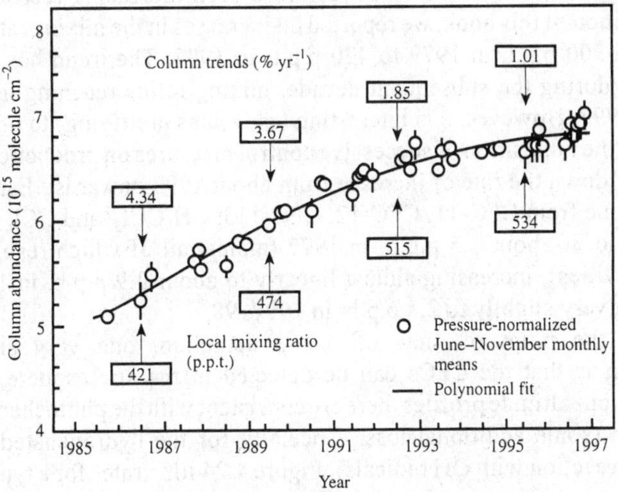

Fig. 4.23. The column amount of CF_2Cl_2 (measured over Jungfraujoch) during the period 1985 to 1996; column trends (% yr^{-1}) and local mixing ratios (p.p.t.: parts per 10^{12}) are shown for four of the years. From *Scientific assessment of ozone depletion: 1998*, World Meteorological Organization, Geneva, 1999.

In 1973, Jim Lovelock[a] and his collaborators reported the presence of halogenated hydrocarbons in the troposphere. It soon became apparent that the quantities of the CFCs were equal, within experimental error, to the total amount ever manufactured. Tropospheric inertness of the CFCs was thus confirmed, and lifetimes of up to hundreds of years were indicated. Only one escape route is possible for the compounds: transport to the stratosphere followed by ultraviolet photolysis

$$CF_2Cl_2 + h\nu \rightarrow CF_2Cl + Cl \qquad (4.58)$$

is the photolysis process for CFC-12. Chlorine generated by the space shuttle was in the minds of atmospheric scientists (see Section 4.6.3). Mario Molina and Sherry Rowland were quick to see[b] that the chlorine atoms from reaction (4.58) were a much more serious threat to the ozone layer, since the CFC source emissions were known to exist, and yielded far more Cl than the shuttle. Tropospheric measurements have since confirmed the presence of all the man-made CFCs, as well as chlorinated compounds such as carbon tetrachloride (CCl_4) and 1,1,1-trichloroethane (methyl chloroform, CH_3CCl_3) which are almost certainly solely of anthropogenic origin. Figure 4.23 shows how

[a]Lovelock, J.E., Maggs, R.J., and Wade, R.J. Halogenated hydrocarbons in and over the Atlantic *Nature, Lond.* **241**, 194 (1973).

[b]Molina, M.J. and Rowland, F.S. Stratospheric sink for chlorofluoromethanes: chlorine atom-catalysed destruction of ozone *Nature, Lond.* **249**, 810 (1974).

tropospheric concentrations of CF_2Cl_2 have been increasing recently. In the second edition of this book, we reported on increases in the mixing ratio for this CFC from 300 p.p.t. in 1979 to 420 p.p.t. in 1987. The trend has evidently continued during the subsequent decade, mixing ratios reaching nearly 540 p.p.t. by 1997. However, it is interesting (as well as gratifying) to observe the effect that the imposition of successive control measures on production has had in slowing down the rate of increase from about 1993 onwards. Estimates of total chlorine from CFC-11, CFC-12, CFC-113, CH_3CCl_3, and CCl_4 suggest a mixing ratio of about 1.5 p.p.b. in 1977 (almost all of which resulted from man's activities), increasing almost linearly to about 2.9 p.p.b. in 1993, and decreasing very slightly to 2.8 p.p.b. in 1997–98.

From the point of view of ozone depletion, one very significant observation is that the CFCs can be detected in the stratosphere, and the concentration–altitude profiles there are consistent with the photochemical loss mechanism (some additional loss, especially for the hydrogenated species, occurs *via* reaction with OH radicals). Figure 4.24 illustrates for a typical CFC how the concentration remains virtually constant throughout the troposphere, where the compound is inert and well mixed, but drops suddenly beyond the tropopause as photolysis destroys it and releases active chlorine. Measurements of the vertical distribution of CH_3Cl (the most important natural chlorine-bearing species) show that chlorine of anthropogenic origin now predominates in the stratosphere. The threat to the ozone layer is thus real and present, and

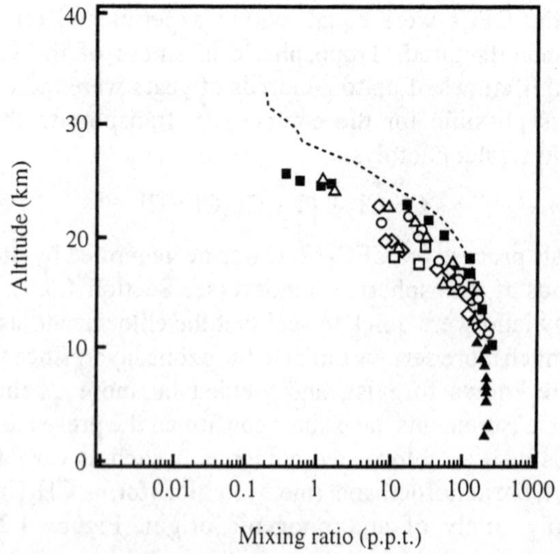

Fig. 4.24. Vertical distribution of $CFCl_3$ in the Northern Hemisphere. The different symbols are used to identify particular experiments performed on different days over the period 1987–90. From *Scientific assessment of ozone depletion: 1994*, World Meteorological Organization, Geneva, 1995.

some more detailed examination both of the catalytic chemistry and of the release of chlorinated hydrocarbons is in order.

Cycle 4

$$Cl + O_3 \rightarrow ClO + O_2 \quad (4.19)$$

$$ClO + O \rightarrow Cl + O_2 \quad (4.20)$$

Net $$O + O_3 \rightarrow O_2 + O_2$$

is the major loss cycle for odd oxygen if ClO_x alone is present as catalyst. At least three other cycles appear to be significant if HO_x and NO_x are also present

Cycle 13

$$Cl + O_3 \rightarrow ClO + O_2 \quad (4.19)$$

$$OH + O_3 \rightarrow HO_2 + O_2 \quad (4.15)$$

$$ClO + HO_2 \rightarrow HOCl + O_2 \quad (4.32)$$

$$HOCl + h\nu \rightarrow OH + Cl \quad (4.59)$$

Net $$O_3 + O_3 \rightarrow O_2 + O_2 + O_2$$

Cycle 14

$$Cl + O_3 \rightarrow ClO + O_2 \quad (4.19)$$

$$ClO + NO \rightarrow Cl + NO_2 \quad (4.24)$$

$$NO_2 + O \rightarrow NO + O_2 \quad (4.18)$$

Net $$O + O_3 \rightarrow O_2 + O_2$$

Cycle 15

$$Cl + O_3 \rightarrow ClO + O_2 \quad (4.19)$$

$$NO + O_3 \rightarrow NO_2 + O_2 \quad (4.17)$$

$$ClO + NO_2 + M \rightarrow ClONO_2 + M \quad (4.34)$$

$$ClONO_2 + h\nu \rightarrow Cl + NO_3 \quad (4.60)$$

$$NO_3 + h\nu \rightarrow NO + O_2 \quad (4.38)$$

Net $$O_3 + O_3 \rightarrow O_2 + O_2 + O_2$$

Each of these cycles involves species from the different HO_x, NO_x, and ClO_x families, so that it is difficult to attribute the odd-oxygen loss to a particular family. Reaction (4.18), for example, is the rate-limiting step in the simple NO_x

catalytic process, cycle 3 (p. 166), as well as in cycle 14. In the contemporary atmosphere, with a Cl mixing ratio of 3×10^{-9}, reaction (4.24) is responsible for 20 per cent of NO oxidation at most; but in a perturbed atmosphere containing four times as much total chlorine, the reaction would become as important as the $NO + O_3$ path, reaction (4.17). Odd-oxygen loss rates due to the four cycles depend on altitude. Above 20 km, cycle 4 is always the major contributor, although the other cycles are operative. At lower altitudes, cycles 13–15 become dominant.

Bromine-containing compounds can also interfere with stratospheric ozone. Source gases are mainly CH_3Br and brominated CFCs (the 'halons'). Methyl bromide is of both natural and anthropogenic origin; we shall return shortly to a fuller discussion of its sources. The halons are of entirely synthetic origin and are used for several purposes, especially as fire extinguishing agents and fire retardants. The compounds halon-1202 (CF_2Br_2), halon-1211 (CF_2BrCl), halon-1301 (CF_3Br), and halon-2402 (CF_2BrCF_2Br) have all been detected in the atmosphere. Of these, the most abundant are halon-1211 and halon-1301. Emissions peaked in the 1980s at roughly 12.5×10^6 kg and 4.5×10^6 kg for halon-1211 and halon-1301, respectively. Regulation was imposed on halon production (see later), but even by 1996 estimated release rates had decreased by less than a factor of two, and the amounts remaining 'banked' in equipment were enough to continue release at these rates for several decades. In that year, atmospheric mixing ratios had reached about 3.5 p.p.t. and 2.3 p.p.t. for the two halons, and appeared to be continuing to increase at a roughly linear rate of 0.16 p.p.t. and 0.12 p.p.t. a year.

Methyl bromide is a most important potential source of Br atoms in the stratosphere, but there remain several puzzles about its sources and budgets in the troposphere. Perhaps the most striking difficulty is that the budget seems to be out of balance: identified sources provide about 122×10^6 kg yr^{-1} to the atmosphere, but identified sinks remove 205×10^6 kg yr^{-1}, so that there is a shortfall of about 83×10^6 kg yr^{-1}. Yet current mean global concentrations are about 9 to 10 p.p.t., and there is certainly no evidence for a strong downward trend. Indeed, there do not seem to have been significant upward or downward trends over the last few years, although concentrations of CH_3Br in the atmosphere may have *increased* by as much as 25 per cent since the beginning of the twentieth century. The conclusion to be drawn is that either a large source term in the budget is missing or the estimates of the magnitude of the sinks are grossly in error. The current best estimate of the tropospheric lifetime of CH_3Br is 0.7 yr, a value calculated from the estimates of chemical losses within the atmosphere and the losses to the ocean and to soils. However, this figure is clearly subject to revision in the light of the uncertainties about the sink rates.

Natural methyl bromide appears to derive mainly from the oceans, with a source strength of about 56×10^6 kg yr^{-1}. Fumigation (especially of soils) accounts for some 41×10^6 kg yr^{-1} of man-made CH_3Br; emissions from automobiles that used leaded gasoline may have made a substantial contribution in the past, although emissions are now probably below 5×10^6 kg yr^{-1} today. Biomass burning is more important, and is responsible for the release of about 20×10^6 kg yr^{-1} of CH_3Br to the atmosphere. These numbers thus indicate that the natural and anthropogenic sources of CH_3Br are very roughly comparable.

Besides methyl bromide, there are several other naturally occurring organic bromine compounds, such as CH_2Br_2, $CHBr_3$, CH_2ClBr, and so on, that are found in the atmosphere. Tropospheric mixing ratios are typically about $5–10 \times 10^{-12}$ (5–10 p.p.t.), but may be higher in the marine environment than elsewhere. Estimates of fluxes to the troposphere suggest a dominant input from natural biogenic sources, and the total bromine flux from such compounds could be comparable to that arising from the combined emissions of CH_3Br and the halons. However, the tropospheric lifetimes are substantially shorter than those of CH_3Br or the halons, so that less reaches the stratosphere. Nevertheless, the impact on catalytic ozone removal may well be significant; at the time of writing, the role of short-lived organic bromine compounds in the (lower) stratosphere is yet to become well defined.

Much of the last few paragraphs has been aimed at quantifying man's contribution to the budget of bromine compounds that might reach the stratosphere. The particular concern with bromine lies in the very high efficiency with which it can destroy ozone in the various catalytic cycles, so that relatively small additional releases can have a disproportionate effect on stratospheric ozone levels. According to one estimate, bromine is 58 times more effective than chlorine at catalysing the destruction of ozone. As explained in Section 4.4.2, the most important factor in determining this high activity is the much 'shallower' reservoirs for Br and BrO than for Cl and ClO. Thus, the bromine analogue of cycle 4 can more effectively destroy ozone than does the chlorine-based cycle. Further, there may be be additional catalytic processes from bromine such as that presented in cycle 9. Also significant is a synergism between the Br- and Cl-catalysed processes. As a result, increases in Br loading in the atmosphere may potentiate Cl, and *vice versa*.

The chlorine–bromine synergism with respect to ozone depletion arises because of a coupling reaction between BrO and ClO that produces Br and Cl atoms which react with ozone, and complete a catalytic cycle

Cycle 16:

$$BrO + ClO \rightarrow Br + Cl + O_2 \qquad (4.61a)$$

$$Br + O_3 \rightarrow BrO + O_2 \qquad (4.62)$$

$$Cl + O_3 \rightarrow ClO + O_2 \qquad (4.19)$$

$$\text{Net} \qquad O_3 + O_3 \rightarrow O_2 + O_2 + O_2$$

With a mixing ratio of 2×10^{-11} of bromine in the stratosphere, the depletion of ozone by the CFCs is enhanced by 5 to 20 per cent. At increased bromine levels (mixing ratios > 10^{-10}), odd-oxygen loss increases rapidly because the fast reaction

$$BrO + BrO \rightarrow Br + Br + O_2 \tag{4.63}$$

is second order in BrO.

There are two further channels for the BrO + ClO interaction that compete with reaction (4.61a)

$$BrO + ClO \rightarrow Br + OClO \tag{4.61b}$$

$$BrO + ClO \rightarrow BrCl + O_2. \tag{4.61c}$$

The branching ratios into the three channels are 0.45, 0.43, and 0.12 for reactions (4.61a), (4.61b), and (4.61c), respectively, at room temperature, although the OClO channel is favoured at low temperatures. Photolysis of OClO regenerates O, so that reaction (4.61b) does not lead to loss of odd oxygen. The BrCl produced in reaction (4.61c) acts as a temporary reservoir for Br but, like the other Br reservoirs, it is rapidly photolysed during the day. Reaction (4.61b) is the only known stratospheric source of OClO, so that observations of the molecule provide useful diagnostic information about stratospheric bromine chemistry.

4.6.5 Halocarbons: loading and ozone depletion potentials

Several factors can influence the impact that releases of particular halocarbons might have on stratospheric ozone. They include the detailed chemistry of the halocarbon once in the stratosphere, the fraction of the emission that reaches the stratosphere, and the absolute amount of the material released.

Although there are some remaining uncertainties in stratospheric models, they do indicate an important feature of perturbation of the stratosphere by CFCs. Since the relative importance of different chlorine-related cycles depends on altitude, the vertical distribution of chlorine sources from different halocarbons must be considered. Destruction of relatively small fractions of ozone at around 15–20 km are more important on an absolute scale than larger fractional destructions at higher altitudes. Ozone is also less rapidly replenished at lower altitudes than at higher ones. Some halocarbons are photolysed at shorter wavelengths, and thus at higher altitudes, than others. Other things being equal, substitution of chlorine by fluorine shifts absorptions to shorter wavelengths. For equal concentrations, $CFCl_3$ has a maximum photolysis rate at ~ 25 km, CF_2Cl_2 at 32 km, while $CClF_2CF_3$ does not produce its maximum contribution until 40 km. It follows that the more heavily chlorinated halocarbons are more active in destroying ozone for two reasons. First, they are photolysed at lower altitudes where their absolute impact is greater. Secondly, they can release more chlorine atoms per molecule to the catalytic cycle.

Atom for atom, bromine is potentially far more destructive towards stratospheric ozone than is chlorine. Photolysis of the source gases occurs lower in the stratosphere, while Br possesses only inefficient reservoirs and participates in synergistic reactions as described at the end of the last section and in Section 4.4.2. Both these factors lead to higher rates of ozone destruction with Br than with Cl. Estimates of the catalytic activity of Br relative to that of Cl, referred to henceforth as α, vary rather widely. Values of α obtained depend not only on the detailed chemistry of the models employed, but also on exactly what comparison of activity is being made: for example, they are functions of altitude, latitude, and season. As stated earlier, recent consensus indicates that α may be as high as 58, although it is also widely accepted that α appropriate to Antarctic ozone loss lies between 40 and 50.

It is obviously important, for control and legislation if for nothing else, to compare the effects of the release of different natural and anthropogenic halocarbons. For simplicity, we introduce some of the key ideas in terms of chlorine alone, and then subsequently extend the concepts to include bromine compounds. In the first instance, it is relatively straightforward to estimate a *chlorine loading potential*, CLP (or, of course, an equivalent *bromine loading potential*, BLP) for the halocarbon molecules in the troposphere. These potentials are effectively the increased tropospheric load of the relevant halogen brought about by the release from the surface of a fixed mass of the halocarbon, measured relative to the loading brought about by the release of the same amount of a reference compound, which is always CFC-11($CFCl_3$) in these studies. The total tropospheric burden of a halocarbon is, as described on p. 137, the rate of its release multiplied by its tropospheric lifetime. In turn, the tropospheric lifetime is determined by a combination of the rates of all loss processes, including transport to the stratosphere, chemical loss in the troposphere, and deposition to the ground and oceans. Photolysis in the stratosphere (and possibly the upper troposphere) constitutes one important sink for the halocarbons, but care must be taken to include in the calculation of these lifetimes the possible return from the stratosphere to troposphere of any molecules that are not removed photochemically.

In terms of tropospheric lifetimes, τ, and relative molecular masses, M, CLP is given simply by the relation

$$\mathrm{CLP} = \frac{M_{\mathrm{CFC\text{-}11}}}{M_{\mathrm{X}}} \cdot \frac{n}{3} \cdot \frac{\tau_{\mathrm{X}}}{\tau_{\mathrm{CFC\text{-}11}}}, \qquad (4.64)$$

where the subscripts X and CFC-11 refer to the 'test' compound X and the reference CFC, respectively; n is the number of chlorine atoms in X, which is compared with the three chlorine atoms in $CFCl_3$. The second column of Table 4.6 gives estimated lifetimes for several compounds of interest in the present context, and, from these values, the CLP can immediately be assessed through equation (4.64). The lifetimes themselves reflect the ideas about reactivity just presented. The hydrogenated species CHF_2Cl and CH_3CCl_3 are tropospherically

Table 4.6 Effect of different halocarbons on atmospheric ozone

Halocarbon	Atmospheric lifetime (yr)[a]	Ozone depletion potential[a, b]	Release rate[a] (10^6 kg yr^{-1})		Contribution to ozone loss[c]	
			1990	1998	1990	1998
$CFCl_3$	53	[1.0]	250	103	22.6	21.0
CF_2Cl_2	100	0.82	371	186	27.5	31.2
$CF_2ClCFCl_2$	83	0.90	236	44	19.2	8.1
CHF_2Cl	11.8	0.04	178	282	0.6	2.3
CF_3Br	69	12	4	3	4.3	7.3
CF_2BrCl	36	5.1	10	7	4.6	7.3
CH_3CCl_3	4.8	0.12	718	103	7.8	2.5
CCl_4	46	1.2	63	24	6.8	5.9
CH_3Br	0.7	0.34	206	206	6.3	14.3

[a] *Scientific assessment of stratospheric ozone: 1998*, World Meteorological Organization, Geneva, 1999. Lifetimes from CSIRO model.
[b] 2-D model results. ODP for CH_3Br scaled for $\tau = 0.7$ yr.
[c] Percentages, calculated from only those halogen compounds listed in the table.

much less inert than the other CFCs, and their lifetimes correspondingly small. Measured concentrations of CH_3CCl_3 in 1990 were about three times lower than those of CF_2Cl_2 and twice those of $CFCl_3$, even though the release rates of CH_3CCl_3 were much higher in that year (fourth column of Table 4.6). The effect of unsaturated bonds on tropospheric reactivity is even more dramatic. Both CCl_2CCl_2 and $CHClCCl_2$ were emitted in quantities larger than those of $CFCl_3$ in 1990, yet the concentrations were much lower, and substantially constant: the atmospheric lifetimes for these compounds are only 5 months and 8 days, respectively.

By analogy with the CLP, it seems reasonable to express the ability of any halocarbon to deplete stratospheric ozone in terms of an *Ozone Depletion Potential* (ODP), specified relative to the depletion by $CFCl_3$ on a kilogram-for-kilogram basis. This measure has, indeed, frequently been adopted to provide a comparison of the damage that might be done to the ozone layer by different halocarbons, although it is now recognized that the concept must be treated with caution. The ODP is equal to the CLP multiplied by an *efficiency factor* that incorporates factors such as the fraction of the halocarbon species, f_X, that is photodissociated compared to that of CFC-11, $f_{CFC\text{-}11}$. In the simplest form, eqn (4.64) for CLP is modified to

$$\text{ODP} = \frac{f_X}{f_{CFC\text{-}11}} \cdot \frac{M_{CFC\text{-}11}}{M_X} \cdot \frac{n}{3} \cdot \frac{\tau_X}{\tau_{CFC\text{-}11}}. \tag{4.65}$$

The ratio $f_X/f_{CFC\text{-}11}$ can, in principle, be obtained from direct measurements. In reality, the efficiency factor has also to take into account the altitudes of photolysis and the altitude dependence of the impact (see earlier) so that recourse must be had to a suitable model of stratospheric chemistry and transport.

At this point, it becomes necessary to consider how the concept of ODP needs to be modified if it is to cover bromine-containing compounds in addition to chlorine-containing ones. The loading potential for bromine (BLP) can be calculated in exactly the same way as the loading potential for chlorine (CLP) through eqn (4.64), but the much greater ozone-destroying power of Br compared with Cl means that eqn (4.65) does not provide a sensible picture of the ODP of bromine compounds. The simplest expedient is to multiply the expression for ODP by the factor α introduced earlier to represent the ratio of activity of Br compared to Cl. It will be remembered that one value of α that has found favour recently is 58, so that this number could well be used in first estimates of ODPs for the bromine compounds. In reality, of course, ODPs for such compounds are derived from model simulations of ozone depletion.

The third column in Table 4.6 gives some ODPs obtained using a 2-D model, and at least shows a very clear distinction between the impact of short- and long-lived halocarbons. Taking these figures at face value, and using the release rates provided in the next two columns of the table, allows the percentage contribution of each of the compounds to halogen-mediated ozone loss to be assessed, and the results are given in the final two columns. The two pairs of columns are for the years 1990 and 1998, and show how the legislation to be discussed in Section 4.6.6 has altered the distribution of ozone loss to be attributed to each compound. Probably more important, the total release of these compounds has decreased from 2036 ktonne (10^6 kg) yr^{-1} in 1990 to 958 ktonne yr^{-1} in 1998, while the ODP-weighted emissions have been reduced from 1105 ktonne yr^{-1} to 489 ktonne yr^{-1}. We see from the table that, for every kilogram of halocarbon released, the ozone destroyed is greater the more chlorine there is in the molecule: $CFCl_3$ is more efficient at destroying ozone than CF_2Cl_2, and CCl_4 is the most efficient of all. Hydrogenated compounds have much lower ODPs, but brominated halocarbons possess much greater potential for damage, as highlighted in our earlier discussion. As anticipated by the original theory, $CFCl_3$ and CF_2Cl_2 make the most important contribution of all to ozone loss. Surprisingly, however, the non-fluorinated halocarbons CCl_4 and CH_3CCl_3 contribute several per cent to the overall loss. Although the CFCs were first seen as the potential culprits in anthropogenic ozone destruction, it is now obvious that the enormous release of less inert species such as carbon tetrachloride and methyl chloroform can also have its effect.

It must be emphasized that exact values of these ODPs (as well as the atmospheric lifetimes) depend on the particular model used for their assessment. In addition, there is a potentially serious problem in that the ODP is based on steady-state concepts, comparison being made with the long-lived $CFCl_3$, while short-term releases of highly reactive compounds puts a transient burden on the atmosphere. Put another way, ODPs give a measure of the total number of molecules of ozone that each unit mass of halocarbon can destroy over an infinite time-scale. Long-lived halocarbons will exert their effects over a similarly long time-scale, while the destruction of ozone by an injection of a short-lived compound necessarily occurs over a smaller period of time. For the same ODP, the same number of molecules of ozone will be removed overall for the same mass of halocarbon injected, but if they are removed over a shorter period, they will represent a larger fraction of the total ozone present at that time. The peak ozone depletions in such cases may exceed those predicted on the basis of ODPs.

Despite the difficulties, ODPs do appear to give a reasonable pointer to which compounds are likely to be least damaging. Various improvements have been proposed that attempt to deal with the problem of time-scale by defining a *time-dependent ozone depletion potential*. An allowance can then also be made for the time lag, τ_{lag}, between release of the halocarbon at the surface and its arrival in the stratosphere. A typical formulation modifies equation (4.65) to read

$$\mathrm{ODP}\,(t) = \alpha \cdot \frac{f_X}{f_{CFC-11}} \cdot \frac{M_{CFC-11}}{M_X} \cdot \frac{n}{3} \cdot \frac{\tau_X}{\tau_{CFC-11}} \cdot \frac{\int e^{-(t-\tau_{lag})/\tau_X}\,dt}{\int e^{-(t-\tau_{lag})/\tau_{CFC-11}}\,dt}, \tag{4.66}$$

where the limits of integration are τ_{lag} (i.e. when the halocarbon first enters the stratosphere) and t, the time for which the ODP is being estimated. The factor α has been included directly in this form of the equation to make it immediately applicable to bromine as well as chlorine compounds. Such time-dependent ODPs illustrate nicely the nature of the problem: a short-lived molecule such as CH_3Br (0.7 yr: see Table 4.6) might show an ODP at least ten times higher on a time-scale of one year than the entry in Table 4.6, which is appropriate to 100 yr or more, might suggest. Since the ozone layer is believed to respond relatively rapidly to changes in halogen loading, time-dependent ODPs provide an appropriate measure of the expected response of ozone to changing inputs of source gas. On the other hand, the possible long-term biological impacts of ozone changes may still be better associated with the steady-state ODPs.

ODPs are essentially concerned with quantifying the impacts on stratospheric ozone of individual halocarbon compounds. Another way of viewing the impacts of the *mixture* of halocarbons released by nature and by man has become increasingly popular. In this approach, the starting point is the evaluation of an *Equivalent Effective Stratospheric Chlorine* (EESC) loading of all the compounds of interest. The EESC is the sum of atmospheric mixing ratios of the constituent compounds, each suitably weighted by the number of halogen atoms it contains, the normalized 'release factor' f_X/f_{CFC-11}, and, in the case of bromine compounds, by α. Formally, the function is

$$\mathrm{EESC} = \sum_X n_X \cdot \frac{f_X}{f_{CFC-11}} \cdot x_X + \alpha \cdot \sum_Y n_Y \cdot \frac{f_Y}{f_{CFC-11}} \cdot x_Y, \tag{4.67}$$

where n, f, and α all have the meanings ascribed to them earlier, and x is the atmospheric (stratospheric) mixing ratio of the halocarbon. Subscripts X refer to chlorine-containing species, and Y to the bromine-containing compounds. The mixing ratios, x, might be obtained by direct atmospheric measurement, but are usually calculated by integrating historical and projected release rates of each hydrocarbon, using the measured lifetimes and taking into account the lag time, τ_{lag}. Figure 4.25 shows one such set of calculations, with the future release rates reflecting one of the 'protocol scenarios' that will be described in the next section. The figure indicates the contribution made by each of a number of individual compounds or classes of compounds. Of these, the *hydrochlorofluorocarbons* (HCFCs) are intended as (temporary) alternatives to CFCs as also described in the next section. It will be seen immediately that by far the largest contribution to the EESC is that made by CFCs, and that will remain true until the end of the twenty-first century.

The next step is to relate the EESC to the resultant ozone depletions. Models could be used, as described previously, although then it could be argued that evaluating EESCs was only using a roundabout method of estimating the same ODPs as previously. An entirely different approach is to use observations of trends in both EESC and O_3 (Section 4.8) to predict changes of O_3 in the future. For simplicity, it may be assumed that there is a linear relation between changes in ozone column amount and EESC. The year 1980 has been used as the reference base in many of these calculations, and we may then write

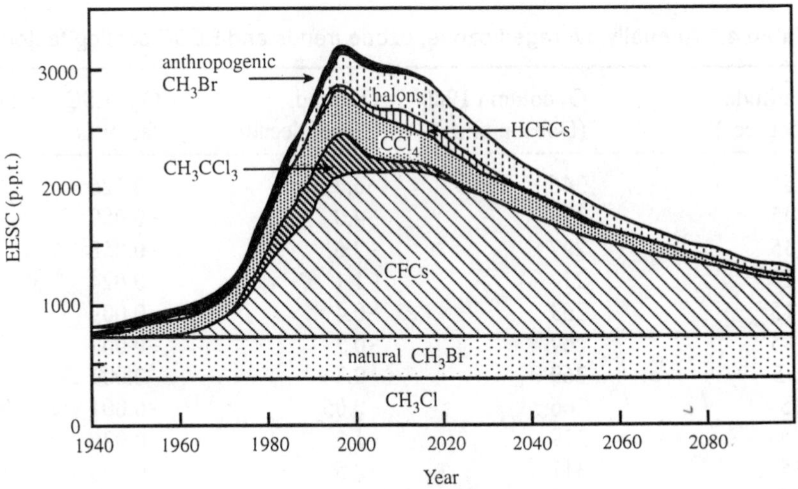

Fig. 4.25. Contributions of different substances or classes of substances to the Equivalent Effective Stratospheric Chlorine. Future emissions reflect one particular control scenario (see Section 4.6.6). From *Scientific assessment of ozone depletion: 1998*, World Meteorological Organization, Geneva, 1999.

$$\frac{O_3(t) - O_3(1980)}{O_3(1980)} = s\left(\frac{EESC(t) - EESC(1980)}{EESC(1980)}\right) \quad (4.68)$$

with the scaling factor, s, to be obtained from trends measured since 1980. One of the great advantages of this procedure is that, while EESC is defined as a global and annual average value, ozone exhibits strong seasonal and geographical variations. The scaling factors can be determined on a latitudinal and seasonal basis. Annual means of s can be calculated if so desired. Table 4.7 presents some calculations of the annually averaged values of s, and the experimental data on which they are based. For EESC, the base value in 1980 is taken to be 1.895 p.p.b., with an average rate of increase of 0.860 p.p.b. per decade. With these scaling factors, it is now possible to convert projected EESC loadings, such as those shown in Fig. 4.25 or those from any other chosen scenario, to predicted amounts of stratospheric ozone in the future. Such predictions form the basis for legislation for control of emissions, and it is to that subject that we turn next.

4.6.6 Halocarbons: control, legislation, and alternatives

So long as the release rates are unaltered, a steady-state concentration of halocarbon must ultimately be reached. Both the time taken to reach the steady state and the value of the limiting concentration depend on the lifetime. The figures given in Table 4.6 suggest a steady-state mixing ratio for CF_2Cl_2 of

Table 4.7 Annually averaged ozone, ozone trends and EESC scaling factors[a]

Latitude (degrees)	O_3 column 1980 (Dobson units)	O_3 trend (% per decade)	O_3/EESC scaling factor (s)
-55	346	-5.7	-0.126
-45	325	-4.0	-0.089
-35	297	-2.6	-0.058
-25	274	-1.1	-0.024
-15	263	-0.4	-0.009
-5	267	-0.3	-0.007
5	269	-0.3	-0.007
15	266	-0.05	-0.001
25	279	-1.2	-0.026
35	311	-2.5	-0.056
45	350	-3.0	-0.067
55	371	-2.7	-0.059

[a]From *Scientific assessment of ozone depletion: 1998*, World Meteorological Organization, Geneva, 1999.

1.8×10^{-9} would be reached if 1990 release rates were maintained, a value very approximately four times the 1990 atmospheric burden shown in Fig. 4.23.

Constant release, leading to steady-state atmospheric concentrations, is only one of many scenarios that can be imagined for the CFCs. For a start, historical release rates have been anything but constant. From a few hundred thousand kilograms of CF_2Cl_2 produced in 1931, an all-time high of 4.74×10^8 kg were produced in 1974. (The cumulative production up to 1990 was about 10^{10} kg.) Similar figures apply to $CFCl_3$. The Western nations used to produce 5–10 times as much of the CFCs as the rest of the world, but since concern about the ozone problem was expressed in 1974, production has been steady and then falling. Some legislation was passed in the 1970s forbidding certain uses of the CFCs, especially in the US, and production in 1981 was more than 20 per cent less than it had been in 1974. By the late 1980s, however, production had begun to creep up again, to reach the rates given in Table 4.6. A much more wide-ranging control is embodied in the 'Montreal Protocol on Substances that Deplete the Ozone Layer' that was agreed in September 1987 and entered into force in January 1989. Each party to the Protocol was to freeze, and then reduce according to an agreed timetable, its production and consumption of CFCs 11, 12, 113, 114, and 155; it was also to freeze consumption and production of the halons. However, it soon became apparent that, even with the 'Montreal' controls, concentrations of the CFCs would continue to increase, albeit perhaps only half as fast as without the controls. A series of amendments (London, 1990; Copenhagen, 1992; Vienna, 1995; Montreal, 1997) have therefore been proposed, and ratified by many, but not all, of the producing countries. These controls called for the 'developed' countries to have phased out production of the halons by 1994 and the five CFCs just listed by 1996; 'developing' countries have until 2010 to completely phase out their production of these compounds. Similar requirements are placed on most other halogen-containing anthropogenic *ozone depleting substances*, including methyl bromide. The projections for future EESC loadings shown in Fig. 4.25 are, in fact, based on these requirements of the Protocol incorporating the 1997 Montreal amendments.

Several uncertainties may require some modifications to be made to the predicted EESC values. For example, the reduction and phase-out dates for developing countries start in 1999 for CFCs, and a few years later for other substances, so that the base levels to which the controls refer also lie partly in the future. Another issue is that emissions of CCl_4 and CH_3CCl_3 are likely to be different from those foreseen in the Protocol scenario: production of CH_3CCl_3 ceased by 1998, even though some production was permitted in the Protocol. Perhaps even more seriously, there are grave uncertainties in the likely release from, or destruction of, the various 'banks' (such as existing equipment) that contain large amounts of CFCs and halons that have already been manufactured. One of the most imponderable unknowns concerns the illegal production of the controlled substances, and the smuggling of them from countries where they can still be produced legally. Recent estimates suggest that as much as 10 ktonnes of CFCs are smuggled annually into the EU, and globally the problem may be three times larger. Whatever precise alterations have to be made to Fig. 4.25, the projections will surely still show that EESCs will not return to the values they had in 1980 until well into the 22nd century, and they will also show that the CFCs themselves are then by far the largest contributors to the EESC.

Reductions in the use of CFCs can be achieved by avoiding their use altogether where possible, or by improving recovery and recycling of them. However, there are instances where halocarbons offer outstanding advantages in particular applications, such as use as refrigerants. There has been an intensive search, therefore, for halocarbons that are 'alternatives' to the conventional ones. The requirements are that the alternative compound should have a low ODP (which means, in essence, that it should have a short tropospheric lifetime) while retaining the desired physical and thermodynamic properties (such as boiling point or heat of vaporization). Two classes of compound that have received particular attention are the *hydrochlorofluorocarbons* (HCFCs) and the *hydrofluorocarbons* (HFCs). Example replacements are CF_3CHCl_2 (ODP $\cong$ 0.013) for $CFCl_3$ (ODP = 1) as a foam-blowing agent; CF_3CH_2F (ODP $\cong$ 0) for CF_2Cl_2 (ODP = 0.9) as a refrigerant, and CH_2ClCF_2Cl (ODP < 0.05) for $CF_2ClCFCl_2$ (ODP = 0.9) as a solvent in the electronics industry. Because both HCFCs and HFCs contain hydrogen in place of one or more of the chlorine atoms in the CFCs, they are subject to attack (hydrogen abstraction) by OH in the troposphere (see Chapter 5). The lifetimes are consequently much shorter than those of the CFCs, so that much smaller amounts reach the stratosphere. Furthermore, the HFCs contain no chlorine at all, so that chlorine-based catalytic cycles for the removal of ozone cannot operate. Although there has been some discussion about catalytic chains involving F- or CF_3-based radicals, it seems that they cannot give rise to much removal of stratospheric ozone. Aspects of the use of HCFCs and HFCs that have given rise to concern are their toxicities and the possible adverse impact of their degradation products in the troposphere. For example, CF_3COOH, a known neurotoxin, might well be formed in the degradation of compounds containing the CF_3 moiety (see Section 5.10.8 for a fuller discussion).

Of the 'alternative' compounds, the HCFCs, in particular, are seen only as *transitional substances*, in the sense that they, too, will be phased out ultimately. They destroy stratospheric ozone, even though to a lesser extent than the CFCs themselves, so that unregulated production is potentially disastrous. For the developed countries, HCFC consumption was frozen at its contemporary level in 1996, and successive reductions are required from 2004 onwards that will lead to complete phase-out by 2030. Developing countries have until 2016 to commence the freeze (which gives rise to uncertainty in base levels), and must phase out the consumption of HCFCs by 2040.

4.6.7 Halocarbons: future ozone depletions

Projected changes in atmospheric loadings of various halocarbons, such as those described in the last section, can be used in conjunction with numerical models of the atmosphere to predict how atmospheric ozone concentrations might alter in the future. Different models, possessing different physical formulations and making different assumptions, produce different predictions. The predicted extent of ozone depletion caused by CFCs has, of course, changed as the chemical input data to the models have been refined, and it will doubtless continue to change in the future. Although the predicted magnitude of the effect has altered, the sense has always been the same: ozone is lost.

As explained on pp. 226–8, there is now an alternative to using direct modelling for the prediction of future ozone depletions. The observational database is now sufficiently extensive for historical changes in stratospheric ozone concentrations to be linked to changes in stratospheric halogen loading (EESC). The results given in Table 4.7 are representative of a comparison of this kind, and provide an 'experimental' sensitivity factor, *s*, that can be used in conjunction with projections of future EESCs to predict the accompanying ozone concentrations. Figure 4.25 has already been introduced to indicate EESCs expected from one specific scenario, and it is then a simple matter to obtain a rough estimate of ozone levels that might be found at any time in the future, assuming of course that changes in stratospheric ozone levels will continue to be dominated by changes in the halogen content of the stratosphere.

Depletions of ozone, whether predicted by models or by the pseudo-experimental method, depend on altitude, season, and latitude. Integrated column ozone concentrations are generally depleted more at high latitudes than at the equator, and the high-latitude depletion is greatest during winter. However, the changes in column abundances are all generally much smaller than the local changes at high altitudes. Some models show depletions for high latitudes of more than 50 per cent at altitudes of 45 km, where the effects of CFC release are at a maximum. Depletion of ozone in the upper stratosphere appears to be partially compensated by increased penetration of solar ultraviolet radiation, which leads to an increased rate of generation of ozone in the lower stratosphere. This compensation is sometimes referred to as 'self-healing'. A further complication arises because changes in concentrations of other atmospheric gases, such as N_2O, CH_4, and CO_2, also affect both the distribution and the total column density of ozone. We shall see in Sections 4.6.8 and 4.6.9 how these gases acting alone or in combination might influence stratospheric ozone.

Regardless of the exact quantitative link between ozone and halogen levels in the stratosphere, it is abundantly obvious from Fig. 4.25 that, without the controls imposed in 1996, ozone depletions could have reached catastrophic proportions in the next century. A simple extrapolation of the EESC values shows that the Montreal Protocol of 1987 would apparently have permitted a doubling of halogen loading in each decade. As discussed in Section 4.6.1, one of the main concerns about reductions in stratospheric ozone has been the consequential increase in UV-B radiation reaching the Earth's surface. It is therefore interesting to end the present examination of the impact of the halocarbons on atmospheric ozone by looking at how UV doses might change for various emission and control scenarios. Figure 4.26 indicates the results of one such set of estimations: the scenarios are for no control whatever, for the Montreal Protocol as envisaged in 1987, and for the Copenhagen amendments of 1992.

4.6.8 Nitrous oxide (N_2O): agriculture

Nitrous oxide from the troposphere is the principal source of stratospheric NO_x. Possible perturbations to its concentration because of human activities could therefore affect stratospheric ozone concentrations. Although NO_x is a catalyst in its own right, and can thus contribute to ozone loss, it also sequesters ClO_x because it forms the reservoir species $ClONO_2$. Furthermore, heterogeneous

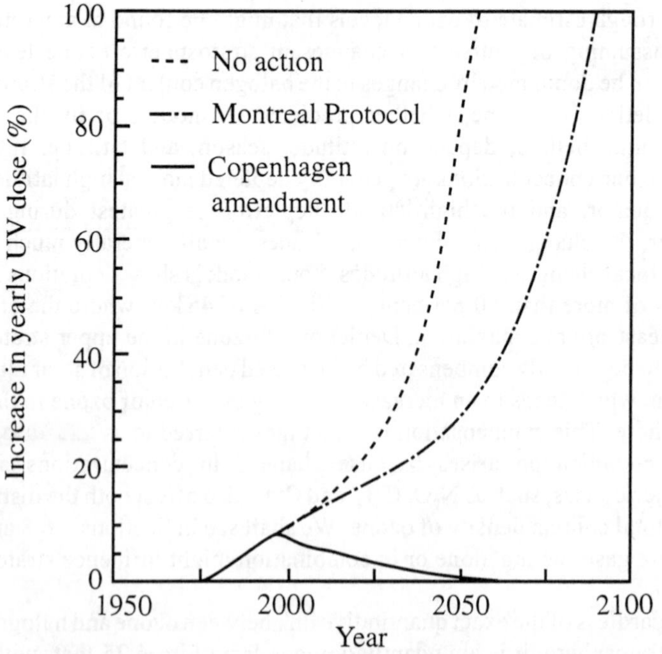

Fig. 4.26. Increases in effective UV dose at a latitude of 40°N for three different scenarios. From *Scientific assessment of ozone depletion: 1998*, World Meteorological Organization, Geneva, 1999.

reactions involving N_2O_5 and $ClONO_2$ are also of major importance, so that particle loading in the stratosphere has an influence on the effect of enhanced NO_x levels. These different influences have already been discussed elsewhere, and especially in Section 4.6.2, and our present enquiry is directed to the anthropogenic contribution to stratospheric N_2O. Nitrous oxide concentrations in the atmosphere are known to be increasing slowly, from an estimated 275 p.p.b. in the pre-industrial era, to 299 p.p.b. in 1976, and 312 p.p.b. in 1996. In general, it appears that changes in NO_y associated with the long-term rise in N_2O will have only a small effect on O_3 on time-scales of tens of years, but might well have a significant role in regulating O_3, and the radiative state of the stratosphere as well, on time-scales of hundreds of years. The atmospheric lifetime of N_2O seems to be more than one hundred years, so that any deleterious effects due to increased N_2O release will manifest themselves decades after the perturbation is first applied.

Intensive use of fertilizers could lead to increased N_2O production in the biosphere, the nitrogen 'fixed' artificially in the manufacture of the fertilizer being returned to the atmosphere through nitrification or denitrification processes (Section 1.5.3). According to recent appraisals of the sources of N_2O, at least 40 per cent of the atmospheric input of the gas is subject to anthropogenic influence, with the production of fertilizer having the largest effect, but with biomass burning, industrial processes, and formation in catalytic convertors for automobile exhaust (see pp. 434–6) all making a contribution. While perturbations due to the use of SSTs or halocarbons may be regarded as discretionary, agricultural use of fertilizers may well be essential if populations continue to grow. The mean trend of increase of N_2O was about 0.25 per cent annually during the 1980s, and early estimates suggested that, with a world population of 6.5×10^9 and a grain consumption of 400 kg per person per year (about 12 times the present value), N_2O production in the biosphere could double by the turn of the century. However, it is now perfectly evident that these predictions were wide of the mark. Indeed, the growth rate of N_2O appears to have been *decreasing* over the last decade: the rate was 1 p.p.b. per year in 1991, but only 0.5 p.p.b. per year in 1993 and 0.6 p.p.b. per year in 1995. The causes of this tailing off are unknown: one difficulty in assigning a single influence is that the eruption of Mount Pinatubo in 1991 (Section 4.5.5) may have had a marked impact on N_2O production. Nevertheless, global *reduction* in the use of nitrogenous fertilizer remains another plausible explanation.

Until recently, it was thought that anaerobic bacterial denitrification (reduction of NO_3^-) was the microbial source of N_2O, a few per cent of the nitrogen converted being released as the oxide. Lowered pH increases the fraction released as N_2O, but decreases the overall rate of denitrification. Denitrification is completely inhibited below a pH of 5.5. Natural N_2O production was assumed to be augmented in proportion to the amount of fixed nitrogen added to the soil as fertilizer. However, normal agricultural practice is to drain and aerate the soil, particularly that soil to which fertilizer is applied. Intensification of agriculture might, according to these ideas, decrease, rather than increase, N_2O output. Incidentally, acid rain (a common form of pollution: see Section 5.10.6) could inhibit N_2O formation *via* denitrification since the pH of raindrops saturated by CO_2 is already near the limiting value of 5.5.

Observations in oceans, rivers, ponds, and soils have provided direct evidence for a quantitative relation between oxidation of organic material and production of N_2O. That is, *nitrification* (oxidation of NH_4^+) rather than denitrification is implicated as the major source of natural N_2O. The fractional yield of N_2O from nitrification of fertilizer nitrogen appears to be linear at low rates of application (10^{-3} moles N_2O per mole of NH_4^+) but increases rapidly at high rates. Where they are heavily loaded with reduced nitrogen, aquatic and soil systems both appear to exhibit non-linear amplification of the N_2O flux in response to increased loading. For both systems, also, enhanced N_2O production is triggered by oxygen depletion. It seems clear, then, that fertilization of soils with ammonium or organic nitrogen, and disposal of human and animal wastes, strongly stimulate nitrification, and are likely to lead to globally significant efflux of N_2O to the atmosphere.

4.6.9 Combined influences: gases, particles, and climate

Trace gases can affect stratospheric ozone concentrations both by direct involvement in catalytic cycles and by indirect alteration of stratospheric chemistry. Changed stratospheric loading of particles, such as sulphate aerosol, is also likely to influence the detailed chemistry. Any model that attempts to estimate future stratospheric ozone concentrations, for example in response to future loadings of halogen compounds, must necessarily consider the influence of all the projected changes simultaneously.

Methane affords an excellent first example of a compound exerting several types of influence, since it can be a source of catalytically active hydroxyl radicals, it can lead to *production* of ozone, it can play a part in the formation of reservoir compounds, and, in turn, its concentration in the troposphere can influence the amounts of other trace gases that reach the stratosphere.

We have already seen (Section 4.4.3) that reaction of excited atomic oxygen, $O(^1D)$, with both H_2O and CH_4 is a source of stratospheric OH, and, further, that oxidation of CH_4 accounts for at least some of the stratospheric H_2O. Hydroxyl radicals enter into several stratospheric cycles, both as participants in catalytic ozone destruction (Section 4.4.1), and in determining the partitioning of other catalysts between active and reservoir forms (Section 4.4.2). Methane also acts in another significant way by converting Cl atoms to the reservoir HCl.

Not only may methane influence the detailed chemistry of the catalytic cycles for ozone destruction, but it may also act as a source of ozone in the lower stratosphere. The chemistry concerned is described in connection with tropospheric ozone production (Section 5.3.3). The essential idea is that peroxy radicals derived from the CH_4 can convert NO to NO_2. Nitrogen dioxide can be photolysed at relatively long wavelengths to yield atomic oxygen, and is thus a source of odd oxygen in altitude regions where there is only feeble penetration of radiation able to dissociate O_2. It is evident, then, that increased $[CH_4]$ could increase $[O_3]$ in the lower stratosphere both by sequestering active chlorine and by converting NO to NO_2.

The concentrations of many species in the stratosphere are in large part determined by processes occurring in the troposphere. As we shall see in Chapter 5, reaction with OH is the major tropospheric sink for important trace gases such as CO, CH_4, H_2, CH_3Cl, NO_2, and many others. Any decrease in tropospheric [OH] would lead to increased global concentrations of naturally occurring species such as CH_4 and CH_3Cl, as well as of anthropogenic substances such as CH_3CCl_3, known to be released in large quantities (cf. pp. 224–5 and Table 4.6). Lowered tropospheric [OH] could thus have an indirect impact on stratospheric ozone through enhanced catalytic destruction. Reactions with carbon monoxide and methane

$$OH + CO \rightarrow H + CO_2 \qquad (4.42)$$

$$OH + CH_4 \rightarrow CH_3 + H_2O \qquad (4.69)$$

are the major sinks for OH in the troposphere (Section 5.3.2), so that increases of either [CO] or $[CH_4]$ can act to reduce [OH]. The detail is complicated, because increased concentration of either carbon species will decrease [OH], and thus lead to increased concentrations of the other species; furthermore, CO is itself one end product of the oxidation of CH_4 (see Sections 1.5.1 and 5.3.2).

Until the 1990s, both $[CH_4]$ and [CO] were increasing in the troposphere, by very roughly one per cent per year (for CO the increase is higher in the Northern Hemisphere and negligible in the Southern). More recently, however, the rate of increase in CH_4 has declined (to about one-third of the earlier value during the period 1992–98), and the trend in CO may even have reversed. As for N_2O, the eruption of Mount Pinatubo may have been responsible, at least in part, for this altered behaviour. Isotopic data suggest that over half the 'new' methane is of fossil origin (e.g. from combustion, natural gas leakage, oil extraction, and possibly from frozen tundra bogs) and less than half from biogenic sources (e.g. rice agriculture and animal husbandry).

Loadings of aerosols and other particles do not seem susceptible to sensible prediction. The volcanic input of sulphate particles into the stratosphere has been unusually high during the past two decades. As explained in earlier sections of this chapter, there are many indicators of enhanced chemical processing on this aerosol material. In addition, we have just seen that an eruption such as that of Mount Pinatubo may affect concentrations of trace gases such as N_2O, CH_4, and CO. Polar stratospheric clouds (PSCs), whose dramatic effect on chemistry is discussed in the next section, depend for their formation on condensation of H_2O, possibly together with HNO_3 and H_2SO_4, at very low temperatures. Changed stratospheric temperature structure (see next paragraphs), may therefore have a marked effect on the abundance of PSCs, and thus on stratospheric chemical behaviour. Changes in concentrations of the precursor gases N_2O and CH_4 exert an influence, and increased aircraft emissions of NO_x and H_2O favour PSC formation.

Carbon dioxide concentrations are increasing in the atmosphere largely as a result of the burning of fossil fuels. Aspects of this increase are discussed in more detail in Sections 9.5 and 9.6. The potential effect on ozone comes about because CO_2 is the principal atmospheric constituent that contributes to stratospheric cooling. Infra-red radiation from CO_2 escapes from the stratosphere (where the CO_2 is optically thin) rather than being trapped as in the lower atmosphere (Section 2.2.4). Increased CO_2 thus leads to lower stratospheric temperatures, which can alter chemical reaction rates and atmospheric dynamics. Similar remarks apply to changes in concentration of all other 'greenhouse' gases such as CO, N_2O, CH_4, and halocarbons that may be present.

A temperature decrease in the stratosphere is converted to an increase in ozone concentration. As we pointed out in Sections 4.4.1 and 4.4.6, the simple Chapman $O + O_3$ process for the loss of odd oxygen, reaction (4.4), has a relatively large activation energy. In the absence of catalytic chains, therefore, loss rates decrease—and O_3 concentrations increase—with lowered temperature. The indirect catalytic routes for odd-oxygen loss are often effective precisely because they have small activation energies. To that extent, catalytic destruction of ozone is less affected by temperature changes than is the direct, oxygen only, route. Nevertheless, cooling the lower stratosphere increases the ozone there by reducing ozone loss resulting from reactions with OH and HO_2. Doubling the present-day CO_2 content would lead to a maximum temperature decrease of 7 to 10 K near 40 km, and a corresponding local ozone increase in the unpolluted stratosphere of 20 to 30 per cent. Column abundances would be increased by ~ 6 per cent. However, because, for example, the CFCs already make a substantial contribution to ozone loss, the increase of ozone due to increased CO_2 is much smaller. The NO_x catalytic chain is affected in a more subtle way by temperature changes. Loss processes for NO in the upper stratosphere include reaction with atomic nitrogen, and the atom concentration increases with decreasing temperature. Lowered temperatures therefore lead to lower NO_x concentrations, less catalytic loss of odd oxygen, and higher ozone concentrations.

The situation with regard to atmospheric temperatures is further complicated by the properties of ozone itself. The heating that leads to the very existence of the stratosphere is provided by the absorption of solar ultraviolet radiation by ozone. A decrease in ozone concentration will thus translate into a decrease in temperature, but there is evidently a (negative) feedback loop in operation, since reduced temperatures themselves lead to an increase in ozone concentrations.

Two-dimensional models can include the alterations to dynamical processes caused by changes in atmospheric temperature structure. Global ozone increases of about 9 per cent are predicted for an approximate doubling of atmospheric CO_2 content for the 'natural' stratosphere, in rough agreement with the 1-D results quoted earlier. Greatest column changes occur at high latitudes, as seems to be indicated for all perturbations of the ozone layer.

Bringing together the indirect influences we have suggested in this section, we can see that the greenhouse gases can influence tropospheric chemistry *via* temperature changes, for example by increasing evaporation of surface water. Increased infra-red opacity due to CO_2 or CFC increases can alter the average H_2O content of the troposphere, and hence the OH concentration. Trace constituent concentrations will then be affected as we described earlier in this section.

A study of the perturbations produced by a single source allows examination of the details of the perturbing influence and of the corresponding feedback processes. The single-source scenario does not, however, reflect reality, and we have already seen that the various perturbations are most surely not additive. Even over the past few decades, CO_2, N_2O, NO_x from subsonic aircraft, and the halocarbons have all increased in concentration. Aerosol loadings are very variable, and have also been unusually high because of volcanic eruptions, but by 1997 were *below* the pre-Mount Pinatubo levels. To interpret long-term ozone trends over the same period, and to predict the effects of future perturbations, we must clearly evaluate how the perturbations operate when they are present simultaneously.

Carbon dioxide cooling (see previous page) has a major effect in offsetting ozone destruction due to release of catalytically active pollutants. Combined scenario models have made it very apparent that carbon dioxide growth rates must be included in any realistic attempt at predicting ozone changes. The CO_2-related increase in O_3 is not algebraically additive with CFC-related decreases, because of the reduced temperature sensitivity of catalytic processes (cf. p. 236). Nevertheless, inclusion of CO_2 effects in models greatly reduces the impact of a given CFC release. When two catalytic species are considered simultaneously, the dominant cycle alone seems to affect the total ozone, although the altitude distribution of depletion may be shifted by the less important cycle.

One key question that has to be answered by the models is how stratospheric ozone values will recover in response to the reduced halogen loadings as the control protocols take effect. Here, the importance of using the combined scenario models is paramount. 2-D models show that increases in future CH_4 will shorten the recovery period, while increases in N_2O or sulphate aerosol will increase it. Inclusion of projected increases in CO_2 shortens the recovery period somewhat. However, with the expected future climate and levels of N_2O, CH_4, and aerosol, some models suggest that O_3 levels may never return to the values that they had in 1979, when Cl_y concentrations were relatively low (see Fig. 4.25). These interactions between different influences make quantitative prediction of ozone depletions more difficult than ever, particularly since the atmospheric trends of species such as CH_4 and N_2O have been recognized only relatively recently, and the reasons for the increases are not yet fully understood. In particular, if the trends in [CH_4] are less than anticipated, or those in [N_2O] are greater, then greater ozone destruction will occur than models currently predict.

4.7 Polar ozone holes

4.7.1 Discovery of abnormal depletion

In 1985, it did not seem as though man's activities should *yet* have had a large effect on stratospheric ozone concentrations, although it was recognized that the build-up of CFCs in the future might lead to substantial ozone depletions (Section 4.6.7). In general terms, models based on known chemistry and the dynamics of the atmosphere seemed to explain stratospheric chemical behaviour well. Indeed, the 'Montreal Protocol' (Section 4.6.6) that sought to control CFC emissions was based on exactly such models. It is thus ironic that the Protocol was being established in the autumn of 1987, just as experimental evidence was confirming extremely large annual depletions of ozone in the Antarctic, and atmospheric scientists had become sure of the connection between these depletions and man's release of the CFCs.

The first intimation of something unexpected in the behaviour of ozone over Antarctica came from scientists of the British Antarctic Survey (BAS), who had been measuring ozone concentrations regularly from their base at Halley Bay at 76°S for many years. The BAS team believed in 1982 that they had detected a decline in (Southern Hemisphere) springtime ozone concentrations since 1977, and by October 1984 they were sure, with something like 30 per cent total ozone loss, and the depletion apparently increasing over the years 1982, 1983, and 1984. Joe Farman and his colleagues published their conclusions in *Nature* in 1985.[a] It is now apparent that this thinning of the ozone layer in the Antarctic spring had already been going on for more than a decade. One worrying feature was that rather more sophisticated ozone instruments (see Section 4.4.6 and Fig. 4.9) such as the Total Ozone Mapping Spectrometer (TOMS) and the Solar Backscattered UltraViolet (SBUV) instrument on the Nimbus 7 satellite had apparently not detected the ozone depletion. However, it subsequently emerged that ozone concentrations as low as those observed by BAS were being rejected from the satellite data on the grounds that they lay outside what was thought to be a 'reasonable' range.

[a]Farman, J.C., Gardiner, B.G., and Shanklin, J.D. *Nature, Lond.*, **315**, 207 (1985).

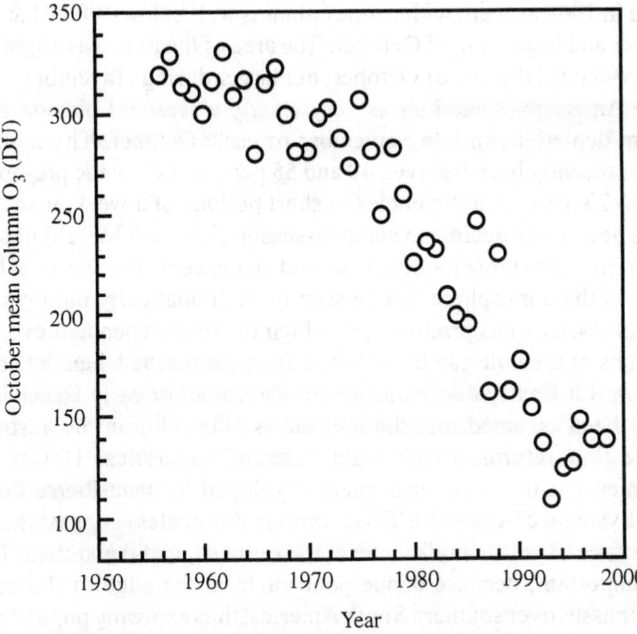

Fig. 4.27. Decline in mean October ozone levels over Halley Bay during the period 1957–1998. Updated from *Stratospheric ozone 1996*, UK Stratospheric Ozone Review Group, Department of the Environment, London, 1996.

When the error was recognized, the satellite results were reprocessed, and they then confirmed the BAS findings. The satellite TOMS, and now GOME, results (Section 4.4.6) give a better overall view of ozone over the region, but it is instructive that the depletion of ozone in the ozone hole was recognized by the British scientists working with relatively simple equipment from the ground. Figure 4.27 shows how the mean October ozone concentrations measured from Halley Bay have evolved since they were first started. There is no room to doubt that something quite dramatic has happened since the mid-1970s, and it is the absence of much of the normal stratospheric ozone that has been called the 'ozone hole'.

The TOMS and GOME data show that the depleted region has grown much deeper, and covered greater areas of the Earth, since 1979. The hole has been defined as an ozone column of 220 Dobson units (DU) or less: bear in mind that the pre-hole column amounts were typically 300–350 DU at Halley Bay during October. Defined in this way, the hole was first larger than the Antarctic continent in 1987. A record for extent was reached 19 September 1998, when the hole covered 27.3 million square kilometres. On 30 September of the same year, the lowest ozone column observed was 90 DU, nearly as low as the lowest value (88 DU) ever measured (28 September 1994). Similar depletions were observed in 1999 as well, with ozone columns well below 100 DU at the end of September and beginning of October. The area of the hole was slightly smaller than in 1998 until the end of October, but larger during November.

The Antarctic ozone hole is very clearly a seasonal phenomenon, with maximum depletions in late September or early October. The monthly total ozone has recently been between 40 and 55 per cent below the pre-hole values, with up to 70 per cent deficiencies in short periods of a week or so. Although there has been some decline in summer-season (January–March) ozone levels, the losses are much more modest at around 10 per cent. The seasonal behaviour of ozone in the atmosphere can be seen most dramatically in moving picture images that NASA has produced, in which the time-dependent evolution and fluctuations of the hole can be followed from the period when the hole begins to form until it finally dissipates in late November or early December. It has generally been assumed that the hole starts to develop in the austral spring, when the Sun returns to the main body of Antarctica. However, recent measurements with a new instrument developed by Jean-Pierre Pommereau (SAOZ: Système d'Analyse d'Observations Zénithales) suggest that the hole really starts each year in mid-winter at the sunlit edge of Antarctica. This result may be important, because ozone-poor air from the edge of the ozone hole regularly passes over southern South America, thus exposing populated regions to elevated doses of UV radiation.

The use of the word 'hole' to describe what happens to ozone in the Antarctic hardly seems an exaggeration. Up to 80 per cent of the ozone in the lower stratosphere (12–20 km) has been lost in recent years in October, and at some altitudes virtually all the ozone is lost, as Fig. 4.28 illustrates. This figure compares a vertical profile of ozone measured over the Antarctic station of Syowa on 6 October 1997 with a long-term average measured for October in the years 1968–80. The profile for the earlier years, when ozone depletion was only just becoming noticeable, shows the normal ozone layer centred on an altitude of about 17 km. In October 1997, the ozone in much of this critical region had entirely vanished. As Susan Solomon, one of the scientists closely connected with interpretation of the new phenomenon, has said, a period of complacency about the ozone layer thus came to an end with the discovery of unprecedented and completely unanticipated depletions of atmospheric ozone over the Antarctic. The magnitudes of the changes were not predicted in 1985 by any model of future stratospheric composition, even for 50 or 100 years ahead. The depletions were the more startling because they occurred in the present-day atmosphere. The seasonal nature of the depletions heightened the mystery, but demonstrated that the effect was real, rather than being a result of slow degradation of the instruments used to measure ozone in the atmosphere.

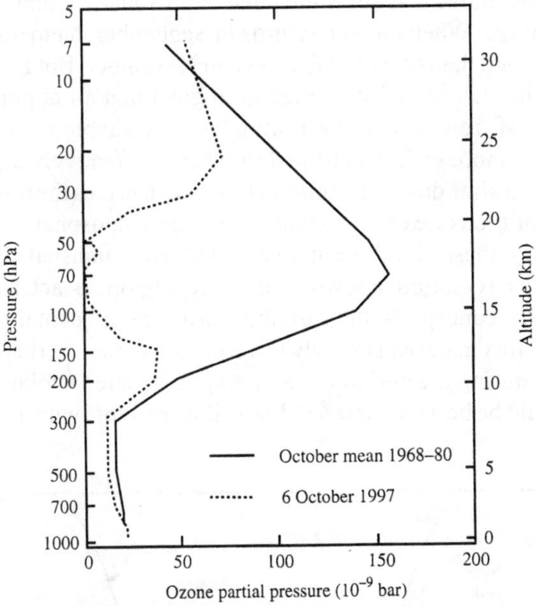

Fig. 4.28. Ozone concentrations over Syowa Station on 6 October 1997, compared with the long-term October average profile for 1968–1980. From *Scientific assessment of ozone depletion: 1998*, World Meteorological Organization, Geneva, 1999.

4.7.2 Special features of polar meteorology

There are two features of polar stratospheric meteorology and dynamics that appear to have a close bearing on the interpretation of polar ozone loss. The very low temperatures (below −80 °C in the Antarctic) lead to the formation of polar stratospheric clouds (PSCs). Section 4.7.4 describes the composition and formation of PSCs, but, for the time being, we note that PSCs are much commoner in the colder Antarctic than in the Arctic regions. There is clear evidence that the polar stratospheric clouds are involved in polar ozone destruction, and it is significant that years when stratospheric temperatures are particularly low are also those in which ozone depletions are greatest. For example, lower stratospheric temperatures were unusually low (by up to 5 °C) in southern middle and polar latitudes in 1998, and this was a year in which the hole was exceptionally wide and deep.

The second, and related, feature of polar meteorology is that a vortex forms as air cools and descends during the winter. A westerly circulation is set up as illustrated in Fig. 4.29, with very high wind speeds of perhaps 100 metres per second or more by spring. The vortex develops a core of very cold air, and it is these low temperatures that allow the polar stratospheric clouds to form in the lower stratosphere. When the Sun returns in September, temperatures rise, the winds weaken, and the vortex breaks down in November. But in the winter and early spring, the stability of the vortex is so great that air at polar latitudes is *almost* sealed off from that at lower latitudes. One simple view is that the air over the pole is more or less confined to what is effectively a giant reaction vessel. There is a slow downward circulation which drives the polar air through the cold core of the vortex that contains the polar stratospheric clouds. As we shall see shortly, these clouds seem to be involved in unusual chemistry, with the downward circulation allowing the core region to act as a 'chemical processor'. The concept is thus of the vortex as a 'containment vessel'. However, in reality material is slowly lost from the vortex, perhaps at about one per cent a day during the period of early August to late October. In that case, the vortex would be best characterized as a 'flow reactor' with a continual slow

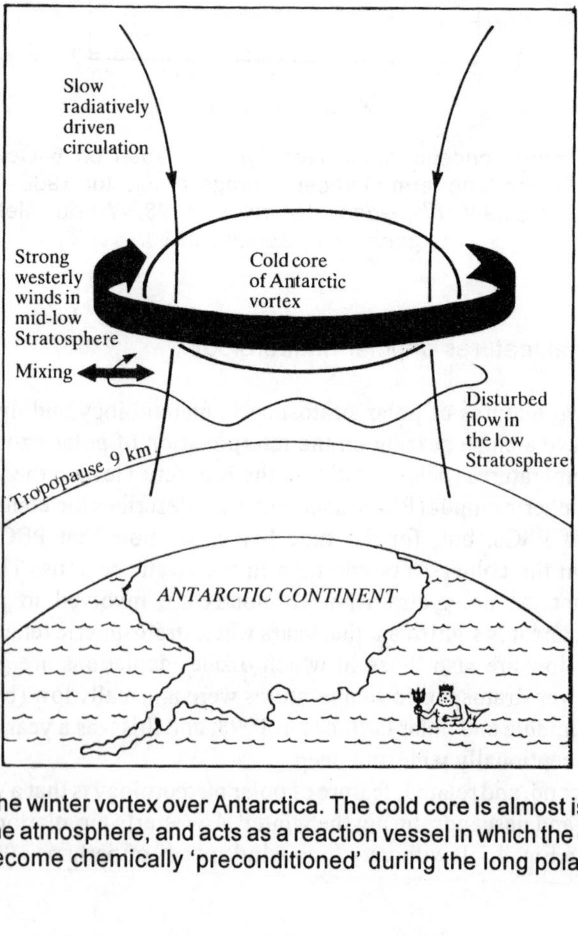

Fig. 4.29. The winter vortex over Antarctica. The cold core is almost isolated from the rest of the atmosphere, and acts as a reaction vessel in which the constituents may become chemically 'preconditioned' during the long polar night.

supply into it of reactants and a flow out of products. A further modification to the simple view is also necessary because the Antarctic vortex is sufficiently extensive that its edges are exposed to sunlight even at the winter solstice. As noted on p. 240, there is evidence for ozone depletion starting in mid-winter that probably originates from such regions. Whatever the details of the processes, the low temperatures, the presence of polar stratospheric clouds, and the unusual dynamics and meteorology of the vortex most assuredly provide the backdrop to some unexpected and surprising chemistry.

4.7.3 Anomalous chemical composition

The confirmation of the British Antarctic Survey observations of ozone depletion naturally generated a great deal of scientific activity. An ozone expedition was mounted in 1986, and it was repeated, in a greatly expanded form, in 1987. The 1987 campaign involved not only ground- and satellite-based measurements and balloons, but two 'airborne laboratories' which flew into the depleted region several times to obtain detailed information about the extent of the perturbed region, and about the chemistry in it. The instrumented aircraft used in the 1987 campaign were a DC8 and the stratospheric ER-2 (see p. 184). The airborne laboratories were able to measure concentrations of ozone; particulate matter, aerosols, and condensation nuclei; the oxides of nitrogen and of the halogens; water vapour concentrations; and temperatures, pressures, and other meteorological parameters. Many of the measurements were obtained by two or more methods. An enormous collaborative effort was involved. Some 150 scientists and support staff from 19 organizations and four nations co-ordinated their efforts in a most effective manner. The wealth of data obtained has permitted much interpretation of the ozone hole phenomenon. Several more campaigns have been mounted subsequently, but it was the discoveries in 1987 that provided the basis for current interpretations of the ozone hole.

The stratospheric aircraft measurements of 1987 confirmed, beyond all doubt, that there is anomalous chemistry going on within the vortex region. First consider the concentrations of ozone and of catalytically active ClO radicals. Figure 4.30 shows measurements obtained at altitudes of around 18 km in late August. Concentrations of ozone are normal. But the chlorine oxide concentrations show a sharp rise at a latitude of about 65 °S. The concentration goes up by a factor of 10 over a few hundred kilometres. In the perturbed region, the concentrations are, in fact, more than 100 times greater than they are at lower latitudes. It is worth remembering that almost all the chlorine in the Earth's atmosphere is a consequence of man's release of compounds containing it! Only a few weeks later, in mid-September, there is a dramatic change in the behaviour of ozone (Fig.4.31). This is a time at which the hole has developed fully. Ozone concentrations decrease over exactly the latitude range where the

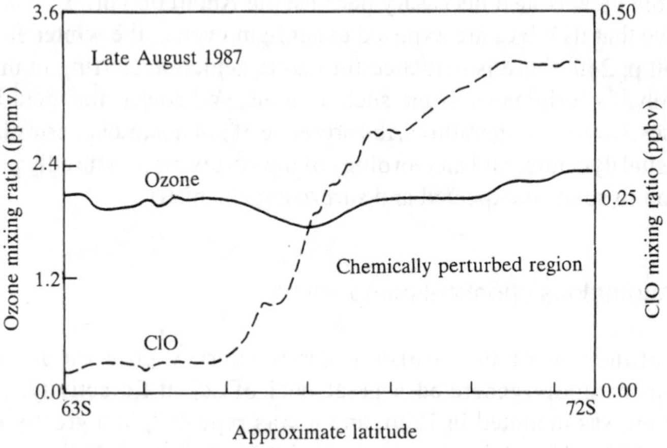

Fig. 4.30. Latitude dependence of ozone and chlorine monoxide (ClO) on entering the chemically perturbed region: late August 1987.

chlorine monoxide concentration increases on entering the chemically perturbed region within the polar vortex. The strong anti-correlation between the ozone and chlorine oxide concentrations is a strong indication that chlorine chemistry is somehow responsible for the ozone depletion. This anti-correlation is exhibited not only by the large-scale changes but also by the smaller fluctuations in concentration seen on traversing the critical latitude region.

Other measurements, summarized in Fig. 4.32, show that the stratosphere within the disturbed region is abnormally dry, and that it is highly deficient in

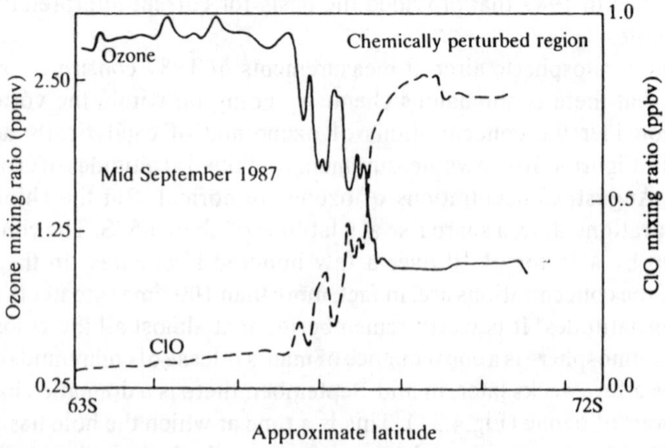

Fig. 4.31. Latitude dependences as for Fig. 4.30, but now in mid-September.

the oxides of nitrogen, the drop from normal stratospheric concentrations occurring at just those latitudes where the chlorine monoxide concentration increases and the ozone concentrations decrease. The dehydration and denitrification are explained by the condensation of water and the conversion of the oxides of nitrogen to nitric acid in the polar stratospheric clouds, the particles of which may become large enough to undergo sedimentation over appreciable distances. If the PSCs contain HNO_3 (see Section 4.7.4), NO_y may be removed irreversibly: up to 90 per cent of the available reactive nitrogen has been observed to be lost in this way in the Antarctic (and the Arctic) vortex. It is this removal of the oxides of nitrogen which leads to the anomalous chlorine chemistry. Only the core of the vortex is cold enough for the larger particles to form, so we see again that it is the combination of low temperature and special dynamics in the atmosphere that set up the conditions needed for perturbed chemistry.

Dehydration may be brought about by a similar sedimentation of PSC particles containing large fractions of water-ice. As discussed in the next section, the formation of such particles requires very low temperatures; for this reason dehydration is observed in the Antarctic, but not in the Arctic where temperatures are higher. Although intense dehydration has not been observed without intense denitrification, the processes are quite possibly independent.

4.7.4 Polar stratospheric clouds

Polar stratospheric clouds are so central to accepted explanations of the ozone-hole phenomenon that their formation and composition warrant some examination. Figure 4.33 summarizes the information that will now be presented.

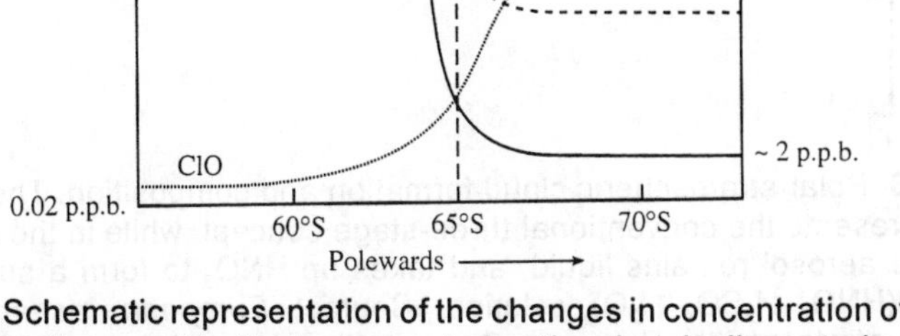

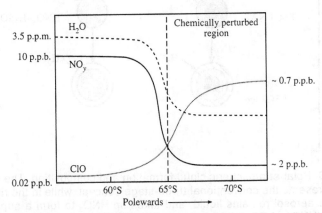

Fig. 4.32. Schematic representation of the changes in concentration of some of the species discussed in the text on entering the chemically perturbed region.

Their optical properties suggest two main classes of PSCs. Type I PSCs are small (< 1 μm diameter) HNO_3-rich particles, and have a *mass* mixing ratio of about 10 p.p.b.m. (parts per billion by mass). Type II PSCs are larger (from 10 μm to perhaps more than 1 mm diameter), are composed primarily of H_2O-ice together with minor amounts of HNO_3 as hydrates, and can constitute up to 1000 p.p.b.m. of the stratosphere when they are present. Several hydrates of the acids may be present in the PSC particles. The hydrates include nitric acid dihydrate (NAD: $HNO_3 \cdot 2H_2O$), nitric acid trihydrate (NAT: $HNO_3 \cdot 3H_2O$), and sulphuric acid tetrahydrate (SAT: $H_2SO_4 \cdot 4H_2O$). Type I PSCs often appear to

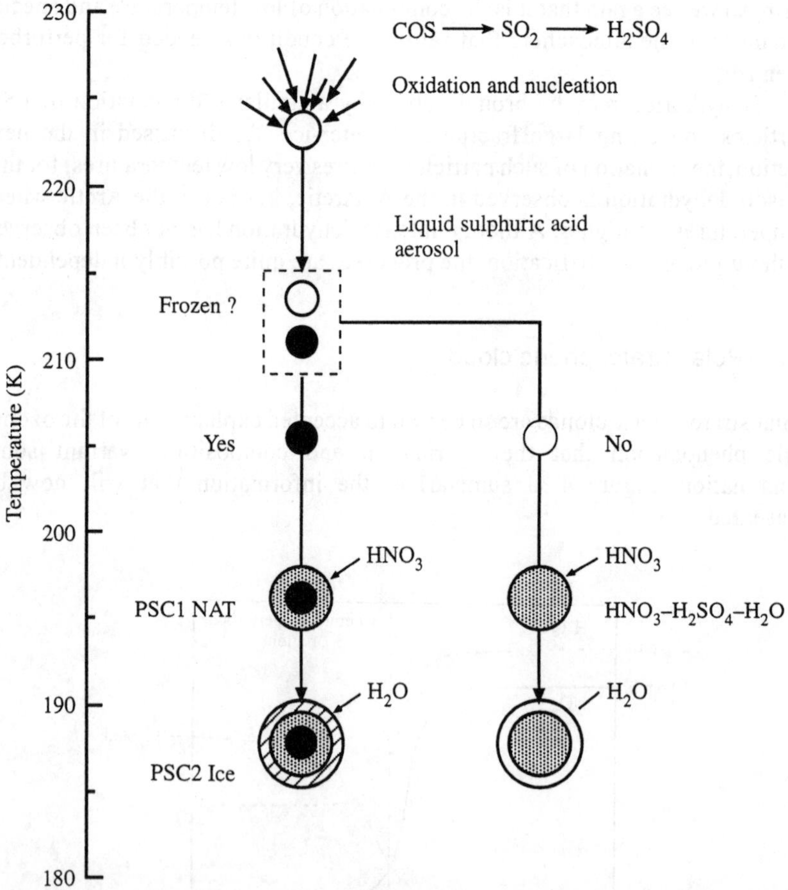

Fig. 4.33. Polar stratospheric cloud formation and composition. The left-hand path represents the conventional three-stage concept, while in the right-hand path, the aerosol remains liquid, and takes up HNO_3 to form a supercooled ternary (HNO_3–H_2SO_4–H_2O) solution. Source: *European research in the stratosphere*, European Communities, Luxembourg, 1997.

belong to one of two sub-categories, Type Ia solid particles consisting of nearly pure NAT and Type Ib particles, which are supercooled *liquid* ternary solutions of HNO_3–H_2SO_4–H_2O.

Figure 4.33 shows that Type I PSCs are formed at substantially higher temperatures—by 5 to 10 K— than the Type II PSCs. Both physical size and chemical composition are thus temperature-dependent. The extent of denitrification (and dehydration) will be affected by the rate of sedimentation, itself obviously more rapid for larger particles. It is worth noting here a difference between 'denoxification' (introduced in Section 4.5.5) and 'denitrification'. NO_x removed by formation of HNO_3 that is incorporated as NAT or as a ternary mixture in Type I PSCs may ultimately be available for release again. The efficacy of the HNO_3 reservoir has been enhanced: NO_x is (temporarily) lost, but NO_y is not, and the atmosphere is denoxified. Subsidence of NAT–ice mixtures to the troposphere in Type II PSCs, on the other hand, removes NO_y permanently, and the atmosphere is denitrified. The chemical consequences of denoxification and denitrification are different, a point to which we shall return in Section 4.7.5.

Some of the heterogeneous reactions introduced in Section 4.4.4

$$N_2O_5 + H_2O \rightarrow 2HNO_3 \tag{4.28}$$

$$ClONO_2 + H_2O \rightarrow HOCl + HNO_3 \tag{4.50}$$

$$ClONO_2 + HCl \rightarrow Cl_2 + HNO_3 \tag{4.51}$$

$$HOCl + HCl \rightarrow H_2O + Cl_2 \tag{4.52}$$

are known to proceed with efficiencies that depend on the nature of the surface. For example, the hydrolysis reactions (4.28) and (4.50) are fast on Type II PSCs (and on supercooled SSA: see p. 215), but slow on Type I PSCs (and on frozen SSA). On the other hand, reactions (4.51) and (4.52) are probably fast on both Type I and Type II PSCs, but the rates may depend on the relative humidity since that will determine the solubility of the HCl reagent in the aerosol particles. Once again, then, reaction probabilities will show an unusually strong dependence on temperature that may be more related to surface composition than to the ordinary considerations of reaction kinetics (Section 3.4.1).

4.7.5 Perturbed chemistry

Figure 4.34 provides, in diagrammatic form, a summary of the chemistry that explains the appearance of the Antarctic ozone holes, and it should be used to map the description that now follows. The central feature of the perturbed chemistry of the polar stratosphere is the conversion of reservoir compounds (Section 4.4.2) to catalytically active species (or their precursors) on the surface of the polar stratospheric clouds (PSCs). Most of the chlorine in the stratosphere is usually bound up in the reservoir molecules hydrogen chloride and chlorine nitrate, as a result of the reactions

$$Cl + CH_4 \rightarrow CH_3 + HCl \tag{4.30}$$

$$ClO + NO_2 + M \rightarrow ClONO_2 + M. \tag{4.34}$$

Liberation of the active chlorine from the reservoirs is normally rather slow. But the two reservoir molecules can react together on PSC particles, as explained in Section 4.4.4,

$$ClONO_2 + HCl \rightarrow Cl_2 + HNO_3. \tag{4.51}$$

The outcome is that molecular chlorine is released as a gas, and the nitric acid remains in the ice particles (as hydrates such as NAT), which can ultimately transport water and nitric acid out of the vortex, and perhaps even to the troposphere if the temperature is low enough. The molecular chlorine is photodissociated to atoms

$$Cl_2 + h\nu \rightarrow Cl + Cl \tag{4.70}$$

if sunlight is present, even at very low intensities. The surface reaction (4.51) has now been characterized in many laboratory studies. The PSCs disturb the balance between active and reservoir chlorine in two related ways. They provide surfaces on which unusual chemical change can occur, and they also

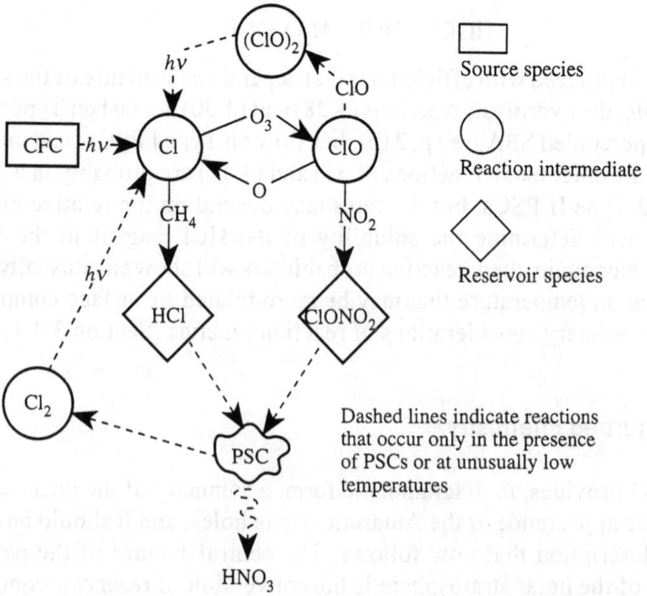

Fig. 4.34 Schematic diagram of chemical conversions in the ClO_x-catalysed decomposition of ozone in the presence of PSCs. Source: *European research in the stratosphere*, European Communities, Luxembourg, 1997.

transport active nitrogen out of the stratosphere, in the form of HNO_3, reducing the amount of $ClONO_2$ reservoir that can be formed in the first place. Denitrification (permanent loss of NO_y) requires lower temperatures than denoxification (loss of NO_x, possibly temporarily), as explained in Section 4.7.4. Low temperatures thus particularly favour reduction of $[ClONO_2]$.

Surface reactions such as reaction (4.51) can obviously play an important role in atmospheric chemistry whenever particles are present (Section 4.4.4), but their involvement in atmospheric chemical transformations was frequently neglected before 1985. Ignorance of reaction (4.51) is a major reason why the ozone hole was not predicted by any atmospheric model. Other surface reactions that have subsequently been demonstrated to occur in the laboratory, and that may be involved in polar chemistry, were introduced in Section 4.4.4, and mentioned again in Section 4.7.4. As examples, the reactions

$$ClONO_2 + H_2O \rightarrow HOCl + HNO_3 \qquad (4.50)$$

$$N_2O_5 + H_2O \rightarrow 2HNO_3. \qquad (4.28)$$

also both produce HNO_3, and reaction (4.28) converts another important reservoir molecule, N_2O_5. The reactions occur on ice particles (with dissolved HCl in the second case), and HNO_3 remains in the ice particle after reaction. HOCl, the gas-phase product of reaction (4.50), is readily photolysed by near-ultraviolet and visible light to yield ClO_x radicals.

Because the conversion can occur on the surface of the polar stratospheric clouds, release of molecular chlorine from the major reservoir molecules can continue in the chemically perturbed region throughout the polar winter and early spring. The vortex largely isolates the air within it, so that this part of the stratosphere can become chemically altered, or 'pre-conditioned', over the long polar night.

Consider, as an example, the molecular chlorine released by reaction (4.51). As we have noted already, when the Sun finally returns again in the Spring, the Cl_2 generated from the reservoir gases as a result of the pre-conditioning is rapidly split into chlorine atoms which can destroy ozone, and at the same time liberate chlorine monoxide

$$Cl_2 + h\nu \rightarrow Cl + Cl \qquad (4.70)$$

$$Cl + O_3 \rightarrow ClO + O_2. \qquad (4.19)$$

The beginnings of an explanation for enhanced concentrations of chlorine monoxide accompanying ozone depletions are already apparent. However, on their own, reactions (4.70) followed by (4.19) cannot lead to much ozone loss. It will be recalled from Section 4.4.1 that, in the mid-stratosphere, a *chain* process (Cycle 4)

$$Cl + O_3 \rightarrow ClO + O_2 \qquad (4.19)$$

$$ClO + O \rightarrow Cl + O_2 \qquad (4.20)$$

destroys large numbers of ozone molecules for each chlorine atom made available. Something of the sort must be going on in the perturbed Antarctic stratosphere, but it cannot be this chain, because the concentration of oxygen atoms in the lower stratosphere is far too small.

Alternative catalytic cycles are required to explain substantial ozone depletion in the polar stratosphere in early spring. The most important of these cycles involves the ClO dimer, $(ClO)_2$, formed in the self-reaction of ClO. These dimers have been shown experimentally to be readily photolysed to yield, by an indirect route, two free chlorine atoms. The cycle is thus

Cycle 17:

$$ClO + ClO + M \rightarrow (ClO)_2 + M \quad (4.71)$$

$$(ClO)_2 + h\nu \rightarrow Cl + ClOO \quad (4.72)$$

$$ClOO + M \rightarrow Cl + O_2 + M \quad (4.73)$$

$$2(Cl + O_3 \rightarrow ClO + O_2) \quad 2 \times (4.19)$$

$$\text{Net} \quad 2O_3 + h\nu \rightarrow 3O_2$$

Dimers such as $(ClO)_2$ are only formed at low temperatures, so that, once again, the low Antarctic polar temperatures are an essential component of another part of the perturbed chemistry. High concentrations of chlorine monoxide also favour the formation of the dimer, and such high concentrations are a feature of the chemically perturbed region of the Antarctic vortex. Recent experiments have shown that, for conditions appropriate to the polar stratosphere, dimer formation in reaction (4.71) is the only channel of importance in the self-reaction of ClO. The photolysis of $(ClO)_2$ in reaction (4.72) was another process completely unknown when the ozone hole was first found. Experimental proof that the reaction liberates atomic chlorine gives some confidence in the validity of the explanation for the Antarctic ozone hole.

The influence of temperature on the chemistry of PSCs was discussed in Section 4.7.4. Here, we return briefly to the subject. The general conclusion is that all the chlorine-activation reactions proceed more rapidly on Type II PSCs than they do on Type I PSCs. The time-scales for the activation processes are thought to range from about ten days at $T = 198$ K on Type I PSCs to one day on Type II PSCs at their threshold temperature (ca. 188 K), to just a few hours at temperatures of 180 K and below. Bearing in mind, as well, the much greater extent of denitrification at the lower temperatures, it becomes evident why stratospheric temperatures are so intimately connected with the depth and extent of the Antarctic ozone holes.

Although the early (pre-1990) atmospheric observations provided sufficient evidence for the basic understanding of ozone-hole chemistry, subsequent developments of *in-situ* and remote-sensing measurements have both confirmed the underlying ideas and provided further information for more detailed interpretations. Outstanding amongst these have been campaigns in the Arctic coordinated by the European Commission and by NASA, and the launch of the NASA Upper Atmosphere Research Satellite (UARS, described on p. 181), which can measure species such as ClO, $ClONO_2$, HCl, HNO_3, H_2O, and several long-lived tracers simultaneously. Some of the results obtained in the Arctic campaigns will be presented in Section 4.7.7. Here, we employ some of the newer data to provide a pictorial summary of the photochemical and dynamical features of polar ozone depletion in Fig. 4.35 The upper panel represents the conversion of chlorine from inactive to active forms during winter, and the formation of the inactive reservoirs again in the spring. Evolution of the polar vortex is indicated in the lower panel, where the temperature scale is meant to indicate changes in minimum temperatures in the lower polar stratosphere. The surface conversion of the reservoirs $ClONO_2$ and HCl to active chlorine occurs rapidly, and virtually to completion, in early winter; the reservoirs begin to reappear in late winter as temperatures become too high for further chemical processing on PSC surfaces. It is interesting that, during the 'normal' part of the year (late spring, summer, and early autumn), $ClONO_2$ is a slightly more abundant reservoir than HCl. Even more interesting is the observation that, on recovery, $ClONO_2$ initially overshoots its 'normal' level by a factor of up to two. It seems that the chlorine rapidly returns to the reservoirs during the recovery period, but that it takes time to re-establish the steady-state partitioning *between* the reservoirs. This behaviour arises because reaction (4.34) forming $ClONO_2$ is much more rapid than reaction (4.30) that yields HCl. While the overshoot shown in the figure is temporal, a similar spatial effect is seen for $ClONO_2$, and for the same reasons. A *collar region* of enhanced $ClONO_2$ concentration is found surrounding the vortex in late winter; in this region, PSC processing is limited or infrequent, and sunlight is available to produce NO_2 from the photolysis of HNO_3.

Other cycles may supplement cycle 17 in the catalytic destruction of ozone in the polar spring. It has been argued that one branch of the coupled reactions of BrO with ClO

$$BrO + ClO \rightarrow Br + Cl + O_2 \tag{4.61a}$$

could make a significant contribution to depletion of ozone in the Antarctic stratosphere *via* cycle 16 (shown on p. 221). Measurements of BrO (which is probably the major bromine-containing species within the perturbed region) establish the abundance at 4–310 p.p.t., and suggest that the BrO–ClO cycle probably only contributes 5–10 per cent to the depletion of ozone. A noteworthy aspect of reaction (4.61a), and cycle 16 in general, is that halogen oxides are recycled to the corresponding halogen atoms *without* the need for sunlight. Other branches of the interaction between BrO and ClO (Section 4.6.4) produce OClO (reaction (4.61b)) and BrCl (reaction (4.61c)). BrCl regenerates Br and Cl, but only on photolysis. Reaction (4.61b) is the only confirmed stratospheric source of OClO, so that the presence of OClO is a good

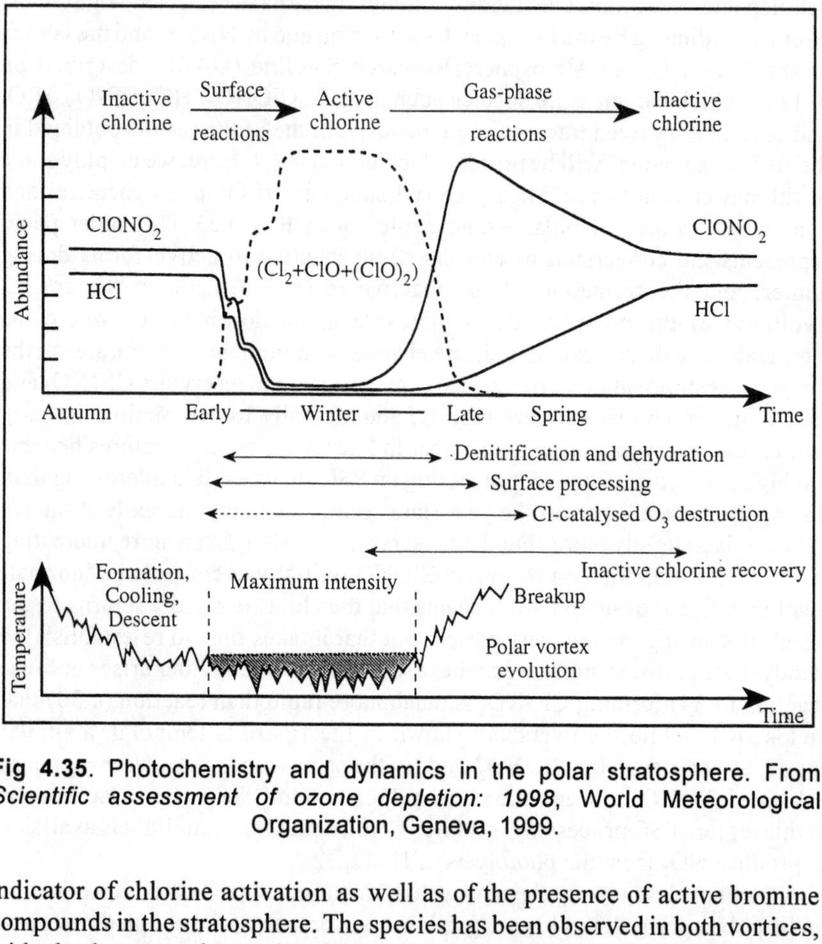

Fig 4.35. Photochemistry and dynamics in the polar stratosphere. From *Scientific assessment of ozone depletion: 1998*, World Meteorological Organization, Geneva, 1999.

indicator of chlorine activation as well as of the presence of active bromine compounds in the stratosphere. The species has been observed in both vortices, with the largest column abundances being found in the Antarctic, and the quantities are broadly consistent with the expectations from model simulations. As noted earlier, enhancements in OClO were observed after the eruption of Mount Pinatubo. The increases occurred before temperatures were low enough for PSC processing, and imply additional ozone destruction due to chlorine (and bromine) activation on volcanic sulphate aerosol.

Because of the denitrification of the Antarctic stratosphere, concentrations of HO_x species are elevated. Yet another cycle (cycle 13, p. 219) then becomes possible based on the reactions

$$ClO + HO_2 \rightarrow HOCl + O_2 \tag{4.32}$$

$$HOCl + h\nu \rightarrow OH + Cl. \tag{4.59}$$

Recent calculations suggest, however, that this cycle contributes much less than the others to the net ozone loss.

4.7.6 Origin of chlorine compounds; dynamics

All the unusual chemistry that leads to the Antarctic ozone hole depends on the presence of chlorine compounds, and most of those chlorine compounds have been released by man in the form of the chlorofluorocarbons. The concentration of the CFCs in the lower atmosphere near the South Pole is almost exactly the same as it is in rural areas of Britain or the United States. In the Antarctic, the compounds have been transported in from populated regions of the Earth. However, at higher altitudes over Antarctica, as the stratosphere is reached, the concentrations of CFCs drop very abruptly, as shown already in Fig. 4.24 for one of the compounds. The absence of CFCs in the stratosphere means, of course, that the chlorine atoms have already almost all been released by photochemical decomposition, and are available to destroy ozone. This release may occur largely over mid-latitudes and the tropics, the chlorine being transported (mainly in the form of the reservoir gases) to polar regions.

The records of polar stratospheric ozone show a considerable variability in the losses from year to year, as even a cursory glance at Fig. 4.27 will show. Within any year, too, there is much variability from day to day in extent and depth of the ozone hole, even though the progression through formation to recovery follows the trends that we have described. Variability is largely consequent on the dynamical features of the vortex, and the random nature of the forces that act on it. Wave activity near the vortex produces much of this variability, and the interaction of waves with the tropospheric weather system. Polar wave activity can also be influenced by low-latitude phenomena such as the QBO (Section 4.5.3), and to a much lesser extent by El Niño (Section 4.5.4). In 'westerly' QBO years (those in which the winds in the equatorial lower stratosphere are from the west), the polar vortices are colder and more intense in both hemispheres than in 'easterly' years.

Features that have the greatest impact on ozone are changes in the spatial and temporal extent of low temperatures, and how long the vortex endures into the spring. Formation of PSCs requires temperatures below a certain threshold, so that fluctuations of a few degrees in temperature can produce substantial changes in the extent of processing inside the vortex. Furthermore, similar small changes greatly affect the extent of denitrification, and thus, in turn, how quickly the reservoir $ClONO_2$ can be re-formed after the Sun returns. If chlorine remains activated when sunlight is present, ozone loss rates increase substantially. A similar influence can be seen at work when wave activity distorts the vortex from a symmetric polar flow into one that transports chemically processed air into lower latitudes that are sunlit. Volcanic eruptions (Section 4.5.5) further add to the variability of the polar ozone in several ways. Increased stratospheric aerosol, for example, leads to a cooling in the lower polar stratosphere, as well as enhanced background chemical processing and additional condensation nuclei for the formation of PSCs.

Superposed on the factors that lead to inter-annual and intra-annual variability seem to be others that provide a trend for a deepening and widening of the Antarctic ozone hole, at least if the evidence of Fig. 4.27 and similar results is taken at face value. One major influence must be the increase in anthropogenic inorganic chlorine that has been reaching the stratosphere at least until the late 1990s. There may even be a non-linear effect, because the concentration of the chlorine monoxide dimer, a key species in ozone destruction, depends on the square of the active chlorine concentration. A positive feedback between ozone concentrations and temperatures may also have an amplifying influence. Reduced ozone in the stratosphere implies less solar heating, so that the vortex remains cold, and it and PSC processing persist longer. In turn, these factors will lead to increased ozone depletion. Long-term declines in ozone are thus expected to cause larger ozone holes each year, and there is certainly observational evidence that the vortices in both hemispheres have been lasting longer in recent years. Changes in tropospheric source gas concentrations may have similar effects. Increased CO_2 is expected to *reduce* lower-stratospheric temperatures and thus increase the frequency and extent of PSC formation; similarly, increased CH_4 burdens are likely to have the same effect.

4.7.7 The Arctic stratosphere

It is of obvious interest to discover if the anomalous chemistry seen in the Antarctic atmosphere could also be of significance in the Arctic. Because the winter stratosphere is generally warmer over the Arctic than over Antarctic regions (the temperatures are roughly 10 °C higher), polar stratospheric clouds are far less abundant and persistent in the north than in the south polar regions. The Arctic vortex is generally smaller, less stable, and shorter lived than the Antarctic vortex. Furthermore, the variabilities of stratospheric ozone within one winter–spring period and from year-to-year, discussed in the previous section, are particularly severe in the Arctic, and it has proved difficult to establish clearly the signal of chemically induced seasonal ozone loss.

Despite the formidable difficulties, ozone losses have now been detected that are comparable with those seen in the Antarctic, and that can thus properly be regarded as 'Arctic Ozone Holes'. For example, in the spring of 1995, 1996, 1997, and 1998 ozone losses were detected for the first time in the Arctic that rivalled in extent those seen in the southern hemisphere roughly a decade earlier. For example, the satellite TOMS (p. 181) instruments observed a record ozone low of 219 DU as the spring Sun dawned over the North Pole on 24 March 1997, and the average 1997 levels were 40 per cent lower than the March averages for the years 1979–82. Beginning in early March, sunlit regions within the Arctic circle (e.g. central and eastern Siberia) experienced ozone thicknesses in the range 240–260 DU, when the normal monthly averaged values for March are 460–500 DU in northern polar latitudes.

The years 1995–97 were characterized by particularly low polar stratospheric temperatures. The winter polar vortex of 1996/1997 was unusually strong and persistent into March. It is thus highly probable that the signature of Arctic ozone depletion has become sharper as polar stratospheric temperatures have dropped in recent years, and as the anthropogenic halogen burden has built up to its maximum level. In contrast, losses were relatively small in the winter of 1998/1999 (less than five per cent loss in the vortex), and this was a winter in which there was a warming in December, the vortex was perturbed, and no PSCs were seen after mid-January 1999. As explained in Section 4.7.6, the lower temperatures of the preceding winters may themselves be a consequence of global losses of ozone (see Section 4.8). If this explanation is correct, then it seems entirely possible that seasonal large losses of ozone over the Arctic could continue, and perhaps even become more severe, even after the halogen loading of the stratosphere begins to decline in response to the control protocols, because average stratospheric temperatures could continue to fall and their recovery lag the halogen reduction.

Temperature sensitivity of ozone depletion has been emphasized by the evidence presented in the last few pages, and it clearly explains why behaviour in the Arctic is so variable. The same understanding opens the way to an enquiry about the effects of future increases in greenhouse gases. Although the build-up of such species in the troposphere leads to global warming near the Earth's surface, where radiation is trapped, at the same time it causes cooling in the stratosphere, where the infra-red absorption is 'optically thin', and energy is radiated to space. Quite small changes in concentrations of radiative gases might tilt the balance between the production of PSCs and their absence. One recent model shows that temperature and wind changes induced by increased concentrations of greenhouse gases would alter the propagation of planetary waves. Such waves would then no longer break up the Arctic vortex as often as they do at present. Sudden stratospheric warmings (pp. 79–80) would occur less frequently. As a consequence of these changes, the Arctic vortex would be more stable, and itself produce significantly lower stratospheric temperatures, adding yet further to the greenhouse *cooling* of the stratosphere. According to one model, the combination of cooling effects might be so great as to double ozone loss in the Arctic by the year 2020. The model indicates that, given a peak in halogen-compound concentration in 2000, Arctic ozone depletion would be greatest during the decade 2010–19, showing a delay of 10–15 years after the maximum stratospheric chlorine levels are reached. In the worst years, up to two-thirds of the Arctic ozone column could be lost in the most severely depleted regions.

The detection of Arctic ozone depletion has required a concerted effort in order to overcome the problems imposed by the strong meteorological variability shown by ozone column amounts. Several campaigns have been mounted over the last decade or so, of which the US Airborne Arctic Stratospheric Expeditions (AASE I and AASE II) and the European Arctic

Stratospheric Ozone Experiment (EASOE: 1991–92), Second European Stratospheric Arctic and Mid-latitude Experiment (SESAME: 1994–95), and Third European Stratospheric Experiment on Ozone (THESEO: 1997–2000) are prime examples. The three European campaigns are especially noteworthy in that they represent one of the very first multinational efforts of this kind. Definitive ozone-loss studies became possible only through pan-European efforts that enabled diverse national activities to be woven into collaborative structures. Much greater effectiveness, for example in terms of the reliability of estimations of ozone loss, was achieved than would have been possible through individual national research.

As a result of the national and transnational efforts, there are today several independent methods by which to quantify Arctic ozone loss. They range from satellite-borne instruments aboard UARS (p. 181), such as the Halogen Occultation Experiment (HALOE) and Microwave Limb Sounder (MLS), to network measurements provided by SAOZ (p. 240) as well as the ozonesondes.

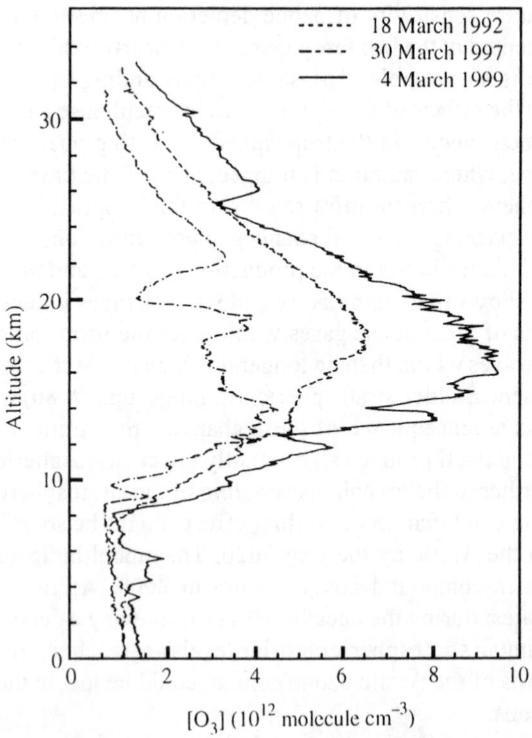

Fig. 4.36. Ozone profiles obtained over Ny-Ålesund (79°N) in March of the three years 1992, 1997, and 1999. Courtesy of P. von der Gathen, Alfred Wegener Institute, Potsdam, May 1999. These results were obtained in part within the EASOE, SESAME, and THESEO campaigns of the European Commission.

Sophisticated analysis is required to separate the signal of chemically induced ozone loss from dynamical effects, and it is therefore difficult to display simply the data from which such depletion may be demonstrated. Nevertheless, even though single profiles cannot be used for quantitative analysis, Fig. 4.36 gives an impression of the amount of chemical ozone destruction in a comparison of the late winters of 1992 and 1997, which illustrates the much greater extent of ozone removal at 17–25 km altitude in the second, colder, winter than in the first. The influence of differing meteorological conditions is then highlighted by the even smaller depletions seen in 1999, a year in which there was a warming the preceding December.

The evidence for perturbed chemistry in the Arctic stratosphere is very similar to that obtained in the Antarctic. The abundance of ClO is much higher than that predicted from gas-phase chemistry alone, and enhancements in [OClO] provide corroboration. Abrupt increases in [ClO] are associated in a number of cases with the edges of PSCs, thus once again implicating cloud processing of precursors. Abundances of $ClONO_2$ were elevated near the edge of the vortex but decreased towards the interior. The ratio of [HCl]/[HF] was reduced by a factor of two or three inside the vortex compared with outside it, indicating chemical conversion of HCl to other chlorine-containing species (the reduction in the perturbed Antarctic atmosphere is even higher, approaching a factor of six).

Extremely low column abundances of NO_2 are found within the Arctic vortex, yet again suggesting that cloud processing brings about denitrification. Measurements of N_2O concentrations can be used to derive the concentrations of NO_y expected under unperturbed conditions. In a typical experiment using this technique, two different methods of measuring N_2O profiles, and a further simultaneous determination of NO_y profiles, were deployed in the Arctic winter of 1995. Over much of the altitude range 15–30 km, measured [NO_y] fell much below that expected without chemical processing. For example, at 20 km and within the vortex, the observed [NO_y] was about 5 p.p.b., while the N_2O measurements predicted about 15 p.p.b. In retrospect, it is now evident that earlier observations, dating back to 1979, had revealed similarly low winter column NO_2 abundances that possessed very steep latitudinal gradients associated with stratospheric flow from polar regions (a phenomenon known as the Noxon 'cliff' after its discoverer). Column concentrations of HNO_3 in the Arctic, on the other hand, are much greater than in the Antarctic. Exceptionally high mixing ratios of 5–6 p.p.b. have been found at 2–3 km above the tropopause, instead of the usual 1–2 p.p.b. Such amounts mean that in the lowermost stratosphere the NO_y concentrations *exceed* those calculated from the N_2O measurements. These and other observations of nitrogen-containing compounds suggest that, although denitrification occurs at altitudes around 20 km, the PSCs may evaporate in the lower stratosphere, thus limiting the reduction in integrated column [HNO_3]. The sedimentation and vaporization of the PSC particles then produces *nitrification* at altitudes around 10–15 km.

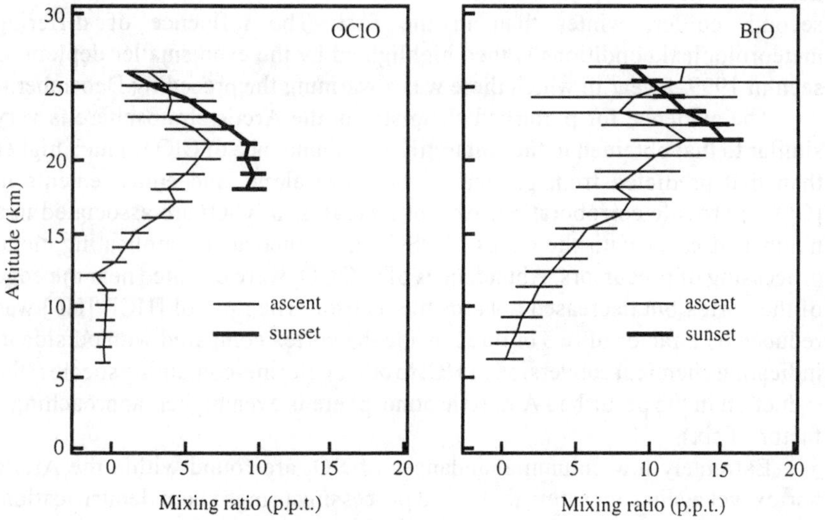

Fig. 4.37. Altitude profiles of OClO and BrO measured by a balloon-borne UV-visible absorption spectrometer. The ascent was at a solar zenith angle (SZA) of 87°; sunset corresponds to 90° SZA. Profiles were obtained on 31 January 1996 within the vortex, at low temperatures (190 K). From *European research in the stratosphere*, European Communities, Luxembourg, 1997.

Another contrast with the Antarctic is the small extent of dehydration accompanying the denitrification. The observations are not yet fully understood, but may indicate that denitrification can occur at higher temperatures than dehydration, as discussed on p. 245.

Many measurements have now been made by several techniques in the Arctic of the radical catalytic-chain carriers, ClO and BrO, both in column amounts and as altitude profiles. Furthermore, profiles of chlorine dioxide, OClO, have also been obtained. Figure 4.37 shows profiles of OClO and BrO obtained by a balloon-borne SAOZ instrument (p. 240), flying within the cold (190 K) Arctic vortex. The observation of OClO is a very clear indicator of both chlorine and bromine activation, as explained on pp. 251–2. The BrO measurements themselves show a peak *concentration* (not mixing ratio) at 17 km, and a mixing ratio above 20 km of 10 p.p.t. The profile is shifted down by roughly 3 km from that calculated from measured Cl_y profiles. Although peak

ClO concentrations are around 1 p.p.b. (i.e. 100 times larger), Br is about 60 times more efficient than chlorine (p. 221) in destroying O_3. The downward shift in the BrO profile thus means that bromine-catalysed ozone destruction could be the dominant loss process in the lowermost stratosphere. A further factor favouring bromine-catalysed destruction in the Arctic compared with the Antarctic is the expected greater efficiency of the combined ClO_x–BrO_x cycle (cycle 16) compared with the ClO_x–ClO_x cycle (cycle 17) at the relatively higher temperatures and lower solar illumination that prevail in the north.

In general, current models of the Arctic polar stratosphere can simulate well the atmospheric observations of the nitrogen compounds (NO_2, HNO_3, HNO_4, N_2O_5, and so on), the reservoirs HCl and $ClONO_2$, and the chain carrier ClO. The measurements do not yet allow for a proper understanding of the partitioning between the activated forms of chlorine (ClO, $(ClO)_2$, and other chlorine radicals). There are also some remaining uncertainties in the models. For example, several models appear to *overestimate* the heterogeneous conversion of NO_2 and N_2O_5 into HNO_3. The timing of the deactivation of chlorine in spring, when $ClONO_2$, and later HCl, are regenerated, does not seem to be correctly captured by the models; associated with this problem are the difficulties of simulating the extent of denitrification and the return of NO_x from HNO_3 in spring. The role of bromine still remains much less certain than that of chlorine, despite evidence that it may be of considerable importance.

The models are certainly well enough developed for the predictions from them to be regarded with confidence. For example, the Cambridge chemical-transport model has been used to simulate ozone profiles such as those shown in Fig. 4.36. In comparisons of the modelled and observed profiles, there is reasonable agreement so long as the model includes all known chemical processes (although the model underestimates the depletion at around 20 km). Without heterogeneous chemical processing, the model generates a profile resembling the experimental one for March 1992, so that the difference between the curves with and without this heterogeneous chemistry provides an estimate of the 'chemical' ozone loss. Cumulative chemical destructions of ozone have been calculated with the same model for winters in different years. Ozone destruction depends on how low temperatures become in the lower stratosphere, in agreement with the observations. By mid-March in the coldest years (e.g. 1996), ozone losses at 20 km altitude, and averaged north of 60°N, reached nearly 25 per cent, again apparently consistent with the observations. The models can, of course, be run for a variety of scenarios. These scenarios can include changes in the chemistry and physics (such as the threshold temperature for permanent denitrification by PSC particles), as well as changes in meteorology and in halogen loading. One dramatic conclusion is that, had the halogen loadings not been limited by the control protocols, we should already have been anticipating Arctic ozone losses as large as those experienced in the Antarctic. Given the distribution of the world's human population, very serious consequences might have followed.

4.7.8 Implications of the polar phenomena

There are two questions to be faced in assessing what the effects of polar ozone depletion might be. First, there are the direct consequences of reduced ozone in those geographical regions where the ozone is lost. Such consequences range from meteorological impacts, including changes of temperature and weather patterns, to the results of increased exposure of living organisms to UV radiation. Secondly, there might be remote consequences on parts of the environment not immediately in or under the ozone-hole region. It is, perhaps, worth noting immediately that, although ozone depletions in the Arctic are much less than those in the Antarctic *so far*, the Arctic is much closer to high densities of population, of humans at least, and the off-pole centering of the Arctic vortex brings the northern hemisphere ozone hole even closer to many of us. For those living in the United Kingdom, March 1996 brought record low values of total ozone. On 3 March, the total ozone over Camborne (50 °N) was 206 DU, and two days later, Lerwick (60 °N) recorded 195 DU. This was the first time that a daily value below 200 DU had been observed over the UK. While meteorological factors contributed to the low ozone values, it is clear that Arctic air in which photochemical ozone loss had *already* occurred was lying over the UK at the time.

A major reason for concern about stratospheric ozone depletion is the biological danger posed by increased solar ultraviolet radiation reaching the Earth's surface (Section 4.6.1). In that context, the creation of polar ozone holes must increase the amount of UV that can penetrate to the ground, especially in the significant UV-B region (λ = 280–320 nm). Spring-time biologically active irradiances in the Antarctic can exceed those experienced in the (unperturbed) midsummer. For example, even by as early as 5 November 1987, the calculated UV-B irradiance averaged over the day exceeded the solstice value by over 50 per cent, and the irradiance at noon was relatively even more enhanced. For the same overhead ozone column, the UV exposure in the Antarctic spring is comparable with typical values at a mid- to low-latitude location such as Miami. A particular threat is posed to indigenous organisms rather than to human visitors to the Antarctic. One area of concern is the possible sensitivity of the spring bloom of phytoplankton living in the surface waters around Antarctica. The UV dose is already greatest in spring, when the ice is much more transparent than in the summer.

Quite apart from the local effects of polar ozone depletion, it is important to know if the processes occurring within the chemically perturbed region can influence ozone concentrations at lower latitudes. Although the vortex is conveniently pictured as a 'containment vessel' that isolates the chemically perturbed region, there is no seal around the polar air, and the containment provided by the vortex is somewhat leaky. Ozone-deficient or chemically processed air can thus be transported to lower latitudes. Later in spring, when the vortex breaks up, more extensive lateral mixing takes place. Export of ozone-poor air from polar regions to middle latitudes might lead to a dilution of ozone that could persist for up to a year because of the relatively slow photochemical replacement of ozone, and transport of processed air rich in active chlorine could enhance the effect. If the deficit lasts from one spring to another, there could be a cumulative, permanent depletion of ozone. There is now clear evidence for a reduction in stratospheric ozone during ozone at mid-latitudes (see Section 4.8), although the causes remain an open question. One potential explanation is erosion of ozone-poor air from the vortices. In the northern hemisphere, frequent extra-vortical PSCs, probably enhanced by lee-wave activity above the Arctic mountain ridges, may also play a role akin to that of PSCs in the vortex itself.

In addition to the 'dilution' effect produced by the transport of ozone-poor air from the polar vortex to middle latitudes, activated chlorine itself might be exported from the vortex (mainly from its edge) to middle latitudes, where ozone loss could then occur. At the time of writing, it seems that such export of activated chlorine does occur, but also it appears as though it is unlikely to have a major impact on mid-latitude ozone. The dilution effect may be more important, although satellite data have indicated that the *centre* of the vortex is generally isolated from mid-latitude air, especially in the Antarctic. There is mixing of air from the edge of the vortex, but there remains controversy about whether or not the extent of such mixing is sufficient to constitute the 'flowing processor' described at the end of Section 4.7.2. Satellite data are consistent with a dilution effect following seasonal Antarctic ozone depletion, although it must be noted that volcanic aerosols and other factors might be responsible for the changes.

The understanding that heterogeneous chemistry is central to Antarctic and Arctic ozone depletion has brought a heightened awareness of the importance of aerosol processes in stratospheric chemistry as well as physics. Laboratory experiments have shown that reactions can proceed on sulphuric acid particles that are similar to those taking place on the surfaces of PSCs. Sulphuric acid aerosol is widely distributed in the stratosphere, and might thus be responsible for ozone depletion in non-polar regions. Such activity is obviously enhanced in periods following volcanic eruptions that deposit sulphate aerosol in the stratosphere. Clear signatures are apparent in the observations of chemical processing following the El Chichón and, especially, the Mount Pinatubo explosions (see Sections 1.3 and 4.5.5) and show that the aerosol was responsible for heterogeneous loss. Ozone measurements exhibit record lows following the eruptions. Total column NO_2 abundances also show a marked decrease, in the same way as in the Antarctic perturbed chemistry. This circumstantial evidence is indicative of chemical processing on aerosol surfaces.

One very important lesson learned from the surprise discovery of the Antarctic ozone hole was that the stratospheric chemistry thought to be essentially 'complete' in the mid-1980s was, in reality, missing a vital part. The possibility that heterogeneous chemistry might significantly affect ozone throughout the stratosphere emphasizes the need to develop atmospheric models that incorporate surface processes with the requisite degree of sophistication. At the same time, it is essential that reliable laboratory experiments provide the appropriate data for use in the models.

4.8 Ozone variations and trends

Section 4.7 introduced the idea of trends in ozone concentrations by examining how ozone concentrations over polar regions have changed during the local spring over the past two decades or so. We now turn to an examination of long-term possible *global* decreases in stratospheric ozone.

Model predictions of present-day ozone concentrations are mainly of interest in so far as they adequately explain experimental observations and thus give credence to the theory on which they are based. Predicted ozone concentrations are not more 'reliable' than measured ones, although there has been some tendency to interpret observations in the light of expectations from some model or other. Increases in atmospheric CO_2, N_2O, and CFCs over the last 10–20 years were sufficient to have had quite marked effects on ozone according to the models. But that is immaterial. The question we really have to ask is: have ozone concentrations actually changed to a detectable and measurable degree?

Quite severe problems interfere with providing a direct answer to our question. Each measurement of ozone concentration is subject to random errors as well as calibration errors that may be regarded as noise. On top of the measurement noise are real random fluctuations of ozone concentration and longer term changes such as those due to seasonal variations. Natural phenomena are known to affect ozone concentrations (Section 4.5). Variations in solar ultraviolet intensities (Section 4.5.2) over the eleven-year solar cycle produce changes of up to two per cent. Column ozone concentrations alter in sympathy with the quasi-biennial oscillation (QBO: Section 4.5.3), typical changes being about 2.7 per cent for mid-latitudes. Volcanoes (Section 4.5.5) can impose a large, but largely unpredictable, signal on top of long-term ozone trends. The observed data series must exist over a long enough period that all the noise and variability factors can be filtered out to reveal possible long-term trends.

Section 4.4.6 contains a brief description of the measurement of ozone in the stratosphere, and suggests some of the advantages and disadvantages of the various types of measurement available. Figure 4.9 provides a diagrammatic summary of the more important techniques that have been used. TOMS (Total Mapping Ozone Spectrometer) devices have been carried by several satellites (Nimbus-7, October 1978 to May 1993; Meteor-3, August 1991 to December 1994; Earth Probe, August 1996 to late 1998, but resuscitated in January 1999), while GOME (Global Ozone Monitoring Experiment) has been in operation on ERS-2 since April 1995 until the time of writing.

Elaborate analyses of the available data have been undertaken several times by the NASA/WMO International Ozone Trends Panel. The first report (1988) exemplifies the methods used for the analysis. The satellite data and ground-based measurements were used to provide cross-checks. For example, the TOMS column ozone values as the satellite passed over each ground station were used for intercomparison and quality control of the ground data. The acceptable ground-based measurements (about 50 to 65 stations in any one month) were then used to back-calibrate the satellite instruments. Allowance was made for increases in tropospheric ozone (which affect the ground instruments more than the satellite ones) and for the effects of pollutants. In this way, the drifts in the SBUV and TOMS instruments were assessed. Finally, the changes in ozone ascribed to natural phenomena (solar cycle and QBO: see above) were removed and long-term trends identified. The analysis revealed that statistically significant downward trends in ozone had already occurred since the 1960s; that losses had taken place at mid-latitudes in the northern hemisphere; and that the losses were larger in winter than in summer. Subsequent assessments have shown much the same pattern.

One clear indication of the overall behaviour can be seen in Fig. 4.38. Global averages of total ozone are shown as a function of month of the year for

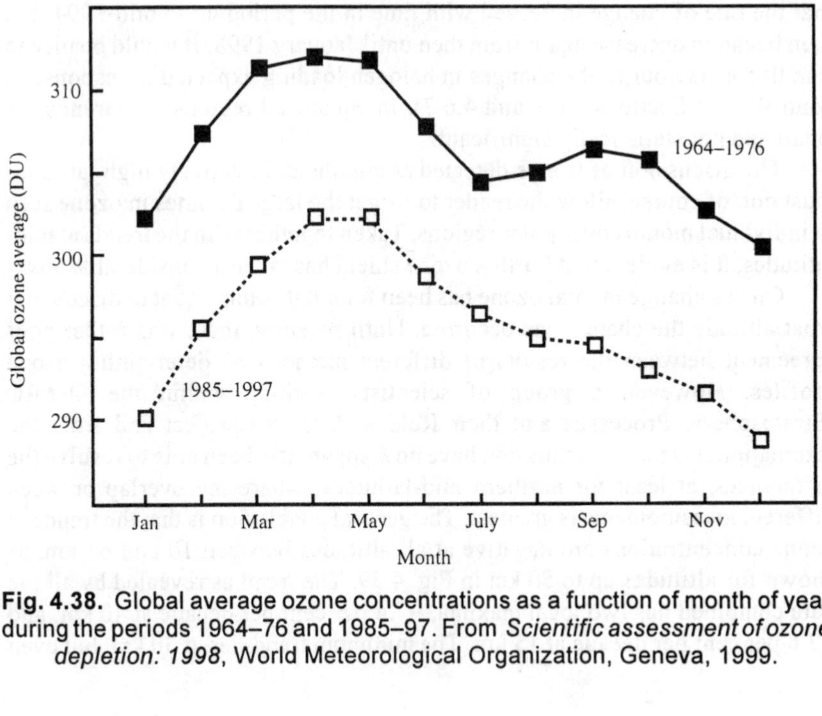

Fig. 4.38. Global average ozone concentrations as a function of month of year during the periods 1964–76 and 1985–97. From *Scientific assessment of ozone depletion: 1998*, World Meteorological Organization, Geneva, 1999.

the baseline years of 1964–76 and for a period 21 years later, 1985–97. Ozone amounts have declined substantially during all months of the year. The strongest declines are in the September–November and February–April periods of the year. These months are those in which the annual ozone loss over Antarctica and over northern middle and high latitudes now occurs. In the earlier years, there was a double maximum in ozone corresponding to the spring maxima over northern and southern middle and high latitudes. In the second set of years, however, the secondary maxima in September–October have disappeared. These years are the ones during which deep ozone holes have generally developed at southern high latitudes during the austral spring. The data summarized in Fig. 4.38 are area-weighted averages from many observing stations. Such stations each record an annual cycle of concentrations. However, for each station at middle and high latitudes, the amplitude of the oscillation has decreased by 15 to 20 per cent, largely as a consequence of the decrease in the annual maximum.

The main conclusion from recent assessments is that there have been large, statistically significant downward trends in total ozone in both hemispheres at middle and high latitudes. These trends are most pronounced in the local winter and spring seasons, being −3 to −6 per cent per decade in the Northern Hemisphere, and reaching as much as −10 per cent per decade in the Southern Hemisphere spring, largely as a result of the influence of the Antarctic ozone hole. In summer and autumn, the trends are smaller, being −1 to −3 per cent per decade in the north and −2 to −5 per cent in the south. There is some evidence that the rate of change increased with time in the period up to mid-1994, but then began to decrease again from then until January 1998. It would be nice to link this behaviour to the changes in halogen loading expected in response to controls (see Sections 4.6.6 and 4.6.7). In equatorial regions, any trends are small and not statistically significant.

The discussion of trends detected at middle and relatively high latitudes must not, of course, allow the reader to forget the large declines in ozone seen in individual months over polar regions. Taken together with the trends at mid-latitudes, it is evident that Earth's ozone shield has taken a considerable blow.

Once a change in total ozone has been found, it is important to discover at what altitude the change has occurred. Until recently, there was rather poor agreement between the results of different methods of determining ozone profiles. However, a group of scientists working within the SPARC (Stratospheric Processes and their Role in Climate) project and from the International Ozone Commission have now apparently been able to resolve the differences, at least for northern mid-latitudes, where the overlap between different measurements is greatest. The general conclusion is that the trends in ozone concentrations are negative at all altitudes between 10 and 65 km, as shown for altitudes up to 50 km in Fig. 4.39. The trend as revealed by all the data combined has two local maxima, −7.4 per cent per decade at 40 km, and −7.6 per cent per decade at 15 km. The minimum trend lies at 30 km, but even

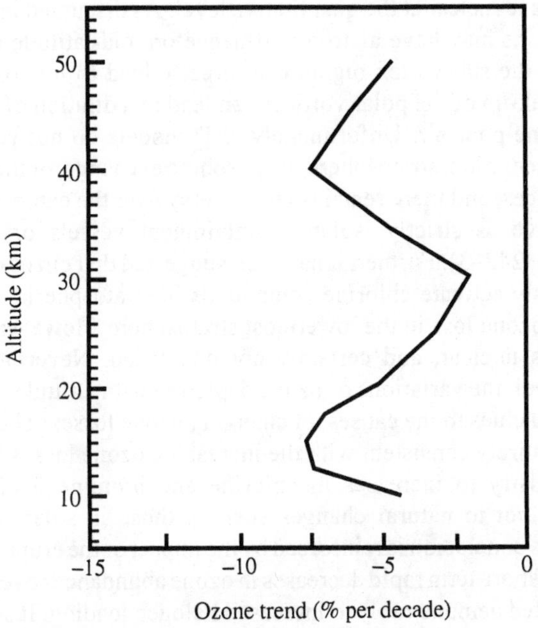

Fig. 4.39. Mean trend of ozone concentration profiles at northern mid-latitudes in the period 1980–96 estimated from combined measurements using a variety of techniques. From *Scientific assessment of ozone depletion: 1998*, World Meteorological Organization, Geneva, 1999.

there it is −2.0 per cent per decade. It is particularly noteworthy that the maximum declines at around 15 km correspond with those regions where the absolute concentrations are highest, so that the impact on total ozone is consequentially greatest. In general, the trends in total ozone amounts and in vertical profile concentrations are now consistent. The magnitude of the trends shows a clear seasonal variation, by a factor of about two, with the largest losses occurring in winter. Although the data are more sparse for southern latitudes, there does not seem to be any significant interhemispheric difference in vertical trends in the data for the upper stratosphere.

Finally in this chapter, it is appropriate to examine briefly the causes of the trend for ozone concentrations to decrease at all altitudes. We have discussed at length in Section 4.7 the reasons for the seasonal loss of ozone in polar regions, and attributed it to the build-up of anthropogenic chlorine and bromine, coupled, perhaps, with a long-term cooling of the lower stratosphere. But what are the causes of the mid-latitude decline of ozone? In the upper stratosphere, the answer again seems clearly to be the increase in anthropogenic halogens. Photochemical models, using the most up-to-date chemistry from laboratory studies, give excellent quantitative agreement with observed reductions in ozone. So far as the lower stratosphere is concerned, the situation is less clear at the quantitative level. As discussed in Section 4.7, the polar regions may have a strong influence on mid-latitude ozone. Polar processing in the sub-vortex region can directly lead to low ozone at mid-latitudes. Break-up of the polar vortices can lead to a dilution of mid-latitude ozone by ozone-poor air. Unfortunately, 2-D models do not yet accurately simulate filamentation, stratosphere–troposphere exchange, or the break-up of the polar vortices, and there remains controversy over the extent to which the vortices behave as strictly isolated containment vessels or as 'flowing processors' (p. 242–3). Further, it has been suggested that cirrus clouds could heterogeneously activate chlorine compounds in stratospheric air, and thus contribute to ozone loss in the lowermost stratosphere. However, the role of cirrus remains unclear, and certainly not quantified. Nevertheless, at the *qualitative* level, the variations of ozone depletion with altitude, latitude, and season all hold clues to the causes of changing ozone losses. The space–time patterns are entirely consistent with the increasing ozone losses being linked through chemistry to increases in chlorine and bromine loading at mid-latitudes, and not to natural changes such as those in solar irradiance or dynamics. This conclusion is reinforced by the impact of the eruption of Mount Pinatubo. The short-term rapid decreases in ozone abundances over a few years can be connected demonstrably to increased halogen loading. It seems fitting to conclude this chapter with the observation that one of the best indicators of man's perturbation of the stratosphere should have been provided by one of nature's most dramatic exhibitions of her power.

Bibliography

Books and articles containing introductions to the study of atmospheric chemistry in general and to the investigation of stratospheric ozone chemistry in particular.

Aeronomy of the middle atmosphere: chemistry and physics in the stratosphere and mesosphere. Brasseur, G. and Solomon, S. 2nd edn. (D. Reidel, Dordrecht, 1986).

Physics and chemistry of the stratosphere. Pommereau, J.P. *Comptes Rendus, Série II, Fasc. A.* **322**, 811 (1996).

Dynamics of the middle atmosphere. Holton, J. R. and Matsuno, T. (eds.) (Terrapub, Tokyo, 1984).

Light, chemical change and life. Coyle, J. D., Hill, R. R., and Roberts, L . R. (eds.) (Open University Press, Milton Keynes, 1982).

These documents from the World Meteorological Organization are major sources of information concerning the chemistry of the stratosphere, models of it, the sources of trace gases, and their concentrations. Much of the quantitative information in Chapter 4 of the present book is gleaned from these publications.

Scientific Assessment of Ozone Depletion: 1998. World Meteorological Organization, Global ozone research and monitoring project: report no. 44. (WMO, Geneva, 1999).

Scientific Assessment of Ozone Depletion: 1994. World Meteorological Organization, Global ozone research and monitoring project: report no. 37. (WMO, Geneva, 1995).

Scientific Assessment of Ozone Depletion: 1991. World Meteorological Organization, Global ozone research and monitoring project: report no. 25. (WMO, Geneva, 1992).

Scientific assessment of stratospheric ozone: 1989. World Meteorological Organization, Global ozone research and monitoring project: report no. 20. (WMO, Geneva, 1990).

Atmospheric ozone 1985. World Meteorological Organization, Global ozone research and monitoring project: report no. 16. (WMO, Geneva, 1986).

The stratosphere 1981. Theory and measurement. World Meteorological Organization, Global ozone research and monitoring project: report no. 11 (WMO, Geneva, 1981).

The reports of the UK Stratospheric Ozone Research Group (SORG) are another important source of information. There have been seven reports up to 1999, of which two examples are cited. The report for 1988 is particularly useful in its coverage of models, source gases, trends in ozone concentrations, and its overview of the Antarctic 'ozone hole' phenomenon.

Stratospheric Ozone 1999. Stratospheric Ozone Research Group (SORG). (Her Majesty's Stationery Office, London, 1999).

Stratospheric Ozone 1988. Stratospheric Ozone Research Group (SORG). (Her Majesty's Stationery Office, London, 1988).

Section 4.2

An ozone climatology based on ozonesonde and satellite measurements. Fortuin, J.P.F. and Kelder, H. *J. geophys. Res.* **103,** 31709 (1998).

Historical introductions to the study of atmospheric ozone

Forty years' research on atmospheric ozone at Oxford: a history. Dobson, G.M.B. *Applied Optics* **7** 387 (1968).

Pioneers of ozone research. A historical survey. Schmidt, M. (Max Planck Institute, Katlenburg–Lindau, 1988).

Sections 4.3 and 4.4

Stratospheric ozone and the influence of catalytic destruction cycles

Catalytic destruction of stratospheric ozone. Lary, D.J. *J. geophys. Res.* **102**, 21515 (1997).

Balloon-borne measurements of stratospheric radicals and their precursors: implications for the production and loss of ozone. Osterman, G.B., Salawitch, R.J., Sen, B., Stachnik, R.A., Pickett, H.M., Toon, G.C., and Margitan, J.J. *Geophys. Res. Letts.* **24**, 1107 (1997).

Removal of stratospheric O_3 by radicals: *in situ* measurements of OH. Wennberg, P. O. and 18 others. HO_2, NO, NO_2, ClO, and BrO. *Science* **266,** 398 (1994).

Halogen-catalysed methane oxidation. Lary, D.J. and Toumi, R. *J. geophys. Res.* **102,** 23421 (1997).

Chemistry of the stratosphere. Thrush, B. A. *Rep. Prog. Phys.* **51**, 1341 (1988).

The photochemistry of the stratosphere. Turco, R. P. in *The photochemistry of atmospheres*. Levine, J. S. (ed.) (Academic Press, Orlando, 1985).

Photochemistry and dynamics of the ozone layer. Prinn, R. G., Alyea, F. M., and Cunnold, D. M. *Annu. Rev. Earth & planet. Sci.* **6**, 43 (1978).

A review of the reactions of halogen oxide radicals, which are amongst the most significant of the catalytic 'families'

Halogen oxides: radicals, sources and reservoirs in the laboratory and in the atmosphere. Wayne, R.P. (ed.) (European Commission, Brussels, 1995); *Atmos. Environ.* **29**, 2675 (1995)

Section 4.4.3

Budgets of catalytically active species

General aspects

Normalization of correlations for atmospheric species with chemical loss. Plumb, I.C., Vohralik, P.F., and Ryan K.R. *J. Geophys. Res.* **104**, 11723 (1999).

Chemical perturbation of the lowermost stratosphere through exchange with the troposphere. Lelieveld, J., Arnold, F., Bregman, B., Bürger, V., Crutzen, P., Fischer, H., Siegmund, P., van Velthoven, P., and Waibel, A. *Geophys Res. Lett.* **24**, 603 (1997).

Evaluation of source gas lifetimes from stratospheric observations. Volk, C.M., Elkins, J.W., Fahey, D.W., Dutton, G.S., Gilligan, J.M., Loewenstein, M., Podolske, J.R., Chan, K.R., and Gunson, M.R. *J. geophys. Res.* **102**, 25543 (1997).

Interrelationships between mixing ratios of long-lived stratospheric constituents. Plumb, R.A. and Ko, M.K.W. *J. geophys. Res.* **97**, 10145 (1992).

Water vapour

Trends in stratospheric humidity and the sensitivity of ozone. Evans, S.J., Toumi, R., Harries, J.E., Chipperfield, M.P., and Russell, J.M., III *J. geophys. Res.* **103**, 8715 (1998).

The dry stratosphere: A limit on cometary water influx. Hannegan, B., Olsen, S., Prather, M., Zhu, X., Rind, D., and Lerner, J. *Geophys. Res. Letts.* **25,** 1649 (1998).

Mass fluxes of O_3, CH_4, N_2O and CF_2Cl_2 in the lower stratosphere calculated from observational data. Gettelman, A., Holton, J.R., and Rosenlof, K.H. *J. geophys. Res.* **102**, 19149 (1997).

Increases in middle atmospheric water vapor as observed by the Halogen Occultation Experiment and the ground-based Water Vapor Millimeter-wave Spectrometer from 1991 to 1997. Nedoluha, G.E., Bevilacqua, R.M., Gomez, R.M., Siskind, D.E., Hicks, B.C., Russell, J.M., III, and Connor, B.J. *J. geophys. Res.* **103**, 3531 (1998).

Implications of satellite OH observations for middle atmospheric H_2O and ozone . Summers, M.E., Conway, R.R., Siskind, D.E., Stevens, M.H., Offerman, D., Riese, M., Preusse, P., Strobel, D.F., and Russell, J.M., III. *Science* **277**, 1967 (1997).

The hydrogen budget of the stratosphere inferred from ATMOS measurements of H_2O and CH_4. Abbas, M.M., Gunson, M.R., Newchurch, M.J., Mitchelsen, H.A., Salawitch, R.J., Allen, M., Abrams, M.C., Chang, A.Y., Goldman, A., Irion, F.W., Moyer, E.J., Nagaraju, R., Rinsland, C.P. Stiller, G.P. and Zander, R. *Geophys. Res. Letts.* **23**, 2405 (1996).

Seasonal variations of water vapor in the lower stratosphere inferred from ATMOS/ATLAS-3 measurements of H_2O and CH_4. Abbas, M.M., Michelsen, H.A., Gunson, M.R., Abrams, M.C., Newchurch, M.J., Salawitch, R.J., Chang, A.Y., Goldman, A., Irion, F.W., Manney, G.L., Moyer, E.J., Nagaraju, R., Rinsland, C.P., Stiller, G.P., and Zander, R. *Geophys. Res. Letts.* **23**, 2401 (1996).

An atmospheric tape-recorder - the imprint of tropical tropopause temperatures on stratospheric water vapor. Mote, P.W., Rosenlof, K.H., McIntyre, M.E., Carr, E.S., Gille, J.C., Holton, J.R., Kinnersley, J.S., Pumphrey, H.C., Russell, J.M. and Waters, J.W. *J. geophys. Res.* **101**, 3989 (1996).

The annual cycle of stratospheric water vapor in a general circulation model. Mote, P.W. *J. geophys. Res.* **100**, 7363 (1995).

Mechanisms controlling water vapor in the lower stratosphere: "A tale of two stratospheres". Dessler, A.E., Hintsa, E.J., Weinstock, E.M., Anderson, J.G., and Chan, K.R. *J. geophys. Res.* **100,** 23167 (1995).

Seasonal variations of water vapor in the tropical lower stratosphere. Mote, P.W., Rosenlof, K.H., Holton, J.R., Harwood, R.S., and Waters, J.S. *Geophys. Res. Letts.* **22**, 1093 (1995).

Methane

Effects of production and oxidation processes on methane emissions from rice fields. Khalil, M.A.K., Rasmussen, R.A., and Shearer, M.J. *J. geophys. Res.* **103,** 25233 (1998).

A dramatic decrease in the growth rate of atmospheric methane in the Northern Hemisphere during 1992. Dlugokencky, E.J., Masarie, K.A., Lang, P.M., Tans, P.P., Steele, L.P., and Nisbet, E.G. *Geophys. Res. Letts.* **21**, 45 (1994).

The growth rate and distribution of atmospheric methane. Dlugokencky, E.J., Steele, L.P., Lang, P.M., and Masarie, K.A. *J. geophys. Res.* **99**, 17021 (1994).

Oxides of nitrogen

Climatology and small-scale structure of lower stratospheric N_2O based on *in situ* observations. Strahan, S.E., Loewenstein, M., and Podolske, J.R. *J. geophys. Res.* **104,** 2195 (1999).

The tropospheric gas-phase degradation of NH_3 and its impact on the formation of N_2O and NO_x. Kohlmann, J.P., and Poppe, D. *J. atmos. Chem.* **32,** 397 (1999).

Stable isotope enrichment in stratospheric nitrous oxide. Rahn, T. and Whalen, M. *Science* **278,** 1776 (1997).

Potential atmospheric sources and sinks of nitrous oxide. 2. Possibilities from excited O_2 "embryonic" O_3 and optically pumped excited O_3. Prasad, S.S. *J. geophys. Res.* **102,** 21527 (1997).

Evidence for an additional source of atmospheric N_2O. McElroy, M.B. and Jones, D.B.A. *Global Biochem. Cycles* **10,** 651 (1996).

Natural atmospheric sources and sinks of nitrous oxide. 1. An evaluation based on 10 laboratory experiments. Prasad, S.S. *J. geophys. Res.* **99,** 5285 (1994).

Effect of lightning on the concentration of odd nitrogen species in the lower stratosphere – an update. Kotamarthi, V.R., Ko, M., Weisenstein, D.K., Rodriguez, J.M., and Sze, N.D. *J. geophys. Res.* **99,** 8167 (1994).

Halogens: mainly chlorine

Reactive Halogen Compounds in the Atmosphere. Fabian, P. and Singh, O.N. (eds.) (Springer-Verlag, Heidelberg, 1998).

Rethinking reactive halogen budgets in the mid-latitude lower stratosphere. Dvortsov, V.L., Geller, M.A., Solomon, S., Schauffler, S.M., Atlas, E.L., and Blake D.R. *Geophys. Res. Letts.* **26,** 1699 (1999).

Natural emissions of chlorine-containing gases: Reactive Chlorine Emissions Inventory. Khalil, M.A.K., Moore, R.M., Harper, D.B., Lobert, J.M., Erickson, D.J., Koropalov, V., Sturges, W.T., and Keene, W.C. *J. geophys. Res.* **104,** 8333 (1999).

Composite global emissions of reactive chlorine from anthropogenic and natural sources: Reactive Chlorine Emissions Inventory. Keene, W.C., Khalil, M.A.K., Erickson, D.J., McCulloch, A., Graedel, T.E., Lobert, J.M., Aucott, M.L., Gong, S.L., Harper, D.B., Kleiman, G., Midgley, P., Moore, R.M., Seuzaret, C., Sturges, W.T., Benkovitz, C.M., Koropalov, V., Barrie, L.A., and Li, Y.F. *J. geophys. Res.* **104**, 8429 (1999).

Atmospheric chloroform. Khalil, M.A.K. and Rasmussen, R.A. *Atmos. Environ.* **33,** 1151 (1999).

Atmospheric methyl chloride. Khalil, M.A.K. and Rasmussen, R.A. *Atmos. Environ.* **33,** 1305 (1999).

Natural formation of chloroform and brominated trihalomethanes in soil. Hoekstra, E.J., de Leer, E.W.B., and Brinkman, U.A.T. *Env. Sci. Technol.* **32,** 3724 (1998).

Measurements of bromine containing organic compounds at the tropical tropopause. Schauffler, S.M., Atlas, E.L., Flocke, F., Lueb, R.A., Stroud, V., and Travnicek, W. *Geophys. Res. Letts.* **25**, 317 (1998).

Ocean–atmosphere exchange of methyl chloride: Results from NW Atlantic and Pacific Ocean studies. Moore, R.W., Groszko, W., and Niven, S.J. *J. Geophys. Res.* **101**, 28529 (1996).

Increase in levels of stratospheric chlorine and fluorine loading between 1985 and 1992. Gunson, M.R., Abrams, M.C., Lowes, L.L., Mahieu, E., Zander, R., Rinsland, C.P., Ko, M.K.W., Sze, N.D., and Weisenstein, D.K. *Geophys. Res. Letts.* **21**, 2223 (1994).

On the relation between stratospheric chlorine / bromine loading and short-lived tropospheric source gases. Ko, M.K.W., Sze, N.D., Scott, C.J., and Weisenstein, D.K. *J. geophys. Res.* **102**, 25507 (1997).

Hydrochloric acid and the chlorine budget of the lower stratosphere. Webster, C. R. and 11 others. *Geophys. Res. Letts.* **21**, 2575 (1994).

Chemical budgets of the stratosphere. Crutzen. P. J. and Schmailzl, U. *Planet. Space Sci.* **31**, 1009 (1983).

The photochemical time constants of minor constituents and their families in the middle atmosphere. Shimazaki, T. *J. atmos. terr. Phys.* **46**, 173 (1984).

Production of odd nitrogen in the stratosphere and mesosphere. An intercomparison of source strengths. Jackman, C. H., Frederick, J. E., and Stolarski, R. S. *J. geophys. Res.* **85**, 7495 (1980).

The role of NO and NO_2 in the chemistry of the troposphere and stratosphere. Crutzen, P.J. *Annu. Rev. Earth & planet. Sci.* **7**, 443 (1979).

Release of active chlorine from NaCl particles: see also Section 5.7

Formation of molecular chlorine from the photolysis of ozone and aqueous sea-salt particles. Oum, K.W., Lakin, M.J., DeHaan, D.O., Brauers, T., and Finlayson-Pitts, B.J. *Science* **279**, 74 (1998).

A mechanism for halogen release from sea-salt aerosol in the remote marine boundary layer. Vogt, R., Crutzen, P.J., and Sander, R. *Nature* **383,** 327 (1996).

Formation of chemically active chlorine compounds by reactions of atmospheric NaCl particles with gaseous N_2O_5 and $ClONO_2$. Finlayson-Pitts, B. J., Ezell, M. J., and Pitts, J. N., Jr. *Nature, Lond.* **337**, 241 (1989).

Bromine is a particularly active catalyst, so that great interest attaches to methyl bromide as a source. There remains uncertainty about how much CH_3Br is natural and how much anthropogenic.

Gas phase atmospheric bromine photochemistry. Lary, D.J. *J. geophys. Res.* **101**, 1505 (1996).

Heterogeneous atmospheric bromine chemistry. Lary, D.J., Chipperfield, M.P., Toumi, R., and Lenton, T. *J. geophys. Res.* **101**, 1489 (1996).

Detection of HBr, and upper limits for HOBr: Bromine partitioning in the stratosphere. Johnson, D.G., Traub, W.A., Chance, K.V. and Jucks, K.W. *Geophys. Res. Letts.* **22,** 1373 (1995).

Is OBrO present in the stratosphere? Renard, J.B., Pirre, M., Robert, C., and Huguenin, D. *Comptes Rendus* **A325**, 921 (1997).

Bromine-containing source gases during EASOE. Fabian, P., Borchers, R., and Kourtidis, K. *Geophys. Res. Letts.* **21,** 1219 (1994).

Scientific uncertainties in the budget of atmospheric methyl bromide. Butler, J.H. *Atmos. Environ.* **30**, R1 (1996).

The Methyl Bromide Issue. Bell, C., Price, N., and Chakrabarti, B. (eds.) (John Wiley and Sons, Ltd., Chichester, 1996).

A source or a sink of methyl bromide? Pilinis, C., King, D.B., and Saltzman, E.S. *Geophys. Res. Letts.* **23,** 817 (1996).

Methyl bromide: its atmospheric science, technology and economics – Montreal protocol assessment supplement. Albritton, D.L., Watson, R.T., Anderson, S.O., and Lee-Bapty, S. (United Nations Environment Programme, Nairobi, 1992).

The potential role of the ocean in regulating atmospheric CH_3Br. Butler, J.H. *Geophys.Res Lett.* **21,** 185-188 (1994).

On the degradation of methyl bromide in sea water. Jeffers, P.M. and Wolfe, N.L. *Geophys. Res. Letts.* **23**, 1773 (1996).

Automobile emissions of methyl bromide. Baker, J.M., Reeves, C.E., Nightingale, P.D., and Penkett, S.A. *Geophys. Res. Letts.* **25,** 2405 (1998).

Measurements of bromine-containing organic compounds at the tropical tropopause. Schauffler, S.M., Atlas, E.L., Flocke, F., Lueb, R.A., Stroud, V., and Travnicek, W. *Geophys. Res. Letts.* **25**, 317 (1998).

Ocean–atmosphere exchange of methyl bromide: NW Atlantic and Pacific Ocean studies. Groszko, W. and Moore, R.M. *J. geophys. Res.* **101,** 16737 (1998).

Vertical distribution of methyl bromide in the stratosphere. Kourtidis, K., Borchers, R., and Fabian, P. *Geophys. Res. Letts.* **25**, 505 (1998).

Seasonal variations of tropospheric methyl bromide concentrations: Constraints on anthropogenic input. Wingenter, O.W., Wang, C.J.-L., Blake, D.R., and Rowland, F.S. *Geophys. Res. Letts.* **25**, 2797 (1998).

Investigation of stratospheric bromine and iodine oxides using the SAOZ balloon sonde. Pundt, I., Pommereau, J.P., and Lefèvre, F. *Proc. XVIII Quad. Ozone Symp.*, 575 (1998).

A global 3-D coupled atmosphere-ocean model of methyl bromide. Lee, J.M., Doney, S.C., Brasseur, G.P., and Müller, J.-F. *J. geophys. Res.* **103**, 16039 (1998).

Green plants: A terrestrial sink for atmospheric CH_3Br. Jeffers, P.M. and Wolfe, N.L. *Geophys Res. Lett.* **25**, 43 (1998).

Bromine emissions from leaded gasoline. Thomas, V.M., Bedford, J.A., and Cicerone, R.J. *Geophys. Res. Letts.* **24**, 1371 (1997).

The potential effect of oceanic biological degradation on the lifetime of atmospheric CH_3Br. Yvon-Lewis, S.A. and Butler, J.H. *Geophys. Res. Letts.* **24**, 1227 (1997).

Bacterial oxidation of methyl bromide in Mono Lake, California. Connell, T.L., Joye, S.B., Miller, L.G., and Oremland, R.S. *Environ. Sci. Technol.* **31**,1489 (1997).

Marine bacterial degradation of brominated methanes. Goodwin, K.D., Lidstrom, M.E., and Oremland, R.S. *Environ. Sci. Technol.* **31**, 3188 (1997).

Methyl halide emissions from savanna fires in southern Africa. Andreae, M.O., Atlas, E., Harris, C.W., Helas, G., De-Kock, A., Koppmann, R., Maenhaut, W., Manö, S., Pollock, W.H., Rudolph, J., Scharffe, D., Schebeske, G., and Welling, M. *J. geophys. Res.* **101**, 23603 (1996).

An improved estimate of the oceanic lifetime of atmospheric CH_3Br. Yvon-Lewis, S.A. and Butler, J.H. *Geophys. Res. Letts.* **23**, 53 (1996).

Biomass burning emissions and vertical distribution of atmospheric methyl halides and other reduced carbon gases in the South Atlantic region. Blake, N.J., Blake, D.R., Sive, B.C., Chen, T.-Y., Rowland, F.S., Collins, J.E. Jr., Sachse, G.W., and Anderson, B.E. *J. geophys. Res.* **101**, 24151 (1996).

Dibromomethane (CH_2Br_2) measurements at the upper troposphere and lower stratosphere. Kourtidis, K., Borchers, R., and Fabian, P. *Geophys. Res. Letts.* **23**, 2581 (1996).

Methyl bromide: Ocean sources, ocean sinks, and climate sensitivity. Andar, A.D., Yung, Y.L., and Chavez, F.P. *Global Biogeochem. Cycles* **10**, 175 (1996).

Methyl bromide under scrutiny, Nature. Butler, J.H. **376**, 469 (1995).

Emission of methyl bromide from biomass burning. Manö, S. and Andreae, M.O. *Science* **263**, 1255 (1994).

Atmospheric methyl bromide: Trends and global mass balance. Khalil, M.A.K., Rasmussen, R.A. , and Gunawardena, R. *J. geophys. Res.* **98**, 2887 (1993).

Iodine is another potentially important catalyst that has received attention recently.

On the role of iodine in ozone depletion. Solomon, S., Garcia, R.R., and Ravishankara, A.R. *J. geophys. Res.* **99** 20491 (1994).

The concurrent observation of methyl iodide and dimethyl sulphide in marine air; implications for sources of atmospheric methyl iodide. Bassford, M.R., Nickless, G., Simmonds, P.G., Lewis, A.C., Pilling M.J., and Evans M.J. *Atmos. Environ.* **33**, 2372 (1999).

Upper limit of iodine oxide in the lower stratosphere. Pundt, I., Pommereau, J.-P., Phillips, C., and Lateltin, E. *J. atmos. Chem.* **30**, 173 (1998).

The atmospheric column abundance of IO: Implications for stratospheric ozone. Wennberg, P.O., Brault, J.W., Hanisco, T.F., Salawitch, R.J., and Mount, G.H. *J. geophys. Res.* **102**, 8887 (1997).

Volatilization of methyl iodide from the soil-plant system. Muramatsu, Y. and Yoshida, S. *Atmos. Environ.* **29**, 21 (1995).

Section 4.4.4

Particles and aerosols are an important component of the stratosphere, and may influence its chemistry. These papers offer introductory reviews of the subject. Additional references can be found in connection with Aerosols and Volcanic activity at the end of Section 1.3. See also the entries for Section 4.5.5 for some specific references to the influence of volcanic aerosol on stratospheric ozone, and entries for Section 4.7.4 for the effects of polar stratospheric clouds.

Estimation of the contribution of heterogeneous reactions on the surface of aerosol particles to the chemistry of the atmosphere. Storozhev, V.B. *J. atmos. Sci.* **26**, 1179 (1995).

Pinatubo aerosol and stratospheric ozone reduction–observations over Central Europe. Ansmann, A., Wagner, F., Wandinger, U., Mattis, I., Gorsdorf, U., Dier, H.D., and Reichardt, J. *J. geophys. Res.* **101**, 18775 (1996).

The role of aerosol variations in anthropogenic ozone depletion at northern midlatitudes. Solomon, S., Portmann, R.W., Garcia, R.R., Thomason, L.W., Poole, L.R., and McCormick, M.P. *J. geophys. Res.* **101**, 6713 (1996).

Heterogeneous reactions on and in sulfate aerosols: Implications for the chemistry of the midlatitude tropopause region. Hendricks, J., Lippert, E., Petry, H., and Ebel, A. *J. geophys. Res.* **104**, 5531 (1999).

Activation of chlorine in sulfate aerosol as inferred from aircraft observations. Kawa, S.R., Newman, P.A., Lait, L.R., Schoeberl, M.R., Stimpfle, R.M., Kohn, D.W., Webster, C.R., May, R.D., Baumgardner, D., Dye, J.E., Wilson, J.C., Chan, K.R., and Loewenstein, M. *J. geophys. Res.* **102**, 3921 (1997).

On the relationship between stratospheric aerosols and nitrogen dioxide. Mills, M. J., Langford, A. O., O'Leary, T. J., Arpag, K., Miller, H. L., Proffitt, M. H., Sanders, R. W., and Solomon, S. *Geophys. Res. Letts.* **20**, 1187 (1993).

Laboratory Studies of Atmospheric Heterogeneous Chemistry. Kolb, C.E., Worsnop, D.R., Zahniser, M.S., Davidovits, P., Keyser, L.F., Leu, M.-T., Molina, M.J., Hanson, D.R., Ravishankara, A.R., Williams, L.R., and Tolbert, M.A. in *Progress and Problems in Atmospheric Chemistry*, Barker, J.R. (ed.) (World Scientific Publishing, Singapore, 1995).

Balloon observations of organic and inorganic chlorine in the stratosphere: The role of $HClO_4$ production on sulphate aerosols. Jaeglé, L., Yung, Y.L., Toon, G.C., Sen, B., and Blavier, J.-F. *Geophys. Res. Letts.* **23**, 1749 (1996).

Aerosol-mediated partitioning of stratospheric Cl_y and NO_y at temperatures above 200 K. Michelsen, H.A., Spivakovsky, C.M., and Wofsy, S.C. *Geophys. Res. Letts.* **26**, 299 (1999).

Observations of stratospheric aerosol using CPFM polarized limb radiances. McLinden, C.A., McConnell, J.C., McElroy, C.T., and Griffioen, E. *J. atmos. Sci.* **56**, 233 (1999).

Aerosol silica as a possible candidate for the heterogeneous formation of nitric acid hydrates in the stratosphere. Bogdan, A. and Kulmala, M. *Geophys. Res. Letts.* **26**, 1433 (1999).

High-resolution modeling of size-resolved stratospheric aerosol. Koziol, A.S. and Pudykiewicz, J. *J. atmos. Sci.* **55**, 3127 (1998).

Airborne aerosol measurements in the tropopause region and the dependence of new particle formation on preexisting particle number concentration. de Reus, M., Strom, J., Kulmala, M., Pirjola, L., Lelieveld, J., Schiller, C., and Zoger, M. *J. Geophys. Res.* **103**, 31255 (1998).

Evidence of heterogeneous bromine chemistry on cold stratospheric sulphate aerosols. Erle, F., Grendel, A., Perner, D., Platt, U., and Pfeilsticker, K. *Geophys. Res. Letts.* **25**, 4329 (1998).

Stratospheric aerosol-sulfuric acid: First *in situ* and on line measurements using a novel balloon-based mass spectrometer apparatus. Arnold, F., Curtius, J., Spreng, S., and Deshler, T *J. atmos. Chem.* **30**, 3 (1998).

Stratospheric aerosol measurements by the Lidar in Space Technology Experiment. Osborn, M.T., Kent, G.S., and Trepte, C.R. *J. Geophys. Res.* **103**, 11447 (1998).

Lidar measurements of stratospheric aerosol over Mauna Loa Observatory. Barnes, J.E. and Hofmann, D.J. *Geophys. Res. Letts.* **24**, 1923 (1997).

Effects of interannual variation of temperature on heterogeneous reactions and stratospheric ozone. Tie, X., Granier, C., Randel, W., and Brasseur, G.P. *J. geophys. Res.* **102**, 23519 (1997).

Atmospheric aerosols: biogeochemical sources and role in atmospheric chemistry. Andreae, M.O. and Crutzen, P.J. *Science* **276**, 1052 (1997).

Modeling the composition of liquid stratospheric aerosols. Carslaw, K.S., Peter,T., and Clegg, S.L. *Rev. Geophys.* **35**, 125 (1997).

Heterogeneous chlorine chemistry in the tropopause region. Solomon, S., Borrmann, S., Garcia, R.R., Portmann, R., Thomason, L., Poole, L.R., Winker, D., and McCormick, M.P. *J. geophys. Res.* **102**, 21411 (1997).

On the occurrence of ClO in cirrus clouds and volcanic aerosol in the tropopause region. Borrmann, S., Solomon, S., Avallone, L., Toohey, D., and Baumgardner, D. *Geophys. Res. Letts.* **24**, 2011 (1997).

Real time kinetics of the uptake of HOBr and $BrONO_2$ on ice and in the presence of HCl in the temperature range 190 to 200K. Allanic, A., Oppliger, R., and Rossi, M.J. *J. Geophys. Res.* **102,** 23529 (1997).

Uncertainties in reactive uptake coefficients for solid stratospheric particles—1. Surface chemistry. Carslaw, K.S. and Peter, T. *Geophys. Res. Letts.* **24,** 1743 (1997).

Heterogeneous $BrONO_2$ hydrolysis: Effect on NO_2 columns and ozone at high latitudes in summer. Randeniya, L.K., Vohralik, P.F., Plumb, I.C., and Ryan, K.R. *J. geophys. Res.* **102**, 23543 (1997).

The heterogeneous reaction of HONO, and HBr on ice, and on sulfuric acid. Seisel, S. and Rossi, M.J. *Ber. Bunsenges. Phys. Chem.* **101**, 943 (1997).

Heterogeneous reaction in sulfuric acid: HOCl and HCl as a Model System. Hanson, D.R. and Lovejoy, E.R. *J. phys. Chem.* **100**, 3697 (1996).

Reaction of $BrONO_2$ with H_2O on submicron sulfuric acid aerosol, and the implication for the lower stratosphere. Hanson, D.R., Ravishankara, A.R. and Lovejoy, E.R. *J. geophys. Res.* **101**, 9063 (1996).

Observations of large reductions in the NO/NO_y ratio near the mid-latitude tropopause and the role of heterogeneous chemistry. Keim, E.R., *et al. Geophys. Res. Letts.* **23**, 3223 (1996).

A physical absorption model of the dependence of $ClONO_2$ heterogeneous reactions on relative humidity. Henson, B.F., Wilson, K.R., and Robinson, J.M. *Geophys. Res. Letts.* **23**, 1021 (1996).

Stratospheric aerosols - Formation, properties, effects. Pueschel, R.F. *J. Aerosol Sci.* **27**, 383 (1996).

Carbon aerosols and atmospheric photochemistry. Lary, D.J., Lee, A.M., Toumi, R., Newchurch, M.J., Pirre, M., and Renard, J.B. *J. geophys. Res.* **102**, 3671 (1996).

The potential of cirrus clouds for heterogeneous chlorine activation. Borrmann, S., Solomon, S., Dye, J.E., and Luo, B. *Geophys. Res. Letts.* **23**, 2133 (1996).

Heterogeneous reaction in liquid sulfuric acid: HOCl + HCl as a model system. Hanson, D.R. and Lovejoy, E.R. *J. phys. Chem.* **100**, 6397 (1996).

Latitudinal distribution of black carbon soot in the upper troposphere and lower stratosphere. Blake, D.F. and Kato, K. *J. geophys. Res.* **100**, 7195 (1995).

Heterogeneous chemistry of bromine species in sulfuric acid under stratospheric conditions. Hanson, D.R. and Ravishankara, A.R. *Geophys. Res. Letts.* **22**, 385 (1995).

Stratospheric aerosol effective radius, surface area and volume estimated from infrared measurements. Grainger, R. G., Lambert, A., Rodgers, C. D., Taylor, F. W., and Deshler, T. *J. geophys. Res.*, **100**, 16507 (1995).

A re-analysis of carbonyl sulfide as a source of stratospheric background sulfur aerosol. Chin, M. and Davis, D.D. *J. Geophys. Res.* **100**, 8999 (1995).

Particle formation in the upper tropical troposphere: a source of nuclei for the stratospheric aerosol. Brock, C.A., Hamill, P., Wilson, J.C., Jonsson, H.H., and Chan, K.R. *Science* **270**, 1650 (1995).

The reaction of $ClONO_2$ with submicrometer sulfuric acid aerosol. Hanson, D.R., and Lovejoy, E.R. *Science* **267**, 1326 (1995).

Origin of condensation nuclei in the springtime polar stratosphere. Zhao, J., Toon, O.B., and Turco, R.P. *J. geophys. Res.* **100**, 5215 (1995).

Heterogeneous chemistry of bromine species in sulfuric acid under stratospheric conditions. Hanson, D.R. and Ravishankara, A.R. *J. phys. Chem.* **22**, 385 (1995).

FTIR studies of low-temperature sulfuric-acid aerosols. Anthony, S.E., Tisdale, R.T., Disselkamp, R.S., Tolbert, M.A., and Wilson, J.C. *Geophys. Res. Letts.* **22**, 1105 (1995).

Sulfuric acid monohydrate: Formation and heterogeneous chemistry in the stratosphere. Zhang R., Leu, M.T., and Keyser, L.F. *J. geophys. Res.* **100**, 18845 (1995).

A model for studying the composition and chemical effects of stratospheric aerosols. Tabazadeh, A., Turco, R.P., and Jacobson, M.Z. *J. geophys. Res.* **99**, 12897 (1994).

Reactive uptake of $ClONO_2$ onto sulfuric acid due to reaction with HCl and H_2O. Hanson, D.R. and Ravishankara, A.R. *J. phys. Chem.* **98**, 5728 (1994).

Heterogeneous reaction of HOBr with HBr and HCl on ice surface at 228 K. Abbatt, J.P.D. *Geophys. Res. Letts.* **21**, 665 (1994).

Heterogeneous reactions in sulfuric acid under stratospheric conditions. Hanson, D.R., Ravishankara, A.R., and Solomon, S. *J. geophys. Res.* **99**, 3615 (1994).

In situ measurements of the ClO/HCl ratio: heterogeneous processing on sulfate aerosols and polar stratospheric clouds. Webster, C. R., May, R. D., Toohey, D. W., Avallone, L. M., Anderson, J. G., and Solomon, S. *Geophys. Res. Letts.* **20**, 2523 (1993).

In situ measurements constraining the role of sulfate aerosols in midlatitude ozone depletion. Fahey, D. and 20 others. *Nature, Lond.* 363, 509 (1993).

A global three-dimensional model of the stratospheric sulfuric acid layer. Golombek, A. and Prinn, R.G. *J. atmos. Chem.* **16**, 179 (1993).

Physical chemistry of the H_2SO_4/H_2O binary system at low temperatures: stratospheric implications. Zhang, R., Wooldridge, P. J., Abbatt, J. P. D., and Molina, M. J. *J. phys. Chem.* **97**, 7351 (1993).

Stratospheric aerosols: Observations, trends and effects. Jäger, H. *J. Aerosol Sci.* **22**, suppl. 1 S517 (1991).

Future changes in stratospheric ozone and the role of heterogeneous chemistry. Brasseur, G., Granier, C., and Walters, S. *Nature, Lond.* **348,** 626 (1990).

Stratospheric aerosols: observation and theory. Turco, R. P., Whitten, R. C., and Toon, O. B. *Rev. Geophys. & Space Phys.* **20**, 233 (1982).

Particles above the tropopause. Toon, O. B., and Farlow, N. H. *Annu. Rev. Earth & planet. Sci.* **9**, 19 (1981).

Possible effects of volcanic eruption on stratospheric minor constituent chemistry. Stolarski, R. S. and Butler, D. M. *Pure & appl. Geophys.* **117**, 486 (1978).

Increases in the stratospheric background sulfuric acid aerosol mass in the past ten years. Hofmann, D.J. *Science* **248**, 996 (1990).

Section 4.4.6

Observations of ozone and of minor constituents. The papers are representative of a large number describing in situ, satellite, and ground-based investigations. See also the papers described under Section 4.7.3.

Remote sensing of the Earth's atmosphere from space with high-resolution Fourier-transform spectroscopy – development and methodology of data-processing for the Atmospheric Trace Molecule Spectroscopy experiment. Abrams, M.C., Gunson, M.R., Chang, A.Y., Rinsland, C.P., and Zander, R. *Appl. Optics* **35**, 2774 (1996).

Analysis of halogen occultation experiment HF versus CH_4 correlation plots—chemistry and transport implications. Luo, M., Cicerone, R.J., and Russell, J.M. *J. geophys. Res.* **100**, 13927 (1995).

Comparison of ClO measurements by airborne and spaceborne microwave radiometers in the Arctic winter stratosphere 1993. Crewell, S., Fabian, R., Kunzi, K., Nett, H., Wehr, T., Read, W., and Waters, J. *Geophys. Res. Letts.* **22**, 1489 (1995).

Experimental methods

Ozone measuring instruments for the stratosphere. Grant, W. B. (ed.) (Optical Society of America, Washington, DC, 1989).

Remote sounding of atmospheres. Houghton, J. T., Taylor, F. W., and Rodgers, C. D. (Cambridge University Press, Cambridge, 1984).

Minor constituents in the stratosphere and mesophere. Hudson, R. D. *Rev. Geophys. & Space Phys.* **17**, 467 (1979).

The middle atmosphere as observed from balloons, rockets and satellites. *Phil. Trans. R. Soc.* **A296**, 1, (1980).

In situ measurements of stratospheric trace constituents. Ehhalt, D. H. *Rev. Geophys. & Space Phys.* **16**, 217 (1978).

Infrared absorption spectroscopy applied to stratospheric profiles of minor constituents. Louisnard, N., Fergant, G., Girard, A., Gramont, L., Lado-Bordowsky, O., Laurent, J., Le Boiteux, S., and Le Maitre, M. P. *J. geophys. Res.* **88**, 5365 (1983).

Remote sensing by IR heterodyne spectroscopy. Kostiuk, T. and Mumma, M. J. *Applied Optics* **22**, 2644 (1983).

Current and future passive remote sensing techniques used to determine atmospheric constituents. Burrows, J.P. in *Approaches to scaling of trace gas fluxes in ecosystems.* Bouwman, A.F. (ed.) (Elsevier, Amsterdam, 1999).

Remote sensing of the lower atmosphere. Stephens, G.L. (Oxford University Press, Oxford, 1994).

Aircraft Laser Infrared Absorption Spectrometer (ALIAS) for *in situ* stratospheric measurements of HCl, N_2O, CH_4, NO_2 and HNO_3. Webster, C.R., May, R.D., Trimble, C.A., Chave, R.G., and Kendall, J. *Appl. Opt.* **33**, 454 (1994).

Improved Limb Atmospheric Spectrometer (ILAS) for stratospheric ozone layer measurements by solar occultation technique. Sasano, Y., Suzuki, M., Yokota, T., and Kanzawa, H. *Geophys. Res. Letts.* **26**, 197 (1999).

Investigation of pole-to-pole performances of spaceborne atmospheric chemistry sensors with the NDSC. Lambert, J.C., Van Roozendael, M., De Maziere, M, Simon, P.C., Pommereau, J.P., Goutail, F., Sarkissian, A., and Gleason, J.F. *J. atmos. Sci.* **56**, 176 (1999).

The UARS and EOS microwave limb sounder (MLS) experiments. Waters, J.W., Read, W.G., Froidevaux, L., Jarnot, R.F., Cofield, R.E., Flower, D.A., Lau, G.K., Pickett, H.M., Santee, M.L., Wu, D.L., Boyles, M.A., Burke, J.R., Lay, R.R., Loo, M.S., Livesey, N.J., Lungu, T.A., Manney, G.L., Nakamura, L.L., Perun, V.S., Ridenoure, B.P., Shippony, Z., Siegel, P.H., Thurstans, R.P., Harwood, R.S., Pumphrey, H.C., and Filipiak, M.J. *J. atmos. Sci.* **56**, 194 (1999).

Ozone profiles from GOME satellite data: Algorithm description and first validation. Hoogen, R., Rozanov, V.V., and Burrows, J.P. *J. Geophys. Res.* **104**, 8263 (1999).

The global ozone monitoring experiment (GOME): Mission concept and first scientific results. Burrows, J.P., Weber, M., Buchwitz, M., Rozanov, V., Ladstatter Weissenmayer, A., Richter, A., De Beek, R., Hoogen, R., Bramstedt, K., Eichmann, K.U., and Eisinger, M. *J. atmos. Sci.* **56**, 151 (1999).

SCIAMACHY: Mission objectives and measurement modes. Bovensmann, H., Burrows, J.P., Buchwitz, M., Frerick, J., Noel, S., Rozanov, V.V., Chance, K.V., and Goede, A.P.H. *J. atmos. Sci.* **56**, 127 (1999).

Validation of ground-based visible measurements of total ozone by comparison with Dobson and Brewer spectrophotometers. Van Roozendael, M., Peeters, P., Roscoe, H.K., De Backer, H., Jones, A.E., Bartlett, L., Vaughan, G., Goutail, F., Pommereau, J.P., Kyro, E., Wahlstrom, C., Braathen, G., and Simon, P.C. *J. atmos. Chem.* **29**, 55 (1998).

Automated ground-based star-pointing UV-visible spectrometer for stratospheric measurements. Roskoe, H.K., Taylor, W.H., Evans, J.D., Tait, A.M., Freshwater, R., Fish, D., Strong, E.K., and Jones, R.L. *Appl. Optics* **36**, 6069 (1997).

Global CF_2Cl_2 measurements by UARS cryogenic limb array etalon spectrometer: validation by correlative data and a model. Nightingale, R.W., Roche, A.E., Kumer, J.B., Mergenthaler, J.L., Gille, J.C., Massie, S.T., Bailey, P.L., Edwards, D.P., Gunson, M.R., Toon, G.C., Sen, B., Blavier, J.-F., and Connell, P.S. *J. geophys. Res.* **101**, 9711 (1996).

Validation of measurements of water vapor from the Halogen Occultation Experiment (HALOE). Harries, J.E., Russell, J.M., III, Tuck, A.F., Gordley, L.L., Purcell, P., Stone, K., Bevilacqua, R.M., Gunson, M.R., Nedoluha, G., and Traub, W.A. *J. geophys. Res.* **101**, 10205 (1996).

Validation of nitric oxide and nitrogen dioxide measurements made by the Halogen Occultation Experiment for UARS platform. Gordley, L.L., Russell, J.M., III, Mickely, L.J., Frederick, J.E., Park, J.H., Stone, K.A., Beaver, G.M., McInerney, J.M., Deaver, L.E., Toon, G.C., Murcray, F.J. Blatherwick, R.D., Gunson, M.R., Abbatt, J.P.D., Mauldin, R.L., III, Mount, G.H., Sen, B., and Blavier, J.-F. *J. geophys. Res.* **101**, 10241 (1996).

The Atmospheric Trace Molecule Spectroscopy (ATMOS) experiment: deployment on the ATLAS Space Shuttle mission. Gunson, M.R., Abbas, M.M., Abrams, M.C., Allen, M., Brown, L.R., Brown, T.L., Chang, A.Y., Goldman, A., Irion, F.W., Lowes, L.L., Mahieu, E., Manney, G.L., Michelsen, H.A., Newchurch, M.J., Rinsland, C.P., Salawitch, R.J., Stiller, G.P., Toon, G.C., Yung, Y.L., and Zander, R. *Geophys. Res. Letts.* **23**, 2333 (1996).

Airborne Raman lidar. Heaps, W.S. and Burris, J. *Appl. Optics* **35**, 7128 (1996).

Validation of hydrogen fluoride measurements made by the HALOE Experiment from the UARS platform. Russell, J.M., III, Deaver, L.E., Luo, M., Cicerone, R.J., Park, J.H., Gordley, L.L., Toon, G.C., Gunson, M.R., Traub, W.A., Johnson, D.G., Jucks, K.W., Zander, R., and Nolt, I. *J. geophys. Res.* **101**, 10163 (1996).

Validation of hydrogen chloride measurements made by HALOE from the UARS platform. Russell, J.M., III, Deaver, L.E., Luo, M., Park, J.H., Gordley, L.L., Tuck, A.F., Toon, G.C., Gunson, M.R., Traub, W.A., Johnson, D.G., Jucks, K.W., Murcray, D.G., Zander, R., Nolt, I., and Webster, C.R. *J. geophys. Res.* **101**, 10151 (1996).

Stratospheric and mesospheric observations with ISAMS. Taylor, F.W., Ballard, J., Dudhia, A., Gosscustard, M., Kerridge, B.J., Lambert, A, Lopez Val Verde, M., Rodgers, C.D., and Remedios, J.J. *Advan. Space Res.* **14**, 41 (1994).

The Halogen Occultation Experiment. Russell, J.M., III, Gordley, L.L., Park, J.H., Drayson, S.R., Hesketh, D.H., Cicerone, R.J., Tuck, A.F., Frederick, J.E., Harries, J.E., and Crutzen, P.J. *J. geophys. Res.* **98**, 10,777 (1993).

Stratospheric photochemical studies with Atmospheric Trace Molecule Spectroscopy (ATMOS) measurements. Natarajan, M. and Callis, L.B. *J. geophys. Res.* **96**, 9361 (1991).

Representative measurements

Balloon profiles of stratospheric NO_3 and HNO_3 for testing the heterogeneous hydrolysis of N_2O_5 on sulfate aerosols. Webster, C. R., May, R. D., Allen, M., Jaegle, L., and McCormick, M. P. *Geophys. Res. Letts.* **21**, 53 (1994).

In situ observations of NO_y, O_3 and the NO_y/O_3 ratio in the lower stratosphere. Fahey, D.W., Donelly, S.G., Keim, E.R., Gao, R.-S., Wamsley, R.C., Del Negro, L.A., Woodbridge, E.L., Proffitt, M.H., Rosenlof, K.H., Ko, M.K.W., Weisenstein, D.K., Scott, C.J., Nevison, C., Solomon S., and Chan, K.R. *geophys. Res. Letts.* **23**, 1653 (1996).

Six years of UARS Microwave Limb Sounder HNO3 observations: Seasonal, interhemispheric, and interannual variations in the lower stratosphere. Santee, M.L., Manney, G.L., Froidevaux, L., Read, W.G. and Water, J.W. *J. Geophys. Res.* **104**, 8225 (1999).

Intercomparison of ILAS and HALOE ozone at high latitudes. Lee, K.M., McInerney, J.M., Sasano, Y., Park, J.H., Choi, W., and Russell, J.M. *Geophys. Res. Letts.* **26**, 835 (1999).

A climatology of total ozone mapping spectrometer data using rotated principal component analysis. Eder, B.K., Le Duc, S.K., and Sickles, J.E. *J. Geophys. Res.* **104**, 3691 (1999).

Space-time patterns of trends in stratospheric constituents derived from UARS measurements. Randel, W.J., Wu, F., Russell, J.M., and Waters, J. *J. Geophys. Res.* **104**, 3711 (1999).

Diurnal variations in the middle atmosphere observed by UARS. Sassi, F. and Salby, M. *J. geophys. Res.* **104**, 3729 (1999).

Balloon-borne *in situ* measurements of stratospheric H_2O, CH_4 and H_2 at midlatitudes. Zoger, M., Engel, A., McKenna, D.S., Schiller, C., Schmidt, U., and Woyke, T. *J. geophys. Res.* **104**, 1817 (1999).

Observations of middle atmosphere CO from the UARS ISAMS during the early northern winter 1991/92. Allen, D.R., Stanford, J.L., Lopez Valverde, M.A., Nakamura, N., Lary, D.J., Douglass, A.R., Cerniglia, M.C., Remedios, J.J., and Taylor, F.W. *J. atmos. Sci.* **56**, 563 (1999).

Observation of the stratospheric NO_2 latitudinal distribution in the northern winter hemisphere. Pfeilsticker, K., Erle, F., and Platt, U. *J. atmos. Chem.* **32**, 101 (1999).

Ozone profiles from GOME satellite data: Algorithm description and first validation. Hoogen, R., Rozanov, V.V., and Burrows, J.P. *J. geophys. Res.* **104,** 8263 (1999).

Selected science highlights from the first 5 years of the Upper Atmosphere Research Satellite (UARS) program. Dessler, A.E., Burrage, M.D., Grooss, J.U., Holton, J.R., Lean, J.L., Massie, S.T., Schoeberl, M.R., Douglass, A.R., and Jackman, C.H. *Rev. Geophys.* **36**, 183 (1998).

Measurements of reactive nitrogen in the stratosphere. Sen, B., Toon, G.C., Osterman, G.B., Blavier, J.-F., Margitan, J.J., Salawitch, R.J., and Yue, G.K. *J. geophys. Res.* **103**, 3571 (1998).

Evidence of transport in water vapor profiles at mid latitudes from Stratospheric Aerosol and Gas Experiment (SAGE II) measurements. Kar, J. and Mahajan, K.K. *J. geophys. Res.* **103**, 31057 (1998).

Stratospheric HBr concentration profile obtained from far-infrared emission spectroscopy. Nolt, I. G., Ade, P.A.R., Alboni, F., Carli, B., Carlotti, M., Cortesi, U., Epifani, M., Griffin, M.J., Hamilton, P.A., Lee, C., Lepri, G., Mencaraglia, F., Murray, A.G., Park, J.H., Park, K., Raspollini, P., Ridolfi, M., and Vanek, M.D. *Geophys. Res. Letts.* **24**, 281 (1997).

Nitrogen partitioning in the middle stratosphere as observed by the Upper Atmosphere Research Satellite. Morris, G.A., Considine, D.B., Dessler, A.E., Kawa, S.R., Kumer, J., Mergenthaler, J., Roche, A., and Russell, J.M., III. *J. geophys. Res.* **102**, 8955 (1997).

HALOE observations of a slowdown in the rate of increase of HF in the lower mesosphere. Considine, G.D., Deaver, L., Remsberg, E.E., Russell, J.M., III, *Geophys. Res. Letts.* **24**, 3217 (1997).

Mid-latitude observations of the seasonal variation of BrO. 1. Zenith-sky measurements. Aliwell, S.R., Jones, R.L., and Fish, D.J. *Geophys. Res. Letts.*, **24**, 1195 (1997).

ATMOS / ATLAS-3 measurements of stratospheric chlorine and reactive nitrogen partitioning inside and outside the November 1994 Antarctic vortex. Rinsland, C.P., Gunson, M.R., Salawitch, R.J., Michelsen, H.A., Zander, R., Newchurch, M.J., Abbas, M.M., Abrams, M.C., Manney, G.L., Chang, A.Y., Irion, F.W., Goldman, A., and Mahieu, E. *Geophys. Res. Letts.* **23**, 2365 (1996).

A twenty-five year record of stratospheric hydrogen chloride. Wallace, L. and Livingstone, W. *Geophys. Res. Letts.* **24**, 2363 (1997).

Vertical profile of night-time stratospheric OClO. Renard, J.B., Lefevre, F., Pirre, M., Robert, C., and Huguenin, D. *J. atmos. Chem.* **26**, 65 (1997).

Airborne heterodyne measurements of stratospheric ClO, HCl, O_3, and N_2O during SESAME 1 over northern Europe. de Valk, J.P., Goede, A.P.H., de Jonge, A.R.W., Mees, J., Franke, B., Crewell, S., Kullmann, H., Urban, J., Wohlgemuth, J., Chipperfield, M.P., and Lee, A.M. *J. geophys. Res.* **102**, 1391 (1997).

Comparison of HF and HCl vertical profiles from ground-based high-resolution infrared solar spectra with Halogen Occultation Experiment observations. Liu, X., Murcray, F.J., Murcray, D.G., and Russell, J.M., III. *J. geophys. Res.* **101**, 10175 (1996).

$ClONO_2$ total vertical column abundances above the Jungfraujoch Station 1986–1994: Long-term trend and winter–spring enhancements. Rinsland, C.P., Zander, R., Demoulin, P., and Mahieu, E. *J. geophys. Res.* **101**, 3891 (1996).

Measurements of stratospheric chlorine monoxide (ClO) from ground based FTIR observations. Bell, W., Patonwalsh, C., Gardiner, T.D., Woods, P.T., Swann, N.R., Martin, N.A., Donohoe, L., and Chipperfield, M.P. *J. atmos. Chem.* **24**, 285 (1996).

Balloon-borne observations of mid-latitude fluorine abundance. Sen, B., Toon, G.C., Blavier, J.-F., Fleming, E.L., and Jackman, C.H. *J. geophys. Res.* **101**, 9045 (1996).

Doppler detection of hydroxyl column abundance in the middle atmosphere. Iwagami, N., Inomata, S., Murata, I., and Ogawa, T. *J. atmos. Chem.* **20**, 1 (1995).

Midlatitude observations of the diurnal variation of stratospheric BrO. Fish, D.J., Jones, R.L., and Strong, E.K. *J. geophys. Res.* **100**, 18,863 (1995).

In situ measurements of BrO during AASE II. Avallone, L.M., Schauffler, S.M., Pollock, W.H., Heidt, L.E., Atlas, E.L., and Chan, K.R. *Geophys. Res. Letts.* **22**, 831 (1995).

Column abundance measurements of atmospheric hydroxyl at 45S. Wood, S.W., Keep, D.J., Burnett, C.R., and Burnett, E.B. *Geophys. Res. Letts.* **21**, 1607 (1994).

Microwave observations and modeling of O_3, H_2O, and HO_2 in the mesosphere. Clancy, R.T., Sandor, B.J., Rusch, D.W., and Muhleman, D.O. *J. geophys. Res.* **99**, 5465 (1994).

Submillimeter heterodyne measurements of stratospheric ClO, HCl, and H_2O: First results. Stachnik, R.A., Hardy, J.C., Tarsala, J.A., Waters, J.W., and Erickson, N.R. *Geophys. Res. Letts.* **19**, 1931 (1992).

Ozone depletion in the upper stratosphere estimated from space shuttle data. Hilsenrath, E., Cebula, R.P., and Jackman, C.H. *Nature* **358**, 131 (1992).

Representative photochemical models of the stratosphere are presented in the references that follow. Some further models are described in references given in connection with Section 4.6.

Buffering interactions in the modeled response of stratospheric O_3 to increased NO_x and HO_x. Nevison, C.D., Solomon, S., and Gao, R.S. *J. Geophys. Res.* **104**, 3741 (1999).

Modelling the chemistry and microphysics of the cold stratosphere. Peter, Th. and Crutzen, P.J. in *Low temperature chemistry of the atmosphere.* Moortgat, G.K., Barnes, A.J., Le Bras, G., and Sodeau, J.R. (eds.) (Springer-Verlag, Berlin, 1994).

A three-dimensional view of the evolution of midlatitude stratospheric intrusions. Bithell, M., Gray, L.J.,and Cox, B.D. *J. atmos. Sci.* **56**, 673 (1999).

Assimilation of total ozone satellite measurements in a three- dimensional tracer transport model. Jeuken, A.B.M., Eskes, H.J., vanVelthoven, P.F.J., Kelder, H.M., Holm, E.V. *J. Geophys. Res.* **104**, 5551 (1999).

Multiannual simulations with a three-dimensional chemical transport model. Chipperfield, M.P. *J. geophys. Res.* **104**, 1781 (1999).

Toward understanding of the nonlinear nature of atmospheric photochemistry: Multiple equilibrium states in the high-latitude lower stratospheric photochemical system. Konovalov, I.B., Feigin, A.M., and Mukhina, A.Y. *J. Geophys. Res.* **104**, 3669 (1999).

The UIUC three-dimensional stratospheric chemical transport model: Description and evaluation of the simulated source gases and ozone. Rozanov, E.V., Zubov, V.A., Schlesinger, M.E., Yang, F.L., and Andronova, N.G. *J. Geophys. Res.* **104**, 11755 (1999).

Development of a chemistry module for GCMs: first results of a multiannual integration. Steil, B., Dameris, M., Bruhl, C., Crutzen, P.J., Grewe, V., Ponater, M., and Sausen, R. *Ann. Geophys.* **16**, 205 (1998).

A three dimensional simulation on the evolution of the middle latitude winter ozone in the middle stratosphere. Douglas, A.R., Rood, R.B., Kawa, S.R., and Allen, D.J. *J. geophys. Res.* **102**, 19217 (1997).

The effect of uncertainties in kinetic and photochemical data on model predictions of stratospheric ozone depletion. Fish, D.J. and Burton, M.R. *J. geophys. Res.* **10**, 25537 (1997).

Three-dimensional chemical forecasting: A methodology. Lee, A.M., Carver, G.D., Chipperfield, M.P., and Pyle, J.A. *J. geophys. Res.* **102**, 3905 (1997).

Evaluating chemical transport models: Comparison of effects of different CFC-11 emission scenarios. Hartley, D.E., Kindler, T., Cunnold, D.C., and Prinn, R.G. *J. geophys. Res.* **101**, 14381 (1996).

A three-dimensional circulation model with coupled chemistry for the middle atmosphere. Rasch, P.J., Boville, B.A., and Brasseur, G.P. *J. geophys. Res.* **100**, 9041 (1995).

A 3D chemical transport model study of chlorine activation during EASOE. Chipperfield, M.P., Cariolle, D., and Simon, P. *Geophys. Res. Letts.* **21**, 1467 (1994).

Chemistry of the 1991–1992 stratospheric winter: Three-dimensional model simulations. Lefevre, F., Brasseur, G.P., Folkins, I., Smith, A.K., and Simon, P. *J. Geophys. Res.* **99**, 8183 (1994).

A new numerical model of the middle atmosphere. 2. Ozone and related species. Garcia, R.R. and Solomon, S. *J. geophys. Res.* **99**, 12937 (1994).

Comparisons between experiment and model, and analysis of results

An examination of the total hydrogen budget of the lower stratosphere. Dessler, A.E., Weinstock, E.M., Hintsa, E.J., Anderson, J.G., Webster, C.R., May, R.D., Elkins, J.W., and Dutton, G.S. *Geophys. Res. Letts.* **21**, 2563 (1994).

A comparison of observations and model simulations of NOx/NOy in the lower stratosphere. Gao, R.S., Fahey, D.W., Del Negro, L.A., Donnelly, S.G., Keim, E.R., Neuman, J.A., Teverovskaia, E., Wennberg, P.O., Hanisco, T.F., Lanzendorf, E.J., Proffitt, M.H., Margitan, J.J., Wilson, J.C., Elkins, J.W., Stimpfle, R.M., Cohen, R.C., McElroy, C.T., Bui, T.P., Salawitch, R.J., Brown, S.S., Ravishankara, A.R., Portmann, R.W., Ko, M.K.W., Weisenstein, D.K., and Newman, P.A. *Geophys. Res. Letts.* **26**, 1153 (1999).

Closure of the total hydrogen budget of the northern extratropical lower stratosphere. Hurst, D.F., Dutton, G.S., Romashkin, P.A., Wamsley, P.R., Moore, F.L., Elkins, J.W., Hintsa, E.J., Weinstock, E.M., Herman, R.L., Moyer, E.J., Scott, D.C., May, R.D., and Webster, C.R. *J. geophys. Res.* **104**, 8191 (1999).

Analysis of seasonal variation of stratospheric nitric acid. Gruzdev, A.N. *Advan. Space. Res.* **22**, 1521 (1999).

Twilight observations suggest unknown sources of HO_x. Wennberg, P.O., Salawitch, R.J., Donaldson, D.J., Hanisco, T.F., Lanzendorf, E.J., Perkins, K.K., Lloyd, S.A., Vaida,V., Gao, R.S., Hintsa, E.J., Cohen, R.C., Swartz, W.H., Kusterer, T.L., and Anderson, D.E. *Geophys. Res. Letts.* **26**, 1373 (1999).

Analysis of BrO measurements from the Global Ozone Monitoring Experiment. Chance, K. *Geophys. Res. Letts.* **25**, 3335 (1998).

Implications of satellite OH observations for middle atmospheric H_2O and ozone. Summers, M.E., Conway, R.R., Siskind, D.E., Stevens, M.H., Offerman, D., Riese, M., Preusse, P., Strobel, D.F., and Russell J.M., III. *Science* **277,** 1967 (1997).

Model overestimates of NO_y in the upper stratosphere. Nevison, C.D., Solomon, S., and Garcia, R.R. *Geophys. Res. Letts.* **24**, 803 (1997).

Measurements of NO_y – N_2O correlation in the lower stratosphere: Latitudinal and seasonal changes and model comparisons. Keim, E.R., Loewenstein, M., Podolske, J.R., Fahey, D.W., Gao, R.S., Woodbridge, E.L., Wamsley, R.C., Donnelly, S.G., Del Negro, L.A., Nevison, C.D., Solomon, S., Rosenslof, K.H., Scott, C.J., Ko, M.K.W., Weisensten, D., and Chan, K.R. *J. geophys. Res.* **102**, 13193 (1997).

Photochemical evolution of ozone in the lower tropical stratosphere. Avallone, L.M. and Prather, M.J. *J. geophys. Res.* **101**, 1457 (1996).

Simultaneous measurements of stratospheric HO_x, NO_x and Cl_x: comparison with a photochemical model. Chance, K., Traub, W.A., Johnson, D.G., Jucks, K.W., Ciarpallini, P., Stachnik, R.A., Salawitch, R.J., and Michelsen, H.A. *J. geophys. Res.* **101**, 9031 (1996).

Ozone production and loss rate measurements in the middle stratosphere. Jucks, K.W., Johnson, D.G., Chance, K.V., Traub, W.A., Salawitch, R.J., and Stachnik, R.A. *J. geophys. Res.* **101**, 28785 (1996).

NO_y correlation with N_2O and CH_4 in the midlatitude stratosphere. Kondo, Y., Schmidt, U., Engel, A., Koike, M., Aimedieu, P., Gunson, M.R., and Rodriguez, J. *Geophys. Res. Letts.* **23**, 2369 (1996).

The total hydrogen budget in the Arctic winter stratosphere during the European Arctic Stratosphere Ozone Experiment. Engel, A., Schiller, C., Schmidt, U., Borchers, R., Ovarlez, H. , and Ovarlez, J. *J. geophys. Res.* **101,** 14495 (1996).

A test of the partitioning between ClO and $ClONO_2$ using simultaneous UARS measurements of ClO, NO_2 and $ClONO_2$. Dessler, A.E., Kawa, S.R., Douglass, A.R., Considine, D.B., Kumer, J.B., Roche, A.E., and Mergenthaler, J.L. *J. geophys. Res.* **101**, 12515 (1996).

Comparison of measured stratospheric OH with prediction. Pickett, H.M. and Peterson, D.B. *J. geophys. Res.* **101**, 16789 (1996).

Stratospheric chlorine partitioning: Constraints from shuttle-borne measurements of [HCl], [$ClNO_3$], and [ClO]. Michelsen, H.A., Salawitch, R.J., Gunson, M.R., Aelling, C., Kämpfer, N., Abbas, M.M., Abrams, M.C., Brown, T.L., Chang, A.Y., Goldman, A., Irion, F.W., Newchurch, M.J., Rinsland, C.P., Stiller, G.P., and Zander, R. *Geophys. Res. Letts.* **23**, 2361 (1996).

The 1994 northern midlatitude budget of stratospheric chlorine derived from ATMOS / ATLAS-3 observations. Zander, R., Mahieu, E., Gunson, M.R., Abrams, M.C., Chang, A.Y., Abbas, M.M., Aellig, C., Engel, A., Goldman, A., Irion, F.W., Kampfer, N., Michelsen, H.A., Newchurch, M.J., Rinsland, C.P., Salawitch, R.J., Stiller, G.P., and Toon, G.C. *Geophys. Res. Letts.* **23**, 2357 (1996).

A re-evaluation of the ozone budget with HALOE UARS data: no evidence for the ozone deficit. Crutzen, P.J., Grooß, J.-U., Brühl, C., Müller, R., and Russell, J.M., III. *Science* **268,** 705 (1995).

Stratospheric trace constituents simulated by a three-dimensional general circulation model: comparison with UARS data. Eckman, R.S., Grose, W.L., Turner, R.E., Blackshear, W.T., Russell, J.M., III, Froidevaux, L., Waters, J.W., Kumer, J.B., and Roche, A.E. *J. geophys. Res.* **100**, 13951 (1995).

Photodissociation of O_2 and H_2O in the middle atmosphere: Comparison of numerical methods and impact on model O_3 and OH. Siskind, D.E., Minschwaner, K., and Eckman, R.S. *Geophys. Res. Letts.* **21**, 863 (1994).

Seasonal evolutions of N_2O, O_3, and CO_2 — 3-dimensional simulations of stratospheric correlations. Hall, T.M. and Prather, M.J. *J. geophys. Res.* **100**, 16699 (1995).

Trajectory mapping and applications to data from the upper-atmosphere research satellite. Morris, G.A., Schoeberl, M.R., Sparling, L.C., Newman, P.A., Lait, L.R., Elson, L., Waters, J., Suttie, R.A., Roche, A., Kumer, J., and Russell, J.M., III. *J. geophys. Res.* **100**, 16491 (1995).

Lagrangian transport calculations using UARS data. 1. Passive tracers. Manney, G.L., Zurek, R.W., Lahoz, W.A., Harwood, R.S., Gille, J.C., Kumer, J.B., Mergenthaler, J.L., Roche, A.E., O'Neill, A., Swinbank, R., and Waters, J.W. *J. atmos. Sci.* **52,** 3049 (1995).

Lagrangian transport calculations using UARS data. 2. Ozone. Manney, G.L., Zurek, R.W., Froidevaux, L., Waters, J.W., O'Neill, A., and Swinbank, R. *J. atmos. Sci.* **52**, 3069 (1995).

Upper atmosphere models and research. Ryecroft, M. J., Kasting, G. M., and Rees, D. (eds.) *Adv. Space Res.* **10**, No. 6 (1990).

Effect of recent rate data revisions on stratospheric modeling. Ko, M. K. W. and Sze, N. D. *Geophys. Res. Lett.* **10**, 341 (1983).

The seasonal and latitudinal behavior of trace gases and O_3 as simulated by a two-dimensional model of the atmosphere. Ko, M. K. W., Sze, N. D., Livshits, M., McElroy, M. B., and Pyle, J. A. *J. atmos. Sci.* **41**, 2381 (1984).

Chemistry and transport in a three-dimensional stratospheric model: chlorine species during a simulated stratospheric warming. Kaye, J. A., and Rood, R. B. *J. geophys. Res.* **94**, 1057 (1989).

A three-dimensional model of chemically active trace species in the middle atmosphere during disturbed winter conditions. Rose, K. and Brasseur, G. *J. geophys. Res.* **94**, 16387 (1989).

The roles of dynamical and chemical processes in determining the stratospheric concentration of ozone in one-dimensional and two-dimensional models. Ko, M. K. W., Sze, N. D., and Weisenstein, D. *J. geophys. Res.* **94**, 9889 (1989).

Section 4.5

Nuclear war might be expected to severely damage the Earth's ozone shield.

Photochemical war on the stratosphere. Hampson, J. *Nature, Lond.* **250**, 189 (1974).

Possible ozone depletions following nuclear explosions. Whitten, R. C., Borucki, W. J., and Turco, R. P. *Nature, Lond.* **257**, 38 (1975).

Ozone concentrations may have been affected by nuclear tests in the past.

Effects of atmospheric nuclear explosions on total ozone. Bauer, E. and Gilmore, F. R. *Rev. Geophys. & Space Phys.* **13**, 451 (1975).

The atmospheric nuclear tests of the 50s and 60s: a possible test of ozone depletion theories. Chang, J. S., Duewer, W. H., and Wuebbles, D. J. *J. geophys. Res.* **84**, 1755 (1979).

An additional atmospheric effect of nuclear war that has been identified concerns the smoke emitted by wide-ranging fires. Indirect effects on ozone concentrations might follow.

The atmosphere after a nuclear war: twilight at noon. Crutzen, P. J. and Birks, J. W. *Ambio* **11**, 114 (1982).

Atmospheric effects of a nuclear war. Birks, J. B. and Crutzen, P. J. *Chem. Br.* **19**, 927 (1983).

Beyond Armageddon. Raloff, J. *Science News* **124**, 314 (1983).

Nuclear winter: global consequences of multiple nuclear explosions. Turco, R. P., Toon, O. B., Ackerman, T. P., Pollock, J. B., and Sagan, C. *Science* **222**, 1293 (1983).

Global atmospheric effects of massive smoke injections from a nuclear war: results from general circulation model simulations. Covey, C., Schneider, S. H., and Thompson, S. L. *Nature, Lond.* **308**, 21 (1984).

Section 4.5.1.

Solar protons, arriving at Earth, enhanced during a 'solar proton event', may produce oxides of nitrogen in the atmosphere. The papers listed describe measurements of the associated depletion of ozone.

An overview of the early November 1993 geomagnetic storm. Knipp, D.J., Emery, B.A., Engebretson, M., Li, X., McAllister, A.H., Mukai, T., Kokubun, S., Reeves, G.D., Evans, D., Obara, T., Pi, X., Rosenberg, T., Weatherwax, A., McHarg, M.G., Chun, F., Mosely, K., Codrescu, M., Lanzerotti, L., Rich, F.J., Sharber, J., and Wilkinson, P. *J. geophys. Res.* **103**, 26197 (1998).

Two-dimensional and three-dimensional model simulations, measurements and interpretation of the influence of the October 1989 solar proton events on the middle atmosphere. Jackman, C.H., Cerniglia, M.C., Nielsen, J.E., Allen, D.J., Zawodny, J.M., McPeters, R.D., Douglas, A.R., Rosenfield, J.E., and Rood, R.B. *J. Geophys. Res.* **100**, 11641 (1995).

Energetic particle influences on NO_y and ozone in the middle atmosphere. Jackman, C.H. *Geophys. Monogr. IUGG.* **15**, 131 (1993).

Possible composition and climate changes due to past intense energetic particle precipitation. Hauglustaine, D. and Gérard, J.-C. *Ann. Geophys.* **8**, 87 (1990).

Effect of solar proton events in 1978 and 1979 on the odd nitrogen abundance in the middle atmosphere. Jackman, C. H. and Meade, P. E. *J. geophys. Res.* **93**, 7084 (1988).

Effect of solar proton events in the middle atmosphere during the past two solar cycles as computed using a two-dimensional model. Jackman, C.H., Douglass, A.R., Rood, R.B., and McPeters, R.D. *J. geophys. Res.* **95**, 7417 (1990).

Solar proton events: stratospheric sources of nitric oxide. Crutzen, P. J., Isaksen, I. S. A., and Reid, G. C. *Science* **189**, 457 (1975).

Section 4.5.2

Solar UV irradiance and solar-cycle variability

An 11-year solar cycle in tropospheric ozone from TOMS measurements. Chandra, S., Ziemke, J.R., and Stewart, R.W. *Geophys. Res. Letts.* **26**, 185 (1999).

Effects of short-term solar UV variability on the stratosphere. Hood, L.L. *Nature, Lond.* **61**, 45 (1999).

Solar variability and its implications for the human environment. Reid, G.C. *Nature, Lond.* **61**, 3 (1999).

The signal of the 11 year solar cycle in the global stratosphere. Van Loon, H. and Labitzke, K. *J. Atm. Terr. Phys.* **61**, 53 (1999).

Solar cycle variability, ozone, and climate. Shindell, D., Rind, D., Balachandran, N., Lean, J., and Lonergan, P. *Science* **284**, 305 (1999).

The solar cycle variation of total ozone : Dynamical forcing in the lower stratosphere. Hood, L.L. *J. geophys. Res.* **102**, 1355 (1997).

The signal of the 11-year sunspot cycle in the upper troposphere–lower stratosphere. Labitzke, K and van Loon, H. *Space Sci. Rev.* **80**, 393 (1997).

Solar activity–total column ozone relationships: Observations and model studies with heterogeneous chemistry. Zerefos, C.S., Tourpali, K., Bojkov, B.R., Balis, D.S., Rognerud, B., and Isaksen, I.S.A. *J. geophys. Res.* **102**, 1561 (1997).

Apparent solar cycle variations of upper stratospheric ozone and temperature: Latitudinal and seasonal dependences. McCormack, J.P. and Hood, L.L. *J. geophys. Res.* **101**, 20933 (1996).

Solar activity forcing of the middle atmosphere. Mohanakumar, K. *Ann. Geophysicae* **13**, 879-885 (1995).

Modeling the effects of UV variability and the QBO on the troposphere-stratosphere system. Part II: The Middle Atmosphere. Rind, D. and Balachandran, N.K. *J. Climate* **8**, 2058 (1995).

The solar cycle variation of ozone in the stratosphere inferred from Nimbus 7 and NOAA 11 satellites. Chandra, S. and McPeters, R.D. *J. geophys. Res.* **99**, 20665 (1994).

The Sun as Variable Star: Solar and Stellar Variations. Pap, J.M. (ed.) (Cambridge University Press, Cambridge, 1994).

Quasi-decadal variability of the stratosphere: Influence of long term solar ultraviolet variations. Hood, L.L., Jirikowic, J.L., and McCormac, J.P. *J. atmos. Sci.* **50**, 3941 (1993).

The response of the middle atmosphere to long-term and short-term solar variability: a two dimensional model. Brasseur, G.P. *J. geophys. Res.* **98**, 23079 (1993).

A numerical response of the middle atmosphere to the 11-year solar cycle. Garcia, R. R., Solomon, S., Roble, R. G., and Rusch, D. W. *Planet. Space Sci.* **32**, 411 (1984).

Section 4.5.3

The latitudinal structure of the quasi-biennial oscillation. Haynes, P.H. *Quart J. Royal Meteorol. Soc.* **124**, 2645 (1988).

Seasonal asymmetry of the low- and middle-latitude QBO circulation anomaly. Kinnersley, J.S. *J. atmos. Sci.* **56**, 1140 (1999).

Simulation of the quasi-biennial oscillation in a general circulation model. Takahashi, M. *Geophys. Res. Letts.* **26**, 1307 (1999).

Investigation of the effect of the quasi biennial oscillation on the southern polar vortex. Shimmin, P., Harries, J.E., and Jackson, D.R. *Advan. Space. Res.* **22**, 1493 (1999).

Global QBO circulation derived from UKMO stratospheric analyses. Randel, W.J., Wu, F., Swinbank, R., Nash, J., and O'Neill, A. *J. atmos. Sci.* **56**, 457 (1999).

Seasonal cycles and QBO variations in stratospheric CH_4 and H_2Oobserved in UARS HALOE data. Randel, W.J., Wu, F., Russell, J.M., III, Roche, A., and Waters, J.W. *J. atmos. Sci.* **55**, 163 (1998).

Effects of an imposed quasi-biennial oscillation in a comprehensive troposphere-stratosphere-mesosphere general circulation model. Hamilton, K. *J. atmos. Sci.* **55**, 2393 (1998).

The influence of the equatorial QBO on the Northern Hemisphere winter circulation of a GCM. Niwano, M. and Takahashi, M. *J. Meteor. Soc. Japan* **76**, 453 (1998).

An analytical study of ozone feedbacks on Kelvin and Rossby gravity waves: Effects on the QBO. Cordero, E.C., Nathan, T.R., and Echols, R.S. *J. atmos. Sci.* **55**, 1051 (1998).

The influence of the equatorial quasi-biennial oscillation on the global circulation at 50mb. Holton, J.R. and Tan, H.-C. *J. atmos. Sci.* **37**, 2200 (1998).

The first simulation of an ozone QBO in a general circulation model. Nagashima, T., Takahashi, M., and Hasebe, F. *Geophys. Res. Letts.* **25**, 3131 (1998).

An estimation of the dynamical isolation of the tropical lower stratosphere using UARS wind and trace gas observations of the quasi- biennial oscillation. Schoeberl, M.R., Roche, A.E., Russell, J.M.III, Ortland, D., Hays, P.B., and Waters, J.W. *Geophys. Res. Letts.* **24**, 53 (1997).

Modeling the quasi-biennial oscillation's influence on tracer transport in the tropics. O'Sullivan, D., Chen, P. *J. Geophys. Res.* **101**, 6811 (1996).

On the relationship between the quasi-biennial oscillation, total chlorine and the Antarctic ozone hole. Butchart, N. and Austin, J. *Quart. J. Royal Meteorol. Soc.* **122**, 183 (1996).

Isolation of the ozone QBO in SAGE II data by singular-value decomposition. Randel, W.J. and Wu, F. *J. atmos. Sci.* **53**, 2546 (1996).

Simulation of the stratospheric quasi-biennial oscillation using a general circulation model. Takahashi, M. *Geophys. Res. Letts.* **23**, 661 (1996).

Observational study of the quasi-biennial oscillation in ozone. Hollandsworth, S.M., Bowman, K.P., and McPeters, R.D. *J. geophys. Res.* **100**, 7347 (1995).

A three-dimensional modeling study of the extra-tropical quasi-biennial oscillation in ozone. Hess, P.G. and O'Sullivan, D. *J. atmos. Sci.* **52**, 1539 (1995).

The quasi-biennial oscillation of ozone in the tropical middle stratosphere: a one dimensional model. Ling, X.-D. and London, J. *J. atmos. Sci.* **51**, 729 (1994).

Quasi-biennial oscillations of ozone and diabatic circulation in the equatorial stratosphere. Hasebe, F. *J. atmos. Sci.* **51**, 729 (1994).

A two-dimensional model study of the QBO signal in SAGE II NO_2 and O_3. Chipperfield, M.P., Gray, L.J., Kinnersly, J.S., and Zawodny, J. *Geophys. Res. Letts.* **21**, 589 (1994).

Section 4.5.4

Spectral comparison of ENSO and stratospheric zonal winds. Kane, R.P. *Int. J. Climatol.* **18**, 1195 (1998).

Principle modes of the total ozone on the southern oscillation timescale and related temperature variations. Kayano, M. *J. geophys. Res.* **102**, 25797 (1997).

On the relative importance of quasi-biennial oscillation and El Niño/Southern Oscillation in the revised Dobson total ozone records. Zerefos, C.S., Bais, A.F., Ziomas, T.C., and Bojkov, R.D. *J. geophys. Res.* **97**, 10135 (1992).

Annual, quasi-biennial and El Niño–Southern Oscillation time scale variations in equatorial total ozone. Shiotani, M. *J. geophys. Res.* **97**, 7625 (1992).

Section 4.5.5

Stratospheric effects of Mt. Pinatubo aerosol studied with a coupled two-dimensional model. Rosenfield, J., Considine, D.B., Meade, D.E., Bacmeister, J.T., Jackman, C.H., and Schoeberl, M.R. *J. geophys. Res.* **102**, 3649 (1997).

Ground-based remote sensing of the decay of the Pinatubo eruption cloud at three Northern Hemisphere sites. Jäger, H., Uchino, O., Nagai, T., Fujimoto, T., Freudenthaler, V., and Homburg, F. *Geophys. Res. Letts.* **22**, 607 (1996).

Nitrogen species in the post-Pinatubo stratosphere: Model analysis utilizing UARS measurements. Danilin, M.Y., Rodriguez, J.M., Hu, W.J., Ko, M.K.W., Weisenstein, D.K., Kumer, J.B., Mergenthaler, J.L., Russell, J.M., III, Koike, M., Yue, G.K., Jones, N.B., and Johnston, P.V. *J. geophys. Res.* **104**, 8247 (1999).

Early evolution of a stratospheric volcanic eruption cloud as observed with TOMS and AVHRR. Schneider, D.J., Rose, W.I., Coke, L.R., Bluth, G.J.S., Sprod, I.E., and Krueger, A.J. *J. geophys. Res.* **104**, 4037 (1999).

Impact of El Chichón and Pinatubo on ozonesonde profiles in north extratropics. Angell J.K. *Geophys. Res. Letts.* **25**, 4485 (1998).

Stratospheric aerosols observed by lidar over northern Greenland in the aftermath of the Pinatubo eruption. diSarra, A., Bernardini, L., Cacciani, M., Fiocco, G., and Fua, D. *J. geophys. Res.* **103**, 13873 (1998).

Ozone depletion at mid-latitudes: Coupling of volcanic aerosols and temperature variability to anthropogenic chlorine. Solomon, S., Portmann, R.W., Garcia, R.R., Randel, W., Wu, F., Nagatani, R., Gleason, J., Thomason, L., Poole, L.R., and McCormick, M.P. *Geophys. Res. Letts.* **25**, 1871 (1998).

Stratospheric aerosol following Pinatubo, comparison of the north and south mid latitudes using *in situ* measurements. Deshler, T., Liley, J.B., Bodeker, G., Matthews, W.A., and Hoffmann, D.J. *Advan. Space Res.* **20**, 2089 (1997).

Estimated impacts of Agung, El Chichon and Pinatubo volcanic eruptions on global and regional total ozone after adjustment for the QBO. Angell, J.K. *Geophys. Res. Letts.* **24**, 647 (1997).

Effect of Pinatubo aerosols on stratospheric NO. Kondo, Y., Sugita, T., Salawitch, R.J. Koike, M., and Deshler, T. *J. geophys. Res.* **102**, 1205 (1997).

Lower stratospheric chlorine partitioning during the decay of the Mt. Pinatubo aerosol cloud. Dessler, A.E., Considine, D.B., Rosenfield, J.E., Kawa, S.A., Douglass, A.R., and Russell, J.M., III. *Geophys. Res. Letts.* **24,** 1623 (1997).

Observations of the impact of volcanic activity on stratospheric chemistry. Coffey, M.T. *J. geophys. Res.* **101**, 6767 (1996).

Evolution of the stratospheric aerosol in the northern hemisphere following the June (1991 volcanic eruption of Mount Pinatubo : Role of tropospheric-stratospheric exchange and transport. Jonsson, H.H., Wilson, J.C., Brock, C.A., Dye, J.E., Ferry, G.V., and Chan, K.R. *J. geophys. Res.* **101**, 1553 (1996).

The effect of volcanic aerosols on ultraviolet radiation in Antarctica. Tsitas, S.R. and Yung, Y.L. *Geophys. Res. Letts.* **23**, 157 (1996).

Recentvolcanic signals in the ozone layer. Zerefos, C.S., Tourpali, K., and Bais, A.F., in *The Mount Pinatubo Eruption: Effect on the Atmosphere and climate.* Fiocco, G., Fua, D., and Visconti, G. (eds.) NATO-ASI series **142**, (1996).

Temporal changes of Mount Pinatubo aerosol characteristics over northern midlatitudes derived from SAGE II extinction measurements. Anderson, J. and Saxena, V.K. *J. geophys. Res.* **101**, 19455 (1996).

Correlations between ozone loss and volcanic aerosol at altitudes below 14 km over McMurdo Station, Antarctica. Deshler, T., Johnson, B.J., Hofmann, D.J., and Nardi, B. *Geophys. Res. Letts.* **23**, 2931 (1996).

Sulfur emissions to the stratosphere from explosive volcanic eruptions. Pyle, D.M., Beattie, P.D., and Bluth, G.J.S. *Bull. Volcan.* **57**, 663 (1996).

The response of stratospheric ozone to volcanic eruptions: Sensitivity to atmospheric chlorine loading. Tie, X.X. and Brasseur, G. *Geophys. Res. Letts.* **22**, 3035 (1995).

Ozone and temperature changes in the stratosphere following the eruption of Mount Pinatubo. Randel, W.J., Wu, F., Russell, J.M. III, Waters, J.W., and Froidevaux, L. *J. geophys. Res.* **100**, 16753 (1995).

Further studies on possible volcanic signal to the ozone layer. Zerefos, C.S., Tourpali, K., and Bais, A.G. *J. geophys. Res.* **99**, 25,741 (1994).

Decay of Mount Pinatubo volcanic perturbation. Goodman, J., Snetsinger, K.G., Pueschel, R.F., and Verma, S. *Geophys. Res. Letts.* **21**, 1129 (1994).

Two-dimensional simulation of Pinatubo aerosol and its effect on stratospheric ozone. Tie, X., Brasseur, G.P., Briegleb, B., and Granier, C. *J. geophys. Res.* **99**, 20545 (1994).

Stratospheric chlorine injection by volcanic eruptions — HCl scavenging and implications for ozone. Tabazedeh, A. and Turco, R.P. *Science* **260**, 1082 (1993).

Anomalous Antarctic ozone during 1992: Evidence for Pinatubo volcanic aerosol effects. Hofmann, D.J. and Oltmans, S.J. *J. geophys. Res.* **98**, 18,555 (1993).

Balloon borne measurements of Pinatubo aerosol during 1991 and 1992 at 41°N, vertical profiles, size distribution and volatility. Deshler, T., Johnson, B.J., and Rozier, W.R. *Geophys. Res. Letts.* **20**, 1435 (1993).

Catastrophic loss of stratospheric ozone in dense volcanic clouds. Prather, M.J. *J. geophys. Res.* **97**, 10187 (1992).

Mount Pinatubo aerosols, chlorofluorocarbons, and ozone depletion. Brasseur, G., and Granier, C. *Science* **257**, 1239 (1992).

Ozone destruction through heterogeneous chemistry following the eruption of El Chichon. Hofmann, D.J. and Solomon, S. *J. Geophys. Res.* **94**, 5029 (1989).

Section 4.6

Ozone in the stratosphere is an essential filter of ultraviolet radiation that protects life on the Earth's surface. One effect of reduced ozone concentrations could therefore be an increase in biologically damaging radiation, a subject explored in the references that follow.

Environmental Effects of Ozone Depletion: 1998 Assessment. Van der Leun, J.C., Tang, X., and Tevini, M. (United Nations Environment Programme, Nairobi, 1998).

Environmental Effects of Ozone Depletion—1994 Update. Van der Leun, J.C., Tang, X., and Tevini, M. (United Nations Environmental Programme, Nairobi, 1994).

Environmental Effects of Stratospheric Ozone Depletion—1991 Update. Van der Leun, J.C., Tevini, M., and Worrest, R.C. (United Nations Environmental Programme, Nairobi, 1991).

The atmosphere and UV-B radiation at ground level, Environmental UV photobiology. Madronich, S. and Young, A.R. (Plenum Press, New York, 1993).

Human Exposure to Ultraviolet Radiation: Risks and Regulations. Passchier, W.R. and Bosnajakovich, B.M.F. (eds.) (Elsevier, Amsterdam, 1987).

Living with our sun's ultraviolet rays. Giese, A. C. (Plenum Press, New York, 1976).

Stratospheric ozone reduction, solar ultraviolet radiation, and plant life. Worrest, W. C., and Calwell, M. M. (Springer–Verlag, Berlin, 1986).

Ozone depletion, related UVB changes and increased skin cancer incidence. Kane, R.P. *Int. J. Climatol.* **18**, 457 (1998).

Effects of increased solar ultraviolet radiation on terrestrial ecosystems. Caldwell, M.M., Bjorn, L.O., Bornman, J.F., Flint, S.D., Kulandaivelu, G., Teramura, A.H., and Tevini, M. *J. Photochem. Photobiol.* **B46**, 40 (1998).

The effects of UV radiation A and B on diurnal variation in photosynthesis in three taxonomically and ecologically diverse microbial mats. Cockell, C.S. and Rothschild, L.J. *Photochem. Photobiol.* **69**, 203 (1999).

The impact of ultraviolet-B radiation on the motility of the freshwater epipelic diatom Nitzschia lineariz. Moroz, A.L., Ehrman, J.M., Clair, T.A., Gordon, R.J., and Kaczmarska, I. *Global Change Biology* **5**, 191 (1999).

Ozone depletion and increased UV-B radiation: is there a real threat to photosynthesis? Allen, D.J., Nogues, S., and Baker, N.R. *J. exp. Bot.* **49**, 1775 (1998).

Effects of the ultraviolet-B radiation (UV-B) on conifers: a review. Laakso, K. and Huttunen, S. *Env. Pollut.* **99**, 319 (1998).

Changes in biologically active ultraviolet radiation reaching the Earth's surface. Madronich, S., McKenzie, R.L., Bjorn, L.O., and Caldwell, M.M. *J. Photochem. Photobiol.* **B46,** 5 (1998).

Modelling the ozone depletion, UV radiation and skin cancer rates for Australia. Koken, P.J.M., Willems, B.A.T., Vrieze, O.J., and Frick, R.A., *Environmetrics* **9**, 15 (1998).

Effects of enhanced ultraviolet-B radiation on plant nutrients and decomposition of spring wheat under field conditions. Yue, M., Li, Y., Wang, X.L. *Env. Exp. Biol.* **40**, 187 (1998).

UV-radiation can affect depth-zonation of Antarctic macroalgae. Bischof, K., Hanelt, D., and Wiencke, C. *Marine Biology* **131**, 597 (1998).

Ultraviolet-absorbing/screening substances in cyanobacteria, phytoplankton and macroalgae. Sinha, R.P., Klisch, M., Groniger, A., and Hader, D.P. *J. Photochem. Photobiol.* **B47**, 83 (1998).

The impact of UVB radiation on marine plankton. Davidson, A.T. *Mutation Res.* **422**, 119 (1998).

Effect of stratospheric ozone variations on UV radiation and on tropospheric ozone at high latitudes. Taalas, P., Damski, J., Kyro, E., Ginzberg, M., and Talamoni, G. *J. geophys. Res.* **102**, 1533 (1997).

Increases in solar UV radiation with altitude. Blumthaler, M., Ambach, W., and Ellinger, R. *J. Photochem. Photobiol.* **B39**, 130 (1997).

The effects of clouds and haze on UV-B radiation. Estupinan, J.G., Raman, S., Crescenti, G.H., Streicher, J.J., and Barnard, W.F. *J. geophys. Res.* **101**, 16,807 (1996).

Measurements of biological effective UV doses, total ozone abundance and cloud effects with multi-channel moderate bandwidth filter instruments. Dahlback, A. *Appl. Opt.* **35**, 6514 (1996).

Estimates of ozone depletion and skin cancer incidence to examine the Vienna Convention achievements. Slaper, H., Velders, G.J.M., Daniel, J.S., de Gruijl, F.R., and van der Leun, J.C. *Nature, Lond.* **384**, 256 (1996).

The relationship between solar UV irradiance and total ozone from observations over southern Argentina. Bojkov, R.D., Fioletov, V.E., and Diaz, S.B. *Geophys. Res. Letts.* **22**, 1249 (1995).

Effects of clouds and stratospheric ozone depletion on ultraviolet radiation trends. Lubin, D. and Jensen, E.H. *Nature, Lond.* **377**, 710 (1995).

Effect of uv-radiation on phytoplankton. Smith, R.C. and Cullen, J.J. *Rev. Geophys.* **33**, 1211 (1995).

Action spectroscopy in complex organisms: Potentials and pitfalls in predicting the impact of increased environmental UVB. Sutherland, B.M. *J. Photochem. Photobiol.* **B31**, 29 (1995).

Changes in ultraviolet radiation reaching the Earth's surface. Madronich, S., McKenzie, R.L., Caldwell, M., and Bjorn, L.O. *Ambio* **24**, 143 (1995).

Effect of ozone depletion on atmospheric CH_4 and CO concentrations. Bekki, S., Law, K.S., and Pyle, J.A. *Nature* **371**, 595 (1994).

Stratospheric ozone depletion between 1979 and 1992: Implication for biologically active ultraviolet-B radiation and non-melanoma skin cancer incidence. Madronich, S. and de Gruijl, F.R. *Photochem. Photobiol.* **59**, 541 (1994).

Ultraviolet radiation, ozone depletion, and marine photosynthesis. Cullen, J.J. and Neale, P.J. *Photosynthesis Research* **39**, 303 (1994).

Ultraviolet Radiation and Biological Research in Antarctica. Booth, R.C. and Madronich, S. (American Geophysical Union Antarctic Research Series, Washington, DC, 1994).

Effects of reductions in stratospheric ozone on tropospheric chemistry through changes in photolysis rates. Fuglestvedt, J.S., Jonson, J.E., and Isaksen, I.S.A. *Tellus* **46B**, 172 (1994).

Skin cancer and UV radiation. Madronich, S. and de Gruijl, F.R. *Nature, Lond.* **366**, 23 (1993).

Evidence for large upward trends in ultraviolet-B radiation linked to ozone depletion. Kerr, J.B. and McElroy, C.T. *Science* **262**, 1032 (1993).

The relationship between erythemal UV and ozone, derived from spectral irradiance measurements. McKenzie, R.L., Matthews, W.A., and Johnston, P.V. *Geophys. Res. Letts.* **18**, 2269 (1991).

Indication of increasing solar ultraviolet-B radiation flux in alpine regions. Blumthaler, M. and Ambach, W. *Science* **248**, 206 (1990).

The budget of biologically active radiation in the earth-atmosphere system. Frederick, J. E., and Lubin, D. *J. geophys. Res.* **93**, 3825 (1988).

Possible long-term changes in the biologically active ultraviolet radiation reaching the ground. Frederick, J. E., and Lubin, D. *Photochem. Photobiol.* **47**, 571 (1988).

Biological effects of ultraviolet radiation. Beddard, G. S., in *Light, chemical change and life* Coyle, J. D., Hill. R. R., and Roberts, D. R. (eds.) (Open University Press, Milton Keynes, 1982).

Possible ozone reductions and UV changes at the Earth's surface. Pyle, J. A. and Derwent, R. G. *Nature, Lond.* **286**, 373 (1980).

Biologically damaging radiations amplified by ozone depletions. Gerstl, S. A. W., Zardecki, A., and Wiser, H. L. *Nature, Lond.* **294**, 352 (1981).

Many activities of man can potentially affect stratospheric ozone concentrations. The first two papers in this set consider the influence that man may have on atmospheric chemistry in general, while the others are devoted more specifically to the problem of stratospheric ozone. The book by Dotto and Schiff describes the political and scientific controversies surrounding the question of ozone depletion.

Changing composition of the global stratosphere. McElroy, M. B. and Salawitch, R. J. *Science* **243**, 763 (1989).

Atmospheric chemistry: response to human influence. Logan, J. A., Prather, M. J., Wofsy, S. C., and McElroy, M. B. *Phil. Trans. R. Soc.* **A290**, 187 (1978).

Pollution of the stratosphere. Johnston, H. S. *Annu. Rev. phys. Chem.* **26**, 315 (1975).

New evidence for ozone depletion in the upper stratosphere. Claude, H., Schönenborn, F., Steinbrecht, W., and Vandersee, W. *Geophys. Res. Letts.* **21** 2409 (1994).

Man and stratospheric ozone. Bower, F. A. and Ward. R. B. (eds.) (CRC, Cleveland, 1980).

Causes and effects of stratospheric ozone reduction: an update. (National Academy Press, Washington, DC, 1982).

Causes and effects of changes in stratospheric ozone: update 1983. (National Academy Press, Washington, DC, 1984).

A catalog of perturbing influences on stratospheric ozone, 1955–1975. Bauer, E. *J. geophys. Res.* **84**, 6929 (1979).

Transport processes and ozone perturbations. Solomon, S., Garcia, R. R., and Stordal, F. *J. geophys. Res.* **90**, 12981 (1985).

The ozone war. Dotto, L. and Schiff, H. I. (Doubleday, Garden City, New York, 1978).

Ozone crisis: the 15-year evolution of a sudden global emergency. Roan, S. L. (John Wiley, Chichester, 1989).

Some specific perturbing influences are now considered, following the order of subdivision of Section 4.6. Section 4.6.2 brings together references on the atmospheric influences of aircraft operations; effects on both stratosphere and troposphere are considered here, as explained in the text.

Section 4.6.2

European scientific assessment of atmospheric effects of aircraft emissions. Brasseur, G., Amantidis, G.T., and Angeletti, G. *Atmos. Environ.* **32**, 2327 (1998).

Aviation and the global atmosphere. Intergovernmental Panel on Climate Change (IPCC), (World Meteorological Organization, Geneva, 1999).

Atmospheric effects of subsonic aircraft: Interim assessment report of the advanced subsonic technology program. Friedl, R.,(ed.) (NASA, Goddard Space Flight Center, Greenbelt, MD, 1997).

Impact of NO_x emissions from aircraft upon the atmosphere at flight altitudes 8-15 km (AERONOX). Schumann, U. (ed.) (European Commission, Brussels, 1995).

1995 Scientific assessment of the Atmospheric Effects of Stratospheric Aircraft. Stolarski, R.S. (ed.) NASA Reference Publication 1381 (NASA, Washington, DC, 1995).

Atmospheric Effects of Stratospheric Aircraft. Stolarski, R. and Wesoky, H.L. (eds.) NASA Reference Publication 1313, (NASA, Washington, DC, 1993).

The possible role of organics in the formation and evolution of ultrafine aircraft particles. Yu, F.Q., Turco, R.P., and Karcher, B. *J. geophys. Res.* **104**, 4079 (1999).

Aircraft emissions: A comparison of methodologies based on different data availability. Romano, D., Gaudioso, D., De Lauretis, R. *Env. Monitoring Assess.* **56**, 51 (1999).

Impact of future subsonic aircraft NO_x emissions on the atmospheric composition. Grewe, V., Dameris, M., Hein, R., Kohler, I., and Sausen, R. *Geophys. Res. Letts.* **26,** 47 (1999).

Direct deposition of subsonic aircraft emissions into the stratosphere. Gettelman, A.and Baughcum, S.L. *J. geophys. Res.* **104**, 8317 (1999).

Factors influencing ozone chemistry in subsonic aircraft plumes. Das-Moulik, M. and Milford, J.B. *Atmos. Environ.* **33**, 869 (1999).

Dilution of aircraft exhaust plumes at cruise altitudes. Schumann, U., Schlager, H., Arnold, F., Baumann, R., Haschberger, P., and Klemm, O. *Atmos. Environ.* **32**, 3097 (1998).

First direct sulfuric acid detection in the exhaust plume of a jet aircraft in flight. Curtius, J., Sierau, B., Arnold, F., Baumann, R., Busen, R., Schulte, P., and Schumann, U. *Geophys. Res. Letts.* **25**, 923 (1998).

In-flight measurement of aircraft non-methane hydrocarbon emission indices. Slemr, F., Giehl, H., Slemr, J., Busen, R., Schulte, P., and Haschberger, P. *Geophys. Res. Letts.* **25**, 321 (1998).

Interactions between sulfur and soot emissions from aircraft and their role in contrail formation. 2. Development. Andronache, C. and Chameides, W.L. *J. geophys. Res.* **103,** 10787 (1998).

Corona-producing ice clouds: a case study of a cold mid-latitude cirrus layer. Sassen, K., Mace, G.G., Hallett, J., and Poellot M.R. *Appl. Optics* **37**, 1477 (1998).

Impact of aircraft NO_x emissions on tropospheric and stratospheric ozone. Part II: 3-D model results. Dameris, M., Gerwe, V., Køhler, I., Sausen, R., Bruhl, C., Groos, J.-U., and Steil, B. *Atmos. Environ.* **32**, 3185 (1998).

Constraining the heterogeneous loss of O_3 on soot particles with observations in jet engine exhaust plumes. Gao, R.S., Karcher, B., Keim, E.R., and Fahey, D.W. *Geophys. Res. Letts.* **25**, 3323 (1998).

The formation and evolution of aerosols in stratospheric aircraft plumes: Numerical simulations and comparisons with observations. Yu, F.Q. and Turco, R.P. *J. geophys. Res.* **103**, 25915 (1998).

Physicochemistry of aircraft-generated liquid aerosols, soot, and ice particles—1. Model description. Karcher, B. *J. geophys. Res.* **103**, 17111 (1998).

Impact of aircraft emissions on tropospheric and stratospheric ozone. Part I: Chemistry and 2-D model results. Grooss, J.U., Bruhl, C., and Peter, T. *Atmos. Environ.* **32**, 3173 (1998).

Effect of aircraft on ultraviolet radiation reaching the ground. Plumb, I.C. and Ryan, K.R. *J. geophys. Res.* **103**, 31231 (1998).

On the occurrence of ClO in cirrus clouds and volcanic aerosol in the tropopause region. Borrmann, S., Solomon, S., Avallone, L., Toohey, D., and Baumgardner, D. *Geophys. Res. Letts.* **24**, 2011 (1997).

On the possible role of aircraft-generated soot in the middle latitude ozone depletion. Bekki, S. *J. geophys. Res.* **102**, 10751 (1997).

In situ observations of air traffic emission signatures in the North Atlantic flight corridor. Schlager, H., Konopka, P., Schulte, P., Schumann, U., Ziereis, H., Arnold, F., Klemm, M., Hagen, D.E., Whitefield, P.D., and Ovarlez, Z. *J. geophys. Res.* **102**, 10739 (1997).

The passive transport of NO_x emissions from aircraft studied with a hierarchy of models. VanVelthoven, P.F.J., Sausen, R., Johnson, C.E., Kelder, H., Kohler, I., Kraus, A.B., Ramaroson, R., Rohrer, F., Stevenson, D., Strand, A., and Wauben, W.M.F. *Atmos. Environ.* **31,** 1783 (1997).

Interactions between sulfur and soot emissions from aircraft and their role in contrail formation.1. Nucleation. Andronache, C. and Chameides, W.L. *J. geophys. Res.* **102,** 21443 (1997).

The impact of nitrogen oxides emissions from aircraft upon the atmosphere at flight altitudes – Results from the AERONOX project. Schumann, U. *Atmos. Environ.* **31,** 1723 (1997).

The impact of aircraft nitrogen oxide emissions on tropospheric ozone studied with a 3D Lagrangian model including fully diurnal chemistry. Stevenson, D.S., Collins, W.J., Johnson, C.E., and Derwent, R.G. *Atmos. Environ.* **31,** 1837 (1997).

A 3-D chemistry transport model study of changes in atmospheric ozone due to aircraft NO_x emissions. Wauben, W.M.F., Van Velthoven, P.F.J., and Kelder, H. *Atmos. Environ.* **31,** 1819 (1997).

Contributions of aircraft emissions to the atmospheric NO_x content. Kohler, I., Sausen, R., and Reinberger, R. *Atmos. Environ.* **31,** 1801 (1997).

On the formation, and persistence of subvisible cirrus clouds near the tropical tropopause. Jensen, E.J., Toon, O.B., Selkirk, H.B., Spinhirne, J.D. and Schoeberl, M.R. *J. geophys. Res.* **101,** 21361 (1996).

Atmospheric impact of NO_x emissions by subsonic aircraft: A three-dimensional model study. Brasseur, G., Müller, J.-F., and Granier, C. *J. geophys. Res.* **101,** 1423 (1996).

A 5 year simulation of supersonic aircraft emission transport using a three-dimensional model. Weaver, C.J., Douglass, A.R., and Considine, D.B. *J. geophys. Res.* **101,** 20975 (1996).

In-situ observations of particles in jet aircraft exhausts and contrails for different sulfur-containing fuels. Schumann, U., Ström, J., Busen, R., Baumann, R., Gierens, K., Krautstrunk, M., Schröder, F.P., and Stingl, J. *J. geophys. Res.* **101,** 6853 (1996).

Particulate-emissions in the exhaust plume from commercial jet aircraft under cruise conditions. Hagen, D.E., Whitefield, P.D., and Schlager, H. *J. geophys. Res.* **101,** 19551 (1996).

Potential impact of SO_2 emissions from stratospheric aircraft on ozone. Weisenstein, D.K., Ko, M.K.W., Sze, N.D., and Rodriguez, J.M. *Geophys. Res. Letts.* **23,** 161 (1996).

Three-dimensional model studies of the effect of NO_x emissions from aircraft on ozone in the upper troposphere over Europe and the North-Atlantic. Flatøy, F. and Hov, Ø. *J. geophys. Res.* **101,** 1401 (1996).

The potential of cirrus clouds for heterogeneous chlorine activation. Borrmann, S., Solomon, S., Dye, J.E., and Luo, B. *Geophys. Res. Letts.* **23,** 2133 (1996).

On the possible role of aircraft-generated soot in middle latitude ozone depletion. Bekki, S. *J. geophys. Res.* **100,** 7195 (1995).

Sensitivity of two-dimensional model predictions of ozone response to stratospheric aircraft: An update. Considine, D.B., Douglass, A.R., and Jackman, C.H. *J. geophys. Res.* **100,** 3075 (1995).

The distribution of hydrogen, nitrogen and chlorine radicals: Implications for changes in O_3 due to emission of NO_y from supersonic aircraft. Salawitch, R.J., and 35 others. *Geophys. Res. Letts.* **21,** 2547 (1994).

North Atlantic air traffic within the lower stratosphere: Cruising times and corresponding emissions. Hoinka, K.P., Reinhardt, M.E., and Metz, W. *J. geophys. Res.* **98**, 23113 (1993).

The impact of aircraft NO_x emissions on tropospheric ozone and global warming. Johnson, C., Henshaw, J., and McInnes, G. *Nature, Lond.* **355**, 69 (1992).

Two-dimensional assessment of the impact of aircraft sulphur emissions on the stratospheric sulphate aerosol layer. Bekki, S. and Pyle, J.A. *J. geophys. Res.* **97**, 15,839 (1992).

Aircraft sulphur emissions. Hofmann, D.J. *Nature, Lond.* **349**, 659 (1991).

Emission measurements of the Concorde supersonic aircraft in the lower stratosphere. Fahey, D. W. and 25 others. *Science,* **270,** 70 (1995).

Nitrogen oxides from high-altitude aircraft: an update on potential effects on ozone. Johnston, H. S., Kinnison, D. E., and Wuebbles, D. J. *J. geophys. Res.* **94**, 16351 (1989).

Section 4.6.3

In-situ measurement of Cl_2 and O_3 in a stratospheric solid rocket motor exhaust plume. Ross, M.N., Ballenthin, J.O., Gosselin, R.B., Meads, R.F., Zittel, P.F., Benbrook, J.R., and Sheldon, W.R. *Geophys. Res. Letts.* **24**, 1755 (1997).

In situ measurement of the aerosol size distribution in stratospheric solid rocket motor exhaust plumes. Ross, M.N., Whitefield, P.D., Hagen, D.E., and Hopkins, A.R. *Geophys. Res. Letts.* **26**, 819 (1999).

A global modeling study of solid rocket aluminum oxide emission effects on stratospheric ozone. Jackman, C.H., Considine, D.B., and Fleming, E.L. *Geophys. Res. Letts.* **25**, 907 (1998).

Observation of stratospheric ozone depletion in rocket exhaust plumes. Ross, M.N., Benbrook, J.R., Sheldon, W.R., Zittel, P.F., and McKenzie , D.L. *Nature, Lond.* **390**, 62 (1997).

Local effects of solid rocket motor exhaust on stratospheric ozone. Ross, M. *Journal of Spacecraft and Rockets.* **33**, 144 (1996).

Space Shuttle's impact on the stratosphere: an update. Jackman, C.H., Considine, D.B., and Fleming, E.L. *J. geophys. Res.* **101**, 12523 (1996).

An assessment of the total ozone mapping spectrometer for measuring ozone levels in a solid rocket plume. Syage, J.A. and Ross, M.N. *Geophys. Res. Letts.* **23**, 3227 (1996).

On the stratospheric impact of launching the Ariane 5 rocket. Jones, A.E., Bekki, S., and Pyle, J.A. *J. geophys. Res.* **100**, 20969 (1995).

The impact of emissions from space transport systems on the state of the atmosphere. Jackman, C.H. in *Proceedings of an International Scientific Colloquium on Impact of Emissions from Aircraft and Spacecraft upon the Atmosphere.* Schumann, U. and Wurzel, D. (eds.) (German Aerospace Research Establishment, 1994).

Atmospheric Effects of Chemical Rocket Propulsion. Report of an AIAA Workshop held June 28-29 1991 in Sacramento, AIAA, 1991).

Section 4.6.4

The nature of the problem; sources, lifetimes, and chemistry of halogenated hydrocarbons

Chlorofluoromethanes in the environment. Rowland, F. S. and Molina, M. J. *Rev. Geophys. & Space Phys.* **13**, 1 (1975).

Halogens in the atmosphere. Cicerone, R. J. *Rev. Geophys. & Space Phys.* **19**, 123 (1981).

Chlorofluoromethanes in the environment: the aerosol controversy. Sugden, T. M. and West, T. F. (eds.) (John Wiley, New Jersey, 1980).

Chlorocarbon emission scenarios: potential impact on stratospheric ozone. Wuebbles, D. J. *J. geophys. Res.* **88**, 1433 (1983).

How have the atmospheric concentrations of halocarbons changed? Prinn, R. G. in *The changing atmosphere*, Rowland, F. S. and Isaksen, I. S. A. (eds.) (John Wiley, Chichester, 1988).

Atmospheric chemistry of CH_2BrCl, $CHBrCl_2$, $CHBr_2Cl$, $CF_3CHBrCl$ and CBr_2Cl_2. Bilde, M., Wallinton, T.J., Ferranto, C., Orlando, J.J., Tyndall, G.S., Estupiñan, E., and Haberkorn, S. *J. phys. Chem.* **102**, 1976 (1998).

Growth of fluoroform (CHF_3, HFC-23) in the background atmosphere. Oram, D.E., Sturges, W.T., Penkett, S.A., McCulloch, A., and Fraser, P.J. *Geophys. Res. Letts.* **25**, 35 (1998).

Global trends and emission estimates of carbon tetrachloride (CCl_4) from *in situ* background observations from July 1978 to June 1996. Simmonds P.G., Cunnold, D.M., Weiss, R.F., Prinn, R.G., Fraser, P.J., McCulloch, A., Alyea, F.N., and O'Doherty, S. *J. geophys. Res.* **103**, 16017 (1998).

Distribution of Halon-1211 in the upper troposphere and lower stratosphere and the 1994 total bromine budget. Wamsley, P.R., Elkins, J.W., Fahey, D.W., Dutton, G.S., Volk, C.M., Myers, R.C., Montzka, S.A., Butler, J.H., Clarke, A.D., Fraser, P.J., Steele, L.P., Lucarelli, M.P., Atlas, E.L., Schauffler, S.M., Blake, D.R., Rowland, F.S., Sturges, W.T., Lee, J.M., Penkett, S.A., Engel, A., Stimpfle, R.M., Chan, K.R., Weisenstein, D.K., Ko, M.K.W., and Salawitch, R.J. *J. geophys. Res.* **103**, 1513 (1998).

Measurements of bromine containing organic compounds at the tropical tropopause. Schauffler, S.M., Atlas, E.L., Flocke, F., Lueb, R.A., Stroud, V., and Travnicek, W. *Geophys. Res. Letts.* **25**, 317 (1998).

Carbon tetrachloride lifetimes and emissions determined from daily global measurements during 1978-1985. Simmonds, P.G., Cunnold, D.M., Alyea, F.N., Cardelino, C.A., Crawford, A.J., Prinn, R.G., Fraser, P.J., Rasmussen, R.A., and Rosen, R.D. *J. atmos. Chem.* **7**, 35 (1998).

Production, sales and atmospheric release of fluorocarbons through 1995. AFEAS (Alternative Fluorocarbon Environmental Acceptability Study). (S. & P.S., Inc., Washington, DC, 1997).

Coupled aerosol-chemical modeling of UARS HNO_3 and N_2O_5 measurements in the Arctic upper stratosphere. Bekki, S., Chipperfield, M.P., Pyle, J.A., Remedios, J.J., Smith, S.E., Grainger, R.G., Lambert, A., Kumer, J.B., and Mergenthaler, J.L. *J. geophys. Res.* **102**, 8977 (1997).

Satellite confirmation of the dominance of chlorofluorocarbons in the global stratospheric chlorine budget. Russell, J.M., III, Luo, M., Cicerone, R.J., and Deaver, L.E. *Nature, Lond.* **379**, 526 (1996).

Trends of OCS, HCN, SF_6, $CHClF_2$ (HCFC-22) in the lower stratosphere from 1985 and 1994. Atmospheric Trace Molecule Spectroscopy experiment measurements near 30°N latitude. Rinsland, C.P., Mahieu, E., Zander, R., Gunson, M.R., Salawitch, R.J., Chang, A.Y., Goldman, A., Abrams, M.C., Abbas, M.M., Newchurch, M.J., and Irion, F.W. *Geophys. Res. Letts.* **23**, 2349 (1996).

Global CF_2Cl_2 measurements by UARS cryogenic limb array etalon spectrometer: Validation by correlative data and a model. Nightingale, R.W., Roche, A.E., Kumer, J.B., Mergenthaler, J.L., Gille, J.C., Massie, S.T., Bailey, P.L., Edwards, D.P., Gunson, M.R., Toon, G.C., Sen, B., Blavier, J.-F., and Connell, P.S. *J. geophys. Res.* **101**, 9711 (1996).

Effect of natural tetrafluoromethane. Harnish, J., Borchers, R., Fabian, P., Gaggeler, H.W., and Schotterer, U. *Nature, Lond.* 384, 32 (1996).

Global stratospheric distribution of halocarbons. Fabian, P., Borchers, R., Leifer, R., Subbaraya, B.H., Lal, S., and Boy, M. *Atmos. Environ.* **308**, 1787-1796 (1996).

The production and global distribution of emissions of trichloroethene, tetrachloroethene and dichloromethane over the period 1988-1992. McCulloch, A. and Midgley, P.M. *Atmos. Environ.* **30**, 601 (1996).

Satellite confirmation of the dominance of chlorofluorocarbons in the global stratospheric chlorine budget. Russell, J.M., III, Luo, M., Cicerone, R.J., and Deaver, L.E. *Nature, Lond.* **379**, 526 (1996).

Recent tropospheric growth rates and distribution of HFC-134a (CF_3CH_2F). Oram, D.E., Reeves, C.E., Sturges, W.T., Penkett, S.A., Fraser, P.J., and Langenfelds, R.L. *geophys. Res. Letts.* **23**, 1949 (1996).

Observations of HFC-134a in the remote troposphere. Montzka, S.A., Myers, R.C., Butler, J.H., Elkins, L.W., Lock, L.T., Clarke, A.D., and Goldstein, A.H. *Geophys. Res. Letts.* **23**, 169 (1996).

The production and global distribution of emissions to the atmosphere of 1,1,1-trichloroethane (methyl chloroform). Midgley, P.M. and McCulloch, A. *Atmos. Environ.* **29**, 1601 (1995).

Measurements of HCFC-142b and HCFC-141b in the Cape Grim air archive: 1978-1993. Oram, D.E., Reeves, C.E., Penkett, S.A., and Fraser, P.J. *Geophys. Res. Letts.* **22**, 2741 (1995).

Bromine-containing source gases during EASOE. Fabian, P., Borchers, R., and Kourtides, K. *Geophys. Res. Letts.* **21**, 1219 (1994).

Estimate of total organic and inorganic chlorine in the lower stratosphere from *in situ* and flask measurements during AASE II. Woodbridge, E.L., Elkins, J.W., Fahey, D.W., Heidt, L.E., Solomon, S., Baring, T.J., Gilpin, T.M., Pollock, W.H., Schauffler, S.M., Atlas, E.L., Loewenstein, M., Podolske, J.R., Webster, C.R., May, R.D., Gilligan, J.M., Montzka, S.A., Boering, K.A., and Salawitch, R.J. *J. Geophys. Res.* **100**, 3057 (1995).

Observed stratospheric profiles and stratospheric lifetimes of HCFC-141b and HCFC-142b. Lee, J.M., Sturges, W.T., Penkett, S.A., Oram, D.E., Schmidt, U., Engel, A., and Bauer, R. *Geophys. Res. Letts.* **22**, 1369 (1995).

Distribution of emissions of chlorofluorocarbons (CFCs) 11, 12, 113, 114 and 115 among reporting and non-reporting countries in 1986. McCulloch, A., Midgley, P.M., and Fisher, D.A. *Atmos. Environ.* **28**, 2567 (1994).

Uncertainties in the calculation of atmospheric releases of chlorofluorocarbons. Fisher, D.A. and Midgley, P.M. *J. geophys. Res.* **99**, 16643 (1994).

Secular evolution of the vertical column abundances of $CHClF_2$ (HCFC -22) in the Earth's atmosphere inferred from ground based IR solar observations at the Jungfraujoch and at Kitt Peak, and comparison with model calculations. Zander, R., Mahieu, E., Demoulin, Ph., Rinsland, C.P., Weisenstein, D.K., Ko, M.K.W., Sze, N.D., and Gunson M.R. *J. atmos. Chem.* **18**, 129 (1994).

Report on Concentrations, Lifetimes and Trends of CFCs, Halons and Related Species. Kaye, J.A., Penkett, S.A., and Ormond, F.M. (eds.) NASA Reference Publication 1339 (NASA, Washington DC, 1994).

Decrease in the growth rates of atmospheric chlorofluorocarbons 11 and 12. Elkins, J.W. and Thompson, T.M. *Nature, Lond.* **364**, 780 (1993).

Chlorine catalyzed destruction of ozone: Implications for ozone variability in the upper stratosphere. Chandra, S., Jackman, C.H., Douglass, A.R., Fleming, E.L., and Considine, D.B. *Geophys. Res. Letts.* **20**, 351 (1993).

The production and release to the atmosphere of chlorodifluoromethane (HCFC-22) Midgley, P.M. and Fisher, D.A. *Atmos. Environ.* **27A**, 2215 (1993).

The production and release to the atmosphere of CFCs 113, 114, and 115. Fisher, D.A. and Midgley, P.M. *Atmos. Environ.* **27**, 271 (1993).

Measurements of halogenated organic compounds near the tropical tropopause. Schauffler, S.M., Heidt, L.E., Pollock, W.H., Gilpin, T.M., Vedder, J.F., Solomon, S., Lueb, R.A., and Atlas, E.L. *Geophys. Res. Letts.* **20**, 2567 (1993).

Stratospheric ozone depletion and future levels of atmospheric chlorine and bromine Prather, M. J. and Watson, R. T. *Nature, Lond.* **344**, 729 (1990).

Even anaesthetics may have ozone-depleting effects.

Volatile anaesthetics and the atmosphere: atmospheric lifetimes and atmospheric effects of halothane, enflurane, isoflurane, desflurane and sevoflurane. Langbein, T., Sonntag, H., Trapp, D., Hoffmann, A., Malms, W., Roth, E.P., Mors, V., and Zellner, R. *Brit. J. Anaesth.* **82**, 66 (1999).

Tropospheric lifetimes of halogenated anaesthetics. Brown, A.C., Canosa-Mas, C.E., Parr, A.D., Pierce, J.M.T., and Wayne, R.P. *Nature, Lond.* **341**, 635 (1989).

Halothane and the atmosphere. Brown, A.C., Canosa-Mas, C.E., Parr, A.D., Pierce, J.M.T., and Wayne, R.P. *Lancet*, **II**, 279 (1989).

Section 4.6.5

Ozone depletion potentials, global warming potentials, and future chlorine/bromine loading. Chapter 13 in *Scientific Assessment of Ozone Depletion: 1994*, United Nations Environmental Program; World Meteorological Organisation Report No.37, (WMO, Geneva, 1995).

Updated evaluation of ozone depletion potentials for chlorobromomethane (CH_2ClBr) and 1-bromo-propane ($CH_2BrCH_2CH_3$). Wuebbles, D.J., Kotamarthi, R., and Patten, K.O. *Atmos. Environ.* **33**, 1641 (1999).

Composite global emissions of reactive chlorine from anthropogenic and natural sources: Reactive Chlorine Emissions Inventory. Keene, W.C., Khalil, M.A.K., Erickson, D.J., McCulloch, A., Graedel, T.E., Lobert, J.M., Aucott, M.L., Gong, S.L., Harper, D.B., Kleiman G., Midgley, P., Moore, R.M., Seuzaret, C., Sturges, W.T., Benkovitz, C.M., Koropalov, V., Barrie, L.A., and Li, Y.F. *J. geophys. Res.* **104**, 8429 (1999).

Timescales in atmospheric chemistry: CH_3Br, the ocean and ozone depletion potentials. Prather, M.J. *Global Biogeochem. Cycles.* **11**, 393 (1997).

On the age of stratospheric air and ozone depletion potentials in the polar regions. Pollock, W.A. *J. geophys. Res.* **97**, 12993 (1992).

Section 4.6.6

Montreal Protocol on substances that may deplete the ozone layer–final act. United Nations Environment Programme (UNEP, Nairobi, Kenya, 1987).

Alternatives to CFCs and their behaviour in the atmosphere. Midgley, P.M. in *Volatile organic compounds in the atmosphere*. Hester, R.E. and Harrison, R.M. (eds.) (Royal Society of Chemistry, Cambridge, 1995).

Potential chlorofluorocarbon replacements : OH reaction rate constants between 250 and 315 K and infrared absorption spectra. Garland, N.L., Medhurst, L.J., and Nelson, H.H. *J. geophys. Res.* **98**, 23107 (1993).

Future emission scenarios for chemicals that may deplete stratospheric ozone. Hammitt, J. K., Camm, F., Connell, P. S., Mooz, W. E., Wolf, K. A., Wuebbles, D. J., and Bemazai,A. *Nature, Lond.* **330**, 711 (1987).

Alternative fluorocarbon environmental acceptability study (AFEAS); Volume II of *Scientific assessment of stratospheric ozone: 1989*. World Meteorological Organization, Global ozone research and monitoring project: report no. 20. (WMO, Geneva, 1990).

Model calculations of the relative effects of CFCs and their replacements on stratospheric ozone. Fisher, D. A., Hales, C. H., Filkin, D. L., Ko, M. K. W., Sze, N. D., Connell, P. S., Wuebbles, D. J., Isaksen, I. S. A., and Stordal, F. *Nature, Lond.* **344**, 508 (1990).

The CFC-ozone issue: progress on the development of alternatives to CFCs. Manzer, L.E. *Science* **249**, 31 (1990).

Section 4.6.7

Industrial emissions of trichloroethene, tetrachloroethene and dichloromethane: Reactive Chlorine Emissions Inventory. McCulloch, A., Aucott, M.L., Graedel, T.E., Kleiman, G., Midgley, P.M., and Li, Y.F. *J. geophys. Res.* **104,** 8417 (1999).

Composite global emissions of reactive chlorine from anthropogenic and natural sources: Reactive Chlorine Emissions Inventory. Keene, W.C., Khalil, M.A.K., Erickson, D.J., McCulloch, A., Graedel, T.E., Lobert, J.M., Aucott, M.L., Gong, S.L., Harper, D.B., Kleiman, G., Midgley, P., Moore, R.M., Seuzaret, C., Sturges, W.T., Benkovitz, C.M., Koropalov, V., Barrie, L.A., and Li, Y.F. *J. geophys. Res.* **104,** 8429 (1999).

Present and future trends in the atmospheric burden of ozone-depleting halogens. Montzka, S.A., Butler, J.H., Elkins, J.W., Thompson, T.M., Clarke, A.D., and Lock, L.T. *Nature, Lond.* **398,** 690 (1999).

Stratospheric trends of CFC-12 over the past two decades: Recent observational evidence of declining growth rates. Engel, N., Schmidt, U., and McKenna, D. *Geophys. Res. Letts.* **25**, 3319 (1998).

Decline in the tropospheric abundance of halogen from halocarbons: implications for stratospheric ozone depletion. Montzka, S.A., Butler, J.H., Meyers, R.C., Thompson, T.M., Swanson, T.H., Clarke, A.D., Lock, L.T., and Elkins, J.W. *Science* **272**, 1318 (1996).

Report on Concentrations, Lifetimes and Trends of CFCs, Halons and Related Species. Kaye, J.A., Penkett, S.A., and Ormond, F.M. NASA Reference Publication 1339 (NASA, Washington, DC, 1994).

Sources of hydrochlorofluorocarbons, hydrofluorocarbons and fluorocarbons and their potential emissions during the next 25 years. McCulloch, A. *Environ. Mon. Assess.* **31**, 167 (1994).

Atmospheric trend and lifetime of chlorodifluoromethane (HCFC-22) and the global tropospheric OH concentration. Miller, B.R., Huang, J., Weiss, R.F., Prinn, R.G., and Fraser, P.J. *J. geophys. Res.* **103**, 13237 (1998).

Present and future trends in the atmospheric burden of ozone-depleting halogens. Montzka, S.A., Butler, J.H., Elkins, J.W., Thompson, T.M., Clarke, A.D., Lock, L.T. *Nature, Lond.* **398**, 690 (1999).

Section 4.6.8

Potential atmospheric sources and sinks of nitrous oxide. 2. Possibilities from excited O_2, 'embryonic' O_3, and optically pumped excited O_3. Prasad, S.S. *J. geophys. Res.* **102,** 21527 (1997).

Potential atmospheric sources and sinks of nitrous oxide. 3. Prasad, S.S., Zipf, E.C., and Zhao, X. *J. geophys. Res.* **102,** 21537 (1997).

A re-examination of the impact of anthropogenically fixed nitrogen on atmospheric N_2O and the stratospheric O_3 layer. Nevison, C.D. and Holland, E.A. *J. geophys. Res.* **102,** 25519 (1997).

Increase in the atmospheric nitrous oxide concentration during the last 250 years. Machida, T., Nakazawa, T., Fujii, Y., Aoki, S., and Watanabe, O. *Geophys. Res. Letts.* **22,** 2921 (1995).

Nitrous oxide emissions from fossil fuel combustion. Linak, W.P., McSorley, J.A., Hall R.E., Ryan, J.V., Srivastava, R.K., Wendt. J.O.L., and Mereb, J.B. *J. geophys. Res.* **95,** 7533 (1990).

Nitrous oxide: trends and global mass balance over the last 3000 years. Khalil, M. A. K. and Rasmussen, R. A. *Ann. Glac.* **10,** 73 (1988).

Ozone calculations with large nitrous oxide and chlorine changes. Kinnison, D. E., Johnston, H. S., and Wuebbles, D. J. *J. geophys. Res.* **93,** 14165 (1988).

Factors influencing the loss of fertilizer nitrogen into the atmosphere as N_2O. Conrad, R., Seiler, W., and Burse, G. *J. geophys. Res.* **88,** 6709 (1983).

Effects of nitrogen fertilizers and combustion on the stratospheric ozone layer. Crutzen, P. J. and Ehhalt, D. H. *Ambio* **6,** 112 (1977).

The nitrogen cycle: perturbations due to man and their impact on atmospheric N_2O and O_3. McElroy, M.B., Wofsy, S. C., and Yung, Y. L. *Phil. Trans. R. Soc.* **B277,** 159 (1977).

Section 4.6.9

Seasonal, semiannual, and interannual variability seen in measurements of methane made by the UARS Halogen Occultation Experiment. Ruth, S.L., Kennaugh, R., Gray, L.J. and Russell, J.M., III. *J. geophys. Res.* **102,** 16189 (1997).

The radiative–dynamical effects of high cloud on global ozone distribution. Williams, V. and Toumi, R. *J. atmos. Chem.* **33,** 1 (1999).

Cooling of the Arctic and Antarctic polar stratospheres due to ozone depletion. Randel, W.J. and Wu, F. *J. Climate* **12,** 1467 (1999).

Time scales in atmospheric chemistry: Coupled perturbations to N_2O, NO_y, and O_3. Prather, M.J., *Science* **279,** 1339 (1998).

Climate change and the middle atmosphere. Part IV: Ozone photochemical response to doubled CO_2. Shindell, D.T., Rind, D., and Lonergan, P. *J. Climate* **11,** 895 (1998).

Doubled CO_2 effects on NO_y in a coupled 2D model. Rosenfield, J.E. and Douglass, A.R. *Geophys. Res. Letts.* **25,** 4381 (1998).

Climate change and the middle atmosphere. Part III: The doubled CO_2 climate revisited. Rind, D., Shindell, D., Lonergan, P., and Balachandran, N.K. *J. Climate* **11**, 876 (1998).

Increased polar stratospheric ozone losses and delayed eventual recovery due to increasing greenhouse gas concentrations. Shindell, D.T., Rind, D., and Lonergan, P. *Nature, Lond.* **392**, 589 (1998).

Effect of interannual meteorological variability on midlatitude O_3. Hadjinicolaou, P., Pyle, J.A., Chipperfield, M.P., and Kettleborough, J.A. *Geophys. Res. Letts.* **23**, 2993 (1997).

Possible links between ozone and temperature profiles. Paul, J., Fortuin, F., and Kelder. H. *Geophys. Res. Letts.* **23**, 1517 (1996).

Changes in CH_4 and CO growth rates after the eruption of Mt. Pinatubo and their link with changes in tropical tropospheric UV flux. Dlugokencky, E.J., Dutton, E.G., Novelli, P.C., Tans, P.P., Masarie, K.A., Lantz, K.O., and Madronich, S. *Geophys. Res. Letts.* **23,** 2761 (1996).

Consistency between variations of ozone and temperature in the stratosphere. Finger, R.G., Nagatani, R.M., Gelman, M.E., Long, C.S., and Miller, A.J. *Geophys. Res. Letts.* **22**, 3477 (1995).

The growth rate and distribution of atmospheric CH_4. Dlugokencky, E.J., Steele, L.P., Lang, P.M. and Mesarie, K.A. *J. geophys. Res.* **99**, 17021 (1994).

Recent changes in atmospheric carbon monoxide. Novelli, P.C., Masari, K.A., Tans, P.P., and Lang, P.M. *Science* **263,** 1587 (1994).

A two-dimensional model with coupled dynamics, radiative transfer, and photochemistry 2. Assessment of the response of stratospheric ozone to increased levels of CO_2, N_2O, CH_4 and CFC. Schneider, H.R., Ko, M.K.W., Shia, R., and Sze, N. *J. geophys. Res.* **98**, 20441 (1993).

Ozone response to a CO_2 doubling: Results from a stratospheric circulation model with heterogeneous chemistry. Pitari, G., Palermi, S., Visconti, G., and Prinn, R.G. *J. Geophys .Res.* **97,** 5953 (1992).

Possibility of an Arctic ozone hole in a doubled-CO_2 climate. Austin, J., Butchart, N., and Shine, K.P. *Nature* **360**, 221 (1992).

Carbon monoxide in the Earth's atmosphere: indications of a global increase. Khalil, M. A. K. and Rasmussen, R. A. *Nature, Lond.* **332**, 242 (1988).

Continuing worldwide increases in tropospheric methane, 1978–1987. Blake, D. R. and Rowland, F. S. *Science* **239**, 1129 (1988).

Increased carbon dioxide and stratospheric ozone. Groves, K. S., Mattingley, S. R., and Tuck, A. F. *Nature, Lond.* **273**, 711 (1978).

Impact of coupled perturbations of atmospheric trace gases on Earth's climate and ozone. Nicoli, M. P. and Visconti, G. *Pure & appl. Geophys.* **120**, 626 (1982).

Effect of coupled anthropogenic perturbations on stratospheric ozone. Wuebbles, D. J., Luther, F. M., and Penner, J. E. *J. geophys. Res.* **88**, 1444 (1983).

On the relationship between the greenhouse effect, atmospheric photochemistry and species distribution. Callis, L. B., Natarajan, M., and Boughner, R. E. *J. geophys. Res.* **88**, 1401 (1983).

Influence of stratospheric cooling from CO_2 on the ozone layer. Isaksen, I. S. A., Hesstvedt, E., and Stordal, F. *Nature, Lond.* **283**, 189 (1980).
Effect of water vapour on the destruction of ozone in the stratosphere perturbed by Cl_x or NO_x pollutants. Lin, S. C., Donahue, T. M., Cicerone, R. J., and Chameides, W. L. *J. geophys. Res.* **81**, 3111 (1976).

Section 4.7

The bibliography for Section 4.7 starts with some general introductions to the phenomenon of polar ozone depletion.

The hole in the sky. Gribbin, J. (Corgi Books, London, 1988).
Punching a hole in the stratosphere. Wayne, R. P. *Proc. Royal Inst.* **61**, 13 (1990).
Polar ozone. Chapter 3 in *Scientific Assessment of Ozone Depletion: 1994*, United Nations Environmental Program; World Meteorological Organization Report No.37. (WMO, Geneva, 1995).
The Antarctic ozone hole, Stolarski, R. S. *Sci. Amer.* **258**, 20 (Jan 1988).
The Antarctic ozone hole. Gardiner, B. G. *Weather*, **44**, 291 (1989).
Ozone depletion at the poles: the hole story emerges. *Physics Today*, p.17 (July 1988).
The mystery of the Antarctic ozone 'hole'. Solomon, S. *Rev. Geophys.* **26**, 131 (1988).
Halocarbons in the Arctic and Antarctic atmosphere. Sturges, W.T. in *The Tropospheric Chemistry of Ozone in the Polar Regions.* Niki, H. and Becker, K.-H. (eds.) (Springer-Verlag, Berlin, 1993).

Proceedings of a conference covering most aspects of the polar ozone phenomena

Polar stratospheric ozone 1997. Harris, N.R.P., Kilbane-Dawe, I., and Amanatidis, G.T. (eds.) (European Communities, Luxembourg, 1998).

Section 4.7.1

Essential characteristics of the Antarctic-spring ozone decline: Update to 1998. Uchino, O., Bojkov, R.D., Balis, D.S., Akagi, K., Hayashi, M., and Kajihara, R. *Geophys. Res. Letts.* **26**, 1377 (1999).
Two-year (1996/1997) ozone DIAL measurement over Dumont d'Urville (Antarctica). Santacesaria, V., Stefanutti, L., Morandi, M., Guzzi, D., and MacKenzie, A.R. *Geophys. Res. Letts.* **26**, 463 (1999).
Decadal evolution of the Antarctic ozone hole. Jiang, Y.B., Yung, Y.L., Zurek, R.W. *J. geophys. Res.* **101**, 8985 (1996).
Ten years of ozonesond measurements at the south pole : Implications for recovery of spring time Antarctic ozone. Hofmann, D.J., Oltmans, S.J., Harris, J.M., Johnson, B.J., and Lathrop, J.A. *J. geophys. Res.* **102**, 8931 (1997).

Record low ozone at the South Pole in the spring of 1993. Hofmann, D.J., Oltmans, S.J., Lathrop, J.A., Harris, J.M., and Vömel, H. *Geophys. Res. Letts.* **21**, 421 (1994).

Midwinter start to Antarctic ozone depletion: Evidence from observations and models. Roscoe, H.K., Jones, A.E., and Lee, A.M. *Science* **278**, 93 (1997).

Annual variation of total ozone at Syowa Station, Antarctica. Chubachi, S. *J. geophys. Res.* **102**, 1349 (1997).

A UARS study of lower stratospheric polar processing in the early stages of northern and southern winters. Yudin, V.A., Geller, M.A., Khattatov, B.V., Douglass, A.R., Cerniglia, M.C., Waters, J.W., Elson, L.S., Roche, A.E., and Russell, J.M. *J. geophys. Res.* **102**, 19137 (1997).

Recovery of Antarctic ozone hole. Hofmann, D.J. *Nature, Lond.* **384**, 222 (1996).

Development of the Antarctic ozone hole. Schoeberl, M.R., Douglass, A.R., Kawa, S.R., Dessler, A.E., Newman, P.A., Stolarski, R.S., Roche, A.E., Waters, J.W., and Russell, J.M. *J. geophys. Res.* **101**, 20909 (1996).

Recovery of ozone in the lower stratosphere at the South Pole during the spring of 1994. Hofmann, D.J., Oltmans, S.J., Johnson, B.J., Lathrop, J.A., Harris, J.M., and Vömel, H. *Geophys. Res. Letts.* **22**, 2493 (1995).

Continued decline of total ozone over Halley, Antarctica, since 1985. Jones, A.E. and Shanklin, J.D. *Nature* **376**, 409 (1995).

Evidence for midwinter chemical ozone destruction over Antarctica. Vomel, H., Hofmann, D.J., Oltmans, S.J., and Harris, J.M. *Geophys. Res. Letts.* **22**, 2381 (1995).

Section 4.7.2

The polar vortexes

The structure of the polar vortex. Schoeberl, M.R., Lait, L.R., Newman, P.A., and Rosenfield, J.E. *J. geophys. Res.* **97**, 7859 (1992).

Transport, radiative, and dynamical effects of the Antarctic ozone hole—a GFDL SKYHI model experiment. Mahlman, J.D., Pinto, J.P., and Uumscheid, L.J. *J. Atmos. Sci.* **51**, 489 (1994).

A Lagrangian study of the Antarctic polar vortex. Paparella, F., Babiano, A., Basdevant, C., Provenzale, A., and Tanga, P. *J. geophys. Res.* **102**, 6765 (1997).

Modified Lagrangian-mean diagnostics of the stratospheric polar vortices. 2. Nitrous oxide and seasonal barrier migration in the cryogenic limb array etalon spectrometer and SKYHI general circulation model. Nakamura, N. and Ma, J. *J. geophys. Res.* **102**, 25721 (1997).

Elliptical diagnostics of stratospheric polar vortices. Waugh, D.W. *Q. J. Roy. Meteorol. Soc.* **123** 1725 (1997).

An objective determination of the polar vortex using Ertel's potential vorticity. Nash, E.R., Newman, P.A., Rosenfield, J.E., and Schoeberl, M.R. *J. geophys. Res.* **101**, 9471 (1996).

Vortex dynamics and the evolution of water-vapor in the stratosphere of the southern hemisphere. Lahoz, W.A., O'Neill, A., Heaps, A., Pope, V.D., Swinbank, R., Hanwood, R.S., Froidevaux, L., Read, W.G., Waters, J.W., and Peckham, G.E. *Q. J. Roy. Meteorol. Soc.* **122**, 423 (1996).
The stratospheric polar vortex and sub-vortex—fluid-dynamics and midlatitude ozone loss. McIntyre, M.E. *Phil. Trans. Roy. Soc. London, Ser A.* **352**, 227 (1995).
Air-mass exchange across the polar vortex edge during a simulated major stratospheric warming. Gunther, G. and Dameris, M. *Ann. Geophys.* **13**, 745 (1995).
Descent of long-lived trace gases in the winter polar vortex. Bacmeister, J.T., Schoeberl, M.R., Summers, M.E., Rosenfield, J.R. and Zhu, X. *J. geophys. Res.* **100**, 11669 (1995).
The stratospheric polar vortex and subvortex: fluid dynamics and midlatitude ozone loss. McIntyre, M.E. *Proc. Roy. Soc. London* **A352**, 227 (1995).
On the motion of air through the stratospheric polar vortex. Manney, G.L., Zurek, R.W., O'Neill, A., and Swinbank, R. *J. atmos. Sci.* **51**, 2973 (1994).
CF_4 and the age of mesospheric and polar vortex air. Harnisch, J., Borchers, R., Fabian, P., and Maiss, M. *Geophys. Res. Letts.* **26**, 295 (1999).
On the magnitude of transport out of Antarctic polar vortex. Wauben, W.M.F., Bintanja, R., Van Velthoven, P.F.J., and Kelder, H. *J. geophys.Res.* **102**, 1229 (1997).
Meteorology of the polar vortex: Spring 1997. Coy, L., Nash, E.R., and Newman, P.A. *Geophys. Res .Lett.* **24**, 2693 (1997).

Section 4.7.3

Reconstruction of the constituent distribution and trends in the Antarctic polar vortex from ER-2 flight observations. Schoeberl, M.R., Lait, L.R., Newman, P.A., Martin, R.L., Proffitt, M.H., Hartmann, D.L., Loewenstein, M., Podolske, J., Strahan, S.E., Anderson, J., Chan, K.R., and Gary, B. *J. geophys. Res.* **94**, 16815 (1989).
Chlorine deactivation in the lower stratospheric polar-regions during late winter – results from UARS. Santee, M.L., Froidevaux, L., Manney, G.L., Read, W.G., Waters, J.W., Chipperfield, M.P., Roche, A.E., Kumer, J.B., Mergenthaler, J.L., and Russell, J.M. *J. geophys. Res.* **101**, 18835 (1996).
ATMOS measurements of $H_2O + 2CH_4$ and total reactive nitrogen in the November 1994 Antarctic stratosphere – dehydration and denitrification in the vortex. Rinsland, C.P., Gunson, M.R., Salawitch, R.J., Newchurch, M.J., Zander, R., Abbas, M.M., Abrams, M.C., Manney, G.L., Michelsen, H.A., Chang, A.Y., and Goldman, A. *Geophys. Res. Letts.* **23**, 2397 (1996).
Missing chemistry of reactive nitrogen in the upper stratospheric polar winter. Kawa, S.R., Kumer, J.B., Douglass, A.R., Roche, A.E., Smith, S.E., Taylor, F.W., and Allen, D.J. *Geophys. Res. Letts.* **22**, 2629 (1995).
Observations of denitrification and dehydration in the winter polar stratospheres. Fahey, D.W., Kelly, K.K., Kawa, S.R., Tuck, A.F., Loewenstein, M., Chan, K.R., and Heidt, L.E. *Nature* **344**, 321 (1990).

In-situ measurements of total reactive nitrogen, total water, and aerosol in a polar stratospheric cloud in the Antarctic. Fahey, D.W., Kelly, K.K., Ferry, G.V., Poole, L.R., Wilson, J.C., Murphy, D.M., Loewenstein, M., and Chan, K.R. *J. geophys. Res.* **94**, 11299 (1989).

Section 4.7.4

Microphysics and heterogeneous chemistry of polar stratospheric clouds. Peter, T. *Annu. Rev. Phys. Chem.* **48**, 785 (1997).

Chemical analysis of polar stratospheric cloud particles. Schreiner, J., Voigt, C., Kohlmann, A., Arnold, F., Mauersberger, K., and Larsen, N. *Science* **283**, 968 (1999).

Thermodynamic stability of PSC particles. Koop, T., Carslaw, K.S., and Peter, T. *Geophys. Res. Letts.* **24**, 2199 (1997).

Trajectory studies of polar stratospheric cloud lidar observations at Sodankyla (Finland) during SESAME: Comparison with box model results of particle evolution. Rizi, V., Redaelli, G., Visconti, G., Masci, F., Wedekind, C., Stein, B., Immler, F., Mielke, B., Rairoux, P., Woste, L., del Guasta, M., Morandi, M., Castagnoli, F., Balestri, S., Stefanutti, L., Matthey, R., Mitev, V., Douard, M., Wolf, J.P., Kyro, E., Rummukainen, M., and Kivi, R. *J. atmos. Chem.* **32**, 165 (1999).

Stratospheric clouds over England. Hervig, M. *Geophys. Res. Letts.* **26**, 1137 (1999).

Type I PSC-particle properties: Measurements at ALOMAR 1995 to 1997. Mehrtens, H., von Zahn, U., Fierli, F., Nardi, B., and Deshler, T. *Geophys. Res. Letts.* **26**, 603 (1999).

HALOE observations of polar stratospheric clouds in the Antarctic in October 1993. Chan, A.H.Y., Jackson, D.R., and Harries, J.E. *Advan. Space. Res.* **22**, 1529 (1999).

Analysis of Polar stratospheric cloud particles. Kondratev, K.Y. *Earth Obs. Remote Sens.* **15**, 333 (1999).

Polar stratospheric clouds of liquid aerosol: An experimental determination of mean size distribution. Pantani, M., Del Guasta, M., Guzzi, D., and Stefanutti, L. *J. Aerosol. Sci.* **30**, 559 (1999).

Estimation of polar stratospheric cloud volume and area densities from UARS, stratospheric aerosol measurement, and polar ozone and aerosol measurement extinction data. Massie, S.T., Baumgardner, D., and Dye, J.E. *J. Geophys. Res.* **103**, 5773 (1998).

Evaluating the role of NAT, NAD, and liquid $H_2SO_4/H_2O/HNO_3$ solutions in Antarctic polar stratospheric cloud aerosol: Observations and implications. Del Negro, L.A., et al. *J. geophys. Res.* **102**, 13255 (1997).

Physical state and composition of polar stratospheric clouds inferred from airborne lidar measurements during SESAME. David, C., Godin, S., Mégie, G., Emery, Y., and Flesia, C. *J. atmos. Chem.* **27**, 1 (1997).

Differences in the reactivity of Type I polar stratospheric clouds depending on their phase. Ravishankara, A.R. and Hanson, D.R. *J. geophys. Res.* **101**, 3885 (1996).

Polar stratospheric cloud due to vapor enhancement: HALOE observations of the Antarctic vortex in 1993. Hervig, M.E., Carslaw, K.S., Peter, T., Deshler, T., Gordley, L.L., Redaelli, G., Biermann, U., and Russell, J.M., III. *J. geophys. Res.* **102**, 28185 (1997).

New evidence of size and composition of polar stratospheric cloud particles. Goodman, J., Verma, S., Pueschel, R.F., Hamill, P., Ferry, G.V., and Webster, D. *Geophys. Res. Letts.* **24**, 615 (1997).

Observations of Antarctic polar stratospheric clouds by POAM II: 1994-1996. Fromm, M.D., Lumpe, J.D., Bevilacqua, R.M., Shettle, E.P., Hornstein, J., Massie, S.T., and Fricke, K.H. *J. geophys. Res.* **102**, 23659 (1997).

A global climatology of stratospheric aerosol surface area density deduced from Stratospheric Aerosol and Gas Experiment II measurements: 1984-1994. Thomason, L.W., Poole, L.R., and Deshler, T. *J. geophys. Res.* **102**, 8967 (1997).

Model study of polar stratospheric clouds and their effect on stratospheric ozone. De Rudder, A.D., Larsen, N., Tie, X., Brassuer, G.P., and Granier, C. *J. Geophys. Res.* **101**, 12567 (1996).

Role of aerosol variations in anthropogenic ozone depletion in the polar regions. Portmann, R.W., Solomon, S., Garcia, R.R., Thomason, L.W., Poole, L.R., and McCormick, P. *J. geophys. Res.* **101**, 22991 (1996).

The role of aerosol trends and variability in anthropogenic ozone depletion at northern midlatitudes. Solomon, S., Portmann, R.W., Garcia, R.R., Thomason, L.W., Poole, L.R., and McCormick, M.P. *J. geophys. Res.* **101**, 6713 (1996).

Formation mechanisms of polar stratospheric clouds. Peter, T. in *Nucleation and Atmospheric Aerosols 1996.* Kulmala, M. and Wagner, P.E. (Elsevier, Oxford, 1996).

Observational constraints on the formation of Type 1a polar stratospheric clouds. Tabazadeh, O., Toon, O.B., Gary, B.L., Bacmeister, J.T., and Schoeberl, M.R. *Geophys. Res. Letts.* **23**, 2109 (1996).

Model study of polar stratospheric clouds and their effect on stratospheric ozone. 2. Model results. Tie, X.X., Brassuer, G.P., Granier, C., Derudder, A., and Larsen, N. *J. geophys. Res.* **101**, 12575 (1996).

Melting of particles upon cooling: implications for polar stratospheric clouds. Koop, T. and Carslaw, K. S. *Science,* **272,** 1638 (1996).

Formation of polar stratospheric clouds on preactivated background aerosols. Zhang, R., Leu, M.T., and Molina, M.J. *Geophys. Res. Letts.* **23,** 1669 (1996).

Role of aerosol variations in anthropogenic ozone depletion in the polar regions. Portmann, R.W., Solomon, S., Garcia, R.R., Thomason, L.W., Poole, L.R., and McCormick, M.P. *J. geophys. Res.* **101**, 22991 (1996).

Polar clouds and sulfate aerosols. Tolbert, M. A. *Science* **272**, 1597 (1996).

Unusual PSCs observed by lidar in Antarctica. Stefanutti, L., Morandi, M., Del Guasta, M., Godin, S., and David, C. *Geophys. Res. Letts.* **22**, 2377 (1995).

Nucleation: measurements, theory, and atmospheric applications. Laaksonen, A., Talanquer, V., and Oxtoby, D.W. *Annu. Rev. Phys. Chem.* **46**, 489 (1995).

On the polar stratospheric cloud formation potential of the northern stratosphere. Pawson, S., Naujokat, B., and Kabitzke, K. *J. geophys. Res.* **100**, 23215 (1995).

Do stratospheric aerosol droplets freeze above the ice frost point? Koop, T., Biermann, U.M., Raber, W., Luo, B.P., Crutzen, P.J., and Peter, T. *Geophys. Res. Letts.* **22**, 917 (1995).

Spectroscopic evidence against nitric acid trihydrate in polar stratospheric clouds. Toon, O. B. and Tolbert, M. A. *Nature, Lond.* **375**, 218 (1995).

Stratospheric aerosol growth and HNO_3 gas phase depletion from coupled HNO_3 and water uptake by liquid particles. Carlsaw, K. S., Luo, B. P., Clegg, S. L., Peter, Th., Brimblecombe, P., and Crutzen, P. J. *Geophys. Res. Letts.* 21, 2479 (1994).

Spectroscopic studies of PSCs. Tolbert, M.A., Middlebrook, A.M., and Koehler, B.G. in *Low temperature chemistry of the atmosphere.* Moortgat, G.K., Barnes, A.J., Le Bras, G., and Sodeau, J.R. (eds.) (Springer-Verlag, Berlin, 1994).

A study of Type I polar stratospheric cloud formation. Tabazadeh, A., Turco, R.P., Drdla, K., and Toon, O.B. *Geophys. Res. Letts.* **21**, 1619 (1994).

A climatology of stratospheric aerosol. Hitchman, M.H., McKay, M., and Trepte, C.R. *J. geophys. Res.* **99**, 20689 (1994).

Changes in the character of polar stratospheric clouds over Antarctica in 1992 due to the Pinatubo volcanic aerosol. Deshler, T., Johnson, B.J., and Rozier, W.R. *Geophys. Res. Letts.* **21**, 273 (1994).

Chlorine chemistry on polar stratospheric cloud particles in the Arctic winter. Webster, C. R. and 9 others. *Science* **261**, 1130 (1993).

Heterogeneous chemistry on polar stratospheric clouds. Molina, M. J. *Atmos. Environ.* **25A**, 2535 (1991).

An analysis of lidar observations of polar stratospheric clouds. Toon, O.B., Browell, E.V., Kinne, S., and Jordan, J. *Geophys. Res. Letts.* **17**, 393 (1990).

Physical processes in polar stratospheric ice clouds. Toon, O.B., Turco, R.P., Jordan, J., Goodman, J., and Ferry, G. *J. geophys. Res.* **94**, 11359-11380 (1989).

Polar stratospheric clouds and the Antarctic ozone hole. Poole, L.R. and McCormick, M.P. *J. geophys. Res.* **93**, 8423 (1988).

Section 4.7.5

A reconstructed view of polar stratospheric chemistry. Coffey, M.T., Mankin, W.G., and Hannigan, J.W. *J. Geophys. Res.* **104**, 8295 (1999).

Progress towards a quantitative understanding of Antarctic ozone depletion. Solomon S. *Nature, Lond.* **347**, 347 (1990).

Chemistry of the Antarctic stratosphere. McElroy, M. B., Salawitch, R. J., and Wofsy, S. C. *Planet. Space Sci.* **36**, 73 (1988).

Chlorine activation and ozone destruction in the northern lowermost stratosphere. Lelieveld, J., Bregman, A., Scheeren, H.A., Strom, J., Carslaw, K.S., Fischer, H., Siegmund, P.C., and Arnold, F. *J. Geophys. Res.* **104**, 8201 (1999).

Influence of Antarctic denitrification on two-dimensional model NO_y/N_2O correlations in the lower stratosphere. Nevison, C.D., Solomon, S., Garcia, R.R., Fahey, D.W., Keim, E.R., Loewenstein, M., Podolske, J.R., Gao, R.S., Wamsley, R.C., Donnelly, S.G., and Del Negro, L.A. *J. geophys. Res.* **102**, 13183 (1997).

Reaction mechanism of chlorine nitrate with HCl on ice surface. Xu, S.C., Zhao, X.S. *Acta Physico-chimica Sinica* **14**, 5 (1998).

Evolution and stoichiometry of heterogeneous processing in the Antarctic stratosphere. Jaegle, L., Webster, C.R., May, R.D., Scott, D.C., Stimpfle, R.M., Kohn, D.W., Wennberg, P.O., Hanisco, T.F., Cohen, R.C., Proffitt, M.H., Kelly, K.K., Elkins, J., Baumgardner, D., Dye, J.E., Wilson, J.C., Pueschel, R.F., Chan, K.R., Salawitch, R.J., Tuck, A.F., Hovde, S.J., and Yung, Y.L. *J. geophys. Res.* **102**, 13235 (1997).

Re-formation of chlorine reservoirs in southern hemisphere polar spring. Grooss, J.U., Pierce, R.B., Crutzen, P.J., Grose, W.L., and Russell, J.M. *J. geophys. Res.* **102**, 13141 (1997).

Limits on heterogeneous processing in the Antarctic spring vortex from a comparison of measured and modeled chlorine. Shindell, D.T. and de Zafra, R.L. *J. geophys. Res.* **102**, 1441 (1997).

Observations of nitric acid pertubations in the winter Arctic stratosphere: Evidence for PSC sedimentation. Arnold, F., Burger, V., Gollinger, K., Roncossek, M., Schneider, J., and Spreng, S. *J. atmos. Chem.* **30**, 49 (1997).

Volcanic perturbation of the atmosphere in both polar regions—1991–1994. Herber, A., Thomason, L.W., Dethloff, K., Viterbo, P., Radionov, V.F., and Leiterer, U. *J. geophys. Res.* **101**, 3921 (1996).

Bromine–chlorine coupling in the Antarctic ozone hole. Danilin, M.Y., Sze, N.D., Ko, M.K.W., Rodriguez, J.M., and Prather, M.J. *Geophys. Res. Letts.* **23**, 153 (1996).

What role do type I polar stratospheric cloud and aerosol parameterizations play in modelled lower stratospheric chlorine activation and ozone loss? Sessler, J., Good, P., MacKenzie, A.R., and Pyle, J.A. *J. geophys. Res.* **101**, 28817 (1996).

Correlated observations of HCl and $ClONO_2$ from UARS and implications for stratospheric chlorine partitioning. Dessler, A.E., Considine, D.B., Morris, G.A., Schoeberl, M.R., Russell, J.M., III, Roche, A.E., Kumer, J.B., Mergenthaler, J.L. Waters, J.W., Gille, J.C., and Yue, G.K. *Geophys. Res. Letts.* **22**, 1721 (1995).

Investigation of the Reactive and Nonreactive Processes Involving $ClONO_2$ and HCl on Water and Nitric Acid Doped Ice. Hanson, D.R. and Ravishankara, A.R. *J. phys. Chem.* **96**, 2682 (1992).

Heterogeneous physicochemistry of the polar ozone hole. Turco, R. P., Toon, O. B., and Hamill, P. *J. geophys. Res.* **94**, 16493 (1989).

Antarctic stratospheric chemistry of chlorine nitrate, hydrogen chloride, and ice: release of active chlorine. Molina, M. J., Tso, T. L., Molina, L. T., and Wang, F. C. Y. *Science* **238**, 1253 (1987).

The role of chlorine chemistry in Antarctic ozone loss: implications of new kinetic data. Rodriguez, J. M., Ko, M. K. W., and Sze, N. D. *Geophys. Res. Lett.* **17**, 255 (1990).

Stratospheric nitric acid vapour measurements in the cold arctic vortex–implications for nitric acid condensation, Arnold, F. and Knop, G. *Nature, Lond.* **338**, 746 (1989).

Nitric acid cloud formation in the cold Antarctic stratosphere: a major cause for the springtime 'ozone hole'. Crutzen, P. J. and Arnold, F. *Nature, Lond.* **324**, 651 (1986).

Heterogeneous reactions on nitric acid trihydrate. Moore, S. B., Keyser, L. F., Leu. M.-T., Turco, R. P., and Smith R. H. *Nature, Lond.* **345**, 333 (1990).

Heterogeneous reactions of N_2O_5 with H_2O and HCl on ice surfaces: implications for Antarctic ozone depletion. Hanson, D. and Mauersberger, K. *Geophys. Res. Lett.* **15**, 855 (1988).

Heterogeneous chemical reaction of chlorine nitrate and water on sulfuric acid surfaces at room temperature. Rossi, M. R., Malhotra, R., and Golden, D. M. *Geophys. Res. Lett.* **14**, 127 (1987).

Heterogeneous interactions of $ClONO_2$, HCl, and HNO_3 with sulfuric acid surface at stratospheric temperatures. Tolbert, M. A., Rossi, M. J., and Golden, D. M. *Geophys. Res. Lett.* **15**, 851 (1988).

Antarctic ozone depletion chemistry: reactions of N_2O_5 with H_2O and HCl on ice surfaces. Tolbert, M. A., Rossi, M.J., and Golden, D. M. *Science* **240**, 1018 (1988).

The stability and photochemistry of dimers of the ClO radical and implications for Antarctic ozone depletion. Cox, R. A. and Hayman, G. D. *Nature, Lond.* **322**, 796 (1988).

Role of the ClO dimer in polar stratospheric chemistry: rate of formation and implications for ozone loss. Sander, S. P., Friedl, R. J., and Yung, Y. K. *Science* **245**, 1095 (1989).

Direct ozone depletion in springtime Antarctic lower stratospheric clouds. Hofmann, D. J. *Nature, Lond.* **337**, 447 (1989).

Section 4.7.6

Effects of fluid-dynamical stirring and mixing on the deactivation of stratospheric chlorine. Tan, D.G.H., Haynes, P.H., MacKenzie, A.R., and Pyle, J.A. *J. geophys. Res.* **103**, 1585 (1998).

Interannual variability of the Antarctic ozone hole in a GCM. Part I: The influence of tropospheric wave variability. Shindell, D.T., Wong, S., and Rind, D. *J. atmos. Sci.* **54** 2308 (1997).

A three-dimensional simulation of the Antarctic ozone hole: Impact of anthropogenic chlorine on the lower stratosphere and upper troposphere. Brasseur, G.P., Tie, X.X., Rasch, P.J., and Lefevre, F. *J. geophys. Res.* **102**, 8909 (1997).

Dynamics of the ozone distrubution in the winter stratosphere: Modelling the interhemispheric differences. Cariolle, D., Amodei, M., and Simon, P. *J. atmos. terr. Phys.* **54**, 627 (1992).

Effects of initial active chlorine concentration on the Antarctic ozone spring depletion. Henderson, G. S., Evans, W. F. J., and McConnell, J. C. *J. geophys. Res.* **95**, 1899 (1990).

Section 4.7.7

Chemical ozone depletion during Arctic winter 1997/98 derived from ground based millimeter-wave observations. Langer, J., Barry, B., Klein, U., Sinnhuber, B.M., Wohltmann, I., and Kunzi, K.F. *Geophys. Res. Letts.* **26**, 599 (1999).

Chemical ozone loss in the Arctic winter 1994/95 as determined by the Match technique. Rex, M., von der Gathen, P., Braathen, G.O., Harris, N.R.P., Reimer, E., Beck, A., Alfier, R., Kruger,Carstensen, R., Chipperfield, M., de Backer, H., Balis, D., O'Connor, F., Dier, H., Dorokhov, V., Fast, H., Gamma, A., Gil, M., Kyro, E., Litynska, Z., Mikkelsen, S., Molyneux, M., Murphy, G., Reid, S.J., Rummukainen, M., Zerefos, C. *J. atmos. Chem.* **32**, 35 (1999).

Partitioning of NO_y species in the summer Arctic stratosphere. Osterman, G.B., Sen, B., Toon, G.C., Salawitch, R.J., Margitan, J.J., Blavier, J.F., Fahey, D.W., and Gao, R.S. *Geophys. Res. Letts.* **26**, 1157 (1999).

Second European Stratospheric Arctic and Mid-Latitude Experiment (SESAME): Part II. Farman, J.C., Harris, N.R.P., Isaksen, I., Megie, G., Peter, T., Pyle, J.A., and Simon, P. *J. atmos. Chem.* **32**, R7 (1999).

The vertical distribution of ClO at Ny-Alesund during March 1997. Ruhnke, R., Kouker, W., Reddmann, T., Berg, H., Hochschild, G., Kopp, G., Krupa, R., and Kuntz, M. *Geophys. Res. Letts.* **26**, 839 (1999).

NO_y–N_2O correlation observed inside the Arctic vortex in February 1997: Dynamical and chemical effects. Kondo, Y., Koike, M., Engel, A., Schmidt, U., Mueller, M., Sugita, T., Kanzawa, H., Nakazawa, T., Aoki, S., Irie, H., Toriyama, N., Suzuki, T., and Sasano, Y. *J. geophys. Res.* **104**, 8215 (1999).

Ozone depletion at the edge of the Arctic polar vortex. Hansen, G. and Chipperfield, M.P. *J. geophys. Res.* **104**, 1837 (1999).

Chemical ozone loss in the Arctic vortex in the winter 1995-96: HALOE measurements in conjunction with other observations. Muller, R., Grooss, J.U., McKenna, D.S., Crutzen, P.J., Bruhl, C., Russell, J.M., Gordley, L.L., Burrows, J.P., and Tuck, A.F. *Ann. Geophys.* **17**, 101 (1999).

Polar stratospheric descent of NO_y and CO and Arctic denitrification during winter 1992-1993. Rinsland, C.P., Salawitch, R.J., Gunson, M.R., Solomon, S., Zander, R., Mahieu, E., Goldman, A., Newchurch, M.J., Irion, F.W., and Chang, A.Y. *J. geophys. Res.* **104**, 1847 (1999).

The temporal evolution of the ratio HNO_3/NO_y in the Arctic lower stratosphere from January to March 1997. Schneider, J., Arnold, F., Curtius, J., Sierau, B., Fischer, H., Hoor, P., Wienhold, F.G., Parchatka, U., Zhang, Y.C., Schlager, H., Ziereis, H., Feigl, C., Lelieveld, J., Scheeren, H.A., Bujok, O. *Geophys. Res. Letts.* **26**, 1125 (1999).

Arctic ozone loss due to denitrification. Waibel, A.E., Peter, T., Carslaw, K.S., Oelhaf, H., Wetzel, G., Crutzen, P.J., Poschl, U., Tsias, A., Reimer, E., and Fischer, H. *Science* **283**, 2064 (1999).

First direct simultaneous HCl and $ClONO_2$ profile measurements in the Arctic vortex. Payan, S., Camy-Peyret, C., Jeseck, P., Hawat, T., Durry, G., and Lefevre, F. *Geophys. Res. Letts.* **25**, 2663 (1998).

Localized rapid ozone loss in the northern winter stratosphere: An analysis of UARS observations. Nair, H., Allen, M., Froidevaux, L., and Zurek, R.W. *J. geophys. Res.* **103**, 1555 (1998).

Ground based millimeter-wave observations of Arctic chlorine activation during winter and spring 1998/97. Raffalski, U., Klein, U., Franke, B., Langer, J., Sinnhuber, B.M., Trentmann, J., Kunzi, K.F., and Schrems, O. *Geophys. Res. Letts.* **25**, 3331 (1998).

Ozone depletion in and below the Arctic vortex for 1997. Knudsen, B.M., Larsen, N., Mikkelsen, I.S., Morcrette, J.J., Braathen, G.O., Kyro, E., Fast, H., Gernandt, H., Kanzawa, H., Nakane, H., Dorokhov, V., Yushkov, V., Hansen, G., Gil, M., and Shearman, R.J. *Geophys. Res. Letts.* **25**, 627 (1998).

Ozone loss rates in the Arctic stratosphere in the winter 1991/92: Model calculations compared with Match results. Becker, G., Muller, R., McKenna, D.S., Rex, M., Carslaw, K.S. *Geophys. Res. Letts.* **25**, 4325 (1998).

In situ measurements of stratospheric ozone depletion rates in the Arctic winter 1991/1992: A Lagrangian approach. Rex, M., von der Gathen, P., Harris, N.R.P., Lucic, D., Knudsen, B.M., Braathen, G.O., Reid, S.J., De Backer, H., Claude, H., Fabian, R., Fast, H., Gil, M., Kyro, E., Mikkelsen, I.S., Rummukainen, M., Smit, H.G., Stahelin, J., Varotsos, C., and Zaitcev, I. *J. geophys. Res.* **103**, 5843 (1998).

Dehydration and denitrification in the Arctic polar vortex during the 1995-1996 winter. Hintsa, E.J., Newman, P.A., Jonsson, H.H., Webster, C.R., May, R.D., Herman, R.L., Lait, L.R., Schoeberl, M.R., Elkins, J.W., Wamsley, P.R., Dutton, G.S., Bui, T.P., Kohn, D.W., and Anderson, J.G. *Geophys. Res. Letts.*.**25**, 501 (1998).

Ground based millimeter-wave observations of Arctic ozone depletion during winter and spring of 1996/97. Sinnhuber, B.M., Langer, S., Klein, U., Raffalski, U., and Kunzi, K. *Geophys. Res. Letts.* **25**, 3327 (1998).

Ozone depletion in the late winter lower Arctic stratosphere: Observations and model results. Bregman, A., van den Broek, M., Carslaw, K.S., Muller, R., Peter, T., Scheele, M.P., and Lelieveld, J. *J. geophys. Res.*, **102**, 10815 (1997).

HALOE observations of the Arctic vortex during the 1997 spring: Horizontal structure in the lower stratosphere. Pierce, R.B., Fairlie, T.D., Remsberg, E.E., Russell, J.M., III, and Grose, W.L. *Geophys. Res. Letts.* **24**, 2701 (1997).

Dehydration and sedimentation of ice particles in the Arctic stratospheric vortex. Vomel, H., Rummukainen, M., Kivi, R., Karhu, J., Turunen, T., Kyro, E., Rosen, J., Kjome, N., and Oltmans, S. *Geophys. Res. Letts.* **24**, 795 (1997).

An investigation of C1O photochemistry in the chemically perturbed arctic vortex. Pierson, J.M., McKinney, K.A., Toohey, D.W., Margitan, J., Schmidt, U., Engel, A. and Newman, P.A. *J. atmos. Chem.* **32**, 61 (1999).

MLS observations of ClO and HNO_3 in the 1996-97 Arctic polar vortex. Santee, M.L., Manney, G.L., Froidevaux, L., Zurek, R.W., and Waters, J.W. *Geophys. Res. Letts.* **24**, 2713 (1997).

Evidence of substantial ozone depletion in winter 1995/96 over Northern Norway. Hansen, G., Svenoe, T., Chipperfield, M.P., Dahlback, A., and Hoppe, U.-P. *Geophys. Res. Letts.* **24**, 799 (1997).

The temporal evolution of the ratio HNO_3/NO_y in the Arctic lower stratosphere from January to March 1997. Schneider, J., Arnold, F., Curtius, J., Sierau, B., Fischer, H., Hoor, P., Wienhold, F.G., Parchatka, U., Zhang, Y.C., Schlager, H., Ziereis, H., Feigl, C., Lelieveld, J., Scheeren, H.A., and Bujok, O. *Geophys. Res. Letts.* **26,** 1125 (1999).

Anomalously low ozone over the Arctic. Newman, P., Gleason, J.F., McPeters, R.D., and Stolarski, R.S. *Geophys. Res. Letts.* **24**, 2689 (1997).

MLS observations of Arctic ozone loss in 1996-1997. Manney, G.L., Froidevaux, L., Santee, M.L., Zurek, R.W., and Waters, J.W. *Geophys. Res. Letts.* **24**, 2697 (1997).

Simultaneous observations of polar stratospheric clouds and HNO_3 over Scandinavia in January 1992. Massie, S.T., et al. *Geophys. Res. Letts.* **24**, 595 (1997).

Evidence of substantial ozone depletion in winter 1996 over northern Norway. Hansen, G., Svenoe, T., Chipperfield, M., Dahlback, A., and Hoppe, U.P. *Geophys. Res. Letts.* **24**, 799 (1997).

HALOE observations of the vertical structure of chemical ozone depletion in the Arctic vortex during winter and early spring 1996-1997. Müller, R., Grooss, J.-U., Brühl, C., McKenna, D.S., Crutzen, P.J., Brühl, C., Russell, J.M., III, and Tuck, A.F. *Geophys. Res. Letts.* **24**, 2717 (1997).

Model studies of chlorine deactivation and formation of $ClONO_2$ collar in the Arctic polar vortex. Chipperfield, M.P., Lutman, E.R., Kettleborough, J.A., Pyle, J.A., and Roche, A.E. *J. geophys. Res.* **102**, 1467 (1997).

Prolonged stratospheric ozone loss in the 1995-96 Arctic winter. Rex, M., Harris, N.R.P., von der Gathen, P., Lehmann, R., Braathen, G.O., Reimer, E., Beck, A., Chipperfield, M.P., Alfier, R., Allart, M., O'Conner, F., Dier, H., Dorokhov, V., Fast, H., Gil, M., Kyrö, E., Litynska, Z., St. Mikkelsen, I., Molyneux, M.G., Nakane, H., Notholt, J., Rummukainen, M., Viatte, P., and Wenger, J. *Nature, Lond.* **389**, 835 (1997).

Severe chemical ozone loss in the Arctic during the winter of 1995-96. Müller, R., Crutzen, P.J., Grooss, J.-U., Brühl, C., Russell, J.M., III, Gernandt, H., McKenna, D.S., and Tuck, A.F. *Nature, Lond.* **389**, 709 (1997).

Chemical loss of polar vortex ozone infrerred from UARS MLS measurements of ClO during the Arctic and Antarctic late winters of 1993. MacKenzie, I.A., Harwood, R.S., Froidevaux, L., Read, W.G., and Waters, J.W. *J. geophys. Res.* **101**, 12505 (1996).

Partitioning between chlorine reservoir species deduced from observations in the Arctic winter stratosphere. Engel, A., Schmidt, U., and Stachnik, R.A. *J. atmos. Chem.* **27**, 107 (1997).

Evidence of the removal of gaseous HNO_3 inside the Arctic polar vortex in January 1992. Höpfner, M., Blom, C.E., Blumenstock, T., Fischer, H., and Gulde, T. *Geophys. Res. Letts.* **22**, 149 (1996).

Chlorine activation and ozone depletion in the Arctic vortex: observations by the halogen occultation experiment on the upper atmosphere research satellite. Müller, R., Crutzen, P.J., Grooß, J.-U., Brühl, C., Russell, J.M., III , and Tuck, A.F. *J. geophys. Res.* **101,** 12531 (1996).

Observations of high concentrations of total reactive nitrogen (NO_y) and nitric acid (HNO_3) in the lower Arctic stratosphere during the stratosphere–troposphere experiment by aircraft measurements (STREAM) II campaign in February 1995. Fischer, H., Waibel, A.E., Welling, M., Wienhold, F.G., Zenker, T., Crutzen, P.J., Arnold, F., Burger, V., Schneider, J., Bregman, A., Lelieveld, J., and Siegmund, P.C. *J. geophys. Res.* **102**, 23559 (1997).

Polar vortex conditions during the 1995-1996 Arctic winter. Manney, G.L., Froidevaux, L., Santee, M.L., Waters, J.W., and Zurek, R.W. *Geophys. Res. Letts.* **23**, 3203 (1996).

The effect of small-scale inhomogeneities on ozone depletion in the Arctic. Edouard, S., Legras, B., Lefevre, F., and Eymard, R. *Nature* **384**, 444 (1996).

Evidence for Arctic ozone depletion in late February and early March 1994. Manney, G.L., Zurek, R.W., Froidevaux, L., Waters, J.W. *Geophys. Res. Letts.* **22**, 2941 (1995).

The variation of available chlorine, Cl_y, in the Arctic polar vortex during EASOE. Schmidt, U., Bauer, R., Engel, A., Borchers, R., and Lee, J. *Geophys. Res. Letts.* **21**, 1215 (1994).

Spectroscopic measurement of bromine oxide and ozone in the high Arctic during Polar Sunrise Experiment 1992. Hausmann, M. and Platt, U. *J. geophys. Res.* **99**, 24399 (1994).

Record low total ozone during northern winters of 1992 and 1993. Bojkov, R.D., Zerefos, C.S., Balis, D.S., Ziomas, I.C., and Bais, A.F. *Geophys. Res. Letts.* **20**, 1351 (1993).

Balloon observations of nitric acid aerosol formation in the Arctic stratosphere: I. Gaseous nitric acid. Schlager, H., Arnold, F., Hofmann, D.J., and Deshler, T. *Geophys. Res. Letts.* **17**, 1275 (1990).

Second European Stratospheric Arctic and Midlatitude Experiment campaign: Correlative measurements of aerosol in the northern polar atmosphere. Brogniez, C., Lenoble, J., Ramananaherisoa, R., Fricke, K.H., Shettle, E.P., Hoppel, K.W., Bevilacqua, R.M., Hornstein, J.S., Lumpe, J., Fromm, M.D., and Krigman, S.S. *J. geophys. Res.* **102**, 1489 (1997).

Observations of denitrification and dehydration in the winter polar stratospheres. Fahey, D. W., Kelley, K. K., Kawa, S. R., Tuck, A. F., Loewenstein, M., Chan, K. R., and Heidt, L. E. *Nature, Lond.* **344**, 321 (1990).

Stratospheric clouds and ozone depletion in the Arctic during January 1989. Hofmann, D. J., Aimedieu, P., Matthews, W. A., Johnston, P. V., Kondo, Y., Sheldon, W. R., and Byrne, G. J. *Nature, Lond.* **340**, 117 (1989).

Observations of stratospheric NO_2 and O_3 at Thule, Greenland. Mount, G. H., Solomon, S., Sanders, R. W., Jakoubek, R. O., and Schmeltekopf, A. L. *Science* **242**, 555 (1988).

Results from the 1989 Airborne Arctic Stratospheric Expedition are summarized in a special issue of Geophys. Res. Lett.

Geophys. Res. Lett. **17**, 313–564 (April 1990).

Section 4.7.8

Polar ozone depletion – a 3-dimensional chemical modeling study of its long-term global impact. Eckman, R.S., Grose, W.L., Turner, D.E., and Blackshear, W.T. *J. geophys. Res.* **101,** 22977 (1996).

Measurements of polar vortex air in the midlatitudes. Newman, P.A., et al. *J. geophys. Res.* **101,** 12879 (1996).

Observational studies of the role of polar regions in midlatitude ozone loss. Jones, R.L. and MacKenzie, A.R. *Geophys. Res. Letts.* **22,** 3485 (1995).

Ozone loss in middle latitudes and the role of the Arctic polar vortex. Pyle, J.A. *Phil. Trans. Roy. Soc. London* **A352,** 241 (1995).

Global impact of the Antarctic ozone hole: chemical propagation. Prather, M. and Jaffe, A.H. *J. geophys. Res.* **95,** 3473 (1990).

Ultraviolet levels under sea ice during the Antarctic spring. Trodahl, H. J. and Buckley, R. G. *Science* **245,** 194 (1989).

A general circulation model simulation of the springtime Antarctic ozone decrease and its impact on mid latitudes. Cariolle, D., Laserre-Bigorry, A., Royes, J.-F., and Geleyn, J. -F. *J. geophys. Res.* **95,** 1883 (1990).

Global impact of the Antarctic ozone hole: dynamical dilution with a three-dimensional chemical transport model. Prather, M., Garcia, M. M., Suozzo, R., and Rind, D. *J. geophys. Res.* **95,** 3449 (1990).

Global impact of the Antarctic ozone hole: chemical propagation. Prather, M. and Jaffe, A. H. *J. geophys. Res.* **95,** 3473 (1990)

Antarctic ozone hole: possible implications for ozone trends in the southern hemisphere. Sze, N. D., Ko, M. K. W., Weisenstein, D. K., Rodriguez, J. M., Stolarski, R. S., and Schoeberl, M. R. *J. geophys. Res.* **94,** 11521 (1989).

Section 4.8

Ozone variability and trends. Chapter 4 of *Scientific assessment of ozone depletion: 1998*. World Meteorological Organization, Global ozone research and monitoring project: report no. 44. (WMO, Geneva, 1999).

Trends in stratospheric and free tropospheric ozone. Harris, N.R.P., and 12 others. *J. geophys. Res.* **102,** 1571 (1997).

The Ozone Trends Panel Reports present the WMO assessments. Examples include

SPARC/IOC/GAW assessment of trends in the vertical distribution of ozone. World Meteorological Organization, Global ozone research and monitoring project: report no. 43. Harris, N., Hudson, R., and Phillips, C. (eds.) (WMO, Geneva, 1998).

Ozone Trends Panel Report, 1988. World Meteorological Organization, Global ozone research and monitoring project: report no. 18. Watson, R. T. (ed.) (WMO, Geneva, 1990).

Other papers deal with the trends, and with some of the factors leading to ozone variability and with the methods for extracting trends from the data.

Stratospheric ozone change. Kaye, J.A. and Jackman, C.H. in *Global atmospheric chemical change.* Hewitt, C.N and Sturges, W.T. (eds.) (Chapman and Hall, London, 1994).

European research in the stratosphere: the contribution of EASOE and SESAME to our current understanding of the ozone layer. Amanatidis, G.T. (ed.) (European Commission, Luxembourg, 1997).

The changing stratosphere. McElroy, M. B., Salawitch, R. J., and Minschwaner, K. *Planet. Space Sci.* **40,** 373 (1992).

The role of dynamics in total ozone deviations from their long-term mean over the Northern Hemisphere. Petzoldt, K. *Ann. Geophys.* **17,** 231 (1999).

Atmospheric chemistry—Uncertain road to ozone recovery. Fraser, P.J. and Prather, M.J. *Nature, Lond.* **398,** 663 (1999).

Characteristics of the ozone decline in the Northern polar and middle latitudes during the winter–spring. Bojkov, R.D., Balis, D.S., and Zerefos, C.S. *Meteorol. Atmos. Phys.* **69,** 119 (1998).

Decadal evolution of total ozone decline: Observations and model results. Tourpali,K., Tie, X.X., Zerefos, C.S., and Brasseur, G. *J. geophys. Res.* **102,** 23955 (1997).

Changes of the lower stratospheric ozone over Europe and Canada. Bojkov, R.D. and Fioletov, V.E. *J. geophys. Res.* **102,** 1337 (1997).

Trends in stratospheric and free tropospheric ozone. Harris, N.R.P., Ancellet, G., Bishop, L., Hoffman, D.J., Kerr, J.B., McPeters, R.D., Prendez, M., Randel, W.J., Staehelin, J., Subbaraya, B.H., Volz-Thomas, A., Zawodny, J., and Zerefos, C.S. *J. geophys. Res.* **102,** 1571 (1997).

Spatial and seasonal characteristics of recent decadal trends in the northern hemispheric troposphere and stratosphere. Kodera, K and Koide, H. *J. geophys. Res.***102,** 19433 (1997).

Long term ozone decline over the Canadian Arctic to early 1997 from ground-based and balloon observations. Fioletov, V.E., Kerr, J.B., Wardle, D.I., Davies, J., Hare, E.W., McElroy, C.T., and Tarasick, D.W. *Geophys. Res. Letts.* **24,** 2705 (1997).

An investigation of dynamical contributions to midlatitude ozone trends in winter. Hood, L.L., McCormack, J.P., and Labitzke, K. *J. geophys. Res.* **102** 13079 (1997).

On the origin of midlatitude ozone changes: Data analysis and simulations for 1979-1993. Gallis, L.B., Natarajan, M., Lambeth, J.D., and Boughner, R.E. *J. geophys. Res.* **102,** 1215 (1997).

A critical analysis of Stratospheric Aerosol and Gas Experiment ozone trends. Wang, H.J., Cunnold, D.M., and Bao.X. *J. geophys. Res.* **101,** 12495 (1996).

Past, present and future modeled ozone trends with comparisons to observed trends. Jackman, C.H., Fleming, E.L., Chandra, S., Considine, D.B., and Rosenfield, J.E. *J. geophys. Res.* **101,** 28753 (1996).

Recent trends in ozone in the upper stratosphere: Implications for chlorine chemistry. Chandra, S., Jackman, C.H., and Fleming. E.L. *Geophys. Res. Letts.* **22,** 843 (1995).

Ozone trends deduced from combined Nimbus 7 SBUV and NOAA 11 SBUV/2 data. Hollandsworth, S.M. et al. *Geophys. Res. Letts.* **22**, 905 (1995).

Estimating the global ozone characteristics during the last 30 years. Bojkov, R.D. and Fioletov, V.E. *J. geophys. Res.* **100**, 16537 (1995).

Seasonal trend analysis of published ground-based and TOMS total ozone data through 1991. Reinsel, G.C., Tiao, G.C., Wuebbles, D.J., Kerr, J.B., Miller, A.J., Nagatani, R.M., Bishop, L., and Ying, L.H. *J. geophys. Res.* **99**, 5449 (1994).

Trends in the vertical distribution of ozone: An analysis of ozone sonde data. Logan, J.A. *J. geophys. Res.* **99**, 25553 (1994).

Chlorine catalysed destruction of ozone: implications for ozone variability in the upper stratosphere. Chandra, S., Jackman, C.H., Douglass, A.R., Fleming, E.L., and Considine, D.B. *Geophys. Res. Letts.* **20**, 351 (1993).

Measured trends in stratospheric ozone. Stolarski, R., Bojkov, R., Bishop, L., Zerefos, C., Staehelin, J., and Zawodny, J. *Science* **256**, 342 (1992).

A critical analysis of SAGE ozone trends. Wang, H.J., Cunnold, D.M., and Bao, X. *J. geophys. Res.* **101**, 12495 (1996).

A statistical trend analysis of revised Dobson data over the Northern Hemisphere. Bojkov, R., Bishop, L., Hill, W.J., Reinsel, G.C., and Tiao, G.C. *J. geophys. Res.* **95**, 9785 (1990).

An analysis of the 7-year record of SBUV satellite ozone data: global profile features and trends in total ozone. Reinsel, G. C., Tiao, G. C., Ahn, S. K., Pugh, M., Basu, S., DeLuisi, J. L., Mateer, C. L., Miller, A. J., Connell, P. S., and Wuebbles, D. J. *J. geophys. Res.* **93**, 1689 (1988).

5 The Earth's troposphere

5.1 Introduction

Hydroxyl radicals dominate the daytime chemistry of the troposphere in the same way that oxygen atoms and ozone dominate the chemistry of the stratosphere. High reactivity of the OH radical with respect to a wide range of species leads to oxidation and chemical conversion of most trace constituents that have an appreciable physical lifetime in the troposphere. Free-radical chain reactions oxidize hydrogen, methane, and other hydrocarbons, and carbon monoxide to CO_2 and H_2O. The reactions thus constitute a low-temperature combustion system. Other species, notably NO_x and sulphur compounds, participate in the reactions to modify the course of the combustion processes.[a] Species that survive both physical loss and chemical conversion in the troposphere (e.g. N_2O, some CH_4, and CH_3Cl) are transported to the stratosphere where they yield the NO_x, HO_x, and ClO_x radicals that destroy ozone in the catalytic cycles discussed in Chapter 4. Hydroxyl radical chemistry in the troposphere provides an efficient chemical scavenging mechanism for both natural and man-made trace constituents, and has a major influence not only on tropospheric composition, but also on stratospheric behaviour.

At night, the nitrate radical, NO_3, takes over from OH as the dominant oxidant in the troposphere. Although NO_3 is generally much less reactive than OH, its peak tropospheric concentration is higher, so that it plays an important role in atmospheric chemical transformations. The diurnal impact of OH and NO_3 is complementary, because OH is generated photochemically only during the day, while NO_3 is readily photolysed, and so can survive only at night.

About 90 per cent of the total atmospheric mass resides in the troposphere, and the bulk of the minor trace-gas burden is also found there. The Earth's surface acts as the main source of the trace gases, although some NO_x and CO may be produced in thunderstorms. Surface emissions include natural and 'pollution' sources, with the latter being concentrated in urban and industrial areas of the Northern Hemisphere. Natural sources are more evenly distributed, but since the land area is twice as great in the Northern as in the Southern Hemisphere, some asymmetry remains.

Radical-chain processes in the troposphere are photochemically driven, although stratospheric ozone limits the solar radiation at the Earth's surface to wavelengths longer than 280 nm (Chapter 4). In 1961, P.A. Leighton wrote his classic work on atmospheric chemistry. Its title, *The photochemistry of air pollution* (see Bibliography) hardly does justice to its coverage, which goes far beyond pollution matters. Leighton noted the need to characterize carefully the flux of solar photons as a function of wavelength as these photons pass through the atmosphere and are absorbed and scattered.

[a]See footnote *a* on p. 166 for a definition of the representations NO_x and NO_y.

The most important species that are photochemically labile at $\lambda > 280$ nm are O_3, NO_2, and HCHO (formaldehyde). As we shall see in Section 5.3, all three of these species can indirectly lead to OH (or HO_2) formation, and thus initiate the oxidation chains. Ozone photolysis is a critical step, since the other photolytic processes owe either their origin or their importance to it. Although only 10 per cent of the total atmospheric ozone is found in the troposphere, all *primary* initiation of oxidation chains in the natural atmosphere depends on that ozone. The origin of tropospheric ozone is therefore of great interest. Mechanisms exist in tropospheric chemistry that potentially can generate O_3 at low altitudes, but their efficiencies are related to absolute and relative concentrations of hydrocarbons and, especially, of NO_x. Ozone is also transported, *via* stratospheric–tropospheric exchange, from the stratospheric ozone layer. The relative importance of stratospheric and locally produced ozone continues to be a controversial subject.

5.2 Sources, sinks, and transport

Some of the natural and artificial sources of various trace species have already been identified in Chapters 1 and 4, and we shall consider further compounds as the present chapter develops. Table 5.1 is a summary[a] of the most important natural and man-made sources of tropospheric trace gases, which include many *volatile organic compounds*, or VOCs. Our immediate interest is to follow the general path of trace components from their sources to their sinks. Physical removal processes are divided into *dry deposition* in which the species are absorbed irreversibly on soil, water, or plant surfaces, and *wet deposition* in which the constituents are incorporated into precipitation elements (clouds, rain droplets, and aerosols), processes that we shall explore more fully in Section 5.2.1. In between release to and removal from the atmosphere, the species may be transported horizontally and vertically. The source–transport–sink sequence defines a *physical lifetime* for the constituent. However, chemical changes may occur on a time-scale comparable with, or smaller than, this lifetime. The *chemical lifetime* may then determine for how long the compound maintains its chemical identity after release, as discussed at much greater length in Section 3.6.1. Degradation of one trace gas frequently generates another which survives long enough to have a separate existence. For example, released methane is not rapidly removed by physical processes, and is oxidized slowly (with a lifetime of several years). Formaldehyde is an intermediate in the oxidation, but is quickly photolysed (lifetime of a few hours). The products are hydrogen and carbon monoxide, which have lifetimes of several years and several months, respectively.

[a]Cox, R.A. and Derwent, R.G. *Gas-phase chemistry of the minor constituents of the troposphere. Gas kinetics and energy transfer*, Specialist Periodical Reports Chem. Soc. **4**, 189 (1981).

Table 5.1 Natural and man-made sources of the minor trace gases of the troposphere

Compound	Natural sources	Man-made sources
Carbon-containing trace gases		
Carbon monoxide (CO)	Oxidation of natural methane, natural C_5, C_{10} hydrocarbons; oceans; forest fires	Oxidation of man-made hydrocarbons; incomplete combustion of wood, oil, gas, and coal, especially in motor vehicles, industrial processes; blast furnaces
Carbon dioxide (CO_2)	Oxidation of natural CO; destruction of forests; respiration by plants	Combustion of oil, gas, coal, and wood; limestone burning
Methane (CH_4)	Enteric fermentation in wild animals; emissions from swamps, bogs, etc.; natural wet land areas; oceans	Enteric fermentation in domesticated ruminants; emissions from paddies; natural gas leakage; sewerage gas; colliery gas; combustion sources
Light alkanes, C_2–C_6	Aerobic biological source	Natural gas leakage; motor vehicle evaporative emissions; refinery emissions
Alkenes, C_2–C_6		Motor vehicle exhaust; diesel engine exhaust
Aromatic hydrocarbons		Motor vehicle exhaust; evaporative emissions; paints; petrol; solvents
Semiterpenes, C_5H_8 Terpenes, $C_{10}H_{16}$ Diterpenes, $C_{20}H_{32}$	Trees, broadleaves, and conifers; plants	
Nitrogen-containing trace gases		
Nitric oxide (NO)	Forest fires; anaerobic processes in soil; electric storms	High-temperature combustion (oil, gas, coal)
Nitrogen dioxide (NO_2)	Forest fires; electric storms	High-temperature combustion (oil, gas, coal); atmospheric transformation of NO
Nitrous oxide (N_2O)	Emissions from denitrifying bacteria in soil; oceans	Combustion of oil and coal

Table 5.1 *(continued)*

Compound	Natural sources	Man-made sources
Peroxyacetyl nitrate (PAN)	Degradation of isoprene	Degradation of hydrocarbons
Ammonia (NH_3)	Aerobic biological source in soil; Breakdown of amino acids in organic waste material	Coal and fuel-oil combustion; waste treatment
Sulphur-containing trace gases		
Sulphur dioxide (SO_2)	Oxidation of H_2S; volcanic activity	Combustion of oil and coal; roasting sulphide ores
Hydrogen sulphide (H_2S)	Anaerobic fermentation; volcanoes and fumaroles	Oil refining; animal manure; Kraft paper mills; rayon production; coke-oven gas
Carbon disulphide (CS_2)	Anaerobic fermentation	Viscose rayon plants; brick making; fish-meal processing
Carbonyl sulphide (COS)	Oxidation of CS_2; slash-and-burn agriculture; volcanoes and fumaroles	Oxidation of CS_2; brick making; effluent from Kraft mills; blast-furnace gas; coke-oven gas; shale and natural gas
Sulphur trioxide (SO_3)		Combustion of S-containing fuel
Methyl mercaptan (CH_3SH)	Anaerobic biological sources	Animal rendering; animal manure; pulp and paper mills; brick manufacture; oil refining
Dimethyl sulphide (CH_3SCH_3)	Aerobic biological sources	Animal rendering; animal manure; pulp and paper mills
Dimethyl disulphide (CH_3SSCH_3)		Animal rendering; fish-meal processing
Other organic sulphur compounds: C_2–C_4 mercaptans; dialkyl disulphides; dimethyl trisulphide; alkyl thiophenes; benzothiophenes	Anaerobic biological sources	Animal rendering; fish-meal processing; brick making

Table 5.1 (*continued*)

Compound	Natural sources	Man-made sources
Halogen-containing trace gases		
Hydrogen fluoride (HF)		Atmospheric degradation of HCFCs and HFCs
Hydrogen chloride (HCl)	Volcanoes and fumaroles; degradation of CH_3Cl	Coal combustion; degradation of chlorocarbons, CFCs and HCFCs
Methyl chloride (CH_3Cl)	Slow combustion of organic matter; microbial action in oceans; algae	PVC manufacture and degradation; tobacco consumption
Methyl bromide (CH_3Br)	Aerobic biological sources	Fumigation of soil and grain; fire retardant
Methyl iodide (CH_3I)	Aerobic biological sources	
Methylene dichloride (CH_2Cl_2)		Solvent
Chloroform ($CHCl_3$)		Pharmaceuticals; solvent; combustion of petrol; wood-pulp bleaching; degradation of $CHClCCl_2$
Carbon tetrachloride (CCl_4)		Solvent; fire extinguishers; degradation of CCl_2CCl_2
Carbon tetrafluoride (CF_4)		Aluminium industry
Sulphur hexafluoride (SF_6)		Electrical insulator, especially in transformers
Methyl chloroform (CH_3CCl_3)		Solvent; degreasing agent
Trichloroethylene ($CHClCCl_2$)		Solvent; dry-cleaning agent; degreasing agent
Tetrachloroethylene (CCl_2CCl_2)		Solvent; dry-cleaning agent; degreasing agent
CFCs, HCFCs, HFCs (see Sections 4.6.4 and 4.6.6): e.g., CF_2Cl_2, $CFCl_3$, CF_3CHCl_2 , CF_3CH_2F		Refrigerants; aerosol propellants; foam-blowing agents (uses now restricted: see Section 4.6.6)
Other minor trace gases		
Hydrogen (H_2)	Oceans; soils; oxidation of methane, isoprene, and terpenes *via* formaldehyde	Motor vehicle exhaust; oxidation of methane *via* formaldehyde
Water vapour (H_2O)	Evaporation, especially from oceans	
Ozone (O_3)	Stratosphere; natural $NO \rightarrow NO_2$ conversion	Man-induced $NO \rightarrow NO_2$ conversion

Updated from Cox, R.A. and Derwent, R.G. *Specialist Periodical Reports Chem. Soc.* **4**, 189 (1981).

5.2.1 Dry and wet deposition

Dry deposition refers to the removal on the Earth's surface of gases and particles by a direct transfer process, and without the involvement of any precipitation. Wet deposition, on the other hand, is a term used for all deposition processes in which the gases or particles are carried to the surface dissolved or entrapped in water. Fog, rain, hail, or snow may thus all be agents of wet precipitation.

Dry deposition acts efficiently only where a specific chemical or biological interaction is available, and even then only when the trace gases are close to the surface. Of the species listed in Table 5.1, only a handful undergo direct deposition. Dry deposition mechanisms exist for SO_2, O_3, CO_2, and SO_3, while microbiological sinks are known for soil removal of CO and H_2. Wet deposition of gas-phase species requires that the compounds are water soluble. Nitrous oxide, carbonyl sulphide (COS), and the chlorofluorocarbons are sufficiently inert, physically and chemically, to survive exchange with the stratosphere. All other species undergo tropospheric photo-oxidation with OH, or direct photolysis (e.g. O_3, NO_2, HCHO). The end of the atmospheric physical and chemical sequence of events then comes when the products of the reactions are finally lost by wet or dry deposition.

Three separate steps can be envisaged in the dry deposition process. First, the species must be transported through the atmosphere to some region in close proximity to the surface, then they must cross to the surface itself (by diffusive or Brownian mechanisms), and finally there must be uptake on the surface. Turbulent atmospheric motions near the Earth's surface effect the first step, whose efficiency is thus determined by the degree of turbulence. Uptake of the gas or particle on the surface is governed by a variety of factors, including shape and smoothness of the surface, the amount of moisture on it, and, of course, the solubility of the depositing species and any specific chemical or biological interactions. Moderately soluble compounds such as SO_2 and O_3 are reversibly absorbed, so that the dampness of the surface is critical, while highly soluble species such as HNO_3 absorb rapidly and irreversibly on virtually any surface.

Rates of dry deposition are often described in terms of a semi-empirical equation

$$F = -u_d C, \tag{5.1}$$

where F is the flux of dry deposition of a species whose concentration is C at some specified height above the surface. The equation then defines the *deposition velocity*, u_d. An assumption is that F remains constant up to the specified reference height; in reality, u_d and F are likely to depend on height above the surface. The equation is often used as a convenient boundary condition for the diffusion equation in atmospheric models in order to take into account dry deposition. Its convenience should not, however, obscure the way in which the parameter u_d lumps together all of the factors in the three steps leading to deposition.

Experimental measurement of dry-deposition velocities has generally proved quite difficult, although it is evident that natural surfaces, such as plant leaves, act as relatively efficient sinks. For example, chamber experiments give velocities of deposition of ethene to spinach, grass, bare soil, and water as 74, 150, 27, and 1.6 $\mu m\,s^{-1}$, respectively, results that show clearly the much greater efficiency of grass compared with sea-water as a surface for dry deposition. The different surfaces afford a variation in the type of plant canopy and the leaf area, and also allow estimates to be made of the relative importance of deposition to the plant canopy and the soil. Different hydrocarbons behave in somewhat different ways; for example, deposition velocities for an alkane (*n*-hexane) were 43, 49, 22, and 3 $\mu m\,s^{-1}$ for the four surfaces in the same order, while for an alkyne (acetylene) the values were 2, 65, 16, and 1.6 $\mu m\,s^{-1}$. Some incompatibility between acetylene and spinach evidently exists! Regardless of the exact numerical values, all these deposition velocities are extremely small, and the experiments confirm the expectation that dry deposition of hydrocarbons is not important in the atmosphere.

Wet deposition by incorporation into falling precipitation ('*washout*') or cloud droplets ('*rainout*') is only significant for those gas-phase species that are water soluble. Particles, on the other hand, can serve as condensation nuclei and be incorporated into a droplet, or they can collide with a pre-existing droplet. A gas or particle taken up by a droplet may, of course, not reach the ground if the droplet evaporates during its descent.

Chemical conversions may play an essential part in the overall scavenging process for gases by generating some new compound that can be removed physically. A tendency exists for natural trace gases of biogenic origin to be reduced (CH_4, terpenes, H_2S, etc.) or only partially oxidized (CO, N_2O). These species are also of only modest solubility in water. By way of contrast, man-made species, which frequently involve combustion sources, are often more highly oxidized (CO_2, NO_2, SO_2) and somewhat more soluble, or at least more readily hydrolysed. Chemical transformations may be precursors of deposition even for the latter species. For example, washout removal times for sulphur dioxide in moderate rainfall are estimated as several hours. Most of the sulphate found in precipitation, however, is a result of rainout of hygroscopic cloud condensation nuclei (Sections 1.3 and 2.5) involving SO_3 or H_2SO_4: that is, the SO_2 has already been oxidized before nucleation occurs. Acids such as HCl, HF, and HNO_3 are readily soluble, as are NH_3, SO_2, and NO_x after conversion to aerosol species: wet deposition is possible.

Figure 5.1 illustrates the many separate steps that go to make up various forms of wet deposition. As pointed out earlier, the chemical species may either enter the cloud or fog that is the precursor of the precipitation, or it may enter the aqueous phase as the precipitation is falling. Direct interception of a cloud (or fog) by the Earth is sometimes recognized as a modification to the wet deposition process. Wet deposition can be thought of as occurring in several discrete steps. First, the atmospheric condensed water ('*hydrometeor*') and the species to be removed must encounter one another. Secondly, the species must be taken up by the hydrometeor. Lastly, the hydrometeor must reach the Earth's surface. Virtually all the processes are reversible, and, at each step, a species may undergo chemical reaction, thus further adding to the complexity of the system.

Quantitative mathematical description of wet deposition is obviously a difficult task in the circumstances described. One of the problems is that the several contributing steps occur on vastly different scales of distance, ranging from molecular mean-free paths of 10^{-10} m, through droplet physical processes occurring over 10^{-6} m, to atmospheric motions over distances of up to 10^3 km. Formulation of the mathematics is made yet more daunting by the need to treat several physical phases (solids, liquids, and gases).

Although substantial progress has been made in the mathematical treatments, the complexity has recommended an altogether simpler approach to many atmospheric scientists. A semi-empirical equation equivalent to eqn (5.1) is written for the rate of transfer of the chemical species into the hydrometeor, with the deposition velocity being replaced by an *uptake* (or *scavenging*) coefficient, γ, as described in Section 3.4.5, and defined in eqn (3.48). Rates of transfer of the chemical species from the cloud to rain together with the rate of below-cloud scavenging go to make up the overall flux of wet deposition. Several assumptions are implicit in this approach, two of the most serious being that the simple equation is applicable only if uptake is irreversible and if γ is independent of concentration of the species already contained in the condensed phase. Even within these limitations, γ remains a quantity that depends on the detailed meteorology and cloud physics pertaining to a particular place and time.

5.2.2 The boundary layer

The lower limit of the troposphere is often envisaged not as the Earth's surface, but rather as a *boundary layer* of the atmosphere. Some hint of this idea has appeared in earlier chapters, but for simplicity was left unexplained. The boundary layer is the region of highly turbulent mixing next to the ground; it is generally confined to the first 0.5–2 km by day, and less at night. Turbulent eddies are generated in part mechanically from strong shear forces as air flow adjusts to avoid slip at the ground. As a result, budgets of trace gases (and momentum and internal energy) for an individual air parcel can no longer be treated without reference to turbulent exchanges with the environment of that parcel, in distinction to the situation for the *free troposphere*. In the free troposphere, most of the kinetic energy is associated with large-scale disturbances and seasonal variations, motions occurring with

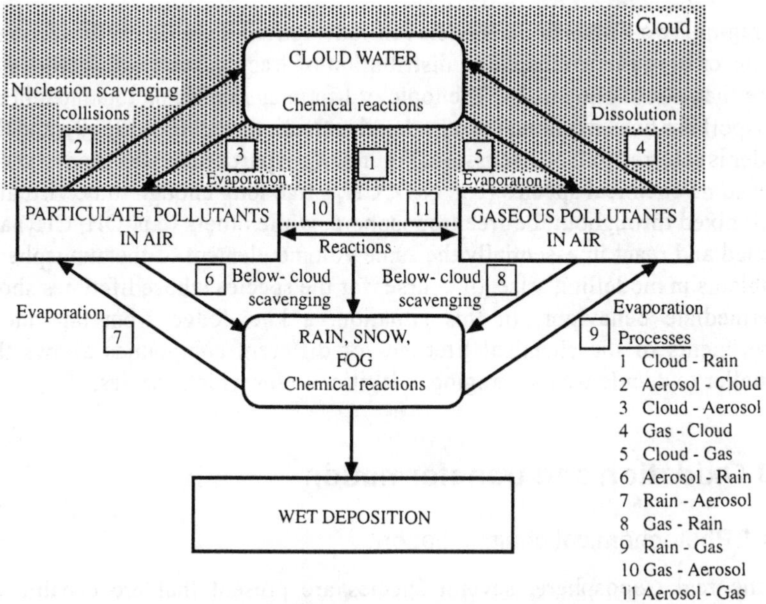

Fig. 5.1. Processes involved in wet deposition. From Seinfeld, J.H. and Pandis, S.N. *Atmospheric chemistry and physics*, John Wiley, 1998.

periods of more than a day. However, inside the boundary layer, as much as half the kinetic energy is concentrated in motions with periods of the order of minutes.

By repeatedly bringing air parcels in contact with the surface, deposition of any local source gas, or its oxidation products, is greatly enhanced. Over a time-scale of days, any remaining constituents are incorporated into the free troposphere, and transported by the atmospheric circulation. Once in the free troposphere, the trace gases move over much larger distances than in the boundary layer. One consequence of the existence of a boundary layer is that local surface releases (natural or artificial) of highly reactive species tend to be geographically confined to the source region.

5.2.3 Transport in the troposphere

In the free troposphere, and at mid-latitudes, wind velocities are typically 10–30 m s^{-1} in the E–W direction, but much less in both N–S and vertical directions. An air parcel is therefore moved around a circle of latitude in a few tens of days, while motions in the N–S direction are much slower. Transfer between hemispheres across the equator is inhibited (cf. p. 77) and occurs only at certain seasons of the year, N → S in the upper troposphere and S ← N in the lower troposphere; the time-scale is about one year. Because of the E–W averaging, two-dimensional models (altitude, latitude) are generally adequate for describing the tropospheric distribution of trace gases with lifetimes of more than about one month. The topic of lifetimes in relation to atmospheric transport has been explored in some detail in Section 3.6.1 (pp. 138–9), and the reader is referred to that discussion. Figure 3.9, in particular, was used to show how some chemical species (e.g. N_2O, CH_4) live long enough to be virtually well mixed throughout the free troposphere, while others (e.g. OH, CH_3) are formed and react in essentially the same volume element of the atmosphere. Problems in modelling arise, of course, for the species whose lifetimes show intermediate behaviour. In this situation, a knowledge from laboratory experiments of the chemical lifetimes of different compounds allows the modeller to decide what averaging is legitimate for which species.

5.3 Oxidation and transformation

5.3.1 Photochemical chain initiation

In the real troposphere, several species are present that are capable of absorbing solar radiation and hence initiating radical-chain oxidation. More insight may be gained if we take as our starting point the artificial situation where no CH_4 has yet been oxidized. Ozone is then the photochemical precursor of hydroxyl radicals. As we discussed at some length in Sections 3.2.1 and 3.2.2, ozone is photolysed at wavelengths less than ~ 310 nm to yield an excited, ^1D, oxygen atom that is energetically capable of reacting with water vapour to yield OH

$$O_3 + h\nu \rightarrow O^*(^1D) + O_2^*(^1\Delta_g) \tag{5.2}$$

$$O^*(^1D) + H_2O \rightarrow OH + OH. \tag{5.3}$$

Since H_2O is itself a minor component of the atmosphere, reaction in process (5.3) is a minor fate of $O^*(^1D)$ atoms compared with quenching

$$O^*(^1D) + M \rightarrow O + M \qquad M = N_2, O_2. \tag{5.4}$$

However, virtually all ground-state O atoms will regenerate ozone

$$O + O_2 + M \rightarrow O_3 + M, \tag{5.5}$$

so that quenching does not constitute a loss of odd oxygen. Although the recycling of atomic oxygen maintains the mass balance of odd oxygen, the rate of OH production and of chain initiation does depend on the relative rates of reactions (5.3) and (5.4), and, in a similar way, on the quantum yield for $O(^1D)$ production in reaction (5.2) as a function of wavelength.

Ozone of stratospheric origin (see Section 5.1) can well be that needed for $O(^1D)$ production, but if NO_2 is also present in the troposphere, then NO_2

photolysis (at $\lambda < 400$ nm)

$$NO_2 + h\nu \rightarrow O + NO, \tag{5.6}$$

followed by the combination reaction (5.5), can provide a tropospheric source. As we shall see shortly, NO itself can be oxidized back to NO_2, so that the formation of O_3 is not stoicheiometrically limited by the number of NO_2 molecules initially present.

5.3.2 Oxidation steps

Hydroxyl radicals formed in reaction (5.3) react mainly with CO and with CH_4

$$OH + CO \rightarrow H + CO_2, \tag{5.7}$$

$$OH + CH_4 \rightarrow CH_3 + H_2O. \tag{5.8}$$

Roughly 70 per cent of the OH reacts with CO, and 30 per cent with CH_4, in the unpolluted atmosphere. The reaction of OH with CO is kinetically and mechanistically unusual, and we shall return to it in Section 5.3.5. For the time being, we note that in both processes an active species is formed that is capable of adding molecular oxygen to produce a peroxy radical

$$H + O_2 + M \rightarrow HO_2 + M \tag{5.9}$$

$$CH_3 + O_2 + M \rightarrow CH_3O_2 + M. \tag{5.10}$$

In tropospheric regions where NO concentrations are very low, the peroxy radicals HO_2 and CH_3O_2 are consumed mainly in the reactions

$$HO_2 + HO_2 \rightarrow H_2O_2 + O_2 \tag{5.11}$$

$$CH_3O_2 + HO_2 \rightarrow CH_3OOH + O_2 \tag{5.12}$$

(Self-reaction of two CH_3O_2 radicals is rather slow at ambient temperatures.) One fate of the hydrogen peroxide (H_2O_2) and the methyl hydroperoxide (CH_3OOH) is that they can dissolve in cloud droplets and be removed from the troposphere in the form of rain. In this respect, therefore, reactions (5.11) and (5.12) are loss or terminating steps, although alternative fates of the peroxides, such as photolysis or reaction with OH, may regenerate radicals.

If oxides of nitrogen are present in the atmosphere, then a quite different course of events can follow the formation of the peroxy radicals in reactions (5.9) and (5.10). We know already from our discussion of stratospheric chemistry (Section 4.4) that HO_2 reacts rapidly with NO, and the analogous reaction occurs with methylperoxy radicals

$$HO_2 + NO \rightarrow OH + NO_2 \tag{5.13}$$

$$CH_3O_2 + NO \rightarrow CH_3O + NO_2. \tag{5.14}$$

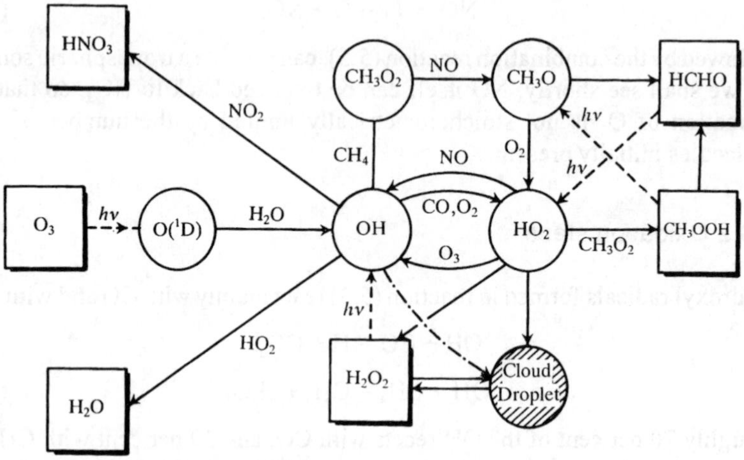

Fig. 5.2. Chemistry of the troposphere. This schematic diagram emphasizes the processes that create and destroy the OH and HO_2 radicals that are key intermediates in the oxidation steps.

Reaction (5.13) regenerates OH, while (5.14) produces a methoxy radical that can in turn react with O_2,

$$CH_3O + O_2 \rightarrow HCHO + HO_2, \tag{5.15}$$

to yield formaldehyde. Formaldehyde itself is photochemically labile; the major photolytic pathway at $\lambda < 338$ nm produces two radical fragments

$$HCHO + h\nu \rightarrow H + HCO, \tag{5.16}$$

and both radicals re-enter the HO_x chain, *via* reaction (5.9) for H, or, in the case of HCO, *via* the reaction

$$HCO + O_2 \rightarrow CO + HO_2. \tag{5.17}$$

The essential feature is the conversion of NO to NO_2 while preserving active radicals capable of oxidizing further molecules of CO and CH_4 or converting more NO to NO_2. This formation of NO_2 is essential to the formation of ozone in the troposphere, as we shall discuss in Section 5.3.3.

Figure 5.2 presents in the form of a 'flow diagram' the chemistry discussed so far, with the emphasis laid on processes creating and destroying the radical intermediates OH and HO_2.

The oxidation steps that we have written for methane obviously have their analogues for higher hydrocarbons, but in all cases depend on the switch between RO_2 (peroxy-) radicals and RO (oxy-) radicals in the interaction with NO, represented for the general case as

$$RO_2 + NO \rightarrow RO + NO_2. \qquad (5.18)$$

We turn now to a rather more detailed consideration of the oxidation of simple alkanes and alkenes. To keep the discussion reasonably compact, we shall assume that the initial attack is always by hydroxyl radicals, although it must be understood that attack by NO_3 is important at night (Section 5.3.6), and that O_3 (Section 5.3.7), HO_2, and O can play a minor role.

Reaction with OH is, in fact, almost exclusively the loss process for alkanes in the troposphere, with H-atom abstraction from the C–H bond yielding an alkyl, R, radical. The radicals react initially with O_2

$$R + O_2 \rightarrow RO_2, \qquad (5.19)$$

a third body probably being needed in the case of $R = CH_3$ [cf. reaction (5.10)]. For ethane and propane, the subsequent steps are probably

$$RO_2 + NO \rightarrow RO + NO_2 \qquad (5.18)$$

$$RO \rightarrow R' + R''CHO \qquad (5.20)$$

$$RO + O_2 \rightarrow R'R''CO + HO_2, \qquad (5.21)$$

where R′ and R″ are daughter alkyl radicals or groups. The aldehydes and ketones (R″CHO, R′R″CO) can be photolysed, as we have seen above for formaldehyde in reaction (5.16), or become further oxidized in thermal reactions. For example, OH abstracts H from acetaldehyde to form CH_3CO, which itself adds oxygen to yield the acetylperoxy radical, $CH_3CO.O_2$

$$CH_3CHO + OH \rightarrow CH_3CO + H_2O, \qquad (5.22)$$

$$CH_3CO + O_2 \rightarrow CH_3CO.O_2. \qquad (5.23)$$

As we shall see in Section 5.3.4, one important reaction of $CH_3CO.O_2$ is addition of NO_2, but the usual [cf. reactions (5.13) and (5.18)] oxidation of NO is possible, and in that case the resulting $CH_3CO.O$ radical can fragment

$$CH_3CO.O_2 + NO \rightarrow CH_3CO.O + NO_2 \qquad (5.24)$$

$$CH_3CO.O \rightarrow CH_3 + CO_2. \qquad (5.25)$$

In the reactions described, the higher alkyl radicals are degraded to lower ones, and ultimately, *via* the reactions of CH_3, to CO and CO_2.

Ethene and propene, C_2H_4 and C_3H_6, appear to react with OH predominantly by an addition mechanism

$$OH + C_2H_4 \rightarrow C_2H_4OH \qquad (5.26)$$

$$OH + CH_3CHCH_2 \rightarrow CH_3CHCH_2OH. \qquad (5.27)$$

For ethene, at least, the reaction appears to require stabilizing collisions with a third body, and shows intermediate-order kinetics (see Section 3.4.2). Both reactions show a negative temperature coefficient of rate, characteristic of an addition process, and a collisionally stabilized adduct has been explicitly observed in the case of reaction (5.27). The formation of the OH–ethene adduct is ~ 134 kJ mol^{-1} exothermic, and elimination of an H atom is endothermic by ~ 30 kJ mol^{-1}, so that the only favourable decomposition path for the adduct is back to the reactants. For propene, the excess energy in the newly formed adduct is similar to that in the ethene case; here, elimination of CH_3 is thermochemically possible, although it seems not to be important for the CH_3CHCH_2OH isomer.

The alkene–OH adducts appear to undergo reactions analogous to those of alkyl radicals. Thus, for terminal addition of OH to propene, the major atmospheric oxidation steps are

$$CH_3CHCH_2OH + O_2 \rightarrow CH_3CH(O_2)CH_2OH, \quad (5.28)$$

$$CH_3CH(O_2)CH_2OH + NO \rightarrow CH_3CH(O)CH_2OH + NO_2, \quad (5.29)$$

$$CH_3CH(O)CH_2OH \rightarrow CH_3CHO + CH_2OH, \quad (5.30)$$

which should be compared with reactions (5.19), (5.18) and (5.20) for the R, RO_2, and RO radicals. In the atmosphere, the CH_2OH reacts exclusively with O_2 to yield $HCHO + HO_2$

$$CH_2OH + O_2 \rightarrow HCHO + HO_2, \quad (5.31)$$

so that the products from attack of OH on propene are the aldehydes HCHO and CH_3CHO, which are ultimately oxidized or photolysed as described earlier. The HO_2 radical is also a product. and it is interesting to add reaction (5.13)

$$HO_2 + NO \rightarrow OH + NO_2 \quad (5.13)$$

to the sequence of reactions (5.27) to (5.31) to emphasize that the oxidation scheme is a chain process. Chain carriers are not consumed overall, even though C_3H_6 has been oxidized to CH_3CHO and HCHO. Of course, two molecules of NO are converted to NO_2 for each OH radical cycle, and that is a matter of importance for ozone production, as we shall now see in the next section.

5.3.3 Tropospheric ozone production

Photolysis of NO_2 is the only known way of producing ozone in the troposphere. Reactions (5.13), (5.14) or (5.18) achieve not only the RO_2 to RO radical conversions, but also the oxidation of NO to NO_2. They thus provide the link required for catalytic generation of O_3 *via* reactions (5.6) and (5.5). One cyclic process for tropospheric ozone production can now be written

$$NO_2 + h\nu \rightarrow O + NO \quad (5.6)$$

$$O + O_2 + M \rightarrow O_3 + M \quad (5.5)$$

$$OH + CO \rightarrow H + CO_2 \quad (5.7)$$

$$H + O_2 + M \rightarrow HO_2 + M \quad (5.9)$$

$$HO_2 + NO \rightarrow OH + NO_2 \quad (5.13)$$

$$\text{Net} \quad CO + 2O_2 + h\nu \rightarrow CO_2 + O_3$$

Very similar chain reactions can be written that involve the RO_2 (e.g. CH_3O_2) species. In all cases, NO_x must be present.

Below a certain critical value of the ratio $[NO]/[O_3]$, ozone loss through the sequence

$$HO_2 + O_3 \rightarrow OH + 2O_2 \quad (5.32)$$

$$OH + CO \rightarrow H + CO_2 \quad (5.7)$$

$$H + O_2 + M \rightarrow HO_2 + M \quad (5.9)$$

$$\text{Net} \quad CO + O_3 \rightarrow CO_2 + O_2$$

dominates over the generation sequence. Current models suggest that production (*via* HO_2 and RO_2 oxidation of NO) balances or exceeds loss of O_3 for atmospheric mixing ratios of $NO \geq 3 \times 10^{-11}$. Regions of the Earth characterized by extremely low concentrations of NO, such as the remote Pacific, are thus likely to provide a net photochemical sink for odd oxygen, while the continental boundary layer at mid-latitudes, where concentrations of NO are relatively high, is likely to provide a net source. So long as NO_x is available, production of ozone in the troposphere is ultimately limited by the supply of CO, CH_4, and other hydrocarbons. Each molecule of CO can generate one molecule of O_3, while it is estimated that as many as 3.5 O_3 molecules could be formed from the oxidation of each CH_4.

Tropospheric loss processes for O_3 include photolysis in the visible and ultraviolet regions, and reaction with species such as NO_2, HO_2, and unsaturated hydrocarbons. Model calculations suggest that the globally averaged tropospheric sources and sinks for ozone are roughly in balance. This view of the *in-situ* balance is consistent with estimates that the surface sink is equal to the stratospheric injection rate.

The tropospheric abundance of ozone is obviously critically important in determining the oxidizing capacity of the troposphere, because O_3 is the primary source of OH radicals (as well as being an oxidizing species in its own

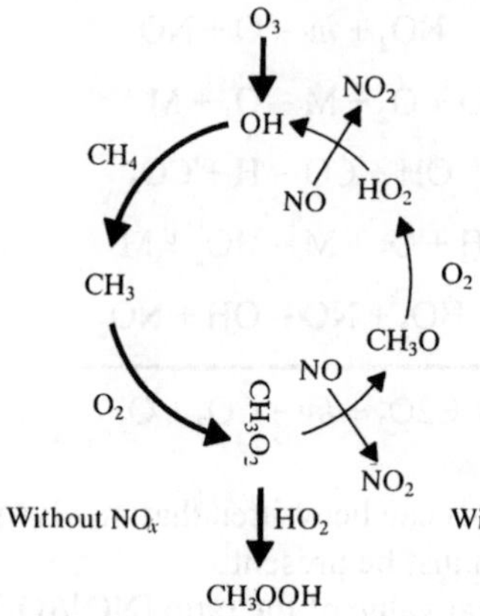

Fig. 5.3. Essential steps in tropospheric methane oxidation. The heavier arrows on the left-hand side of the diagram indicate the steps that can occur in the absence of NO_x. With NO_x present, the processes on the right-hand side can close a loop, with regeneration of OH and oxidation of NO to NO_2. From a diagram devised by M.E. Jenkin.

right). Background concentrations of NO_x and of the peroxy radicals derived from CO, CH_4, and other hydrocarbons determine the rate of formation of ozone. Future trends of tropospheric oxidizing capacity will thus be tied closely to future atmospheric burdens of all these species.

5.3.4 The importance of NO_x

The discussion of the previous two sections has shown how important NO_x, and especially NO, is to the overall oxidation rates and to the ozone distribution in the troposphere. The influence of NO_x is emphasized in Fig. 5.3, which is a restatement of the essential steps of Fig. 5.2 in another form. The left-hand semicircle shows the oxidation steps that do not require the presence of NO. In this case, the sequence of transformations following the heavy arrows leads from O_3 to CH_3OOH. If, however, NO is present, then the transformations of the right-hand semicircle close the loop, with regeneration of OH and the concomitant production of two molecules of NO_2 (and thus, potentially, of O_3).

Model estimates of tropospheric ozone production obviously depend critically on the altitude–concentration profiles adopted for NO_x, so that we must consider the atmospheric budgets for these species. Natural sources of NO_x include microbial actions in the soil, which produce NO as well as N_2O (cf. Sections 1.5.3 and 4.6.8). Controlled laboratory experiments suggest that NO might even be a slightly more abundant initial product than N_2O, but NO has a far shorter chemical lifetime. Oxidation of biogenic NH_3, initiated by OH radicals, would be another significant natural source of NO_x. Lightning appears to be responsible for less than 10 per cent of the total NO_x budget, while biomass burning and high-temperature combustion (i.e. 'anthropogenic' sources) are perhaps responsible for 50 per cent globally. The ratio of $[NO]/[NO_2]$ depends on the rate of photolysis (reaction (5.5)) and the rates of NO oxidation in processes (5.13), (5.14), and (5.18), together with the reaction

$$NO + O_3 \rightarrow NO_2 + O_2. \tag{5.33}$$

This latter reaction has a significant activation energy (~ 13 kJ mol^{-1}) so that $[NO]/[NO_2]$ is larger at higher tropospheric altitudes where the temperature is lower.

It is important to note that all the sources mentioned, except lightning, release NO_x to the boundary layer. Since the major loss for NO_x involves nitric acid formation,

$$OH + NO_2 + M \rightarrow HNO_3 + M, \tag{5.34}$$

followed by wet deposition, the NO_x release to the free troposphere may be appreciably smaller than the surface source strengths suggest. Some workers have argued that injection of NO_x from the stratosphere could provide a more important source for the upper troposphere than the surface emissions. Over the Pacific Ocean, where there is no surface source, mixing ratios of NO in the lower troposphere are as low as 4×10^{-12}, which would suggest that the stratospherically derived NO_x is alone important and that the upper troposphere is the dominant region for photochemical production of ozone. However, several other observations in the remote troposphere indicate mixing ratios in the range 10^{-11} to 2×10^{-10}, and ozone production in the lower (<5 km) troposphere would then be important. The question remains in dispute at the time of writing.

Measurements in regions remote from man-made sources have generally been thought to be representative of NO_x in the 'natural' atmosphere, because the chemical lifetime of NO_x is small compared with the times taken for geographical redistribution. The situation is rather more involved than appears at first sight, because a reservoir species for NO_x is now known that can extend the lifetime of NO_x and provide a source when others are absent. This reservoir species is an adduct of the peroxyacetyl radical, $CH_3CO.O_2$, with nitrogen dioxide, of formula $CH_3CO.O_2NO_2$. It is known universally as peroxyacetyl nitrate, or PAN.[a] PAN has been recognized for many years as a component of photochemical air pollution (Section 5.10.7), but is now known to be present also in the clean air of remote oceanic regions. Concentrations of 10 to 400

[a]A proper systematic name for PAN might be ethane peroxoic nitric anhydride as the compound is a mixed anhydride of two acids; the compound is not an ester of HNO_3 so the 'nitrate' in the name is not really appropriate.

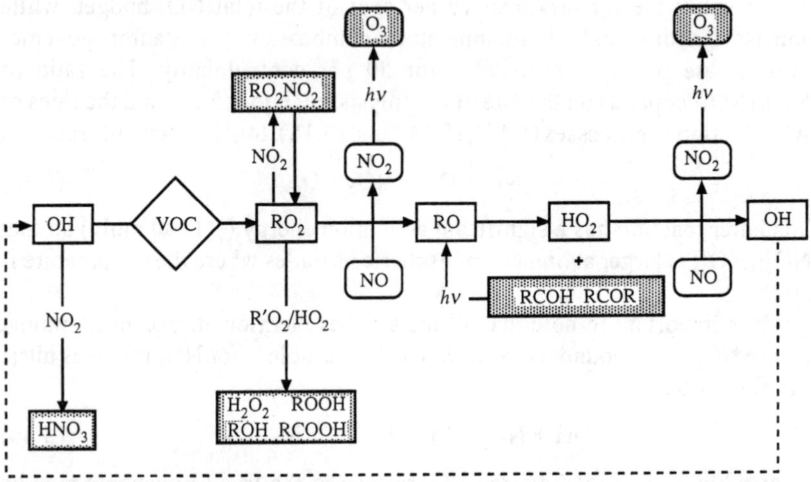

Fig. 5.4. Oxidation of a volatile organic compound (VOC) during the daytime. Oxidation is initiated by attack of OH radicals on the VOC, followed by the addition of O_2 to form a peroxy radical, RO_2. From Le Bras, G. (ed.) *Chemical processes in atmospheric oxidation*, Springer-Verlag, Berlin, 1997.

parts in 10^{12} of PAN have been observed over the Pacific Ocean, where NO_2 itself constitutes less than 30 parts in 10^{12} of the air. The importance of PAN is that it is in thermal equilibrium with its precursors,

$$CH_3CO.O_2 + NO_2 \rightleftharpoons CH_3CO.O_2NO_2, \tag{5.35}$$

and that the equilibrium is shifted to the right-hand side at lower temperatures. Above the boundary layer, temperatures are sufficiently low for PAN to be relatively stable, but the molecule is unstable close to the surface. PAN will release NO_2 as it is transferred from cooler to warmer regions.

Formation of PAN, and of other peroxyacyl nitrates, requires the initial generation of the peroxyacyl radicals. As we saw in Section 5.3.2, many hydrocarbons, both saturated and unsaturated, produce aldehydes as oxidation intermediates, and it is attack of OH on these carbonyl compounds that yields the acyl, and ultimately the peroxyacyl, radicals. Several other oxidation mechanisms can lead to alkoxy radicals, RO. So long as R contains more than two carbon atoms, oxidation can yield a carbonyl fragment R′CO and thence, *via* reactions analogous to (5.22) and (5.23), a peroxyacyl species. PAN is therefore a compound expected when hydrocarbons are oxidized in the presence of NO_2. It is now apparent that the species must be included with NO, NO_2, and HNO_3 when considering NO_x or NO_y budgets and transport in the troposphere.

The many reactions discussed so far are summarized in the flow diagram of Fig. 5.4, which is a more sophisticated form of Fig. 5.3. It illustrates how a generalized volatile organic compound (VOC) undergoes oxidation during the daytime, and includes processes such as the formation of PAN, the photolysis of carbonyl compounds, and the other key steps.

5.3.5 The reaction OH + CO

Oxidation of CO by OH,

$$OH + CO \rightarrow H + CO_2, \tag{5.7}$$

is one of the most important reactions in tropospheric chemistry. The process exhibits some interesting kinetic features that must be taken into account in constructing tropospheric models. First, the reaction shows a temperature dependence that deviates markedly from the Arrhenius form (p. 119). At temperatures below ~ 300 K, the activation energy is near zero. Secondly, the overall rate of reaction is pressure-dependent, the apparent rate coefficient increasing with pressure.

A possible explanation of the kinetic behaviour is that the reaction involves discrete steps with an HOCO intermediate

$$OH + CO \rightleftharpoons HOCO^* \tag{5.36}$$

$$HOCO^* \rightarrow H + CO_2 \tag{5.37}$$

$$HOCO^* + M \rightleftharpoons HOCO + M \tag{5.38}$$

$$HOCO + O_2 \rightarrow HO_2 + CO_2. \tag{5.39}$$

Reaction (5.38) followed by (5.39) offers a pressure-dependent pathway that does not lead to the $H + CO_2$ products of reaction (5.7), but that particular path can operate only when O_2 is present, which it certainly is in the Earth's atmosphere! The reversible association reaction (5.36) will be favoured at low temperatures, and thus account for the slight negative temperature-dependence observed for the overall rate.

5.3.6 The nitrate radical

Radical intermediates in chemical reactions are most positively identified through their optical absorption spectra. The nitrate radical, NO_3, was observed in this way over 100 years ago, in 1881, and may thus have been the first transient radical species to be detected directly. Over the past two decades, it has become evident that the NO_3 radical plays a significant part in chemical transformations in both the stratosphere (see Section 4.4.2) and the troposphere of the Earth.

In the Earth's atmosphere, NO_3 is formed by the reaction

$$NO_2 + O_3 \rightarrow NO_3 + O_2 \tag{5.40}$$

in both stratosphere and troposphere. Dissociation of N_2O_5

$$N_2O_5 + M \rightarrow NO_3 + NO_2 + M \tag{5.41}$$

is apparently an additional source, but since N_2O_5 is formed by the reaction

$$NO_3 + NO_2 + M \rightarrow N_2O_5 + M \tag{5.42}$$

it is ultimately dependent on the occurrence of reaction (5.40). Dinitrogen pentoxide, N_2O_5, is an important product in its own right, since it can react heterogeneously with H_2O (Section 5.8) to yield HNO_3, and thus contribute to atmospheric acidification (cf. Section 5.10.6). Incidentally, NO_3 does not itself appear to react with H_2O, although it does react with aqueous negative ions to yield nitrate ions and other products.

During the day, the NO_3 radical is rapidly photolysed: the product channels may be $NO + O_2$ or $NO_2 + O$, the relative and absolute yields being dependent on wavelength

$$NO_3 + h\nu \rightarrow NO + O_2 \tag{5.43a}$$

$$NO_3 + h\nu \rightarrow NO_2 + O. \tag{5.43b}$$

In the stratosphere, therefore, reaction (5.40) followed by photolysis contributes to the detailed balance of the chemistry of odd-nitrogen compounds. Reactions of NO_3 in the troposphere present a different picture because of the multitude of organic compounds available. Although the hydroxyl radical is usually the main agent of attack on organic species during the day, the nitrate radical may be the most important oxidizing species in the troposphere at night. For certain species, such as CH_3SCH_3 and some terpenes, the reaction at night with NO_3 may even dominate over the daytime reaction with OH. Two main kinds of initial step can be envisaged: hydrogen abstraction, and addition to unsaturated bonds, as typified by the reactions

$$NO_3 + RH \rightarrow HNO_3 + R \tag{5.44}$$

$$NO_3 + {>}C{=}C{<} \rightarrow {>}C(ONO_2)\dot{C}{<}. \tag{5.45}$$

Nitric acid is thus a direct product of hydrogen abstraction by the radical. Furthermore, the radical produced in reaction (5.44) is likely to add O_2 in air to form a peroxy radical, RO_2. In the special case that RH is formaldehyde, HCHO, the HCO radical will ultimately yield the HO_2 radical; for other aldehyde precursors, the acyl products of reaction (5.44) yield acylperoxy radicals, $R.CO.O_2$, and are thus potential sources of peroxyacylnitrates.

The initial adduct formed in reaction (5.45) can eliminate NO_2 to yield an epoxide. However, the species is a radical, and in the presence of air it is therefore expected to add oxygen. For example, for the radical derived from

propene, the reaction is

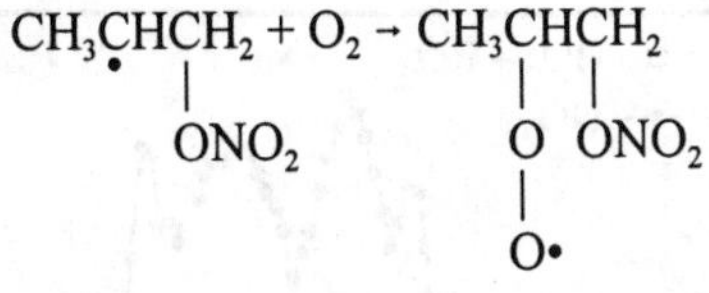

(5.46)

Several products are observed in the system when O_2 and NO_x are present, including CH_3CHO, HCHO, 1,2-propanediol dinitrate (PDDN), nitroxyperoxypropyl nitrate (NPPN), and α-(nitrooxy)acetone

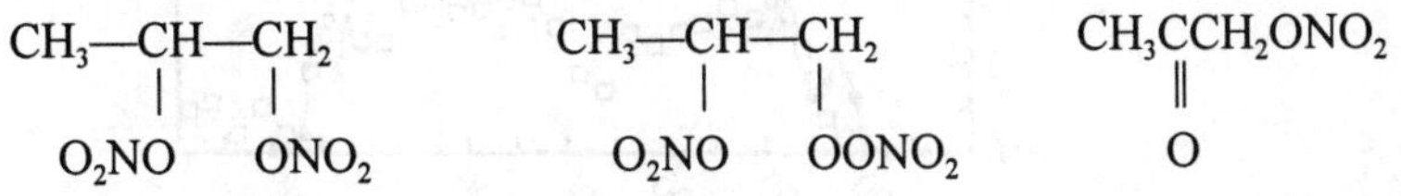

PDDN NPPN α-(nitrooxy)acetone

The formation of the nitrated acetone from the peroxy radical product of reaction (5.46) is illustrated by the sequence of processes

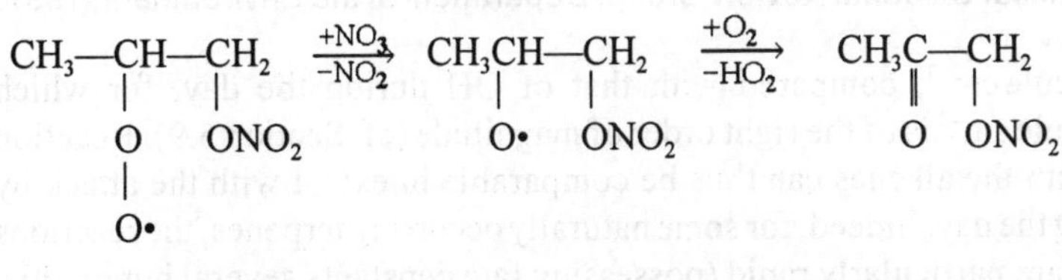

(5.47)

which is clearly the analogue of reactions (5.18) followed by (5.21), but with NO_3 rather than NO as the reactant in the first step (see footnote on next page). The possible formation of α-(nitrooxy)acetone in ambient air is also of interest, since the compound has been reported to be a mutagen. The other products are also of concern because dinitrates have been shown to produce several adverse health reactions. In the general case, the steps that make up reaction (5.47) have the form

$$RO_2 + NO_3 \rightarrow RO + NO_2 + O_2 \quad (5.48)$$

$$RO + O_2 \rightarrow R'R''CO + HO_2. \quad (5.21)$$

An important additional point is now made clear: an HO_2 radical is formed in the second step, so that NO_3-initiated night-time chemistry is a source of both RO_2 and HO_2 radicals. We return to this point shortly.

Rate constants for the reactions of NO_3 with hydrocarbons are generally much lower than those for the reactions of OH. Typical relative reactivities of OH and NO_3 towards hydrocarbons such as butane and 1-butene are roughly 10^5 and 3000 at ambient temperature. Against these intrinsic reactivities must be set the greater concentration of NO_3 during the night, which may be about

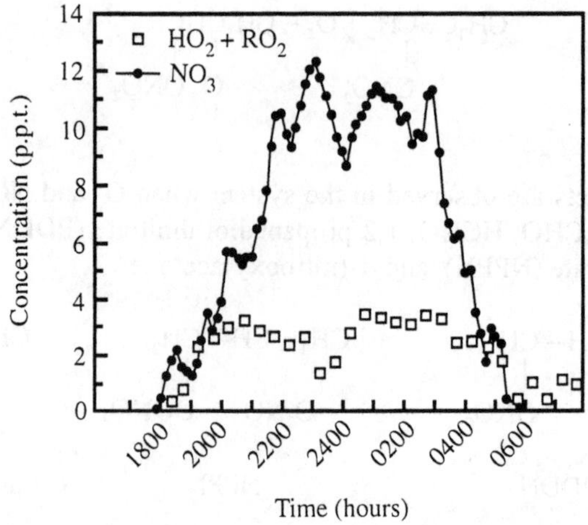

Fig. 5.5. Concentrations of NO_3 and peroxy radicals measured on the Norfolk coast. Modified from data presented in *Ozone in the United Kingdom*, UK Photochemical Oxidants Review Group, Department of the Environment, 1997.

10^9 molecule cm^{-3}, compared with that of OH during the day, for which 10^6 molecule cm^{-3} is of the right order of magnitude (cf. Section 5.9). Reaction of NO_3 with the alkenes can thus be comparable in extent with the attack by OH during the day. Indeed, for some naturally occurring terpenes, the reactions with NO_3 are particularly rapid (possessing rate constants several hundredths of those for the corresponding reactions with OH), so that the night-time NO_3 processes dominate. We shall see in Section 5.6 that NO_3 may play a similarly important role in the oxidation of some sulphur compounds. Reaction of NO_3 with alkanes is probably never more than a minor contributor to the total diurnally averaged oxidation; from the point of view of HNO_3 formation, however, reaction (5.44) may add substantially to other sources, such as hydrolysis of N_2O_5.

Although the reactions of NO_3 with alkanes and simple alkenes are relatively slow, the same is not true of the reactions with radical species. Radicals such as RO_2 and HO_2 persist at night and, as we have just seen, are even generated in the reactions with NO_3. Interactions of NO_3 with such radicals may thus be of atmospheric significance, particularly since NO, so important in the reactions of the peroxy radicals in reactions (5.13) and (5.18), usually falls to low concentrations during the night.[a] Indeed, it is now apparent that the reactions might even constitute a night-time source of OH radicals,

[a] If tropospheric O_3 concentrations are abnormally low, then some NO may persist at night. These are conditions that will lead to a suppression of night-time NO_3 chemistry, since NO_3 requires O_3 for its formation, and it is also destroyed in a fast reaction with NO.

species that were hitherto regarded as of daytime photochemical origin. Two reactions can convert the HO_2 product of reaction (5.21) to OH

$$HO_2 + O_3 \rightarrow OH + 2O_2 \qquad (5.32)$$

$$HO_2 + NO_3 \rightarrow OH + NO + O_2, \qquad (5.49)$$

the relative importance of each depending on the amount of NO_x present, which determines the ratio of $[NO_3]$ to $[O_3]$. Taking reaction (5.32) as the source of OH as an example, the complete cycle consists of the steps

$$NO_2 + O_3 \rightarrow NO_3 + O_2 \qquad (5.40)$$

$$NO_3 + \text{organic compound} \rightarrow R + \text{products} \qquad (5.50)$$

$$R + O_2 \rightarrow RO_2 \qquad (5.19)$$

$$RO_2 + NO_3 \rightarrow RO + NC_2 + O_2 \qquad (5.48)$$

$$RO + O_2 \rightarrow R'R''CO + HO_2 \qquad (5.21)$$

$$HO_2 + O_3 \rightarrow OH + 2O_2. \qquad (5.32)$$

The sequence of reactions makes it clear that OH radicals are generated at the expense of ozone (and whatever NO_x is lost in reaction (5.50)).

This unexpected potential source of HO_x is highlighted by the particular case where the radical, RO_2, is the acetylperoxy radical, formed in the reaction

$$CH_3CO + O_2 \rightarrow CH_3CO.O_2 \qquad (5.23)$$

and partly stored in the reservoir PAN, generated in the reversible process

$$CH_3CO.O_2 + NO_2 \rightleftharpoons CH_3CO.O_2NO_2. \qquad (5.35)$$

If NO_3 is present, a modification of reaction (5.48) occurs

$$CH_3CO.O_2 + NO_3 \rightarrow CH_3 + CO_2 + NO_2 + O_2 \qquad (5.51)$$

and the CH_3 radicals then act as any other alkyl radical, R, in reactions (5.19), (5.48), and (5.21). Thus, the presence of NO_3 promotes the decomposition of acetylperoxy radicals by offering channel (5.51) in competition with the forwards path of reaction (5.35); CH_3O_2, and ultimately HO_2 and even OH radicals are the products.

Evidence is accumulating to confirm the validity of the concepts just outlined. Significant concentrations of peroxy radicals have been detected at night, and, depending on the exact nature of the observations, there is support both for the formation of RO_2 initiated by the reactions of NO_3, and for the interaction between the RO_2 and NO_3. For example, Fig. 5.5 shows how concentrations of peroxy radicals ($[RO_2]$ + $[HO_2]$) correlate well with simultaneous measurements of $[NO_3]$ made on the east coast of England. Sunset was between 1800 and 1900 hrs, and sunrise between 0500 and 0600 on the occasion of these measurements, and it is quite clear that concentrations

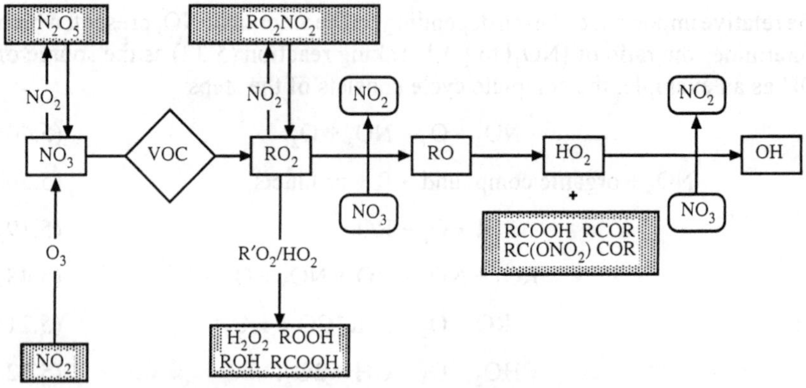

Fig. 5.6. Oxidation of a volatile organic compound at night. Oxidation is initiated by attack of NO_3 radicals on the VOC, followed by the addition of O_2 to form a peroxy radical, RO_2 (cf. Fig. 5.4). From Le Bras, G. (ed.) *Chemical processes in atmospheric oxidation*, Springer-Verlag, Berlin, 1997.

of both peroxy and nitrate radicals rise sharply at sunset, and fall sharply again at sunrise, thus strongly supporting the idea that NO_3 plays a major role in generating RO_2 and HO_2.

Figure 5.6 provides a pictorial summary of night-time oxidation of a VOC in a manner identical to that given for the day in Fig. 5.4.

5.3.7 Reactions with ozone

Emphasis has been laid so far on the initiation of the atmospheric oxidation of volatile organic compounds (VOCs) by hydroxyl radicals during the day, and by nitrate radicals at night. Both these radicals indirectly owe their existence to ozone in the troposphere. However, unsaturated VOCs such as alkenes and dienes may be attacked in the atmosphere not only by OH and NO_3, but by O_3 itself. Organic radicals, and even OH, may be formed as products, and a chain oxidation follow the initial step. In circumstances when solar intensity is low (resulting in low [OH]) or NO_x levels are small (resulting in low [NO_3]), the direct reactions with O_3 can make a significant contribution to the loss of VOC.

The relative importance of initiation of oxidation by OH, NO_3, and O_3 can be judged from the lifetimes of different VOCs with respect to reaction with the three attacking agents. Estimates of such lifetimes for rural sites in the UK are shown in Table 5.2. Absence of an entry indicates that attack by that species is negligible, so that it immediately emerges that only unsaturated compounds are susceptible to atmospheric attack by O_3. For many of those compounds, however, it is evident that reaction with O_3 makes a significant contribution to the overall reaction. The more alkyl-substituted alkenes react most rapidly of all with O_3; 2-butene and 2-methyl-2-butene have *shorter* lifetimes with respect to reaction with O_3 than with OH. It is worth noting, however, that for these two compounds, and a few others, notably isoprene (Section 5.4) and dimethyl sulphide (Section 5.6), reaction with NO_3 at night is the most important of all.

Figure 5.7 shows in diagrammatic form the essential steps in the 'ozonolysis' of a simple alkene. Reaction starts by the addition of O_3 to the

Table 5.2 Chemical lifetimes of selected VOCs with respect to attack by OH, O_3, and NO_3.[a]

VOC	Lifetime OH	O_3	NO_3
Methane	3.0 years		
Ethane	29 days		65 years
Propane	6.3 days		5.6 years
Butane	2.9 days		1.9 years
2-Methyl butane	1.9 days		11 months
Ethene	20 hours	9.7 days	5.2 months
Propene	6.6 hours	1.5 days	3.5 days
1-Butene	5.5 hours	1.6 days	2.5 days
2-Butene	2.9 hours	2.4 hours	2.1 hours
2-Methyl-2-butene	2.0 hours	0.9 hours	0.09 hours
Isoprene	1.7 hours	1.2 days	1.2 hours
Benzene	5.7 days		
Toluene	1.2 days		1.3 years
o-Xylene	12 hours		2.9 months
m-Xylene	7.1 hours		4.7 months
p-Xylene	12 hours		2.4 months
Dimethyl sulphide	1.5 days		0.7 hours[b]

[a]Data from *Ozone in the United Kingdom*, Department of the Environment, London, 1997. Lifetimes are for UK rural sites at assumed ambient levels of oxidant concentration (OH $= 1.6 \times 10^6$, $O_3 = 7.5 \times 10^{11}$, and $NO_3 = 3.5 \times 10^8$ molecule cm^{-3}).

[b]In the marine troposphere, where DMS oxidation is most important, concentrations of NO_3 might be three times less than those used for the calculations; the lifetime of DMS would then be roughly two hours.

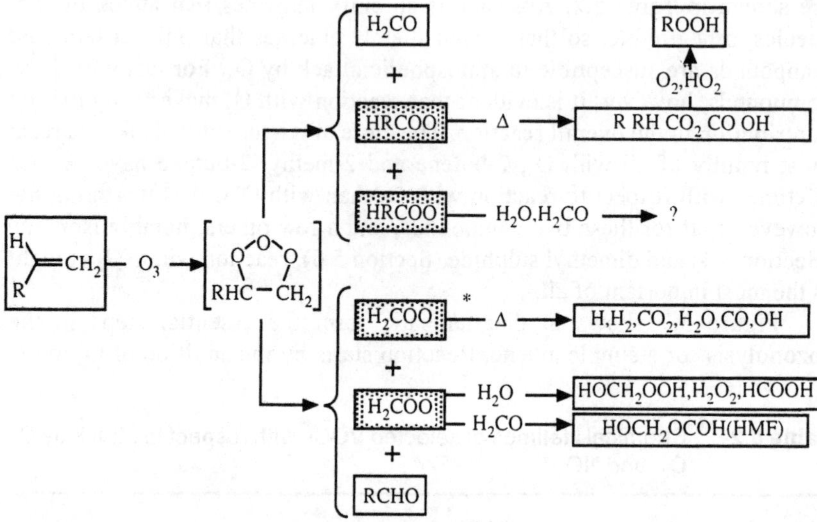

Fig. 5.7. Flow diagram showing the reactions in the ozonolysis of a simple alkene. From Le Bras, G. (ed.) *Chemical processes in atmospheric oxidation*, Springer-Verlag, Berlin, 1997.

double bond to form an ozonide. This ozonide is energy-rich, and rapidly decomposes. One suggestion is that a biradical (*Criegee radical*) is formed, together with a carbonyl compound, as illustrated by the reaction of O_3 with propene

$$O_3 + CH_2{=}CHCH_3 \rightarrow CH_3\dot{C}HO\dot{O} + HCHO \qquad (5.52a)$$

$$\rightarrow \dot{C}H_2O\dot{O} + CH_3CHO. \qquad (5.52b)$$

The Criegee biradicals are also formed energy-rich, and can undergo decomposition, illustrated here for the product of reaction (5.52a)

$$CH_3\dot{C}HO\dot{O} \rightarrow CH_3 + CO + OH \qquad (5.53a)$$

$$\rightarrow CH_3 + CO_2 + H \qquad (5.53b)$$

$$\rightarrow CH_4 + CO_2. \qquad (5.53c)$$

Thus the ozonolysis reactions may produce organic free radicals that can form RO_2 by addition of O_2, as well as HO_x (either OH, or HO_2 after addition of O_2 to H). Laboratory experiments demonstrate the formation of OH in the interaction of ozone with alkenes, and the yield of OH increases with increasing alkyl substitution of the double bond. Thus propene, 2-butene, and 2-methyl-2-butene yield 30, 60, and 90 per cent of OH on reaction. Since these are just the alkenes that also react most rapidly with O_3 (see Table 5.2), there is evidently the potential for significant HO_x production in the troposphere when such alkenes are present. The formation of organic radicals, ultimately in the form of RO_2, also has interesting consequences. We know already that such radicals are agents in tropospheric O_3 production (Section 5.3.3), so that, depending on the ambient conditions, initiation of oxidation by the ozone interaction can lead to net ozone *production.*

It is the *energy-rich* Criegee radicals that decompose in the three channels of reaction (5.53). However, a substantial fraction of the excited radicals are stabilized in the atmosphere as a result of collisional removal of their energy. The most important atmospheric reactions of these stabilized radicals seem to be isomerization and interaction with water vapour

$$CH_3\dot{C}HO\dot{O} \rightarrow CH_3CO.OH \quad (5.54)$$

$$CH_3\dot{C}HO\dot{O} + H_2O \rightarrow CH_3CH(OH)OOH \quad (5.55a)$$

$$\rightarrow CH_3CHO + H_2O_2 \quad (5.55b)$$

to form carboxylic acids, aldehydes, and hydroperoxides.

5.4 Biogenic volatile organic compounds

Section 5.3 examined tropospheric oxidation of volatile organic compounds (VOCs) in terms of the radical-chain oxidation steps, as well as the initiation of those chains as a result of interactions of the VOCs with OH, NO_3, or O_3. In the present section and the next three, we turn our attention to the reactants themselves. It will have become evident by now that the troposphere contains an amazing mixture of VOCs, some of them released by nature ('biogenic') and some as a result of man's activities ('anthropogenic'). Our immediate interest is in the 'natural' atmosphere, but we shall touch on man's pollution at appropriate points, and then examine the issue in more depth in Section 5.10.

VOCs generated naturally are often hydrocarbons, but smaller quantities of partially oxidized VOCs, such as alcohols, aldehydes, ketones, and acids are released. These compounds may also, of course, be formed *in situ* in the atmosphere as intermediates in the oxidation of the primary biogenic releases. Sulphur compounds (Section 5.6) and some halogenated hydrocarbons (Section 5.7) are further components of the inventory of natural organic chemicals.

Of the biogenic hydrocarbons, methane is by far the most abundant, which is why this compound is used to exemplify the oxidation steps described in Sections 5.3.1 and 5.3.2. The quantities of *non-methane hydrocarbons* (NMHCs) and of partially oxidized VOCs released to the atmosphere are even larger than those of methane. These compounds are much more reactive in the atmosphere than is methane, so that the concentrations of individual species are much smaller. However, the high reactivity also means that the biogenic VOCs make a significant, as well as fascinating, contribution to atmospheric chemistry.

5.4.1 Methane

Methane mixing ratios in the contemporary atmosphere lie at just over 1.7 p.p.m. Until the mid-1990s, concentrations had been rising at a rate of roughly one per cent every year, although the rate of increase has dropped more recently to about 0.3 per cent per year (see p. 235). Ice-core samples suggest that the mixing ratio was nearly constant, at about 0.7 p.p.m., for the 800 years before the industrial revolution, implicating man as the cause of the increases.

Table 5.3 shows annual emissions from the major biogenic and anthropogenic sources of methane, as estimated in the 1994 assessment of the Intergovernmental Panel on Climate Change (IPCC). A grand total of 535 Tg (10^{12} g $\equiv 10^9$ kg) of CH_4 is emitted annually, with a fraction of 160/535 ascribed to natural sources and 375/535 to all identified anthropogenic sources. Of the many uncertainties, one obvious one is how to attribute the release of natural fossil-gas methane between natural and man-related budgets. Total fossil-carbon releases of CH_4 can be estimated as roughly 20 per cent from the isotopic ratio ^{14}C:^{12}C in the atmosphere, as described on p. 23. Similar analysis of the isotope ratios suggests that, of the 'new' methane that has contributed to the rises of mixing ratios in recent years, more than half is of fossil origin (see p. 235). Another most striking feature highlighted by the table is the *dominant* contribution (275/535) from biospheric processes that are nevertheless associated with man's activities.

As befits one of the simplest of organic compounds, the *primary* step in the atmospheric chemistry of methane is very straightforward. Hydrogen abstraction by the OH radical

$$OH + CH_4 \rightarrow CH_3 + H_2O \tag{5.8}$$

yields methyl radicals, which then participate in the oxidation and degradation steps outlined in Section 5.3.2. Reactions of CH_4 with either NO_3 or O_3 are far too slow to be of any atmospheric importance whatever, so that the daytime

Table 5.3 Sources of atmospheric methane: source strengths in units of 10^9 kg yr^{-1}

Natural		Anthropogenic (biospheric)		Anthropogenic (fossil fuels)	
Wetlands	115	Enteric fermentation	85	Natural gas	40
Termites	20	Rice paddies	60	Coal mining	20
Oceans	10	Biomass burning	40	Coal burning	25
Other	15	Wastes, landfill, sewage	90	Petroleum industry	15
Total	160	Total	275	Total	100

Based on a table presented in Seinfeld, J.H and Pandis, S.N. *Atmospheric chemistry and physics*, John Wiley, 1998, itself derived from the 1994 assessment of the Intergovernmental Panel on Climate Change.

reaction with OH represents the only tropospheric chemical loss process for this most abundant of hydrocarbons.

Methane is relatively unreactive even towards OH, the rate constant for reaction (5.8) being 6.6×10^{-15} cm^3 molecule^{-1} s^{-1} at $T = 290$ K. Taking this rate constant in conjunction with the OH concentration used in compiling Table 5.2 (1.6×10^6 molecule cm^{-3}) gives, in a simplified calculation, the lifetime of 3.0 years entered in the table. This lifetime is clearly more than an order of magnitude longer than the lifetimes of any of the other hydrocarbons for which values are given, yet the reaction is responsible for the oxidation of 445×10^9 kg yr^{-1} of CH_4. A further 40×10^9 kg yr^{-1} are exchanged to the stratosphere, and 30×10^9 kg yr^{-1} deposited to the surface, thus making up total sinks of 515×10^9 kg yr^{-1}. The shortfall between these losses and the production rate (535×10^9 kg yr^{-1}) is the reason for the slow rise in atmospheric mixing ratios of methane.

5.4.2 Non-methane hydrocarbons and other compounds

Although many of the non-methane hydrocarbons and partially oxidized VOCs found in the present-day atmosphere are of anthropogenic origin, nature also makes its own contribution to these compounds. Vegetation is known to release a wide range of hydrocarbons to the atmosphere, some of them of a fairly exotic nature. The list of natural VOCs known to enter the atmosphere now extends to well over 1000 compounds. Most of these compounds are emitted primarily from terrestrial plants. Except for methane (and dimethyl sulphide: Section 5.6), animals, soils, sediments, and the oceans are much weaker sources. Soils and sediments possess a high capacity for microbial oxidation of organic species (it is estimated that only 10 per cent of the methane formed escapes such oxidation), which accounts in part for the weakness of their emissions of VOCs. Forests cover more than 40 per cent of all land surfaces, and the release of VOCs, and especially NMHCs, from them plays an important role in shaping tropospheric chemistry. Table 5.4 gives an indication of the sources and emission strengths of important biogenic VOCs. The table shows that very substantial quantities of isoprene and the terpenes are produced by plants. The terpenes are isoprenoid structures. Isoprene (C_5H_8) itself, and the related compound 2-methylbut-2-en-3-ol (MBO: C_5H_9OH) are classed as hemiterpenes; monoterpenes ($C_{10}H_{16}$) and sesquiterpenes ($C_{15}H_{24}$) are the volatile or semivolatile terpenes emitted by plants. Figure 5.8 shows some typical structures of these NMHCs. The monoterpene family contains about 1000 members, and the sesquiterpenes more than 6500, so the potential number of minor VOCs emitted to the atmosphere is very great indeed.

Oxygenated compounds include the C_1 family (CH_3OH, HCHO, HCOOH) and the corresponding C_2 species (C_2H_5OH, CH_3CHO, CH_3COOH). The only C_3 compound known to be released is acetone (CH_3COCH_3), but biogenic

Terpenes

Table 5.4 Sources, emission strengths, and lifetimes of biogenic VOCs[a]

Compound	Main natural sources	Annual emission (10^9 kg C yr^{-1})	Scale of lifetime
Methane	See Table 5.3	160	Years
Dimethyl sulphide	See Section 5.6	15–30	Hours to days
Ethylene (ethene)	Plants, soils, oceans	8–25	Days
Isoprene	Plants	175–503	Hours
Terpenes	Plants	127–480	Hours
Other reactive VOCs[b]	Plants	~260	Hours
Less reactive VOCs[c]	Plants, soils	~260	Days to months

[a]based on a table presented by Fall, R. in Hewitt, C.N. (ed.) *Reactive hydrocarbons in the atmosphere*, Academic Press, San Diego , 1999.

[b]For example, acetaldehyde, 2-methylbut-3-en-2-ol (MBO), hexenal family.

[c]For example, methanol, ethanol, formic acid, acetic acid, acetone.

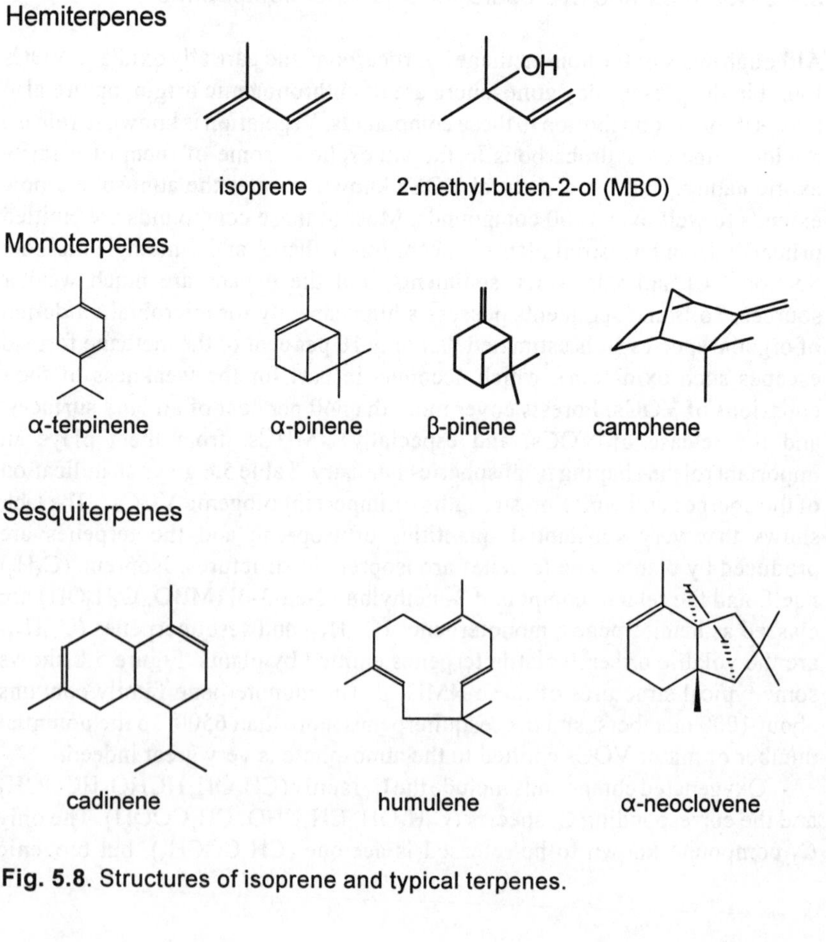

Fig. 5.8. Structures of isoprene and typical terpenes.

formation of this ketone is of particular interest, because surprisingly high concentrations (0.5 p.p.b.) have been found in the free troposphere at northern mid-latitudes, with rather less (0.2 p.p.b.) at southern latitudes. Concentrations as high as these could contribute to the formation of HO_x and to the sequestering of NO_x. As well as being attacked by OH, acetone in the troposphere can be photolysed

$$CH_3COCH_3 + h\nu \rightarrow CH_3CO + CH_3, \qquad (5.56)$$

and addition of O_2 to the radicals yields the peroxy radicals CH_3O_2 (reaction (5.10), p. 331) and $CH_3CO.O_2$ (reaction (5.23), p. 333). Thus the usual sequence of reactions (5.14)–(5.17) (pp. 331–2) generates HO_2, while reaction (5.35) (p. 338) produces PAN and removes NO_2.

Volatile C_6 compounds produced include several isomers of hexenol and hexenal; (3*Z*)-hexenyl acetate is also liberated from some plants. A mixture of these compounds is said to give rise to the characteristic odour of green tea, and the 'green' smell of cut grass or damaged leaves is also attributed to combinations of some of the aldehydes and alcohols. Indeed, (2*E*)-hexenal is sometimes called 'leaf aldehyde' and (3*Z*)-hexenol referred to as 'leaf alcohol'. Hexenyl acetate is one of the few volatile esters shown to be liberated by vegetation; another is methyl salicylate, which is generated by diseased tobacco plants.

Plants thus have the metabolic potential to generate and emit a large variety of VOCs. Figure 5.9 illustrates the concept in the form of a hypothetical 'VOC tree' that emits all the most important VOCs that we have discussed. The probable tissues and compartments for the generation of different compounds are indicated in the boxes to each side of the tree, together with a suggestion of the associated production mechanism. There is clearly great diversity in both how and where the biogenic VOCs are formed. The metabolic pathways are complex and fascinating, but this is not the place to delve into the topic. A similar diversity characterizes the way in which the VOCs leave the plant. Some, such as isoprene and methanol and many monoterpenes, escape smoothly, and perhaps even in a regulated manner, through stomatal pores in the leaves. Others, including certain monoterpenes, depart from external storage structures. Mint plants possess such structures ('trichomes') on their leaves; normally, monoterpene emission is controlled by conductance through the cuticle of the trichome, but crushing ruptures the cuticle, facilitating volatilization, and greatly enhancing the odour for a human observer. Many other VOCs are emitted without any control by the plant, as, for example, are the monoterpenes and oxygenated VOCs released after physical wounding or damage to the plant.

Although our 'VOC tree' emits many of the known VOCs, in reality different plants and plant families emit different combinations of VOCs. For example, all North American oaks are high emitters of isoprene, but many European oak species do not emit isoprene at all. Several pine species emit the

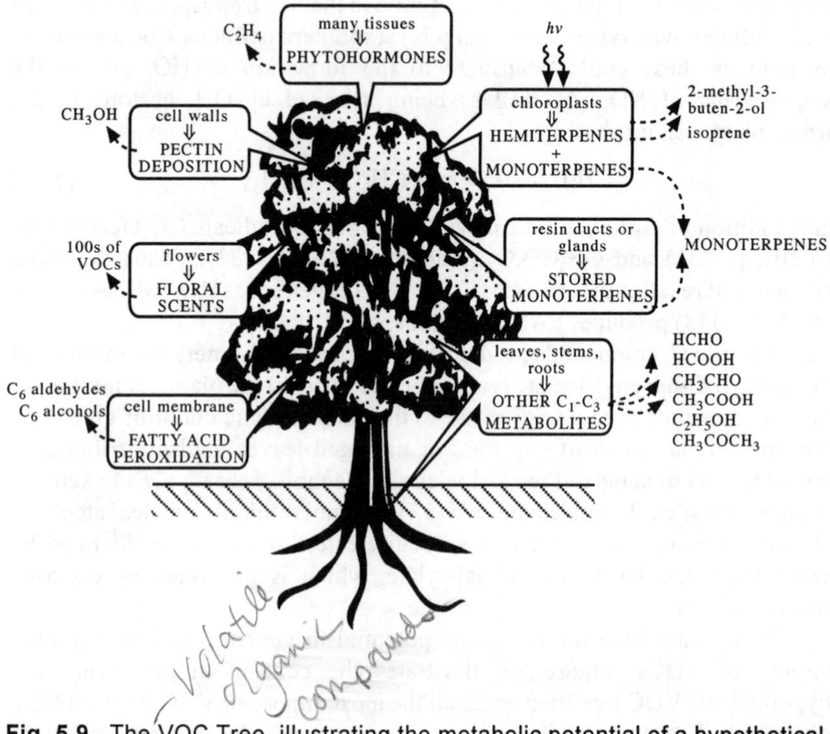

Fig. 5.9. The VOC Tree, illustrating the metabolic potential of a hypothetical tree to produce all the main known VOCs emitted by plants. Based on a diagram of Fall, R. presented in Hewitt, C.N. (ed.) *Reactive hydrocarbons in the atmosphere*, Academic Press, San Diego, 1999.

C_5 alcohol MBO, which may be an important atmospheric precursor of acetone above pine forests. Acetone itself is emitted from the buds of conifers, and it is generated when dried leaves, grass, and conifer needles are wetted. Conifers, mints, and citrus plants are particularly prolific producers of monoterpenes, as the smells of the crushed needles or leaves of these plants demonstrate. Sesquiterpenes are emitted by a wide variety of plants, accounting for up to nine per cent of the non-methane VOCs emitted from some vegetation sites. They are released by oilseed rape (a commercial crop), and are implicated in the irritant effect that this crop can produce in humans. Sesquiterpenes are also synthesized by fungi, and are associated with the dry-rot of potatoes.

Why do plants produce VOCs on a large scale and in wide variety? Strongly emitting plants typically convert one or two per cent of the CO_2 that they have fixed by photosynthesis into the VOCs, and the biochemical energy required to convert the CO_2 into VOC is considerable. It would seem likely that there would be some advantage to the plant in expending this energy. Floral scents are, perhaps, fairly obviously designed to attract pollinating insects, and other VOCs play the same role. Attracting predators of herbivores would be another similar use, and some VOCs are repellants of the herbivores themselves. Many of the VOCs that have been identified also possess antibacterial, antiviral, or antifungal activity. The C_6 alcohols and aldehydes, for example, possess antibiotic properties, and their rapid formation—within less than a minute—after leaf injury probably reflects the plant's attempt to prevent infection by bacteria that reside on the leaf surfaces.

Ethylene (ethene) has been shown to play a most important role in controlling plant growth and development, and it behaves as a volatile plant hormone. Seed germination, flowering, fruit ripening, senescence of flowers and leaves, and sex determination are all affected by exposure to ethylene. Furthermore, ethylene production is greatly enhanced in response to stress: infection, physical wounding, and exposure to some chemicals can all increase the production rate by up to 400-fold. The effect of chemical stress is particularly interesting in the context of atmospheric chemistry, since species such as ozone, PAN, and sulphur dioxide can trigger the response and cause dramatic increases in the rate of release of ethylene to the atmosphere, and thus in turn affect atmospheric chemistry itself. In general, it seems that stress ethylene serves to induce protective responses in affected plants.

The function or functions of isoprene, which is probably the VOC emitted by plants in greatest quantity, is rather uncertain. One suggestion is that isoprene interferes with interactions between the host plant and competing plant species, or between the plant and insects. Another idea is that isoprene serves as a volatile mediator of membrane fluidity, in order to protect the photosynthetic apparatus of the plant from thermal damage. One piece of evidence that supports this view is that leaves of plants grown at low temperature produce little isoprene, but rapidly generate isoprene when exposed to high temperatures.

Monoterpenes have several ecological roles that have been established with some certainty. They provide direct defence against herbivores and pathogens; they attract pollinators and predators of herbivores; and they interfere with competing plants. One interesting role of monoterpenes in conifers is evident in its outward manifestations to anyone who has stood in a pine forest. A wounded conifer releases *oleoresin*, a pleasant smelling sticky substance that is made up of monoterpenes and diterpene acids, both of which are toxic to insects and to pathogenic fungi. The monoterpene mixture (collectively called *turpentine*) acts as a solvent for the acids. Evaporation of the turpentine then leaves behind a hardened barrier of *rosin* (a substance used by string-instrument musicians to treat the hairs of their bows) that seals the wound. The biological role of the sesquiterpenes is less clear. They are emitted from plants in response to leaf injury, and they may act as *signalling molecules*: their release may guide a predator to the herbivore, say a caterpillar, that is causing the injury. The concept of volatile molecules being able to convey information may even extend to 'communication' between one plant and another. Methyl salicylate, for example, may be such a signal molecule, eliciting a defensive response from plants near a diseased, emitting, plant.

Emissions of biogenic VOCs by plants generally exhibit a strong diurnal cycle and a very strong seasonal cycle. The hemiterpenes (isoprene and MBO) show a marked dependence on exposure of the plant to sunlight, as do *some* monoterpene emissions (see the top left-hand box in Fig. 5.9). In many ways, the emission of isoprene parallels the rate of photosynthesis, even exhibiting a similar light-saturation phenomenon (a levelling off in the production rate at high light intensities). However, withdrawal of CO_2 does not lead to decreased isoprene emissions, so that the effect of light may not *directly* relate to photosynthesis. Whatever the internal leaf mechanism, the consequence is that isoprene is not emitted from most plants during the night. It has, however, been suggested that there is a sharp *rise* in the rate of production just before the rapid fall at night. Leaf temperature, too, has a strong influence on isoprene production rates. In the willow leaf, for example, emission rates increase by a factor of ten in raising the temperature from 20 °C to 40 °C, and then decline rapidly at higher temperatures. An enzyme involved in isoprene production, isoprene synthase, obtained from the same leaf shows similar variations in activity. For many monoterpenes, and for the sesquiterpenes, light does not have an influence, and emission is regulated only by temperature. These compounds are therefore emitted throughout the night, but with a smaller flux than during the warmer day. Seasons characterized by high solar flux also experience the highest temperatures, sometimes with a time lag, so that season has a particularly marked influence on all emissions.

Having examined the production of biogenic VOCs, we now turn to a consideration of their behaviour once they reach the atmosphere. One noteworthy feature is the involvement of several compounds in the formation of aerosols. Oxidation of biogenic compounds initiated by reaction with OH, NO_3, or O_3 often leads to the formation of oxygenated or nitrated products with vapour pressures lower than those of the starting reactants. The product compounds may then partition to the aerosol phase. For example, a major product of the oxidation of α-pinene is pinonic acid. Pinonic acid has only a small vapour pressure, and therefore undergoes gas-to-particle formation to yield an organic aerosol. The volatility of some of the sesquiterpenes may be so low that they could condense as the emitted gases rise and encounter lower temperatures, even without any chemical conversion. Organic aerosols can influence climate by scattering solar radiation back to space (net cooling) and by absorption of thermal radiation from the surface of the Earth (net warming). There is some speculation about the possible role of biogenic VOC emissions as natural precursors of cloud condensation nuclei, thus providing an analogue for continental areas to the cloud-formation mechanism postulated for dimethyl sulphide in marine zones (see Sections 1.6 and 5.6).

The 'blue hazes' that are formed in summer over certain forested areas, such as the Smoky Mountains (USA) or the Blue Mountains (Australia), probably owe their origin to VOCs emitted by the vegetation. Reaction with ozone originating in the stratosphere (p. 164) or generated in the troposphere (Sections 5.3.3) and other oxidants will lead to the formation of aerosol in a kind of natural photochemical smog (Section 5.10.7). Indeed, parts of California experienced something like photochemical smog long before the invention of the motor car, and it may well be that terpenes from the citrus groves interacted with ozone to produce the hazes. A rather fine laboratory demonstration mimics this behaviour. A piece of orange peel is squeezed near a glass jar containing ozonized oxygen: as fine jets of the orange oils encounter the ozone, a multitude of white streamers develop in the jar.

The gas-phase atmospheric chemistry of the biogenic VOCs proceeds along the lines already explained in Section 5.3. With ethylene, $CH_2{=}CH_2$, the initial reactions have been set out in Sections 5.3.2, 5.3.6, and 5.3.7 for attack by OH, NO_3, and O_3, respectively; the normal oxidation steps follow. The more complex VOCs undergo essentially similar reactions, but the presence of different functional groups attached to each end of the unsaturated bond, and possibly the existence of more than one such double bond, increase the number of distinguishable reaction sites and thus reaction pathways. For example, isoprene, whose formula we shall represent as $CH_2{=}C(CH_3)CH{=}CH_2$ for the present purposes, has four chemically different carbon atoms attached to the double bonds. Thus attack by, say, OH can yield four different radical adducts

$$\dot{C}H_2C(OH)(CH_3)CH{=}CH_2 \ \text{(I)} \qquad CH_2{=}C(CH_3)C(OH)\dot{C}H_2 \ \text{(II)}$$

$$\overline{CH_2\dot{C}(CH_3)CH}CH_2(OH) \ \text{(III)} \qquad CH_2(OH)\overline{C(CH_3)\dot{C}HCH_2} \ \text{(IV).}$$

It might be supposed that, if each of these radicals were represented as R, the next stages could be represented by four sequences of the type

$$R \xrightarrow{+O_2} RO_2 \xrightarrow{+NO} RO \xrightarrow{+O_2} HO_2 + HCHO + \text{carbonyl.} \qquad (5.57)$$

However, the second pair of initial adducts, (III) and (IV), are allylic radicals, and the free electron of the radical adduct is delocalized over two bonds, leading to isomerization in the allyl species. Each of these radicals thus produces *two* peroxy radicals, RO_2, on addition of oxygen, making a total of six subsequent pathways. Remember, also, that RO_2 can react in the atmosphere not only with NO, as suggested by eqn (5.57), but also with HO_2 and all other RO_2 species present, and the complexities of the chemistry begin to emerge. Figure 5.10 will provide a guide to the description that follows; it is a simplified diagram of the steps that can occur in OH-initiated oxidation of isoprene.

The products observed in the laboratory include, in addition to formaldehyde, HCHO, the unsaturated carbonyl compounds methacrolein, $CH_2{=}C(CH_3)CHO$, and methyl vinyl ketone, $CH_3COCH{=}CH_2$. Pathways can readily be written to methyl vinyl ketone from radicals (I) and (IV), and to

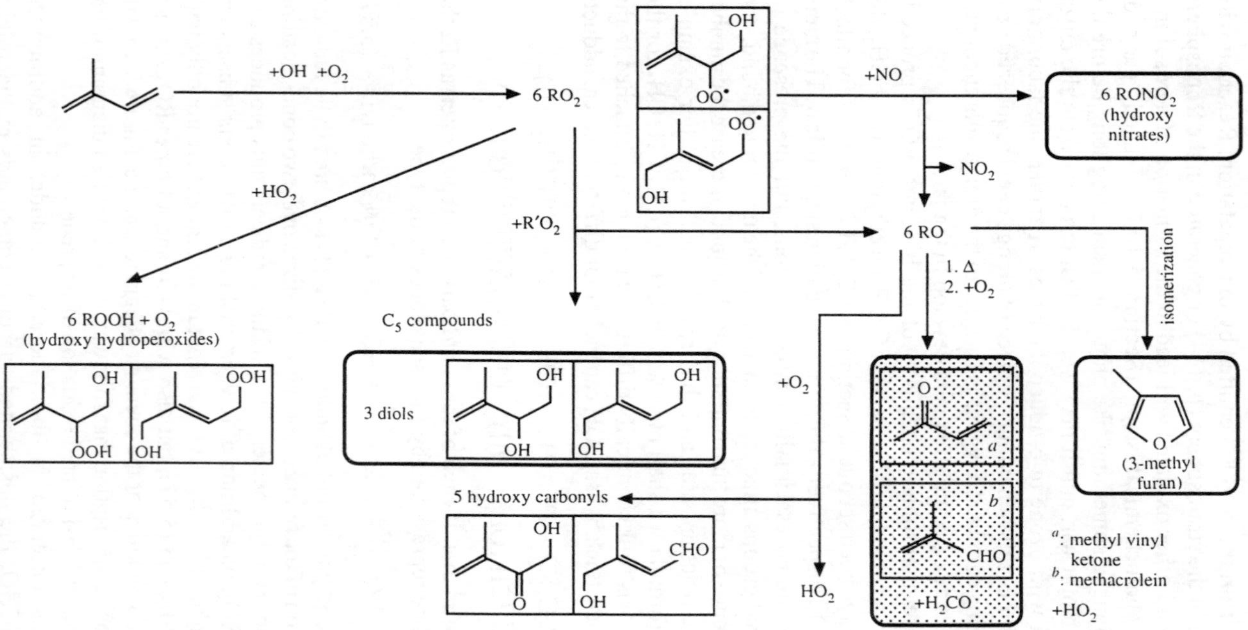

Fig. 5.10. Simplified mechanism for the OH-initiated oxidation of isoprene. Numbers written in front of a product indicate the number of different compounds that can be expected, disregarding geometrical isomers: thus there are six possibilities for RO_2 formed in the initial steps, as discussed in the text. Products enclosed in light rectangular boxes are examples of the types of product formed; those in heavy rounded boxes are identified in laboratory studies (unshaded) or identified in laboratory and field studies (shaded). From Le Bras, G. (ed.) *Chemical processes in atmospheric oxidation*, Springer-Verlag, Berlin, 1997.

methacrolein from radicals (II) and (III). For example, radicals (I) and (II) add O_2 and then react with NO to yield the alkoxy radicals

$$\dot{O}CH_2C(OH)(CH_3)CH{=}CH_2 \text{ (V)} \qquad CH_2{=}C(CH_3)C(OH)CH_2\dot{O} \text{ (VI)},$$

respectively. These radicals undergo elimination of the CH_2O moiety and react further in the presence of oxygen

$$\dot{O}CH_2C(OH)(CH_3)CH{=}CH_2 \xrightarrow{+O_2} HCHO + CH_2{=}C(CH_3)CHO + HO_2 \tag{5.58}$$

$$CH_2{=}C(CH_3)C(OH)CH_2\dot{O} \xrightarrow{+O_2} HCHO + CH_3COCH{=}CH_2 + HO_2. \tag{5.59}$$

Both methacrolein and methyl vinyl ketone themselves react with OH, NO_3, and O_3, and they are also photolysed in the atmosphere, so that the degradation chemistry continues even after the steps already outlined. One especially interesting process involves the peroxy radical derived from methacrolein, $CH_2{=}C(CH_3)C(O)\dot{O}_2$, which can add to NO_2

$$CH_2{=}C(CH_3)C(O)\dot{O}_2 + NO_2 \rightarrow CH_2{=}C(CH_3)C(O)\dot{O}_2NO_2 \tag{5.60}$$

to form peroxymethacryloyl nitrate (PMAN, but sometimes abbreviated MPAN). This compound is a peroxynitrate analogous to PAN, and has been detected in the troposphere at concentrations as high as 10–30 per cent of those of PAN itself. In the atmosphere, it is formed almost solely from isoprene by the reactions outlined, and its presence is thus an excellent indicator of recent ozone production from *biogenic*, and not anthropogenic, hydrocarbons.

Although we have looked at initiation of oxidation by attack of OH on isoprene, inspection of Table 5.2 shows that reaction with NO_3 during the night might be rather more important, and that the reaction with O_3 also makes a contribution. For the terpenes themselves, reactions with NO_3 and O_3 may even be dominant. A good example is afforded by the compound terpinoline, for which lifetimes have been calculated of 49 minutes towards OH, 17 minutes towards O_3, and just 7 minutes towards NO_3, with OH and NO_3 concentrations at typical day and night 12-hour average values, and O_3 at a reasonable 24-hour average concentration. At face value, these numbers mean that 64 per cent of terpinoline in the atmosphere reacts with NO_3, 27 per cent with O_3, and only nine per cent with OH. It is worth pausing at this point to recollect that the lifetime of methane with respect to attack by OH was of the order of *years*, and to reflect how the presence of the double bonds in the terpene have enhanced its reactivity. Other terpenes show different absolute and relative reactivities towards the oxidants OH, NO_3, and O_3. However, whatever the detail, it is quite evident that an understanding of the atmospheric chemistry of the biogenic VOCs must encompass an unravelling of the reaction paths for attack by NO_3 and O_3 as well as that by OH.

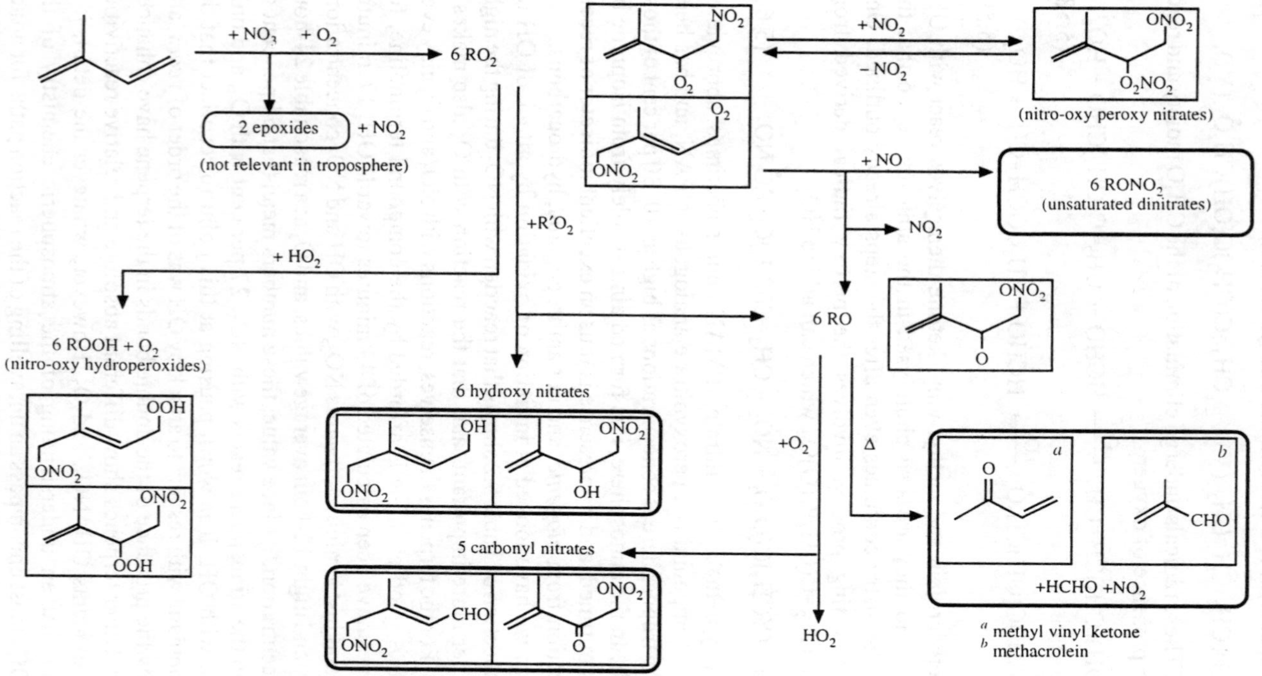

Fig. 5.11. Simplified mechanism for the NO_3-initiated oxidation of isoprene. The meaning of the numbers and types of boxes is the same as described for Fig. 5.9. From Le Bras, G. (ed.) *Chemical processes in atmospheric oxidation*, Springer-Verlag, Berlin, 1997

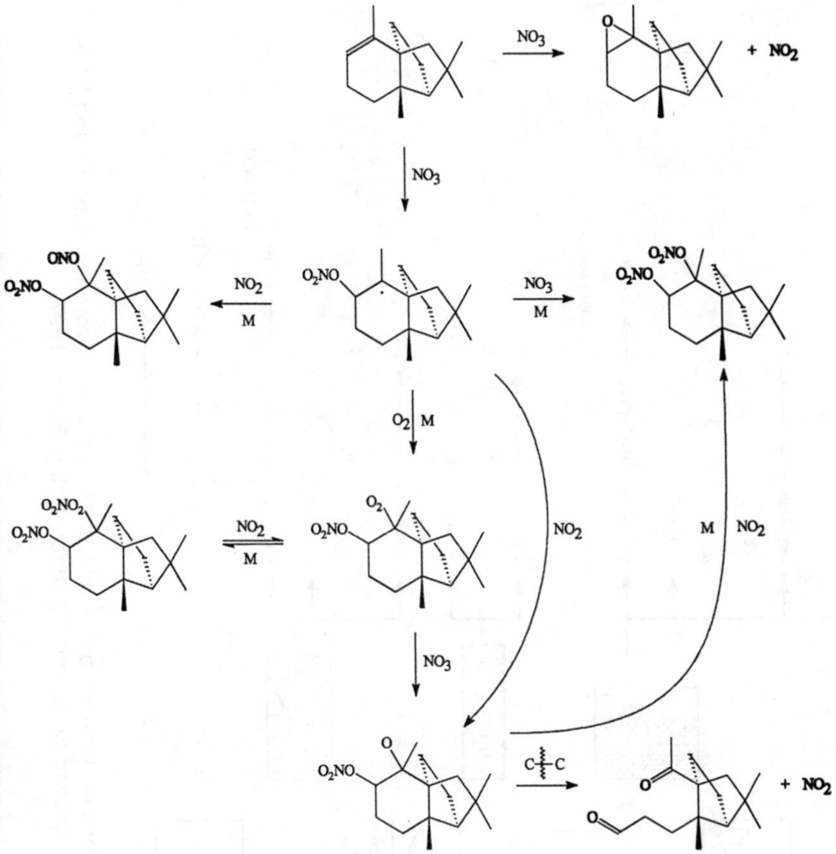

Fig. 5.12. Mechanism for the night-time atmospheric oxidation of α-neoclovene. From King, M.D., D.Phil. Thesis, University of Oxford, 1998.

Figure 5.11 provides the simplified mechanism for NO_3-initiated oxidation of isoprene in the same format as Fig. 5.10. Attack by the NO_3 radical proceeds according to the principles set out in Section 5.3.6. The products of the reaction of the nitrate radical with isoprene in the presence of O_2 include, as for attack by OH, formaldehyde, methacrolein, and methyl vinyl ketone, and, in addition, various nitrato-substituted C_4 and C_5 carbonyl compounds. In principle, initial attack could be on any of the carbons attached to double bonds, to yield the analogues of each of the radicals (I) to (IV), but with $-ONO_2$ groups in place of the $-OH$ groups. However, addition appears to be

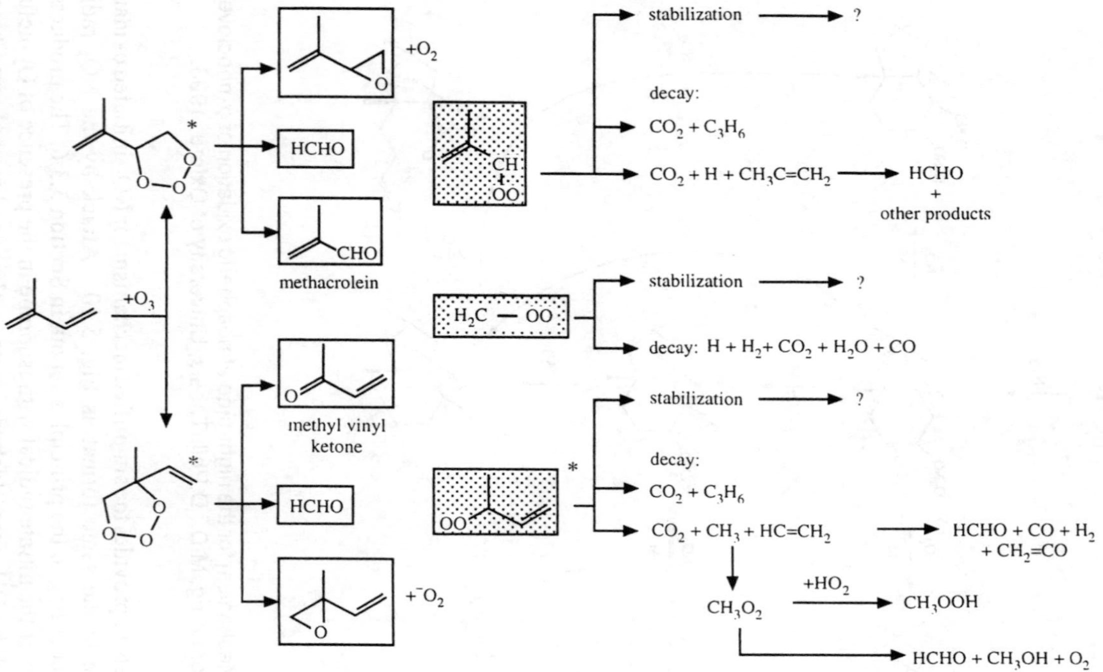

Fig. 5.13. Simplified proposed mechanism for the gas-phase ozonolysis of isoprene. From Le Bras, G. (ed.) *Chemical processes in atmospheric oxidation,* Springer-Verlag, Berlin, 1997.

largely to the terminal carbons (positions 1 and 4), which yields the analogues of the allyl radicals (III: position 4) and (IV: position 1). Products such as 4-nitroxy-2-methyl-1-butan-3-one and methacrolein are consistent with such attack. Addition at position 1 (carbon nearest the $-CH_3$ substituent) apparently occurs preferentially. That means that the most important primary radical will be the allyl species $CH_2(ONO_2)\overline{C(CH_3)\dot{C}HCH_2}$. One of the two peroxy radicals derived from this radical, $CH_2(ONO_2)C(CH_3)=CHCH_2\dot{O}_2$, can then be seen to be the precursor of one of the main observed products in the reaction of NO_3 with isoprene, 3-methyl-4-nitroxy-2-butenal, $CH_2(ONO_2)C(CH_3)=CHCHO$. A particularly important question that remains is the extent to which the initial adduct (which might eliminate NO_2 to form an epoxide) is actually converted to the nitro-oxy-peroxy radicals in the atmosphere.

There is no doubt that the chemistry in these systems is highly complex, and further laboratory studies will be needed before all possible steps in the reaction sequences are understood and characterized, especially with some of the more complicated VOCs such as the terpenes themselves. To give an idea of what is involved, Fig. 5.12 illustrates the most probable pathways for reaction of the NO_3 radical with a sesquiterpene, in this case α-neoclovene.

Reactions of ozone with biogenic VOCs are similarly rather complex. In the case of isoprene, O_3 adds initially to both double bonds to yield two primary ozonides. These two ozonides *may* then each decompose to form carbonyl compounds and Criegee biradicals, as described in Section 5.3.7. As for reaction with OH and with NO_3, the main products are formaldehyde, methacrolein, and methyl vinyl ketone, together with small amounts of 1,2-epoxymethyl butene. Hydroxyl radicals are formed (cf. reactions (5.52) and (5.53)) in a yield approaching 30 per cent of the number of O_3 molecules lost. Figure 5.13 shows a proposed, but still simplified, scheme for the gas-phase ozonolysis of isoprene. The several question marks highlight some of the current uncertainties. The fates of the Criegee radicals —and even if they are formed at all—are the subject of some disagreement; the matter is of importance because of the potential generation of OH at night.

5.5 Aromatic compounds

Aromatic hydrocarbons, and especially benzene, toluene, and the isomers of xylene, represent a major class of organic compound associated with the urban environment. Model calculations suggest that the aromatic species might contribute more than 30 per cent to the formation of photochemical oxidant in urban areas. The chemical behaviour of these compounds in the atmosphere shows some distinct differences from that of the aliphatic species considered so far, and it therefore deserves explicit consideration.

Motor vehicles are the dominant contributors to the emissions of aromatic VOCs, at least in the USA and densely populated parts of Europe. These VOCs are released in part because of the incomplete combustion of fuel, and in part from losses through fuel vaporization. There is evidence for a dramatic reduction between 1970 and 1980 in the ambient concentrations of benzene and toluene around the southern coast of the USA, the decline probably resulting from the introduction of catalytic converters on motor vehicles during this period (see end of Section 5.10.7). There are strong diurnal and seasonal variations in the concentrations of toluene, benzene, and the other aromatic compounds, with a clear minimum developing during the afternoon in many places. The aromatic compounds are removed by chemical reaction after a morning build-up, and are also removed by the prevailing meteorological conditions. Benzene persists, but at much lower concentrations, into the oceanic free troposphere, and even in remote regions of the Southern Hemisphere. Toluene, however, usually remains below the detection limits, the difference between benzene and toluene reflecting the greater chemical reactivity of toluene towards attack by OH (see Table 5.2). Table 5.5 gives an indication of present-day summertime concentrations of the most important aromatic VOCs at a rural site in the UK, together with estimates of emissions from transport and non-transport sources.

The initial step in the atmospheric oxidation of aromatic hydrocarbons is mainly attack by OH during the daytime. The process involves primarily an addition to the ring, although, if the molecule possesses an alkyl side-chain, H-abstraction from the alkyl group can occur as a minor process. Thus, for toluene, we may write the two possible pathways as

$$OH + C_6H_5CH_3 \rightarrow \overset{\bullet}{\overline{C_6H_5}}(OH)CH_3 \quad (5.61a)$$

$$\rightarrow C_6H_5\dot{C}H_2 + H_2O \quad (5.61b)$$

with channel (5.61a) contributing about 90 per cent to the overall reaction. Writing the equations without showing the aromatic rings is convenient in the

Table 5.5 Summertime mixing ratios at a rural site and emission estimates for important aromatic VOCs in the UK[a]

Hydrocarbon	Mixing ratio p.p.b.	Emissions in 10^6 kg yr^{-1} Transport	Non-transport	Total
Benzene	0.24	50	6	56
Toluene	0.46	75	54	129
Ethylbenzene	0.09	28	0.1	28
m-, *p*-Xylene	0.21	38	47	85
o-Xylene	0.14	9	24	33
Total	1.14	200	131	331

[a]Compiled from data presented in *Ozone in the United Kingdom*, Department of the Environment, London, 1997.

case of reaction (5.61a), since the $\overset{\bullet}{C_6H_5}(OH)CH_3$ radical fragment (hydroxymethyl cyclohexadienyl) can have, in principle, the OH attached at positions *ortho*, *meta*, or *para* to the methyl group. Calculations, backed up by some experimental evidence, indicate that the favoured structure is with the OH at the *ortho* position. Laboratory data show that, under atmospheric conditions, the adducts will react predominantly with O_2, to form methyl cyclohexadienylperoxy radicals as an addition product, although a minor abstraction channel also exists, so that, with the aromatic structures now shown, the major and minor channels are

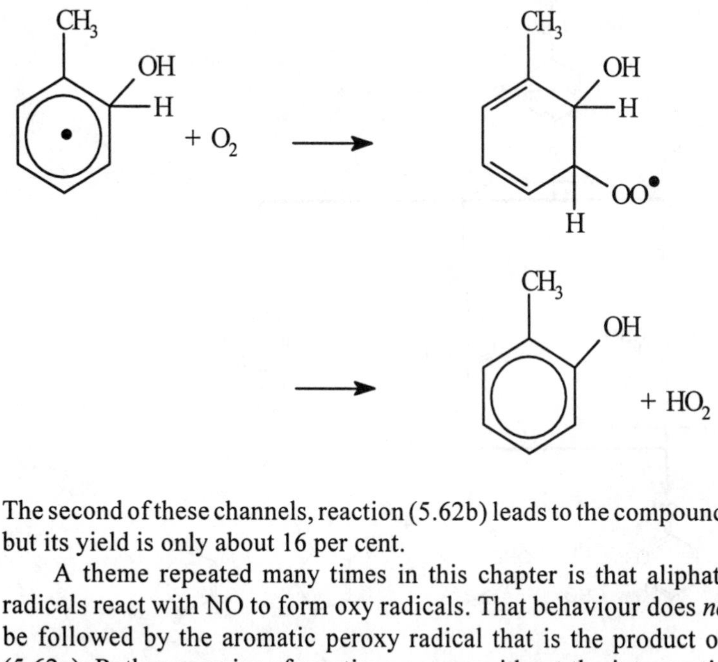

(5.62a)

(5.62b)

The second of these channels, reaction (5.62b) leads to the compound *o*-cresol, but its yield is only about 16 per cent.

A theme repeated many times in this chapter is that aliphatic peroxy radicals react with NO to form oxy radicals. That behaviour does *not* seem to be followed by the aromatic peroxy radical that is the product of reaction (5.62a). Rather, a series of reactions occurs without the intervention of NO_x that leads to several dicarbonyl and unsaturated dicarbonyl compounds. Figure 5.14 is an attempt to show one current view of how the observed products, labelled (1) to (6) in the diagram, might be formed.

Product formation in the system seems best explained by the direct formation of muconaldehydes (such as hexa-2,4-diene-1,6-dial) from the peroxy radical products of reaction (5.62a). Several of these muconaldehydes have been synthesized, and on reaction with OH they yield unsaturated 1,4-dicarbonyl compounds, glyoxal, methyl glyoxal, and maleic anhydride, which are those products also seen in the reaction of OH with toluene in the presence of O_2 (see Fig. 5.14). Many of the aromatic oxidation products, and especially the unsaturated 1,6-dicarbonyl species, are known to be toxic, with both carcinogenic and mutagenic properties.

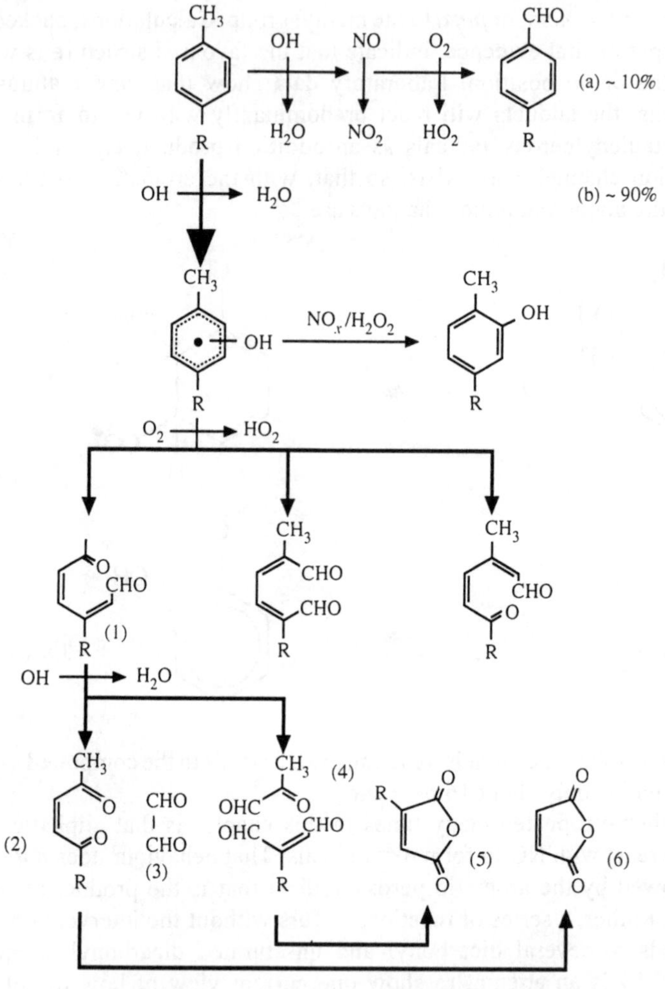

Products	R = H	R = CH$_3$
1	6-oxo-2,4,-heptadienal	3-methyl-6-oxo-2,4-heptadienal
2	4-oxo-2-pentenal	3-hexene-2,5-dione
3	glyoxal	glyoxal
4	methylglyoxal	methylglyoxal
5, 6	maleic anhydride	methyl maleic anhydride

Fig. 5.14. Proposed mechanism for the atmospheric oxidation of aromatic hydrocarbons (toluene: R = H; *p*-xylene: R = CH$_3$). (a) Abstraction channel; (b) addition channel. From Le Bras, G. (ed.) *Chemical processes in atmospheric oxidation*, Springer-Verlag, Berlin, 1997.

Several experiments suggest a short lifetime for the aromatic peroxy radicals, such as the methyl cyclohexadienylperoxyl product of reaction (5.62a), and it seems likely that it is for this reason that they are unable to oxidize NO to NO_2. Various routes from the peroxy radicals have been proposed, including one starting with cyclization to form a bicyclic radical, but there seems little hard evidence in support or rebuttal of the hypothesis. A further puzzle is that the reaction of the OH-adduct with O_2 produces HO_2 radicals with high yield on a short time-scale. If this is the case, then the peroxy radical formed in reaction (5.62a) seems necessarily to be the precursor of HO_2, but what the mechanism might be remains unclear. The process could, of course, be of some importance in that it impinges on the balance of HO_x in the system. There are yet further uncertainties concerning the way in which NO_x behaves in the atmospheric chemistry of aromatic hydrocarbons. Laboratory experiments show that cresol formation becomes significant if NO_x is added to the OH–toluene system; high concentrations of H_2O_2 likewise enhance cresol yields. These observations are reflected in Fig. 5.14 (horizontal path in the second row down), but a convincing explanation is still awaited. Further experiments will doubtless soon shed light on the interesting questions that have arisen.

5.6 Compounds of sulphur

In Section 1.5.4, we presented the main features of the natural biogeochemical cycle of sulphur in the atmosphere. Biogenic processes generally emit sulphur species in the reduced forms of H_2S, CS_2, COS, and the organic compounds CH_3SH (methyl mercaptan), CH_3SCH_3 (dimethyl sulphide, DMS) and CH_3SSCH_3 (dimethyl disulphide, DMDS). The importance of these compounds in producing aerosols in the troposphere (Section 1.6) and stratosphere (Sections 1.5.4, 4.5.5 and 4.7.8) has been mentioned already; oxidation products act either as the main components or as condensation nuclei in the aerosols. Carbonyl sulphide, COS, is rather uniformly distributed throughout the troposphere, at high concentrations (mixing ratios = $5–6 \times 10^{-10}$). Each of these observations is attributable to the long atmospheric lifetime, as discussed later. Carbonyl sulphide, from natural or anthropogenic sources, can therefore reach the stratosphere and there be oxidized to the stratospheric sulphuric acid or sulphate aerosol (see Chapter 4, and especially Sections 4.4.4, 4.5.5, 4.7.4, and 4.7.8). Oceanic emissions of biogenic sulphur gases provide an important flux to the atmosphere, and constitute a source that is believed to be responsible for the background levels of SO_2, methane sulphonic acid (CH_3SO_3), and non-sea-salt sulphate. The dominant sulphur compound released from the oceans (> 90 per cent) is DMS, which is produced by metabolic processes in certain algae, as described later. Oceanic surface waters are supersaturated with DMS with respect to atmospheric concentrations and

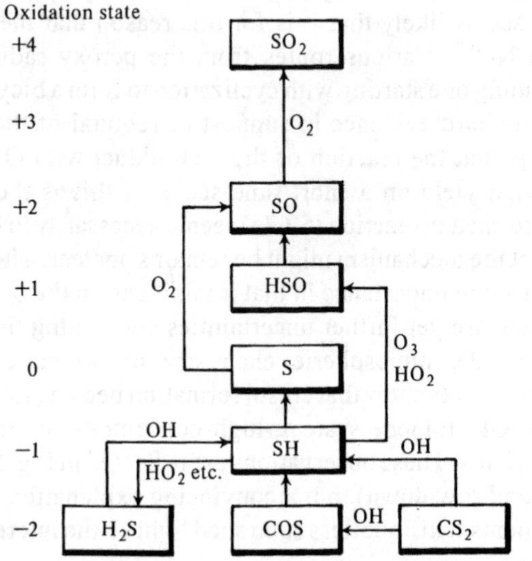

Fig. 5.15. Conversions of sulphur-containing species in the troposphere, showing the progression towards more oxidized compounds.

thus supply a net flux to the atmosphere. Sulphur dioxide, SO_2, may be a product (perhaps minor) of the oxidation of the biogenic sulphur compounds, but man makes an especially large contribution to the atmospheric sulphur burden in urban and industrial regions, almost entirely in the form of SO_2. For this reason, it is convenient to divide sulphur chemistry into two parts. First, we treat in this section the oxidation of the reduced compounds, released naturally, through perhaps six oxidation states to SO_2. Then, in Section 5.10.4, we consider the oxidation of anthropogenic and natural SO_2 to H_2SO_4.

Considerable uncertainties still exist in estimates of fluxes of sulphur compounds to the atmosphere. Oceanic biological activity is thought to contribute 12–58 × 10^9 kg S yr^{-1}, and the land biosphere only 0.1–7 × 10^9 kg S yr^{-1}. Volcanoes provide a flux of 3–9 × 10^9 kg S yr^{-1}. Man's activities release much more sulphur: combustion of fossil fuels is responsible for 70–100 × 10^9 kg S yr^{-1} (and biomass burning for a further 1–4 × 10^9 kg S yr^{-1}). Despite the uncertainties, it is obvious that the oceanic DMS constitutes the most important biogenic flux of sulphur compounds. Biogenic inputs are dominant for most latitude bands in the Southern Hemisphere, but anthropogenic sulphur is dominant for all latitudes in the Northern Hemisphere. It is worth noting that anthropogenic SO_2 is rapidly oxidized and removed from the atmosphere in rainfall and in deposition of particles, so that the large contributions of man to the sulphur budget may nevertheless have

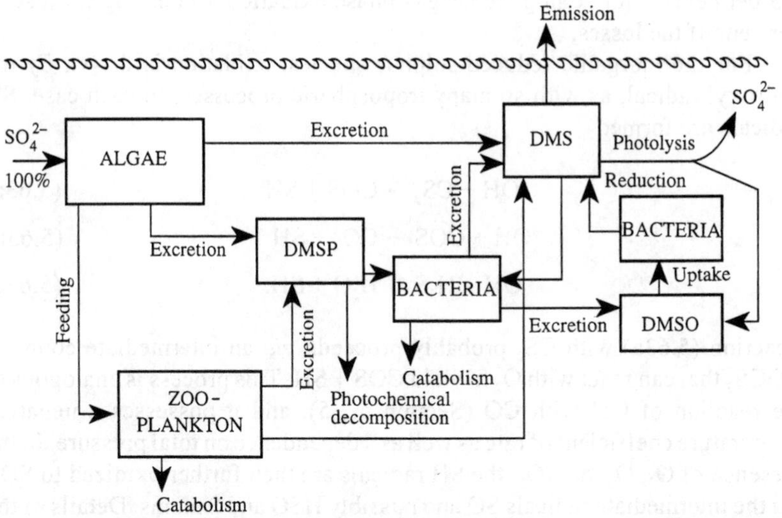

Fig. 5.16. Processes that regulate the interactions of DMS and related sulphur species in the surface ocean. From Andreae, M.O. and Crutzen, P.J. *Science* **276**, 1052 (1997).

their main impacts on the regional scale. The longer lived biogenic compounds, on the other hand, may have a global influence.

Examination of source strengths for the compounds provides only part of the picture, since the atmospheric concentrations of column burdens may be more significant from the point of view of processes affected by atmospheric sulphur species present in the gas phase or in aerosols. Source strengths and concentrations are connected through atmospheric lifetimes, which may, in turn, be closely related to chemical reactivity. Carbonyl sulphide, COS, for example is the most abundant sulphur compound in the atmosphere, the total burden being about 15 times that of SO_2, yet the source flux of COS is probably roughly 100 times *less* than that of SO_2. These values imply an atmospheric lifetime for COS some 1500 times greater than that of SO_2. Current estimates of the global lifetime of COS are about seven years; SO_2 in the boundary layer is removed in about one day by dry deposition, and in one week by gas-phase chemistry in the free troposphere. Thus the lifetimes are apparently compatible with the observed fluxes and concentrations for the two species. The low gas-phase reactivity of COS means that it can be transported to the stratosphere, and oxidized there to H_2SO_4, a mechanism that is generally considered to be the dominant source of the stratospheric sulphate aerosol during periods of volcanic quiescence. There is thus very considerable interest in COS for this reason alone. Oxidation of CS_2 (see next paragraph) seems to be responsible for nearly one-half of the source flux of COS, and the oceans for one-fifth. Major sinks include plant uptake (48 per cent) and soil uptake (33 per cent), with homogeneous gas-phase oxidation accounting for just 13 per cent of the losses.

For the inorganic reduced-sulphur species, oxidation is driven by the hydroxyl radical, as with so many tropospheric processes. In each case, SH radicals are formed

$$OH + CS_2 \rightarrow COS + SH \tag{5.63a}$$

$$OH + COS \rightarrow CO_2 + SH \tag{5.63b}$$

$$OH + H_2S \rightarrow H_2O + SH. \tag{5.63c}$$

Reaction (5.63a) with CS_2 probably proceeds *via* an intermediate complex $HOCS_2$ that can react with O_2 to yield COS + SH. This process is analogous to the reaction of OH with CO (Section 5.3.5), and it possesses a negative temperature coefficient of rate as well as a dependence on total pressure. In the presence of O_2, O_3, or NO_2, the SH radicals are then further oxidized to SO_2, *via* the intermediate radicals SO and possibly HSO and S atoms. Details of the mechanism are not known, but the oxidation steps are probably of the type

$$SH + O_2 \rightarrow OH + SO, \tag{5.64}$$

$$SH + OH \rightarrow S + H_2O, \tag{5.65}$$

$$S + O_2 \rightarrow SO + O, \tag{5.66}$$

$$SO + O_2 \rightarrow SO_2 + O. \tag{5.67}$$

The disproportionation reaction

$$SH + SH \rightarrow H_2S + S \tag{5.68}$$

creates one S atom that can be oxidized in reaction (5.66) followed by (5.67), but also provides a pathway for the formation of H_2S from the more oxidized SH radicals. Reaction of SH with HO_2, HCHO, H_2O_2, and CH_3OOH likewise generates H_2S, so there exist photochemical pathways that link CS_2 and COS to H_2S. Figure 5.15 illustrates the main transformations that lead to SO_2 from reduced inorganic sulphur compounds.

The species that result from the atmospheric oxidation of DMS are of considerable importance. It has been proposed (see Sections 1.5.4 and 1.6) that DMS may be a link in an ecosystem regulation of cloudiness over Earth. Sulphate aerosols are very efficient cloud condensation nuclei. If oxidation of DMS leads to SO_2, then SO_4^{2-} can be formed in the aqueous phase (see Section 5.10.4), and the activity of the marine biota could influence cloud cover. In turn, backscattering by clouds influences both illumination and temperature at the surface, and thus affects the rate at which DMS is generated. The elements of a 'planetary thermostat' are thus in place (cf. Section 1.6). Quite apart from this regulatory mechanism, oxidation of DMS could contribute as much as 25 per cent to total acid deposition (Section 5.10.6), at least over Europe.

The potential for large-scale climate change and the awareness that DMS possesses the largest atmospheric flux of the biogenic gases have between them stimulated much effort into unravelling the biogeochemistry of DMS and its precursors. DMS is produced in seawater by the breakdown of dimethyl sulphoniopropionate (DMSP) into DMS and acrylic acid

$$(CH_3)_2S^+CH_2CH_2COO^- \rightarrow (CH_3)_2S + CH_2{=}CHCOOH. \qquad (5.69)$$

Reaction (5.69) is catalysed both inside and outside the cells of certain phytoplankton and bacteria by the enzyme DMSP-lyase. DMSP itself is made by a variety of algae in sea-water to provide osmoregulation, to afford cryoprotection, and to act as a methyl donor in a variety of biochemical processes. The intimate link between DMSP (and thence DMS) production and marine phytoplankton suggests that the potential might exist for the estimation of DMS concentrations from measured distributions of chlorophyll (used as a measure of plant biomass), but there appears in reality to be a rather poor correlation. The factors that regulate the production of DMSP and DMS in oceanic waters, and the final emission of DMS to the atmosphere, are evidently very complex. Figure 5.16 attempts to show some of the processes that regulate the interactions.

Oxidation of the organic sulphur compounds can be initiated by attack either of OH or of NO_3. For CH_3SCH_3 (DMS), a typical atmospheric lifetime against attack by NO_3 at night is less than 0.7 hr, as against 1.5 days for daytime attack by OH radicals (Table 5.2). CH_3SSCH_3 seems to react predominantly with OH, although NO_3 still contributes about 10 per cent to the loss of the compound.

The types of oxidation process that the organic sulphides undergo can be illustrated with reference to oxidation of DMS initiated by reaction with hydroxyl radicals. Figure 5.17 is intended to provide a simplified guide to the description of the reactions that follows. Products observed are dimethyl sulphoxide (DMSO), COS, SO_2, H_2SO_4, methane sulphonic acid (MSA), and dimethyl sulphone ($DMSO_2$). Initial attack of OH on DMS proceeds *via* two parallel steps

$$OH + CH_3SCH_3 \rightarrow H_2O + CH_2SCH_3 \qquad (5.70a)$$

$$OH + CH_3SCH_3 \rightarrow CH_3S(OH)CH_3, \qquad (5.70b)$$

of which one involves hydrogen abstraction to yield a radical that behaves much as an alkyl radical, and the other addition of OH to form an adduct. The figure shows that these two pathways lead to different final products, and it is noteworthy that the branching ratio is very temperature-dependent: abstraction is favoured at higher temperatures and addition at lower. Thus, at $T = 285$ K,

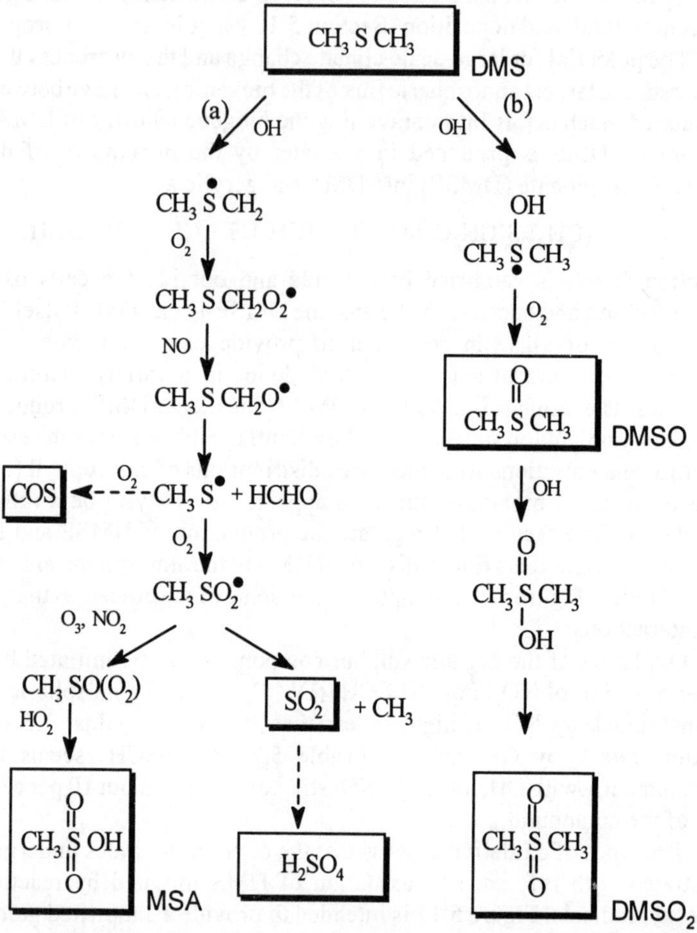

Fig. 5.17. Simplified pathways for the atmospheric oxidation of dimethyl sulphide (DMS) initiated by reaction with OH. Species in boxes are observed atmospheric products: DMSO = dimethyl sulphoxide; $DMSO_2$ = dimethyl sulphone; MSA = methyl sulphonic acid.

the two channels are of equal importance, but by $T = 298$ K channel (5.70a) makes up about 80 per cent of the reaction. Relative product yields will therefore be dependent on atmospheric temperatures.

In the presence of air, initial products from both channels react further with O_2. The peroxy radical derived from channel (5.70a), $CH_3SCH_2O_2$, reacts with NO as though it were a simple alkylperoxy radical, the oxy radical fragmenting to a methyl thiyl radical (CH_3S) and formaldehyde. As with other RO_2 radicals, if NO_x is low (as it may be in the marine boundary layer) other reactions with HO_2 compete (cf. reaction (5.12)).

A series of competing reactions now follows. First, the CH_3S radical may react with O_2

$$CH_3S + O_2 \rightarrow CH_3SO_2 \qquad (5.71a)$$

$$\rightarrow CH_2S + HO_2 \qquad (5.71b)$$

in two parallel channels. The CH_2S product of reaction (5.71b) can form COS, an observed product, in a sequence of steps. Furthermore, there are several pathways for the reaction of CH_3SO_2

$$CH_3SO_2 \rightarrow CH_3 + SO_2 \qquad (5.72)$$

$$CH_3SO_2 + O_3 \rightarrow CH_3SO(O_2) + O_2 \qquad (5.73)$$

$$CH_3SO_2 + NO_2 \rightarrow CH_3SO(O_2) + NO. \qquad (5.74)$$

Reaction (5.72) is a source of SO_2, which itself can undergo oxidation to the sulphate ion (or sulphuric acid), as will be explained in more detail in Section 5.10.4. Reactions (5.73) and (5.74), on the other hand, lead *via* reaction with HO_2 to methane sulphonic acid, CH_3SO_3H (MSA), shown at the bottom left of Fig. 5.17. The radical decomposition process (5.72) is likely to possess a relatively large activation energy, while the two bimolecular steps of reactions (5.73) and (5.74) will probably have small activation energies. It is therefore interesting that there is a temperature dependence of the atmospheric ratio of non-sea-salt sulphate to MSA, which reflects the contributions of reaction (5.72) relative to those of reactions (5.73) + (5.74). The ratio increases from 2.5 in polar regions to nearly 12 near the equator.

The addition channel of the primary interaction, reaction (5.70b), is responsible for the formation of yet other sulphur-containing species found in the atmosphere, such as dimethyl sulphoxide (DMSO) and dimethyl sulphone ($DMSO_2$). One plausible channel leads directly from the adduct to DMSO

$$CH_3S(OH)CH_3 + O_2 \rightarrow CH_3S(O)CH_3 + HO_2. \qquad (5.75)$$

and DMSO adds a further OH radical to form yet another adduct that is the precursor of $DMSO_2$.

The reaction of NO_3 with DMS is more rapid than many interactions of this radical with organic compounds, and it possesses a small negative temperature coefficient, which is indicative of an addition mechanism. However, HNO_3 is rapidly eliminated from the adduct first formed, so that the products *appear* to be those that would arise from simple abstraction

$$CH_3SCH_3 + NO_3 \rightarrow CH_3S(NO_3)CH_3 \rightarrow CH_3SCH_2 + HNO_3. \qquad (5.76)$$

The subsequent oxidation steps are thus identical to those described for the abstraction channel of OH-initiated oxidation of DMS. The rapid reaction of NO_3 may be especially important during high-latitude winter, even in remote regions where NO_x is relatively low. It is not at present clear if reactions of

NO_3 are responsible for observed cyclic seasonal variations in the ratio of non-sea-salt sulphate to MSA, which show a pronounced maximum in winter.

Field measurements, laboratory studies, and models all currently indicate that SO_2 is the major oxidation product from DMS under most atmospheric conditions, and that DMS is the main source of non-sea-salt sulphate. This conclusion is consistent with the hypothesis introduced earlier that seeks to find DMS to be a significant source of cloud-condensation nuclei (CCN), and thus potentially of climate forcing. For DMS to act in this way, it must increase the *number* of CCN particles, and not just the total mass of sulphate aerosol. Such an increase could arise either by the creation of new particles or by causing the growth of pre-existing smaller particles to a size at which they can act as CCN. Direct evidence for the contribution to particle and CCN formation is provided, for example, by measurements made over the South Atlantic. Strong correlations exist between the marine DMS emission flux and the concentrations both of total aerosol particles and of CCN.

5.7 Natural halogen-containing species

Halogenated compounds are found in wide variety in the atmosphere, many of them being of anthropogenic origin (see, for example, Sections 5.10.8 and 5.10.10). A few halogenated compounds are of natural origin. For example, it has long been recognized that the methyl halides, CH_3Cl, CH_3Br, and CH_3I, are present: they are produced predominantly in the oceans. Since the 1980s, however, there has been increasing awareness of the possibility that *inorganic* chlorine might participate in tropospheric chemistry.

Vast quantities of inorganic halides are present as particles in the atmosphere. Wave action in marine areas generates small airborne droplets of sea-water that either remain as concentrated aqueous solutions, or evaporate to leave suspended particles of the solids. Sea-salt itself contains 55.7 per cent of Cl, 0.19 per cent of Br, and 2×10^{-5} per cent of I by weight. Salt flats, and inland bodies of salt water, such as the Dead Sea, provide additional sources of salt-containing particles as they are generated by wind action and erosion of crusts. Some volcanic eruptions, such as that of El Chichón, generate a brief pulse of salt emissions. The immediate question is, then, if significant amounts of these halides can be activated in such a way that the halogens can play a role in the chemistry of the troposphere. The somewhat surprising answer is that inorganic chlorine and bromine do, indeed, contribute to such atmospheric chemistry. The answer was unexpected, because processes sufficiently energetic to liberate the reactive species to the gas phase were unrecognized until the stimulus of research on *stratospheric* halogen chemistry drew attention to the importance of heterogeneous processes (see Sections 4.4.4 and 4.7). We return shortly to possible mechanisms that would release reactive halogen species from inorganic salts, but first look at some atmospheric processes and observations.

Free halogen atoms in the troposphere might have several significant effects on the course of chemical behaviour. Amongst the most important processes might be included: (i) direct attack on organic substrates to initiate oxidation chains; (ii) formation of organo-halogen compounds; and (iii) involvement in catalytic cycles that would destroy tropospheric ozone. We examine each of these effects in turn.

Atomic chlorine is very highly reactive towards many organic compounds, especially those containing abstractable hydrogen or those possessing unsaturated bonds to which the chlorine may add. Rate constants may be one or two orders of magnitude larger for the Cl reactions compared with those for the corresponding OH processes. For example, in the comparable reactions

$$OH + CH_4 \rightarrow CH_3 + H_2O \quad (5.8)$$

$$Cl + CH_4 \rightarrow CH_3 + HCl \quad (5.77)$$

the rate coefficient for the second process is about 130 times that for the first at ambient temperature. Secondary reactions of CH_3 follow as described in Section 5.3.2, regardless of the source of the radicals. As a result, chlorine atoms in the troposphere could initiate oxidation, even if present at concentrations orders of magnitude smaller than those of OH. A further detail that is of significance is that Cl-atom concentrations would be at a peak near dawn, while OH reaches its highest concentration around local noon, so that a comparison of rates of oxidation has to address time-dependent effects as well as those averaged over a day. Atomic bromine is generally much less reactive than Cl towards organic substrates. Although any sea-salt source would provide much less flux of Br than of Cl (molar ratio about 1:660 in sea salt), concentrations of Br could nevertheless be relatively high in some circumstances because of the longer lifetime with respect to the gas-phase organic sinks.

Quite apart from the contribution that atomic halogens, and chlorine in particular, might make to the initiation of oxidation, these halogens could also lead to the formation of halogenated products through addition and subsequent secondary reactions. In this way, *natural* hydrocarbons such as terpenes and *natural* halogens could be the precursors of substantial quantities of halogen-containing derivatives. An example should show why discovery of such pathways might be important. Significant concentrations of chlorinated acetic acids have been found in pine needles in several locations. One obvious explanation is that the degradation products of anthropogenic organo-chlorine compounds have been incorporated into the vegetation. On the other hand, the existence of a biogenic route to similar compounds means that sensible policy decisions depend on proper quantification of the natural contribution, to assess if any strategy controlling man's activities could have a perceptible impact.

The great significance of halogen-catalysed destruction of ozone in the stratosphere (Sections 4.4.1, 4.6.4 and 4.7) makes a search for similar processes in the troposphere an evident priority. Bromine would be a prime candidate for tropospheric ozone destruction. We have already seen in Chapter 4 (p. 221) that bromine is an even more active catalyst in the stratosphere than is chlorine. We also know from the discussion of the preceding paragraph that more Br than Cl might be available to react with O_3. In addition, there are some special features in bromine chemistry that make catalytic chain removal of ozone plausible. The Br-based equivalent of 'cycle 4' (p. 166) would employ the catalytic pair of reactions

$$Br + O_3 \rightarrow BrO + O_2 \qquad (5.78)$$

$$BrO + O \rightarrow Br + O_2, \qquad (5.79)$$

but, in the troposphere, O-atom concentrations are far too low to make the cycle effective. However, for BrO (but *not* for ClO) another reaction that requires neither light nor atomic oxygen

$$BrO + BrO \rightarrow Br + Br + O_2 \qquad (5.80)$$

can regenerate the halogen atom. Heterogeneous processes may, in fact, be even more important than reaction (5.80) in recycling halogen atoms, as we shall see later. Considerable evidence has now accumulated to show the existence of tropospheric ozone depletion that is probably catalysed in this way by free halogen atoms. In the mid-1980s, a surprising phenomenon was discovered from ground-based measurements in the troposphere. Ozone was found to be depleted rapidly in Arctic regions during April (at polar sunrise) over periods of hours to days, the removal being quite extensive both vertically and horizontally. Reports on episodes in the Norwegian Arctic showed almost complete O_3 loss from ground level to altitudes of ca. 2 km, and evidence was obtained linking the depletion to the liberation of free atomic bromine. The field studies provide evidence for the presence of photochemical precursors of both Cl and Br before and during Arctic sunrise, and the concentrations of the atoms at the time of the episodes have been estimated to be in the range 10^3–10^5 atom cm^{-3} for Cl and two to three orders of magnitude larger for Br.

A singularly telling piece of evidence that links elevated halogen atom levels with the Arctic tropospheric ozone depletions is reproduced in Fig. 5.18. Concentrations of *i*-butane, *n*-butane, and propane were measured as they decayed as a result of chemical reaction, and in the absence of new emissions or physical dilution. Referring to the three hydrocarbons as *i*, *n*, and *p*, the rate constants for reaction with OH are roughly in the ratios $i{:}n{:}p = 2{:}2{:}1$, while for reaction with Cl the rate-constant ratios are approximately $i{:}n{:}p = 1{:}2{:}1$. If, then, loss of the hydrocarbons is primarily in reaction with OH, it might be expected that $[i]/[n]$ would remain constant, while $[i]/[p]$ decreased. On the other hand, if reaction with Cl were largely responsible for chemical loss of the hydrocarbons, $[i]/[p]$ would remain constant while $[n]/[i]$ would decrease. The figure shows that the experimental data divide themselves into two distinct

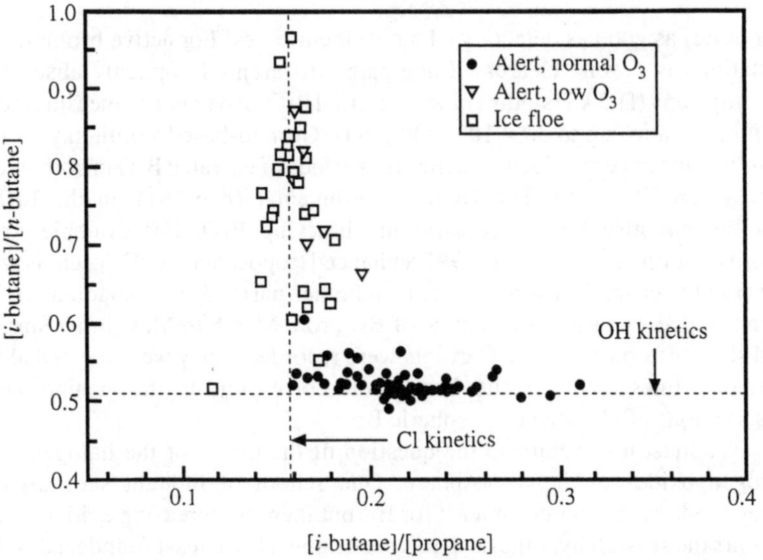

Fig. 5.18. Relative concentrations of propane, *n*-butane, and *i*-butane in the Arctic troposphere at Alert (Canada) and on an ice floe 150 km north of Alert. As presented by Finlayson-Pitts, B.J. and Pitts, J.N., Jr. *Science* **276**, 1045 (1997).

groups. When there is a normal ozone concentration at Alert, the behaviour is that of OH-controlled concentrations, while when O_3 is low at Alert or on the ice floe, the loss of hydrocarbon is essentially that expected if losses were dominated by reaction with Cl. There is thus good evidence that Cl concentrations are elevated at the times of O_3 loss, and it does not require too great an act of faith to believe that the elevated Cl, and species associated with it, might itself be responsible for the O_3 destruction.

Other researchers have adopted the tactic of utilizing the differing patterns of reactivities of groups of chemical compounds towards Cl and OH in order to estimate concentrations of Cl. For example, by measuring the differences in concentrations of selected groups of organic compounds, peak concentrations of Cl atoms at dawn over the Atlantic and Pacific oceans were evaluated to be in the range 10^4 to 10^5 atom cm^{-3}. Emission estimates and measurements of the concentrations of tetrachloroethylene have been used to infer a globally and annually averaged concentration of Cl atoms of less than 10^3 atom cm^{-3}. Such low averaged values are, of course, not necessarily inconsistent with the much higher estimates of peak concentrations.

All the evidence presented so far for the existence of free halogen atoms and halogen oxides in the troposphere has been circumstantial. There is, however, some more direct information. Atomic chlorine has not, so far, been detected explicitly, but *molecular* chlorine has been identified and measured in a marine area by a specially adapted form of mass spectrometry. Since Cl_2 is without question photolysed to Cl atoms, the discovery of Cl_2 is almost (but not quite) as good as detecting Cl atoms themselves! For active bromine, the situation is even clearer. Long-path differential optical absorption spectroscopy (DOAS) studies have identified BrO, and obtained measurements of mixing ratios up to 30×10^{-12} (30 p.p.t). Ground-based zenith-sky DOAS experiments have provided evidence for periods of elevated BrO in both Arctic and Antarctic Spring. The GOME instrument (see p. 181) on the ERS-2 satellite has also been successful in observing BrO. For example, from February until the end of May 1997, enhanced tropospheric BrO columns were observed over the Hudson Bay area, and other parts of the Canadian Arctic, which suggest a large local source of Br. From March to May, other smaller and shorter tropospheric BrO events were recorded. They were located along the coast lines of the Arctic Ocean, and correspond to the ground-based observations of elevated tropospheric BrO.

We must now return to the question of the origin of the halogens and halogen oxides in the troposphere. One school of thought sees organic compounds as the source, at least for Br, but there is increasing evidence that favours the sea-salt hypothesis. It has been known for at least four decades that sea-salt particles are deficient in chloride and bromide ions relative to sodium, suggesting that reactions with atmospheric gases (acids, oxides of nitrogen, or other oxidants) may have converted condensed-phase halides into gas-phase halogen species. Reactions with acids such as HNO_3

$$NaCl_{(cond)} + HNO_{3\,(g)} \rightarrow HCl_{(g)} + NaNO_{3\,(cond)} \tag{5.81}$$

$$2H^+ + NO_3^-$$

$$\uparrow H_2O$$

$$N_2O_5\,(gas) \longrightarrow N_2O_5\,(liq) \rightleftharpoons NO_2^+ + NO_3^-$$

$$H_2O \downarrow \qquad\qquad \downarrow Cl^-$$

$$2H^+ + 2NO_3^- \qquad ClNO_2\,(liq.)$$

$$\downarrow h\nu$$

$$ClNO_2\,(gas) \longrightarrow Cl + NO_2$$

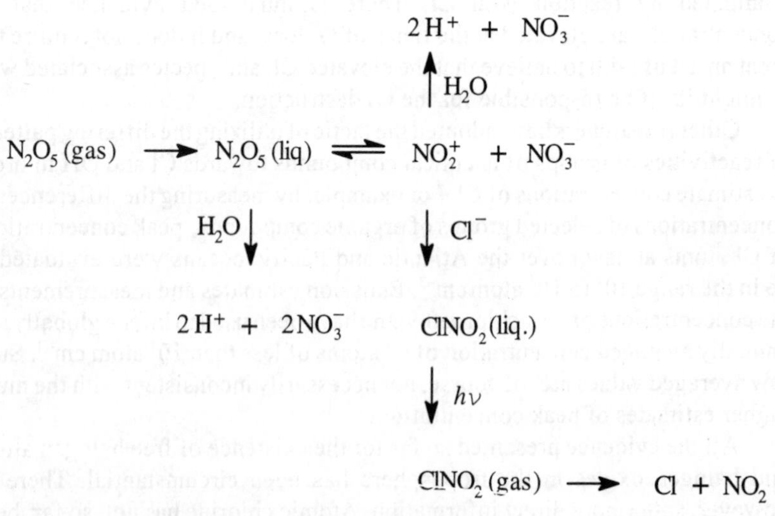

Fig 5.19. Proposed reaction scheme to explain the formation of $ClNO_2$, and then Cl, from N_2O_5 and droplets of NaCl solution.

are well known. However, subsequent liberation of free atomic chlorine by reaction with, for example, OH

$$OH + HCl \rightarrow Cl + H_2O \tag{5.82}$$

is probably too slow to produce the apparent atmospheric concentrations. Instead, other heterogeneous reactions have been invoked by analogy with those that are important in the polar stratosphere (p. 177 and pp. 247–9)

$$NaCl_{(s)} + N_2O_{5\,(g)} \rightarrow ClNO_{2\,(g)} + NaNO_{3\,(s)} \tag{5.83}$$

$$NaCl_{(s)} + ClONO_{2\,(g)} \rightarrow Cl_{2\,(g)} + NaNO_{3\,(s)}. \tag{5.84}$$

Both $ClNO_2$ and Cl_2 are readily photolysed to yield Cl atoms. Reaction (5.83) appears to be the major one operative in the troposphere, and it proceeds in several steps involving ionic reactions within the droplet. Figure 5.19 represents one suggested scheme for the formation of $ClNO_2$ and its subsequent photolysis to yield Cl.

Analogous reactions of NaBr and KBr give photolysable Br species, and the NO_3 radical appears to react with both NaCl and KBr to yield halogen atoms. There are thus a number of atmospherically reasonable reactions that have been proven in the laboratory to form Cl or Br atoms.

A further interesting atmospheric source of halogen atoms involves the photolysis of ozone in the presence of wet sea-salt particles. Several pieces of direct and indirect evidence from the laboratory indicate that the reaction indeed produces Cl_2 or Cl, although the mechanism remains to be elucidated.

As remarked earlier, tropospheric ozone depletion in polar regions may depend on heterogeneous processes to achieve the recycling. We now wish to show that such reactions may not only regenerate the chain carriers for ozone decomposition, but may also liberate additional active halogen from the inorganic salt. One sequence involves the steps

$$BrO + HO_2 \rightarrow HOBr + O_2 \tag{5.85}$$

$$HOBr + Cl^- + H^+ \rightarrow BrCl + H_2O. \tag{5.86}$$

The second step may occur on the surface of sea ice. BrCl is photolysed at wavelengths that penetrate into the troposphere

$$BrCl + h\nu \rightarrow Br + Cl, \tag{5.87}$$

so that the Br atom is regenerated, and can re-enter the cycle at reaction (5.78). More than this, a Cl atom has been liberated into the gas phase from the sea-salt particle, so that the process is autocatalytic in terms of free halogen. A process exactly analogous to reaction (5.86) can be written with Br^- as the negative ion, and Br_2 as the product. However, the process probably involves several individual steps. According to current opinion, it would then be possible to write an autocatalytic cyclic destruction of ozone in the form

$$2(Br + O_3 \rightarrow BrO + O_2) \qquad 2 \times (5.78)$$

$$2(BrO + HO_2 \rightarrow HOBr + O_2) \qquad 2 \times (5.85)$$

$$HOBr + Cl^- + H^+ \rightarrow BrCl + H_2O \qquad (5.86)$$

$$BrCl + Br^- \rightarrow Br_2Cl^- \rightarrow Br_2 + Cl^- \qquad (5.88a)$$

$$Br_2 + h\nu \rightarrow Br + Br \qquad (5.88b)$$

$$\text{Net} \quad 2HO_2 + H^+ + 2O_3 + Br^- + h\nu \rightarrow HOBr + 4O_2 + H_2O$$

There remain a great many speculations and uncertainties about this area of atmospheric chemistry, and there are doubtless other surprises to come in what promises to be an exciting field.

Organic halogen compounds in the atmosphere have been more widely studied. We confine ourselves here to biogenic species (see also Sections 5.10.8 and 5.10.10), Methyl chloride and methyl bromide both have tropospheric lifetimes toward attack by OH in excess of one year, so that they are able to reach the stratosphere and play a role there as catalytic species in ozone chemistry (Section 4.4). The behaviour of methyl iodide is, in some ways, more interesting. Unlike the chlorine and bromine compounds, CH_3I can be photolysed by the radiation that penetrates to the troposphere, and it generates iodine atoms

$$CH_3I + h\nu \rightarrow CH_3 + I \qquad (5.89)$$

which could participate in tropospheric chemistry. For example, the reactions

$$I + O_3 \rightarrow IO + O_2 \qquad (5.90)$$

$$IO + IO \rightarrow 2I + O_2 \qquad (5.91)$$

provide a catalytic route to ozone destruction that might play a minor part in ozone loss in the troposphere. However, the rather low tropospheric abundances of CH_3I and the existence of alternative channels for the interaction of two IO radicals (to yield $I_2 + O_2$ or an IO dimer) limit the importance of the chain. It has also been suggested that IO radicals might influence NO_x and HO_x chemistry in the troposphere by forming $IONO_2$ and HOI (species to be compared with the analogous 'reservoir' compounds of the stratosphere, $ClONO_2$ and HOCl, Section 4.4.2). Long-path differential optical absorption spectroscopy (LP-DOAS) has proved successful at detecting the IO radical. For example, at Mace Head, Ireland, IO has been measured at concentrations of up to 6 p.p.t. (6×10^{-12}). The IO follows solar radiation intensity very closely, with concentrations below the level of detectability during the night. This correlation suggests a photochemical origin of the radicals, most probably from the precursors CH_2I_2 and CH_2BrI which were observed at elevated concentrations. The observed amounts of IO are quite sufficient to have a substantial influence on boundary-layer photochemistry: a mixing ratio of 6 p.p.t. can increase the rate of removal of O_3 by as much as 70 per cent over the ocean and 10 per cent over land. Although the measurements of IO refer to rather special conditions, and the participation of IO in the 'mainstream' chemistry of the troposphere is uncertain at present, the oxidizing potentiality of the radical should clearly be kept under review.

5.8 Heterogeneous processes and cloud chemistry

Although the Earth's atmosphere is made up mostly of gases, it has suspended in it liquid and solid particles that significantly influence the chemistry that occurs. These particles also affect the way in which ultraviolet, visible, and infrared radiation is distributed within the atmosphere; in turn, the alterations to the radiation field can regulate both climate and the atmospheric chemical processes themselves.

The subject of particles in the atmosphere was introduced in Section 1.3, and some of the kinetic considerations associated with *heterogeneous* and *multiphase* processes were presented in Sections 3.4.3–3.4.5. Our more detailed discussions of the stratosphere and the troposphere will have demonstrated very clearly the range of influence of what are often rather loosely called 'heterogeneous reactions'.[a] Sections in Chapter 4 that are specially relevant included 4.4.4 (heterogeneous reactions in general, and their influence on reservoir species), 4.5.5 (volcanoes), and 4.6.2 (aircraft). Almost the whole of Section 4.7, which deals with polar ozone depletion, turns around the theme of chemical processing on particles, but Section 4.7.4 (polar stratospheric clouds) is particularly intimately involved with heterogeneous reactions. In the present chapter, too, we have seen many examples. Section 5.3.6 considers the influence of heterogeneous processes on N_2O_5 and NO_3; Section 5.4 describes the formation of particles in the oxidation of biogenic organic compounds; Section 5.6 has as one of its themes the formation of cloud condensation nuclei in the oxidation of natural sulphur compounds, and Section 5.10.4 will return to the theme of oxidation of SO_2 on particles; and halogens may be released to the troposphere in processes involving sea-salt particles, as we have just seen in Section 5.7. Yet other ways in which particles are important in atmospheric chemistry will emerge as this book develops. Our

[a]The processes of interest are chemical changes occurring within liquid droplets, and on the surfaces of solid and liquid aerosol particles. Heterogeneous chemical reactions are usually thought of as those occurring at the interface between two phases. Some important atmospheric processes, however, involve transfer of reactants from the gaseous to a condensed phase followed by homogeneous chemical change within the condensed phase. For our purposes, therefore, it is convenient to include in the category of heterogeneous reactions all processes that involve more than one phase.

Table 5.6 Heterogeneous processes of importance in the troposphere[a]

Substrate / Reactant	diesel or aircraft soot	mineral dust	organic matter (secondary)	sulphates bisulphates	sea salt (deliquescent)	sulphuric acid	cloud droplets ice particles
	→ Increasingly hygroscopic or water containing →						
OH	reactive loss[b]	unknown	reactive loss[b]	unknown	reactive loss[b]	uptake	uptake
HO_2	reactive loss[b]	unknown	unknown	unknown	uptake, reaction[b]		uptake and reaction
RO_2	reactive loss[b]	unknown	uptake, reaction[b]		uptake		solubility limited
O_3	reactive loss surface ageing synergisms[b]	unknown	unknown		importance of direct uptake to be found		solubility limited uptake reactive loss
NO_2	chemisorption reduction to NO HONO formed	HONO formed[b]	nitration (more likely by N_2O_5 / NO_2^+)		ClNO formation on dry NaCl		HONO formation on liquid water[b] or on ice[b]
NO_3	reactive loss[b]	unknown	reaction (e.g. with aromatic species)	unknown	limited solubility reaction with I^- etc		low solubility
N_2O_5	hydrolysis or reaction as NO_2^+, depending on substrate reaction probabilities are very variable formation of particulate nitrate or gaseous HNO_3				formation of $ClNO_2$ and other halogen compounds	hydrolysis	
SO_2	catalytic oxidation	catalytic oxidation[b]			oxidation in polluted marine atmosphere	unimportant	oxidation by H_2O_2, O_3, etc
Other processes	adsorption synergisms[b]	solubilization of catalytic metals	aerosols formed from biogenic VOC	pH-dependent uptake of acidic or alkaline gases		uptake of NH_3	aerosol processing

[a]From a figure presented by Schurath, U. at *EUROTRAC Symposium, 1998*, Garmisch-Partenkirchen, March 1998, and reproduced by his kind permission.
[b]uncertain.

purpose here is to look at some of the more general aspects of heterogeneous chemistry in the troposphere, and to extend the discussion elsewhere with some additional examples.

A useful starting point is provided by Table 5.6, in which an attempt is made to summarize our current state of knowledge about heterogeneous processes on different types of solid and liquid aerosol. The substrate materials are arranged so that they are increasingly hygroscopic, and then water containing, as the table is read from left to right. Although much has been discovered in recent years, it is striking how much uncertainty, or even complete lack of knowledge, remains.

As the table illustrates, the substrate surfaces on which atmospheric heterogeneous reactions occur are represented by a wide range of physical and chemical types. Condensed water is the predominant form of suspended matter in the troposphere, and clouds the most abundant form of condensed water. Liquid-water droplets, rather than ice, are favoured only in the lowest part of the troposphere. The cirrus clouds formed in the middle and upper troposphere are made up of ice crystals. More than 50 per cent of the surface of the Earth is usually covered by clouds at some altitude, and roughly seven per cent of the volume of the troposphere contains clouds. A fairly dense cloud will contain about 0.5 millionth of its volume as liquid water. If all the droplets possess a diameter of 10 μm, there will be almost 1000 droplets in every cm^3 of air, and the total surface area of water will be about 3×10^{-3} cm^2. The total surface area of water within the troposphere, and on which surface reactions can occur, is thus enormous. As to the other substrates, they are much less abundant, but nevertheless significant. Sulphuric acid and sulphates are quite abundant, and have natural sources, as we saw in Section 5.6, as well as the anthropogenic ones to be discussed in Section 5.10.4. Sea-salt is widespread, especially in marine areas. The organic component remains rather poorly characterized at the time of writing. Some, at least, is of biogenic origin, presumably resulting from the oxidation of terpenes and other biogenic VOCs (Section 5.4.2). This biogenic source is strongly implicated in heavy organic aerosol loading in the Amazon basin during the wet season, when there is little biomass burning. One of the puzzles is the presence of 'black carbon' outside the burning season, since this material is often thought to be soot carbon from combustion. It seems possible that some of this material may be *primary* biogenic aerosol, such as particles of bacteria and fungi, plant debris (waxes and leaf fragments), and humic matter.

Four simple categories can be recognized for the chemistry that creates aerosols and droplets, and occurs on and in them:

(i) condensation of a single component;
(ii) reaction of more than one gas to form a new particle;
(iii) reaction of gases on a pre-existing particle; and
(iv) reactions within the particles themselves.

Category (i), the condensation of a single gaseous component to form a new suspended particle is *homogeneous, homomolecular* nucleation, and it is obviously central to cloud formation. The thermodynamics of this process were discussed in Section 2.5. The most obvious example is the aggregation of sufficient H_2O molecules from the gas phase to produce a droplet of liquid, or of solid, water. Category (ii) is the analogous process for reacting species involving two or more gases to form a condensable product species in a *homogeneous, heteromolecular* process. A typical example is the reaction

$$NH_{3\,(g)} + HNO_{3\,(g)} \rightarrow NH_4NO_{3\,(s)}, \tag{5.92}$$

which can form NH_4NO_3 that ultimately produces particles. Reaction (5.92) occurring on a particle that already exists—that is to say, a *heterogeneous, heteromolecular* reaction of type (iii)—may be a more important route to aerosol NH_4NO_3 than the direct formation of a new solid particle. Another example of the heterogeneous, heteromolecular process is afforded by a reaction we have encountered already, that of NO_2 or HNO_3 with sea-salt particles (NaCl)

$$NaCl_{(cond)} + HNO_{3\,(g)} \rightarrow HCl_{(g)} + NaNO_{3\,(cond)} \tag{5.81}$$

to form $NaNO_3$. Heteromolecular reactive condensation of gas-phase molecules on pre-existing particles is sometimes called *aerosol scavenging*. It can have an impact on bulk tropospheric chemistry by providing a sink for nitrogen and hydrogen species such as HNO_3, NO_3, N_2O_5, H_2O_2, and HO_2, as well as organic nitrates and peroxides. Clouds and raindrops have a major effect on gas-phase species through the scavenging mechanism. Rainout (removal of gases by cloud droplets) is believed to be more important than washout (removal of gases by raindrops) because of the longer lifetime and greater surface area of cloud droplets compared with raindrops. Water-soluble species, such as the acids, acid anhydrides, and peroxides are obviously particularly susceptible to removal by these mechanisms.

Finally, category (iv) includes chemical reactions that occur within the aerosol itself to form particles of changed composition, as in the oxidation of SO_2 to sulphate ions in clouds. These are multiphase processes, since they involve transfer from (and perhaps back to) the gas phase, and the condensed phase is almost always liquid, since diffusion rates in solids are too slow to permit significant extents of reaction. Discussion of the particular case of SO_2 oxidation is deferred until Section 5.10.4, but similar considerations apply to the chemistry and photochemistry of background air. As well as providing a liquid phase in which soluble gases can be dissolved, clouds also offer an active chemical medium for aqueous-phase reactions that can affect the distribution of active species in the troposphere. Despite the relatively small fractional volume of the troposphere occupied by cloud droplets or particles (see p. 381), the direct chemical influence of clouds on the tropospheric reactants may be more important in atmospheric photochemistry than the scattering and reflection of actinic ultraviolet radiation. It is therefore appropriate to consider now how that influence is exerted.

The general principles that give rise to the differences between gas-phase and condensed-phase reactions were set out in Section 3.4.4. The ideas revolve around cage effects, consequential diffusion- and activation-control of reaction, solvation, and so on. The disproportionate influence of the aqueous-phase reactions derives in part from the concentration of the reagents in solution, which especially enhances the rate of second-order reactions. Added to the concentration effect, the solution reactions often have lower activation energies than the corresponding gas-phase processes, thus further increasing the rates. But a special feature in the tropospheric reactions is high solubility of certain key compounds such as HO_2 or N_2O_5; the reactions of these species within the droplets ensure that dissolution is irreversible. The partitioning into the aqueous phase is thus strongly favoured, and partially offsets the small relative volume of the water droplets. Because NO is relatively insoluble, HO_2 and NO are separated and prevented from participating in the reaction

$$HO_2 + NO \rightarrow OH + NO_2 \qquad (5.13)$$

that converts NO to NO_2 (and thus promotes ozone formation) in the gas phase. The inhibition of this reaction in the liquid phase has a most pronounced effect on the chemistry and oxidizing capacity of the atmosphere in the vicinity of clouds. Photochemical processes may also be different in condensed phases, because of changes in absorption spectra, the opening of new photolytic channels, and alterations of actinic flux (see p. 386).

Some of the most important atmospheric reactions exemplify the principles very nicely. One case is the formation of HNO_3 from N_2O_5

$$[N_2O_5 + H_2O \rightarrow HNO_3 + HNO_3]_{aq}, \qquad (5.93)$$

the square brackets written here being intended to demonstrate that the reaction occurs *within* a water droplet. The gas-to-aqueous phase transfer of N_2O_5 is limited by gas-phase diffusion and transfer through the interface, while reaction (5.93) within the aqueous phase is so fast as to be essentially 'instantaneous', so that the dissolution of N_2O_5 is irreversible. The reaction between N_2O_5 and H_2O is slow in the gas phase, and the difference between the gas-phase and liquid-phase reactivity is a consequence of interactions with the solvent opening up lower energy ionic reaction pathways in solution.

The gas-phase processes involved in the formation of N_2O_5 are the steps (Section 5.3.6)

$$NO_2 + O_3 \rightarrow NO_3 + O_2 \qquad (5.40)$$

$$NO_3 + NO_2 + M \rightarrow N_2O_5 + M \qquad (5.42)$$

$$N_2O_5 + M \rightarrow NO_3 + NO_2 + M, \qquad (5.41)$$

but NO_3 is rapidly photolysed during the day, so that N_2O_5 is formed only at night. Reaction (5.93) should thus have a particularly noticeable effect during the dark winter period of the year, and at high latitudes. N_2O_5, NO_2, and NO_3 are essentially in thermal equilibrium, which is maintained by the opposing reactions (5.42) and (5.41). Since reaction (5.93) removes N_2O_5, it prevents regeneration of NO_2 in reaction (5.41). High altitudes accentuate the effect, because N_2O_5 is more stable at lower temperatures when the decomposition reaction (5.41) becomes slower. Observations confirm large losses in tropospheric NO_x (by up to 90 per cent) under such conditions.

Cloud reactions further influence the oxides of nitrogen by removal of NO_3 radicals (Section 5.3.6). Because the radicals have a long lifetime against removal in the gas-phase, they can be incorporated into cloud water as $(NO_3)_{aq}$ Although NO_3 is not particularly soluble in pure water, the dissolved radical reacts rapidly with the chloride ion

$$[NO_3 + Cl^- \rightarrow NO_3^- + Cl]_{aq} \qquad (5.94)$$

present in the sea-salt aerosols of the marine boundary layer to remove NO_3 permanently. Conversion times are a few tens of seconds, and NO_x is thus removed from the gas phase quite rapidly. At the relatively higher temperatures and lower NO_x concentrations experienced in the marine boundary layer, the reaction may be more efficient in removing NO_x from ('denitrifying') the atmosphere than is the route involving hydrolysis of N_2O_5, since reaction (5.42), which generates N_2O_5, requires NO_2, and its rate *decreases* with increase of temperature.

The effect of reaction (5.94) is not only to remove oxides of nitrogen from the gas phase, but also both to enhance the concentration of NO_3^- dissolved in the cloud water and to generate atomic chlorine. Thus further oxidation may be initiated in the liquid phase (indeed OH can be formed from NO_3^-: see, for example, p. 386), and Cl can be exchanged to the gas phase, so that free halogen atoms can be formed without the high degree of acidification implied by reaction (5.83).

Heterogeneous (multiphase) loss of N_2O_5 in reaction (5.93) is essentially irreversible; the equilibrium implied by reactions (5.42) and (5.41) is disturbed, and NO_2 and NO_3—neither of which is very soluble in water—are removed indirectly by the presence of cloud droplets. A similar sequence is responsible for the removal of PAN (pp. 337–8), another important component of the NO_y system (see, again, the footnote on p. 166 for an explanation of NO_y). PAN is another compound that is not highly soluble in water, and clouds do not affect its concentration directly either by dissolving it or acting as a medium for chemical reaction. PAN and its precursors, it will be remembered, are connected by the equilibrium

$$CH_3CO.O_2 + NO_2 \rightleftharpoons CH_3CO.O_2NO_2. \qquad (5.35)$$

Peroxy radicals, such as $CH_3CO.O_2$, *are* highly soluble, so that the equilibrium is shifted to the left in the presence of clouds, and the atmosphere is 'renoxified': that is, NO_y is converted to NO_x.

The discussion of the last few paragraphs has indicated how heterogeneous chemistry can bring about changes in concentration of gas-phase NO_x. These changes can in turn have a considerable impact on both O_3 and OH in the troposphere. One set of model predictions, for example, suggests a depletion, as far towards the equator as 25 °N, of O_3 by 25 per cent and of OH by 20 per cent as a result of the occurrence of the heterogeneous reaction (5.93). The effects were even more pronounced at higher latitudes, for the reasons explained earlier. The models show that O_3 production is suppressed because of the lowered $[NO_2]$, and that [OH] is diminished in two ways. First, the loss of ozone itself results in lowered OH generation in steps (5.2) and (5.3) (Section 5.3.1). Secondly, the all-important reaction

$$HO_2 + NO \rightarrow OH + NO_2 \qquad (5.13)$$

is less able to convert HO_2 to OH both because there is less NO available, and because of the highly favoured partitioning of HO_2, but not NO, into the droplets, as explained on p. 383.

One very interesting result of the model simulations just described was that sulphate aerosol turned out to be the main agent in the reduction of atmospheric oxidizing efficiency, and that sea-salt particles played a relatively small part in this aspect of heterogeneous tropospheric chemistry. One reason for the predominance of processing on sulphate particles is that anthropogenic emissions of sulphur oxides and nitrogen oxides tend to be concentrated in the same geographical locations. Much less NO_x is found in the marine boundary layer (MBL), which is where the sea-salt particles are largely found, so that sea-salt aerosol processing is limited to coastal regions immediately downwind of urban and industrial areas. It goes without saying, of course, that these considerations apply just to the one multiphase process, reaction (5.93), and that chemistry involving the heterogeneous chemistry of halogen atoms or of NO_3 radicals may be most important of all within the MBL.

Another heterogeneous reaction can convert HO_2 to H_2O_2 by producing first the superoxide ion which then reacts with more HO_2

$$[HO_2 \rightarrow H^+ + O_2^-]_{aq} \qquad (5.95)$$

$$[HO_2 + O_2^- \xrightarrow{H^+} H_2O_2]_{aq}. \qquad (5.96)$$

Evaporation of clouds containing high concentrations of dissolved H_2O_2 could represent a net source of H_2O_2 to the atmosphere. However, the H_2O_2 can also act with OH in the aqueous phase

$$[OH + H_2O_2 \rightarrow H_2O + HO_2]_{aq}, \qquad (5.97)$$

so that the sequence of reactions (5.95) + (5.96) + (5.97) constitutes a free radical sink

$$[OH + HO_2 \rightarrow H_2O + O_2]_{aq}. \tag{5.98}$$

Thus these processes provide yet another way in which the overall oxidizing capacity of the troposphere can be strongly influenced by cloud chemistry. The reduction in HO_x radical concentrations is compounded both by the destruction of hydrated HCHO and by inhibition of its photolysis to give HO_2 in reactions (5.16) and (5.17). Indeed, there are even aqueous-phase reactions that destroy ozone itself, such as

$$[O_2^- + O_3 + H_2O \rightarrow 2O_2 + OH + OH^-]_{aq}. \tag{5.99}$$

Various reaction sequences can be written, of which the most obvious is the pair of reactions (5.95) + (5.99), which then acts as the aqueous-phase analogue of the gas-phase sink reaction (5.32) between HO_2 and O_3. Despite the relatively poor solubility of O_3 in water, reaction (5.99) also turns out to be the dominant source of dissolved hydroxyl radicals.

We mentioned on p. 386 that, as well as thermal reactions, photochemical processes can be modified by the presence of particles and droplets in the atmosphere. There are several reasons for the altered behaviour. Light intensities may be much higher near the tops of thick clouds than they are in cloud-free regions, but very much lower beneath the clouds. These changed light intensities can affect photolysis rates both in the gas phase and within liquid droplets. One estimate suggests that the gas-phase rate of photolysis of NO_2, for example, can be fivefold greater near the top of a cloud than below it. Absorption spectra may also be different in different phases; spectral shifts, altered transition probabilities, and the formation of new absorbers are all known processes. The absorption of ozone illustrates the effect of a shift of the spectrum of aqueous solutions to longer wavelengths compared with the gas phase. Because the absorption of O_3 falls in a spectral region where solar intensities in the troposphere increase extremely rapidly with increasing wavelength, a relatively small *bathochromic shift* (shift to longer wavelengths) of the spectrum can lead to a disproportionately large increase in the rate of photolysis. Formation of new absorbers in solution is exemplified by the reactions of N_2O_5. Photodissociation of this compound in the gas phase is not a significant loss process in the troposphere. However, in the presence of cloud droplets, there is strong partitioning to aqueous HNO_3 because of the occurrence of reaction (5.93), as emphasized so many times before. The NO_3^- anion, to which HNO_3 dissociates almost completely in aqueous solution, absorbs at wavelengths that penetrate to the troposphere, and photodissociates to O^-, which then rapidly protonates to form the OH radical

$$NO_3^- + h\nu \rightarrow NO_2 + O^- \tag{5.100}$$

$$O^- + H_2O \rightarrow OH + OH^-, \tag{5.101}$$

once again altering the oxidizing capacity of the atmosphere, and providing a new source of oxidation within the cloud droplet.

A quantitative interpretation of heterogeneous chemistry requires a method for introducing the coupling of the chemistry that occurs in the two phases, as well as a knowledge of the solubilities, interfacial mass-transport kinetics, and the reaction kinetics of the species in the condensed phase. The theoretical background to the coupling needs further development, and the whole study of heterogeneous reactions in particles and droplets is in its infancy. However, in view of the likely involvement of reactions in and on surfaces in both the troposphere and the stratosphere, this field is one that merits the most detailed attention. The past decade has seen great progress, with very considerable effort expended in both experimental and theoretical research, but enormous uncertainties remain.

Most of the examples discussed in the present section have referred to particles belonging to the categories in the three right-hand columns of Table 5.6: that is to say, particles that are aqueous or hygroscopic. Important new goals will be to understand better the origins and behaviour of primary and secondary organic aerosol, and of mineral particles. Biogenic organic aerosol may have a far-reaching influence on both atmospheric chemistry and climate. Such species act as effective cloud condensation nuclei, and tropical continental clouds—which play a major role in planetary heat redistribution—are strongly influenced by the presence of the particles. Furthermore, there is some speculation that organic particles might be the true cloud condensation nuclei in the marine troposphere as well, but that, since they can also act as surfaces for the deposition of H_2SO_4 from the oxidation of DMS (Section 5.6), they might be mistaken for cloud condensation nuclei of marine biogenic origin. Mineral aerosols, too, are expected to have a part to play in tropospheric chemistry, although quantitative information is sparse. Experiments have demonstrated, for example, that minerals can increase the rate of photochemical degradation of adsorbed organic compounds, and that the reaction products are often different from those evolving from gas-phase reactions. There could be significant atmospheric consequences, especially if the products turn out to be more toxic than the reactants. The presence of water on minerals associated with atmospheric particles can have a significant influence on rates of reaction, especially in relation to the transfer of transition metal ions as a result of the formation of organic complexes. We should note that transition metals such as iron, manganese, and cobalt are now known to be very important catalysts in the atmospheric oxidation of sulphur (IV) compounds (Section 5.10.4).

Understanding of heterogeneous chemistry has evidently reached the stage where its complexity and probable importance in tropospheric chemistry are apparent, but where basic information on composition and reaction channels and mechanisms is sorely needed.

5.9 Models, observations, and comparisons

Although the broad outline of tropospheric behaviour now seems clear enough, there remain many problems of detail. The variability in concentration of several species is one cause of uncertainty. Relatively short atmospheric lifetimes mean that geographical distributions are frequently determined by local release patterns and meteorological conditions. Furthermore, tropospheric measurements of concentrations often prove to be much more difficult and imprecise than those made in the stratosphere. Chemical reactions are themselves more complex, involving a wide variety of relatively large molecules, with the additional possibility of heterogeneous processes on surfaces on the planet or in the atmosphere. Is it possible, with the uncertainties and complexities, to compare the predictions of models with field observations of concentrations in the troposphere? Such comparisons, it will be remembered, provide one basis for the validation of stratospheric models (Section 4.4.6). The present section seeks to answer this question; it starts out with a brief description of tropospheric models, then examines methods of measuring atmospheric concentrations of trace species, and ends by comparing observations with models for some of the most important chemical species.

Two obvious candidate species for study are the radical OH, which is central to tropospheric oxidation chemistry, and CO, which is both formed and destroyed in OH-mediated reactions (Section 5.3.2). Since tropospheric OH plays such a large part in controlling the oxidizing capacity of the atmosphere, accurate representation of OH in models is prerequisite. Reliable values for OH concentrations are also needed to estimate tropospheric lifetimes for many trace gases, including CO and CH_4, the sources of which are extremely difficult to quantify. The analysis is complicated since OH itself is lost mainly in reaction with these two gases (see Section 5.3.2), both of whose tropospheric concentrations have been, at least until recently, slowly increasing (see p. 235, Section 4.6.9). It is evident that verification of the theories of tropospheric OH would benefit from good measurements of the spatial and temporal distribution of concentrations of the radical. Validation of a time-dependent model requires simultaneous determination not only of [OH] itself, but also of all the parameters that control the local OH chemistry.

Since OH is very reactive, its chemical lifetime is so short that atmospheric transport plays only a minor role in determining concentrations and distributions of the radical (see Fig. 3.9). Comparison of model predictions with experimental data is thus possible with a simple zero-dimensional model that takes into account only the chemical part of the budget. Without agreement at this level, there is little point in evaluating models of higher dimensionality and complexity that incorporate the same chemistry.

Both the chemistry and the representation of transport in models is provided by simulations of the latitudinal dependence and the seasonal cycle of CO. The lifetime of this species allows transport over hundreds of kilometres, again as indicated in Fig. 3.9, so that the CO distribution may be expected to be determined to a large extent by atmospheric flow patterns, on which will be superimposed variations in latitudinal and seasonal chemistry. Species of intermediate lifetime, such as HCHO or PAN, may be used in tests of 1-D or 2-D models.

Ozone itself is an interesting case. Much research on tropospheric chemistry is directed towards understanding ozone concentrations in natural and polluted environments. Although there are many localized measurements of ozone concentrations in the troposphere, monitoring of the global distributions and seasonal behaviour of O_3 have, up to now, been generally inadequate. Assessment of the global ozone budget in the troposphere therefore currently relies strongly on models of chemistry and transport, making clear once again the need for proper validation of the models.

5.9.1 Tropospheric models

The general principles in constructing models of tropospheric chemistry follow those set out in Section 3.6. The models are not so very different in concept from those used for the stratosphere, but they may present much greater difficulty in implementation. Sophisticated modelling of transport is essential, and clouds and other particles need to be accommodated by the physics and the chemistry. Furthermore, the transport of trace constituents in the atmosphere is intimately related to the hydrological cycle, so that a high-quality simulation of the hydrological cycle is a necessary prerequisite for successful modelling of trace gases in the free troposphere.

Three-dimensional (3-D) chemistry–transport models are being employed increasingly widely for the representation of the complex interactions between the main precursors of O_3, such as NO_x and CO, and the highly variable meteorological conditions encountered in the troposphere. A first requirement is that they properly reflect transport properties. One way of evaluating these properties is through tracer experiments. A tracer particularly well suited to this purpose is ^{222}Rn, a radioactive isotope of radon. The gas is emitted from soils, and since the half-life for radioactive decay is 5.5 days, ^{222}Rn distributions are quite sensitive to relatively small-scale meteorological processes. In general, the tracer tests show that many current models perform well in representing the effects of synoptic weather systems, but are much less good at simulating behaviour in the upper troposphere, suggesting that the effects of deep convection are not being correctly modelled. Since NO_x and other short-lived species in the free troposphere are influenced by vertical exchange processes, the poor performance of the models in this respect indicates one direction in which improvement must be made.

An interesting development over the past decade has been the pooling of modelling resources in the coordinated efforts of several research groups. This approach was exemplified within the project *EUROTRAC* (*European Transport and Chemical Transformation of Pollutants*). In one subproject, the main focus was the development of a 3-D Eulerian model system for the simulation of air pollution in the troposphere over Europe, and its application to scientific problems and environmental planning. The development of models on the European scale was complemented by work on smaller scale chemical transport models, and by sensitivity as well as process studies using a hierarchy of models. The hierarchy was assembled in a way that allowed zooming of chemical transport calculations to smaller areas embedded in the larger European domain.

5.9.2 Tropospheric measurements of trace species

This book is not the place to provide an in-depth presentation of the many techniques now available for making measurements of trace gases in the atmosphere. However, a few remarks must be made about the subject, since 'field measurements' represent one of the main strands of atmospheric chemistry, the study of which seeks to explain the nature and concentrations of the components of the atmosphere.

Figure 5.20 is similar in intent to Fig. 4.9 provided for stratospheric measurements of ozone. Both *in-situ* and remote-sensing experiments are represented, and the advantages and disadvantages of each class remain the same as set out in Table 4.3. Tropospheric measurements, especially of boundary layer components, have historically relied rather heavily on localized *in-situ* methods. The very first experiments in atmospheric chemistry naturally examined the composition of air that could be sampled in the immediate vicinity. Stratospheric chemistry, in contrast, was initially reliant on remote sounding, because suitable platforms did not exist. These distinctions are slowly disappearing, as remote-sensing techniques are developed for tropospheric use, and as measuring devices can be carried into the stratosphere more easily. Satellite methods, in particular, are now being adapted widely to make tropospheric determinations of composition and concentration.

Tropospheric *in-situ* methods are themselves sometimes perceived as falling into the two categories of 'spectroscopic' and 'chemical', although the borderline between the two is often blurred. Christian Friedrich Schönbein is often regarded as the 'father of air chemistry' because, soon after his discovery of ozone in the laboratory in 1839, he demonstrated the presence of ozone in the atmosphere of several European cities. For this purpose, he employed test papers coated with potassium iodide. As Schönbein had found out already, ozone liberates iodine from the iodide ions, and the iodine can interact with the starch in the paper (or deliberately added) to form a dark-blue complex. By comparing the depth of colour with tints on a calibration chart after the test papers had been exposed to air for a fixed period, it was possible to reach an

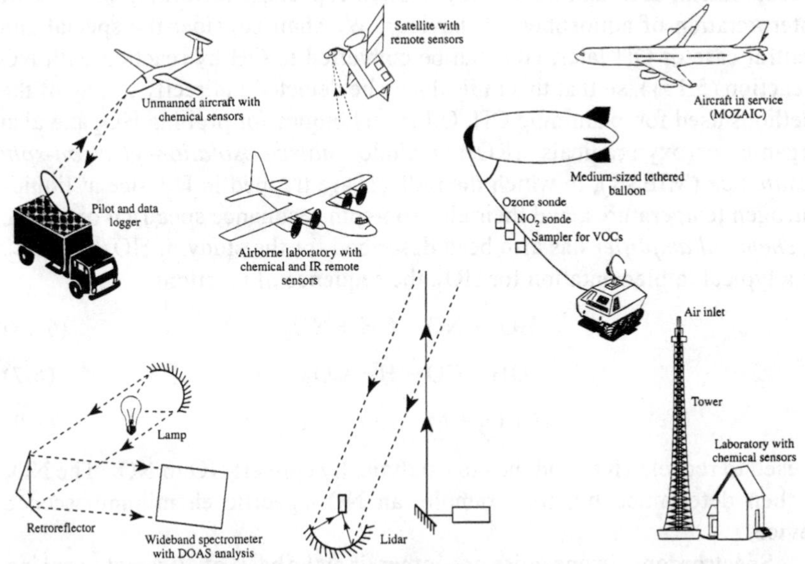

Fig. 5.20. Some platforms used for measurements of tropospheric trace gases. After Roscoe, H.K. and Clemitshaw, K.C. *Science* **276**, 1065 (1997).

estimate of the atmospheric ozone concentration. Many problems beset the use of 'Schönbein papers', but variants of the method of liberating iodine from iodide long remained one of the staples of atmospheric ozone measurements. Schönbein's experiments thus seem to have been the first application of chemical methods to the identification and determination of concentration of a trace component in the atmosphere.

Present-day chemical sensors include gas chromatographs, mass spectrometers, and all the armoury of modern analytical chemistry. One method now of wide application is that of *chemiluminescence*. Certain chemical reactions can release part of their energy in the form of light. Two of these involve atmospheric species. Nitric oxide, NO, reacts with ozone (reaction (5.33)) to emit red light; photoelectric measurement of the intensity emitted in the presence of a known amount of O_3 provides a quick, convenient, and specific way of determining [NO]. The system can be run in reverse, with NO added to the air sample, in order to measure [O_3]. Nitrogen dioxide, NO_2, undergoes a chemiluminescent reaction with the organic compound luminol, this time emitting blue light, so that this process affords a method for determining [NO_2]. A variety of schemes exist for converting NO_2, and other members of the NO_y family, such as HONO, N_2O_5, or PAN, to NO, so that the chemiluminescent detector employing O_3 can provide access to concentrations of all these species.

Free-radical species offer considerable challenges in detection and measurement, although their study is often especially rewarding in terms of interpretation of atmospheric behaviour. We shall consider the special and central case of OH later. HO_2 can be converted to OH by reaction with NO (reaction (5.13)), so that this radical may be detected indirectly by any of the methods used for examining OH. Other techniques for probing HO_2 and also organic peroxy radicals (RO_2) include *matrix isolation–electron-spin resonance* (MIESR), in which the radicals are trapped in D_2O-ice at liquid-nitrogen temperatures, and their electron-spin resonance spectrum obtained. A *chemical amplifier* has also been described for the study of HO_2 and RO_2. In a typical implementation for HO_2, the sequence of reactions

$$HO_2 + NO \rightarrow OH + NO_2 \qquad (5.13)$$

$$OH + CO \rightarrow H + CO_2 \qquad (5.7)$$

$$H + O_2 + M \rightarrow HO_2 + M \qquad (5.9)$$

is used to recycle HO_2, and thus to catalytically convert NO to NO_2. The NO_2 is then determined by, for example, an NO_2-specific chemiluminescence device.

Spectroscopy, in one guise or another, is at the heart of all remote-sensing experiments. It is, however, also very well suited to sensitive and specific *in-situ* applications. Ultraviolet, visible, and infrared regions are all widely used. *Diode-array detectors* are often used for wavelengths shorter than about 1100 nm, since they permit simultaneous examination of the whole of the spectrum of interest, in contrast to the point-by-point data gathering of older scanning spectrometers. At longer wavelengths, *Fourier-transform infrared* (FTIR) spectroscopy serves a similar purpose.

Samples may be collected for off-line analysis in the usual way, but a more direct method is to measure the optical absorption in the atmosphere itself. For this purpose, observations may be made over straight horizontal paths of up to tens of kilometres in length, or the absorption beam can be reflected back to the location of the source, as illustrated schematically in Fig. 5.20. Another application is *lidar* (light detection and ranging) which is the optical analogue of radar; it is also represented in the figure. In lidar, absorption is observed in a laser beam that has been scattered downward by the atmosphere. A short pulse of laser radiation is used, and the scattered light is monitored as a function of time after the pulse. Since the time is proportional to distance travelled, the technique allows vertical concentration profiles to be determined.

One problem that besets optical measurements in the ultraviolet and visible regions is the scattering of the source radiation by both aerosol particles and by molecules (see Section 2.6), especially when long optical paths are employed. A great advance has been achieved in atmospheric studies by the adoption of the technique of *differential absorption optical spectroscopy* (DOAS). The spectrum is obtained in the ordinary way, but it then undergoes additional computer processing. Many small molecules present in the atmosphere exhibit an absorption with sharp and discrete spectral features that result from the electronic, vibrational, and rotational quantum states that are populated. These absorptions are superposed on the spectral characteristics of the source radiation, and altered by atmospheric scattering. The DOAS procedure is to generate a second spectrum by computer that is deliberately degraded; the discrete features are smoothed out, but the general underlying shape is left behind. The difference between the original spectrum and this new one is then the true absorption of the gas-phase atmospheric molecules, which is relatively immune to the existence and characteristics of the interfering scattering. In the ideal situation, where the source output itself contains no sharp features, the procedure also allows for the spectral distribution of the source, and even the spectral response of the spectrometer. DOAS is capable of very considerable sensitivity, and several species can be detected simultaneously with a modern wide-band instrument. Not only can relatively long-lived molecules such as O_3, HCHO, SO_2, HONO, and NO_2 be examined at concentrations down to a few p.p.b., but radical species such as OH, NO_3, and BrO can be detected at much lower concentrations (0.005 p.p.b. for OH).

Measurement of OH concentrations in the atmosphere is particularly important, for reasons that we have already established, and we now return to a discussion of how the radical can be examined. Three methods currently exist for the time-resolved measurement of local hydroxyl radical concentrations. *Laser-induced fluorescence* (LIF) is a modification of the resonance-fluorescence method described in Section 4.4.6, in which lasers are used as the source of radiation resonant in wavelength with the OH ($A^2\Sigma^+ \leftarrow X^2\Pi$) transition at $\lambda \simeq 308$ nm. With continuous-wave (CW) lasers, molecular and aerosol scattering limit the sensitivity, because the scattered light is also at the same wavelength, and tends to mask the true fluorescence. One way round the problem is to use a pulsed laser. The scattered light follows the decay of the laser pulse, while the fluorescence decays with a longer lifetime that, in a vacuum, would be the radiative lifetime of the $A^2\Sigma^+ \leftarrow X^2\Pi$ transition. However, in air at atmospheric pressure, the fluorescence lifetime is much less than the radiative lifetime because of deactivating collisions, and the emission decays almost as fast as any available laser pulse. This difficulty has itself been overcome by expanding the sampled gas through a nozzle from 1 atmosphere (ca. 1 bar) to, say, 1 mbar, thus increasing the fluorescence lifetime, and permitting time discrimination from the exciting pulse. The technique is *fluorescence analysis with gas expansion* and is given the acronym FAGE.

Although the weight of evidence is against it, there exists the possibility of interference with the measurements with high-intensity laser sources. Artificial OH can be produced from atmospheric H_2O and O_3 by the laser pulse itself. However, recent studies suggest that only insignificant amounts of OH are, in reality, created in this way, and comparisons of FAGE measurements with the results of other techniques have also shown excellent agreement.

Long-path absorption spectroscopy of OH has also been remarkably successful, especially when DOAS methods are used. Some results using DOAS will be shown later. In a recent high-resolution instrument, several absorption features of OH in the ultraviolet region are monitored simultaneously in order to overcome interference problems. The method gives absolute [OH] without the need for calibration.

A further technique for determining [OH] relies on ^{14}C isotopic measurements of CO. ^{14}CO is a natural constituent of the atmosphere, and has been used to obtain [OH] as a global average and to calculate its distribution in a more detailed way. Cosmic-ray bombardment of atmospheric nitrogen is the source of new ^{14}C. Natural biospheric processes liberate and consume a small but significant amount of ^{14}CO, but human combustion processes do not, since they burn ancient fossil carbon in which the radioactive ^{14}C component has decayed. The assumption is that, without the variable anthropogenic component, the source strength can be estimated. Losses are largely controlled by the rapid reaction with OH (see Section 5.3.5). The concentrations of ^{14}CO are extremely small, always being less than about 30 molecule cm^{-3}. The derived global mean $[OH] = 6.5 \pm 2.5 \times 10^5$ molecule cm^{-3}, which corresponds reasonably well with the values obtained by other techniques. Although the calculations were first conducted in the early 1980s, a re-evaluation nearly a decade later suggested that the average [OH] derived from the ^{14}CO measurements was unaffected by the improved data for kinetic and other factors, probably as a result of fortuitous cancellation of effects!

5.9.3 Comparison of measurements and model predictions

Figure 5.21 introduces this section by showing the results of one comparison between calculated and measured OH concentrations. The field observations were made at two rural sites in Germany, and the predictions come from a simple box model which, as explained earlier, should be perfectly adequate for a short-lived species such as OH. According to the figure, there is evidently an excellent correlation between the observed and predicted concentrations over an [OH] range that spans a factor of five. In absolute terms, the model overpredicts concentrations by about 20 per cent, but that discrepancy can be readily accommodated by the *combined* uncertainties in all the rate constants that are needed for the reactions in the model. The qualitative and quantitative

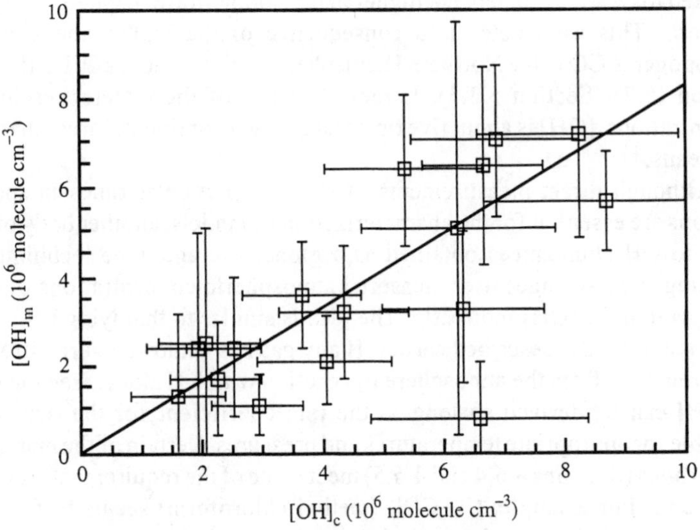

Fig. 5.21. Correlation of measured and observed OH concentrations at two rural sites in Germany. From Poppe, D., Zimmermann, J., and Kuhn, M. in Ebel, A., Friedrich, R., and Rodhe, H. (eds.) *Tropospheric modelling and emission estimation*, Springer-Verlag, Berlin, 1997.

behaviour of the model thus gives considerable confidence in the basic chemical scheme that underlies our current understanding of tropospheric chemistry, and illustrates why measurements of OH are so important in achieving this kind of validation.

Daytime [OH] ranges from 1 to 10 × 10^6 molecule cm^{-3}, in line with the general predictions of realistic models of the troposphere; winter-time values of 1–5 × 10^6 molecule cm^{-3} being realistic, and 5–10 × 10^6 molecule cm^{-3} being generated during the summer. At night, concentrations drop to < 5 × 10^6 molecule cm^{-3}. Recent *in-situ* observations show that, especially in the pollution-free atmosphere, OH concentrations are controlled by O_3 photolysis and the subsequent reaction of $O(^1D)$ with water vapour (see Section 5.3.1). Under more polluted conditions, the presence of NO_x both recycles HO_2 and increases O_3 concentrations; in turn, OH concentrations then become enhanced. Natural and anthropogenic non-methane hydrocarbons (NMHCs) remove OH. On the other hand, intermediate products of NMHC degradation, such as acetone (cf. pp. 351–2), can act as agents for the *production* of OH. Formaldehyde (see reaction (5.16)) and, to a lesser extent, H_2O_2, can act in the same way, and formaldehyde produced in the oxidation of NMHCs could be an important photolytic source of free radicals in the early morning.

Models predict that the highest OH concentrations of all will arise in the tropics, where high humidities and large light intensities lead to rapid photolysis of O_3, and thus rapid production of $O(^1D)$. Concentrations are also expected to be about 20 per cent higher in the Southern Hemisphere than in the Northern. This asymmetry is a consequence of the higher abundance of anthropogenic CO in the Northern Hemisphere, which removes OH in the rapid reaction (5.7) (Section 5.3.5). Largely because of the differences in NO_x concentrations, [OH] is about five times larger over continental areas than over the oceans.

Although direct measurements of [OH] at particular times in specific locations are essential for the characterization of models, another body of data refers to OH abundances obtained as a global average. One technique for obtaining such averages uses measured atmospheric concentrations of some trace gas to infer [OH] indirectly. The idea is similar to that lying behind the use of natural ^{14}CO described earlier. If any gas has a known source strength, and is removed from the atmosphere by reaction with OH alone, then the value of [OH] can be derived so long as the rate coefficient for the reaction is available for appropriate temperatures and pressures. Certain of the man-made halocarbons (Sections 4.6.4 and 4.6.5) meet some of the requirements for a test compound. For example, CH_3CCl_3 (methyl chloroform) seems to fulfil the criteria quite well, the global emission rates probably being known to within 5 per cent. Uncertainties in the absolute atmospheric concentrations of CH_3CCl_3 pose one of the greatest obstacles to accurate derivation of [OH]. According to values accepted at the time of writing, the atmospheric lifetime of CH_3CCl_3 is 4.8 years (cf. Table 4.6), and the corresponding globally averaged [OH] is about 9×10^5 molecule cm^{-3}. The data confirm the model predictions that [OH] is somewhat higher in the Northern Hemisphere than in the Southern. Halocarbons with a shorter lifetime than CH_3CCl_3 potentially provide a more sensitive indicator of latitudinal and seasonal gradients of OH concentration. Chemically suitable halocarbons include 1,2-dichloroethane (CH_2ClCH_2Cl) or tetrachloroethene (CCl_2CCl_2), but unfortunately the absolute and seasonally varying source strengths are too poorly known to allow reliable conclusions about the OH field to be drawn.

One major objective of obtaining globally averaged values of [OH] has been to assess the rate of tropospheric scavenging of man-made pollutants. For example, the determination of the impact of halocarbons on stratospheric ozone (Section 4.6.5) first requires calculation of the atmospheric lifetime, a quantity strongly dependent on OH concentrations. However, the use of a global average of [OH] in the calculations is of dubious validity, first because of the lack of uniformity of the abundances both of the OH itself and of the molecules with which it is to react, and secondly because of differences in temperature dependences of the rates of the reactions being compared. One procedure that has been used to avoid some of these difficulties is to calculate initially a three-dimensional tropospheric OH field that allows for observed trace-gas mixing ratios, and climatological factors such as temperatures and sunlight intensities. Relatively simple chemistry is used, but the predicted and

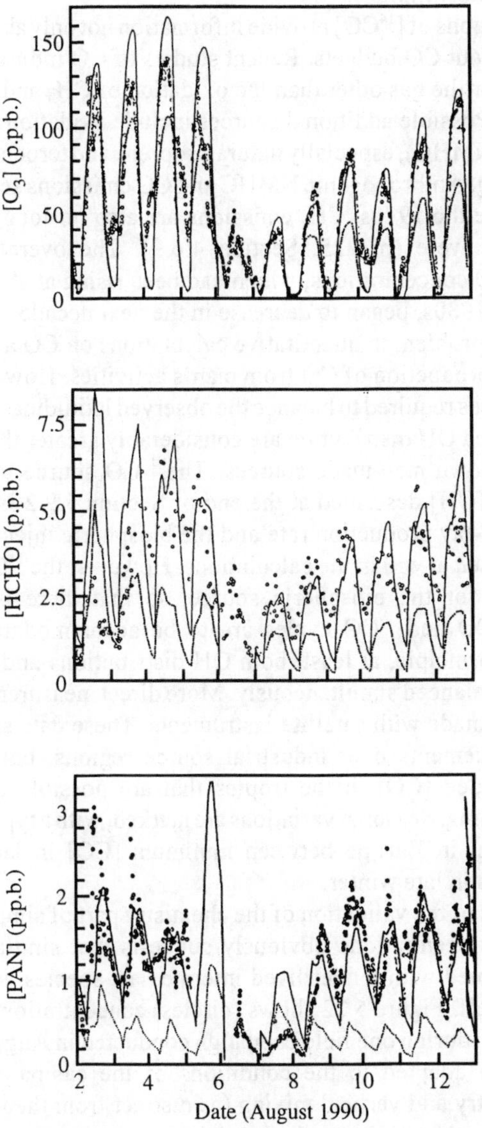

Fig. 5.22. Diurnal cycles of O_3, HCHO, and PAN in a campaign in Germany during August 1990. The circles represent experimental measurements, and the lines are the results of model calculations: the dotted line is from a model with hydrocarbon contributions from anthropogenic sources only, while the solid line includes biogenic isoprene. From Poppe, D., Zimmermann, J., and Kuhn, M., in Ebel, A., Friedrich, R., and Rodhe, H. (eds.) *Tropospheric modelling and emission estimation*, Springer-Verlag, Berlin, 1997.

observed mixing ratios for CH_3CCl_3 are subsequently matched by scaling the absolute OH concentrations.

Determinations of [^{14}CO] provide information not only about tropospheric OH, but also about CO budgets. Recent studies of CO indicate that there are large sources for the gas other than the oxidation of CH_4 and the combustion of fossil fuels. Possible additional sources include oxidation of non-methane hydrocarbons (NMHC), especially natural isoprene and terpenes, together with biomass burning. Anthropogenic NMHC and CO emissions seem to have been decreasing since the 1970s. CH_4 emissions are also no longer increasing as rapidly as they were (p. 225, Section 4.6.9). The overall result is that tropospheric CO concentrations, which had been rising at about one per cent a year until the 1980s, began to decrease in the next decade.

Part of the problem in quantitative calculations on CO arises because of large localized production of CO from man's activities. However, the carbon monoxide releases required to balance the observed latitudinal CO distribution and the calculated OH distribution are considerably greater than the injection rate from all known man-made sources. The ^{14}CO source term used in the calculations of [OH] described at the end of Section 5.9.2 is the sum of the (known) cosmic-ray production rate and the biospheric injection. This latter term can be chosen at will in the calculations. However, the ^{14}C to ^{12}C isotopic ratio is known for the biospheric source, so that there is an additional constraint if ^{14}CO and ^{12}CO cycles are to be accommodated in the same calculation. In principle, at least, both OH distributions and biospheric CO sources can be balanced simultaneously. More direct measurements of the CO field have been made with satellite instruments. These data show clearly the expected enhancements over industrial source regions, but there are also regions of elevated [CO] in the tropics that are possibly associated with agricultural burning. Seasonal variations are marked, with a typical modulation of ± 25 per cent in Europe between minimum [CO] in late summer and maximum values in late winter.

The success of the validation of the chemistry part of simple models with measured OH concentrations obviously suggests that similar comparisons might be performed with longer lived intermediate species, such as HCHO, PAN, and O_3 itself. Figure 5.22 shows (circles) concentrations of these three species measured during one field campaign conducted in August 1990. A 1-D model was then adapted to the conditions of the campaign. This model included chemistry and vertical mixing (as distinct from the box model with chemistry only used for OH); it calculated the concentrations of all short-lived and all secondary pollutants. Local concentrations of anthropogenic compounds were measured and used as input to the model. One interesting conclusion drawn from the calculations was that concentrations of O_3, peroxides, HCHO, and PAN near the ground were almost equally influenced by chemistry and by vertical transport. The results of two sets of calculation are shown as lines on the figure. The dotted line allows for an input of hydrocarbons from anthropogenic sources only, while the solid line includes an estimated biogenic source of isoprene. Although the general diurnal fluctuations of all three compounds are quite well represented by both lines, there is obviously a significantly better fit in time, and a much better fit in concentration, if the isoprene contribution is taken into account. These experiments thus show what an important part biogenic hydrocarbons play in tropospheric chemistry (Section 5.4.2). However, the match between experiment and theory is by no means perfect, and the discrepancies probably point to a need to improve the isoprene degradation scheme that was used, and to quantify the source strengths somewhat better.

Ozone is one of the most important trace gases in the troposphere since it is the ultimate initiator of photochemical chain oxidation (Section 5.3.1). Tropospheric ozone is produced both by the *in-situ* chemistry described in Section 5.3.3 and by transfer from the stratosphere (Section 2.3.3). Model calculations suggest that sources and sinks for ozone within the troposphere are nearly balanced on a global scale, and other estimates indicate that the stratospheric injection is roughly equal to the surface sinks. However, both observations and analysis of regional budgets show that $[O_3]$ is increasing in the Northern Hemisphere, probably because of a shift in the balance of *in-situ* production and loss. The balance depends on the abundance of NO_x in the background atmosphere (see Section 5.3.4). The rate of conversion of NO to NO_2 is also a key factor, and it is directly influenced (Section 5.3.2) by the concentrations of peroxy radicals that are a consequence of oxidation of CO, CH_4, and the non-methane hydrocarbons.

Stratosphere–troposphere exchange (STE: Section 2.3.3) of O_3 adds a very considerable complication in the interpretation of tropospheric ozone concentrations and distributions, because it provides an additional source term that may be difficult to quantify. A few sets of experimental evaluations of the contribution exist. One such study put the global mean flux of O_3 to the stratosphere at $510 \times 10^9\,kg\,yr^{-1}$. However, most estimates of the extent of STE depend not on atmospheric experiment, but on 3-D chemistry–transport models. Most of these models indicate an STE flux of $> 400 \times 10^9\,kg\,yr^{-1}$. The contribution of STE to O_3 in the lower troposphere remains the subject of debate. Some researchers suggest that STE events can sometimes penetrate towards the surface, but most believe that only elevated locations (including some of the high-mountain ozone monitoring sites) experience the direct influence of STE. Nevertheless, after transport of O_3 to the upper troposphere by STE, further downward transport and mixing can bring the stratospheric O_3 to lower levels. Models suggest that, on a global scale, STE can account for about 40 per cent of the total ozone found in the troposphere. In the Southern Hemisphere, up to 45 per cent of the *surface* O_3 may have arrived by STE. In the Northern Hemisphere, the anthropogenic contribution to O_3 generation is sufficiently large to reduce the STE part of the budget to only 30 per cent, Southern. STE-derived ozone makes the largest fractional contribution at

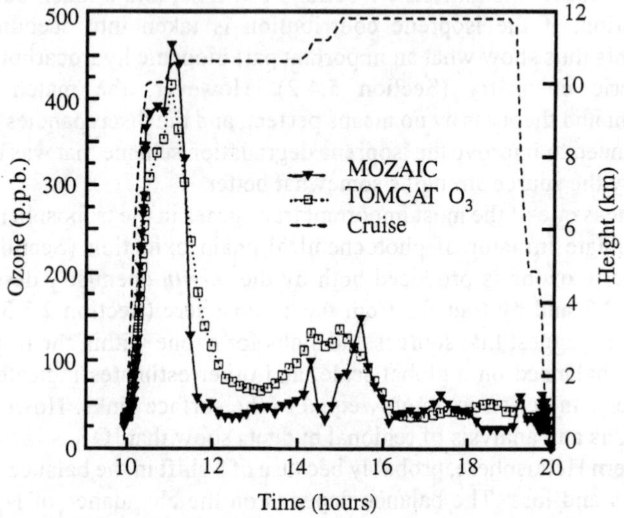

Fig. 5.23. Measured (filled triangles) and modelled (open squares) concentrations of O_3 along a flight track from Paris to Caracas on 3 March 1995. From Law, K.S., Plantevin, P.H., Shallcross, D.E., Rogers, H.L., Pyle, J.A., Grouhel, C., Thouret, V., and Marenco, A. *J. geophys. Res.* **103**, 25721 (1998).

middle and high latitudes during winter; conversely, it has only a very small influence in the tropical troposphere.

Recent 3-D models estimate that 3500–4500 × 10^9 kgyr^{-1} of ozone is generated photochemically within the troposphere itself, and that chemical destruction provides a sink for 100–600 × 10^9 kgyr^{-1} less than the production rate. If STE adds an additional 500 × 10^9 kgyr^{-1}, that means that balance can be maintained only if dry deposition removes 600–1100 × 10^9 kgyr^{-1}. It may even be that there is no true balance between production and loss, although it currently seems difficult to establish if the troposphere is a net source or sink of O_3. As expected, the models show that if NMHC are included in the chemistry, predicted rates of both production and destruction of ozone increase by 30 to 40 per cent, with a corresponding predicted increase in ozone concentration of 15 to 20 per cent (cf. ozone curve in Fig 5.22).

Ozone measurements are a central aim of a new type of experiment that is typified by the MOZAIC (Measurement of Ozone and water vapour by Airbus in-service Aircraft) programme. In these studies, commercial aircraft (in the first instance, Airbus A340s) are suitably instrumented for the measurement of species such as O_3, CO, and NO_y. The aim is to provide the

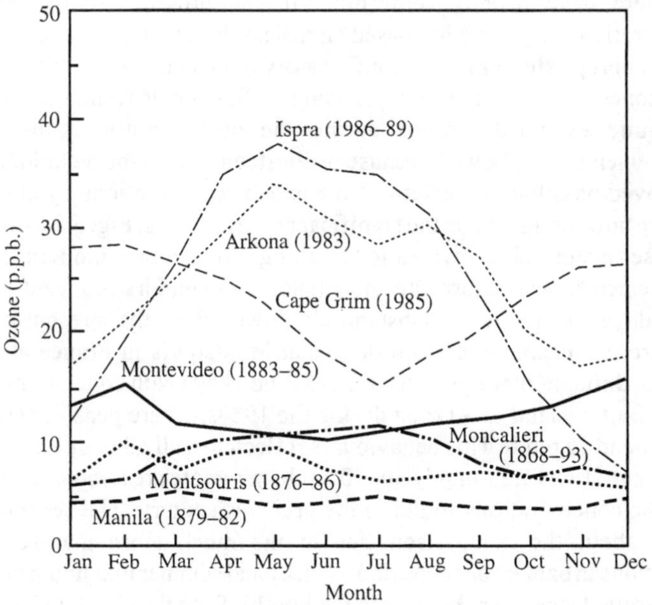

Fig. 5.24. Average monthly concentrations of ozone at several sites in the 19th century compared with others during the 20th century. From Hov, Ø. (ed.) *Chemistry in the atmosphere*, Commission of the European Communities, Brussels, 1993.

database for the validation of chemistry–transport models rather than to detect the direct effects of aircraft emissions on the ozone budget. The data provide detailed information about the 9–12 km altitude region, which is still imperfectly described in existing models. Since the aircraft flights are widely spread geographically, the observations provide a coverage for which it would be virtually impossible to obtain funding for a purely scientific venture. To provide but one example of the application of the MOZAIC studies, Fig. 5.23 shows measured O_3 concentrations (filled diamonds) along a flight track from Paris to Caracas (flight of 3 March 1995). For comparison, the output of a model calculation (TOMCAT: open squares) is shown. As the aircraft left Paris, it entered a low-altitude (ca. 5 km) region of high stratospheric O_3, which was shown to be the result of a stratospheric intrusion bringing O_3-rich air to very low altitudes. The model reproduces this feature well. Later in the flight, a tongue of stratospheric air was encountered, which separates into two separate peaks at times of 1415 and 1530 UT. The model captures the structure, although it appears smeared out, producing an overestimate of O_3. The general agreement between model prediction and experiment is generally excellent, but there is evidently still room for improvement in the prediction of absolute concentrations.

There have been several hints in the present section that ozone concentrations may have increased significantly over the past century, and it is now appropriate to present confirmatory evidence. Several sets of data of ozone concentrations were obtained using the Schönbein technique (pp. 390–1) over quite extended periods during the past century. Although these measurements were flawed because of shortcomings in the technique itself, it has proved possible to recalculate the ozone concentrations by allowing for what are now thought to be the significant interferences. Fig. 5.24 shows some of these historical re-evaluations, along with more modern series of measurements. Two features are immediately evident. First, concentrations one hundred years ago were substantially lower than the present-day ones. Compare, for example, concentrations at Montsouris in France at less than 10 p.p.b. throughout the year in the 1870s and 1880s with concentrations at the remote Baltic island of Arkona during the 1980s, where peak concentrations were around 35 p.p.b. This behaviour is reflected at all other sites, and surface O_3 concentrations are roughly 30–40 p.p.b. in remote areas, some three to four times the concentrations found in the previous century. The second obvious feature about the comparison lies in the much more marked seasonal fluctuations in ozone concentration, with a local summer maximum (June–July in the North, December–January in the South). Both the absolute increase and the emphasized seasonal oscillations can be attributed to enhanced NO_x emissions associated with the combustion of fossil fuels (the NO_x, of course, acting as the photochemical precursor of O_3). The point is an important one, since changes in O_3 are necessarily accompanied by changes in the *oxidizing capacity of the troposphere*. In turn, a changed oxidation capacity has impacts on climate (through the lifetime of CH_4 and the formation of O_3), on photochemical oxidants (through the concentrations and geographical distribution of O_3), on acidification and eutrophication of ecosystems (through the spatial pattern of deposition of sulphur and nitrogen compounds), and on stratospheric ozone concentrations (through the tropospheric lifetimes of biogenic and anthropogenic substances that yield catalysts for ozone destruction in the stratosphere).

We have emphasized, in this section, determinations of [OH], [CO], and [O_3] to illustrate the possibilities and problems of tropospheric measurements. Much effort is being put into the investigation of other trace gases, and the interpretation of secular, seasonal, and geographical trends of concentrations will further our understanding of tropospheric chemistry. Man's agricultural and industrial activities may be altering tropospheric chemistry on a global scale, as concentrations of important trace species, and the tropospheric oxidizing capacity itself, undergo change. It is not yet clear what effects these global changes might have. On a more local scale, the consequences of pollution of the atmosphere by man is all too evident, and it is with such pollution that the rest of this chapter is concerned.

5.10 Air pollution

5.10.1 Clean and polluted air

Air is never perfectly 'clean' in the sense of containing only N_2, O_2, CO_2, H_2O, and the inert gases. Even before man existed, there were many sources of the trace-constituent species, some of which would now be classified as pollutants. 'Natural air pollution' is caused by volcanic eruptions, breaking waves, pollens, dust, and terpenes from plants, windblown dust, and forest fires. Man has accentuated the frequency and intensity of release of some of these pollutants. Agricultural practices, in particular, have been a source of additional pollutant species ever since man has been on Earth. 'Slash-and-burn' clearing of land adds greatly to the fire source; the removal of natural vegetation cover itself increases the rate of dust erosion.

Nitrogen fixed as fertilizer is in part returned to the atmosphere as N_2O, and alterations in the types and utilization of crops also affect the N_2O budget. Increased population of livestock and of rice-paddy cultivation imply an increase in CH_4 production. Destruction of forests, and their replacement by grasslands, may provide suitable habitats for termites that are a potentially significant source of atmospheric methane. Human-accentuated forms of pollution, although significant, may seem minor compared with the releases of man-made pollutants in industrial societies. Man seems capable of releasing —by accident or design—almost all known gaseous species to the atmosphere. Even substances such as lead, not commonly regarded as gaseous, can be produced as aerosols by man's activities. It is not our purpose to catalogue here all the pollutants that have been, or might be, released. Rather, we wish to concentrate on pollution processes which are of rather widespread occurrence, with respect both to geographical location and to frequency. We have already seen, in Chapter 4, that some trace gases whose release is due to man (e.g. CF_2Cl_2), or accentuated by him (e.g. N_2O, CH_4), can have global consequences because of stratospheric interactions. In the troposphere, the most important forms of pollution seem to be related to the combustion of fossil fuels. It is these types of pollution that will provide the emphasis in the remaining sections of this chapter.

Combustion of fossil fuels affects the atmosphere in three main ways: by the formation of carbon dioxide, by the release of substances such as sulphur dioxide and partially oxidized or unburnt fuel, and by the high-temperature fixation of atmospheric N_2 and O_2 to yield oxides of nitrogen. The carbon dioxide problem arises because, in the course of a few hundred years, appreciable fractions of the Earth's fuel deposits have been burnt to release carbon dioxide, thus reversing the chemistry that photosynthesis took millions of times as long to achieve. At present, atmospheric CO_2 concentrations are increasing by about one part in 300 a year, corresponding to about one-fifth of the annual release of carbon dioxide from the combustion of fossil fuels (see

Fig. 1.5). The potential climatic effects of increases in a 'greenhouse' gas are considered in relation to the evolving atmosphere (Chapter 9); we have already seen that the changes in stratospheric temperature consequential on CO_2 increase can affect ozone chemistry (Section 4.6.9). Coal and petroleum products contain appreciable quantities of sulphur compounds (up to 2.5 per cent of coal by weight, and as much as 1.8 per cent in residual fuel oil). Combustion then leads to sulphur dioxide release. Particulate matter (e.g. soot), carbon monoxide, and unburnt hydrocarbons are also important minor by-products of the combustion process. We shall discuss sulphur dioxide chemistry in Section 5.10.4, and examine some consequences of SO_2 release in the following two sections. Sulphur dioxide and the oxides of nitrogen, formed when combustion is supported by air, contribute to the problems of acid rain (Section 5.10.6), and the oxides of nitrogen also play a special part in 'photochemical air pollution' (Section 5.10.7).

Pollutants released to the boundary layer may be quite rapidly removed by wet or dry deposition, and so degrade air quality only near the source. Soot, SO_2, and malodorous compounds emitted from factories often fall into this category, and contribute to *local air pollution*. More widespread dispersal within the troposphere can lead to *regional pollution* of quite large geographical areas. If the entire atmosphere is affected, as is the case for trends discussed at the end of Section 5.9 or for the long-term CO_2 increases, then the pollution is *global* in extent. Attempts to reduce the local impact of released pollutants can sometimes backfire by increasing the impact on the larger scale. For example, by building very tall chimney stacks for factories, local pollution can be much reduced. However, the removal of pollutants at higher altitudes may be a far slower process than near the surface, and on the regional scale the pollution problem may be exacerbated.

5.10.2 Effects of pollution

Air pollution is seen as a growing threat to our welfare, and especially to human health, as ever-increasing emission of contaminants is made into the constant-sized atmosphere. An average adult male gets through about 13.5 kg of air a day, compared with about 1.2 kg of food and 2 kg of water. The quality of the air breathed is therefore at least as important as the cleanliness of our food and water. Air pollutants can exert an influence on health in two ways. First, they may be physiologically toxic, and secondly they may possess a nuisance value. Thus pollution can, by being smelly or dirty, have a health effect far beyond that of simple poisoning.

One of the major difficulties we face in assessing the impact of a pollutant on *human* health concerns the experiments that are possible. It is obviously out of the question to perform clinical tests of human response to any substance at concentrations that could conceivably be toxic. We have, therefore, to fall back either on the effects of accidental exposure of small groups to high doses, or to statistical correlations between mortality or morbidity and (chronic) exposure to low doses: such studies form the basis of *epidemiology*. By far the most hazardous pollution is that inflicted on self (and neighbours) by tobacco smokers, and the difficulties of air-pollution epidemiology are compounded because of this factor. Indeed, the differences in virtually all health indicators are so large as between smokers and non-smokers that it is often impossible to detect whether other forms of air pollution are presenting a health hazard. Cigarette smoke contains, in addition to carcinogens and nicotine, sufficient carbon monoxide to clearly worsen coronary heart disease and respiratory problems. Toxicological trials on animals are of only limited relevance, because we cannot be sure that human responses will match those of the species tested, even after due allowance has been made for the size of the species.

We shall consider the health effects of individual pollutants in subsequent sections, but there are some comments of general applicability which can be made here. The main route of entry into the body of air pollutants is *via* the respiratory system, although some solid material deposited there may subsequently be swept to the gastrointestinal system. Because of the mode of access, respiratory symptoms are the most common response produced by pollutants. Low concentrations of oxidants such as O_3 or NO_2 produce nasal and throat irritation, while slightly higher concentrations impair mechanical lung function. In healthy subjects, the change is reversible within a few hours. So long as the change in function is unaccompanied by symptoms or by decreased work capacity, the exposure probably does not correspond to a real impairment of health. Of course, in persons with underlying respiratory illness, the situation is altered, since the *normal* function may already limit activity. Further increases in irritant concentration (say > 0.3 p.p.m. of ozone) lead to sufficient discomfort to restrict normal activity. Even in the healthy, the effects experienced depend on whether the subjects are exercising or at rest. Presumably, when exercising the expiratory flow rates are increased, and breathing is through the mouth, so that the pollutant is delivered deeper within the respiratory tree.

Particulate matter in the atmosphere can obviously present a challenge to the respiratory system. Recent epidemiological studies suggest that particles of diameter less than 10 μm (identified as PM10) can increase premature mortality by one per cent if their concentration increases by as little as $10\,\mu g\,m^{-3}$. The relation seems to hold in all geographical locations where the studies were conducted, regardless of whether the local pollution was predominantly of SO_2 and primary particles, or of the *precursors* of photochemical oxidants and secondary particles (see Section 5.10.3). Two possible interpretations of these observations are either that there are common toxic or mutagenic species found in most particles, or that inhalation leads to a general inflammatory response independent of the specific composition.

Smaller particles can reach the lungs even more easily, and a PM2.5 standard is under discussion in the USA. The smallest particles are generally those produced in combustion and by gas-to-particle conversion (see Fig. 1.1). If particle size is of great significance in determining the health hazard, then pollution control strategies may have to be examined particularly carefully. For example, there has been a great effort to reduce the emissions of particles from diesel motors, visible as black smoke from older engines. Although improvements in design have led to a substantial reduction in emissions measured as total mass and of numbers of particles in the size range of 0.05 to 1 μm, these reductions may have been gained at the expense of greatly *increased* numbers of emitted particles of diameters less than about 0.05 μm. As a result, the reduction in atmospheric pollutant load (as well as perceived 'dirtiness' of the engine) may be somewhat illusory in terms of benefits to the health of the population.

Living species other than man may be at risk from man-made pollutants. As we shall see in Section 5.10.6, fish seem to be highly sensitive to the pH of their environment, and release of pollutants tending to increase acidity can have serious consequences for freshwater fish populations. Acute and chronic injury to the leaves of trees and plants can be caused by a number of pollutant gases. Sulphur dioxide, and the oxidants O_3 and NO_2, all cause damage: they are said to be *phytotoxic*. Combinations of the gases seem particularly harmful. Economic consequences in terms of lost foodstuffs or forests can be very great indeed. The indirect effects of reduced nitrogen fixation and CO_2 turnover, and of interference with entire ecosystems, could be of even greater long-term importance.

Economic loss can also be sustained by damage to inanimate materials. Many organic substances are susceptible to attack by oxidants such as ozone, especially if they contain double bonds. Elastomers such as natural and synthetic rubber, paints, and dyes are attacked. Chemical measures introduced to prevent damage may themselves constitute a major cost in production. Structural damage can be caused to buildings by species such as sulphur dioxide or sulphuric acid, which attack carbonates present in the stone. Compounds of larger volume are formed, which lead to flaking of the stonework. Disfigurement by blackening of stone surfaces may be a result of simple deposition of particles, but it may also involve a more complex organic process that first requires the absorption of SO_2.

Aerosols reduce visibility in their own right; in addition, particles may act as condensation nuclei to cause or aggravate water fogs. Apart from the 'nuisance' aspect of reduced visibility, it is also evident that limited visual range is a contributing cause of automobile and aircraft accidents.

Release of species such as CO or CH_4, or alteration by man of their natural production rates, seems to exert its main effect in an indirect way. Of course, carbon monoxide is toxic at high concentrations, and may be implicated in traffic accidents at rush-hour periods when urban atmospheric concentrations

Table 5.7 Global man-made and natural emissions of various species

Species	Emission estimate (10^9 kg yr^{-1})	
	Man-made	Natural
CO_2	2.6×10^7	5.5×10^8
CH_4	375	160
CO	1815	430
SO_2	146	~ 15
H_2S	4	1–2
N_2O	16	26
NO_x (as NO)	85	24
NH_3	30	15

Based on tables presented in Seinfeld, J.H and Pandis, S.N. *Atmospheric chemistry and physics*, John Wiley, 1998.

Natural CO_2 emission estimates are assumed to be the primary production from land (2.2×10^{17} kg yr^{-1}) and exchanged from the oceans (3.3×10^{17} kg yr^{-1}).

(up to 85 p.p.m.) exceed the amounts known to degrade behavioural performance. However, average tropospheric concentrations (e.g. ~ 0.2 p.p.m. even in the industrialized Northern Hemisphere) are too low to elicit a direct response. Much more important is the reduction in tropospheric [OH] that is likely to follow increased CO emission rates. Reaction (5.7) between CO and OH (cf. Section 5.3.5) is a major loss process for OH (~ 70 per cent), so that [OH] is very sensitive to [CO]. Since the remaining 30 per cent of OH reacts with CH_4 in the 'clean' atmosphere, and CO is the product of oxidation (Section 5.3.2), increases in atmospheric CH_4 also reduce hydroxyl radical levels in the troposphere. Smaller tropospheric [OH] in turn has consequences for the concentrations of a wide variety of natural and man-made trace species, since reaction with OH is often the principal scavenging mechanism. As we have pointed out frequently, the possible responses then include not only changes in the troposphere but also modifications to stratospheric chemistry and global climate.

5.10.3 Primary and secondary pollutants

Pollutants may be grouped into two categories: primary and secondary. *Primary pollutants* are the chemical species emitted directly from identifiable sources. *Secondary pollutants*, on the other hand, are species formed from the primary pollutants by chemical transformation. Adverse effects of pollution are often associated more with the secondary than with the primary pollutants. For example, although atmospheric sulphur dioxide has itself many harmful effects, the sulphuric acid formed as a secondary pollutant by oxidation of SO_2 is even more damaging to the environment.

Table 5.8 Global emissions of man-made species (1985–1995)

Source	Emission estimate (10^9 kg yr^{-1})						
	Particles	SO_2	N_2O	NO_x (as NO)	CO	CH_4	NMHC
Fuel combustion	5	140 (fuel combustion and industry combined)	–	51	890	100 (fuel combustion and industry combined)	115
Industry	40		3.7	–	425		18
Biomass burning	60	6	1.4	17	500	40	–
Cultivated soils	–	–	10	13	–	60	–
Other	200	–	1.1	4	–	175	10

Based on tables presented in Seinfeld, J.H. and Pandis, S.N. *Atmospheric chemistry and physics*, John Wiley, 1998.

Table 5.7 shows estimates of some man-made emissions. For comparison, a rough idea is given of the natural source strengths for the chemical species. (We note here that different assumptions about natural sources lead to very different values: the figures quoted in Table 5.7 are intended to be consistent with those for the man-made emissions. Man-accentuated processes are not included, so that the total contribution of man's activities to species such as CH_4, CO, or NO, may be much higher.)

Examination of Table 5.7 shows immediately that man releases very large quantities of material to the atmosphere. The units are 10^9 kg yr^{-1}—that is, millions of tonnes annually. For most species, the 'anthropogenic' emissions are a minor, but significant, contribution to the global total budget. In the case of sulphur dioxide, man produces more than the natural sources. Overall sulphur emissions include H_2S, COS, CS_2, and the organic sulphides, as well as SO_2. As a world average, man contributes about 40 per cent, the rest coming from biogenic sources, sea-spray, and volcanoes, but in the industrial Northern Hemisphere man dominates over the sum of natural sulphur emissions.

Table 5.8 gives a more detailed breakdown of the emission sources in order to show the importance of combustion-related processes to atmospheric pollution. Industrial operations include chemical manufacturing, metal smelting and refining, and mineral extraction, and thus contain a combustion element. Even without that contribution, fuel combustion is by far the largest source of the oxidized species (SO_x, NO_x, CO). Transport (especially private cars and light duty vehicles using petroleum fuel) is the single most polluting activity in the USA and other heavily urbanized parts of the world; the quantity of CO and unburnt hydrocarbons emitted reveals something about the efficiency of the internal combustion engine! That the sulphur emission is not higher is a result of desulphurization of the fuel. Motor spirit ('gasoline') contains between 0.026 per cent (USA Premium grade) and 0.040 per cent (UK) of sulphur, compared with several per cent in the crude oil.

5.10.4 Sulphur dioxide chemistry

Sulphur dioxide is present at mixing ratios ranging from 0.02×10^{-9} in the free troposphere of remote areas (e.g. the South Pacific) to more than 10^{-9} (1 p.p.b.) in polluted continental air. Concentrations of several hundred p.p.b. are found in heavily polluted urban environments. The results suggest an atmospheric lifetime measured in days or weeks, and a largely anthropogenic source. Natural contributions to sulphur dioxide are generally a result of oxidation of reduced sulphur compounds as discussed in Section 5.6.

In the troposphere, SO_2 is almost all oxidized to H_2SO_4, in the form of aerosol, and the atmospheric sulphur cycle is closed by wet precipitation of the sulphuric acid. The aerosol particles, once formed, are rapidly incorporated into water droplets, and the particles may, indeed, act as condensation nuclei. Gas-phase, aqueous-phase, and surface oxidation steps all seem to contribute to the overall conversion of SO_2 to H_2SO_4. Meteorological conditions such as the relative humidity and the presence or absence of clouds or fogs are likely to control the relative importance of the homogeneous and heterogeneous (Section 5.8) processes. Enhanced oxidation rates at high relative humidities or when condensed water is present indicate the participation of liquid-phase reactions. Rates of wet and dry deposition of SO_2 itself are five to ten times faster than the rate of homogeneous oxidation in the boundary layer, thus suggesting that heterogeneous conversion of SO_2 to H_2SO_4 must normally constitute the major oxidation path. Some reaction may occur on solid aerosol surfaces, but within a cloud most of the oxidation takes place in the liquid phase. However, there are uncertainties about the rates for many of the reactions possible in the gas and liquid phases. In highly polluted environments, species present other than sulphur compounds may act as oxidants, and various metals can act as catalysts. Quantitative estimates of the contributions of the different routes to SO_2 oxidation are not, therefore, available. Instead, we must examine the processes most likely to effect homogeneous and heterogeneous oxidation.

Photodissociation of SO_2 has a threshold wavelength of about 210 nm, corresponding to the O–SO bond energy of ca. 548 kJ mol^{-1}. Radiation of this wavelength does not penetrate to the troposphere, and the process cannot, therefore, play a part in the tropospheric oxidation of SO_2. Significant quantities of sunlight are absorbed to generate excited singlet and triplet states, but physical deactivation is the major fate, and direct photo-oxidation of gaseous SO_2 to H_2SO_4 in air is negligible. The ubiquitous hydroxyl radical seems once again to initiate the dominant oxidation route by addition to SO_2

$$OH + SO_2 + M \rightarrow HOSO_2 + M. \qquad (5.102)$$

The $HOSO_2$ radical is further oxidized to H_2SO_4, although the mechanism is probably more complex than just the addition of a further OH radical. Kinetic studies of the OH–SO_2 reaction, using long-path Fourier transform infrared spectroscopy to follow chemical changes, have begun to help in further elucidation of the oxidation mechanism. In irradiated mixtures of HONO (source of OH), CO, SO_2, NO_x and O_2/N_2, the [OH] is insensitive to the concentration of SO_2 in the mixture, suggesting that SO_2 termination of HO–HO_2-chain reactions is unimportant. The overall reaction seems most nearly represented by the equation

$$OH + SO_2 (+ O_2, H_2O) \rightarrow H_2SO_4 + HO_2. \qquad (5.103)$$

Other experiments have shown that aerosol can be formed when OH reacts with SO_2 in the presence of O_2 and H_2O. The direct reaction of O_2 with $HOSO_2$

$$O_2 + HOSO_2 \rightarrow HO_2 + SO_3 \qquad (5.104)$$

would be followed by rapid hydrolysis of SO_3 to yield H_2SO_4. Reaction (5.104) is probably slightly endothermic, so it may be that O_2 adds to hydrated $HOSO_2$ radicals, $HOSO_2.H_2O$, the enthalpy of hydration then being available to promote reaction to the final products, H_2SO_4 + HO_2. Experimental determinations of the rate constant for reaction (5.104) suggest that the lifetime of the $HOSO_2$ adduct would be only 0.5 μs at the Earth's surface, so that regardless of whether hydration of the adduct occurs before reaction, H_2SO_4 formation seems assured following addition of OH to SO_2. An interesting consequence of HO_2 formation is that the oxidation is catalytic in the presence of NO, since OH is regenerated in the process

$$HO_2 + NO \rightarrow OH + NO_2. \qquad (5.13)$$

As an alternative to the exchange reaction with O_2, it has been suggested that the oxidation of $HOSO_2$ may first involve addition of O_2 to form HSO_5 ($HOSO_2O_2$), which has some of the properties of other peroxy radicals, but which may also become hydrated with n water molecules

$$HOSO_2 + O_2 \overset{M}{\rightleftharpoons} HSO_5 \qquad (5.105)$$

$$HSO_5 + nH_2O \rightarrow HSO_5(H_2O)_n. \qquad (5.106)$$

Hydrated HSO_5 is seen as a strong oxidizing agent that can convert SO_2 to SO_3 in a reaction that is described as 'quasi-heterogeneous' because, with large n, the hydrated radicals are virtually aerosol particles. Sulphuric acid formation is possible, as indicated by the equation

$$HSO_5(H_2O)_n + SO_2 \rightarrow HSO_4(H_2O)_nSO_3$$
$$\rightarrow HSO_4(H_2O)_{n-m} + H_2SO_4(H_2O)_{m-1}, \qquad (5.107)$$

m of the original n molecules of water being used to hydrolyse and hydrate the

SO_3. Reactions analogous to (5.107) can also oxidize NO and HO_2. Nitric acid is a potential product of the reaction with NO

$$HSO_5(H_2O)_n + NO \rightarrow HSO_4(H_2O)_nNO_2$$
$$\rightarrow H_2SO_4(H_2O)_{n-m} + HNO_3(H_2O)_{m-1}, \qquad (5.108)$$

a matter of considerable interest since nitric and sulphuric acids are frequently found together in polluted environments (see Section 5.10.6). The hydrated HSO_4 radical which appears as a product in reaction (5.107) is likely to be a highly reactive species that could react with itself, or with HO_2 or NO_2.

Peroxy radicals may play a minor role in the atmospheric oxidation of SO_2. HO_2 itself reacts rather slowly with SO_2, but the reaction of methylperoxy (CH_3O_2) is relatively rapid, and in the boundary layer, especially of polluted atmospheres, oxidation *via* this route may be significant. Ozone cannot react significantly with SO_2 *in the gas phase*. However, ozone reacts with alkenes to produce 'Criegee biradicals' (Section 5.3.7) such as CH_3CHOO (formed from propene) that are thought to oxidize SO_2

$$CH_3CHOO + SO_2 \rightarrow CH_3CHO + SO_3. \qquad (5.109)$$

These reactions cannot be a major route for SO_2 oxidation under most atmospheric conditions, but they are distinguished from many other SO_2 oxidation mechanisms in that they can occur at night.

We consider now the oxidation from the S(IV) oxidation state of SO_2 to S(VI) of SO_4^{2-} in the aqueous droplets that constitute aerosols, clouds, fogs, and rain. Solution of SO_2 in water proceeds first through hydration of the neutral SO_2 and then by ionization to yield HSO_3^- and SO_3^{2-} ions

$$SO_2 + H_2O \rightleftharpoons H^+ + HSO_3^-, \qquad (5.110)$$
$$HSO_3^- \rightleftharpoons H^+ + SO_3^{2-}. \qquad (5.111)$$

These ions are finally oxidized to SO_4^{2-}. Hydrogen peroxide and ozone dissolved from the gas phase both oxidize HSO_3^-. The reaction with ozone probably proceeds *via* an ionic mechanism that can be expressed as

$$HSO_3^- + OH^- + O_3 \rightarrow SO_4^{2-} + H_2O + O_2. \qquad (5.112)$$

The rate constant for this reaction increases with increasing pH, as does the solubility of the S(IV) reactants, so that oxidation by ozone is probably important only for pH > 4.5. Oxidation of S(IV) by hydrogen peroxide, on the other hand, has a rate coefficient that decreases with increasing pH, which offsets the increasing solubility of S(IV), and makes the production of S(VI) only slightly dependent on pH. The great solubility of H_2O_2 in water droplets means that equilibrium concentrations of H_2O_2 in solution would be about six orders of magnitude greater than those of O_3 expected for atmospheric conditions, thus making H_2O_2 a plausible oxidant for S(IV) in solution.

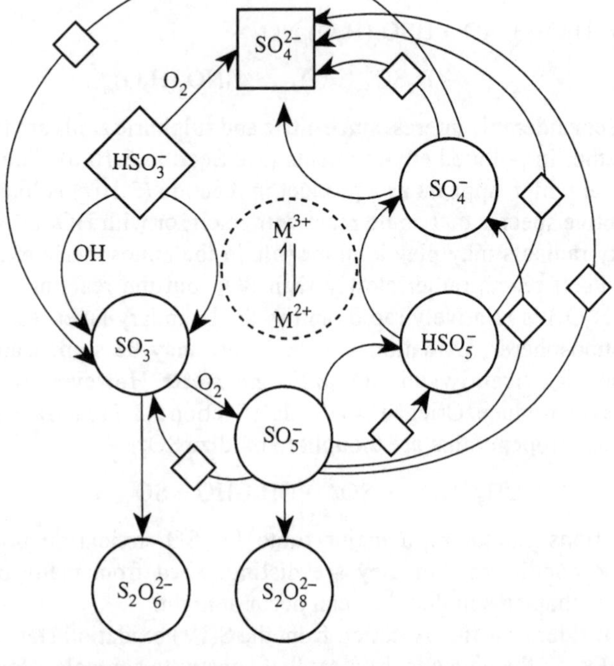

Fig. 5.25. Generalized scheme for the oxidation of sulphur (IV) in aqueous solution. Reaction pathways marked with a diamond shape consume HSO_3^- as one of the reactants. The dotted circle and arrow links in the centre represent the catalytic pathways opened by the presence of transition metal ions. Based on a diagram presented by Warneck, P. (ed.) *Heterogeneous and liquid phase processes*, Springer-Verlag, Berlin, 1996.

Molecular oxygen plays a more equivocal role, since it is not clear if any oxidation occurs at all in the absence of catalysts; several metals have, however, been shown to promote the reaction.

In evaluating different routes for the oxidation of S(IV) to S(VI), consideration must always be given to the availability of sufficient oxidant to effect the conversion. For example, in urban environments, where SO_2 concentrations can be around 100 p.p.b., it seems that, however *efficient* an oxidant H_2O_2 might be, it is almost certainly present in quantities only enough to oxidize a small fraction of the SO_2. Even O_3 concentrations might be inadequate for the purpose. It is in this respect that chain processes, such as the sequence of gas-phase reactions (5.103) + (5.104) + (5.13), look especially attractive. Since cloud droplets enhance the rate of oxidation, an obvious step is to look for similar cyclic processes in the aqueous phase. Free radicals in the liquid phase, such as OH and HO_2, may be important in promoting the reactions. The radicals can be scavenged from the gas phase, or formed by reactions in the droplets, as described in Section 5.8. Aqueous OH reacts with HSO_3^- and SO_3^{2-} to yield SO_4^{2-}; intermediates such as SO_3^-, SO_4^-, and SO_5^- are believed to be involved. As before, the first step is the formation of bisulphite (HSO_3^-) ions in reaction (5.110). A simplified mechanism can then be written in the form

$$OH + HSO_3^- \rightarrow SO_3^- + H_2O \qquad (5.113)$$

$$SO_3^- + O_2 \rightarrow SO_5^- \qquad (5.114)$$

$$SO_5^- + HSO_3^- \rightarrow SO_4^{2-} + SO_4^- + H^+ \qquad (5.115)$$

$$SO_4^- + HSO_3^- \rightarrow SO_4^{2-} + SO_3^- + H^+ \qquad (5.116)$$

$$SO_5^- + SO_5^- \rightarrow 2SO_4^- + O_2 \qquad (5.117)$$

in which the radical-ion intermediates participate. Reactions (5.114), (5.115), and (5.116) constitute a first chain, in which two HSO_3^- ions (S(IV)) are converted to two SO_4^{2-} ions (S(VI)). Inclusion of reaction (5.117) affords yet another way to close a loop. In reality, further reactions involving the species HSO_5^-, formed in the reaction

$$SO_5^- + HSO_3^- \rightarrow HSO_5^- + SO_3^- \qquad (5.118)$$

must also be included, and an entirely similar set of reactions occurs with SO_3^{2-} instead of HSO_3^- ions.

Figure 5.25 is an attempt to present diagrammatically this rather complex chemistry. Ignore initially the large dashed circle in the middle of the diagram. The diamond shape contains the starting S(IV) compound, HSO_3^-, while the square contains the oxidized S(VI) product, SO_4^{2-}; the circles contain the various intermediates, as well as two other terminal ions, $S_2O_6^{2-}$ and $S_2O_8^{2-}$. The lines are the available chemical pathways. Several cyclic paths can be constructed, although a critical link seems to be between SO_4^- and SO_3^- ions, corresponding to reaction (5.116). A diamond shape on a line means that SO_3^- is the co-reactant in the process, so that each of the seven reactions so marked allows yet another S(IV) compound to enter the oxidation chain.

Much evidence has accumulated to show that transition metals catalyse the oxidation of S(IV) compounds, and such metals are often abundant in atmospheric particles present in just those polluted environments where SO_2 concentrations are also high. Catalytic pathways in the radical-ion oxidation mechanism have now been discovered, and they are indicated schematically as the inner part of Fig. 5.25, where the variable-valence metal ions are shown within the dotted circle, and the arrows show the species linked through the catalyst. The first step of the uncatalysed scheme, reaction (5.113) may be replaced by the step

$$FeOH^{2+} + SO_3^- \rightarrow [Fe(OH)HSO_3]^+ \rightarrow Fe^{2+} + SO_3^- + H_2O \qquad (5.119)$$

and the pair of reactions

$$Fe^{2+} + SO_5^- + H_2O \rightarrow FeOH^{2+} + HSO_5^- \quad (5.120)$$

$$Fe^{2+} + HSO_5^- \rightarrow FeOH^{2+} + SO_4^- \quad (5.121)$$

both supplement reaction (5.115) and regenerate the catalyst $FeOH^{2+}$. One most important feature of this scheme is that the initial formation of SO_3^- no longer requires the presence of dissolved OH radicals.

Transition metals other than iron are also apparently effective catalysts. Manganese participates in an oxidation mechanism very similar to that proposed for iron, and cobalt and copper exhibit catalytic activity. It is therefore tempting to conclude that all one-electron metal oxidants operate similarly, but that hypothesis remains unproven at present. The common feature is that a metal ion or complex in a higher oxidation state is needed to initiate the formation of the SO_3^- radical-ion.

In the real atmosphere, a combination of the different proposed mechanisms may bring about the oxidation of SO_2, and each may be of dominant importance under different conditions. Whatever the oxidation route, and be it gas-phase or aqueous-phase, H^+ ions become dissolved in cloud water and lower its pH. Precipitation thus brings with it acidity as well as dissolved sulphate. Any SO_2 that is not oxidized remains to exert its own irritant and phytotoxic effects. It is clear that SO_2 is a particularly noxious pollutant, and effects associated with it are discussed in the next two sections.

5.10.5 Smoke and sulphur pollution

Air pollution from the burning of coal has been a problem for centuries. At the beginning of the fourteenth century, Edward I forbade the use of coal because of the smell and smoke it produced: it has been said that a violator of that law was executed![a] London, although not unique in suffering from the combination of smoke and SO_2 pollution produced by coal combustion, has had severe problems well into the twentieth century. London 'pea-souper' fogs were a recurrent nightmare in Victorian London (of course, the Hollywood version includes Sherlock Holmes and Jack the Ripper prowling through the nearly impenetrable gloom). The word *smog* was coined in 1905 to describe the combination of smoke and fog that was so disastrous.[b] Almost all heavily industrialized cities suffered to some extent. Probably the special place of London in smog history arose because of the size of the city and the scale of industrialization at a time when possible control measures were unknown.

[a] Professor Peter Brimblecombe, who has made a special study of historical records of pollution, thinks that this story is an eighteenth- or nineteenth-century exaggeration. Contemporary documents hint at 'grievous ransoms', which probably means a stiff fine.

[b] *Smog* is now used to describe any smoky or hazy pollution of the atmosphere, and includes the conditions encountered in Los Angeles (Section 5.10.7). A suitable qualifier, such as 'London', 'classical', or 'Los Angeles, 'photochemical', is often used to provide greater clarity of expression.

British bituminous coal is high in sulphur content, and the tars and hydrocarbons make for a high smoke yield. Such a combination is dramatically effective in fog nucleation in a climate already humid and possibly supersaturated. In 1952, a tragic air pollution episode occurred in London, as a result of which more than 4000 people died.

The deaths that followed acute smog episodes usually involved those with pre-existing heart and respiratory problems, primarily the elderly. It is recognized, however, that human health can suffer as a result of slower subtle effects, especially on the respiratory system, produced at lower air pollution levels. Sulphur dioxide is itself a respiratory irritant, the effects appearing at concentrations above about 1 p.p.m., substantially larger than the largest *now* recorded, even for highly industrialized urban areas. Below about 25 p.p.m., irritation is confined to the upper respiratory tract. The situation is greatly altered if particles (e.g. soot) are also present in the pollution. The lower part of the respiratory tract may then be involved at the lower concentration levels. A three- or fourfold potentiation of the irritant response to SO_2 results from the presence of particulate matter. Chronic and acute bronchitis, pleurisy, and emphysema are all produced by SO_2-containing smoke such as that generated in the combustion of bituminous coal. Part of the potentiation may be caused by the delivery and retention in the respiratory system of substances, including SO_2, absorbed on soot particles. In addition, as we saw in Section 5.10.4, SO_2 oxidation to H_2SO_4 (or SO_3) occurs efficiently on aerosol surfaces; in heavily polluted atmospheres, the presence of metals may enhance catalytic effects.

Shortly after the 1952 London smog disaster, decisive action was taken in Britain to alleviate pollution. Controls were placed on the type of fuel burnt and the kinds of smoke that might be emitted. A ban on all but 'smokeless' fuels in urban areas has been particularly effective in reducing the emissions of particulate matter, if not of sulphur dioxide. Although there were a few further serious episodes in the period immediately following the legislation, pollution disasters related to smog now appear to have become a thing of the past in London, and there seems to be a continuing downward trend in emissions. During the period 1974–1994, SO_2 concentrations measured in Stepney (E. London) decreased by a factor of roughly six, and smoke by a factor as large as 10. Most other cities and regions report a similar success with smoke control. For example, averaged estimated aggregate emissions of SO_2 in the United States and Canada fell about 12 per cent from 1980 to 1982, remained roughly level for a decade, and then fell approximately 15 per cent in 1992–1995. However, as the coal sources of SO_2 and smoke has diminished, emissions of pollutants, including particles, from road traffic have increased. Catalytic converters (see Section 5.10.7) have greatly decreased the emissions of gaseous species from motor vehicles, but the increased market share of diesel cars—which also has a beneficial effect on gas-phase emissions—may have led to less improvement in the emissions of particles. There is the further complication of the influence of particle size on health effects (see pp. 405–6).

Despite the great improvements made in emissions in North America and Europe, in particular, sulphur dioxide remains a problem, as the emission inventories of Tables 5.7 and 5.8 show. Developing nations may now be major contributors to the atmospheric load of SO_2 (and other pollutants), and the situation might well worsen. The most serious consequences may be due in part to the acidity of precipitation, and we turn now to this question.

5.10.6 Acid rain

Natural precipitation—rain and snow—is slightly acid, in part because carbon dioxide is dissolved in the falling droplets. Indeed, the acidity forms part of the geochemical weathering cycle, as discussed in Chapter 1. However, 'carbonic acid' is a weak acid, so that the pH in water saturated with CO_2 is limited to minimum values of ~ 5.6. With the presence of other naturally generated acidic gases, such as NO_x and SO_2, and in the absence of any neutralizing ions, the pH of natural rainwater is probably about 5.0, an estimate supported by examination of rain in remote and unpolluted regions. Over the last few decades, rainwater of much greater acidity (lower pH) has been of widespread occurrence in industrial areas, and the acids involved have mostly been 'strong' ones such as sulphuric, nitric, and hydrochloric. Acidic rain is potentially damaging to the environment, the two most serious influences appearing to be on freshwater fish and on forest ecology. Strong acids such as H_2SO_4 and HNO_3 have their origin in gaseous SO_2 and NO_2, while HCl may be produced by the reaction of H_2SO_4 with atmospheric NaCl of marine origin. We shall show shortly that most of the load of strong acid is a consequence of fossil fuel combustion.

Two regions have been particularly badly affected by the acid rain problem. They are the north-eastern United States and neighbouring parts of Canada, and the Scandinavian countries, particularly Sweden and southern Norway. 'Fossil' precipitation is sometimes preserved as glacial ice, and it indicates that the pH was generally greater than 5 outside urban areas prior to about 1930. Direct records in north-west Europe have been kept for five decades now, and show that precipitation has become increasingly acid and that this acidity is more widespread geographically. In some parts of Scandinavia, the H^+ concentration in precipitation has increased by a factor of more than 200 over the last two decades. Figure 5.26 is a map of the global pattern of precipitation pH as it appeared at the end of the 1980s, and it exhibits clearly the widespread nature of rainwater acidification, as well as the especially high acidity in the two specific regions just mentioned. Average pH levels in the north-eastern United States are now between 4.0 and 4.2, but values as low as 2.1 have been recorded for individual storms. The greatest increase in acidity of precipitation in the USA appears to have taken place some time between 1930 and 1950. On the other hand, acid rain is no new phenomenon. In the early part of this century, acidity was recognized in the

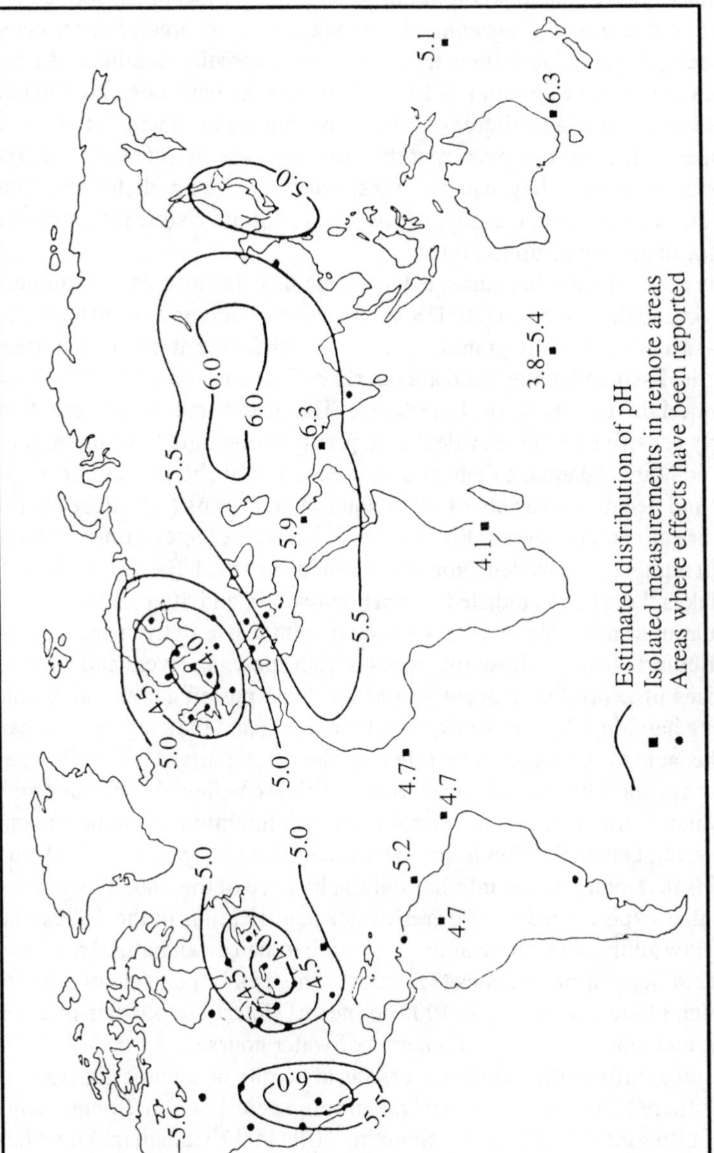

Fig. 5.26. Global pattern of acidity of precipitation. Redrawn from Seinfeld, J.H. and Pandis, S.N. *Atmospheric chemistry and physics*, John Wiley, Chichester, 1998.

rainwater of the industrial northern cities of England. What is new is the commonplace nature of acid precipitation in regionally widespread and remote areas. The increase and geographical extension of acid precipitation came during a period when local industrial pollution generally declined. As we pointed out at the end of Section 5.10.1, smoke-stacks have been heightened to disperse pollutants and thereby reduce concentrations locally at ground level. Dispersal means, however, that the pollutants are delivered to the free troposphere, in which they can be transported over long distances. One infamous stack in Ontario, Canada, is over 400 m tall: its visible plume can be detected for distances of up to 200 km.

Let us now consider the damage done by acidity. Large areas of southern Scandinavia and the north-eastern USA—those very regions most affected by acid rain—are underlain by granite-type rocks. Surface waters in such areas contain little dissolved matter, and are poorly buffered (in distinction to those waters underlain by chalk or limestone). The lakes and rivers are thus particularly sensitive to the prevalent acid precipitation, and have undergone extensive ecological damage. Fish are sensitive to acidity, both as a direct toxic response, and because a variety of other aquatic organisms (e.g. algae) in the food web are adversely altered. Freshwater fish have become extinct, or have declined in number, in Sweden, Norway, Canada, and the USA. About 10 000 Swedish lakes have been acidified to a pH below 6.0, and 5000 to below 5.0. Fish populations have been seriously affected, with losses of trout and salmon being particularly heavy. Those two types of fish spawn in rivers and streams where pulses of acidity may occur just at the same period as the vulnerable stage of egg hatching. Ion separation in the freezing and thawing process can lead to the acidity being concentrated in the first portion of melt-water liberated in spring. The direct toxic effects on fish are believed to be due to the low pH by itself, and to the enhanced solubility of aluminium from the stream-bed and the adjacent soil. Aluminium salts accumulate on specialized cells on the fish gills that maintain the internal sodium balance of the fish. Many of the countries that experienced severe freshwater acidification in the 1970s and 1980s are now adding limestone or dolomite rather than await natural recovery of pH as acid deposition declines (see later). This is a large-scale operation, with Sweden alone dispersing 200 000 tonnes of limestone powder in some 7500 lakes and many thousand kilometres of water courses.

Additional information about the effects of acidity on aquatic ecosystems has been gathered from an area of natural airborne acidification. Spontaneous burning of bituminous shales at the Smoking Hills in the Canadian Arctic has been in progress for at least several hundreds, and perhaps thousands, of years. Sulphur dioxide, and H_2SO_4 aerosol, are released, the pH of summer rainstorms being as low as 2.0. Of the ponds studied, 30 per cent have a pH in the range 2.5 to 3.5, so that the region provides a valuable opportunity of assessing biological and chemical responses to long-term acid inputs. Other ponds in the region, buffered by the HCO_3^-/CO_2 system, have a water content of much higher pH (6.5 to 10.5), and represent the normal Arctic environment. One major conclusion of the studies is that the acid ponds have become barren, and have then been repopulated. The biota of the alkaline ponds are typically Arctic in character, while those of the acidic ponds resemble the biota found world-wide in highly acidic waters. The acid-dwelling biota are not a subset of the much more diverse biota of alkaline ponds, but rather seem to be pre-adapted invaders.

The other effect of acid rain that we shall consider is that on land vegetation. Interpretation of response is difficult because the gaseous pollutants are themselves phytotoxic. Soil may also be affected, *via* leaching of inorganic ions, reduced nitrogen availability, and decreased soil respiration. Conversely, acid rain may actually supply needed sulphur or nitrogen to soils deficient in these species. Adverse effects can therefore be masked by nutritional benefits. Forest damage, in particular, has given rise to particular concern, above all in Germany, where the possibility of acid rain leading to forest decline excited strong emotions. That this decline is caused mainly by acid precipitation, or the gaseous pollutants, acting alone or in combination, looks increasingly unlikely.

Gaseous pollutants such as SO_2, NO_x, or O_3 at high concentrations have long been recognized as having an adverse effect on plants. However, concentrations of SO_2 and NO_x now found even in urban areas of countries such as the UK or the USA seem too low to exact any large-scale damage to crops or forests. The situation for O_3 is more complex. Concentrations have been increasing over recent decades, and peak concentrations are encountered during summer, and thus during the growing season. Ozone pollution thus has the potential for an adverse economic impact on commercial crops. A mean exposure to 50 p.p.b. of ozone—well within the concentrations sometimes experienced in the UK during summer—can reduce crop yield by 10 per cent. In California, hourly-mean ozone concentrations can reach 300 p.p.b., and O_3 damage is thought to lead to agricultural losses of up to 56 per cent, and to changes in the coastal sage scrub.

Acid precipitation must obviously be distinguished from the gaseous pollutants and, indeed, it does exert its influence on plants in different ways, having direct physical and chemical effects at the leaf surface. Whether the deposition is rain or snow, fog or mist, is also important in determining the impact. Acids themselves can leach nutrients from plant foliage, accelerate cuticular erosion of the leaves, and directly damage the leaves at pH values below 3.5. Germination of seeds is reduced, and the establishment of seedlings made less secure under acid conditions. Finally, other organisms associated with the plant systems (e.g. pathogens) may change in quantity and type in response to increased acidity. In describing the ways in which acid rain could damage plants, it is important not to lose sight of the *beneficial* effects of the components of rain. Sulphate, for example, is essential for plant growth and metabolism, and is not phytotoxic. Indeed, in sulphur deficient regions—and parts of the UK may possess this deficiency—sulphur deposition may be a desirable contributor to plant nutrition, and current pollution control measures may require compensatory increased use of fertilizers for commercial crops.

The initial concern about forest decline followed reports of a disease that yellowed the foliage of trees, and caused its premature loss. Some commentators sought to connect these observations with increased acidity of precipitation. Extensive surveys subsequently showed that the cumulative forest area that could be considered damaged was around 20 per cent in Germany. Other nations in Europe, and the USA, possess forests exhibiting various degrees of damage, but only some symptoms are found in common, and mostly seem to be general rather than specific to any one cause. Perhaps the most widely recorded kind of damage is *chlorosis*, which is associated with magnesium deficiency and consequential impairment of photosynthesis and carbohydrate translocation. Low soil magnesium levels quite probably reflect intensive forestry practice, but they can obviously be exacerbated by leaching of minerals from the soil by acid rain. The effects can be remedied by application of neutralizing or magnesium-containing fertilizers. Enhanced mineral leaching may, in fact, lie behind the apparent implication of the acidification of soil in damage to trees. As it turns out, predictions of forest decline of progressive severity and geographical spread have not been fulfilled. There are even signs of forest recovery in Europe. The majority of forests in North America seem healthy, and acid rain is now considered in the USA as a minor factor affecting the health and productivity of forests (there is, it seems, a contradictory observation concerning red spruce at one high altitude site).

One important question to be answered is whether or not the increased acidity in precipitation is causally connected with combustion-related releases of SO_2 and NO_x. An alternative possibility, which would require quite different control strategies, would be that changes in biogenic sources (perhaps related to fertilizer use) had arisen over the last 20 to 30 years. On the global scale, man contributes more of both SO_2 and NO_x than nature (see Table 5.7). Several pieces of evidence plainly implicate combustion of fossil fuels in the generation of acidity. Potential sources of biogenic sulphur emission, such as the salt marshes along the eastern seaboard of the USA, or in the Baltic sea, generally lie downwind of the areas experiencing highest acidity. Major industrial areas lie upwind, and there is a high correlation between low pH in precipitation and storm tracks that have passed over large emission sources.

Various types of 'tracer' experiment bear out the idea that the acids, or their oxidic precursors, are transported over long distances. Sulphur hexafluoride (SF_6) has been injected into the waste gases leaving power station chimneys, and its dispersion followed by an instrumented aircraft. Release from the east coast of England can be followed well on the way to Scandinavia. Detection of the element vanadium in Arctic air suggests as a source the burning of fuel oil, which is rich in vanadium, in middle latitudes. Manganese is also found, and this element is released by metal processing and coal burning. Measurements of the two elements together pinpoint the source of the metals in the Arctic as the central (former) Soviet Union and western Europe. Similar analysis of the [Mn]/[V] 'signature' for the north-eastern USA indicates sources in the Midwest, where the ratio is high (1–10), rather than the east, where less coal and more oil is burned (ratio 0.1–0.2). Caution must be exercised in extending these data to acid precipitation, since the metals leave the air as deposited aerosols, while the acidic components remain as gases. None the less, trace element analysis, perhaps utilizing isotopic abundances of arsenic and selenium, may prove to be a valuable method for identifying distant pollution sources.

The data presented clearly suggest anthropogenic sources of pollutants, coupled with quite long-range transport, as the cause of acid precipitation. Meteorological factors may lead to deposition, in a small area, of pollutants picked up over widespread densely urbanized and industrialized areas. Such a situation is thought to exist in Scandinavia, where the surface features favour precipitation of moisture from the air. More than 70 per cent of the sulphur in the atmosphere over Sweden is thought to be anthropogenic, and, of this load, 77 per cent may originate outside Sweden. Britain and the Ruhr Valley, together with some former Eastern Bloc countries, are cited as the sources of the foreign sulphur and the associated increase in acid precipitation.

Most of the acidity in the affected rainwater used to be due to H_2SO_4, but the relative concentrations of HNO_3 have increased dramatically over the last few decades. For example, in the period 1955–1985, $[NO_3^-]/[SO_4^{2-}]$ in rainwater at a site in central England increased fairly steadily from 0.2 to 0.7, the change being brought about both by a reduction in SO_4^{2-} and an increase in NO_3^-. SO_2 emissions continue to decrease in the industrialized nations, but perhaps not elsewhere, as explained at the end of Section 5.10.5. After the passage of the Clean Air Act in the USA in 1970, emissions of SO_2 and particulate matter were reduced, and the concentrations of SO_4^{2-} in surface waters in eastern North America declined. However, the relation between SO_2 emission and acidification is apparently complex, because surface waters throughout the region that are low in acid-neutralizing capacity (ANC) have not seen a commensurate recovery in pH or in ANC. Further reductions in emissions and in acid deposition are following 1990 amendments to the Clean Air Act, but the measurements suggest that even if the acidification of ecosystems is reversible, recovery may be very slow. There is thus a major policy implication, since it now seems questionable if the amendments will be adequate to protect surface waters and forest soils against further anthropogenic acidification.

While on the subject of reduction of pollution, control measures, and policy-making, it may be appropriate here to introduce the concept of *critical load*. In the context of acid rain, the critical load may be defined as 'the highest load that will not bring about chemical changes that would produce long-term harmful effects on the most sensitive ecological systems'. It is expressed as a quantity of atmospheric pollutant deposited on unit area. Such a load thus represents a threshold quantity, based on a dose–response relationship. There are obviously attractions in using critical loads for formulating policy decisions, rather than adopting arbitrary concentrations or 'air quality standards' as the basis of legislation.

The importance of sensible legislation is paramount, because the economic, as well as the ecological, costs of acidic precipitation are potentially enormous. Control of SO_2 and NO_x emission is unfortunately also costly. Energy conservation, resulting in reduced fuel consumption, is an attractive strategy, especially in view of the shortage of fossil fuel resources. On the other hand, smaller energy requirements mean a reduction in the supply of material goods, which conflicts with industry's view of the needs of the community. Fuels can be desulphurized, or low-sulphur fuels used. Effluent stack gases can have their SO_2 and NO_x content reduced before release. But these processes are expensive: an estimate has been suggested of \$250 per tonne of SO_2 removed. Reduction by one-half of 146×10^9 kg yr^{-1} (Table 5.7) would cost the staggering sum of about \$$1.8 \times 10^{10}$ annually. Put another way, the cost of electric power would increase by at least 10 per cent, which might prove crippling to the already burdened consumer. Remember that large sums are already expended in achieving the very substantial reductions in SO_2 seen over the past few decades. Alternative energy sources, such as nuclear fuels, are currently unpopular with the general public precisely because of the pollution dangers. Until fusion power becomes commercially available, some conflict seems bound to arise between economic and environmental pressures. Of course, combustion of fossil fuels will last for only a brief period in man's history. No energy conservation plan could reasonably envisage large fractions of our energy needs being met by fossil fuels alone for more than a few hundred years. By that time, an alternative energy source or a completely new life-style must have been found. Until then, acid rain—often produced by one nation to the detriment of another—seems to be one of the most serious problems in atmospheric pollution.

5.10.7 Photochemical ozone and smog

The London type of air pollution, characterized by particulate matter and SO_x (Section 5.10.5) has been recognized for centuries. In the mid-1940s, the effects of a new kind of oxidizing air pollution, which caused eye irritation, plant damage, and visibility degradation, became evident in Los Angeles. The oxidants include ozone, nitrogen dioxide, and peroxyacetyl nitrate (PAN: see p. 337), and they are formed photochemically by the action of solar radiation on mixtures of NO_x and hydrocarbons (HC) in air. Because of the region where it was first observed, and because of its photochemical origin, oxidizing pollution of the kind described is called *Los Angeles Smog* or *Photochemical Smog*. Very many other cities, especially in the south-west USA, but also in

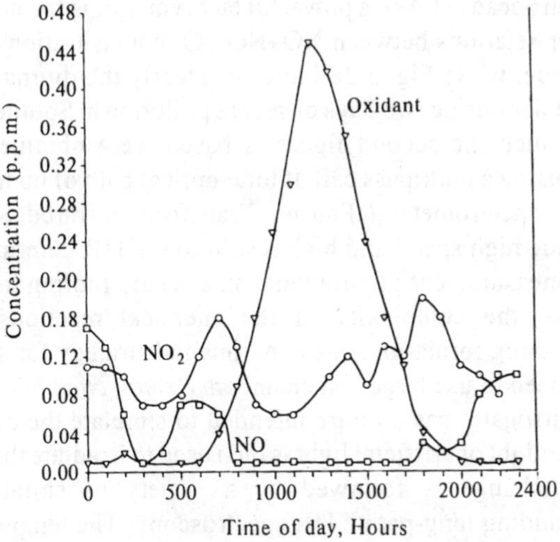

Fig. 5.27. Variations in concentration of oxidant (mainly ozone) and oxides of nitrogen during the course of a smoggy day in Southern California. From Finlayson-Pitts, B.J. and Pitts, J.N., Jr. *Adv. Environ. Sci. Technol.* **7**, 75 (1977).

Greece, Israel, Japan, Australia, and even in the UK, have subsequently been found to suffer from photochemical pollution. Photochemical smog has, however, been a particularly serious problem in Los Angeles and Southern California, and there are several contributing factors. The Los Angeles basin faces the Pacific Ocean to the south-west, and is otherwise almost enclosed by mountain ranges. This topography results in frequent temperature inversions in the boundary layer and lower troposphere, and pollutants are trapped within the basin. Intensely sunny days are frequent, thus promoting photochemical processes. Last, but not least, there is a very high density of automobiles that are thought to be the most important source of primary pollutants. Curiously, it turns out that photochemical 'pollution' is not really a new phenomenon at all! San Pedro Bay was named the 'Bay of Smokes' in 1542, and eye irritation was first recorded in Los Angeles by 1868. The blue haze of the Smoky Mountains—a tourist attraction—probably owes its existence to a biogenic variant of the automobile problem (cf. p. 354 and p. 430). The severity and incidence of pollution have, however, grown out of all recognition from the early precursors of photochemical smog, and the growth has paralleled the vast expansion of use of the internal combustion engine (especially for light motor vehicles) over the last fifty or sixty years.

What is observed on a smoggy day? Nitric oxide (NO) concentrations build up during the night and during the early-morning period of heavy commuter traffic. After dawn, NO becomes replaced by NO_2, and ozone is generated. By noon, there are high concentrations of ozone and nitrogen dioxide in the atmosphere, there is a brown haze because particles are present, and the eyes run because PAN, a powerful lachrymator, is formed. Figure 5.27 shows the inter-relations between NO–NO_2–O_3 concentrations during an air pollution episode, while Fig. 5.28 indicates clearly the diurnal variations of several key species during two days of severe pollution in Southern California. The data on which the second figure is based were obtained by infrared spectroscopy, using a multipass cell of total optical path of up to 1 km, and an interferometric spectrometer ('Fourier Transform Infrared Spectroscopy', FTIR) to provide high speed and high resolution. FTIR determinations have placed *in-situ* measurement of pollutants on a secure footing in recent years.

Because of the complexity of the chemical reactions involved in photochemical smog formation, one experimental strategy for studying smog processes has been to use large test chambers ('*smog chambers*') in which is contained a 'surrogate' gas mixture intended to simulate the composition of polluted air. Sunlight or artificial light is then used to irradiate the mixture, and photochemical change is followed by a variety of suitable analytical techniques, including long-path FTIR spectroscopy. The temporal behaviour of pollutant species in the atmosphere is well-mimicked by the smog chamber experiments. Figure 5.29 demonstrates the effects of irradiating an alkene–NO_x–air mixture with a 25 kW xenon lamp as source. Trends observed in the real pollution episodes (Figs 5.27 and 5.28) show up clearly: NO_2 builds up at the expense of NO, and O_3 at the expense of NO_2, and propene is consumed.

We must now consider the chemistry that gives rise to the observed behaviour. The primary pollutants in automobile exhaust are NO_x (mainly NO) from the high-temperature combustion, carbon monoxide, partially oxidized and unburnt hydrocarbons (HCs), and sulphur dioxide from sulphur-containing fuels. It is these species that then undergo photochemical transformation to ozone, nitrogen dioxide, aldehydes, ketones and acids, PAN, and inorganic and organic aerosols. As we have already emphasized (Section 5.3.3), NO_2 photolysis followed by O + O_2 combination is the only known tropospheric source of O_3

$$NO_2 + h\nu \rightarrow O + NO \tag{5.6}$$

$$O + O_2 + M \rightarrow O_3 + M. \tag{5.5}$$

A route must therefore be found for conversion of the primary species NO to NO_2, before ozone can be formed. Inorganic chemistry on its own seems unable to bring about the oxidation. Three-body reaction

$$NO + NO + O_2 \rightarrow 2NO_2 \tag{5.122}$$

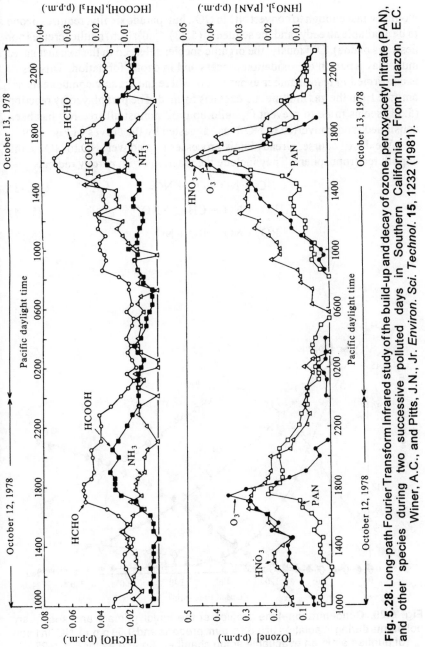

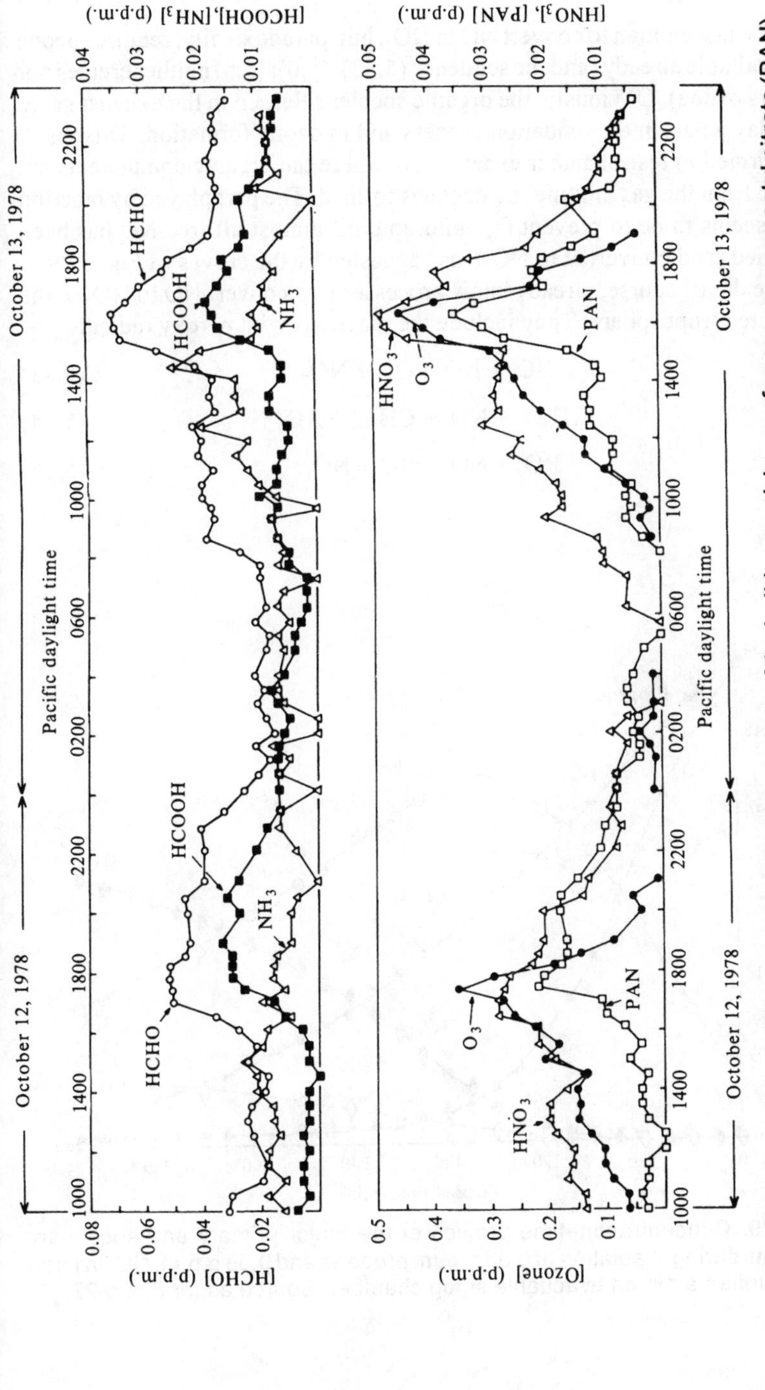

Fig. 5.28. Long-path Fourier Transform Infrared study of the build-up and decay of ozone, peroxyacetylnitrate (PAN), and other species during two successive polluted days in Southern California. From Tuazon, E.C. Winer, A.C., and Pitts, J.N., Jr. *Environ. Sci. Technol.* **15**, 1232 (1981).

is far too slow at the concentrations of NO present (its rate is proportional to $[NO]^2$). Reaction of NO with O_3,

$$NO + O_3 \rightarrow NO_2 + O_2, \qquad (5.33)$$

would be fast enough to convert NO to NO_2, but, paradoxically, requires ozone to be available already (and the sequence (5.33), (5.6), (5.5) neither creates nor destroys ozone). Obviously, the organic species released in the exhaust gases must play a part in the oxidation process and in ozone formation. This result is confirmed in test-chamber experiments where the organic components are omitted from the gas mixture: no ozone is formed. The part played by reaction (5.33) seems to be to prevent O_3 build-up until almost all free NO has been consumed (and converted to NO_2), as suggested by the curves of Fig. 5.29.

We do, of course, already know processes that convert NO to NO_2 in the unpolluted troposphere. They include the reactions with peroxy radicals

$$HO_2 + NO \rightarrow OH + NO_2 \qquad (5.13)$$

$$CH_3O_2 + NO \rightarrow CH_3O + NO_2 \qquad (5.14)$$

$$RO_2 + NO \rightarrow RO + NO_2. \qquad (5.18)$$

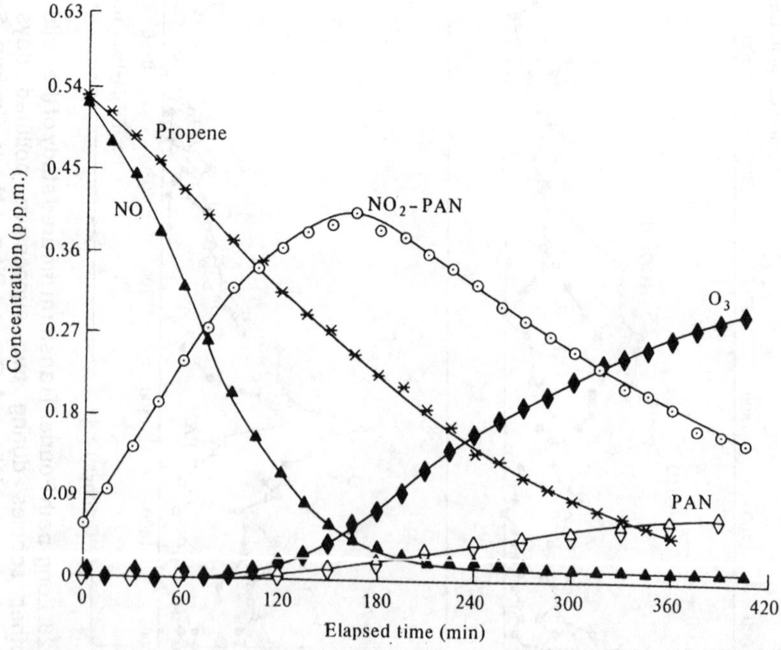

Fig. 5.29. Concentration–time profiles of the major primary and secondary pollutants during irradiation of 0.53 p.p.m. propene and 0.59 p.p.m. NO_x in 1 atm of purified air in an evacuable smog chamber. Source as for Fig. 5.27.

These reactions do, indeed, seem to be the critical ones for oxidizing NO in photochemical smog. Smog chemistry is, then, a grotesquely exaggerated form of the oxidation and transformation chemistry (Section 5.3) of the unperturbed troposphere. Higher concentrations of primary species (e.g. NO_x, HCs) are present in the polluted atmosphere, and perhaps a wider variety of saturated, unsaturated, and aromatic HCs is liberated. But the oxidation chain is still carried by OH, HO_2, and organic oxy- and peroxy-radicals, as in the natural troposphere. We may emphasize the conversion of NO to NO_2 in the chain process by writing the reaction sequence following attack of OH on an alkane, RCH_3 [reactions (5.124) to (5.126) are more detailed forms of eqns (5.19), (5.18), and (5.21)]

$$OH + RCH_3 \rightarrow H_2O + RCH_2 \tag{5.123}$$

$$RCH_2 + O_2 \rightarrow RCH_2O_2 \tag{5.124}$$

$$RCH_2O_2 + NO \rightarrow RCH_2O + NO_2 \tag{5.125}$$

$$RCH_2O + O_2 \rightarrow RCHO + HO_2 \tag{5.126}$$

$$HO_2 + NO \rightarrow OH + NO_2 \tag{5.13}$$

Net $$RCH_3 + 2NO + 2O_2 \rightarrow RCHO + 2NO_2 + H_2O.$$

Attack of OH on RCHO continues the hydrocarbon oxidation by yielding carbonyl radicals

$$OH + RCHO \rightarrow RCO + H_2O, \tag{5.127}$$

that can lead, directly or indirectly, to carbon monoxide and the radical R possessing one less carbon atom than the starting hydrocarbon. Acids, RCOOH, are a minor product from RCO radicals, but they, together with the aldehydes, are found in photochemical smog (Fig. 5.28). The degradation from RCH_2 to R radicals is accompanied by formation of aldehydes and acids with all possible numbers of carbon atoms, down to the first members, HCHO (formaldehyde) and HCOOH (formic acid).

Carbon monoxide is the final oxidation product of the organic chain, but is itself oxidized by OH (Section 5.3.5). The sequence

$$OH + CO \rightarrow H + CO_2 \tag{5.7}$$

$$H + O_2 + M \rightarrow HO_2 + M \tag{5.9}$$

$$HO_2 + NO \rightarrow OH + NO_2 \tag{5.13}$$

Net $$NO + O_2 + CO \rightarrow NO_2 + CO_2$$

can be written to show again how this oxidation of CO can be accompanied by the oxidation of NO to NO_2.

Although we have illustrated the oxidation chain with an alkane as fuel, alkenes (olefins), in fact react with OH radicals even faster, at rates approaching the collision- or diffusion-controlled limit. The initial attack appears to be the addition of OH to the double bond (pp. 333–4), with the major products being the appropriate aldehydes and ketones (e.g. CH_3CHO from C_2H_4, and $C_2H_5CHO + CH_3COCH_3$ from C_3H_6). Aromatic compounds (Section 5.5) constitute a significant fraction of the reactive hydrocarbons in automobile exhaust gases. Hydroxyl radical attack is rapid, and the products are of particular interest because they may include long-chain oxygenated compounds that can be involved in aerosol formation (see p. 340).

The general mechanism of hydrocarbon oxidation, and of NO to NO_2 conversion, in a free-radical chain reaction seems well established. Hydroxyl radicals (and probably HO_2) attack the organic 'fuel' to propagate the chain. We must therefore now consider the origin of radicals in polluted atmospheres. The ozone photochemical source (Section 5.3.1) that is important in the natural troposphere may be supplemented by several other processes. Of these, two are of particular interest, since they involve species detected in photochemical smog. Nitrous acid (HONO) can be formed in the process

$$NO + NO_2 + H_2O \rightarrow 2HONO \qquad (5.128)$$

by either a homogeneous or a heterogeneous route. The molecule is photolysed at relatively long wavelengths ($\lambda < 400$ nm) that reach ground level

$$HONO + h\nu \rightarrow OH + NO. \qquad (5.129)$$

One oxidized molecule, NO_2, is lost in reaction (5.128), but two HONO molecules, and thus potentially two chain-initiating OH radicals, are created. Aldehydes may provide an important entry into the radical chain. One channel for the photodissociation of formaldehyde (at $\lambda < 340$ nm) yields H and HCO radicals. Both these radicals are converted to HO_2 in the presence of O_2, so that the photochemical initiation steps can be represented by the sequence

$$HCHO + h\nu \rightarrow H + HCO \qquad (5.16)$$

$$HCO + O_2 \rightarrow CO + HO_2 \qquad (5.17)$$

$$H + O_2 + M \rightarrow HO_2 + M \qquad (5.9)$$

$$\text{Net} \qquad HCHO + 2O_2 + h\nu \rightarrow 2HO_2 + CO$$

Higher aldehydes are probably less significant sources of chain carriers, since they require shorter wavelength ultraviolet light for photolysis. Formaldehyde itself is emitted directly into the air from automobile exhausts.

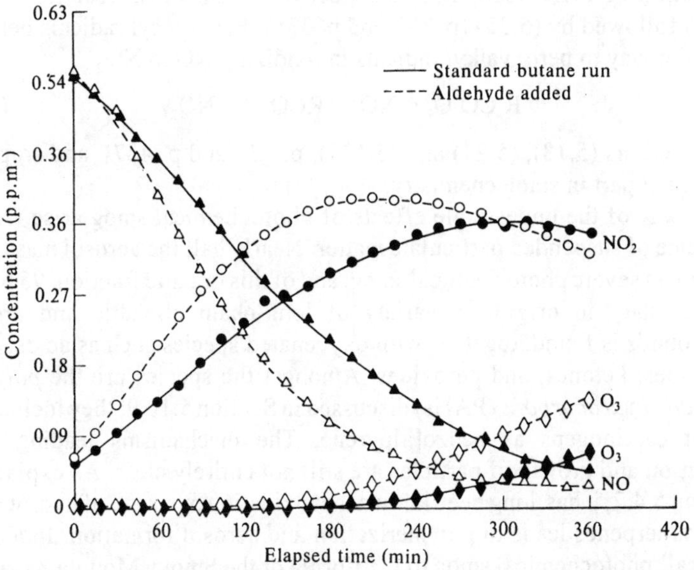

Fig. 5.30. Concentration–time profiles for some primary and secondary pollutants during irradiation of 2.2 p.p.m. *n*-butane and 0.61 p.p.m. NO_x in an evacuable smog chamber without added aldehyde (—) and with 0.13 p.p.m. HCHO added (- - -). Source as for Fig. 5.27.

Smog-chamber experiments support the idea that aldehydes are important. Addition of formaldehyde to a hydrocarbon–NO_x–air mixture causes NO to be converted much more rapidly into NO_2, and ozone appears earlier and at higher concentrations. Concentration–time profiles are shown in Fig. 5.30 for irradiation of butane–NO_x mixtures with and without added aldehyde. Even the relatively small amount of HCHO used has a dramatic effect on the conversion rates.

Aldehydes are also implicated in the formation of peroxyacetyl nitrate, an important component of photochemical smog. Carbonyl compounds, either emitted as primary pollutants, or produced *via* processes such as (5.126) as oxidation intermediates, can be converted to acyl radicals in reaction with OH, and thence to peroxyacyl radicals

$$OH + RCHO \rightarrow RCO + H_2O \tag{5.127}$$

$$RCO + O_2 \rightarrow RCO.O_2. \tag{5.130}$$

Addition of NO_2 to $RCO.O_2$ then yields a peroxyacyl nitrate

$$R.CO.O_2 + NO_2 \rightarrow RCO.O_2.NO_2. \quad (5.131)$$

Formation of PAN itself from acetaldehyde is shown in reactions (5.22)–(5.23) followed by (5.35) (p. 333 and p. 338). Peroxyacyl radicals behave in a similar way to peroxyalkyl radicals in oxidizing NO to NO_2

$$R.CO.O_2 + NO \rightarrow RCO.O + NO_2, \quad (5.132)$$

[cf. reactions (5.18), (5.24) and (5.125), p. 333 and p. 427], and so play an additional part in smog chemistry.

Many of the undesirable effects of photochemical smog arise from the presence of suspended particulate matter. Nearly half the aerosol mass can be organic in severe photochemical smog, and of this organic fraction, 95 per cent is secondary in origin. A variety of long-chain aliphatic and aromatic compounds is found, together with oxygenated species such as acids, esters, aldehydes, ketones, and peroxides. Amongst the species are the *polycyclic aromatic hydrocarbons* (PAHs) discussed in Section 5.10.9; they include such potent carcinogens as benzo[a]pyrene. The mechanisms leading to the formation and growth of particles are still not entirely clear. As explained in Section 5.4.2, it has long been known that oxidation by ozone of hydrocarbons such as terpenes leads to polymerization and aerosol formation. Indeed, the 'natural' photochemical smog of California or the Smoky Mountains, alluded to at the beginning of this section, is ascribed to the reaction between oils from pine forests or citrus groves with ozone naturally present in the troposphere (Section 5.3.3).

Inorganic aerosol in photochemical smog includes sulphate, nitrate, and ammonium ions, as well as a variety of trace metals. Sulphuric acid is formed (Section 5.10.4) from SO_2 released by combustion of sulphur-containing fuels, while nitric acid involves the usual reaction with OH [process (5.34) p. 337]. Ammonia is assigned a prominent role in the neutralization of the acids, especially when there are high local concentrations produced by primary sources (e.g. cattle stations), as in California. The time dependence of NH_3 and HNO_3 concentrations shown in Fig. 5.28 is suggestive of reaction between the two species, with $[NH_3]$ decreasing as $[HNO_3]$ builds up. Sulphur dioxide to H_2SO_4 conversion seems to be accelerated by photochemical smog. Observation shows that, in the Los Angeles area, oxidation of SO_2 may proceed up to 100 times faster than the simple photo-oxidation rate. Recent laboratory investigations have also shown that an intermediate produced in ozone–alkene reactions in air can rapidly oxidize gaseous SO_2 to sulphate (cf. p. 412).

Photochemical air pollution degrades the 'quality' of the environment in all the ways outlined in Section 5.10.2. Human health is affected primarily by the oxidant species such as ozone, but PAN, NO_2, and the aerosols are also harmful. Impairment in physical performance has been demonstrated at oxidant levels (ozone + PAN) above 0.15 p.p.m. Attacks in asthmatics are exacerbated at 0.25 p.p.m. of oxidant (but the same level may have no effect on healthy persons). In general, it seems that discomfort can be perceived—as chest pains, cough, and headache—for concentrations of oxidant beyond 0.25 to 0.3 p.p.m. The World Health Organization (WHO) guidelines for human health, which incorporate safety margins, are 0.076–0.100 p.p.m. for a one-hour exposure and 0.050–0.060 p.p.b. for an eight-hour exposure. The current USA 'air quality standard' quotes a limiting ozone concentration of 0.120 p.p.m. over one hour, and the State of California has set its own more stringent standard of 0.09 p.p.m. over one hour. Reference to Figs 5.27 or 5.28 shows that the ozone concentrations in the episodes represented greatly exceeded the standards throughout the period 10 a.m. to 6 p.m., and that the levels around midday were sufficient to have noticeable effects on health. In Great Britain, meteorological conditions rarely favour serious smog formation, although atmospheric ozone has exceeded the USA air quality standard on several occasions since regular monitoring began in 1970. During the exceptionally hot summer of 1976, photochemical pollution was enhanced, and between 22 June and 17 July rural hourly-mean ozone levels exceeded 0.25 p.p.m. At one rural site, ozone concentrations were in excess of 0.1 p.p.m. over at least eight hours for 18 consecutive days of the 21 days of the episode. More recently, hourly-mean ozone concentrations reached more than 0.100 p.p.m. on several days in the summer of 1995 at both urban and rural monitoring stations in the UK.

Peroxyacetylnitrate, PAN, is a powerful lachrymator, as well as having the effects on the respiratory system of the oxidants. That is, it causes intense irritation of the eye, with consequent tear formation. Irritation increases steadily for oxidant concentrations between 0.1 and 0.45 p.p.m., although ozone on its own is not an eye irritant. We cannot be clear whether eye irritation constitutes a real impairment of public health since it is reversible, and there is no proven association between pollution-induced irritation and chronic eye damage. Nevertheless, the effect is undoubtedly unpleasant, and is perhaps the most obviously perceivable nuisance aspect of being exposed to photochemical air pollution.

Vegetation is easily harmed by photochemical air pollution. Once again, the main agents of damage are PAN, which is one of the most phytotoxic substances known, and ozone. Plants respond to the oxidants by first increasing their cell-membrane permeability. Higher doses lead to cellular and biochemical changes with visible leaf injury, leaf drop, and reduced vigour and growth, and finally death. The classic injury caused by PAN is a glaze, followed by bronzing of the lower leaf surface, although the upper leaf surface is affected in plants such as tobacco. Serious leaf damage to experimental tobacco crops in England was reported in early July 1976, following the June–July pollution episode. Growth experiments in field chambers have shown that there can be up to 50 per cent loss in yield in citrus fruits, grapes, potatoes, and tobacco, and up to 30 per cent loss in cotton. As little as 0.05 p.p.m. of oxidant can cause significant effects, so that the implications for agricultural economy are very serious. The WHO guidelines for exposure of terrestrial vegetation to ozone are 0.100 p.p.m. (one-hour), 0.033 p.p.m. (24-hour), and 0.030 p.p.m. (growing season).

The concept of critical loads, introduced in Section 5.10.5 for pollution by sulphur dioxide and particulate matter, has been extended to a *critical level* in the discussion of the effects of ozone on vegetation. These levels have been formulated using the *accumulated ozone time* (AOT) concept, which defines the measure of ozone exposure through the accumulated time during which the hourly ozone concentration exceeds a given threshold. Thus AOT40 refers to a threshold level of 40 p.p.b., and an exposure of 1 hour at 60 p.p.b. would add 20 p.p.b.-h. to the AOT40. Proposed critical levels have included 3000 p.p.b.-h. for crops and 10 000 p.p.b.-h. for trees. Note that, since damage may increase continuously with concentration, rather than starting at a non-zero threshold, setting critical levels may require some decision about the extent of acceptable damage. For example, the figure just quoted for crops incorporates an 'acceptable' five per cent loss of yield. In considering the application of critical levels for human health, both the United Nations Economic Commission for Europe (UNECE) and the World Health Organization (WHO) have concluded that more information is required than is provided by integration over time of concentration exceeding a defined threshold. A proper assessment of health impacts requires *at least* a knowledge of the frequency distributions of daily maximum ozone concentrations (eight-hour rolling means), so that AOT40 on its own cannot be used to set quality standards.

Can anything be done to reduce and control the formation of photochemical smog? Photochemical ozone concentrations can, in principle, be decreased by reductions in HC and other VOC emissions and by reductions in NO_x. Unfortunately, the most effective control strategy can be elusive, because of a non-linear (and sometimes inverse: see later) dependence of ozone production on precursor emissions. Various photochemical air-quality models have been developed that seek to predict ozone concentrations and their response to different control measures. Such efforts encounter difficulties associated with the variability in place and time of source and sink terms, with the transport and age of the air sample, and with the formulation and detailed kinetics of the chemistry. It is necessary both to assess the impact of a mix of anthropogenic VOCs with widely differing reactivities, and to evaluate that impact in relation to ozone produced by biogenic hydrocarbons.

A first step in understanding the effects of anthropogenic releases is to develop *reactivity scales* that compare different VOCs. Because of the inherent non-linearity of the system, attention is often directed to some measure of *incremental reactivity*, R_i, defined for a specific VOC as

$$R_i = \frac{\text{ozone response in p.p.b}}{\text{incremental increase of reactive hydrocarbon } i \text{ in mass units}}. \quad (5.133)$$

It may be convenient to express the incremental reactivity relative to some

Table 5.9 Contributions to summertime ozone production of some selected hydrocarbons at rural sites in the UK[a]

Hydrocarbon	Mean concentration /p.p.b.	$10^{12}\, k_{OH}$ / $cm^3\, molecule^{-1}\, s^{-1}$	O_3 production rate/p.p.b. hr^{-1}
Iso-butene	0.21	51.4	0.387
Propene	0.27	26.3	0.256
Ethene	0.67	8.52	0.206
Isoprene	0.05	101	0.182
1,2,4-Trimethylbenzene	0.15	32.5	0.176
(*m*- + *p*-) Xylene	0.21	19	0.144
1,3,5-Trimethylbenzene	0.06	57.5	0.128
E-but-2-ene	0.05	64	0.115
Toluene	0.46	5.96	0.099
E-pent-2-ene	0.04	66.9	0.097
1,3-Butadiene	0.04	66.6	0.096
Z-but-2-ene	0.04	56.4	0.081
Iso-pentane	0.52	3.9	0.073
o-Xylene	0.14	13.7	0.069
But-1-ene	0.05	31.4	0.057

[a]From Derwent, R.G. in Hewitt, C.N. (ed.) *Reactive hydrocarbons in the atmosphere*, Academic Press, San Diego, 1999.

chosen compound. The *photochemical ozone creation potential* (POCP) uses ethylene as the standard for comparison, and is defined as

$$POCP = \frac{R_i}{R_{ethylene}} \times 100. \tag{5.134}$$

POCPs can be evaluated using

1. Smog-chamber data
2. Rate-coefficients for the attack of OH on the VOCs
3. Air-quality simulation models
4. Explicit chemical mechanisms.

Each of these methods has seen fairly widespread application. Perhaps the most versatile is the last one: in a recent example of its use, POCPs were determined for 120 reactive hydrocarbons using a highly detailed and explicit chemical mechanism (the 'master chemical mechanism'), and a realistic trajectory model for regional-scale ozone formation across north-west Europe. However, it is much more straightforward to assemble a reactivity scale that is based on the rate coefficients for reaction of OH with the VOC. The problem is that the ability to produce ozone depends not only on the rate of initiation of oxidation chains, but also on the efficiency with which the subsequent steps proceed. Despite these drawbacks, the method provides a first estimate of the POCP. The actual ozone production rates will depend on the concentrations of the individual hydrocarbons present in the atmosphere. Table 5.9 presents selected data for calculated ozone production rates calculated from the rate coefficients and a series of HCs, for which concentration measurements are available for rural sites in the UK. The 'top ranking' 15 hydrocarbons are included, out of the 85 that constituted the original study. Because the rate constants were obtained in laboratory experiments, and the concentrations come from *in-situ* measurements, the table is based entirely on observations. It is noteworthy that alkenes sit in the top four positions, and dominate the list (although aromatic hydrocarbons also make a strong showing). Isoprene is high in the list, not because its concentration is high, but because it is so reactive. The only alkane in this list is iso-pentane: it owes its position to a high ambient concentration rather than to a high intrinsic reactivity.

Armed with this kind of information, it becomes possible to discover which sources of VOCs make the most damaging contribution, and which might therefore be most susceptible to effective control. A further step is then to weight the emissions from different types of activities ('sectors') by the POCPs of the chemical species emitted, to produce a list of sectoral POCPs. In both North America and the UK, vehicle exhausts and fuel evaporation dominate the list, as might be anticipated, but stationary combustion, solvent usage, surface coating, industrial processes and chemical manufacture, petroleum refining and natural gas leakage, agriculture, and biomass burning are all responsible for quite considerable amounts of ozone formation.

Attacking the primary target of motor vehicles, the State of California, one of the worst afflicted areas, has taken the lead in recommending and enforcing legislation. One of the first moves, in 1961, was to require the installation of positive crankcase ventilation on new and used cars, followed by the approval of catalytic converters in 1964–6 to reduce hydrocarbon and carbon monoxide emissions. In 1966, an alternate, 'lean-burn' method of reducing HC and CO emissions was implemented that did not require catalytic devices or afterburners. Unfortunately, this approach led simultaneously to large increases in NO_x emissions. We return to current strategies after looking at the NO_x problem a little more deeply.

Control of NO_x emissions is no less essential than that of HCs, but there is considerable debate as to the degree of control that is appropriate for any particular area. The nature of the problem can be seen from the consequences of the 1966 California legislation, which led to *increased* NO_x levels. The legislation did have the required effect of reducing average ozone levels in downtown Los Angeles, but, unfortunately, oxidant levels downwind of the central area actually increased. Ozone concentrations near the release area were decreased because NO—the main component of the increased NO_x emission—rapidly destroys O_3 [reaction (5.33)]. While the air is being

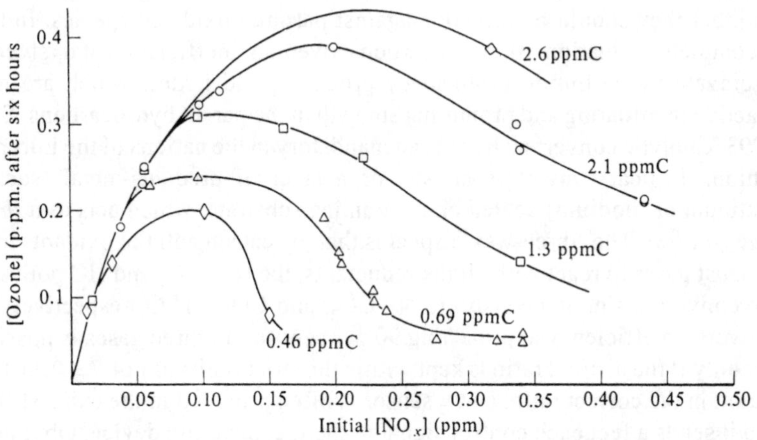

Fig. 5.31. Ozone concentration reached after a 6-h irradiation of NO_x in a surrogate mixture of hydrocarbons simulating ambient air, with varying initial concentrations and ratios of [HC]/[NO_x]. Source as for Fig. 5.27.

transported downwind, the NO_2 product is photolysed, and ultimately produces ozone. Similar pollution patterns to those of the Los Angeles basin were seen in the British smog episodes of 1976. Rural sites downwind of the primary pollutant source (in this case, presumably, Greater London) experienced the highest oxidant concentrations. Ozone measurements made at five different sites during the episode suggest that the Greater London source was augmented by secondary pollutant formation over north-west Europe as a whole, and brought to the UK by the prevailing winds.

The complexities of pollution response to NO_x control are well illustrated by the smog chamber results set out in Fig.5.31. Ozone concentrations were measured for fixed irradiation periods of different mixtures of HC and NO_x with air. Reduced reactive [HC] at constant [NO_x] leads to decreased [O_3] in all cases. However, if one is on the right-hand side of the maximum in the curves (a region typical of urban ambient air), reducing [NO_x] at constant [HC] will raise the maximum ozone level. These results, and the atmospheric observations described in the preceding paragraph, emphasize an important aspect of pollution reduction. That is, the most effective strategy may differ according to whether it is areas upwind or downwind of the primary source that are to be protected; what time of day release occurs; and what the relative and absolute HC and NO_x concentrations are.

The present strategy is to reduce vehicle emissions by catalytic conversion of the exhaust gases. By 1975, catalytic converters were virtually universal in the USA, and stringent standards for HC and CO emissions could be met without greatly impairing engine performance or economy. An essential requirement of catalytic afterburners is that they have a long enough lifetime, and that they should be effective against partially oxidized species. Indeed, incomplete oxidation of hydrocarbons over an inefficient catalyst could aggravate the pollution problem by producing aldehydes, which are more reactive in initiating and promoting smog than the parent hydrocarbons. Since 1993, catalytic converters have been mandatory in the nations of the European Union. Typical converters consist of a layer of precious-metal (such as platinum or rhodium) coated on an alumina substrate, which acts as a *three-way catalyst*. The 'three-way' aspect is that, by causing all the oxidants in the exhaust gases to react with all the reductants, the NO, CO, and HC pollutants are converted simultaneously (to N_2, CO_2, and CO_2 + H_2O, respectively). A conversion efficiency approaching 90 per cent for all three gases is possible, but only if the air:fuel ratio is kept within the strict limits of (14.7 ± 0.1):1. To maintain the correct ratio, an O_2 sensor is often provided in the exhaust flow, and it sends a feedback control signal to the fuel metering device. Obviously, such measures can only work with new vehicles. In 1994, more than 80 per cent of vehicle emissions of VOCs came from cars produced before 1981, even though they accounted for only 40 per cent of the distances travelled. The economic cost for new vehicles is substantial: one estimate puts the sales-weighted average at about $750 for each vehicle for the pollution control measures. That the measures are effective, however, seems in little doubt. HC and NO_x emissions from petrol-engine cars each decreased by a factor of ten for vehicles subject to the European Union directive, and CO emissions decreased by a factor of three. At the time of writing, it is not clear if these reduced emissions have *yet* translated into better air quality. In general, rather scattered results are obtained from European stations reporting ozone concentrations. Only for nine sites in the Netherlands are there strong indications of a downward trend (0.4–1.6 per cent per year between 1981 and 1994). Analysis of trends is obviously extremely important when assessing the outcome of VOC protocols and other measures taken to reduce emissions. In order to remove some of the interfering factors, detailed scrutiny has been given to measurements of ozone made at Mace Head, on the west coast of Ireland. Even here, however, the results are not clear-cut. During the period 1990–1994, ozone concentrations did, indeed, decline, by about 1.6 per cent per year, but during 1995–1996 the background ozone concentrations recovered, to give an overall *upward* trend over the period 1990–1996.

Alternative fuels must, of course, be considered. The emissions of unburnt HC and of CO are markedly better from diesel-powered cars than from those with petrol engines, and NO_x emissions only marginally worse than those from petrol cars with efficient catalytic converters. The real problem is particle emissions, as discussed on p. 407. Exhausts from engines powered by methanol or ethanol, and by natural gas, are potentially much less polluting. Ethanol can be produced from biomass, and Brazil has a large-scale biofuel programme in operation. Pure ethanol from sugar-cane fermentation is used to power about four million light vehicles in the country, and a remaining seven million run on a blend of ethanol and gasoline ('gasohol'). However, formaldehyde and acetaldehyde produced in the oxidation of methanol and ethanol are much more polluting than the parent alcohols, and incomplete combustion could thus negate the apparent advantages. Both aldehydes are toxic and potential carcinogens. Elevated concentrations of acetaldehyde, much higher than in urban areas of other countries, have been found in the air of towns of Brazil. Hydrogen-powered and electric vehicles are cleaner yet on the road, although the pollution audit must include the generation of the hydrogen or the electricity. The economic costs of emission control can be offset by changes in life-style. Small decreases in legal speed limits not only conserve fuel, but significantly reduce NO_x emission with negligible increase in hydrocarbon. Increased use of public transport rather than the private car further reduces total emissions. Use of fuels such as hydrogen eliminates at once the hydrocarbon problem, but it is not yet commercially feasible. The parts of this chapter concerned with pollution will have shown that, in the long term, the cleanliness of the air we breathe, as well as the climate of our planet, will depend on the development of non-combustion energy sources.

5.10.8 Degradation of HFCs and HCFCs

Section 4.6.6 describes the introduction of the hydrofluorocarbons (HFCs) and hydrochlorofluorocarbons (HCFCs) as substitutes for the CFCs. The rationale for the replacement of the CFCs is that the alternative compounds will be removed in the troposphere, and thus be unable to reach the stratosphere. The deliberate release by Man of reactive organic compounds to the troposphere evidently merits examination.

Lifetimes of the HFCs and HCFCs range from about one to 250 years with respect to the dominant loss in reaction with OH. The compounds are thus fairly unreactive compared with the VOCs listed in Table 5.2. Release rates, although large in terms of manufactured compounds, are nevertheless small compared with release rates of biogenic species or the major anthropogenic pollutants. These two factors taken together mean that the HFCs and HCFCs are unlikely to make a significant contribution to ozone production in the troposphere. There are, however, other environmental questions attached to the release of the compounds. Some of them are concerned with generation of halogens or other ozone-destroying catalysts within the stratosphere. These problems are not relevant to the enquiry of this chapter. On the other hand, the tropospheric degradation and oxidation of the compounds might lead to the formation of secondary species that could possess an adverse impact within the

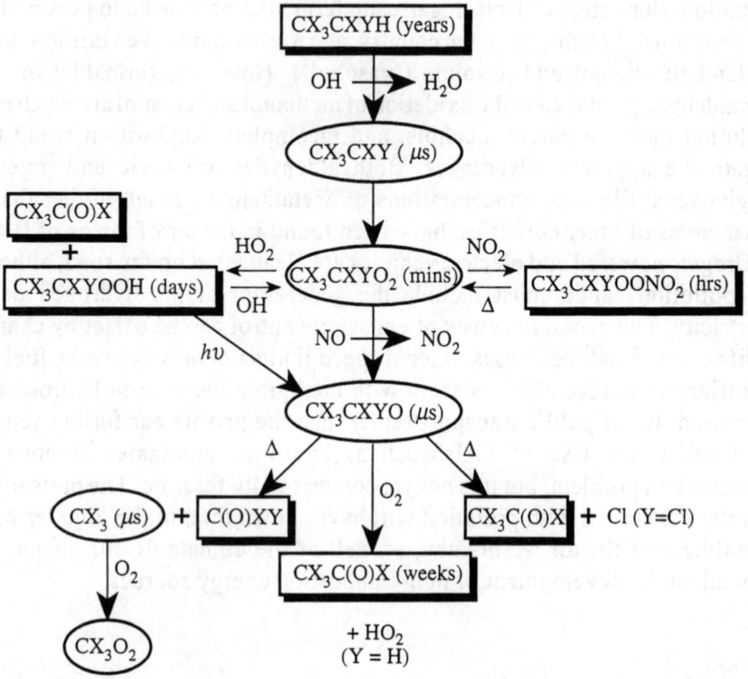

Fig. 5.32. Scheme for the tropospheric oxidation of the C_2-hydrocarbon CX_3CXYH, where X and Y are each Cl or F. Estimated lifetimes of the various species are given in parentheses; closed-shell species are enclosed in rectangles, and radicals in ellipses. From Wallington, T.J. and Nielsen, O.J. in Boule, P. (ed.) *Environmental Photochemistry*, Springer-Verlag, Berlin, 1999.

troposphere or at the Earth's surface, or that might themselves be catalytic species that could reach the stratosphere.

Initial attack by OH consists of H-abstraction in the case of the haloalkanes to yield a haloalkyl radical that adds an O_2 molecule to form a haloalkyl peroxy radical. In almost all respects, this radical will undergo reactions similar to those of simple alkyl radicals, as outlined in Section 5.3.2, although the halogen substitution may provide some different reaction pathways. Figure 5.32 is a generic scheme for the oxidation of a C_2-halocarbon CX_3CXYH (where X and Y can each be either Cl or F). The essence is that peroxy radicals can react, as usual, with NO, NO_2, and HO_2. An alkoxy radical CX_3CXYO may then be generated, most obviously by the NO-route. The atmospheric fate of these radicals is either decomposition or reaction with O_2. Decomposition can occur by C–C bond fission or by elimination of a Cl atom; reaction with O_2 is possible only if an α-hydrogen is present (as, for example, in CF_3CFHO). These three reactions are typified by the processes

$$CF_3CFHO \rightarrow CF_3 + HC(O)F \quad (5.135)$$

$$CF_3CFHO + O_2 \rightarrow CF_3C(O)F + HO_2 \quad (5.136)$$

$$CF_3CFClO \rightarrow CF_3C(O)F + Cl. \quad (5.137)$$

Note that reactions (5.135) and (5.136) represent competing routes for reaction of CF_3CFHO. This alkoxy radical is derived from one of the commercially important HCFCs, CF_3CFH_2 (HFC-134a); in the atmosphere, 7–20 per cent reacts by channel (5.136).

As indicated in Fig. 5.32, the lifetimes of the carbonyl products, such as HC(O)F and $CF_3C(O)F$, are measured in weeks, much longer than those of any of the preceding radical intermediates. Although gas-phase oxidation of species such as $CX_3C(O)H$ and $CF_3C(O)Cl$ is possible, most of the carbonyl compounds are removed primarily by incorporation into water droplets, in which they are hydrolysed. HC(O)F, for example, is hydrolysed to HF and HCOOH, $CF_3C(O)F$ to $CF_3C(O)OH$ and HF, and $CF_3C(O)Cl$ to $CF_3C(O)OH$ and HCl. The acids are then removed through wet deposition (with a lifetime of tens of days) or dry deposition to the surface and the oceans (with a lifetime of the order of years).

Acids such as HF or HCl are produced in quantities small compared with other sources. However, atmospheric breakdown of several of the HFCs and HCFCs produces trifluoroacetyl halides, $CF_3C(O)F$ and $CF_3C(O)Cl$, which in turn form trifluoroacetic acid (TFA), $CF_3C(O)OH$, on hydrolysis. TFA would therefore be precipitated in rain or snow, and these processes are the only known ones capable of producing the compound, a known neurotoxin, in the atmosphere. The environmental fate of TFA is not properly understood, but reports suggest that it may have adverse effects, including the inhibition of plant growth. TFA has already been observed in surface water, rain, and tropospheric air. Curiously, the concentrations are much larger than the breakdown of HFCs and HCFCs could explain, so that there appears to be an additional, and unknown, source of the acid.

It will be noted that reaction (5.135) is a source of the CF_3 radical; addition of O_2 produces CF_3O_2 radicals, and thence the corresponding alkoxy radicals, CF_3O. This reaction, and several similar ones involving HFCs and HCFCs possessing the CF_3 group, can produce CF_3O in both the troposphere and the stratosphere. We have already mentioned speculation that the radical might be a potential catalyst for stratospheric ozone depletion (p. 230), although current evidence is against a significant effect. The possibility arises because of an unusual thermodynamic stability of the radical. Most halo-alkoxy radicals either react with O_2 or eliminate a halogen atom, in analogues of reactions (5.136) or (5.137). Neither of these losses is thermodynamically possible for CF_3O, and the radical might therefore be expected to be rather persistent. Its main atmospheric losses seem to involve reaction with NO (to form $C(O)F_2$ and FNO) or with CH_4 (to yield CF_3OH and CH_3).

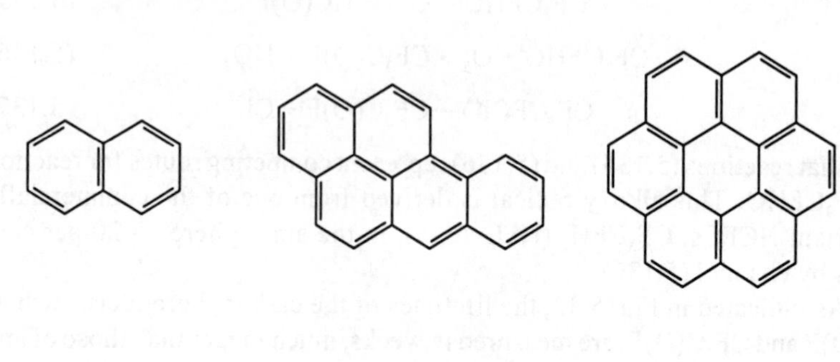

Fig. 5.33. Structures of some atmospherically important polycyclic aromatic hydrocarbons.

5.10.9 Polycyclic aromatic hydrocarbons (PAHs)

Atmospheric constituents that could be carcinogenic or mutagenic, rather than simply toxic, must come under special scrutiny. One of the first classes of atmospheric species that were shown to be carcinogenic were the *polycyclic aromatic hydrocarbons* (PAHs). A small contribution to the atmospheric load of PAHs may come from natural sources such as forest fires or volcanoes, but the predominant sources are anthropogenic.

PAHs are hydrocarbons that consist of two or more benzene rings fused together in a variety of ways as linear, angled, or clustered structures. More than 100 different PAH species have been identified in the atmosphere, ranging from naphthalene (two benzene rings; relative molecular mass 128) to coronene (seven rings; RMM 300). Figure 5.33 illustrates some of the structures. The largest PAHs, with five, six, and seven aromatic rings, are found in the atmosphere predominantly as aerosols, while naphthalene exists exclusively in the gas phase. The compounds may also become adsorbed on the surfaces of other aerosol particles, such as those generated in combustion processes, and products more polar and reactive than the parent PAHs can be formed.

In general, the PAHs are liberated to the atmosphere as a result of incomplete combustion. Anthropogenic sources include motor vehicles (both diesel and petrol), stationary power plants (coal and oil fired), domestic (coal and wood burning, tobacco smoke), as well as deliberate biomass burning (see Section 5.10.10). The compounds are thus ubiquitous in our atmospheric environment. The importance of the compounds resides largely in their impact on human health. As long ago as 1942, it was discovered that the organic extract from particles in the ambient air could produce cancer in experimental animals, and benzo[a]pyrene was subsequently identified as one of the causative agents. Benzo[a]pyrene in tobacco smoke is also often cited as the reason that cigarette smoking produces cancer, although many other carcinogenic compounds, including other PAHs, are implicated in reality. Of course, PAHs that are present as aerosols or adsorbed on other particles—and cigarette smoke is a fine example of that kind of system—offer a special hazard for the reasons outlined on pp. 406–7.

Despite the identification of carcinogenic PAHs in the atmosphere, epidemiological surveys suggested that other, unidentified, carcinogens must be present in some polluted atmospheres. PAHs such as benzo[a]pyrene are *promutagens* that require activation before a cancer can be produced. The activity of some air particles suggests that other species may be present that possess direct, and powerful, mutagenic properties. Considerable evidence now points to mono- and di-nitro derivatives of PAHs as these agents. Several such compounds (for example, 1-nitropyrene and 3-nitrofluoranthene) have been identified in the particles collected from several combustion sources. Even more important may be secondary species, such as 2-nitropyrene and 2-nitrofluoranthene, that are found in the atmosphere of widespread geographical locations, even though they are not emitted significantly by combustion. Atmospheric chemical transformation of primary PAHs is the probable source of these nitro-compounds. Radical reactions in the gas phase effect the conversions when NO_2 is present in the atmosphere. The initiator is

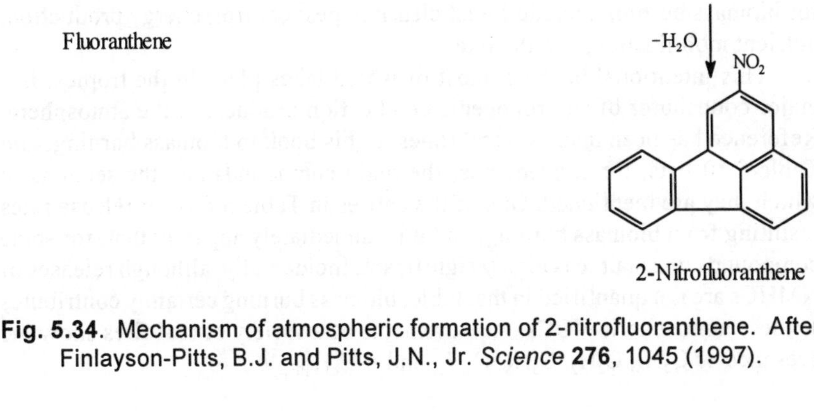

Fig. 5.34. Mechanism of atmospheric formation of 2-nitrofluoranthene. After Finlayson-Pitts, B.J. and Pitts, J.N., Jr. *Science* **276**, 1045 (1997).

Table 5.10 Biomass burning as a source of atmospheric trace constituents: references made in Chapters 4 and 5[a]

Species	Reference
CO	Section 5.9.3
CH_4	Sections 5.4.1, 5.9.3
CH_3Cl	Section 4.6.4
CH_3Br	Section 4.6.4
NO_x	Section 4.3.4
N_2O	Section 4.6.7
S-compounds	Section 5.6
VOC Species	Section 5.10.7
PAH	Section 5.10.9
Particles	Section 5.8

[a]See also entries in Table 5.8.

OH during the day, as usual. However, NO_3 can effect conversion at night, and a high degree of mutagenicity of air downwind of Los Angeles has been attributed to nitronaphthalenes and methylnitronaphthalenes generated by the initial attack of NO_3 on naphthalene and methylnaphthalene.

Figure 5.34 shows the steps envisaged in the atmospheric formation of 2-nitrofluoranthene in the OH-initiated conversion of fluoranthene. Similar atmospheric reactions may be responsible for the formation of nitrolactones (for example, isomers of nitrodibenzopyranone from phenanthrene) that appear to be even more potent mutagens than the nitroarenes.

5.10.10 Biomass burning

Man burns plant matter for a variety of reasons, and *biomass burning* is an accepted aspect of land cultivation in many parts of the world. The motivations for biomass burning include forest clearing, pest control, energy production, nutrient mobilization, and the like.

This intentional burning, most of which takes place in the tropics, is a major contributor of anthropogenic combustion products to the atmosphere. Reference has been made several times in this book to biomass burning, and Table 5.10 lists, for convenience, the main compounds and the sections in which they are mentioned. One of the entries in Table 5.8 is for release rates resulting from biomass burning, and it is immediately apparent that, for some compounds, the source is highly significant. Incidentally, although releases of NMHCs are not quantified in that table, biomass burning certainly contributes to these atmospheric VOCs: measurements of a variety of real fires show the presence of C_2H_6, C_2H_2, C_3H_8, C_3H_6, and n-C_4H_{10}.

Dry plant matter is mainly carbohydrate (empirical formula CH_2O), and burns to CO_2 and H_2O. The CO_2 itself may be of concern (Sections 1.5.1, 2.2, and 9.6), but here we are concerned with minor products of combustion. In addition to the carbohydrate, smaller quantities of other elements such as N, Cl, S, P, and K are present, and combustion allows volatilization of some of the compounds that are formed. Temperatures do not generally become high enough to oxidize atmospheric N_2 (in distinction to the conditions that often obtain in the combustion of fossil fuel). The nitrogen compounds emitted must therefore come from N in the fuel itself. Laboratory test fires show that H_2, CH_3Cl, N_2, COS, particulate matter, and several other minor compounds are emitted. In the early, flaming, stage of combustion, when temperatures are high, compounds of high oxidation state (e.g. CO_2, SO_2, NO_x) are emitted, while in the later, smouldering, stage, less oxidized compounds (e.g. CH_4, N_2O) are produced. According to such experiments, for every one kilogram of dry biomass burnt, emissions are about 0.5 g of N as NO_x, 42 g of C as CO, and 4 g of C as CH_4. Given the enormous amounts of biomass burned each year, these values show immediately why the process is so important. During the dry season, when it usually takes place, biomass burning can be so widespread and intense that it can easily be seen from space. Emissions from the fires can significantly enhance tropospheric ozone concentrations. Large-scale burning of the savanna in Africa, for example, is believed to produce as much regional-scale ozone as urban industrialized regions. The emissions may also be a major source of nitric acid in the tropics. The effects on atmospheric chemistry and climate are critically important.

The sad message with which this chapter must end, therefore, is that Man has many, many ways in which he can pollute his environment, and they are by no means all dependent on industrial activity. The encouraging part, which this chapter should have highlighted, is that scientific understanding of the problems is becoming well enough advanced that sensible policy decisions and control strategies can be formulated.

Bibliography

Leighton's seminal book still provides a satisfactory overview of tropospheric chemical processes in unpolluted as well as polluted atmospheres, despite revisions in the details of the chemistry since it was written (see also p. 148 and p. 321).

Photochemistry of air pollution. Leighton, P.A. (Academic Press, New York, 1961).

Sections 5.1 and 5.2

The bibliography continues with some general introductions to the troposphere, to the compounds found in it, and to the chemistry that occurs there. The formation of ozone in the troposphere is a central and recurring theme.

Reactive hydrocarbons in the atmosphere. Hewitt, C.N. (ed.) (Academic Press, San Diego, 1999).

Microbiology of atmospheric trace gases: sources, sinks and global change processes. Murrell, J.C. and Kelly, D.P. (eds.) (Springer-Verlag, Berlin, 1996).

Volatile organic compounds in the atmosphere. Hester, R.E. and Harrison, R.M. (eds.) (Royal Society of Chemistry, Cambridge, 1995).

Natural and anthropogenic organic compounds in the global atmosphere. Atlas, E.L., Li, S.-M., Standley, L.J., and Hites, R.A. in *Global atmospheric chemical change.* Hewitt, C.N and Sturges, W.T. (eds.) (Chapman and Hall, London, 1994).

Sources, atmospheric lifetimes, and chemical fates of species in the natural troposphere. Chapter 14 in *Atmospheric chemistry.* Finlayson-Pitts, B.J. and Pitts, J.N., Jr. (John Wiley, Chichester, 1986).

Tropospheric ozone research. Hov, Ø. (ed.) EUROTRAC Report Volume 6. (Springer-Verlag, Berlin, 1998).

Ozone in the United Kingdom. UK Photochemical Oxidants Review Group. (Department of the Environment, Transport, and the Regions, 1998).

The global impact of human activity on tropospheric ozone. Levy, H., Kasibhatla, P.S., Moxim, W.J., Klonecki, A.A., Hirsch, A.I., Oltmans, S.J., and Chameides, W.L. *Geophys. Res. Letts.* **24,** 791 (1997).

Ozone in the troposphere. Crutzen, P.J., in *Composition, chemistry, and climate of the atmosphere* Singh, H.B. (ed.) (Van Nostrand, New York, 1995).

Tropospheric ozone: regional and global scale interactions. Isaksen, I.S.A. (ed.) NATO ASI Series C, (D. Reidel, Dordrecht, 1988).

Ozone in the troposphere. Fishman, J., in *Ozone in the free atmosphere*, Whitten, R.C. and Prasad, S.S. (eds.) (Van Nostrand Reinhold, New York, 1985).

Tropospheric ozone: seasonal behavior, trends and anthropogenic influence. Logan, J.A. *J. geophys. Res.* **90**, 10463 (1985).

The chemistry of the atmosphere: oxidants and oxidation in the Earth's atmosphere. Bandy, A.R. (ed.) (Royal Society of Chemistry, London, 1995).

Global tropospheric chemistry. Crutzen, P.J. in *Low temperature chemistry of the atmosphere.* Moortgat, G.K., Barnes, A.J., Le Bras, G., and Sodeau, J.R. (eds.) (Springer-Verlag, Berlin, 1994).

Gas-phase tropospheric chemistry of organic compounds: a review. Atkinson, R. *Atmos. Environ.* **24A**, 1(1990).

Atmospheric chemistry. Finlayson-Pitts, B.J. and Pitts, J.N., Jr. (John Wiley, Chichester, 1986).

Atmospheric chemistry and physics of air pollution. Seinfeld, J.L. (John Wiley, Chichester, 1986).

Chemistry of the natural atmosphere. Warneck, P. (Academic Press, London, 1988).

Tropospheric chemistry: a global perspective. Logan, J.A., Prather, M.J., Wofsy, S.C., and McElroy, M.B. *J. geophys. Res..* **86**, 7210 (1981).

The global troposphere: biogeochemical cycles, chemistry and remote sensing. Levine, J.S. and Allario, F. *Environ. Monitg & Assessm.* **1**, 263 (1982).

Chemistry in the troposphere. Chameides, W.L. and Davis, D.D. *Chem. Engng. News* **60**, No. 40, 39 (1982).

Gas-phase chemistry of the minor constituents of the troposphere. Cox, R.A. and Derwent, R.G. *Gas kinetics and energy transfer*. Specialist Periodical Reports of the Chemical Society **4**, 189 (1981).

Organic chemistry of the Earth's atmosphere. Isidorov, V.A., trans. Kovoleva, E.A. (Springer Verlag, Berlin, 1990).

There is considerable evidence that the concentrations of many trace gases in the troposphere are increasing, probably as a result of man's activities. The oxidizing capacity of the troposphere, and hence the ability of the atmosphere to process added substances, may thus be reduced. See also the references for Sections 1.5 and 5.4 (which describe changes in atmospheric CH_4, for example), and Section 5.9.

Atmospheric change and biodiversity. Maciver, D.C. *Env. Monitoring Assess.* **49**, 177 (1998).

Changes in tropospheric composition and air quality. Tang, X., Madronich, S., Wallington, T., and Calamari, D. *J. Photochem. Photobiol.* **B46**, 83 (1998).

Global atmospheric chemical change. Hewitt, C.N. and Sturgess, W.T. (eds.) (Chapman and Hall, London, 1994).

Global atmospheric chemical change, Hewitt, C.N. and Sturgess, W.T. (eds.) (Elsevier, Amsterdam, 1993).

Concentrations and trends of tropospheric ozone precursor emissions in the USA and Europe, Kley, D., Geiss, H., and Mohnen, V.A. in *The Chemistry of the Atmosphere: Its impact on Global Change,* Calvert, J.G. (ed.) (Blackwell, Oxford, 1994).

The oxidizing capacity of the Earth's atmosphere: Probable past and future changes. Thompson, A.M. *Science* **256**, 1157 (1992).

The changing atmosphere. Rowland, F.S. and Isaksen, I.S.A. (eds.) (John Wiley, Chichester, 1988).

Increased tropospheric concentrations of trace gases can increase radiation trapping and thus influence climate. This topic is explored further in Section 9.6.

Trace gas trends and their possible role in climate change. Ramanathan, V., Cicerone, R.J., Singh, H.B., and Kiehl, J.T. *J. geophys. Res.* **90**, 5547 (1985).

Biogenic sulfur in the environment. Saltman, E.S., and Cooper, W.J. (eds). ACS Symposium Series No. 393, American Chemical Society, Washington, DC, 1989.

Photochemical production of carbonyl sulphide in marine surface waters. Ferek, R.J and Andreae, M.O. *Nature, Lond.* **307**, 148 (1984).

Oxidation of CS_2 and COS: sources for atmospheric SO_2. Logan, J.A., McElroy, M.B., Wofsy, S.C., and Prather, M.J. *Nature, Lond.* **281**, 185 (1979).

The contribution of volcanoes to the global atmospheric sulfur budget. Berresheim, H. and Jaeschke, W. *J. geophys. Res.* **88**, 3732 (1983).

Section 5.2.2

The atmospheric boundary layer. Garrett, J.R. (Cambridge University Press, Cambridge, 1992).

Section 5.2.3

Dynamics of transport processes in the upper troposphere. Mahlman, J.D. *Science* **276**, 1079 (1997).

Tropospheric chemistry and transport. Kley, D. *Science* **276**, 1043 (1997).

Section 5.3

Chemical processes in atmospheric oxidation. Le Bras, G. (ed.) EUROTRAC Report Volume 3. (Springer-Verlag, Berlin, 1997).

Tropospheric ozone research: tropospheric ozone in the regional and sub-regional context. Hov, Ø. (Springer-Verlag, Berlin, 1997).

Gas-phase tropospheric chemistry of organic compounds. Atkinson, R. in *Volatile organic compounds in the atmosphere.* Hester, R.E. and Harrison, R.M. (eds.) (Royal Society of Chemistry, Cambridge, 1995).

Gas-phase tropospheric chemistry of organic compounds: a review. Atkinson, R. *Atmos. Environ.,* **24A**, 1 (1990).

QSAR study on the tropospheric degradation of organic compounds. Gramatica, P., Consonni, V., and Todeschini, R. *Chemosphere* **38,** 1371 (1999).

Hydrogen radicals, nitrogen radicals, and the production of O_3 in the upper troposphere. Wennberg, P.O. and 23 co-authors. *Science* **279,** 49 (1998).

Understanding the production and interconversion of the hydroxyl radical during the tropospheric OH experiment. Eisele, F.L., Mount, G.H., Tanner, D., Jefferson, A., Shetter, R., Harder, J.W., and Williams, E.J. *J. geophys. Res.* **102**, 6457 (1997).

Atmospheric oxidation mechanism of *n*-butane: the fate of alkoxy radicals. Jungkamp, T.P.W., Smith, J.N., and Seinfeld, J.H. *J. phys. Chem.* **101**, 4392 (1997).

Relationships between ozone photolysis rates and peroxy radical concentrations in clean marine air over the Southern Ocean. Penkett, S.A., Monks, P.S., Carpenter, L.J., Clemitshaw, K.C., Ayers, G.P., Gillett, R.W., Galbally, I.E., and Meyer, C.P. *J. geophys. Res.* **101**, 12805 (1997).

Gas-phase tropospheric chemistry of organic compounds. Atkinson, R. *J. phys. Chem. Ref. Data, Monograph 2,* 1 (1994).

Net yield of OH, CO, and O_3 from the oxidation of atmospheric methane. Tie, X., Kao, C.-Y., and Mroz, E.J. *Atrnos. Environ.* **26A**, 125 (1992).

The chemistry of the hydroxyl radical in the troposphere. Ehhalt, D.H., Dorn, H.-P., and Poppe, D. *Proc. Roy. Soc. Edinburgh* **B97**, 17 (1991).

Evaluation of a detailed gas-phase atmospheric reaction mechanism using environmental chamber data. Carter, W.P.L. and Lurmann, F.W. *Atmos. Environ.* **25A,** 2771 (1991).

A detailed mechanism for the gas-phase atmospheric reactions of organic compounds. Carter, W.P.L. *Atmos. Environ.* **24A**, 481 (1990).

Permutation reactions of organic peroxy radicals. Madronich, S. and Calvert, J.G. *J. geophys. Res.* **95**, 5697 (1990).

The atmospheric chemistry of organic nitrates. Roberts, J.M. *Atmos. Environ.* **24A**, 243 (1990).

The photochemistry of the troposphere. Graedel, T.E., in *The photochemistry of atmospheres*, Levine, J.S. (ed.) (Academic Press, Orlando, Fla, 1985).

Photochemistry of biogenic gases. Levine J.S. in *Global ecology: towards a science of the biosphere*, Rambler, M.B., Margulis, L., and Fester, R. (Academic Press Orlando, Fla, 1989).

Atmospheric chemistry of ethane and ethylene. Aikin, A.C. *J. geophys. Res.* **87**, 3105 (1982).

On the mechanism of the interaction between hydroxyl radicals and alkenes, and the detection of a stabilized adduct

Photoionization mass spectrometer studies of the collisionally stabilized product distribution in the reaction of OH radicals with selected alkenes at 298 K. Biermann, H.W., Harris, G.W., and Pitts, J.N., Jr. *J. phys. Chem.* **86**, 2958 (1982).

Section 5.3.2

Global atmospheric chemistry of reactive hydrocarbons. Seinfeld, J.H. in *Reactive hydrocarbons in the atmosphere.* Hewitt, C.N. (ed.) (Academic Press, San Diego, 1999).

Atmospheric degradation of anthropogenic molecules. Wallington, T.J. and Nielsen, O.J. in *Environmental photochemistry.* Boule, P. (ed.) (Springer-Verlag, Berlin, 1999).

Diurnal cycles of short-lived tropospheric alkenes at a north Atlantic coastal site. Lewis, A.C., McQuaid, J.B., Carslaw, N., and Pilling, M.J. *Atmos. Environ.* **33,** 2417 (1999).

On the budget of photooxidants in the marine boundary layer of the tropical South Atlantic. Junkermann, W., and Stockwell, W.R. *J. geophys. Res.* **104,** 8039 (1999).

Identification of organic peroxides in the oxidation of C_1–C_3 alkanes. Qi, B., Zhang, Y.H., Chen, Z.M., Shao, K.S., Tang, X.Y., Hu, M., *Chemosphere* **38,** 1213 (1999).

Atmosphere hydrogen peroxide and organic hydroperoxides: A review. Jackson, A.V. *Crit. Revs. Env. Sci. & Technol.* **29,** 175 (1999).

Methane photooxidation in the atmosphere: Contrast between two methods of analysis. Johnston, H. and Kinnison, D. *J. geophys. Res.* **103,** 21967 (1998).

Photochemistry of formaldehyde during the 1993 tropospheric OH photochemistry experiment. Fried, A., McKeen, S., Sewell, S., Harder, J., Henry, B., Goldan, P., Kuster, W., Williams, E., Baumann, K., Shetter, R., and Cantrell, C. *J. geophys. Res.* **102**, 6283 (1997).

Organic peroxy radicals: kinetics, spectroscopy and tropospheric chemistry. Lightfoot, P.D., Cox, R.A., Crowley, J.N., Destriau, M., Hayman, G.D., Jenkin, M.E., Moortgat, G.K., and Zabel, F. *Atmos. Environ.* **26A**, 1805 (1992).

Section 5.3.3

On the background photochemistry of tropospheric ozone. Crutzen, P.J., Lawrence, M.G., and Poschl, U. *Tellus* **A51,** 123 (1999).

Tropospheric ozone: distribution and sources. Fehsenfeld, F.C. and Liu, S.C. in *Global atmospheric chemical change.* Hewitt, C.N and Sturges, W.T. (eds.) (Chapman and Hall, London, 1994).

The Tropospheric Chemistry of Ozone in the Polar Regions. Niki, H. and Becker, K.-H. (eds.) (Springer-Verlag, Berlin, 1993).

Studies of oxidant production at the Weybourne Atmospheric Observatory in summer and winter conditions. Penkett, S.A., Clemitshaw, K.C., Savage, N.H., Burgess, R.A., Cardenas, L.M., Carpenter, L.J., McFadyen, G.G., and Cape, J.N. *J. atmos. Chem.* **33,** 111 (1999).

An evaluation of chemistry's role in the winter-spring ozone maximum found in the northern midlatitude free troposphere. Yienger, J.J., Klonecki, A.A., Levy, H., Moxim, W.J., Carmichael, G.R., *J. geophys. Res.* **104,** 3655 (1999).

Evidence for photochemical control of ozone concentrations in unpolluted marine air. Ayers, G. Penkett, S.A., Gillett, R.W., Bandy, B., Galbally, I.E., Meyer, C.P., Elsworth, C.M., Bentley, S.T., and Forgan, B.W. *Nature, Lond.* **360**, 446 (1992).

An important question, to which some uncertainty is still attached, is that of how much ozone in the troposphere is of tropospheric origin, and how much has been transported down from the stratosphere. The last papers of this group investigate the influence of tropospheric ozone on climate.

An estimate of the flux of stratospheric reactive nitrogen and ozone into the troposphere. Murphy, D.M. and Fahey, D.W. *J. geophys. Res.* **99,** 5325 (1994).

Tropospheric ozone: the role of transport. Levy, H., II, Mahlman, J.D., Moxim, W.J., and Liu, S.C. *J. geophys. Res.***90**, 3753 (1985).

Vertical profiles of tropospheric gases: chemical consequences of stratospheric intrusions. Bamber, D.J., Healey, P.G.W., Jones, B.M.R., Penkett, S.A., Tuck, A.F., and Vaughan, G. *Atmos. Environ.* **18**, 1759 (1984).

Stratospheric ozone in the lower troposphere. II. Assessment of downward flux and ground level impact. Viezee, W., Johnson, W.B., and Singh, H.B. *Atmos. Environ.* **17**, 1979 (1983).

On the origin of tropospheric ozone. Liu, S.C., Kley, D., McFarland, M., Mahlman, J.D., and Levy, H., II, *J. geophys. Res.* **85**, 7546 (1980).

Effects of anthropogenic emissions on tropospheric ozone and radiative forcing. Berntsen, T., Isaksen, I.S.A., Fuglestvedt, J.S., Myhre, G., Stordal, F., Freckleton, R.S., and Shine, K.P. *J. geophys. Res.* **102**, 28101 (1997).

Tropospheric ozone, its changes and possible radiative effect. Bojkov, R.D., World Meteorological Organization Special Environment Report No. 16 (1983).

Tropospheric ozone and climate. Fishman, J., Ramanathan, P.J., Crutzen, P.J., and Liu, S.C. *Nature, Lond.* **282**, 818 (1979).

Section 5.3.4

The oxides of nitrogen play a central role in tropospheric chemistry. Nitric oxide (NO) converts peroxy radicals to oxy radicals, and is itself oxidized to NO_2*. Nitrogen dioxide (*NO_2*) is a photochemical source of O atoms. The nitrate radical (*NO_3*) provides a potential source of night-time oxidation (see Section 5.3.6). Peroxyacetyl nitrate has been identified as a species that could be involved in the spatial transport of* NO_x*. One aspect that continues to stimulate investigation is the extent to which lightning discharges represent a source of the oxides of nitrogen.*

Reactive odd-nitrogen (NO_y) in the atmosphere, Roberts, J.M. in *Composition, Chemistry, and Climate of the Atmosphere.* Singh, H.B. (ed.) (Van Nostrand Reinhold, New York, 1995).

Ozone and peroxy radical budgets in the marine boundary layer: Modeling the effect of NO_x. Cox, R.A., *J. geophys. Res.* **104,** 8047 (1999).

Three-dimensional study of the relative contributions of the different nitrogen sources in the troposphere. Lamarque, J.-F., Brasseur, G.P., Hess, P.G., and Müller, J.-F. *J. geophys. Res.* **101**, 22955 (1996).

Biogenic nitric oxide from wastewater land application. Rammon, D.A., and Peirce, J.J. *Atmos. Environ.* **33,** 2115 (1999).

Nitric oxide emission from a Norway spruce forest floor. Pilegaard, K., Hummelshoj, P., and Jensen, N.O. *J. geophys. Res.* **104,** 3433 (1999).

On the origin of tropospheric ozone and NO_x over the tropical South Pacific. Schultz, M.G., Jacob, D.J., Wang, Y.H., Logan, J.A., Atlas, E.L., Blake, D.R., Blake, N.J., Bradshaw, J.D., Browell, E.V., Fenn, M.A., Flocke, F., Gregory, G.L., Heikes, B.G., Sachse, G.W., Sandholm, S.T., Shetter, R.E., Singh, H.B., and Talbot, R.W. *J. geophys. Res.* **104,** 5829 (1999).

Characterization of biogenic nitric oxide source strength in the southeast United States. Aneja, V.P., Roelle, P.A., and Robarge, W.P. *Env. Pollut.* **102,** 211 (1998).

NO_y blue ribbon panel. Crosley, D.R. *J. geophys. Res.* 101, 2049 (1996).

Empirical model of global soil-biogenic NO_x emissions. Yienger, J.J. and Levy, H., II. *J. geophys. Res.* **100**, 11447 (1995).

An evaluation of the mechanism of nitrous acid formation in the urban atmosphere. Calvert, J.G., Yarwood, G., and Dunker, A. *Res. Chem. Intermediates* **20,** 463 (1994).

Nitrogen oxides in the troposphere: global and regional budgets. Logan, J.A. *J. geophys. Res.* **88**, 10785 (1983).

Enhanced O_3 and NO_2 in thunderstorm clouds: Convection or production? Winterrath, T., Kurosu, T.P., Richter, A., and Burrows, J.P. *Geophys. Res. Letts.* **26**, 1291 (1999).

Vertical distributions of lightning NO_x for use in regional and global chemical transport models. Pickering, K.E., Wang, Y.S., Tao, W.K., Price, C., and Muller, J.F. *J. geophys. Res.* **103,** 31203 (1998).

NO_x from lightning: Global distribution based on lightning physics. Price, C., Penner, J., and Prather, M. *J. geophys. Res.* **102**, 5929 (1997).

Impact of lightning and convection on reactive nitrogen in the tropical free troposphere. Kawakami, S., Kondo, Y., Koike, M., Nakajima, H., Gregory, G.L., Sachse, G.W., Newell, R.E., Browell, E.V., Blake, D.R., Rodrigues, J.M., and Merrill, J.T. *J. geophys. Res.* **102**, 28367 (1997).

Lightning and atmospheric chemistry: the rate of atmospheric NO production. Lawrence, M.G., Chameides, W.L., Kasibhatla, P.S., Levy, H., and Moxim, W. in *Handbook of atmospheric electrodynamics, Vol. 1.* Volland, H. (ed.) (CRC Press Inc., Boca Raton, 1995).

Nitrogen oxides produced by lightning. Franzblau, E. and Popp, C.J. *J. geophys. Res.* **94,** 11089 (1989).

Production of nitrogen oxides by lightning discharges. Tuck, A.F. *Q. J. R. Meteorol. Soc.* **102**, 749 (1976).

Production of NO and N_2O by soil nitrifying bacteria. Lipschulz, F., Zafiriou, O.C., Wofsy, S.C., McElroy, M.B., Valois, F.W., and Watson, S.W. *Nature, Lond.* **294,** 641 (1981).

Production of nitrous oxide and consumption of methane by forest soils. Keller, M., Goreau, T., Wofsy, S.C., Kaplan, W.A., and McElroy, M.B. *Geophys. Res. Letts.* **10,** 1156 (1983).

The role of NO and NO_2 in the chemistry of the troposphere and stratosphere. Crutzen, P.J. *Annu. Rev. Earth & Planet. Sci.* **7**, 443 (1979).

The atmospheric chemistry of organic nitrates. Roberts, J.M. *Atmos. Environ.* **24A**, 243 (1990).

Physical and chemical influences on PAN concentrations at a rural site. McFadyen, G.G. and Cape, J.N. *Atmos. Environ.* **33,** 2929 (1999).

Simulated global tropospheric PAN: Its transport and impact on NO_x. Moxim, W.J., Levy, H.,II, and Kasibhatla, P.S. *J. geophys. Res.* **101**, 12621 (1996).

The global distribution of peroxyacetyl nitrate. Singh, H.B., Salas, L.J., and Viezee, W. *Nature, Lond.*, **321**, 588 (1986).

Atmospheric measurements of peroxyacetyl nitrate (PAN) in rural southeast England: seasonal variations, winter photochemistry and long range transport. Brice, K.A., Penkett, S.A., Atkins, D.H.F., Sandalls, F.J., Bamber, D.J., Tuck, A.F., and Vaughan, G. *Atmos. Environ.* **18**, 2691 (1984).

Influence of peroxyacetyl nitrate (PAN) on odd nitrogen in the troposphere and lower stratosphere Aikin, A.C., Herrnan, J.R., Maier, E.J.R., and McQuillan, C.J. *Planet. Space Sci.* **31**, 1075 (1983).
Peroxyacetyl nitrate in the free troposphere. Singh, H.B. and Salas, L.J. *Nature, Lond.* **302**, 326 (1983).

Section 5.3.5

The first paper is representative of current thinking about the OH + CO reaction, while the second one looks at the indirect atmospheric consequences of the reaction.

The pressure dependence of the rate constant of the reaction of OH radicals with CO. Paraskevopoulos, G. and Irwin, R.S. *J. chem. Phys.* **80**, 259 (1984).
Correlative nature, of ozone and carbon monoxide in the troposphere: implication for the tropospheric ozone budget. Fishman, J. and Seiler, W. *J. geophys. Res.* **88**, 3662 (1983).

Section 5.3.6

The nitrate radical: physics, chemistry and the atmosphere. Wayne, R.P. (ed.) *Atmos. Environ.* **25A**, 1 (1991).
Experimental study of the nitrate radical. Wayne, R.P. in *N-centered radicals*, Alfassi, Z.B. (ed.) (Wiley, Chichester, 1998).
Observations of the nitrate radical in the marine boundary layer. Allan, B.J., Carslaw, N., Coe, H., Burgess, R.A., and Plane, J.M.C. *J. atmos. Chem.* **33**, 129 (1999).
The vertical distribution of NO_3 in the atmospheric boundary layer. Fish, D.J., Shallcross, D.E., and Jones, R.L. *Atmos. Environ.* **33**, 687 (1999).
Observations of the nitrate radical in the free troposphere at Izaña de Tenerife. Carslaw, N., Plane, J.M.C., Coe, H., and Cuevas, E. *J. geophys. Res.* **102**, 10613 (1997).
Simultaneous observations of nitrate and peroxy radicals in the marine boundary layer. Carslaw, N., Carpenter, L.J., Plane, J.M.C., Allan, B.J., Burgess, R.A., Clemitshaw, K.C., Coe, H., and Penkett, S.A. *J. geophys. Res.* **102**, 18917 (1997).
Kinetics and mechanisms of the gas-phase reactions of the NO_3 radical with organic compounds. Atkinson, R. *J. phys. Chem. Ref. Data* **20**, 459 (1991).
Reactive uptake of NO_3 on pure water and ionic solutions. Rudich, Y., Talukdar, R.K., Fox, R.W., and Ravishankara, A.R. *J. geophys. Res.* **101**, 21023 (1996).
Long-term observation of nitrate radicals at the Tor station, Kap Arkona (Rügen). Heintz, F., Platt, U., Flentje, H., and Dubois, R. *J. geophys. Res.* **101**, 22891 (1996).
A study of nighttime nitrogen oxide oxidation in a large reaction chamber — the fate of NO_2, N_2O_5, HNO_3, and O_3 at different humidities. Mertel, T.F., Bleilebens, D., and Wahner, A. *Atmos. Environ.* **30**, 4007 (1996).

Nitrate radicals in tropospheric chemistry. Platt, U., and Heintz, F. *Israel J Chem.* **34,** 289 (1994).

Section 5.3.7

Gas-phase ozonolysis of alkenes and the role of the Criegee intermediate. Sander, W., Träubel, M., and Komnick, P. in *Chemical processes in atmospheric oxidation.* Le Bras, G. (ed.) EUROTRAC Report Volume 3. (Springer-Verlag, Berlin, 1997).

A mechanistic study on the ozonolysis of ethene. Thomas, W., Zabel, F., Becker, K.-H., and Fink, E.H. in *Tropospheric Oxidation Mechanisms.* Becker, K.-H. (ed.) Report EUR 16171 EN (European Commission, Luxembourg, 1995).

Tropospheric OH formation by ozonolysis of terpenes. Pfeiffer, T., Forberich, O., and Comes, F.J *Chem. Phys. Letts.* **298**, 351 (1998).

Direct observation of OH production from the ozonolysis of olefins. Donahue, N.M., Kroll, J.H., Anderson, J.G., and Demerjian, K.L. *Geophys. Res. Letts.* **25,** 59 (1998).

OH radical formation yields from the gas-phase reactions of O_3 with alkenes and monoterpenes. Chew, A.A. and Atkinson, R. *J. geophys. Res.* **101**, 28649 (1996).

OH radical production from the gas-phase reactions of O_3 with a series of alkenes under atmospheric conditions. Atkinson, R. and Aschmann, S.M. *Environ. Sci. Technol.* **27,** 1357 (1993).

Formation of secondary ozonides in the gas-phase ozonolysis of simple alkenes. Neeb, P., Horie, O., and Moortgat, G.K. *Tetrahedron Letts.* **37,** 9297 (1996).

The nature of the transitory product in the gas-phase ozonolysis of ethene. Neeb, P., Horie, O., and Moortgat, G.K. *Chem. Phys. Letts.* **246,** 150 (1995).

Products of the gas-phase reactions of O_3 with alkenes. Atkinson, R., Tuazon, E.C., and Aschmann, S.M. *Environ. Sci. Technol.* 29, 1860 (1995).

Ozonolysis of trans-2-butenes and cis-2-butenes in low parts-per-million concentration ranges. Horie, O., Neeb, P., and Moortgat, G.K. *Int. J. Chem. Kinet,* **26,** 1075 (1994).

Formation of formic acid and organic peroxides in the ozonolysis of ethene with added water vapor. Horie, O., Neeb, P., Limbach, S., and Moortgat, G.K. *Geophys. Res. Letts.* **21**, 1523 (1994).

Decomposition pathways of the excited Criegee intermediates in the ozonolysis of simple alkenes. Horie, O., and Moortgat, G.K. *Atmos. Environ.* **25A**, 1881 (1991).

Atmospheric ozone–olefin reactions. Niki, H., Maker, P.D., Savage, C.M., and Breitenbach, L.P. *Environ. Sci. Technol.* **17**, 312A (1983).

Section 5.4

See initially the first three citations given for Sections 5.1 and 5.2.

Atmospheric VOCs from natural sources. Hewitt, C.N., Cao, X.-L., Boissard, C, and Duckham, S.C. in *Volatile organic compounds in the atmosphere.* Hester, R.E. and Harrison, R.M. (eds.) (Royal Society of Chemistry, Cambridge, 1995).

Trace gas emissions by plants. Sharkey, T.J., Holland, E.A., and Mooney, H.A. (eds.) (Academic Press, San Diego, 1991).

A global model of natural volatile organic compound emissions. Guenther, A., and 15 others *J. geophys. Res.* 100, 8873 (1995).

Methane

Atmospheric methane: sources, sinks, and role in global change. Khalil, M.A.K. (ed.) (Springer-Verlag, 1993).

Changing concentration, lifetime and climate forcing of atmospheric methane. Lelieveld, J., Crutzen, P.J., and Dentener, F.J. *Tellus* **B50**, 128 (1998).

Effects of production and oxidation processes on methane emissions from rice fields. Khalil, M.A.K., Rasmussen, R.A., and Shearer, M.J. *J. geophys. Res.* **103,** 25233 (1998).

An inverse modeling approach to investigate the global atmospheric methane cycle. Hein, R., Crutzen, P.J., and Heimann, M. *Global Biogeochem. Cycles* **11**, 43 (1997).

Anthropogenic sources of halocarbons, sulfur hexafluoride, carbon monoxide, and methane in the southeastern United States. Bakwin, P.S., Hurst, D.F., Tans, P.P., and Elkins, J.W. *J. geophys. Res.* **102,** 15915 (1997).

Methane in the tropical South Atlantic: Sources and distribution during the late dry season. Bartlett, K.B., Sachse, G.W., Collins, J.E. Jr., and Harriss, R.C. *J. geophys. Res.* **101**, 24139 (1996).

The growth rate and distribution of atmospheric methane. Dlugokencky, E.J., Steele, L.P., Lang, P.M., and Masarie, K.A. *J. geophys. Res.* **99**, 17021 (1994).

A dramatic decrease in the growth rate of atmospheric methane in the Northern Hemisphere during 1992. Dlugokencky, E.J., Masarie, K.A., Lang, P.M., Tans, P.P., Steele, L.P., and Nisbet, E.G. *Geophys. Res. Letts.* **21**, 45 (1994).

Models and observations of the impact of natural hydrocarbons in rural ozone. Trainer, M., Williams, E.T., Parrish, D.D., Buhr, M.P., Allwine, E.J., Westberg, H.H., Fehsenfeld, F.C., and Liu, S.C. *Nature* **329**, 705 (1987).

The influence of termites on atmospheric trace gases: CH_4, CO, $CHCl_3$, N_2O, CO, H_2, and light hydrocarbons. Khalil, M.A.J., Rasmussen, R.A., French, J.R.J., and Holt, J.A. *J. geophys. Res.* **95**, 3619 (1990).

Global production of methane by termites. Rasmussen, R.A. and Khalil, M.A.K. *Nature, Lond.* **301**, 700 (1983).

Models and observations of the impact of natural hydrocarbons in rural ozone. Trainer, M., Williams, E.T., Parrish, D.D., Buhr, M.P., Allwine, E.J., Westberg, H.H., Fehsenfeld, F.C., and Liu, S.C. *Nature* **329**, 705 (1987).

Non-methane hydrocarbons

Biogenic emissions of volatile organic compounds from higher plants. Fall, R. in *Reactive hydrocarbons in the atmosphere.* Hewitt, C.N. (ed.) (Academic Press, San Diego, 1999).

The impact of nonmethane hydrocarbon compounds on tropospheric photochemistry. Houweling, S., Dentener, F., and Lelieveld, J. *J. geophys. Res.* **103,** 10673 (1998).

Biogenic hydrocarbon contribution to the ambient air of selected areas. Arnts, R.R. and Meeks, S.A. *Atmos. Environ.* **15,** 1643 (1981).

Organic material in the global troposphere. Duce, R.A., Mohnen, V.A., Zimmerman, P.R., Grosjean, D., Cautreels, W., Chatfield, R., Jaenicke, R., Ogren, J.A., Pellizzari, E.D., and Wallace, G.T. *Revs. Geophys. Space Phys.* **21,** 921 (1983).

Measurements of atmospheric hydrocarbons and biogenic emission fluxes in the Amazon boundary layer. Zimmerman, P.R., Greenberg, J.P., and Westberg, C.E. *J. geophys. Res.* **93,** 1407 (1988).

Tropospheric reactions of isoprene and oxidation products: kinetic and mechanistic studies, Ruppert, L., Barnes, I, and Becker, K.-H. in *Tropospheric Oxidation Mechanisms.* Becker, K.-H. (ed.) Report EUR 16171 EN (European Commission, Luxembourg, 1995).

The OH radical initiated oxidation of isoprene: mechanism construction. Jenkin, M.E., and Hayman, G.D. in *Tropospheric Oxidation Mechanisms.* Becker, K.-H. (ed.) Report EUR 16171 EN (European Commission, Luxembourg, 1995).

Relationships among isoprene emission rate, photosynthesis, and isoprene synthase activity as influenced by temperature. Monson, R., Jaeger, C., Adams, W., Driggers, E., Silver, G., and Fall, R. *Plant Physiol.* **92** 1175 (1992).

Gas-phase terpene oxidation products: a review. Calogirou, A., Larsen, B.R., and Kotzias, D. *Atmos. Environ.* **33,** 1423 (1999).

Volatile organic compound emission rates from mixed deciduous and coniferous forests in Northern Wisconsin, USA. Isebrands, J.G., Guenther, A.B., Harley, P., Helmig, D., Klinger, L., Vierling, L., Zimmerman, P., and Geron, C. *Atmos. Environ.* **33,** 2527 (1999).

Chemical processing of biogenic hydrocarbons within and above a temperate deciduous forest. Makar, P.A., Fuentes, J.D., Wang, D., Staebler, R.M., and Wiebe, H.A. *J. geophys. Res.* **104,** 3581 (1999).

Effects of ozone concentrations on biogenic volatile organic compounds emission in the Mediterranean region. Penuelas, J., Llusia, J., and Gimeno, B.S. *Env. Pollut.* **105,** 17 (1999).

Ecological and evolutionary aspects of isoprene emission from plants. Harley, P.C., Monson, R.K., and Lerdau, M.T. *Oecologia* **118,** 109 (1999).

Emissions of volatile organic compounds from cut grass and clover are enhanced during the drying process. de Gouw, J.A., Howard, C.J., Custer, T.G., and Fall, R. *Geophys. Res. Letts.* **26,** 811 (1999).

Evolutionary significance of isoprene emission from mosses. Hanson, D.T., Swanson, S., Graham, L.E., and Sharkey, T.D. *Amer. J. Bot.* **86,** 634 (1999).

Emission of reactive terpene compounds from orange orchards and their removal by within-canopy processes. Ciccioli, P., Brancaleoni, E., Frattoni, M., DiPalo, V., Valentini, R., Tirone, G., Seufert, G., Bertin, N., Hansen, U., Csiky, O., Lenz, R., and Sharma, M. *J. geophys. Res.* **104,** 8077 (1999).

Atmospheric formic and acetic acids: An overview. Khare, P., Kumar, N., Kumari, K.M., and Srivastava, S.S. *Rev. Geophys.* **37,** 227 (1999).

Measurements of PAN, PPN, and MPAN made during the 1994 and 1995 Nashville intensives of the Southern Oxidant study: implications for regional ozone production from biogenic hydrocarbons. Roberts, J.M and 18 others. *J. geophys. Res.* **103**, 22473 (1998).

The role of biogenic hydrocarbons in urban photochemical smog: Atlanta as a case study. Chameides, W.L., Lindsay, R.W., Richardson, J., and Kiang, C.S. *Science* **241**, 1473 (1998).

Nighttime isoprene chemistry at an urban-impacted forest site. Starn, T.K., Shepson, P.B., Bertman, S.B., Riemer, D.D., Zika, R.G., and Olszyna, K. *J. geophys. Res.* **103**, 22437 (1998).

Observations of isoprene chemistry and its role in ozone production at a semirural site during the 1995 Southern Oxidants study. Starn, T.K., Shepson, P.B., Bertman, S.B., White, J.S., Splawn, B.G., Riemer, D.D., Zika, R.G., and Olszyna, K. *J. geophys. Res.* **103**, 22425 (1998).

Volatile organic compounds and isoprene oxidation products at a temperate deciduous forest site. Helmig, D., Greenberg, J., Guenther, A., Zimmerman, P., and Geron, C. *J. geophys. Res.* **103**, 22397 (1998).

Nonmethane hydrocarbon measurements during the tropospheric OH photochemistry experiment. Goldan, P.D., Kuster, W.C., and Fehsenfeld, F.C. *J. geophys. Res.* **102**, 6315 (1997).

Formation of organic aerosols from the oxidation of biogenic hydrocarbons. Hoffmann, T., Odum, J.R., Bowman, F., Collins, D., Klokow, D., Flagan, R.C., and Seinfeld, J.H. *J. atmos. Chem.* **26**, 189 (1997).

Distribution and seasonality of selected hydrocarbons and halocarbons over the Western Pacific basin during PEM-West A and PEM West B. Blake, N.J., Blake, D.R., Chen, T.-Y., Collins, J.E. Jr., Sachse, G.W., Anderson, B.E., and Rowland, F.S. *J. geophys. Res.* **102**, 28315 (1997).

Development and evaluation of a detailed mechanism for the atmospheric reactions of isoprene and NO_x. Carter, W.P.L., and Atkinson, R. *Int. J. Chem. Kinet.* **28,** 497 (1996).

Seasonal measurements of nonmethane hydrocarbons and carbon monoxide at the Mauna Loa Observatory during the Mauna Loa Observatory Photochemistry Experiment 2. Greenberg, J.P., Helmig, D., and Zimmerman, P.R. *J. geophys. Res.* **101**, 14581 (1996).

Carboxylic acids in the rural continental atmosphere over the Eastern United States during the Shenandoah Cloud and Photochemistry Experiment. Talbot, R.W., Mosher, B.W., Heilas, B.G., Jacob, D.J., Munger, H.W., Daube, B.C., Keene, W.C., Maben, J.R., and Artz, R.S. *J. geophys. Res.* **100**, 9335 (1995).

Measurements of 3-methyl furan, methyl vinyl ketone, and methacrolein at a rural forested site in the southeastern United States. Montzka, S.A., Trainer, M., Angevine, W.M., and Fehsenfeld, F.C. *J. geophys. Res.* **100** 11393 (1995).

Biogenic emissions in Europe. 1. Estimates and uncertainties. Simpson, D., Guenther, A., Hewitt, C.N., and Steinbrecher, R. *J. geophys. Res.* **100,** 22875 (1995).

Seasonal and diurnal variation of isoprene and its reaction products in a semi-rural area. Yokouchi, Y. *Atmos. Environ.* **28**, 2651 (1994).

High concentrations and photochemical fate of oxygenated hydrocarbons in the global troposphere. Singh, H.B., Kanakidou, M., Crutzen, P.J., and Jacob, D.J. *Nature, Lond.* **378,** 50 (1994).

Isoprene and its oxidation-products, methyl vinyl ketone and methacrolein, in the rural troposphere. Montzka, S.A., Trainer, M., Goldan, P.D., Kuster, W.C., and Fehsenfeld, F.C. *J. geophys. Res.* **98,** 1101 (1993).

A biogenic hydrocarbon emission inventory for the U.S.A. using a simple forest canopy model. Lamb, B., Gay, D., Westberg, H., and Pierce, T. *Atmos. Environ.* **27A**, 1673 (1993).

Acetone in the atmosphere: distribution, sources, and sinks. Singh, H.B., O'Hara, D., Hereth, D., Sachse, W., Blake, D.R., Bradshaw, J.D., Kanakidou, M., and Crutzen, P.J. *J. geophys. Res.* **99**, 1805 (1993).

Atmospheric photooxidation of isoprene Part II: The ozone–isoprene reaction. Paulson, S.E., Flagan, R.C., and Seinfeld, J.H. *Int. J. Chem. Kinet.* **24,** 103 (1992).

Atmospheric photooxidation of isoprene. Part I: The hydroxyl radical and ground state atomic oxygen reactions. Paulson, S.E., Flagan, R.C., and Seinfeld, J.H. *Int. J. Chem. Kinet.,* **24,** 79 (1992).

Biological input to visibility-reducing aerosol particles in the remote arid southwestern United States. Mazurek, M.A., Cass, G.R., and Simoneit, B.R.T. *Environ. Sci. Technol.* **25,** 684 (1991).

Reactions of OH with α-pinene and β-pinene in air: estimate of global CO production from the atmospheric oxidation of terpenes. Hatakeyama, S., Izumi, K., Fukuyama, T., Akimoto, H., and Washida, N. *J. geophys. Res.* **96**, 947 (1991).

Reactions of ozone with α-pinene and β-pinene in air: yields of gaseous and particulate products. Hatakeyama, S., Izumi, K., Fukuyama, T., and Akimoto, H. *J. geophys. Res.* **94**, 13013 (1989).

Volatile atmospheric compounds in the atmosphere of forests. Isodorov, V.A., Zenkevich, I.G., and Ioffe, B.V. *Atmos. Environ.* 19, 1 (1985).

Volatile organic material of plant origin in the atmosphere. Rasmussen, R.A., and Went, F.W. *Proc. Natl. Acad. Sci.* **53,** 215 (1965).

Section 5.5

Oxidation of aromatic compounds. Section 2.7 in *Chemical processes in atmospheric oxidation.* Le Bras, G. (ed.) EUROTRAC Report Volume 3. (Springer-Verlag, Berlin, 1997).

Mechanism of atmospheric photooxidation of aromatics: a theoretical study. Andino, J.M., Smith, J.N., Flagan, R.C., Goddard, W.A., III, and Seinfeld, J.H. *J. phys. Chem.* **100**, 10967 (1996).

Section 5.6

Sulfur in the atmosphere. Berresheim, H., Wine, P.H., and Davis, D.D., in *Composition, chemistry, and climate of the atmosphere*. Singh, H.B. (ed.) (Van Nostrand, New York, 1995).

Biogenic sulfur in the environment. Saltman, E.S. and Cooper, W.J. (eds.) ACS Symposium Series No 393, American Chemistry Society, Washington, DC, 1989.

Oxidation of atmospheric reduced sulphur compounds: perspective from laboratory studies. Ravishankara, A.R., Rudich, Y., Talukdar, R., and Barone, S. *Phil. Trans. Roy. Soc. Lond.* **B352**, 171 (1997).

Reaction of OH with dimethyl sulfide. 2. Products and mechanisms. Turnipseed, A.A., Barone, S.B., and Ravishankara, A.R. *J. phys. Chem.*, **100,** 14703 (1996).

FTIR product study of the OH-initiated oxidation of dimethyl sulphide: observation of carbonyl sulphide and dimethyl sulphoxide. Barnes, I., Becker, K.-H., and Patroescu, I. *Atmos. Environ.* **30,** 1805 (1996).

Role of adducts in the atmospheric oxidation of dimethyl sulfide. Barone, S.B., Turnipseed, A.A., and Ravishankara, A.R. *Faraday Discuss.* **100**, 39 (1995).

Human impact on the atmospheric sulfur balance. Rodhe, H. *Tellus* **A51,** 110 (1999).

Dimethyl sulfide oxidation in the equatorial Pacific: Comparison of model simulations with field observations for DMS, SO_2, H_2SO_4(g), MSA(g), MS, and NSS. Davis, D., Chen, G., Bandy, A., Thornton, D., Eisele, F., Mauldin, L., Tanner, D., Lenschow, D., Fuelberg, H., Huebert, B., Heath, J., Clarke, A., and Blake, D. *J. geophys. Res.* **104,** 5765 (1999).

DMS oxidation in the Antarctic marine boundary layer: Comparison of model simulations and field observations of DMS, DMSO, $DMSO_2$, H_2SO_4 (g), MSA (g) and MSA (p). Davis, D.D., Chen, G., Kasidhatla, P., Jefferson, A., Tanner, D., Eisele, F., Lenshow, D., Neff, W., and Berresheim, H. *J. geophys. Res.* **103**, 1657 (1998).

Sulfur chemistry in the Antarctic troposphere experiment: An overview of project SCATE. Berresheim, H. and Eisele, F.L. *J. geophys. Res.* **103**, 1619 (1998).

Marine sulphur emissions. Liss, P.S., Hatton, A.D., Malin, G., Nightingale, P.D., and Turner, S.M. *Phil. Trans. Royal Soc.* **B352**, 159 (1997).

The interaction between the nitrogen and sulfur cycles in the marine boundary layer. Yvon, S.A., Plane, J.M.C., Nien, C.-F., Cooper, D.J., and Saltman, E.S. *J. geophys. Res.* **101**, 1379 (1996).

Global gridded inventories of anthropogenic emissions of sulfur and nitrogen. Benkovitz, C.M., Scholtz, T.M., Pacyna, J., Rarrason, L., Dignon, J., Voldner, E.C., Spiro, P.A., Logan, J.A., and Graedel, T.E. *J. geophys. Res.* **101**, 29239 (1996).

Sulfur emissions to the stratosphere from explosive volcanic eruptions. Pyle, D.M., Beattie, P.D. and Bluth, G.J.S. *Bull. Volcan.* **57**, 663 (1996).

The tropospheric oxidation of dimethyl sulfide: a new source of carbonyl sulfide. Barnes, I., Becker, K.-H., and Patroescu, I. *Geophys. Res. Letts.* **21**, 2389 (1994).

A global three-dimensional model of the tropospheric sulfur cycle. Langner, J. and Rodhe, H. *J. atmos. Chem.* **13**, 225-263 (1991).

Global ocean–to–atmosphere dimethyl sulfide flux. Erickson, D.J., III, Ghan, S.J., and Penner, J.E. *J. geophys. Res.* **95**, 7543 (1990).

Human influence on the sulphur cycle. Brimblecombe, P., Hammer, C., Rohde, H., Ryaboshapko, A., and Boutron, C.F. SCOPE **39**, 77 (1989).

Are global cloud albedo and climate controlled by marine phytoplankton? Schwartz S.E. *Nature, Lond.* **336**, 441 (1988).

Section 5.7

Heterogeneous chemistry in the troposphere: experimental approaches and applications to the chemistry of sea salt particles. DeHaan, D.O., Brauers, T., Oum, K., Stutz, J., Nordmeyer, T., and Finlayson-Pitts, B.J. *Int. Revs. Phys. Chem.* **18**, No. 3 (1999).

Reactive Halogen Compounds in the Atmosphere. Fabian, P. and Singh, O.N. (eds.) Springer-Verlag, Heidelberg, 1998.

Halogens in the atmospheric environment. Singh, H.B. in *Composition, Chemistry, and Climate of the Atmosphere.* Singh, H.B. (ed.) (Van Nostrand Reinhold, New York, 1995).

Formation of molecular chlorine from the photolysis of ozone and aqueous sea-salt particles. Oum, K.W., Lakin, M.J., DeHaan, D.O., Brauers, T., and Finlayson-Pitts, B.J. *Science* **279**, 74 (1998).

Non-sea-salt-sulphate formation in sea-salt aerosol. Clegg, N.A., and Toumi, R. *J. geophys. Res.* **103,** 31095 (1998).

Production and decay of $ClNO_2$ from the reaction of gaseous N_2O_5 with NaCl: bulk and aerosol experiments. Behnke, W., George, C., Scheer, V., and Zetzsch, C. *J. geophys. Res.* **102**, 3795 (1997).

A mechanism for halogen release from sea-salt aerosol in the remote marine boundary layer. Vogt, R., Crutzen, P.J., and Sander, R. *Nature* **383,** 327 (1996).

Formation of chemically active chlorine compounds by reactions of atmospheric NaCl particles with gaseous N_2O_5 and $ClONO_2$. Finlayson-Pitts, B.J., Ezell, M.J., and Pitts, J.N., Jr. *Nature, Lond.* **337,** 241 (1989).

Uptake coefficients of NO_3 radicals on solid surfaces of sea-salts. Gratpanche, F. and Sawerysyn, J.P. *J. Chim. phys.* **96,** 213 (1999).

Tropospheric budget of reactive chlorine. Graedel, T.E. and Keene, W.C. *Global Biogeochem. Cycles.* **9**, 47 (1995).

The chemistry of halogen compounds in the Arctic troposphere. Platt, U. in *Tropospheric Oxidation Mechanisms.* Becker, K-H. (ed.) Report EUR 16171 EN (European Commission, Luxembourg, 1995).

Tetrachlororthylene as an indicator of low Cl atom concentrations in the troposphere. Singh, H.B., Thakur, A.N., Chen, Y.E., and Kanakidou, M. *Geophys. Res. Letts.* **23**, 1529 (1996).

The budget of ethane and tetrachloroethene: Is there evidence for an impact of reactions with chlorine atoms in the troposphere? Rudolph, J., Koppmann, R., and Plass-Dulmer, C.H. *Atmos. Environ.* **30**, 1887 (1996).

Short-term variations in the C-13/C-12 ratio of CO as a measure of Cl activation during tropospheric ozone depletion events in the Arctic. Rockmann, T., Brenninkmeijer, C.A.M., Crutzen, P.J., and Platt, U. *J. geophys. Res.* **104,** 1691 (1999).

Hydrocarbon measurements during tropospheric ozone depletion events: Evidence for halogen atom chemistry. Ramacher, B., Rudolph, J., and Koppmann, R. *J. geophys. Res.* **104,** 3633 (1999).

Depletion of lower tropospheric ozone during Arctic spring: The Polar Sunrise Experiment, 1988. Bottenheim, J.W., Barrie, L.A., Atlas, E., Heidt, L.E., Wiki, H., Rasmussen, R.A., and Shepson, P.B. *J. geophys. Res.* **95**, 18555 (1990).

Ozone destruction and bromine photochemistry in the Arctic spring. Finlayson-Pitts, B.J., Livingston, F.E., and Berko, H.N. *Nature, Lond.* **343**, 622 (1990).

DOAS measurements of tropospheric bromine oxide in mid-latitudes. Hebestreit, K., Stutz, J., Rosen, D., Matveiv, V., Peleg, M., Luria, M., and Platt, U. *Science* **283,** 55 (1999).

Evidence for bromine monoxide in the free troposphere during the Arctic polar sunrise. McElroy, C.T., McLinden, C.A., and McConnell, J.C. *Nature, Lond.* **397,** 338 (1999).

Global distribution of atmospheric bromine monoxide from GOME on Earth observing satellite ERS-2. Hegels, E., Crutzen, P.J., Klupfel, T., Perner, D., and Burrows, J.P. *Geophys. Res. Letts.* **25,** 3127 (1998).

GOME observations of tropospheric BrO in northern hemispheric spring and summer 1997. Richter, A., Wittrock, F., Eisinger, M., and Burrows, J.P. *Geophys. Res. Letts.* **25,** 2683 (1998).

Analysis of BrO measurements from the Global Ozone Monitoring Experiment. Chance, K. *Geophys. Res. Letts.* **25,** 3335 (1998).

Gas phase atmospheric bromine photochemistry. Lary, D.J. *J. geophys. Res.* **101**, 1505 (1996).

Spring measurements of tropospheric bromine at Barrow, Alaska. Sturges, W.T., Schnell, R.C., Dutton, G.S., Garcia, S.R., and Lind, J.A. *Geophys. Res. Letts.* **20**, 201 (1995).

A possible mechanism for combined chlorine and bromine catalyzed destruction of tropospheric ozone in the Arctic. Le Bras, G. and Platt, U. *Geophys. Res. Letts.* **22,** 599 (1995).

Spectroscopic measurement of bromine oxide and ozone in the high Arctic during Polar Sunrise Experiment 1992. Hausmann, M. and Platt, U. *J. geophys. Res.* **99**, 25399 (1994).

Iodine oxide in the marine boundary layer. Alicke, B., Hebestreit, K., Stutz, J., and Platt, U. *Nature, Lond.* **397**, 573 (1999).

Iodine chemistry and its role in halogen activation and ozone loss in the marine boundary layer: A model study. Vogt, R., Sander, R., Von Glasow, R., and Crutzen, P.J. *J. atmos. Chem.* **32,** 375 (1999).

Volatilization of methyl iodide from the soil-plant system. Muramatsu, Y. and Yoshida, S. *Atmos .Environ.* **29**, 21 (1995).

Short-lived alkyl iodides and bromides at Mace Head, Ireland: Links to biogenic sources and halogen oxide production. Carpenter, L.J., Sturges, W.T., Penkett, S.A., Liss, P.S., Alicke, B., Hebestreit, K., and Platt, U. *J. geophys. Res.* **104,** 1679 (1999).

Potential impact of iodine on tropospheric levels of ozone and other critical oxidants. Davis, D., Crawford, J., Liu, S., McKeen, S., Bandy, A., Thornton, D., Rowland, F., and Blake, D. *J. geophys. Res.* **101,** 2135 (1996).

Iodine: its possible role in tropospheric photochemistry. Chameides, W.L. and Davis, D.D. *J. geophys. Res.* **85,** 7383 (1980).

The dimethyl sulfide reaction with atomic chlorine, and its implications for the budget of methyl chloride. Langer, S., McGovney, B.T., Finlayson-Pitts, B.J., and Moore, R.M. *Geophys. Res. Letts.* **13,** 1661 (1996).

Reactive chlorine: A potential sink for dimethylsulphide and hydrocarbons in the marine boundary layer. Keene, W.C., Jacob, D.J. and Fan, S.-M. *Atmos. Environ.* **30,** (1996).

Aqueous-phase chemical processes in deliquescent sea-salt aerosols — a mechanism that couples the atmospheric cycles of S and sea salt. Chameides, W.L. and Stelson, A.W. *J. geophys. Res.* **97,** 20565 (1992).

Atmospheric methyl chloride. Khalil, M.A.K. and Rasmussen, R.A. *Atmos. Environ.* **33,** 1305 (1999).

Atmospheric chloroform. Khalil, M.A.K. and Rasmussen, R.A., *Atmos. Environ.* **33,** 1151 (1999).

Natural production of chlorinated organic compounds in soil. Hoekstra, E.J. and De Leer, E.W.B. in *Contaminated Soils 1993.* Arendt, F., Annohlee, G.J., Bosman, R., and van de Brink, W.J. (eds.) (Kluwer, Dordrecht, 1993).

Natural production of chloroform by fungi. Hoekstra, E.J., Verhagen, F.J.M., Field, J.A., De Leer, E.W.B., and Brinkman, U.A.T. *Phytochemistry* **49,** 91 (1998).

Section 5.8

Much tropospheric chemistry involves processes occurring at surfaces or within aerosols and raindrops. Further general references to this very important, but rather poorly understood, topic can be found in the lists for Sections 1.3 and 3.4.3–3.4.5.

Heterogeneous and liquid-phase processes. Warneck, P. (ed.) EUROTRAC Report Volume 2. (Springer-Verlag, Berlin, 1997).

Aquatic Chemistry (3rd edn.) Stumm, W. and Morgan, J.J. (Wiley, New York, 1996).

Laboratory Studies of Atmospheric Heterogeneous Chemistry. Kolb, C.E., Worsnop, D.R., Zahniser, M.S., Davidovits, P., Keyser, L.F., Leu, M.-T., Molina, M.J., Hanson, D.R., Ravishankara, A.R., Williams, L.R., and Tolbert, M.A. in *Progress and Problems in Atmospheric Chemistry*, Barker, J.R., (ed.) (World Scientific Publishing, Singapore, 1995).

A chemical aqueous-phase radical mechanism for tropospheric chemistry. Herrmann, H., Ervens, B., Nowacki, P., Wolke, R., and Zellner, R. *Chemosphere* **38,** 1223 (1999).

Atmospheric aerosols: biogeochemical sources and role in atmospheric chemistry. Andreae, M.O. and Crutzen, P.J. *Science* **221**, 744 (1997).

Heterogeneous and multiphase chemistry in the troposphere. Ravishankara, A.R. *Science* **276**, 1058 (1997).

Formation of organic aerosols from the oxidation of biogenic hydrocarbons. Hoffmann, T., Odum, J.R., Bowman, F., Collins, D., Klokow., Flagan, R.C., and Seinfeld, J.H. *J. atmos. Chem.* **26**, 189 (1997).

Carbon aerosols and atmospheric photochemistry. Lary, D.J., Lee, A.M.,Toumi, R., Newchurch, M.J., Pirre, M., and Renard, J.B. *J. geophys. Res.* **102**, 3671 (1997).

Effect of aqueous phase cloud chemistry on tropospheric ozone. Liang, J. and Jacob, D.J. *J. geophys. Res.* **102,** 5993 (1997).

The effects of clouds on aerosol and chemical species production and distribution. 1. Cloud model formulation, mixing, and detrainment. Taylor, G.R., Kreidenweis, S., and Zhang, Y.P. *J. geophys. Res.* **102,** 23851 (1997).

Tropospheric ozone chemistry. The impact of cloud chemistry. Jonson, J.E. and Isaksen, I.S.A. *J. atmos. Chem.* **16,** 99 (1993).

Fine particles in the global troposphere—a review. Heintzenberg, J. *Tellus* **41B**, 149 (1989).

The free radical chemistry of cloud droplets and its impact upon the composition of rain. Chameides, W.L., and Davis, D.D. *J. geophys. Res.* **87,** 4863 (1982).

Organic acids in continental background aerosols. Limbeck, A., and Puxbaum, H. *Atmos. Environ.* **33,** 1847 (1999).

Chemical characteristics of free tropospheric aerosols over the Japan Sea coast: aircraft-borne measurements. Mori, I., Iwasaka, Y., Matsunaga, K., Hayashi, M., and Nishikawa, M. *Atmos. Environ.* **33,** 601 (1999).

Arctic lower tropospheric aerosol trends and composition at Alert, Canada: 1980-1995. Sirois, A. and Barrie, L.A. *J. geophys. Res.* **104,** 11599 (1999).

Characterization of aerosols over ocean from POLDER/ADEOS-1. Deuze, J.L., Herman, M., Goloub, P., Tanre, D., and Marchand, A. *Geophys. Res. Letts.* **26,** 1421 (1999).

Processes determining the relationship between aerosol number and non-sea-salt sulfate mass concentrations in the clean and perturbed marine boundary layer. Van Dingenen, R., Raes, F., Putaud, J.P., Virkkula, A., and Mangoni, M. *J. geophys. Res.* **104,** 8027 (1999).

Nanoparticle formation in marine airmasses: contrasting behaviour of the open ocean and coastal environments. Allen, A.G., Grenfell, J.L., Harrison, R.M., James, J., and Evans, M.J. *Atmos. Res.* **51,** 1 (1999).

Heterogeneous reactions on and in sulfate aerosols: Implications for the chemistry of the midlatitude tropopause region. Hendricks, J., Lippert, E., Petry, H., and Ebel, A. *J. geophys. Res.* **104,** 5531 (1999).

Aerosol chemical composition and distribution during the Pacific Exploratory Mission (PEM) Tropics. Dibb, J.E., Talbot, R.W., Scheuer, E.M., Blake, D.R., Blake, N.J., Gregory, G.L., Sachse, G.W., and Thornton, D.C. *J. geophys. Res.* **104,** 5785 (1999).

Locations and preferred pathways of possible sources of Arctic aerosol. Xie, Y.L., Hopke, P.K., Paatero, P., Barrie, L.A., and Li, S.M. *Atmos. Environ.* **33,** 2229 (1999).

Atmospheric processing of organic aerosols. Ellison, G.B., Tuck, A.F., and Vaida, V. *J. geophys. Res.* **104,** 11633 (1999).

New particle formation in the remote troposphere: A comparison of observations at various sites. Weber, R.J., McMurry, P.H., Mauldin, R.L., Tanner, D.J., Eisele, F.L., Clarke, A.D., and Kapustin, V.N. *Geophys. Res. Letts.* **26,** 307 (1999).

Gas-to-particle conversion in the atmosphere: I. Evidence from empirical atmospheric aerosols. Clement, C.F., and Ford, I.J. *Atmos. Environ.* **33,** 475 (1999).

Is aerosol production within the remote marine boundary layer sufficient to maintain observed concentrations? Capaldo, K.P., Kasibhatla, P., and Pandis, S.N. *J. geophys. Res.* **104,** 3483 (1999).

Optical properties and size distribution of aerosols derived from simultaneous measurements with lidar, a sunphotometer, and an aureolemeter. Hayasaka, T., Meguro, Y., Sasano, Y., and Takamura, T. *Appl. Optics* **38,** 1630 (1999).

Sources of aerosol sulphate at Alert: Apportionment using stable isotopes. Norman, A.L, Barrie, L.A., Toom-Sauntry, D., Sirois, A., Krouse, H.R., Li, S.M., and Sharma, S. *J. geophys. Res.* **104,** 11619 (1999).

Physical and chemical characteristics of aerosols at Spitsbergen in the spring of 1996. Staebler, R., Toom-Sauntry, D., Barrie, L., Langendorfer, U., Lehrer, E., Li, S.M., and Dryfhout-Clark, H. *J. geophys. Res.* **104,** 5515 (1999).

The background aerosol size distribution in the free troposphere: An analysis of the annual cycle at a high-alpine site. Nyeki, S., Li, F., Weingartner, E., Streit, N., Colbeck, I., Gaggeler, H.W., and Baltensperger, U. *J. geophys. Res.* **103,** 31749 (1998).

Atmospheric particles. Physicochemical characteristics. Masclet, P. and Cachier, H. *Analusis* **26,** M11 (1998).

A new source of tropospheric aerosols: Ion-ion recombination. Turco, R.P., Zhao, J.X., and Yu, F.Q. *Geophys. Res. Letts.*, **25** 635 (1998).

Remote sensing of atmospheric aerosols from active and passive optical techniques. Devara, P.C.S. *Int. J. remote Sensing* **19,** 3271 (1998).

The influence of in-cloud chemical reactions on ozone formation in polluted and non-polluted areas. Walcek, C.J., Yuan, H.-H., and Stockwell, W.R. *Atmos. Environ.* **31,** 1221 (1997).

The atmospheric aerosol-forming potential of whole gasoline vapor. Odum, J.R., Jungkamp, T.P.W., Griffin, R.J., Flagan, R.C., and Seinfeld, J.H. *Science* **276,** 96 (1997).

New insights on OH: Measurements around and in clouds. Mauldin, R.L., III, Madronich, S., Flocke, S.J., Eisele, F.L., Frost, G.J., and Prevot, A.S.H. *Geophys. Res. Lett* **24,** 3033 (1997).

Interaction of HNO_3 with water-ice surfaces at temperatures of free troposphere. Abbatt, J.P.D. *Geophys. Res. Letts.* **24,** 1479 (1997).

The role of mineral aerosol as a reactive surface in the global troposphere. Dentener, F.J., Carmichael, G.R., Zhang, Y., Lelieveld, J., and Crutzen, P.J. *J. geophys. Res.* **101,** 22869 (1996).

Unexpected low ozone concentration in mid-latitude tropospheric ice clouds: A case study. Reichardt, J., Ansmann, A., Serwazi, M., Weitkamp, C., and Michaelis, W. *Geophys. Res. Letts.* **23**, 1929 (1996).

Nucleation: measurements, theory, and atmospheric applications. Laaksonen, A., Talanquer, V., and Oxtoby, D.W. *Annu. Rev. Phys. Chem.* **46**, 489 (1995).

Role of deep cloud convection in the ozone budget of the troposphere. Lelieveld, J. and Crutzen, P.J. *Science* **264**, 1759 (1994).

Heterogeneous reactions in sulfuric acid aerosols: A framework for model calculations. Hanson, D.R., Ravishankara, A.R., and Solomon, S. *J. geophys. Res.* **99**, 3615 (1994).

Reaction of N_2O_5 on tropospheric aerosols: impact on the global distribution of NO_3, O_3, and OH. Dentener, F.J., and Crutzen, P.J. *J. geophys. Res.* **98**, 7149 (1993).

The role of clouds in tropospheric photochemistry. Lelieveld, J. and Crutzen, P.J. *J. atmos. Chem.* **12**, 229 (1991).

Influences of cloud photochemical processes on tropospheric ozone. Lelieveld, J. and Crutzen, P.J. *Nature, Lond.* **343**, 227 (1990).

Sensitivity analysis of a chemical mechanism for aqueous-phase atmospheric chemistry. Pandis, S.N., and Seinfeld, J.H. *J. geophys. Res.* **94**, 1105 (1989).

Particulate matter in the atmosphere: primary and secondary particles. Chapter 12 in *Atmospheric chemistry*. Finlayson-Pitts, B.J. and Pitts, J.N., Jr. (John Wiley, Chichester, 1986).

Section 5.9

The bibliography for this section is divided into three main parts: (i) publications principally concerned with models of the atmosphere; (ii) publications describing atmospheric measurements; and (iii) publications describing the link between measurements and models, and the analysis and interpretation of the data.

Models

Tropospheric modelling and emission estimation. Ebel, A., Friedrich, R., and Rohde, H. (eds.) EUROTRAC Report Volume 7. (Springer-Verlag, Berlin, 1997).

Intercomparison of the gas-phase chemistry in several chemistry and transport models. Kuhn, M., Builtjes, P.J.H., Poppe, D., Simpson, D., Stockwell, W.R., Andersson-Sköld, Y., Baart, A., Das, M., Fiedler, F., Hov, Ø., Kirchner, F., Makar, P.A., Milford, J.B., Roemer, M.G.M., Ruhnke, R., Strand, A., Vogel, B., and Vogel, H. *Atmos. Environ.* **32**, 693 (1998).

A new mechanism for regional atmospheric chemistry modeling. Stockwell, W.R., Kirchner, F., Kuhn, M., and Seefeld, S. *J. geophys. Res.* **102**, 25847 (1997).

A global three-dimensional chemical transport model for the troposphere. 1. Model description and CO and ozone results. Berntsen, T.K. and Isaksen, I.S.A. *J. geophys. Res.* **102**, 21239 (1997).

Results from the Intergovernmental Panel on Climatic Change photochemical model intercomparison (Photocomp). Olson, J. and 20 co-authors. *J. geophys. Res.* **102,** 5979 (1997).

Global uncertainty analysis of a regional scale gas-phase chemical mechanism. Gao, D., Stockwell, W.R., and Milford, J.B. *J. geophys. Res.* **101**, 9107 (1996).

First order sensitivity and uncertainty analysis for a regional scale gas-phase chemical mechanism. Gao, D., Stockwell, W.R., and Milford, J.B. *J. geophys. Res.* **100**, 23153 (1995).

IMAGES:A three-dimensional chemical transport model of the global troposphere. Müller, J.-F. and Brasseur, G. *J. geophys. Res.* **100**, 16445 (1995).

Effect of kinetic uncertainties on calculated constituents in a tropospheric photochemical model. Thompson, A.M. and Stewart, R.W. *J. geophys. Res.* **96**, 13089 (1991).

A detailed mechanism for the gas-phase atmospheric reactions of organic compounds. Carter, W.P.L. *Atmos. Environ.* **24A**, 481 (1990).

Measurements

Spectroscopic determinations play a key role in measuring concentrations of trace gases. Fourier Transform Infrared (FTIR) methods have become increasingly popular (see Section 5.10.7), as have Differential Optical Absorption Spectroscopy (DOAS) and Tunable Diode Laser (TDL) methods. The first group of publications describe these and other general aspects of atmospheric measurements.

Current and future passive remote sensing techniques used to determine atmospheric constituents. Burrows, J.P. in *Approaches to scaling of trace gas fluxes in ecosystems.* Bouwman, A.F. (ed.) (Elsevier, Amsterdam, 1999).

Atmospheric monitoring by Differential Optical Absorption Spectroscopy. Plane, J.M.C. and Smith, N. in *Spectroscopy in environmental sciences*. Hester, R.E. and Clark, R.J.H. (eds.) (Wiley, London, 1995).

Differential Optical Absorption Spectroscopy (DOAS). Platt, U. in *Air monitoring by spectroscopic techniques.* Sigrist, M.W. (ed.) (Wiley, London, 1994).

Remote sensing measurements of tropospheric ozone by ground-based thermal emission spectroscopy. Evans, W.F.J., and Puckrin, E. *J. atmos. Sci.* **56,** 311 (1999).

First results from the MOZAIC-1 program (*measurement of ozone and water vapor by Airbus in-service aircraft*). Special section of *J. geophys. Res.* **103**, 25631 (1998).

Measurement techniques in gas-phase tropospheric chemistry: a selective view of the past and future. Roscoe, H.K. and Clemitshaw, K.C. *Science* **276**, 1065 (1997).

Trace gas and aerosol measurements using aircraft data from the North Atlantic Regional Experiment (NARE 1993). Buhr, M., Sueper, D., Trainer, M., Goldan, P., Kuster, B., Fehsenfeld, F., Kok, G., Shillawski, R., and Schanot, A. *J. geophys. Res.* **101**, 29013 (1996).

Ozone

Ozone and C_2–C_5 hydrocarbon observations in the marine boundary layer between 45 degrees S and 77 degrees S. Gros, V., Martin, D., Poisson, N., Kanakidou, M., Bonsang, B., LeGuern, F., and Demont, E. *Tellus* **B50,** 430 (1998).

Direct measurement of tropospheric ozone distributions from space. Munro, R., Siddans, R., Reburn, W.J., and Kerridge, B.J. *Nature, Lond.* **392**, 168 (1998).

The climatological distribution of tropospheric ozone derived from satellite measurements using version 7 of Total Ozone Mapping Spectrometer (TOMS) and Stratospheric Aerosol and Gas Experiment (SAGE) data sets. Fishman, J. and Brackett, V.G. *J. geophys. Res.* **102**, 19275 (1997).

Aircraft measurements of tropospheric ozone over the western Pacific Ocean. Tsutsumi, Y., Makino, Y., and Jensen, J. *Atmos. Environ.* **30**, 1763 (1996).

OH and HO_2

Measurement of free radicals OH and HO_2 in Los Angeles smog. George, L.A., Hard, T.M., and O'Brien, R.J. *J. geophys. Res.* **104,** 11643 (1999).

Measurements of OH during PEM-Tropics A. Mauldin, R.L., Tanner, D.J., and Eisele, F.L. *J. geophys. Res.* **104,** 5817 (1999).

Highly time resolved measurements of OH during POPCORN using laser-induced fluorescence spectroscopy. Holland, F., Aschmutat, U., Hessling, M., Hofzumahaus, A., and Ehhalt, D.H. *J. atmos. Chem.* **31,** 205 (1998).

POPCORN: A field study of photochemistry in North-Eastern Germany. Plass-Dulmer, C., Brauers, T., and Rudolph, J. *J. atmos. Chem.* **31,** 5 (1998).

Intercomparison of tropospheric OH measurements by different laser techniques during the POPCORN campaign 1994. Hofzumahaus, A., Aschmutat, U., Brandenburger, U., Brauers, T., Dorn, H.P., Hausmann, M., Hessling, M., Holland, F., PlassDulmer, C., and Ehhalt, D.H. *J. atmos. Chem.* **31,** 227 (1998).

In-situ measurements of tropospheric hydroxyl radicals by folded long-path laser absorption during the field campaign POPCORN. Brandenburger, U., Brauers, T., Dorn, H.P., Hausmann, M., and Ehhalt, D.H. *J. atmos. Chem.* **31,** 181 (1998).

An overview of the tropospheric OH photochemistry experiment, Fritz Peak / Idaho Hill, Colorado, fall 1993. Mount, G.H. and Williams, E. *J. geophys. Res.* **102,** 6171 (1997).

In situ measurements of OH and HO_2 in the upper troposphere and stratosphere. Wennberg, P.O., Hanisco, T.F., Cohen, R.C., Stimpfle, R.M., Lapson, L.B., and Anderson, J.G. *J. atmos. Sci.* **52,** 3413 (1995).

Tropospheric OH concentration measurements by laser long-path absorption spectroscopy. Hofzumahaus, A., Dorn, H.-P., Callies, J., Platt, U., and Ehhalt, D. H. *Atmos. Environ.* **25A**, 2017 (1991).

Methyl chloroform concentrations provide another way of calculating [OH] in the troposphere (see text). The subsequent papers present general assessments of OH concentrations, and the inter-relation between OH, CH_4, and CO.

Global OH trend inferred from methylchloroform measurements. Krol, M., van Leeuwen, P.J., and Lelieveld, J. *J. geophys. Res.* **103,** 10697 (1998).

Atmospheric trends and lifetime of CH_3CCl_3 and global OH concentrations. Prinn, R.G., Weiss, R.F., Miller, B.R., Huang, J., Alyea, F.N., Cunnold, D.M., Fraser, P.J., Hartley, D.E., and Simmonds, P.G. *Science* **269,** 187 (1995).

The production and global distribution of emissions to the atmosphere of 1,1,1-trichloroethane (methyl chloroform). Midgley, P.M. and McCulloch, A. *Atmos. Environ.* **29,** 1601 (1995).

Oceanic consumption of CH_3CCl_3 : Implications for tropospheric OH. Butler, J.H., Elkins, J.W., Thompson, T.M., Hall, B.D., Swanson, T.H., and Koropalov, V. *J. geophys. Res.* **96,** 22347 (1991).

Simulations of cosmogenic (CO)-C-14 using the three-dimensional atmospheric model MATCH: Effects of C-14 production distribution and the solar cycle. Jockel, P., Lawrence, M.G., and Brenninkmeijer, C.A.M. *J. geophys. Res.* **104,** 11733 (1999).

Changing trends in tropospheric methane and carbon monoxide: a sensitivity analysis of the OH-radical. Van Dop, H. and Krol, M. *J. atmos. Chem.* **25**, 271 (1996).

On the indirect determination of atmospheric OH radical concentrations from reactive hydrocarbon measurements. McKeen, S.A., Trainer, M., Hsie, E.Y., Tallamraj R.K., and Liu, S.C. *J. geophys. Res.* **95,** 7493 (1990).

The tropospheric lifetime of halocarbons and their reactions with OH radicals: an assessment based on the concentrations of ^{14}CO. Derwent, R. and Volz-Thomas, A. *Alternative fluorocarbon environmental acceptability study (AFEAS)* Volume II of *Scientific assessment of stratospheric ozone: 1989.* World Meteorological Organization, Global ozone research and monitoring project: report no. 20. (WMO, Geneva, 1990).

NO_x and NO_y. See also bibliography for Sections 5.3.4 and 5.3.6.

NO_x and NO_y over the northwestern North Atlantic: Measurements and measurement accuracy. Peterson, M.C., and Honrath, R.E. *J. geophys. Res.* **104,** 11695 (1999).

Measurements of tropospheric NO_3 at midlatitude. Aliwell, S.R., and Jones, R.L. *J. geophys. Res.* **103**, 5719(1998).

Ammonia is of interest as the only naturally occurring alkaline trace gas. It can be converted (via OH oxidation) to the oxides of nitrogen.

The concentration of ammonia in southern ocean air. Ayers, G.P. and Gras, J. *J. geophys. Res.* **88**, 10655 (1983).

Hydrocarbons: methane and the non-methane hydrocarbons are of particular significance as the 'fuels' that are oxidized in tropospheric chemical conversions. See also bibliography for Sections 5.4.1 and 5.4.2.

Global distribution of reactive hydrocarbons in the atmosphere. Bonsang, B. and Boissard, C. in *Reactive hydrocarbons in the atmosphere.* Hewitt, C.N. (ed.) (Academic Press, San Diego, 1999).

Measurements of atmospheric CO_2 and CH_4 using a commercial airliner from 1993 to 1994. Matsueda, H. and Inoue, H.Y. *Atmos. Environ.* **30**, 1647 (1996).

Alkyl nitrate and selected halocarbon measurements at Mauna Loa Observatory, Hawaii. Atlas, E., Schauffler, S.M., Merrill, J.T., Hahn, C.J., Ridley, B., Walega, J., Greenberg, J., Heidt, L., and Zimmerman, P. *J. geophys. Res.* **97**, 10331 (1992).

Aldehydes, and especially formaldehyde, are capable of yielding hydroxyl radicals at the wavelengths of light penetrating to the troposphere. Formaldehyde is also an expected product of the oxidation of methane in the natural troposphere. The identification of formaldehyde in air far removed from human sources of pollutants is thus of considerable interest.

Formaldehyde (HCHO) measurements in the nonurban atmosphere. Lowe, D.C. and Schmidt, U. *J. geophys. Res.* 88, 10844 (1983).

Sulphur compounds. See also bibliography for section 5.6.

Tropospheric sulfur dioxide observed by the ERS-2 GOME instrument. Eisinger, M. and Burrows, J.P. *Geophys. Res. Letts.* **25,** 4177 (1998).

Carbon monoxide. See also bibliography for Section 5.3.5.

Measurements of the atmospheric carbon monoxide column with a ground- based length-modulated radiometer. Tolton, B.T. and Drummond, J.R. *Appl. Optics* **38,** 1897 (1999).

Seasonal measurements of nonmethane hydrocarbons and carbon monoxide at the Mauna Loa Observatory during the Mauna Loa Observatory Photochemistry Experiment 2. Greenberg, J.P., Helmig, D., and Zimmerman, P.R. *J. geophys. Res.* **101**, 14581 (1996).

Peroxy radicals and peroxides

Distribution of hydrogen peroxide and methylhydroperoxide over the Pacific and South Atlantic Oceans. O'Sullivan, D.W., Heikes, B.G., Lee, M., Chang, W., Gregory, G.L., Blake, D.R., and Sachse, G.W. *J. geophys. Res.* **104,** 5635 (1999).

An improved chemical amplifier technique for peroxy radical measurements. Cantrell, C.A., Shetter, R.E., Lind, J.A., McDaniel, A.H., Calvert, J.G., Parrish, D.D., Fehsenfeld, F.C., Buhr, M.P., and Trainer, M. *J. geophys. Res.* **98,** 2897 (1993).

Comparisons of measurements and models, and the analysis and interpretation of the data. The order in which the references appear follows that used in the presentation of the measurements themselves.

Chemical compounds in the remote Pacific troposphere: Comparison between MLOPEX measurements and chemical transport model calculations. Brasseur, G.P., Hauglustaine, D.A., and Walters, S. *J. geophys. Res.* **101**, 14795 (1996).

Three-dimensional view of the large-scale tropospheric ozone distribution over the North-Atlantic Ocean during summer. Kasibhatla, P., Levy, H., II, Klonecki, A., and Chameides, W.L. *J. geophys. Res.* **101**, 29305 (1996).

Tropospheric box-modelling and analytical studies of the hydroxyl (OH) radical and related species: Comparison with observations. Grenfell, J.L., Savage, N.H., Harrison, R.M., Penkett, S.A., Forberich, O., Comes, F.J., Clemitshaw, K.C., Burgess, R.A., Cardenas, L.M., Davison, B., and McFadyen, G.G. *J. atmos. Chem.* **33,** 183 (1999).

Sources of upper tropospheric HO_x: A three-dimensional study. Muller, J.F., and Brasseur, G. *J. geophys. Res.* **104,** 1705 (1999).

Measurement of the diurnal variation of the OH radical concentration and analysis of the data by modelling. Forberich, O., Pfeiffer, T., Spiekermann, M., Walter, J., Comes, F.J., Grigonis, R., Clemitshaw, K.C., and Burgess, R.A. *J. atmos. Chem.* **33,** 155 (1999).

OH field measurements: A critical input to model calculations on atmospheric chemistry. Comes, F.J., Forberich, O., and Walter, J. *J. atmos. Sci.* **54,** 1886 (1997).

The influence of human activities on the distribution of hydroxyl radicals in the troposphere. Derwent, R.G. *Phil. Trans. Royal Soc.* **A354**, 501 (1996).

Comparison of measured OH concentrations with model calculations. Poppe, D. and 22 co-authors. *J. geophys. Res.* **99**, 16633 (1994).

Tropospheric OH: local measurements and their interpretation. Comes, F.J., Armerding, W., Grigonis, R., Herbert, A., Spiekermann, M., and Walter, J. *Ber. Bunsenges. phys. Chem.* **96,** 284 (1992).

NO_y partitioning from measurements of nitrogen and hydrogen radicals in the upper troposphere. Keim, E.R., McKeen, S.A., Gao, R.S., Donnelly, S.G., Wamsley, R.C., DelNegro, L.A., Fahey, D.W., Hanisco, T.F., Lanzendorf, E.J., Proffitt, M.H., Margitan, J.J., Hintsa, E.J., Jaegle, L., Webster, C.R., May, R.D., Scott, D.C., Salawitch, R.J., Wilson, J.C., McElroy, C.T., Atlas, E.L., Flocke, F., and Bui, T.P. *Geophys. Res. Letts.* **26,** 51 (1999).

Distribution of reactive nitrogen species in the remote free troposphere: data and model comparisons. Thakur, A.N., Singh, H.B., Mariani, P., Chen, Y., Wang, Y., Jacob, D.J., Brasseur, G., Muller, J.F., and Lawrence, M. *Atmos. Environ.* **33,** 1403 (1999).

Correlative nature of ozone and carbon monoxide in the troposphere: implications for the tropospheric ozone budget. Fishman, J., and Seiler, W. *J. geophys. Res.* **88**, 3662 (1983).

Peroxy radicals from photostationary state deviations and steady state calculations during the tropospheric OH experiment at Idaho Hill, Colorado, 1993. Cantrell, C.A., Shetter, R.E., Calvert, J.G., Eisele, F.L., Williams, E., Baumann, K., Brune, W.H., Stevens, P.S., and Mather, J.H. *J. geophys. Res.* **102**, 6396 (1997).

Effect of peroxy radical reactions on the predicted concentrations of ozone, nitrogenous compounds and radicals. Kirchner, F. and Stockwell, W.R. *J. geophys. Res.* **101**, 21007 (1996).

Peroxy radicals measured during Mauna Loa Observatory Photochemistry Experiment 2: the data and first analysis. Cantrell, C.A., Shetter, R.E., Gilpin, T.M., and Calvert, J.G. *J. geophys. Res.* **101**, 14643 (1996).

Peroxy radicals measured and calculated from trace gas measurements in the Mauna Loa Observatory Photochemistry Experiment 2. Cantrell, C.A., Shetter, R.E., Gilpin, T.M., Calvert, J.G., Eisele, F.L., and Tanner, D.J. *J. geophys. Res.* **101**, 14653 (1996).

Peroxy radicals as measured in ROSE and estimated from photostationary state deviations. Cantrell, C.A. and 10 others. *J. geophys. Res.* **98,** 18355 (1993).

These discussions of trends in tropospheric trace-gas concentrations complement those listed in the references for Sections 5.1 and 5.2; they are generally oriented to explanations of the changes.

Ozone Measurements in Historic Perspective Kley, D., Volz, A., and Mulheims, F. in *Tropospheric Ozone: Regional and Global Scale Interactions.* Isaksen, I.S.A. (ed.) NATO ASI Ser. C. 227 (D. Reidel, Dordrecht, 1998).

Trends of ozone in the troposphere. Oltmans, S.J. and 17 co-authors. *Geophys. Res. Letts.* **25,** 139 (1998).

Changes in tropospheric composition and air quality. Tang, X., Madronich, S., Wallington, T., and Calamari, D. *J. Photochem. Photobiol.* **B46,** 83 (1998).

Trends in stratospheric and free tropospheric ozone. Harris, N.R.P., Ancellet, G., Bishop, L., Hoffman, D.J., Kerr, J.B., McPeters, R.D., Prendez, M., Randel, W.J., Staehelin, J., Subbaraya, B.H., Volz-Thomas, A., Zawodny, J., and Zerefos, C.S. *J. geophys. Res.* **102**, 1571 (1997).

Ozone measurements made in the 19th century: An evaluation of Montsouris series. Volz, A. and Kley, D. *Nature* **332**, 240 (1988).

Concentrations and trends of tropospheric ozone precursor emissions in the USA and Europe, Kley, D., Geiss, H., and Mohnen, V.A. in *The Chemistry of the Atmosphere: Its impact on Global Change,* Calvert, J.G. (ed.) (Blackwell, Oxford, 1994).

Calculation of trends in the tropospheric concentrations of O_3, OH, CO, CH_4, and NO. Isaksen, I.S.A. and Hov, Ø. *Tellus* **39B**, 271 (1987).

Indications and causes of ozone increase in the troposphere. Penkett, S.A., in *The changing atmosphere*. Rowland, F.S. and Isaksen, I.S.A. (eds.) (John Wiley, Chichester, 1988).

Secular increase of the vertical column abundance of methane derived from IR solar spectra recorded at the Jungfraujoch station. Zander, R., Demoulin, P., Ehhalt, D.H., and Schmidt, U. *J. geophys. Res.* **94**, 11029 (1989).

Secular increase of the total vertical column abundance of carbon monoxide above central Europe since 1950. Zander, R., Demoulin, P., Ehhalt, D.H., Schmidt, U., and Rinsland, C.P. *J. geophys. Res.* **94**, 11021 (1989).

Is the oxidizing capacity of the atmosphere changing? Isaksen, I.S.A., in *The changing atmosphere*. Rowland, F.S. and Isaksen, I.S.A. (eds.) (John Wiley, Chichester, 1988).

Sensitivity of tropospheric oxidants to global chemical and climate change. Thompson, A.M., Stewart, R.W., Owens, M.A., and Heruche, J.A. *Atmos. Environ.* **23**, 519 (1989).

Section 5.10

The first references are to general accounts of tropospheric air pollution and the physico-chemical behaviour of pollutants. National bodies, such as the National Academy of Sciences in the USA, issue reports from time to time, as do international organizations such as the European Commission.

Social, economic, and political aspects are dealt with in addition to the chemical ones. Texts more specifically addressing the chemical problems include a series reporting conferences held under the auspices of the European Commission, and entitled Physico-chemical behaviour of atmospheric pollutants.

The first EUROTRAC project, mentioned briefly at the end of Section 5.9.1, started in 1988 and, at its peak, comprised some 250 research groups. The scientific results are summarized in a ten-volume report that represents a major resource in air-pollution studies, and in tropospheric chemistry more generally. The series title and list of volumes follows: certain individual volumes are referenced elsewhere.

Transport and chemical transformation of pollutants in the troposphere. Borrell, P., Borrell, P.M., Cvitaš, T., Kelly, K., and Seiler, W. (series eds.) (Springer-Verlag, Berlin, 1997 &c). *Volume 1: Transport and chemical transformation of pollutants; Volume 2: Heterogeneous and liquid-phase processes; Volume 3: Chemical processes in atmospheric oxidation; Volume 4: Biosphere–atmosphere exchange of pollutants and trace substances; Volume 5: Cloud multi-phase processes and high alpine air and snow chemistry; Volume 6: tropospheric ozone research; Volume 7: Tropospheric modelling and emission estimation; Volume 8: Instrument development for atmospheric research and monitoring; Volume 9: Exchange and transport of air pollutants over complex terrain and the sea; Volume 10: Photo-oxidants, acidification and tools: policy applications of EUROTRAC results.*

These first two books, listed also as introductory texts, are particularly strong on the chemistry of air pollution. The other references provide good overviews of various general aspects of air pollution.

Atmospheric chemistry. Finlayson-Pitts, B.J. and Pitts, J.N., Jr. (John Wiley, Chichester, 1986).

Atmospheric chemistry and physics of air pollution. Seinfeld, J.H. (John Wiley, Chichester, 1986).

Reactive hydrocarbons in the atmosphere at urban and regional scales. Ciccioli, P., Brancaleoni, E., and Frattoni, M. in *Reactive hydrocarbons in the atmosphere.* Hewitt, C.N. (ed.) (Academic Press, San Diego, 1999).

Anthropogenic emissions of volatile organic compounds. Friedrich, R. and Obermeier, A. in *Reactive hydrocarbons in the atmosphere.* Hewitt, C.N. (ed.) (Academic Press, San Diego, 1999).

Air pollution in the United Kingdom. Davison, G. and Hewitt, C.N. (Royal Society of Chemistry, Cambridge, 1997).

Atmospheric environmental research. Möller, D. and Schaller, E. (eds.) (Springer-Verlag, Berlin, 1998).

Air pollution: an introduction. Colls, J. (Spon, London, 1997).

National Air Pollutant Emissions Trends, 1900-1995. (US Environmental Protection Agency, Research Triangle Park, NC, 1996).

Source inventories and control strategies for VOCs. Passant, N.R. in *Volatile organic compounds in the atmosphere.* Hester, R.E. and Harrison, R.M. (eds.) (Royal Society of Chemistry, Cambridge, 1995).

Atmospheric pollution: a global problem. Elsom, D.M. (Blackwell, Oxford, 1992).

The big smoke: a history of air pollution in London since medieval times. Brimblecombe, P. (Routledge, London, 1988).

Hazardous air pollutants (HAPs) and their effects on biodiversity: An overview of the atmospheric pathways of persistent organic pollutants (POPs) and suggestions for future studies. Finizio, A., DiGuardo, A., and Cartmale, L. *Env. Monitoring Assess.* **49**, 327 (1998).

The global impact of human activity on tropospheric ozone. Levy, H., II, Kasibhatla, P.S., Moxim, W.J., Klonecki, A.A., Hirsch, A.I., Oltmans, S.J., and Chameides, W.L. *Geophys. Res. Letts.* **24**, 791 (1997).

Tropospheric air pollution: ozone, airborne toxics, polycyclic aromatic hydrocarbons, and particles. Finlayson-Pitts, B. and Pitts, J.N., Jr. *Science* **276**, 1045 (1997).

Urban air pollution: state of the science. Seinfeld, J.H. *Science* **243**, 745 (1989).

Air pollution by particles. Shaw, R.W. *Scient. Am.* **257** (2), 84 (Aug 1987).

Pollution: causes, effects and control. 2nd edn. Harrison, R.M. (ed.) (Royal Society of Chemistry, London, 1990).

The following paper describes how the abundance ratios of elements provide a 'signature' that can help identify the source of pollution. Specific examples are discussed in Section 5.10.6.

Elemental tracers of distant regional pollution aerosols. Rahn, K.A. and Lowenthal, D.H. *Science* **223**, 132 (1984).

Section 5.10.2

Some effects of air pollution, excluding those on human health. A guide to the literature on health problems can be found in the textbooks on air pollution described earlier.

Plant response to air pollution. Yunus, M. and Iqbal, M. (Wiley, Chichester, 1996).
Particles in our air: concentrations and health effects. Wilson, R. and Spengler, J.D. (Harvard University Press, Cambridge, MA, 1996).
Aerosol effects on UV radiation in nonurban regions. Reuder, J., and Schwander, H. *J. geophys. Res.* **104,** 4065 (1999).
Ozone flux to vegetation and its relationship to plant response and ambient air quality standards. Musselman, R.C., and Massman, W.J. *Atmos. Environ.* **33,** 65 (1999).
Long-term concentrations of ambient air pollutants and incident lung cancer in California adults: Results from the AHSMOG study. Beeson, W.L., Abbey, D.E., and Knutsen, S.F. *Env. Health Perspects.* **106,** 813 (1998).
An association between air pollution and mortality in six U.S. cities. Dockery, D.W., Pope, C.A., III, Xu, X., Spengler, J.D., Ware, J.H., Fay, M.E., Ferris, B.G., Jr., and Speizer, F.E. *N. Engl. J. Med.* **329**, 1753 (1993).
Non-health effects of airborne particulate matter. Lodge, J.P., Jr., Waggoner, A.P., Klodt, D.T., and Crain, C.T. *Atmos. Environ.* 15, 431 (1981).
Effects of Gaseous Air Pollution in Agriculture and Horticulture. Unsworth, M.H. and Ormrod, D.P. (eds.) (Butterworth, London, 1982).
Review: Atmospheric deposition and plant assimilation of gases and particles. Hosker, R.P., Jr. and Lindberg, S.E. *Atmos. Environ.* **16**, 889 (1982).

Section 5.10.3

Sources of air pollutants. Middleton, P., in *Composition, Chemistry, and Climate of the Atmosphere.* Singh, H.B. (ed.) (Van Nostrand Reinhold, New York, 1995).
High concentrations and photochemical fate of oxygenated hydrocarbons in the global troposphere. Singh, H.B., Kanakidou, M., Crutzen, P.J., and Jacob, D.J. *Nature* **378,** 50 (1995).

Sections 5.10.4 and 5.10.5

Sulfur dioxide distribution over the Pacific Ocean 1991-1996. Thornton, D.C., Bandy, A.R., Blomquist, B.W., Driedger, A.R., and Wade, T.P. *J. geophys. Res.* **104,** 5845 (1999).

A study of the spectra and reactivity of oxysulfur-radical anions involved in the chain oxidation of S(IV): a pulse and γ-radiolysis study. Buxton, G.V., McGowan, S., Salmon, G.A., Williams, J.E., and Wood, N.D. *Atmos. Environ.* **30,** 2483 (1996).

Kinetics and mechanism of the oxidation of sulfur (IV) by ozone in aqueous solutions. Botha, C.F., Hahn, J., Pienaar, J.J., and Vaneldik, R. *Atmos. Environ.* **28,** 3207 (1994).

Heterogeneous sulphate production in an urban fog. Pandis, S.N., Seinfeld, J.H., and Pilinis, C. *Atmos. Environ.* **26**, 2509 (1992).

Rate constants for some oxidations of S(IV) by radicals in aqueous solutions. Huie, R.E., and Neta, P. *Atmos. Environ.* **21**, 1743 (1987).

Human influence on the sulphur cycle. Brimblecombe, P., Hammer, C., Rohde, H., Ryaboshapko, A., and Boutron, C.F., in *Evolution of the global biogeochemical sulphur cycle*, Brimblecombe, P. and Lein, A.Y. (eds) SCOPE 39, (John Wiley, Chichester, 1989).

The homogeneous chemistry of atmospheric sulfur. Graedel, T.E. *Rev. Geophys. & Space Phys.* **15**, 421 (1977).

These references are concerned with mechanisms for the oxidation of SO_2 and with the formation of H_2SO_4. Conversion in solution and in cloud droplets is the subject of the later papers.

The mechanism of the HO–SO_2 reaction. Stockwell, W.R. and Calvert, J.G. *Atmos. Environ.* **17**, 2231 (1983).

Mechanism of the homogeneous oxidation of sulfur dioxide in the troposphere. Calvert, J.G., Su, F., Bottenheim, J.W., and Strausz, O.P. *Atmos. Environ.* **12**, 197 (1978).

SO_2 oxidation *via* the hydroxyl radical: atmospheric fate of the HSO_x radicals. Davis, D.D., Ravishankara, A.R., and Fischer, S. *Geophys. Res. Letts.* **6**, 113 (1979).

Kinetic flash spectroscopic study of the CH_3O_2–CH_3O_2 and CH_3O_2–SO_2 reactions. Kan, C.S., McQuigg, R.D., Whitbeck, M.R., and Calvert, J.G. *Int. J. Chem. Kinet.* **11**, 921 (1979).

Formation of sulfuric and nitric acids in acid rain and fogs. Chapter 11 in *Atmospheric chemistry*. Finlayson-Pitts, B.J. and Pitts, J.N., Jr. (John Wiley, Chichester, 1986).

Aqueous oxidation of SO_2 by hydrogen peroxide. Kunen, S.M., Lazrus, A.L.. Kok, G.L., and Heider, B.G. *J. geophys. Res.* **88**, 3671(1983).

Measurements of the oxidation rate of sulfur (IV) by ozone in aqueous solution and their relevance to SO_2 conversion in non-urban tropospheric clouds. Maahs, H.G. *Atmos. Environ.* **17**, 341(1983).

Kinetics and mechanism of the oxidation of S(IV) by ozone in aqueous solution with particular reference to SO_2 conversion in nonurban tropospheric clouds. Maahs, H.G. *J. geophys. Res.* **88**, 10721 (1983).

Section 5.10.6

Acid rain is regarded as one of the most serious problems of tropospheric pollution. The first papers deal with aspects of oxidative transformation and acid formation not specifically concerned with H_2SO_4 chemistry (Section 5.10.4).

Acid generation in the troposphere by gas-phase chemistry. Calvert, J.G. and Stockwell, W.R. *Environ. Sci. Technol.* **17**, 428A (1983).

A modelling study of SO_x–NO_x–hydrocarbon plumes and their transport to the background troposphere. Balko, J.A. and Peters, L.K. *Atmos. Environ.* **17**, 1965 (1983).

The free radical chemistry of cloud droplets and its impact upon the composition of rain. Chameides, W.L. and Davis, D.D. *J. geophys. Res.* **87**, 4863 (1982).

The mechanism of NO_3 and HONO formation in the nighttime chemistry of the urban atmosphere. Stockwell, W.R. and Calvert, J.G. *J. geophys. Res.* **88**, 6673 (1983).

Effects of heterogeneous processes on NO_3, HONO, and HNO_3 chemistry in the troposphere. Heikes, B. and Thomson, Anne M. *J. geophys. Res.* **88**, 10883 (1983).

Aqueous-phase source of formic acid in clouds. Chameides, W.L. and Davis, D.D. and Martin, L.R. *Atmos. Environ.* **17**, 2005 (1983).

Aqueous-phase source of formic acid in clouds. Chameides, W.L. and Davis, D.D. *Nature, Lond.* **304**, 427 (1983).

Naturally acidic water systems exist because of 'natural' pollution, and the ecology of these systems provides useful clues about the possible consequences of anthropogenic acidification.

The smoking hills: natural acidification of an aquatic ecosystem. Havas, M. and Hutchinson, T.C. *Nature, Lond.* **301**, 23 (1983).

The next references are to reports concerned directly with the causes and effects of acid rain.

Acid rain. ApSimon, H. (Earthwatch, 1997).

Environmental acidification. Brimblecombe, P. in *Global atmospheric chemical change*. Hewitt, C.N and Sturges, W.T. (eds.) (Chapman and Hall, London, 1994).

Atmospheric acidity. Radojevic, M. and Harrison, R.M. (Elsevier Applied Science, London, 1992).

Emissions Involved in Acidic Deposition Processes. State-of-Science/Technology Report 1. National Acid Precipitation Assessment Program. Placet, M., Battye, R.E., Fehsenfeld, F.C., and Bassett, G.W. (U.S. Government Printing Office, Washington, DC, 1990).

Regional trends in wet deposition of sulphate in the United States and SO_2 emissions from 1980 through 1995. Shannon, J.D. *Atmos. Environ.* **33**, 807 (1999).

Long-term effects of acid rain: response and recovery of a forest ecosystem. Likens, G.E., Driscoll, C.T., and Buso, D.C. *Science* **272**, 244 (1996).

Global gridded inventories of anthropogenic emissions of sulfur and nitrogen. Benkovitz, C.M., Scholtz, T.M., Pacyna, J., Rarrason, L., Dignon, J., Voldner, E.C., Spiro, P.A., Logan, J.A., and Graedel, T.E. *J. geophys. Res.* **101**, 29239 (1996).

Mathematical modeling of acid deposition due to radiation fog. Pandis, S.N. and Seinfeld, J.H. *J. geophys. Res.* **94,** 12911 (1989).

Acid deposition: unravelling a regional phenomenon. Schwartz S.E. *Science* **243**, 753 (1989).

Quantification of changes in lakewater chemistry in response to acidic deposition. Sullivan, T.J., Charles, D.F., Smol, J.P., Cumming, B.F., Selle, A.R., Thomas, D.R., Bernert, J.A., and Dixit, S.S. *Nature, Lond.* **345**, 54 (1990).

Biological recovery of an acid lake after reductions in industrial emissions of sulphur. Gunn, J.M. and Keller, W. *Nature, Lond.* **345**, 431(1990).

Acid rain on acid soil: a new perspective. Krug, E.C. and Frink, C.R. *Science* **221**, 520 (1983).

Air pollutants and forest decline. Tomlinson, G.H., II, *Environ. Sci. Technol.* **17**, 246A (1983).

Acid deposition and forest decline. Johnson, A.H. and Siccama, T.G. *Environ. Sci. Technol.* **17**, 294A (1983).

Section 5.10.7

The next references discuss the mechanisms of smog formation and possible control strategies in kinetic terms. Despite the apparent geographical restriction implied by the title, the first book provides an excellent description of photochemical oxidants in the atmosphere.

Ozone in the United Kingdom. UK Photochemical Oxidants Review Group. (Department of the Environment, Transport, and the Regions, 1998).

Air pollution by photochemical oxidants. Colbeck, I. and Mackenzie, A.R. (Elsevier, Amsterdam, 1994).

The relation between ozone, NO_x and hydrocarbons in urban and polluted rural environments. Sillman, S., *Atmos. Environ.* **33,** 1821 (1999).

Influence of regional-scale anthropogenic activity in northeast Asia on seasonal variations of surface ozone and carbon monoxide observed at Oki, Japan. Pochanart, P., Hirokawa, J., Kajii, Y., Akimoto, H., and Nakao, M. *J. geophys. Res.* **104,** 3621 (1999).

The global impact of human activity on tropospheric ozone. Levy, H., Kasibhatla, P.S., Moxim, W.J., Klonecki, A.A., Hirsch, A.I., Oltmans, S.J., and Chameides, W.L. *Geophys. Res. Letts.* **24,** 791 (1997).

Ozone precursor relationships in the ambient atmosphere. Chameides, W.L., Fehsenfeld, F., Rodgers, M.O., Cardelino, C., Martinez, J., Parrish, D., Lonneman, W., Lawson, D.R., Rasmussen, R.A., Zimmerman, P., Greenberg, J., Middleton, P., and Wang, T. *J. geophys. Res.* **97**, 6037 (1992).

The role of biogenic hydrocarbons in urban photochemical smog — Atlanta as a case-study. Chameides, W.L., Lindsay, R.W., Richardson, J., and Kiang, C.S. *Science* **241,** 1473 (1988).

Photochemical air pollution: mechanisms of formation and chemical basis of control strategy options for oxidant and gaseous airborne toxic chemicals. Part 5 (Chapters 9 and 10) in *Atmospheric chemistry*. Finlayson-Pitts, B.J. and Pitts, J.N., Jr. (John Wiley, Chichester, 1986).

Air pollution by photochemical oxidants: formation, transport, control, and effects on plants. Guderian, R. (ed.) (Springer-Verlag, Berlin, 1985).

The photochemistry of anthropogenic nonmethane hydrocarbons in the troposphere. Brewer, D.A., Augustsson, T.R., and Levine, J.S. *J. geophys. Res.* **88,** 6683 (1983).

Keys to photochemical smog control. Pitts, J.N., Jr. *Environ. Sci. Technol.* **11**, 456 (1977).

The chemical basis of air quality: kinetics and mechanisms of photochemical air pollution and application to control strategies. Finlayson-Pitts, B.J. and Pitts, J.N., Jr. *Adv. Environ. Sci. Technol.* **7**, 75 (1977).

Photochemistry of the polluted troposphere. Finlayson, B.J. and Pitts, J.N., Jr. *Science* **192**, 111 (1976).

Photochemical oxidant creation potentials (POCPs) and other reactivity scales

Reactive hydrocarbons and photochemical smog. Derwent, R.G. in *Reactive hydrocarbons in the atmosphere.* Hewitt, C.N. (ed.) (Academic Press, San Diego, 1999).

Photochemical ozone creation potentials for oxygenated volatile organic compounds: sensitivity to variations in kinetic and mechanistic parameters. Jenkin, M.E., and Hayman, G.D. *Atmos. Environ.* **33,** 1275 (1999).

Photochemical pollution indicators. Perros, P.E., and Marion, T. *Analusis* **26,** M43 (1998).

Ozone productivity of atmospheric organics. Bowman, F.M., and Seinfeld, J.H. *J. geophys. Res.* **99**, 5309 (1994).

Fundamental basis of incremental reactivities of organics in ozone formation in VOC/NO_x mixtures. Bowman, F.M., and Seinfeld, J.H. *Atmos. Environ.* **28,** 3359 (1994).

Development of ozone reactivity scales for volatile organic compounds. Carter, W.P.L. *J. Air & Waste Manage. Assoc.* **44,** 881 (1994).

The use of radical traps has been suggested as a pollution control technique, but many critics believe the method to be difficult in application, and potentially dangerous.

Control of photochemical smog by diethylhydroxylamine. Heicklen, J. *Atmos. Environ.* **15**, 229 (1981).

Models and computer simulations of smog formation have provided new insights into the chemical processes occurring. One of the first is described in this paper.

Computer simulation of the rates and mechanisms of photochemical smog formation Calvert, J.G. and McQuigg, R.D. *Int. J. chem. Kin. Symp.* **1**, 113 (1975).

Haziness and reduced visibility is an attribute of photochemical smog. Aerosols are formed in the oxidation of substances released by man's combustion of fossil fuels. Aerosols are, however, observed in many environments not polluted by man. They may be produced by processes analogous to those involved in photochemical smog, but using natural hydrocarbons such as terpenes as the precursor material.

New particle formation in the remote troposphere: A comparison of observations at various sites. Weber, R.J., McMurry, P.H., Mauldin, R.L., Tanner, D.J., Eisele, F.L., Clarke, A.D., and Kapustin, V.N. *Geophys. Res. Letts.* **26,** 307 (1999).

Gas-to-particle conversion in the atmosphere: I. Evidence from empirical atmospheric aerosols. Clement, C.F., and Ford, I.J. *Atmos. Environ.* **33,** 475 (1999).

Biological input to visibility-reducing aerosol particles in the remote arid southwestern United States. Mazurek, M.A., Cass, G.R., and Simoneit, B.R.T. *Environ. Sci. Technol.* **25**, 684 (1991).

Natural organic atmospheric aerosols of terrestrial origin. Zenchelsky, S. and Youssefi, M. *Rev. Geophys. & Space Phys.* **17**, 459 (1979).

Blue hazes in the atmosphere. Went, F.W. *Nature, Lond.* **187**, 641 (1960).

A better understanding of problems of composition of photochemical smog, and chemical transformations in it, has been provided by the use of tunable lasers and long-path Fourier Transform Infra-red Spectrometers (see also references for 'measurements' in Section 5.9).

Optical systems unravel smog chemistry. Pitts, J.N., Jr., Finlayson-Pitts, B.J., and Winer, A.M. *Environ. Sci. Technol.* **11**, 568 (1977).

Section 5.10.8

Atmospheric degradation of anthropogenic molecules. Wallington, T.J. and Nielsen, O.J. in *Environmental photochemistry.* Boule, P. (ed.) (Springer-Verlag, Berlin, 1999).

Alternatives to CFCs and their behaviour in the atmosphere. Midgley, P.M. in *Volatile organic compounds in the atmosphere.* Hester, R.E. and Harrison, R.M. (eds.) (Royal Society of Chemistry, Cambridge, 1995).

Atmospheric chemistry of alternative halocarbons. Francisco, J.S. and Maricq, M.M. in *Advances in photochemistry, Vol. 20.* Neckers, D.C., Volman, D.H., and von Bünau, C. (eds.) (John Wiley, New York, 1995).

Degradation of HFCs and HCFCs in the atmosphere. Sidebottom, H. in *Tropospheric Oxidation Mechanisms.* Becker, K.-H. (ed.) Report EUR 16171 EN (European Commission, Luxembourg, 1995).

The environmental impact of CFC replacements — HFCs and HCFCs. Wallington, T.J., Schneider, W.F., Worsnop, D.R., Neilsen, O.J., Sehested, J., de Bruyn, W.J., and Shorter, J.A. *Environ. Sci. Technol.* **28**, 320A (1994).

The atmospheric fate and impact of hydrofluorocarbons and chlorinated solvents. Sidebottom, H. and Franklin, J. *Pure and Appl. Chem.* **68**, 1757(1996).

Phosgene ($COCl_2$) and more complex substituted carbonyl halides are intermediates that may be hydrolysed to acids. A particular problem arises in the degradation of compounds with the $-CF_3$ grouping, since trifluoroacetic acid, a known toxin, may be formed. Chlorinated acetic acids have been discovered in the environment, but there remains doubt as to whether or not their origin is entirely anthropogenic.

The fate of atmospheric phosgene and the stratospheric loading of its parent compounds: CCl_4, C_2Cl_4, C_2HCl_3, CH_3CCl_3 and $CHCl_3$. *J. geophys. Res.* **100**, 1235 (1995).

Uptake of haloacetyl and carbonyl halides by water surfaces. De Bruyn, W.J., Shorter, J.A., Davidovits, P., Worsnop, D.R., Zahniser, M.S., and Kolb, C.E. *Environ. Sci. Technol.* **29**, 1179 (1995).

Production of trifluoroacetic acid from HCFCs and HFCs: A three-dimensional modeling study. Kotamarthi, V.R., Rodriguez, J.M., Ko, M.K.W., Sze, N.D., and Prather, M.J. *J. geophys. Res.* **103**, 5747 (1998).

Trifluoroacetate in the environment. Evidence for sources other than HFC/HCFCs. Jordan, A. and Frank, H. *Env. Sci. Tech.* **33**, 522 (1999).

Environmental trifluoroacetate. Frank, H., Klein, A., and Renschen, D. *Nature* **382**, 34 (1996).

Mass balance of trichloroacetic acid in the soil top layer. Hoekstra, E.J., de Leer, E.W.B., and Brinkman, U.A.T. *Chemosphere* **38,** 551 (1999).

Findings supporting the natural formation of trichloroacetic acid in soil. Hoekstra, E.J., de Leer, E.W.B., and Brinkman, U.A.T. *Chemosphere* **38,** 2875 (1999).

Section 5.10.9

Tropospheric air pollution: ozone, airborne toxics, polycyclic aromatic hydrocarbons, and particles. Finlayson-Pitts, B. and Pitts, J.N., Jr. *Science* **276**, 1045 (1997).

Section 5.10.10

Biomass burning appears to be an important source of atmospheric trace gases

Global biomass burning: atmospheric, climatic, and biospheric impications. Levine, J.S. (ed.) (MIT Press, Cambridge, MA, 1991).

Influence of biomass combustion emissions on the distribution of acidic trace gases over the southern Pacific basin during austral springtime. Talbot, R.W., Dibb, J.E., Scheuer, E.M., Blake, D.R., Blake, N.J., Gregory, G.L., Sachse, G.W., Bradshaw, J.D., Sandholm, S.T., and Singh, H.B. *J. geophys. Res.* **104,** 5623 (1999).

Frequency and distribution of forest, savanna, and crop fires over tropical regions during PEM-Tropics A. Olson, J.R., Baum, B.A., Cahoon, D.R., and Crawford, J.H. *J. geophys. Res.* **104,** 5865 (1999).

Satellite detection of smoke aerosols over a snow/ice surface by TOMS. Hsu, N.C., Herman, J.R., Gleason, J.F., Torres, O., and Seftor, C.J. *Geophys. Res. Letts.* **26,** 1165 (1999).

On springtime high ozone events in the lower troposphere from Southeast Asian biomass burning. Liu, H.Y., Chang, W.L., Oltmans, S.J., Chan, L.Y., and Harris, J.M. *Atmos. Environ.* **33,** 2403 (1999).

Biomass burning as a source of formaldehyde, acetaldehyde, methanol, acetone, acetonitrile, and hydrogen cyanide. Holzinger, R., Warneke, C., Hansel, A., Jordan, A., Lindinger, W., Scharffe, D.H., Schade, G., and Crutzen, P.J. *Geophys. Res. Letts.* **26,** 1161 (1999).

Detection of biomass burning combustion products in Southeast Asia from backscatter data taken by the GOME spectrometer. Thomas, W., Hegels, E., Slijkhuis, S., Spurr, R., and Chance, K.I. *Geophys. Res. Letts.* **25,** 1317 (1998).

Biomass-burning influence on tropospheric ozone over New Guinea and South America. Kim, J.H. and Newchurch, M.J. *J. geophys. Res.* **103**, 1455 (1998).

Hydrogen peroxide, organic hydroperoxide and formaldehyde as primary pollutants from biomass burning. Lee, M., Heikes, B.G., Jacob, D.J., Sachse, G. and Anderson, B. *J. geophys. Res.* **102**, 1301 (1997).

OH, HO_2 and NO in two biomass burning plumes: Sources of HO_x and implications for ozone production. Folkins, I., Wennberg, P.O., Hanisco, T.F., Anderson, J.G., and Salawitch, R.J. *Geophys. Res. Letts.* **24**, 3185 (1997).

Ozone and aerosol distributions and air mass characteristics over the South Atlantic basin during the burning season. Browell, E.V., et. al. *J. geophys. Res.* **101**, 24043 (1996).

Observations of ozone concentrations in the Brazilian cerrado during the TRACE A field expedition. Kirchoff, V.W.J.H., Alves, J.R., da Silva, F.R., and Fishman, J. *J. geophys. Res.* **101**, 24029 (1996).

Biomass burning emissions and vertical distribution of atmospheric methyl halides and other reduced carbon gases in the South Atlantic region. Blake, N.J., Blake, D.R., Sive, B.C., Chen, T.-Y., Rowland, F.S., Collins, J.E. Jr., Sachse, G.W., and Anderson, B.E. *J. geophys. Res.* **101**, 24151 (1996).

Emissions of trace gases and aerosol particles due to vegetation burning in southern hemisphere Africa. Scholes, R.J., Ward, D.E., and Justice C.O. *J. geophys. Res.* **101,** 23677 (1996).

Methyl halide emissions from savanna fires in southern Africa. Andreae, M.O., Atlas, E., Harris, C.W., Helas, G., De Kock, A., Koppmann, R., Maenhaut, W., Manö, S., Pollock, W.H., Rudolph, J., Scharffe, D., Schebeske, G., and Welling, M. *J. geophys. Res.* **101**, 23603 (1996).

The quantity of biomass burned in southern Africa. Scholes, R.J., Kendall, J., and Justice, C.O. *J. geophys. Res.* **101**, 23667 (1996).

Biogenic NO emissions from savanna oils as a function of fire regime, soil type, soil nitrogen and water status. Parsons, D.A.B., Scholes, M.C., Scholes, R.J., and Levine J.S. *J. geophys. Res.* **101**, 23683 (1996).

Biogenic soil emissions of nitric oxide (NO) and nitrous oxide (N_2O) from savannas in South Africa: the impact of wetting and burning. Levine, J.S., Winstead, E.L., Parsons, D.A.B., Scholes, M.C., Scholes, R.J., Cofer W.R., III, Cahoon T., D.R., and Sebacher, D.I. *J. geophys. Res.* **101**, 23689 (1996).

Field study of the emissions of methyl chloride and other halocarbons from biomass burning in western Africa. Rudolph, J. Khedim, A. Koppmann, R., and Bonsang, B. *J. atmos. Chem.* **22**, 67 (1995).

Emission of methyl bromide from biomass burning. Manö, S. and Andreae, M.O. *Science* **263**, 1255 (1994).

Emission of some trace gases from biomass fires. Hegg, D.A., Radke, L.F., Hobbs, P.V., Rasmussen, R.A., and Riggan, P.J. *J. geophys. Res.* **95**, 5669 (1990).

Importance of biomass burning in the atmospheric budgets of nitrogen-containing gases. Lobert, J.M., Scharffe, D.H., Hao, W.M., and Crutzen, P.J. *Nature, Lond.* **346**, 552 (1990).

Enhanced biogenic emissions of nitric oxide and nitrous oxide following surface biomass burning. Anderson, I.C., Levine, J.S., Poth, M.A., and Riggan, P.J. *J. geophys. Res.* **93**, 3893 (1988).

Carbon monoxide and the burning earth. Newell, R.E., Reichle, H.G., Jr., and Seiler, W. *Scient. Am.* **261** (4), 58 (Oct 1989).

Biomass burning as a source of atmospheric gases CO, H_2, N_2O, NO, CH_3Cl and COS. Crutzen, P.J., Heidt, L.E., Krasnec, J.P., Pollock, W.H., and Seiler, W. *Nature, Lond.* **282**, 253 (1979).

6 Ions in the atmosphere

6.1 Electrical charges in the atmosphere

Charged particles represent only a minute fraction of the total mass of the Earth's atmosphere, but they play a crucial part in many geophysical phenomena such as variations in the geomagnetic field, lightning, and auroras. Neutral species may be produced or destroyed in reactions involving ions or electrons, and ion chemistry would be important even in relation to neutral composition and behaviour alone. Communication by radio waves is now regarded as an essential feature of 'civilized' life, so that the reflection, absorption, and propagation through the charged atmosphere of radio-frequency electromagnetic waves is of vital importance. Much of the motivation for atmospheric research on charged species has derived from the need to understand how to provide secure and reliable communication, or even how to prevent it! The military implications of such communications have been regrettably clear, although civil interests are considerable. Electrons are now known to be the particles immediately involved in interactions with radio propagation. Atmospheric ions and their chemistry, which are properly the subjects of our study, play a central role in determining the concentration and distribution of electrons. Our survey of atmospheric chemistry would evidently be incomplete without some discussion of radio propagation, especially as radio experiments have given so much information about atmospheric ions. Section 6.1.3 is devoted to an explanation of the radio studies, and it follows a very brief introduction to two geophysical manifestations of charged species that are of particular historical interest.

Ions and electrons are most abundant at altitudes greater than ~ 60 km (i.e. within the mesosphere or above). Several factors contribute to this high altitude distribution. Most of the Sun's ionizing electromagnetic radiation, a major source of ions as discussed in Section 6.3, is absorbed at levels above about 60 km. Ions and electrons are highly reactive, because their charges can produce long-range attractive forces (Section 3.4.1). The rapid increase in mean free path with increasing altitude in itself favours longer lifetimes for the charged particles. But, in addition, loss processes are more efficient for complex molecular ions than for atomic ions (Section 3.4.2), and the atomic ions are found with greater relative abundance at higher altitudes (Section 6.2). Free electrons at concentrations sufficient to affect radio-wave propagation are certainly only located at levels above 60 km, and this region has conventionally been called the ionosphere. Peak instantaneous electron densities can exceed 10^6 electrons cm^{-3} at ~ 300 km, and drop to near-zero at 60 km. Total charged particle concentrations (positive ions + negative ions + electrons) also fall to a minimum of a few hundred per cm^3 at ~ 60km, but increase a little in the stratosphere. Near ground level, the ion density is of the order of 10^3 cm^{-3}, but, of course, this value corresponds to a tiny mixing ratio (less than, say, 1 part in 10^{16}), while at 300 km several parts per thousand of the atoms and molecules are ionized. Stratospheric positive and negative ions are hard to investigate because of their small fractional abundance. Recent studies have, however, been successful, and some results are discussed in Section 6.5.

6.1.1 Aurora

Nature provides in the high atmosphere a phenomenon so obvious and striking that it has been the subject of scientific study for centuries. In regions not far from the geographic poles, a spectacular display of coloured lights can sometimes be seen at night. Complicated shapes are produced, which often move and change rapidly. The luminosity is given the name *aurora*, or 'dawn', with the qualifiers *borealis* ('northern') or *australis* ('southern') being added according to the polar region in which the display is seen.

Measurements of the angle of elevation of one auroral feature from two places simultaneously showed early on that most of the light comes from altitudes of around 100 km. Such observations thus gave man some knowledge of the atmosphere at heights then inaccessible to *in-situ* experiment. Spectroscopically, the auroral light is not dissimilar to that seen in the laboratory from electric discharges through low-pressure air. Atomic and molecular emission systems of oxygen and nitrogen predominate, the *auroral green line* of atomic oxygen, $O(^1S \rightarrow {}^1D)$, at $\lambda = 557.7$ nm being especially pronounced (see Section 7.4.1 for a further discussion of this line and of excitation mechanisms). The excitation seems to involve bombardment of atmospheric gases by particles from above, followed by ionization. Neutralization by electron capture leads to excitation, and directly or indirectly to emission of visible radiation. If the bombarding particles were charged species (e.g. electrons or protons) projected from the Sun, then they would be deflected as they approached the Earth because of the latter's magnetic field. The charged beam would impinge equally on day and night sides, and would be concentrated in circular regions, one round each pole. Most auroral activity is, in fact, observed in an oval belt around the geomagnetic pole, between 15° and 30° from it. Violent changes in the observed intensity of auroral light, according to this picture, correspond to changes in the primary ionizing stream originating initially from the Sun. There is also a connection between *solar activity* and auroral phenomena. Solar 'activity' is often assessed in terms of the number of sudden 'storms' that disturb the Sun's surface. These storms reveal themselves as intense localized bursts of H-alpha light ($\lambda = 656.3$ nm; the $n = 3 \rightarrow n = 2$ transition). *Solar flares* of this kind are accompanied by the liberation of huge amounts of energy as electromagnetic radiation (especially in the X-ray region) and as kinetic energy of particles (protons mainly, but also α-particles and heavier nuclei). It is generally believed that the energy is produced by the annihilation of magnetic fields associated with sunspots. Sunspots are areas, dark only relative to the solar disc, that appear from time to time on the Sun's surface, and move across the disc in ~ 13.5 days (solar rotational period ~ 27 days) A definite sunspot cycle of the number of spots is known to repeat itself every eleven years, and is thought to represent some important oscillation in the structure of the Sun itself. One of several manifestations in our atmosphere of solar storms is the enhanced auroral activity mentioned earlier. A complication arises in that the ionizing particles are not simply the storm particles emitted by the Sun, partly demonstrated in that there is too great a time delay between the storm and the auroral response. One suggestion is that particles already trapped within the Earth's magnetic field are released and accelerated by the arrival of the storm particles. Whatever the detailed explanation, there is a statistical correlation between solar activity and frequency of auroras

6.1.2 Geomagnetic fluctuations

Studies of the Earth's magnetism provided further early evidence that incident radiation might ionize the atmosphere at great altitudes. Detailed investigations had shown that the apparent direction of the Earth's magnetic field fluctuated regularly throughout a day, and that the fluctuations were smaller in winter than in summer. In 1882, Balfour Stewart suggested that electric currents circulating in the upper atmosphere might be responsible for diurnal changes in field. The atmospheric gases, *if conducting* and made to move across the Earth's magnetic field, would generate and carry such currents, as in a giant dynamo. Currents of the order of tens of thousands of amperes are involved! If solar radiation were responsible for the ionization, then the day-to-night variations would be explicable, as would the seasonal dependence of the magnitude. Solar particles seem a less likely agent of ionization, since they would arrive at the atmosphere near the auroral zones, while the geomagnetic effect is spread over the entire earth. Nevertheless, the diurnal excursions of magnetic field do show a sympathy with the sunspot cycle, the swing being nearly twice as great at sunspot maximum as at minimum. This cyclic behaviour is now interpreted in terms of the increased ionization from solar X-rays at times of high sunspot activity.

As early as 1740, large, rapid, and irregular changes had been noticed as an occasional feature of the geomagnetic field. *Magnetic storms* of this kind are often seen when auroral displays are in progress, and they are most intense within the auroral zones. They are not, however, exclusively confined to the auroral regions, so that the excitation of auroras and the increase of atmospheric conductivity do not have the same *immediate* cause, even though both are somehow connected with the solar cycle.

6.1.3 Radio propagation

Marconi's success, in 1901, in sending signals from England to America is often regarded as heralding the birth of radio as a means of communication. Radio waves are electromagnetic radiation and travel in nearly straight lines, as does light. Diffraction cannot possibly account for the travel of radio waves across the Atlantic, so that Marconi's feat demanded some explanation. An overhead reflector was suggested by Heaviside in England and by Kennelly in America: the hypothetical reflecting layer was called the *Heaviside Layer*. Commercial communication links were set up after Marconi's experiment, and it was soon noticed that signal strengths varied in a regular way throughout the day, the season, and the solar cycle. Magnetic storms were found to be associated with disturbances of the diurnal variability and sometimes with disruption of communication. When radio broadcasting started in the early 1920s, it was found that, *at night*, the signal strengths at distances of about 100 km varied widely over a few minutes, sometimes disappearing completely. This *fading* of the signals was ascribed to interference effects between a reflected *sky wave* and the direct *ground wave*. Reflections of the frequencies used for broadcasting thus seemed possible even for nearly vertical incidence, but only at night. The phenomenon is now very apparent with the large number of transmitting stations. At night, medium-distance (up to a few thousand km) stations can be received only too readily, with resulting interference between stations. During the day, only relatively local signals can be received at the frequencies involved (~ 1 MHz). Such discoveries suggested that radio reflection is caused by ionized layers in the atmosphere, and that the ionization is at least influenced, and probably caused, by solar radiation. As we shall see shortly, ionization at low altitudes leads also to absorption of broadcast frequencies, and prevents reflection from higher altitudes during the day. At night, the absorption is much diminished, and the various fading and interference effects become pronounced.

The early observations of the behaviour of radio waves were of great importance, because they indicated that radio waves could be used as a tool to explore the nature of the upper atmosphere and the ionization in it. Until rockets and satellites were able to enter the ionosphere or to examine it from above, radio studies provided much the most detailed information about our ionosphere. A crucial experiment concerned the polarization of reflected waves produced by the Earth's magnetic field. The results showed that the charged particles causing reflection were electrons and not ions. Sir Edward Appleton discovered that radio waves of frequency greater than 4–5 MHz penetrate the Heaviside layer, but, instead of escaping to space, they are reflected by a higher layer. This layer was named the *Appleton layer* after its discoverer. Appleton himself used the symbols E and F for the electric fields in the lower and upper layers, and suggested that the layers be called by the same letters in order to accommodate other possible layers by neighbouring letters of the alphabet. When discussing the ionosphere, it is universal practice to use this nomenclature, with D, E, and F layers or regions being particularly important. Electron densities in the ionosphere have been most commonly measured from the ground by a device called an *ionosonde*. A pulse of radiation of known frequency is directed vertically upwards, and the time delay of any reflection

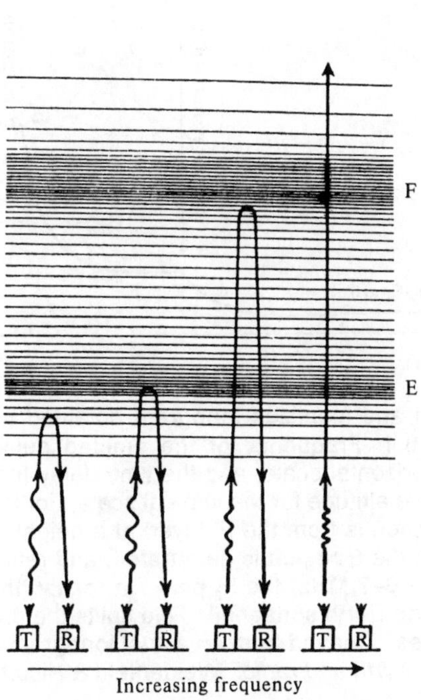

Fig. 6.1. Schematic representation of the operation of an ionosonde. T is the transmitter, and R the receiver. Horizontal shading indicates rate of change of ionization in the ionosphere. The four T–R pairs show, from left to right, what happens to radiowave pulses as the frequency increases. Based on a diagram of Iribarne, J.V. and Cho, H.-R. *Atmospheric physics*, D. Reidel Co., Dordrecht, 1980.

is recorded automatically. The frequency of the pulse train is increased to permit penetration to greater altitudes (see below). Figure 6.1 illustrates the principle of the ionosonde, and Fig. 6.2 is an ionogram obtained by day. At frequencies up to 4.4 MHz, reflection is by the E layer at an altitude of ~100 km. Wave penetration to the undisturbed F layer has a quite sharp onset at 4.4 MHz. Note that the F layer is partially split into two layers labelled F_1 and F_2: this is a day-time phenomenon, the F_1 'ledge' disappearing at night. The D region is more involved with absorption, rather than reflection, of radio waves, and so must be investigated by different techniques. *Topside sounding* by satellite instruments now complements the data provided by ground-based ionosondes.

We have so far discussed the experimental results obtained on penetration, reflection, and absorption without explanation of why the phenomena occur, or why they are dependent on frequency, electron concentration, or altitude. Reflection itself is brought about by a process akin to the total internal

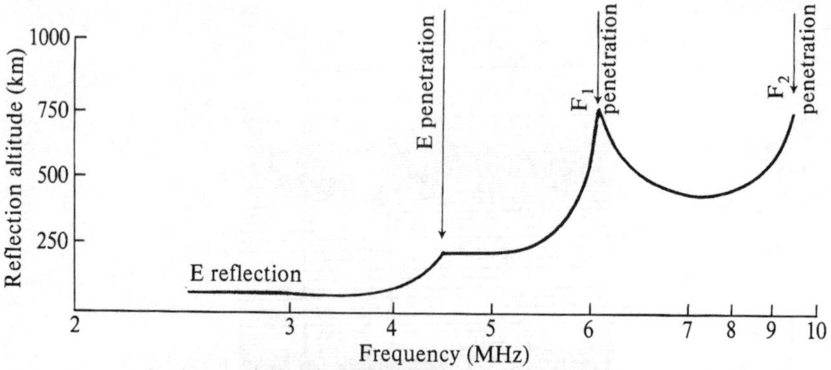

Fig. 6.2. Simplified and idealized ionogram obtained using the ionosonde technique of Fig. 6.1. Frequency of the emitted pulse of radiowaves is displayed on the horizontal scale, and the time delay for the return pulse is converted to effective altitude for the vertical scale. For frequencies less than about 4 MHz, reflection is from the E 'layer' at a height of about 120 km. At higher frequencies, the E region is penetrated, and reflection is from the F_1 ledge, and (above ~ 6–7 MHz) the F_2 peak. In reality, the traces tend to be doubled, because the Earth's magnetic field splits the original wave into two 'characteristic' waves. Derived from an actual ionogram trace reproduced in Ratcliffe, J.A. *Sun, earth and radio*, Weidenfeld & Nicolson, London, 1970.

reflection familiar in optics, although the detail is rather more subtle. Radio waves set atmospheric electrons into oscillation and cause them to radiate secondary wavelets. Since the electrons are randomly placed, the secondary wavelets coherently reinforce each other only in the original direction of propagation of the wave. (The *incoherent* or *Thomson* scatter in other directions produces a very weak wave that is also most valuable in ionospheric studies, as discussed at the end of this section). Although the wavelets constituting the forward-scattered wave have traversed equal paths, the re-radiated wavelets are $\pi/2$ advanced in phase compared with the original driving wave: that is, they are one-quarter of a cycle ahead. As a result, the composite combined wave also leads the driving wave by a phase angle dependent upon the relative contributions of scattered and unscattered radiation. Greater electron densities give rise to larger phase advances. In a very qualitative way, we can also say that higher frequencies are less able to force oscillation in the electrons, so that for the same electron density the phase advance is lower at high frequencies. The phase advance in the composite wave corresponds to an effective increase in the propagation velocity of the wave. An oblique wave is thus swung round and reflected as it enters regions of increasing electron density, since the higher parts of the wave move more rapidly than the lower. An extension of the argument then applies to vertically transmitted waves. The increase in velocity of propagation (see footnote [a], next page) corresponds to a decrease (below unity) of the refractive index, μ, of the medium. Total internal reflection becomes possible for angles of incidence, i, with the vertical such that $\mu = \sin i$. Reflection of vertically transmitted signals is possible if $\mu \to 0$. Theory shows that quantitatively

$$\mu^2 = 1 - (e^2/4\pi\epsilon_0 m \nu^2)n, \tag{6.1}$$

where n is the number density of electrons of charge e and mass m; ϵ_0 is the permittivity of free space, and ν is the frequency of the radio wave. The reflection condition can thus be written in terms of the number density of electrons, n_{reflect}, needed to achieve zero refractive index

$$n_{\text{reflect}} = \text{constant} \times \nu^2 \tag{6.2}$$

$$(\text{constant} = 1.24 \times 10^{-8}\ \text{s}^2\,\text{cm}^{-3} \text{ for } n_{\text{reflect}} \text{ in electrons cm}^{-3}).$$

The equation shows that, if electron densities increase with altitude, then at some critical altitude reflection may occur for a particular frequency. Higher frequencies will, however, penetrate to higher altitudes before reflection occurs. A further important conclusion to be drawn from our discussion and the nature of eqn (6.2) is that *layers* of electrons are not necessary for reflection to occur, but rather only sufficient densities. Nevertheless, the fairly sharp transitions between reflection altitudes at specific frequencies (Fig. 6.2) suggest that electron densities must themselves change rather sharply with altitude at those levels. Figure 6.3 gives some representative electron density profiles based on a variety of measurement techniques. The labels D, E, F (F_1, F_2) are now indeed seen to be associated with regions rather than layers, the boundaries between regions being altitudes where electron densities increase particularly rapidly with height. Day–night variations are shown for the two extremes of the sunspot cycle. Night-time electron densities are orders of magnitude smaller than daytime ones, and the distinction between F_1 and F_2 regions is lost. Electron densities in the D region become near-zero at night (although negative ions continue to be present in this region).

The presence of electrons in the daytime D region is responsible for the attenuation of reflected waves of the medium frequencies (~ 1 MHz) used for broadcasting, as suggested earlier. For the frequencies concerned, eqn (6.2) predicts an electron density for reflection of ~ 10^4 electrons cm^{-3}, which is that found at altitudes of ~ 80 km during the day. However, total gas densities—and hence collision rates—are relatively high in this low part of the ionosphere. Energy is removed from the oscillating electrons by collisions with neutral molecules, so that the radio-frequency signal is absorbed rather than reflected. Any process tending to increase ionization at low altitudes, where collision frequencies are high, tends also to bring with it enhanced absorption of radio

[a]To be more precise, the *phase velocity*; the group of waves in the transmitted pulse moves with the *group velocity*. Individual waves travel faster than the group in the ionosphere, so that when they reach the front of the group they die out and are replaced by new ones at the rear. At the turning point, the group velocity becomes zero for vertical reflection.

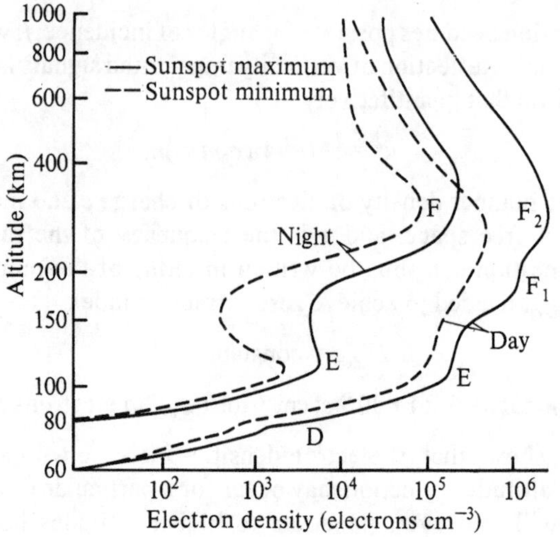

Fig. 6.3. Typical electron density–altitude profiles for the mid-latitude ionosphere. From a wallchart prepared by Swider, W., entitled *Aerospace environment*, Air Force Geophysics Laboratory, Hanscom Air Force Base, Massachusetts.

waves. We have already seen how observations early established an association of solar activity and of magnetic storms with disturbed radio conditions, and we shall return, in passing, to this point several times again.

Mention was made earlier of the use of *incoherent scatter* techniques to probe the ionosphere. An incoherent-scatter spectrum depends on parameters such as electron density and temperature, and ion composition, temperature and velocity. Although it is very difficult to determine all the parameters simultaneously from the scatter measurements themselves, various methods of fixing or modelling some of the parameters allow the others to be extracted. One possibility is thus to use the technique to determine ion composition, a subject of particular interest in our present examination of ionospheric chemistry. One highly successful programme has the acronym EISCAT (European Incoherent **SCAT**ter), and uses VHF and UHF radars situated in Norway, Finland, and Sweden with the overall objective of conducting high-latitude upper-atmosphere research.

6.2 Ion chemistry in the atmosphere

The regions of the ionosphere have been introduced in this chapter in terms of the reflection of radio waves, and the electron densities that cause the reflections. An equally valid division of the ionosphere into its regions can be based on the chemical identity of the dominant ions and the chemical processes occurring at different altitudes.

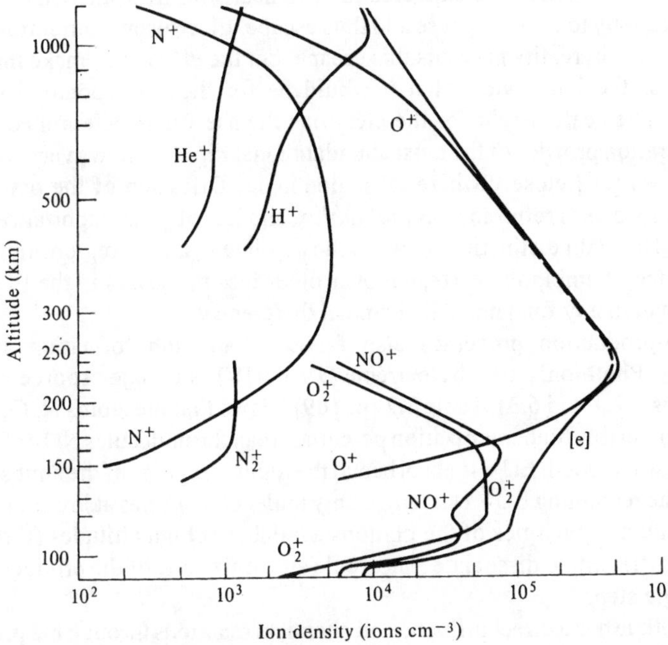

Fig. 6.4. Composition of positive ions in the E and F regions. Ion distributions were obtained by mass spectrometer experiments during daytime and at solar minimum; the data are normalized to the measured electron-density distribution From Johnson, C.Y. *J. geophys. Res.* **71**, 330 (1966).

Figure 6.4 shows ion compositions measured mass spectrometrically above about 100 km, together with total electron densities. The F region (say above 150 km) is characterized by atomic ions: O^+ and N^+ are dominant throughout most of the region, although H^+ and He^+ are more abundant at the very highest altitudes. Molecular ions (NO^+, O_2^+) are the most important ions in the E region (100–150 km), although total concentrations are lower than in the F region. At lower altitudes, in the D region, more complex molecular positive ions dominate, and negative ions are seen for the first time. We shall examine more carefully the composition and chemistry of each region in later sections. Here it is of interest to identify in outline some influences that lead to the compositions observed.

Gravitational separation (Section 2.1) of ionic species and their neutral precursors obviously favours atomic over molecular ions at high levels in the heterosphere. The abundances of the various atomic ions above the F peak of electron density at ~ 250 km fits this expectation well, with O^+ and N^+ being overtaken by the lightest ions H^+ and He^+ beyond about 1000 km. One point of interest concerns the scale heights (Section 2.1) for these ions. Electrons, being so light, could perhaps be expected to have near-infinite scale height, and to diffuse rapidly to the exosphere and thus escape. Electrostatic attraction to the positive ions in reality prevents this escape, but the effect is to make the scale height for the ions twice what it would be for the corresponding neutral species. The scale height for the electrons themselves is determined by the concentration profiles of the most abundant ions: Fig. 6.4 shows very well that [e] follows [O^+] closely where O^+ is dominant. Diffusion of the oppositely charged species is referred to as *ambipolar*, and it is of great importance in the F layer. The main distinction between the daytime F_1 and F_2 regions lies in the dominance of ambipolar diffusion over other loss processes in the F_2 region rather than in any fundamental chemical differences.

Ion-production processes also favour atomic-ion formation at high altitudes. Photoionization by extreme UV (EUV) is a major source of ions (Sections 3.2.1 and 6.3). Table 3.2 (p. 109) shows that the atoms N, O, and H have appreciably higher ionization potentials than the molecules NO or O_2. The shortest wavelength EUV is absorbed in the highest regions of the atmosphere, so that the remaining EUV can ionize only molecular species at lower altitudes (E, D regions). Energies of the photons available at high altitudes (F region) may be sufficient to dissociate a molecule and ionize one of the product atoms in a single step.

Altitude has a direct influence on chemical reactions through the pressure and hence the frequency of collisions. Ion–molecule reactions can convert the ions initially produced to different species that appear as *intermediate* and *terminal* ions. Although ion–molecule reactions are generally quite efficient because of long-range electrostatic attractions (Section 3.4.1), they are obviously faster for a given ion concentration at lower altitudes where the neutral reactant is more abundant. A number of ionic processes of atmospheric importance are brought together in Table 6.1. Some, such as the exchange processes (6.5) and (6.6), have clear counterparts in neutral chemistry, while others, such as charge transfer, (6.3), do not.

Conversion in the ionosphere of atomic ions to molecular ions is of critical importance in permitting charge neutralization, and thus in determining electron and ion concentrations throughout the atmosphere. Charge is lost, in the reactions shown, only in the recombination processes (6.7) to (6.9). As we discussed in Section 3.4.2, the ionization energy must be dissipated. Radiative recombination, process (6.7), is the only way in which atomic ions can be neutralized; it is extremely inefficient. Molecules can undergo dissociative recombination, as in reaction (6.8), where the fragments carry off the ionization energy as translational motion. Similarly, positive and negative molecular ions can interact [reaction (6.9)] to yield neutral particles that disperse the energy released as internal motions as well as kinetic energy of the products. A further mechanism for ion–ion recombination is recognized: a three-body association involving a neutral third-body, M, that can dissipate

Table 6.1 Some types of ionic process of atmospheric importance

Charge transfer			
	$N_2^+ + O_2 \rightarrow N_2 + O_2^+$		6.3
	$H^+ + O \rightarrow O^+ + H$		6.4
Exchanges			
	$N_2 + O^+ \rightarrow N^{(*)} + NO^+$		6.5
	$O_2^+ + N_2 \rightarrow NO + NO^+$		6.6
Recombinations			
	$O^+ + e \rightarrow O + h\nu$	Radiative	6.7
	$NO^+ + e \rightarrow N^{(*)} + O$	Dissociative	6.8
	$NO^+ + NO_2^- \rightarrow NO + NO_2$	Ion–ion	6.9
Attachments			
	$e + O \rightarrow O^- + h\nu$	Radiative	6.10
	$e + O_2 + M \rightarrow O_2^- + M$	Three-body	6.11
	$e + O_3 \rightarrow O^- + O_2$	Dissociative	6.12
Detachments			
	$O_2^- + M \rightarrow O_2 + e$	Collisional	6.13
	$O_2^- + O \rightarrow O_3 + e$	Associative	6.14
	$O^- + h\nu \rightarrow O + e$	Radiative	6.15
Clustering			
	$O_2^+ + O_2 + M \rightarrow O_2^+.O_2 + M$		6.16
	$H^+(H_2O)_n + H_2O\ (+\ M) \rightarrow H^+(H_2O)_{n+1}\ (+\ M)$		6.17
	$NO_3^- + H_2O + M \rightarrow NO_3^-.H_2O + M$		6.18

Note: Products of exothermic processes may possess electronic as well as translational (and, where appropriate, vibrational and rotational) excitation. For example, the atomic nitrogen product of reactions (6.5) and (6.8) can be formed in the excited ^2D state. The radiative detachment process (6.15) is referred to as a *photodetachment*.

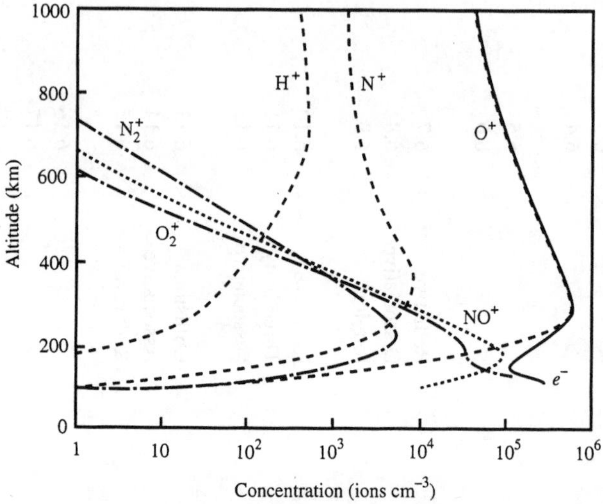

Fig. 6.5. Daytime ionospheric concentration profiles modelled for a 'quiet' ionosphere. From Diloy, P.-Y. *et al. Ann. Geophysicae* **14**, 191 (1996).

excess energy in the usual way. Such ion–ion charge-annihilation processes are of particular importance at atmospheric altitudes less than about 60 km, since almost all the negative charge is carried by ions rather than electrons, although we shall see in Sections 6.4.3, 6.4.4 and 6.5 that the ions involved are considerably more complex than the NO^+ and NO_2 represented in eqn (6.9).

Attachment of electrons to neutral molecules to form negative ions requires the removal of excess energy in the same way that neutralization does. Radiative attachment, reaction (6.10), is once again much less efficient than the three-body or reactive processes, (6.11) and (6.12), and is of little atmospheric significance. Much more important is the three-body reaction, but since its rate depends on the square of the pressure of neutral constituents, it is confined to the D region and below (as is the ozone reaction, (6.12)). Negative ions are consequently only found at low altitudes. Detachment processes, such as reactions (6.13) to (6.15), convert negative ions back to neutrals and electrons, so that the detailed balance between the number densities of negative ions and electrons depends on the competing rates of attachment, detachment, and ion–ion recombination.

The attractive forces between neutral molecules are not usually strong enough to favour formation of long-lived 'Van der Waals' species except at very low temperatures. For ions, however, the interactions with polarizable, or especially dipolar, molecules are much stronger, and 'cluster ions' are a feature of ion chemistry at low altitudes, as we shall see later. Some typical clustering processes are shown in reactions (6.16) to (6.18).

Although the mass-spectrometric data represented in Fig. 6.4 are more than 30 years old, models developed in the latter part of the 1990s using the best available chemical mechanisms and kinetic data reproduce the observations rather well. Figure 6.5, for example, shows the predictions of a model developed in connection with EISCAT studies (see end of Section 6.1.3). The trace for electron concentrations exhibits two peaks characteristic of the E and F regions. These peaks are associated with different ion populations, O_2^+ in the E region, and O^+ in the F region. Between these two peaks, NO^+ is the major ion, reaching a maximum concentration at an altitude of 180 km; O^+ begins to dominate again above 190 km. These results are in good agreement with both the mass spectrometric data and with EISCAT observations. Above the F-region peak, N^+ becomes an important minor ion, the ratio of $[N^+]/[O^+]$ reaching about five per cent at 1000 km. The production region for N^+ is around 350 km, but the N^+ ions diffuse upwards through the O^+. As a result, the scale height for N^+ is very close to that for the plasma, and the ionic composition is nearly constant with altitude, a conclusion consistent with the results of satellite measurements. The concentration of H^+ is about one per cent of that of O^+, according to the model described here.

6.3 Ionization mechanisms

Positive ions are formed in the atmosphere by three principal agencies: (i) solar radiation in the EUV and X-ray wavelength regions; (ii) galactic cosmic rays (GCR) and other galactic radiations; and (iii) precipitating energetic particles of solar origin and from the Earth's radiation belts. In the lower troposphere, radioactive emanation from rocks provides an ionization source at all times.

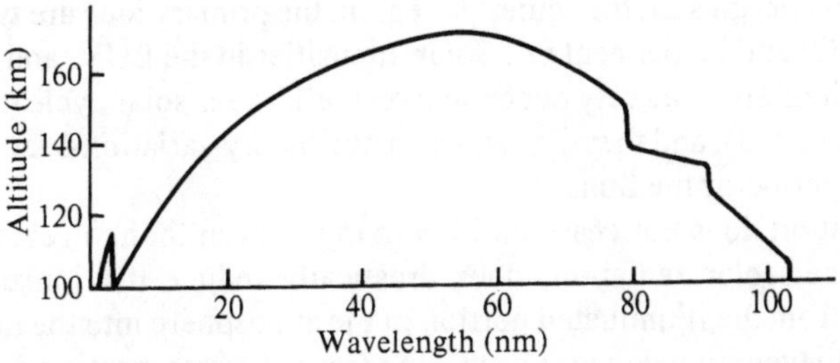

Fig. 6.6. Height at which the remaining overhead atmospheric layer possesses unity optical depth, shown as a function of wavelength. This height corresponds to the altitude at which the solar intensity has been attenuated to a value 1/e times the intensity incident outside the atmosphere. The height also corresponds approximately to the altitude of maximum rate of absorption for any wavelength. Adapted from Rishbeth, H. and Garriott, O.K. *Introduction to ionospheric physics*, Academic Press, New York, 1969.

Photoionization, by EUV and X-rays, is the most important ion source above 60 km during the daytime with a 'quiet' Sun (i.e. in the absence of solar storms). Figure 6.6 shows the altitude by which the atmospheric components afford unity optical depth (p. 55) at different wavelengths: these altitudes correspond to the heights at which the incident solar radiation has been attenuated by a factor e. Radiation of all wavelengths in the range reaches the F_2 region (> 200 km), while the range 20–91 nm probably contributes to F_1 ionization (~ 140–200 km). E region (~ 90–140 km) ionization comes from the more deeply penetrating part of the spectrum, EUV radiation between roughly 80 and 103 nm and X-rays from 1 to 10 nm wavelength. Several strong atomic lines are superposed on the solar continuum in the EUV region: H Lyman-β (102.6nm), C III (97.7 nm) and He I, II (58.4 nm, 30.4 nm) are prominent. Lower down in the atmosphere, in the D region (~ 70–90 km), the only surviving ionizing radiations are EUV at $\lambda > 103$ nm and 'hard' X-rays ($\lambda \sim 0.2$–0.8 nm). The only strong atomic feature is H Lyman-α at $\lambda = 121.6$ nm (≡ 10.19 eV); rather curiously, O_2 has a gap in its spectrum, where absorption is relatively weak, at exactly this wavelength, providing a 'window' that allows the radiation to penetrate more deeply than it would otherwise (cf. p. 109). Primary ions formed depend on the available wavelengths and the neutral precursors at any altitude. In the F region, therefore, O^+, together with some N^+, are the dominant primary ions (see Table 3.2, p. 109, for photoionization limits). Molecular ions become the major primary products in the E region, O_2^+ being generated in particular by Ly-β and C III solar atomic lines. Soft X-rays ionize N_2 to make N_2^+ the second most important primary ion. Only minor constituents such as NO or metal atoms (cf. Table 3.2) can be ionized by the Ly-α that penetrates into the D region. In addition, metastable *excited* O_2, in the $^1\Delta_g$ state, can be ionized in the wave-length range 102.7–111.8 nm (Section 3.2.1). Hard X-rays can ionize O_2 and N_2; cosmic rays can ionize all constituents. Figure 6.7 shows the relative daytime rates of ionization in the D and lower E regions due to photons of different energies. In the 'quiet' D region, the primary ions are typically 80 per cent NO^+ and 20 per cent O_2^+. Solar intensities in the EUV, and especially X-ray, regions are markedly dependent on the 11-year solar cycle (cf. Section 6.1.1 and Fig. 6.3), and they also show an oscillatory variation over the 27-day rotational period of the Sun.

Ionization does not cease entirely at night, even though removal of the main source, solar radiation, does drastically reduce the ionization rate. Radiation from the illuminated portion of the atmosphere into the night sector may be achieved by resonance scattering (resonance absorption followed by fluorescence). In the F region, $\lambda = 58.4$ and 30.4 nm lines scattered by He are a source of O^+; in the E region, $\lambda = 102.6$ nm Ly-β radiation, scattered by H, is an important source; and, in the D region, scattered Ly-α ($\lambda = 121.6$ nm) can ionize NO. Additional sources include starlight (stellar continuum radiation in the spectral interval 91.1–102.6 nm) in the E region, and galactic cosmic rays (which continue to arrive at night, of course) in the D region and below.

Corpuscular ionization—that is, ionization by impact with energetic particles from, or released by, the Sun—is a potential source of night- or day-time ionization. During 'disturbed' solar conditions, especially just after the maximum of the sunspot cycle, solar storms and flares (Section 6.1.1) are fairly frequent. Increases in ionospheric electron density are associated with the storms, and lead to a variety of ionospheric disturbances. *Sudden ionospheric disturbances* (SID) lead to shortwave fadeout of radio signals due to absorption in the enhanced D region (Section 6.1.3). The SID is produced by an intensification of hard X-ray emissions by several orders of magnitude, and is thus a photon effect. Corpuscular effects are generally delayed by at least several hours after the solar flare, and can persist for several days, in distinction to the flare itself that lasts for tens of minutes. One effect involves energetic electrons trapped in the Earth's radiation belts, which are slowly precipitated in the days following the storm to produce increased ionization in latitudes from approximately 45° to 72°. *Solar particle* (or *proton*) *events* (SPE), on the other hand, lead to enhanced ionization at high magnetic latitudes (> 60°), and are due to energetic particles, mainly protons, ejected from the Sun. The delay in arrival of the particles is a result of their having been guided into a long spiral path by the interplanetary magnetic field. Radio blackouts lasting several days can follow SPEs; because the D-region enhancement covers the polar regions only, such radio disturbances are called *polar cap absorption* (PCA) events. In almost all disturbances (SID, electron, and SPE) of the D region, the primary ions are almost 100 per cent O_2^+, since the major constituent O_2 is capable of being ionized (NO^+, it will be remembered, is dominant in the quiet D region).

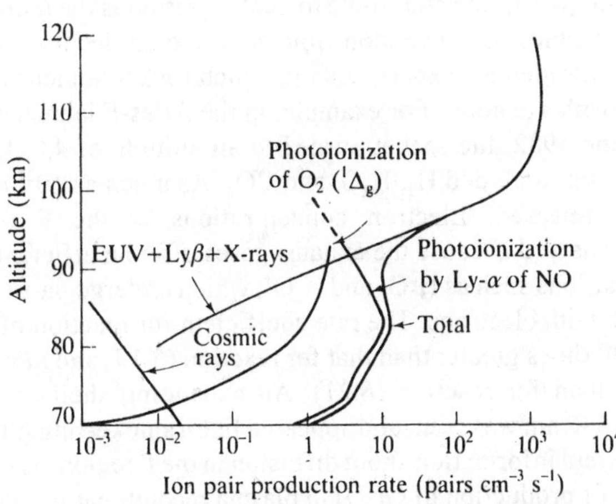

Fig. 6.7. Contribution of several ionization processes to ion pair formation in the 'quiet' daytime E and F regions. From McEwan, M.J. and Phillips, L.F. *Chemistry of the atmosphere*, Edward Arnold, London, 1975.

6.4 Chemistry of specific regions

6.4.1 F-region processes

Chemistry of ions in the F region is relatively simple, and it centres on the conversion of O^+, formed as the primary ion (Section 6.3) to secondary molecular ions that can recombine with electrons (Section 6.2). Two pairs of processes illustrate the conversion and neutralization

$$O^+ + O_2 \rightarrow O + O_2^+ \quad (6.19)$$

$$O_2^+ + e \rightarrow O(*) + O, \quad (6.20)$$

and

$$O^+ + N_2 \rightarrow N + NO^+ \quad (6.21)$$

$$NO^+ + e \rightarrow N(*) + O. \quad (6.8)$$

Reaction (6.19) has the *effect* of charge transfer, although it may proceed partially through an exchange mechanism. Electronic excitation in one of the fragment atoms in reactions (6.20) or (6.8) can help to carry off excitation energy, and it is represented in the equations by the bracketed asterisk. Optical emission from the excited atoms is then sometimes observed as *airglow*, a subject we explore further in the next chapter. For example, O(*) from reaction (6.20) can be in the ^1D state, and the emission (cf. pp. 534–6) is of the forbidden $^1D \rightarrow {}^3P$ atomic line in the red at $\lambda = 630$ nm. At F region altitudes, $[N_2] \gg [O_2]$ because photodissociation of O_2 is almost complete. Thus, although the rate constant for reaction (6.21) is about an order of magnitude less than that for reaction (6.19), the NO^+ route to neutralization is the more important.

The significance of conversion from atomic to molecular ions is nicely illustrated by 'accidental' experiments in which rocket launches produce an artificial 'ionospheric hole'. For example, in the Atlas-F launch of a weather satellite in June 1982, the rocket burned to an altitude of 434 km, ejecting exhaust gases that included H_2, H_2O, and CO_2. As much as 10^{27} molecule s^{-1} of H_2O was released. Electron concentrations in the F region were approximately halved because the exhaust gases offer an efficient route from O^+ to molecular ions such as H_2O^+ and H_3O^+, which undergo particularly rapid recombination with electrons. The rate coefficient for reaction of O^+ with H_2 and H_2O is 100 times greater than that for reaction (6.19) and more than 1000 times greater than for reaction (6.21). An expanding shell of $O(^1D \rightarrow {}^3P)$ airglow at $\lambda = 630$ nm was seen, and appeared like a smoke-ring. Studies of the emission can yield information about diffusion in the F region, as well as about the efficiency of production of $O(^1D)$ in plasma recombination. Indeed, these fortuitous observations have been followed up by deliberate experiments, as in those possessing the highly appropriate acronym RED AIR (Release Experiments to Derive Airglow Inducing Reactions). In these experiments, sounding rockets were used to place equal quantities of CO_2 above and below the maximum of the nocturnal F region. The reaction

$$O^+ + CO_2 \rightarrow O_2^+ + CO \qquad (6.22)$$

possesses an unusually large rate constant, and in the presence of injected CO_2, it converts the primary ion of the F region to O_2^+ (see also Section 8.3.3. for a discussion of the occurrence of this reaction in the Martian ionosphere). The additional O_2^+ is thus able to participate in reaction (6.20) to form $O(^1D)$. Enhanced red airglow was observed, and the plasma was depleted as expected. Detailed analysis of the optical and plasma parameters provided new insights into the excitation mechanism. We return briefly to the topic in Section 7.4.1.

The daytime peak in the F_2 region ion and electron concentrations might appear, at first sight, to be a result of Chapman layer formation (see Section 4.3.2). However, the peak concentrations occur at altitudes (~ 250–300 km) much greater than those (~100–150 km) where the rate of ionization is maximum. The peculiar behaviour is in part a result of the detailed charge neutralization (i.e. loss) mechanism. Since the rate-determining step in the loss processes is conversion of O^+ to molecular ions, mainly in reaction (6.21), the loss rate is proportional to $[N_2]$. Ionization rates are proportional to [O], so that the steady-state electron concentration is proportional to $[O]/[N_2]$. But the concentration of N_2 falls off more rapidly than [O] with increasing altitude (gravitational separation, etc.), so that electron density in this steady-state picture *increases* with height. In reality, the chemical destruction of electrons and ions from any altitude is supplemented by transport loss. Diffusion rates increase with decreasing pressure (increasing altitude) so that electron concentrations ultimately fall. The position of the F_2 peak will thus occur where the chemical and diffusive loss rates are identical.

Above the F_2 peak, the region dominated by O^+ gives way to the *protonosphere* dominated by H^+ (Fig. 6.4), the boundary being strongly influenced by the rapid and near-resonant charge-exchange reaction

$$O^+ + H \rightleftharpoons O + H^+, \qquad (6.23)$$

of which we have represented the reverse step alone in eqn (6.4). Below the F_2 peak, concentrations of NO^+ and O continue to increase, even though O^+ and electron concentrations decrease. In the lower F (or F_1) region, the molecular ion concentrations may even exceed $[O^+]$. The species are not, however, primary ions, but rather the products of reactions (6.21) and (6.19) favoured by the higher N_2 and O_2 concentrations and by the reduced electron densities.

6.4.2 E-region processes

Molecular ions are the primary ionization products in the E region. As discussed in Section 6.3, both O_2^+ and N_2^+ are produced from the major neutral atmospheric constituents. Observed ion distributions (Fig. 6.4) show only a small contribution from N_2^+, so that this ion must be consumed rapidly in secondary reactions. In fact, the reactions that represent losses of N_2^+ are also sources of the terminal ions

$$N_2^+ + O \rightarrow N(*) + NO^+ \qquad (6.5)$$

$$N_2^+ + O_2 \rightarrow N_2 + O_2^+. \qquad (6.3)$$

Exchange with atomic oxygen in the first reaction is favoured at high altitudes, and charge transfer with O_2 at lower levels, following the $[O]/[O_2]$ profile in the atmosphere.

Neutral nitric oxide begins to play an important part in atmospheric chemistry at E region altitudes. Ionic processes can themselves generate the molecule. For example, one source of excited atomic nitrogen, $N(^2D)$, is the dissociative recombination reaction (6.8); the excited atom can react with O_2

$$N(^2D) + O_2 \rightarrow NO + O. \qquad (6.24)$$

Reaction of O_2^+ with N_2

$$O_2^+ + N_2 \rightarrow NO + NO^+ \qquad (6.6)$$

has a rather small rate coefficient, but can be an important source of both NO and NO^+ because of the high values of $[N_2]$ relative to other species. Once present, NO tends to act as a charge-transfer acceptor

$$O_2^+ + NO \rightarrow O_2 + NO^+, \qquad (6.25)$$

because of its low ionization potential (Table 3.2). Reactions such as (6.6) and (6.25) thus tend to increase the NO^+ concentration at the expense of O_2^+. Dissociative recombination of NO^+ in reaction (6.8), rather than of O_2^+ in (6.20), is the most important charge neutralization step in the E region, since $[e] \ll [NO]$, and most O_2^+ ions will collide with NO to transfer charge before they meet an electron.

From time to time, unusual propagation of short-wave radio signals suggests the presence of local areas of increased ionization in the E region. These effects are known as *Sporadic E* phenomena. One manifestation of Sporadic E is long-distance reception of television pictures that have arrived by a reflected path that is usually absent. Narrow (1–3 km) localized layers of metal ions are often associated with sporadic E. Peak metal ion concentrations

---

Fig. 6.8. (*opposite*). Results of some rocket-borne mass spectrometric determinations of ion composition in the D region: (a) above Red Lake, Ontario; the main proton hydrates are shown, and the decrease in concentration of these species above ~ 85 km is clearly marked; (b) above Kiruna, N. Sweden; the transition— here at ~ 90 km—between proton hydrates and NO^+, O_2^+ is even more evident; (c) a mass-spectral scan (from the Kiruna flight of (b)) just below the transition height showing that $H^+ (H_2O)_n$, $n = 3$ to 12, are the dominant ions. From Kopp, E. and Herrman, U. *Ann. Geophysicae* **2**, 83 (1984).

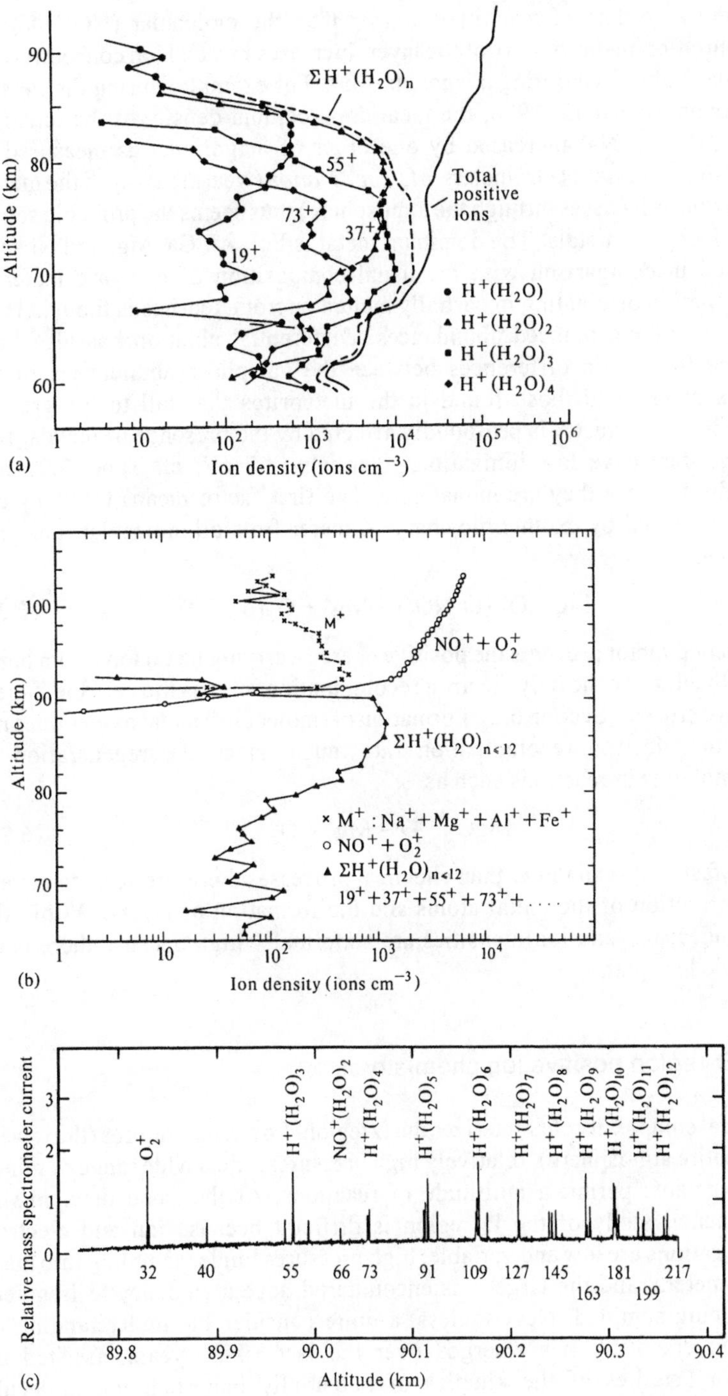

ΣH+(H2O)n
55+
73+
37+
19+
Total positive ions
• H+(H2O)
▲ H+(H2O)2
■ H+(H2O)3
◆ H+ (H2O)4
Altitude (km)
Ion density (ions cm−3)
(a)
M+
NO+ + O2+
ΣH+(H2O)n<12
× M+ : Na+ + Mg+ + Al+ + Fe+
○ NO+ + O2+
▲ ΣH+(H2O)n<12
19+ + 37+ + 55+ + 73+ + ...
Altitude (km)
Ion density (ions cm−3)
(b)
O2+
H+ (H2O)3
NO+(H2O)2
H+ (H2O)4
H+ (H2O)5
H+ (H2O)6
H+(H2O)7
H+ (H2O)8
H+ (H2O)9
H+(H2O)10
H+(H2O)11
H+(H2O)12
32
40
55
66
73
91
109
127
145
163
181
199
217
Relative mass spectrometer current
Altitude (km)
(c)

can be two orders of magnitude higher than the molecular (NO^+, O_2^+) ion concentration in the vicinity of the layer. Increases in metal ion concentrations have been observed during meteor showers. For example, during the Perseid shower of August 12, 1976, the mean total column density of the ions Fe^+, Mg^+, Ca^+, and Na^+ increased by one order of magnitude, as measured by rocket-borne mass spectrometers. *Meteor ablation* (wearing away of the meteor by friction as it passes through the atmosphere) thus seems the probable source of the E-region metals. The dominant metal is Fe^+; Al, Ca, Mg, and Ni seem depleted in comparison with the usual composition of meteoric material. Fractionation originating in partially melted micrometeorites is thought to be the cause of the reduced abundances. Differential ablation has also been invoked to explain differences between the gas-phase abundances of the *neutral* metals and those found in the meteorites that fall to the ground. Ionospheric chemistry is profoundly affected by the presence of metals, first because they have low ionization potentials ($\lesssim 7$ eV: cf. Table 3.2), and secondly because they are monatomic. The first factor means that they can become ionized by exothermic charge transfer from all molecular ions; e.g. with magnesium

$$Mg + O_2^+ \text{ (or } NO^+\text{)} \rightarrow Mg^+ + O_2 \text{ (or NO)}. \tag{6.26}$$

The second factor prevents the positive charge-carrying metal ions from being neutralized, since the only electron recombination route is the very inefficient radiative process (Section 6.2). Formation of (molecular) metal oxides does not assist the electron recombination rate much, since the regeneration of elemental ions in reactions such as

$$MgO^+ + O \rightarrow Mg^+ + O_2 \tag{6.27}$$

is very fast. Physical rather than chemical processes therefore seem to govern the distribution of the metal atoms and the formation of layers. Within the layers, electron (and ion) densities are abnormally high because there is no available loss pathway.

6.4.3 D-region positive ion chemistry

Chemical complexity characterizes the D region! Low temperatures (the lowest in the entire atmosphere), relatively high pressures, and a wide range of minor trace reactants permit a multitude of reactions. At the same time, *in-situ* experimental study of the D region is difficult because ion and electron concentrations are low and variable; high pressures hinder sampling into mass spectrometers; and the large ions encountered have a tendency to fragment while being sampled. Nevertheless, a quite considerable understanding of D-region chemistry has emerged over the last 10–20 years, assisted by laboratory studies of the kinetics of potentially important ion–molecule reactions. Investigations of positive-ion compositions (in the altitude range 64–112 km) date from 1965. These earliest mass-spectrometric studies straight away showed that some unexpected ions were present at the lower altitudes, with mass numbers separated by 18 units: that is, the mass of H_2O. At first, there was much suspicion that contamination or some other artefact was responsible for the mass-spectrometric peaks, but it is now certain that the ions are ionospheric hydrated protons, $H^+(H_2O)_n$, with n ranging from 2 to at least 8, and occasionally even 20: H_3O^+ itself seems, at most, to be of minor importance. Figure 6.8 shows some typical results.

A clear boundary, at ~ 82–85 km, is observed in most measured ion profiles, with the $H^+(H_2O)_n$ water cluster ions being dominant below the boundary, and NO^+ and O_2^+ above it. At the same altitudes, most daytime rocket flights find that the electron density decreases by almost an order of magnitude within a height decrease of a few kilometres. Reduced photoionization rates cannot account for the abrupt decrease in electron densities at this 'ledge', which means that the changes must have been brought about by greatly increased loss rates at lower altitudes. Negative-ion formation accounts for some of the electron loss,

$$e + O_2 + M \rightarrow O_2^- + M, \tag{6.11}$$

and, since it is a three-body process, its rate increases as the square of the pressure. Even more important, the large water cluster ions are exceptionally good at dissipating recombination energy, since several polyatomic fragments are formed

$$H^+(H_2O)_n + e \rightarrow H + nH_2O. \tag{6.28}$$

Dissociative recombination of $H^+(H_2O)_n$, with n ~ 6, is at least an order of magnitude faster than recombination with NO^+, reaction (6.8), at D-region temperatures (~200 K). Reaction (6.28) is not only a loss process for electrons, but also for cluster ions. It is thus at least self-consistent that $H^+(H_2O)_n$ is absent when $[e]$ is relatively large ($\gtrsim$82 km) and that $[e]$ is small when $[H^+(H_2O)_n]$ is relatively large ($\lesssim$ 82 km). But we now have to ask what factors make the water-ion clustering process so altitude-dependent. The easiest way of answering this question is to turn to the results of laboratory experiments.

Many techniques exist for the study of ion–molecule reactions, including ion beam, ion-cyclotron resonance, stationary discharge, high-pressure mass spectrometer, and drift-tube methods. However, the largest contribution to the understanding of ionospheric chemistry has probably come from flow studies in which ions are generated upstream of the reaction region and sampled by a downstream mass spectrometer. The method as applied to studies of ion reactions is almost always referred to as the flowing-afterglow technique. Its development by Eldon Ferguson and co-workers at Boulder, Colorado, has given us an understanding of D-region chemistry formerly lacking, as well as providing detailed knowledge about ionic processes in the other regions.

Variants of the flowing afterglow technique combine it with drift tubes or with upstream mass-spectrometric selection (*selected-ion flow tube*, 'SIFT') to sift the reactant ions and provide kinetic data of even greater sophistication. Studies using conventional cooling techniques to reach D-region temperatures (150–200 K) are of limited applicability when water is a reactant because of the condensation problem. One solution is to rapidly expand gases containing the reactants, thus cooling them. By forcing the gases through an expansion nozzle, a flowing stream of cold gas is established which can be used for flow-kinetic studies. At supersonic flow velocities, temperatures less than 200 K can be attained from a room-temperature starting point: in essence, the three-dimensional translational energy of the molecules has been rearranged to be one dimensional in the direction of flow. With such a system carrying ionized gases, and with a mass spectrometer as detector, D-region ion reactions can be studied in the laboratory at the temperature appropriate to the atmosphere.

With O_2^+ ions present in ultra-dry O_2, the ions that are produced at low temperature are oxygen self-clusters: $O_2^+.O_2$ and even $O_2^+.(O_2)_2$. Addition of a trace of water, however, prevents the formation of $O_2^+.(O_2)_2$, and diminishes the yield of $O_2^+.O_2$. In their place, first the ion $O_2^+(H_2O)$ and then the ion $O_2^+(H_2O)_2$ appear. Further addition of water leads to the conversion of $O_2^+(H_2O)_2$ to $H^+(H_2O)_2$, and then to higher hydrates. Clusters with n up to 78 have been seen in the laboratory: they can almost be regarded as ice surrounding an H^+ ion. The essential reactions in the scheme envisaged to explain the observations are

$$O_2^+ + O_2 + M \rightarrow O_2^+.O_2 + M \qquad \text{Cluster,} \qquad (6.16)$$

$$O_2^+.O_2 + H_2O \rightarrow O_2^+(H_2O) + O_2 \qquad \text{Switch,} \qquad (6.29)$$

$$O_2^+(H_2O) + H_2O \rightarrow O_2^+(H_2O)_2 \qquad \text{Cluster,} \qquad (6.30)$$

$$O_2^+(H_2O)_2 + H_2O \rightarrow H^+(H_2O)_2 + O_2 + OH \qquad \text{Exchange,} \qquad (6.31)$$

$$H^+(H_2O)_2 + H_2O\ (+\ M) \rightarrow H^+(H_2O)_3\ (+\ M) \qquad \text{Cluster,} \qquad (6.32)$$

and so on. An essential feature of the scheme is that entry into the series of reactions leading to $H^+(H_2O)_n$ involves $O_2^+.O_2$, whose formation is a three-body process. That is, the rate of clustering is limited by a process whose rate is proportional to pressure squared, and is thus highly altitude-dependent. Altitude dependence is further magnified by a reaction between atomic oxygen and $O_2^+.O_2$ that destroys the cluster

$$O + O_2^+.O_2 \rightarrow O_2^+ + O_3. \qquad (6.33)$$

Atomic oxygen concentrations increase sharply with altitude, so that, as a consequence of reactions (6.33) and (6.16), $[O_2^+.O_2]$, and hence $[H^+(H_2O)_n]$, increases very rapidly with decreasing altitude.

Molecular oxygen ions, O_2^+, are dominant in the D region only in 'disturbed' conditions (cf. Section 6.3). However, reactions analogous to those just described can be written for the major (80 per cent) ion of 'quiet' conditions, NO^+. Direct addition of water to NO^+ is too slow to account for conversion to $H^+(H_2O)_n$. Rather, NO^+ must first cluster with one of the major neutral gases in a three-body process

$$NO^+ + N_2\,(\text{or } O_2,\, CO_2) + M \rightarrow NO^+.N_2\,(\text{or } O_2,\, CO_2) + M. \qquad (6.34)$$

The resultant ion 'switches' in a fast reaction to form $NO^+(H_2O)$, which can then add a further water molecule: the dihydrate exchanges with H_2O to yield the proton hydrates

$$NO^+(H_2O)_2 + H_2O \rightarrow H^+(H_2O)_2 + HNO_2. \qquad (6.35)$$

Once again, the laboratory experiments show no $H^+(H_2O)_n$ with $n < 2$, in substantial agreement with ionospheric observation. A summary of the reaction schemes is provided diagrammatically in Fig. 6.9.

An essential feature of the scheme presented is the formation of NO^+ cluster ions with N_2, O_2, and CO_2. Such ions are rather weakly bound, $NO^+.N_2$ having a bond dissociation energy of only 20 kJ mol^{-1}. Sampling into rocket-borne mass spectrometers is difficult, but some success has been achieved. Mass-spectrometric peaks have been observed *in situ* that are assigned to the clusters $NO^+.N_2$, $NO^+.CO_2$ and $NO^+.H_2O.CO_2$. The rate of formation of the weakly bound cluster has a very strong *negative* temperature dependence. Accordingly, the conversion of NO^+ ions to water cluster ions is extremely sensitive to temperature changes, and the D-region positive-ion composition is expected to manifest strong seasonal, latitudinal, and even day-to-day variations as atmospheric temperatures fluctuate. Experimentally, such variability in the D region is well known, and the compositions and concentrations of Fig. 6.8 must be regarded only as representative.

Proton hydrates may be involved in the mesospheric phenomenon of *noctilucent clouds*. As the name implies, these clouds are only visible in the dark sky long after sunset: they are similar in appearance to thin cirrus clouds, but much higher and more tenuous. In order that they may still be lit up by the Sun, while the Earth's surface is in darkness, the noctilucent clouds must be situated at great height. The clouds lie at altitudes between 80 and 87 km, with the base most frequently near 82 km. Noctilucent clouds seem to be largely of summer–high latitude occurrence. That is, they are formed in the regions of the atmosphere and the seasons of the year where atmospheric temperatures are the lowest—perhaps even as low as 100–120 K. They are also, it will be noticed, formed at altitudes close to the D-region transition in ionospheric composition and concentration. Various pieces of evidence, including direct rocket sampling, suggest that noctilucent clouds are composed of water-ice. However, even at the low temperatures involved, it is hard to see how ice crystals could be formed by conventional nucleation mechanisms (Section 2.5). Dust particles from meteoric debris might conceivably be capable of acting as condensation nuclei. An alternative hypothesis is that the ionic water-cluster species, known

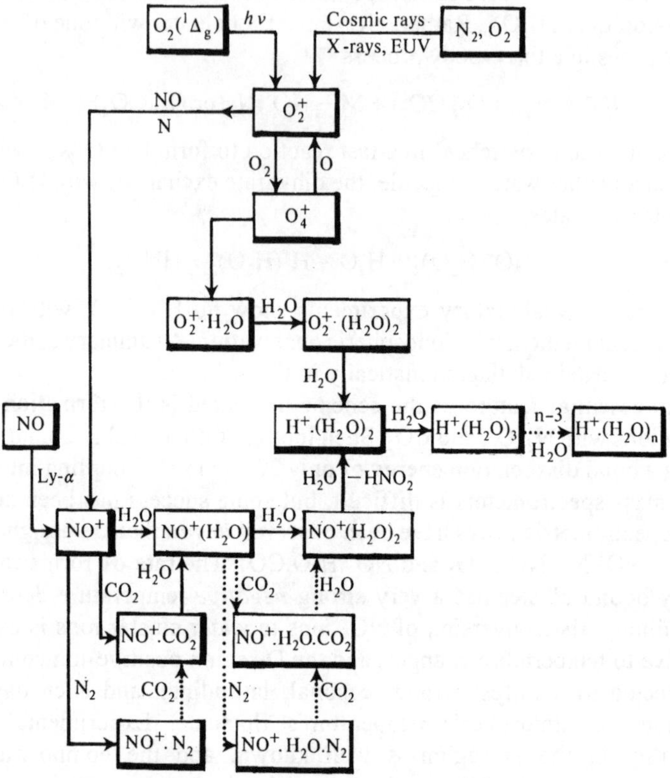

Fig. 6.9. Reaction scheme for positive-ion chemistry in the D region showing the major steps in the conversion of the primary ions, O_2^+ and NO^+, to the terminal ions, $H^+(H_2O)_n$. Adapted from Ferguson, E.E., Fehsenfeld, F.C., and Albritton, D.L. in Bowers, M. T. (ed.) *Gas phase ion chemistry, Vol. 1*, Academic Press, New York, 1979.

to be abundant at the cloud altitudes, are associated with the nucleation process. One possibility is that the dissociative recombination of $H^+(H_2O)_n$ in reaction (6.28) leaves a neutral cluster fragment $(H_2O)_m$ sufficiently large to start uninhibited growth (cf. Section 2.5)

$$H^+(H_2O)_n + e \rightarrow H + (H_2O)_m + (n-m)H_2O. \qquad (6.36)$$

Whatever the detailed formulation, the salient feature of the proton hydrate idea is that the energy barrier that exists to nucleation of neutral molecules can be overcome by the electrostatic ion–dipole attractive forces in the protonic species. Mass-spectrometric rocket measurements of positive-ion composition have shown that there may be a depletion, or even complete disappearance, of the main positive proton hydrates as an intense noctilucent cloud is traversed. There is thus evidence for the attachment of the positive ions to the particles of the cloud.

6.4.4 D-region negative-ion chemistry

Rather few detailed mass-spectrometric measurements exist for D-region negative ions. One reason is that negative ions exist as important species only where total (ion + electron) concentrations are low ($z < 78$ km). Further, the negative-ion population consists of many different chemical species, so that any individual mass-spectrometric peak is weaker than the less distributed positive-ion peaks. Technical problems also interfere with the sampling and detection of negative ions. Notwithstanding the problems, an understanding, at least in outline, of D-region negative-ion chemistry has begun to emerge.

The terminal ions appear to be species such as CO_3^-, HCO_3^-, Cl^-, and the hydrates $NO_3^-(H_2O)_n$ and $CO_3^-(H_2O)_m$, while the primary negative ion is O_2^-, produced by three-body electron attachment

$$e + O_2 + M \rightarrow O_2^- + M. \tag{6.11}$$

Such a process will show a (pressure)^2 dependence of reaction rate, so that negative-ion production is confined to altitudes below 70–80 km. Dissociative attachment of electrons to ozone,

$$e + O_3 \rightarrow O^- + O_2, \tag{6.12}$$

is a minor source of negative ions; as with reaction (6.11), there is a strong altitude dependence because of the rapid increase in $[O_3]$ with decreasing altitude in this region. The transition from ions to electrons as dominant carriers of negative charge is sharpened by the steep increase in [O] above 75–80 km. Atomic oxygen inhibits the growth of stable negative ions, an effect that can be expressed (with some simplification) as a result of the associative detachment reaction

$$O_2^- + O \rightarrow O_3 + e. \tag{6.14}$$

The arguments for a sudden increase in negative-ion concentrations below a boundary altitude are thus very similar to those used to explain $H^+(H_2O)_n$ emergence at the same altitudes, but with the feature of reduced dissociative recombination being absent.

Electron affinities of the neutral precursors of simple atmospheric negative ions are generally rather small, and much less than the ionization potentials of similar neutral species. For O_2 itself, the electron affinity is ~0.44 eV (42.5 kJ mol^{-1}). The negative ions are thus relatively unstable and have a tendency to participate in reactions, such as (6.14), that lead back to the release of electrons. In competition with the loss processes, ion–molecule reactions with minor molecular species can lead to the terminal ions that are stable enough to resist electron detachment. For example, the electron detachment energy of NO_3^- is ~3.9 eV (≡ 380 kJ mol^{-1}), one of the largest known for a simple negative ion.

Photodetachment (the analogue of photoionization) is possible with visible and even infrared radiation for ions with small electron affinity. With O_2 itself, near-infrared radiation can remove the electron

$$O_2^- + h\nu\,(\lambda \lesssim 2800\ \text{nm}) \rightarrow O_2 + e. \tag{6.37}$$

Daytime negative ion-to-electron concentration ratios are thus potentially affected by photodetachment processes, and there seems little doubt that the electron–ion transition boundary is shifted several kilometres to lower altitudes by day. A more detailed investigation of the rate at which the D-region negative ions build up at twilight, and electrons build up at dawn, reveals further subtleties. Effects of illumination are particularly well revealed by absorption of radio waves during PCA events (see Section 6.3). Absorption results from increased free-electron concentrations in the D region. Formation of negative ions therefore reduces the influence of the PCA. Very large day–night modulations of the absorption are observed, with the anomalous effects virtually disappearing at night. At dawn and twilight, however, the changes in electron density seem to require that the radiation causing photodetachment is screened by a layer for 30–40 km above the Earth's surface. Ozone is the obvious screening substance, but it filters out ultraviolet, and not visible radiation. Reaction (6.37) cannot, therefore, be the major detachment process. Instead, atomic oxygen from O_2 and O_3 photolysis (only ultraviolet is active) removes O_2^- by the associative detachment reaction

$$O_2^- + O \rightarrow O_3 + e. \tag{6.14}$$

Ozone photolysis,

$$O_3 + h\nu\,(\lambda \lesssim 310\ \text{nm}) \rightarrow O(^1D) + O_2(^1\Delta_g), \tag{6.38}$$

provides not only atomic oxygen, but also an excited molecular fragment (Section 3.2.1) that is curiously efficient as an electron-detaching partner in the reaction

$$O_2^- + O_2(^1\Delta_g) \rightarrow 2O_2 + e. \tag{6.39}$$

The progression from O_2^- to the terminal ions (NO_3^- and its hydrates) proceeds *via* a series of increasingly complex ions. Two threads can be identified in the conversion, one of which involves the ions O_4^-, CO_4^-, NO_3^{-*} ($O_2^-.NO$) and the other O_3^-, CO_3^-, NO_2^-. Clarification of the sequences can be achieved by writing the ions as though they were clusters, with a dot separating the initial ion and its cluster partner. Charge is, in reality, delocalized over the ion, but the cluster structure emphasizes the differences between, for example, the two nitrate ions

$$NO_3^- \text{ or } \left(O{-}N{<}^{O}_{O}\right)^- \quad \text{and} \quad NO_3^{-*} \text{ or } (O{-}ONO)^-$$

The initial step in each case is an addition, and each conversion represents a switching process with the appropriate neutral molecule

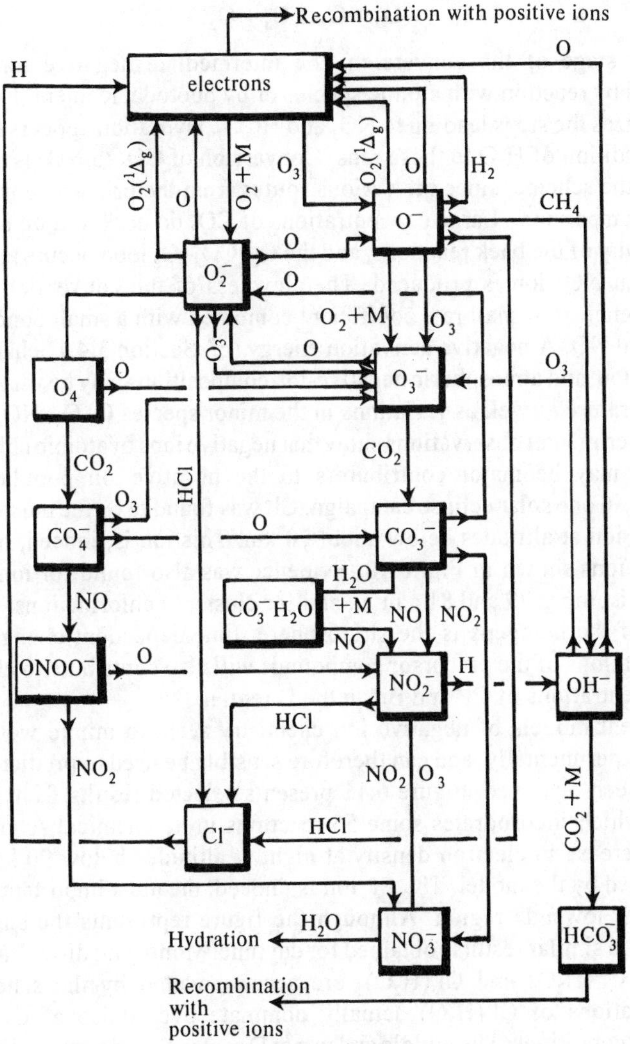

Fig. 6.10. Reaction scheme for negative-ion chemistry in the D region. O_2^- and O^- are the primary ions, while the terminal ions are species such as NO_3^-, CO_3^-, and their hydrates. Source as for Fig. 6.9.

$$O_2^- \xrightarrow{O_2} O_2^-.O_2 \xrightarrow{CO_2} O_2^-.CO_2 \xrightarrow{NO} O_2^-.NO \longrightarrow O^-.NO_2, \tag{6.40}$$

or

$$O^- \xrightarrow{O_2} O^-.O_2 \xrightarrow{CO_2} O^-.CO_2 \xrightarrow{NO} O^-.NO \xrightarrow{NO_2 \text{ or } O_3} O^-.NO_2. \tag{6.41}$$

The O_3^- ion ($O^-.O_2$) can be reached not only from the minor O^- primary species, but also from O_2^-

$$O_2^- + O_3 \rightarrow O_3^- + O_2. \tag{6.42}$$

At each stage of the conversion, the intermediate negative ions can be removed by reaction with atomic species or by photodetachment. Figure 6.10 summarizes the steps leading to NO_3^- and HCO_3^-. Hydration appears to involve direct addition of H_2O to these ions. Conversion of CO_3^- to NO_2^- is the bottleneck in the scheme, since the various routes from the primary ions to O_3^- are fast in comparison. Large concentrations of CO_3^- do not build up because of the rapidity of the back reactions, and the O_2^-–CO_3^-–O_2^- loop occurs many times before an NO_2^- ion is produced. The slowness of the conversion step is a consequence of a small rate coefficient combined with a small concentration of neutral NO. A negative activation energy (cf. Section 3.4.1) characterizes the reaction, and atmospheric negative-ion compositions may be very sensitive to temperature as well as variations in the minor species O, O_3, NO, and H.

Experimental observations show that negative ions of atomic chlorine and bromine may be major contributors to the negative-ion population. For example, in one solar eclipse campaign, Cl^- was found to be the most abundant negative ion at altitudes below about 70 km. This ion is, indeed, one of the terminal ions shown in Fig. 6.10. Evidence was also found for ions of both bromine isotopes 79 and 81 a.m.u., and for clustered chlorine ions. Since the source of the halogens is the stratosphere, time-dependent changes in the concentrations of the precursor compounds will also contribute to changes in the concentrations of Cl^- and Br^- in the D region.

Recent models of negative ion chemistry seem to mimic well what is known experimentally, and can therefore sensibly be used to predict what has not yet been observed. Figure 6.11 presents selected results from one such model, which incorporates some 55 reactions in its chemical scheme. The sharp decrease in electron density at night at altitudes below 90 km is well reproduced by the model. The Cl^- ion is, indeed, the most important negative ion of the lower D region. Although the figure represents the situation at midnight, a similar result is obtained for daytime. Mono- and di-hydrates of the Cl^- ion, $Cl^-(H_2O)$ and $Cl^-(H_2O)_2$ are accommodated by the scheme, and concentrations of $Cl^-(H_2O)$ actually dominate over those of Cl^- for the midnight conditions. The middle and upper D regions are dominated by HCO_3^-.

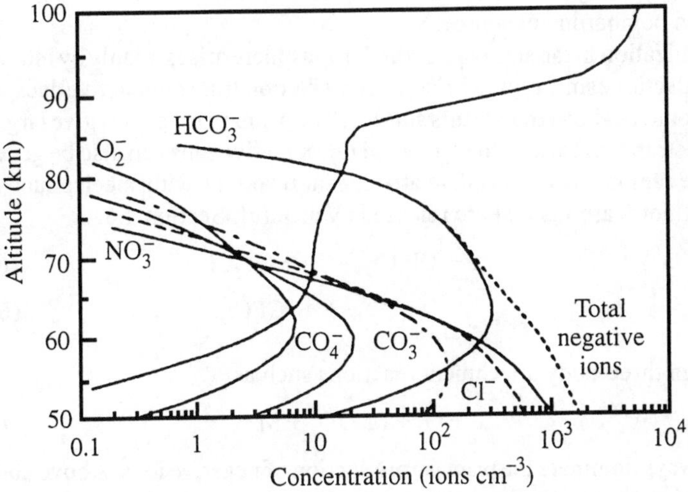

Fig. 6.11. Altitude profiles for the major negative ions modelled for summer mid-latitude conditions at midnight. From Fritzenwaller, J. and Kopp, E. *Advan. Space Res.* **21**, 891 (1998).

6.5 Ions in the stratosphere and troposphere

Although the D region is the bottom part of the 'conventional' ionosphere, speculation about underlying 'C', and lower regions has hardened into fact with the identification of ions in the stratosphere and troposphere. Rocket, balloon, and aircraft measurements have recently become feasible in these regions, and have revealed the presence of rather complex ions. Gaseous ions below the stratopause can play a role in trace-gas processes and in aerosol formation. Long-range forces often make reaction cross-sections ('rate coefficients') large for ion–molecule processes. In addition, a newly recognized class of process, *ion-catalysed reactions*, may occur in the stratosphere. Large cluster ions react with gas-phase species in a manner similar to surface-catalysed reactions: the reactant molecule can be regarded as being adsorbed on the 'surface' of the cluster ion. Ions promote nucleation *via* several processes, of which growth of large clusters (*ion nucleation*) and formation and growth of stable ion pairs by ion–ion recombination (*polyion nucleation*) are the most important. Gaseous ions are of additional interest in the stratosphere and troposphere since they can be used as powerful probes for detection of neutral gases. Selective reactions of naturally occurring ions with neutral trace gases lead to characteristic ionic products. Mass-spectrometric measurements of product and reactant ions allow concentrations of the trace species to be inferred with considerable accuracy. *Passive Chemical Ionization Mass Spectrometry* (PACIMS) of this kind has been supplemented by an active variant (ACIMS) in which ions are created in the stratospheric medium by an electron bombardment source.

Ionization in the stratosphere and troposphere arises mainly by interaction with galactic cosmic rays, with some contribution from radioactive decay in the last kilometre above the Earth's surface. The primary processes give largely O_2^+ and electrons that attach to O_2 to yield O_2^-. Negative ions can also be generated in these regions by dissociative attachment reactions with species such as the chlorofluorocarbons that are released by man (cf. Section 4.6.4)

$$e + CF_2Cl_2 \rightarrow Cl^- + CF_2Cl \qquad (6.43a)$$

$$\rightarrow F^- + CFCl_2, \qquad (6.43b)$$

although three-body attachment reactions such as

$$e + O_2 + M \rightarrow O_2^- + M \qquad (6.11)$$

will always dominate the primary production of negative ions. Above about 35 km, the major terminal positive ions are water clusters, as in the lower D region. A fairly sharp transition is seen at 30–35 km to mixed and non-proton hydrate cluster ions, of which $H^+.(CH_3CN)_l.(H_2O)_{n-l}$ and $H^+.(CH_3CN)_m$ are typical. Evolution of these ions can be envisaged as an extension of the D-region chemistry reviewed in Fig. 6.9. Formation of $H^+(H_2O)_n$ ions is driven by the large proton affinity (~ 711 kJ mol $^{-1}$) of H_2O, and the strong bonding of H_2O molecules to the hydronium ion, H_3O^+. Displacement reactions with a molecule C

$$H^+(H_2O)_n + C \rightarrow H^+C(H_2O)_{n-1} + H_2O, \qquad (6.44)$$

are exothermic if the proton affinity of C is greater than that of water. Acetonitrile, CH_3CN, falls into this category, since its proton affinity is 777 kJ mol^{-1}. Tropospheric sources (perhaps biomass burning: Section 5.10.10) seem most likely for CH_3CN, and the molecule survives transport to the stratosphere because it is neither attacked by OH nor photolysed efficiently (cf. Chapter 5). Other molecules with high proton affinities include CH_3OH and HCHO from methane oxidation (Section 5.3.2), and metal compounds such as NaOH or NaCl from meteor ablation and downward mixing from the mesosphere.

Stratospheric negative ions can be grouped into two families, $NO_3^-.(HNO_3)_p$ ($p \sim 2$–3) and $HSO_4^-.(H_2SO_4)_q.(HNO_3)_r$, ($q \leq 3$), with the former being dominant below about 25–30 km. The larger $HSO_4^-(H_2SO_4)_n$ ions are markedly hydrated, suggesting an increase in the H_2O bond energy for increasing q. Several other negative ions have been observed in smaller concentrations, most consisting of high electron affinity core molecules (e.g. CN, CO_3) and gas-phase ligands of high acidity (e.g. HCl, HNO_2, HOCl, HSO_3, H_2O). The major ions seem to be formed by a mechanism that starts from O_2^- as in the D region (Fig. 6.10). Nitric acid concentrations are sufficient in the stratosphere to displace water from $NO_3^-.(H_2O)_m$, and establish a quasi-equilibrium size distribution of $NO_3^-.(HNO_3)_p$. A second stage in the sequence involves reactions with S-containing gases (mostly H_2SO_4 and HSO_3), leading to HSO_4^- cores. Mixed cluster ions, $HSO_4^-.(H_2SO_4)_q.(HNO_3)_r$, are subsequently formed by the displacement of HNO_3 ligands by H_2SO_4. Strong co-operative bonding between H_2SO_4 and H_2O ligands leads to marked hydration of these mixed cluster ions, and the ions already resemble a small solution droplet composed of an H_2SO_4–H_2O mixture that has a large ΔH of mixing. Implications for stratospheric aerosol formation are apparent (cf. p. 177, p.197, p. 210, p. 213, and the Bibliography to Chapter 4).

The nature of tropospheric ions is at present just beginning to be unravelled, and the formidable problem of ion sampling from high pressures to be overcome. Aircraft flights at levels down to 3 km below the tropopause have returned mass-spectrometric data on positive and negative ions. For example, the major negative ion is identified as $NO_3^-.(HNO_3)_2$ and there are sizeable contributions from $NO_3^-.HNO_3$ and NO_3^- itself. Negative ions up to mass 352 have been observed, with the highest mass ion being identified as $HSO_4^-.(H_2SO_4)_2.HNO_3$. A potential difference between tropospheric and stratospheric ions may arise from the small, and possibly strongly variable, abundances of trace gases that can be depleted by heterogeneous interaction with aerosols. Detailed *in-situ* composition measurements of tropospheric ions will undoubtedly become refined in the near future, and investigation of the part played by ions in tropospheric chemistry will be a fruitful area of research.

Bibliography

Introductory texts dealing with the ionosphere and emphasizing the physical effects of charged particles in the atmosphere.

The solar-terrestrial environment. Hargreaves, J.K. (Cambridge University Press, Cambridge, 1992).

Sun, earth, and radio. Ratcliffe, J.A. Weidenfeld & Nicolson, London (1970).

An introduction to the ionosphere and magnetosphere. Ratcliffe, J.A. (Cambridge University Press, Cambridge, 1972).

Physics and chemistry of the upper atmosphere. Rees, M.H. (Cambridge University Press, Cambridge, 1989).

The Earth's ionosphere. Kelley, M.C. (Academic Press, San Diego, 1989).

Predictions and observations of ionospheric change.

The first real-time worldwide ionospheric predictions network: An advance in support of spaceborne experimentation, on-line model validation, and space weather. Szuszczewicz, E.P., Blanchard, P., Wilkinson, P., Crowley, G., Fuller-Rowell, T., Richards, P., Abdu, M., Bullett, T., Hanbaba, R., Lebreton, J.P., Lester, M., Lockwood, M., Millward, G., Wild, M., Pulinets, S., Reddy, B.M., Stanislawska, I., Vannaroni, G., and Zolesi, B. *Geophys. Res. Letts.* **25,** 449 (1998).

Review of long-term trends in the upper mesosphere, thermosphere and ionosphere. Danilov, A.D. *Advan. Space Res.* **22,** 907 (1998).

Section 6.1.1

The aurora. Bine, N. (Wiley, Chichester, 1996).

The northern light. Brekke, A. and Egeland, A. (Springer-Verlag, Berlin, 1983).

The aurora. Akasofu, S-I., in *Atmospheric phenomena.* Lynch, D.K. (ed.) pp. 141–9. (Freeman, San Francisco, 1980). (Scient. Am., December 1965).

The motion of ions in the auroral ionosphere. Fujii, R., Nozawa, S., Buchert, S.C., Matuura, N., and Brekke, A. *J. geophys. Res.* **103,** 20685 (1998).

VLF remote sensing of high-energy auroral particle precipitation. Cummer, S.A., Bell, T.F., Inan, U.S., and Chenette, D.L. *J. geophys. Res.* **102,** 7477 (1997).

Ion composition and effective ion recombination rate in the nighttime auroral lower ionosphere. Del Pozo, C.F., Hargreaves, J.K., and Aylward, A.D. *J. atmos. solar-terrest. Phys.* **59,** 1919 (1997).

Synoptic auroral distribution: A survey using polar ultraviolet imagery. Liou, K., Newell, P.T., Meng, C.I., Brittnacher, M., and Parks, G. *J. geophys. Res.* **102,** 27197 (1997).

Observation and analysis of NI 520.0 nm auroral emissions. Shepherd, M.G., Gattinger, R.L., and Jones, A.V. *J. atmos. terrest. Phys.* **58,** 579 (1996).

Energetic electron-precipitation during auroral events observed by incoherent-scatter radar. Osepian, A., Kirkwood, S., and Smirnova, N. *Advan. Space Res.* **17,** 149 (1995).

Section 6.1.2

Geomagnetic storm effects at low latitudes. Rastogi, R.G. *Ann. Geophys.* **17,** 438 (1999).

Comparison of models and data at Millstone Hill during the 5-11 June 1991 storm. Pavlov, A.V., Buonsanto, M.J., Schlesier, A.C., and Richards, P.G. *J. atmos. solar-terrest. Phys.* **61,** 263 (1999).

The storm of 10 January 1997: electrodynamics of the high latitude E region from EISCAT data. Schlegel, K., and Collis, P.N. *J. atmos. solar-terrest. Phys.* **61,** 217 (1999).

Forecasting ionospheric structure during the great geomagnetic storms. Cander, L.R. and Mihajlovic, S.J. *J. geophys. Res.* **103,** 391 (1998).

Trends in the ionospheric E and F regions over Europe. Bremer, J. *Ann. Geophys.* **16,** 986 (1998).

Post storm effects in middle and subauroral latitudes. Bremer, J. *Advan. Space Res.* **22,** 837 (1998).

The role of vibrationally excited oxygen and nitrogen in the ionosphere during the undisturbed and geomagnetic storm period of 6–12 April 1990. Pavlov, A.V. *Ann. Geophys.* **16,** 589 (1998).

Midlatitude particle and electric field effects at the onset of the November 1993 geomagnetic storm. Foster, J.C., Cummer, S., and Inan, U.S. *J. geophys. Res.* **103,** 26359 (1998).

A comparative study of global ionospheric responses to intense magnetic storm conditions. Szuszczewicz, E.P., Lester, M., Wilkinson, P., Blanchard, P., Abdu, M., Hanbaba, R., Igarashi, K., Pulinets, S., and Reddy, B.M. *J. geophys. Res.* **103,** 11665 (1998).

Prompt mid!atitude electric field effects during severe geomagnetic storms. Foster, J.C. and Rich, F.J. *J. geophys. Res.* **103,** 26367 (1998).

Variations of total electron content during geomagnetic disturbances: A model/observation comparison. Lu, G., Pi, X., Richmond, A.D., and Roble, R.G. *Geophys. Res. Letts.* **25,** 253 (1998).

Small-scale structure of electron density in the D region during onset phases of an auroral absorption substorm. Ranta, H. and Yamagishi, H. *Advan. Space Res.* **19,** 159 (1997).

Ionospheric storm effects in the nighttime E region caused by neutralized ring current particles. Bauske, R., Noel, S., and Prolss, G.W. *Ann. Geophys.* **15,** 300 (1997).

On the seasonal response of the thermosphere and ionosphere to geomagnetic storms. Fuller-Rowell, T.J., Codrescu, M.V., Rishbeth, H., Moffett, R.J., and Quegan, S. *J. geophys. Res.* **101,** 2343 (1996).

VLF and LF signatures of mesospheric lower ionospheric response to lightning discharges. Inan, U.S., Slingeland, A., Pasko, V.P., and Rodriguez, J.V. *J. geophys. Res.* **101,** 5219 (1996).

Section 6.1.3

Electrons in the ionosphere affect the propagation, absorption, and scattering of radio waves. Observations of radio-wave behaviour have thus provided an important tool for probing the ionosphere. Incoherent scattering measurements, such as those conducted in the EISCAT programme (see text) have recently yielded much important new information. Polar cap absorption is treated in Section 6.3.

Ionospheric D region remote sensing using VLF radio atmospherics. Cummer, S.A., Inan, U.S., and Bell, T.F. *Radio Sci.* **33,** 1781 (1998).

Coherent backscatter cross-section ratio measurements in the midlatitude E region ionosphere. Koehler, J.A., Haldoupis, C., and Schlegel, E. *J. geophys. Res.* **104,** 4351 (1999).

Ion composition measurements and modelling at altitudes from 140 to 350 km using EISCAT measurements. Litvine, A., Kofman, W., and Cabrit, B. *Ann. Geophys.* **16,** 1159 (1998).

The EISCAT Svalbard radar: A case study in modern incoherent scatter radar system design. Wannberg, G., Wolf, I., Vanhainen, L.G., Koskenniemi, K., Rottger, J., Postila, M., Markkanen, J., Jacobsen, R., Stenberg, A., Larsen, R., Eliassen, S., Heck, S., and Huuskonen, A. *Radio Sci.* **32,** 2283 (1997).
Coordinated EISCAT/DMSP measurements of electron density and energetic electron precipitation. Anderson, P.C., McCrea, I.W., Strickland, D.J., Blake, J.B., and Looper, M.D. *J. geophys. Res.* **102,** 7421 (1997).
Incoherent scatter radar and Digisonde observations at tropical latitudes, including conjugate point studies. Scali, J.L., Reinisch, B.W., Kelley, M.C., Miller, C.A., Swartz, W.E., Zhou, Q.H., and Radicella, S. *J. geophys. Res.* **102,** 7357 (1997).
Incoherent scatter radar observations of the F-region ionosphere at Arecibo during January 1993. Zhou, Q.H. and Sulzer, M.P. *J. atmos. solar-terrest. Phys.* **59,** 2213 (1997).
Ionospheric composition measurement by EISCAT using a global fit procedure. Cabrit, B., and Kofman, W. *Ann. Geophys.* **14,** 1496 (1996).
Comparison of F-region electron density observations by satellite radio tomography and incoherent scatter methods. Nygren, T., Markkanen, M., Lehtinen, M., Tereshchenko, E.D., Khudukon, B.Z., Evstafiev, O.V., and Pollari, P. *Ann. Geophys.* **14,** 1422 (1996).
Incoherent scatter radar contributions to high latitude D-region aeronomy. Turunen, E. *J. atmos. terrest. Phys.* **58,** 707 (1996).
EISCAT and the EISCAT data-base — a tool for ionospheric modeling (E- region and D-region). Schlegel, K. *Advan. Space Res.* **16,** 147 (1995).
The high latitude D-region and mesosphere revealed by the EISCAT incoherent scatter radars during solar proton events. Collis, P.N. *Advan. Space Res.* **18,** 83 (1995).

Sections 6.2 and 6.3

The nature and rates of reactions involving charged species are also discussed in Section 3.4.1 and the associated references. See Section 6.4.3 for experimental methods for the laboratory study of such reactions. The first three reviews are excellent general accounts of atmospheric ion chemistry, and that by Thomas also provides an introduction to the chemistry of neutral species in atmospheric regions above the stratosphere. The next two are concerned more particularly with the higher altitude regions (above about 100 km) and the last four with the lower altitudes.

Ions in the terrestrial atmosphere and in interstellar clouds. Smith, D. and Spanel, P. *Mass Spec. Revs.* **14,** 255 (1995).
Neutral and ion chemistry of the upper atmosphere. Thomas, L., in *Handbuch der Physik* Rawer, K. (ed.) XLIX/6, 7 Part VI. (Springer-Verlag, Berlin, 1982).
Upper atmosphere models and research. Ryecroft, M.J., Kasting, G.M., and Rees, D., (eds.) *Adv. Space Res.* **10**, No. 6 (1990).

Ion chemistry of the Earth's atmosphere. Ferguson, E.E., Fehsenfeld, F.C., and Albritton, D.L., in *Gas phase ion chemistry*. Bowers, M.T.(ed.) Vol. 1, pp. 45–82. (Academic Press, New York, 1979).

Ionospheric Chemistry. Torr, D.G. *Rev. Geophys. & Space Phys.* **17**, 510 (1979).

Chemistry of the thermosphere and ionosphere. Torr, D.G. and Torr, M.R. *J. atmos. terr. Phys.* **41**, 797, (1979).

Modelling of the ion composition of the middle atmosphere. Thomas, L. *Ann. Geophysicae* **1**, 61 (1983).

Chemistry of middle atmospheric ionization—a review. Mitra, A.P. *J. atmos. terr. Phys.* **43**, 737 (1981).

On ionization of the lower mesosphere. Chakrabarty, D.K., Chakrabarty, P. and Witt, G. *J. atmos. terr. Phys.* **43**, 23 (1981).

Ion composition and electron and ion loss processes in the Earth's atmosphere. Arnold, F. and Krankowsky, D., in *Dynamical and chemical coupling*. Grandal, B. and Holtet, J.A. (eds.) (D. Reidel, Dordrecht, 1977).

The experimental study in the laboratory of gas-phase ion reactions

Reactions of mass-selected cluster ions in a thermal bath gas. Viggiano, A.A., Arnold, S.T., and Morris, R.A. *Int. Revs. Phys. Chem.* **17**, 147 (1998).

Low-temperature rate studies of ions and radicals in supersonic flows. Smith, M.A. *Int. Revs. Phys. Chem.* **17**, 35 (1998).

Ion–ion chemistry of high-mass multiply charged ions. McLuckey, S.A., and Stephenson, J.L. *Mass Spect. Revs.* **17,** 369 (1998).

Atmospheric chemistry and the flowing afterglow technique. Squires, R.R. *J. mass Spect.* **32,** 1271 (1997).

Ionospheric meteorology

Dynamical coupling processes between the middle atmosphere and lower ionosphere. Hocking, W.K. *J. atmos. terrest. Phys.* **58,** 735 (1996).

Winds in the ionosphere – a review. Titheridge, J.E. *J. atmos. terrest. Phys.* **57,** 1681 (1995).

The ionosphere and atmospheric change

Atmospheric and ionospheric response to trace gas perturbations through the ice age to the next century in the middle atmosphere. 2. Ionization. Beig, G. and Mitra, A.P. *J. atmos. solar-terrest. Phys.* **59,** 1261 (1997).

Ionization processes

Studies of the auroral E region neutral solar cycle: Quiet days. Nozawa, S., and Brekke, A. *J. geophys. Res.* **104,** 45 (1999).

HALOE nitric oxide measurements in view of ionospheric data. Friedrich, M., Siskind, D.E., and Torkar, K.M. *J. atmos. solar-terrest. Phys.* **60,** 1445 (1998).
Solar activity effects in the ionospheric D region. Danilov, A.D. *Ann. Geophys.* **16,** 1527 (1998).
Solar radiation fluxes in the 10–30 nm range from studying the E- region and E–F valley. Nusinov, A.A., Antonova, L.A., and Katyushina, V.V. *Solar Phys.* **177,** 191 (1998).
Definition of disturbance and quietness with topside ionosonde data. Gulyaeva, T.L., Barbatsi, K., Boska, J., Defranceschi, G., Kouris, S.S., Moraitis, S., Pulinetz, S., Radicella, S.M., Stanislawska, I., Xenos, T., and Zhang, M.L. *Advan. Space Res.* **16,** 143 (1995).
Ionospheric electron-densities calculated using different EUV flux models and cross-sections — comparison with radar data. Buonsanto, M.J., Richards, P.G., Tobiska, W.K., Solomon, S.C., Tung, Y.K., and Fennelly, J.A. *J. geophys. Res.* **100,** 14569 (1995).
Photoionization rates in the night-time E- and F-region ionosphere. Strobel, D.F., Opal, C.B., and Meier, R.R. *Planet. Space Sci.* **28,** 1027 (1980).

Polar cap absorption

Structure and occurrence of polar ionization patches. Coley, W.R., and Heelis, R.A. *J. geophys. Res.* **103,** 2201 (1998).
Daytime ionospheric absorption features in the polar cap associated with poleward drifting F-region plasma patches. Nishino, M., Nozawa, S., and Holtet, J.A. *Earth Planets & Space* **50,** 107 (1998).
Structures in ionospheric number density and velocity associated with polar cap ionization patches. Kivanc, O. and Heelis, R.A. *J. geophys. Res.* **102,** 307 (1997).
Twilight anomaly, midday recovery and cutoff latitudes during the intense polar-cap absorption event of March 1991. Ranta, H., Yamagishi, H., and Stauning, P. *Ann. Geophys.* **13,** 262 (1995).

Section 6.4.1

Electron energy balance in the F region above Jicamarca. Aponte, N., Swartz, W.E., and Farley, D.T. *J. geophys. Res.* **104,** 10041 (1999).
Thermalization of $O(^1D)$ atoms in the thermosphere. Shematovich, V., Gerard, J.C., Bisikalo, D.V., and Hubert, B. *J. geophys. Res.* **104,** 4287 (1999).
Intercomparison of physical models and observations of the ionosphere. Anderson, D.N., Buonsanto, M.J., Codrescu, M., Decker, D., Fesen, C.G., Fuller-Rowell, T.J., Reinisch, B.W., Richards, P.G., Roble, R.G., Schunk, R.W., and Sojka, J.J. *J. geophys. Res.* **103,** 2179 (1998).
Comparison of measured high latitude F-region ion composition climatological variability with models. Grebowsky, J.M., Erlandson, R.E., Sojka, J.J., Schunk, R.W., and Bilitza, D. *Advan. Space Res.* **22,** 885 (1998).

The effects of meridional neutral winds on the O^+–H^+ transition altitude over Arecibo. MacPherson, B., Gonzalez, S.A., Bailey, G.J., Moffett, R.J., and Sulzer, M.P. *J. geophys. Res.* **103,** 29183 (1998).

Computer-simulation of electron and ion densities and temperatures in the equatorial F-region and comparison with hinotori results. Watanabe, S., Oyama, K.I., and Abdu, M.A. *J. geophys. Res.* **100,** 14581 (1995).

Some modelling studies of the equatorial ionosphere using the Sheffield University Plasmasphere Ionosphere Model. Bailey, G.J., and Balan, N. *Advan. Space Res.* **18,** 59 (1995).

Atomic oxygen in the thermosphere during the July 13, 1982, solar proton event deduced from far ultraviolet images. Drob, D.P., Meier, R.R., Picone, J.M., Strickland, D.J., Cox, R.J., and Nicholas, A.C. *J. geophys. Res.* **104,** 4267 (1999).

A study of the daytime E–F_1 region ionosphere at mid latitudes. Buonsanto, M.J. *J. geophys. Res.* **95**, 7735 (1990).

Dissociative recombination of ions such as O_2 or NO^+ can frequently lead to electronically excited fragment atoms. The efficiency of the process with O_2^+ is discussed in connection with atomic oxygen airglow in Section 7.4.1 and the associated references. Artificial injection of molecular species dramatically decreases electron densities by providing an efficient recombination route.

Images of transequatorial F region bubbles in 630- and 777-nm emissions compared with satellite measurements. Tinsley, B.A., Rohrbaugh, R.P., Hanson, W.B., and Broadfoot, A.L. *J. geophys. Res.* **102,** 2057 (1997).

Mechanism for the green glow of the upper ionosphere. Guberman, S.L. *Science* **278,** 1276 (1997).

OI 630.0 nm nightglow depletion observations at Watukosek, Indonesia. Shibasaki, K., Kita, K., Iwagami, N., Ogawa, T., and Sripto, A. *Earth Planets & Space* **49,** S197 (1997).

Heterogeneous structure of the ionosphere F region formed by rockets. Nagorskii, P.M. *Geomag. Aeronom.* **38,** 100 (1998).

A study of oxygen 6300 A airglow production through chemical modification of the nighttime ionosphere. Semeter, J., Mendillo, M., Baumgardner, J., Holt, J., Hunton, D.E., and Eccles, V. *J. geophys. Res.* **101,** 19683 (1996).

Early time evolution of a chemically produced electron depletion. Scales, W.A., Bernhardt, P.A., and Ganguli, G. *J. geophys. Res.* **100,** 269 (1995).

Ionospheric hole caused by rocket engine. Rycroft, M.J. *Nature, Lond.* **217**, 537 (1982).

Section 6.4.2

A numerical model of the ionosphere, including the E-region above EISCAT. Diloy, P.Y., Robineau, A., Lilensten, J., Blelly, P.L., and Fontanari, J. *Ann. Geophys.* **14,** 191 (1996).

Model results for the ionospheric E region: Solar and seasonal changes. Titheridge, J.E. *Ann. Geophys.* **15,** 63 (1997).

Comparison of topside and bottomside irregularities in equatorial F region ionosphere. Kuo, F.S., Chou, S.Y., and Shan, S.J. *J. geophys. Res.* **103,** 2193 (1998).

Studies of the E-region ion-neutral collision frequency using the EISCAT incoherent scatter radar. Nygren, T. *Advan. Space Res.* **18,** 79 (1995).

A numerical study of ionospheric profiles for midlatitudes. Zhang, S.R. and Huang, X.Y. *Ann. Geophys.* **13,** 551 (1995).

'Sporadic E'

Sporadic E: current views and recent progress. Mathews, J.D. *J. atmos. solar-terrest. Phys.* **60,** 413 (1998).

Evidence for planetary wave effects on midlatitude backscatter and sporadic E layer occurrence. Voiculescu, M., Haldoupis, C., and Schlegel, K. *Geophys. Res. Letts.* **26,** 1105 (1999).

Large airglow enhancements produced via wave-plasma interactions in sporadic E. Djuth, F.T., Bernhardt, P.A., Tepley, C.A., Gardner, J.A., Kelley, M.C., Broadfoot, A.L., Kagan, L.M., Sulzer, M.P., Elder, J.H., Selcher, C., Isham, B., Brown, C., and Carlson, H.C. *Geophys. Res. Letts.* **26,** 1557 (1999).

Decameter mid-latitude sporadic-E irregularities in relation with gravity waves. Bourdillon, A., Lefur, E., Haldoupis, C., LeRoux, Y., Menard, J., and Delloue, J. *Ann. Geophys.* **15,** 925 (1997).

Ablating meteors are usually thought to provide a source of metal atoms that inhibit recombination, so that meteor showers can enhance electron densities in the E region. Further references to metals in the atmosphere are given in the bibliography for Section 7.4.2.

Mesospheric Na layer at 40 degrees N: Modeling and observations. Plane, J.M.C., Gardner, C.S., Yu, J.R., She, C.Y., Garcia, R.R., and Pumphrey, H.C. *J. geophys. Res.* **104,** 3773 (1999).

Global transport and localized layering of metallic ions in the upper atmosphere. Carter, L.N., and Forbes, J.M. *Ann. Geophys.* **17,** 190 (1999).

Model calculations of the silicon and magnesium chemistry in the mesosphere and lower thermosphere. Fritzenwaller, J. and Kopp, E. *Advan. Space Res.* **21,** 859 (1998).

Diurnal occurrence of thin metallic ion layers in the high-latitude ionosphere. Bedey, D.F. and Watkins, B.J. *Geophys. Res. Letts.* **25,** 3767 (1998).

An ion–molecule mechanism for the formation of neutral sporadic Na layers. Cox, R.M., and Plane, J.M.C. *J. geophys. Res.* **103,** 6349 (1998).

GLO observations of E- and F-region metal atoms and ions. Gardner, J.A., Murad, E., Viereck, R.A., Knecht, D.J., Pike, C.P., and Broadfoot, A.L. *Advan. Space Res.* **21,** 867 (1998).

Differential ablation of cosmic dust and implications for the relative abundances of atmospheric metals. McNeil, W.J., Lai, S.T., and Murad, E. *J. geophys. Res.* **103,** 10899 (1998).

On the abundance of metal ions in the lower ionosphere. Kopp, E. *J. geophys. Res.* **102,** 9667 (1997).

Thin layers of ionization observed by rocketborne probes in equatorial E region. Gupta, S.P. *Advan. Space Res.* **19,** 169 (1997).

Large-scale transport of metallic ions and the occurrence of thin ion layers in the polar ionosphere. Bedey, D.F. and Watkins, B.J. *J. geophys. Res.* **102,** 9675 (1997).

Multiple thin layers of enhanced ionization in the ionospheric E- region derived from VLF wave measurements. Okada, T., Mambo, M., Fukami, T., Nagano, I., and Okumura, K. *J. Geomag. Geoelectr.* **49,** 69 (1997).

A model for meteoric magnesium in the ionosphere. McNeil, W.J., Lai, S.T., and Murad, E. *J. geophys. Res.* **101,** 5251 (1996).

Sporadic metal layers in the upper mesosphere. Gardner, C.S. *Faraday Disc.* **101,** 7917 (1996).

Sporadic neutral metal layers in the mesosphere and lower thermosphere. Clemesha, B.R. *J. atmos. terrest. Phys.* **57,** 725 (1995).

The following paper suggests that sulphur may also be released on meteor ablation, and that SO and SO^+ ions become the principal carriers of sulphur following chemical conversion from S.

Sulfur chemistry in the E region. Swider, W., Murad, E., and Herrmann, U. *Geophys. Res. Letts.* **6**, 560 (1979).

Section 6.4.3

Ion composition

Ions and electrons of the lower-latitude D region. Kull, A., Kopp, E., Granier, C., and Brasseur, G. *J. geophys. Res.* **102,** 9705 (1997).

Silicon molecular ions in the D-region. Kopp, E., Balsiger, F., and Murad, E. *Geophys. Res. Letts.* **22,** 3473 (1995).

Hydrogen constituents of the mesosphere inferred from positive ions: H_2O, CH_4, H_2CO, H_2O_2, and HCN. Kopp, E. *J. geophys. Res.* **95,** 5613 (1990).

First ion composition measurements in the stratopause region using a rocket-borne parachute drop sonde. Pfeilsticker, K. and Arnold, F. *Planet. Space Sci.* **37**, 315 (1989).

Ion composition in the lower ionosphere. Kopp, E. and Herrman, U. *Ann. Geophysicae* **2**, 83 (1984).

Modelling

Empirical D-region modelling, a progress report. Friedrich, M. and Torkar, K.M. *Advan. Space Res.* **22,** 757 (1998).
Comparison between an empirical and a theoretical model of the D-region. Friedrich, M. and Torkar, K.M. *Advan. Space Res.* **21,** 895 (1998).
Modelling of the electron density profile in the lowest part of ionosphere D-region on the basis of radiowave absorption data. 1. Theoretical model. Mukhtarov, P. and Pancheva, D. *J. atmos. terrest. Phys.* **58,** 1721 (1996).
Modelling of the electron density profile in the lowest part of ionosphere D-region on the basis of radiowave absorption data. 2. Seasonal variations. Mukhtarov, P. and Pancheva, D. *J. atmos. terrest. Phys.* **58,** 1729 (1996).

Variability of the D region and ionization mechanisms

Diurnal and seasonal variation of D-region electron density at low latitude. Gupta, S.P. *Advan. Space Res.* **21,** 875 (1998).
Meteoric dust effects on D-region incoherent scatter radar spectra. Cho, J.Y.N., Sulzer, M.P., and Kelley, M.C. *J. atmos. solar-terrest. Phys.* **60,** 349 (1998).
Time-variations of D region electron densities, and comparisons with model computations. Ganguly, S., and Zinn, J. *Studia Geophysica et Geodaetica* **42,** 500 (1998).
Solar activity effects in the ionospheric D region. Danilov, A.D. *Ann. Geophys.* **16,** 1527 (1998).
Possible role of NO and $O_2(^1\Delta_g)$ in the formation of the D-region winter anomaly in ionosphere. Medvedev, V.V. and Zenkin, V.I. *Geomag. Aeronom.* **38,** 156 (1998).

Other aspects of D-region meteorology and behaviour

Contribution of mesospheric ozone variability to the diurnal asymmetry in D-region electron density. Arunamani, T., Somayaji, T.S.N., and Rao, D.N.M. *Current Sci.* **70,** 391 (1996).
Artificial periodic irregularities in the auroral ionosphere. Rietveld, M.T., Turunen, E., Matveinen, H., Goncharov, N.P., and Pollari, P. *Ann. Geophys.* **14,** 1437 (1996).
Breakdown of the neutral atmosphere in the D region due to lightning driven electromagnetic pulses. Rowland, H.L., Fernsler, R.F., and Bernhardt, P.A. *J. geophys. Res.* **101,** 7935 (1996).

Cluster ions are of particular importance in the D region (and below).

Cluster reactions. Castleman, A.W., Jr. and Wei, S. *Annu. Rev. phys. Chem.* **45,** 685 (1994).
Reappraisal of the contribution from $[O_2(H_2O)_n]^+$ cluster ions to the chemistry of the ionosphere. Angel, L. and Stace, A.J. *J. phys. Chem.* A **103,** 2999 (1999).

Nucleation and particle formation in the upper atmosphere. Keesee, R.G. *J. geophys. Res.* **94**, 14683 (1989).

Understanding the middle atmosphere *via* laboratory: ion cluster investigations. Keesee, R.G. and Castleman, A.W., Jr. *Ann. Geophysicae* **1**, 75 (1983).

Studies of ion clusters: relationship to understanding nucleation and solvation phenomena. Castleman, A.W., Jr., in *Kinetics of ion–molecule reactions* Ausloos, P. (ed.) (Plenum Publishing Corp., 1979).

A reconsideration of nucleation phenomena in light of recent findings concerning the properties of small clusters, and a brief review of some other particle growth processes. Castleman, A.W., Jr. *Astrophys. Space Sci.* **65**, 337 (1979).

Cluster ions may be implicated in the nucleation leading to formation of noctilucent clouds.

In situ measurements of the vertical structure of a noctilucent cloud. Gumbel, J. and Witt, G. *Geophys. Res. Letts.* **25**, 493 (1998).

Positive ion depletion in a noctilucent cloud. Balsiger, F., Kopp, E., Friedrich, M., Torkar, K.M., Walchli, U., and Witt, G. *Geophys. Res. Letts.* **23**, 93 (1996).

Noctilucent clouds and the thermal structure near the Arctic mesopause in summer. Lubken, F.-J., Fricke, K.H., and Langer, M. *J. geophys. Res.* **101**, 9489 (1996).

Hydroxyl temperature and intensity measurements during noctilucent cloud displays. Taylor, M.J., Lowe, R.P., and Baker, D.J. *Ann. Geophys.* **13**, 1107 (1995).

Dynamics, radiation, and photochemistry in the mesosphere: implications for the formation of noctilucent clouds. Garcia, R.R. *J. geophys. Res.* **94**, 14605 (1989).

On the diurnal variation of noctilucent clouds. Jensen, E., Thomas, G.E., and Toon, O.B. *J. geophys. Res.* **94**, 14693 (1989).

Noctilucent clouds. Gadsden, M. *Space Sci. Rev.* **33**, 279 (1982).

Noctilucent clouds. Soberrnan, R.K., in *Atmospheric phenomena.* Lynch, D.K. (ed.) (Freeman, San Francisco, 1980). (from *Scient. Am.*, June 1963).

The references presented next discuss some of the laboratory techniques that have proved of particular value in the interpretation of ionospheric chemistry. The first three are reviews, while the remainder trace the development of flowing afterglow, drift, SIFT, and supersonic flow methods, and describe results obtained using the techniques.

Low-temperature rate studies of ions and radicals in supersonic flows. Smith, M.A. *Int. Rev. phys. Chem.* **17**, 35 (1998).

Reactions of mass-selected cluster ions in a thermal bath gas. Viggiano, A.A., Arnold, S.T., and Morris, R.A. *Int. Rev. phys. Chem.* **17**, 147 (1998).

Studies of ion–ion recombination using flowing afterglow plasmas. Smith, D. and Adams, N.G., in *Physics of ion–ion and electron–ion collisions.* Brouillard, F. and McGowan, J.W. (eds.) (Plenum Publishing Corp., New Jersey, 1983).

Recent advances in flow tubes: measurement of ion–molecule rate coefficients and product distributions. Smith, D. and Adams, N.G., in *Gas phase ion chemistry*, Vol. 1 Bowers, M.T. (ed.) (Academic Press, New York, 1979).

Flowing afterglow measurement of ion–neutral reactions. Ferguson, E.E., Fehsenfeld, F.C., and Schmeltekopf, A.L. *Adv. atomic. molec. Phys.* **5**, 1 (1969).

The critical hydration reactions of NO^+ and NO_2^+. Angel, L. and Stace, A.J. *J. chem. Phys.* **109**, 1713 (1998).

Reactions of NO^+ in heterogeneous water clusters. Angel, L., and Stace, A.J. *J. phys. Chem.* A **102,** 3037 (1998).

Calculated thermodynamics of reactions involving NO^+–X complexes (where X=H_2O, N_2 and CO_2). Mack, P., Dyke, J.M., and Wright, T.G. *Chem. Phys.* **218,** 243 (1997).

The effect of clustering on mutual neutralization rates using electrostatically merged ion-beams — $H_3O^+(H_2O)_{n=0-3} + OH(H_2O)_{m=0-3}$. Plastridge, B., Cohen, M.H., Cowen, K.A., Wood, D.A., and COE, J.V. *J. phys. Chem.* **99**, 118 (1995).

Thermal energy ion–neutral reaction rates. I. Some reactions of helium ions. Fehsenfeld, F.C., Schmeltekopf, A.L., Goldan, P.D., Schiff, H.I., and Ferguson, E.E. *J. chem. Phys.* **44**, 4087 (1966).

Ion–molecule reaction studies from 300° to 600° in a temperature controlled flowing afterglow system. Dunkin, D.B., Fehsenfeld, F.C., Schmeltekopf, A.L., and Ferguson, E.E. *J. chem. Phys.* **49**, 1365 (1968).

Temperature dependence of some ionospheric ion–neutral reactions from 300° to 900°K. Lindinger, W., Fehsenfeld, F.C., Schmeltekopf, A.L., and Ferguson, E.E. *J. geophys. Res.* **79**, 4753 (1974).

Flow-drift technique for ion-mobility and ion–molecule reaction rate constant measurements. I. Apparatus and mobility measurements. McFarland, M., Albritton, D.L., Fehsenfeld, F.C., Ferguson, E.E., and Schmeltekopf, A.L. *J. chem. Phys.* **59**, 6610 (1973).

Effects of ion speed distributions in flow-drift studies of ion–neutral reactions. Albritton, D.L., Dotan, I., Lindinger, W., McFarland, M., Tellinghuisen, J., and Fehsenfeld, F.C. *J. chem. Phys.* **66**, 410 (1977).

Product-ion distributions for some ion–molecule reactions. Adams, N.G. and Smith, D. *J. Phys. B, Atom. molec. Phys.* **9,** 1439 (1976).

The selected ion flow tube (SIFT): a technique for studying ion–neutral reactions. Adams, N.G. and Smith, D. *Int. J. Mass Spec. Ion Phys.* **21**, 349 (1976).

Laboratory investigation of the ionospheric O_2^+ ($X^2\Pi_g$, $v \neq 1$) reaction with NO. Lindinger, W. and Ferguson, E.E. *Planet. Space Sci.* **31**, 1181 (1983).

Rate coefficient at 300 K for the vibrational energy transfer reactions from $N_2(v=1)$ to $O_2^+(v=0)$ and $NO^+(v=0)$. Ferguson, E.E., Adams, N.G., Smith, D., and Alge, E. *J. chem. Phys.* **80**, 6095 (1984).

Section 6.4.4

D-region negative ions.

Model calculations of the negative ion chemistry in the mesosphere with special emphasis on the chlorine species and the formation of cluster ions. Fritzenwaller, J., and Kopp, E. *Advan. Space Res.* **21,** 891 (1998).

Chlorine and bromine ions in the D-region. Kopp, E. and Fritzenwaller, J. *Advan. Space Res.* **20,** 2111 (1997).

Section 6.5

Physics and chemistry of atmospheric ions (stratosphere and troposphere). Arnold, F., in *Atmospheric chemistry.* Goldberg, E.D. (ed.) (Springer-Verlag, Berlin, 1982).

On unequal distributions of positive and negative ions in the upper stratosphere. Chakrabarty, D.K. *Advan. Space Res.* **20**, 2191 (1997).

Positive and negative ions in the stratosphere. Arijs, E. *Ann. Geophysicae* **1**, 149 (1983).

Positive ion composition measurements between 33 and 20 km altitude. Arijs, E. Nevejans, D., Ingels, J., and Frederick, P. *Ann. Geophysicae* **1**, 161 (1983).

Modelling of stratospheric ions: a first attempt. Brasseur, G. and Chatel, A. *Ann. Geophysicae* **1**, 173 (1983).

Composition measurements of tropospheric ions. Heitmann, H. and Arnold, F. *Nature, Lond.* **306**, 747 (1983).

Anthropogenic perturbations of tropospheric ion composition. Beig, G. and Brasseur, G. *Geophys. Res. Letts.* **26,** 1303 (1999).

Ion measurements can yield valuable data about the neutral trace gas abundances.

Detection of massive negative chemiions in the exhaust plume of a jet aircraft in flight. Arnold, F., Curtius, J., Sierau, B., Burger, V., Busen, R., and Schumann, U. *Geophys. Res. Letts.* **26,** 1577 (1999).

Stratospheric chemical ionization mass spectrometry: nitric acid detection by different ion molecule reaction schemes. Arijs, E., Barassin, A., Kopp, E., Amelynck, C., Catoire, V., Fink, H.P., Guimbaud, C., Jenzer, U., Labonnette, D., Luithardt, W., Neefs, E., Nevejans, D., Schoon, N., and Van Bavel, A.M. *Int. J. mass Spect.* **181,** 99 (1998).

Upper stratosphere negative ion composition measurements and inferred trace gas abundances. Arnold, F. and Qui, S. *Planet. Space Sci.* **32**, 169 (1984).

Stratosphere *in situ* measurements of H_2SO_4 and HSO_3 vapours during a volcanically active period. Qui, S. and Arnold, F. *Planet. Space Sci.* **32**, 87 (1984).

Ion chemistry in the stratosphere can lead to significant changes in the composition of neutral species.

Effects of intense stratospheric ionisation events. Reid, G.C., McAffee, J.R., and Crutzen, P.J. *Nature, Lond.* **175**, 489 (1978).

Stratospheric ions may also be of importance in nucleation of aerosol growth.

A new source of tropospheric aerosols: Ion–ion recombination. Turco, R.P., Zhao, J.X., and Yu, F.Q. *Geophys. Res. Letts.*, **25** 635 (1998).

Ion nucleation—a potential source for stratospheric aerosols. Arnold, F. *Nature, Lond.* **299**, 134 (1982).

Stratospheric sources of CH_3CN and CH_3OH. Murad, E., Swider, W., Moss, R.A., and Toby, S. *Geophys. Res. Letts.* **11**, 147 (1984).

Implications for trace gases and aerosols of large negative ion clusters in the stratosphere. Arnold, F., Viggiano, A.A., and Schlager, H. *Nature, Lond.* **297**, 371 (1982).

Condensation nuclei events at 30 km and possible influences of solar cosmic rays. Hofmann, D.J. and Rosen, J.M. *Nature, Lond.* **302**, 511 (1983).

7 The airglow

7.1 Optical emission from planetary atmospheres

Sources located outside our atmosphere and within it illuminate the night sky. Moon, stars, and surface lights all contribute, but if the light from these sources were eliminated, the sky would not be completely black. A faint glow would remain that has its origins in atmospheric photochemical processes, and to which the name *airglow* is given. Astronauts see this airglow as an envelope that sheaths the night side of the Earth, but Earth-bound observers are also able to perceive it. The visible radiation is rather feeble, and has been equated in intensity to the light of a candle at a distance of 100 m. In fact, the night sky would appear far brighter were it not that the strongest airglow emission features lie in the near-infrared region, just beyond the response of the human eye. By day, the airglow intensities are orders of magnitude larger, but are not detectable by eye because they are completely dominated by atmospherically scattered sunlight. According to the time of day at which observations are made, the airglow is described as *nightglow* or *dayglow*. When the Sun is below the horizon at ground level, but in view from the upper atmosphere, the emission is termed the twilight glow. Several of the lines and bands are identical to those seen in the aurora (Section 6.1.1), and some distinction needs to be drawn between the two phenomena of airglow and aurora. Airglow occurs continuously, is weak, and is observed at all latitudes. In contrast, auroras are much more intense, but irregular in form and occurrence, and are restricted to a region near the geographical poles. The differences arise from the excitation mechanisms. Airglow is driven by photons from the Sun, whereas auroras are excited by the impact of energetic solar particles.

Airglow emissions are a feature of most planetary atmospheres. The light consists of atomic and molecular line, band, and continuum systems of atmospheric constituents, both neutral and ionized. Because many of the emission features are measurable at the Earth's surface, airglow investigations provided a valuable source of information about the composition of our own as well as other planetary atmospheres long before the era of direct investigations by rocket probes and satellites. Other emission features are absorbed by the terrestrial atmosphere, and thus require instrumentation on space vehicles for their study. Airglow investigations continue to assist interpretations of the photochemistry and dynamics of planetary atmospheres.

Most of the spectral features of the airglow have their origins in transitions from *electronically excited* species (although very important components of the Earth's airglow are due to *vibrationally* excited OH, as described in Section 7.4.3, and NO). Many of the transitions are formally *forbidden* by the electric dipole selection rules (see Section 3.1). Indeed, there may be no lower level to which an electronically excited species can make an allowed transition, in which case the species will have a long radiative lifetime, and is said to be *metastable*. Metastable species can therefore be transported away from the region in which they are formed, and subsequently give up their energy in collisional quenching processes. Excited species can participate in otherwise endothermic reactions (Section 3.2.2). Significant non-local effects are introduced, particularly into thermospheric chemistry, as a result of the formation and transport of metastable atoms and molecules. Whatever the ultimate fate of the excitation energy, the emitting species responsible for the airglow are greatly removed from thermal equilibrium with their surroundings so far as their electronic temperature is concerned. One of our interests will be to look at the physical and chemical processes that can lead to specific excitation of a particular energy level (Section 7.2).

For many of the atoms and molecules of the atmosphere, the first allowed optical transition to and from the ground state (the *resonance* transition) lies in the 'vacuum' ultraviolet, and so by definition does not penetrate the Earth's atmosphere. Resonance emissions from helium ($\lambda = 58.4$ nm), nitrogen (120.0 nm), hydrogen (121.6 nm), and oxygen (130.2, 130.4, 130.6 nm) which might otherwise be expected to be strong, are not observed at the ground, although they are major features of the Earth's airglow as seen from space. We have already alluded to the resonant absorption and re-radiation of sunlight by hydrogen and helium in connection with sources of night-time ionization (Section 6.3). Satellite measurements have an important part to play in airglow studies. For example, the Upper Atmosphere Research Satellite (UARS), first mentioned in connection with stratospheric investigations (p. 181), carries two very interesting experiments designed to examine different aspects of airglow phenomena. These experiments are the *Wind Imaging Interferometer* (WINDII) and the *High-Resolution Doppler Imager* (HRDI): in general, the Doppler shifts of the airglow lines allow information about winds and other motions to be derived, while the line shapes and strengths yield information about atmospheric composition and temperature.

Emissions at wavelengths longer than the atmospheric cut-off may sometimes be removed by specific absorption processes. For example, the strongest feature of the terrestrial dayglow is due to a transition from the first excited state of O_2, the $^1\Delta_g$, to the ground state, $^3\Sigma_g^-$, and is known as the *Infrared Atmospheric Band*. The (0,0) band of this system lies at $\lambda = 1270$ nm, just into the infrared region, and in a part of the spectrum generally free of atmospheric absorptions. The exception, of course, is for the lines of the oxygen bands originating in the $v'' = 0$ level of the ground state. Although the $^1\Delta_g$–$^3\Sigma_g^-$ transition is highly forbidden (note that the electric dipole selection rules involving **S**, Λ, and *g/u* are all broken!), the great optical path of atmospheric oxygen filters out from the surface most of the $\lambda = 1270$-nm line produced at altitudes of around 50 km (Section 7.4.1). Atmospheric absorption of the (0, 1) band (i.e., $^1\Delta_g, v' = 0 \leftarrow {}^3\Sigma_g^-, v'' = 1$) at $\lambda = 1580$ nm is almost non-existent because the population of $v' = 1$ in O_2 is small at atmospheric temperatures. This band can thus be observed from the ground, while rockets, balloons, or aeroplanes must be used to study the (0,0) emission. Cunning experimentation allows the (0,0) transition from the airglows of *other* planets to be observed with ground-based telescopes, even though at first sight such investigations might appear doubly restricted. However, high-resolution spectroscopy (using interferometric methods) can separate out individual rotational lines in the bands. If Earth and the target planet have a sufficiently high relative velocity, then the planetary airglow lines are Doppler-shifted away from the terrestrial absorption lines. Successful investigations of $O_2(^1\Delta_g)$ in the airglows of Mars and Venus have been achieved by the technique.

Doppler shifts are also used to good effect in an application of investigations of terrestrial airglow. The shifts are, of course, dependent on the speed and direction of motion of the emitting particles. High-resolution spectrometers measuring airglow wavelengths can therefore be used to probe the velocities, and such measurements are now used routinely to provide information on temperatures and wind speeds in the upper atmosphere. Some spectrometers even have the capability of examining the entire overhead sky. All-sky camera studies of this kind are also useful for studying *gravity waves*[a] in the atmosphere. The WINDII and HRDI instruments described briefly on the previous page are satellite-borne devices that utilize Doppler shifts, but ground-based spectrometers have been widely employed.

Altitude profiles of the emitting species often afford valuable insight into excitation and deactivation mechanisms of specific airglow features. Early studies used the variation in intensity with viewing angle to estimate the profile, but this rather unreliable method has been replaced almost completely by rocket measurements. The total (integrated) overhead intensity is measured as a function of altitude attained by the rocket, and the profile obtained by numerical differentiation of the result. Figures 7.1 and 7.2 illustrate the difficulties inherent in the method. Data for overhead intensities of the night-glow Infrared Atmospheric Band are given in the first figure (the intensity units are explained in Section 7.3) and the derived $[O_2(^1\Delta_g)]$ profile in the second. The two concentration peaks arise from relatively small changes of slope in Fig. 7.1 (they may well arise from interactions with gravity waves, which we have just mentioned, with vertical wavelengths of about 10 km). Noise on the data could greatly affect the magnitude and position of the peaks, especially in the case of the higher altitude one where the remaining overhead intensity is low. In Section 7.3, we consider what factors determine the absolute intensities and the altitude dependence of airglow emissions. Dozens of features are known in the terrestrial airglow. Our intention is not to catalogue and explain all these features. Instead, we discuss in Section 7.4 a few of the more interesting and important transitions that are observed. First,

[a]Gravity waves are an important component of upper-atmosphere dynamics. They are created within the troposphere by convective thunderstorms and orographic forcing, and they propagate to the mesosphere where they break, depositing significant quantities of energy.

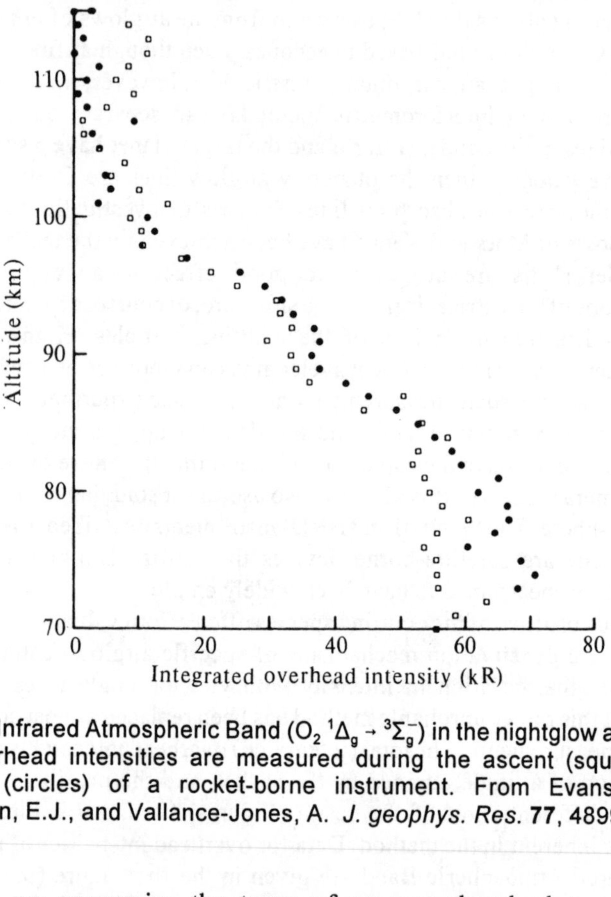

Fig. 7.1. Infrared Atmospheric Band ($O_2\ ^1\Delta_g \rightarrow {}^3\Sigma_g^-$) in the nightglow at $\lambda = 1270$ nm. Overhead intensities are measured during the ascent (squares) and descent (circles) of a rocket-borne instrument. From Evans, W.F.J., Llewellyn, E.J., and Vallance-Jones, A. *J. geophys. Res.* **77**, 4899 (1972).

however, we summarize the types of process that lead to excitation, deliberately using as illustrations some additional features of the airglow.

7.2 Excitation mechanisms

During the day, sunlight can be absorbed directly by atmospheric constituents, to yield electronically excited products. Radiative decay then contributes to the airglow, and the absorption–radiation sequence is *resonance fluorescence* or *resonance scattering*.[a] This is the process we discussed in connection with the residual ionization at night (Section 6.3) caused predominantly by H and He

[a]Properly speaking, the emitted radiation must be at the same wavelength as the absorption for the fluorescence to qualify as a resonant process. However, in molecular systems the concept of resonance is often extended to include emission from and to vibrational levels (of the same electronic states) different from those initially populated.

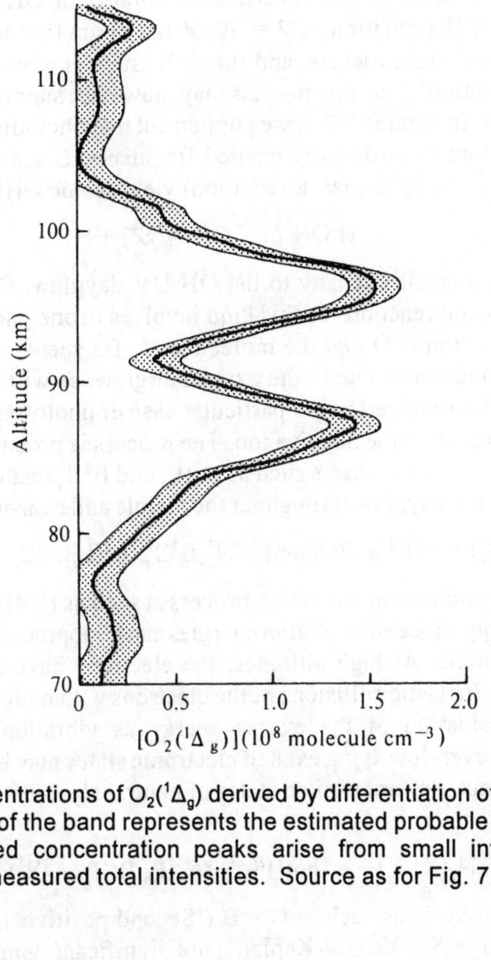

Fig. 7.2. Concentrations of $O_2(^1\Delta_g)$ derived by differentiation of the data of Fig. 7.1. The width of the band represents the estimated probable error. Note how the two marked concentration peaks arise from small inflections in the measured total intensities. Source as for Fig. 7.1.

atoms. Atoms such as H have particular roles to play in resonance scattering, since the hydrogen resonance line is a major component of the solar spectrum and there is thus a match that efficiently couples energy transfer from Sun to Earth. The upper and lower electronic levels involved are the ^2P and ^2S states of H, so that the excitation–emission sequence may be written

$$H(^2S) + h\nu\,(\text{solar; } \lambda = 121.6 \text{ nm}) \rightarrow H(^2P), \tag{7.1}$$

$$H(^2P) \rightarrow H(^2S) + h\nu\,(\text{airglow; } \lambda = 121.6 \text{ nm}). \tag{7.2}$$

Other species, for which there is not a specific solar emission line, can absorb from the solar continuum. Even molecular species such as NO or OH are efficiently excited in this way, the NO 'γ system' ($A^2\Sigma^+ \rightarrow X^2\Pi$) being the strongest dayglow feature in the spectral range 200–300 nm (but observable only by rockets because of atmospheric absorption). The OH resonance band of the $A^2\Sigma^+ \rightarrow X^2\Pi$ transition at $\lambda = 306.4$ nm is another intense dayglow emission observed from rockets, and the $A^2\Sigma^+$ state is mainly populated by resonance absorption. A second process may, however, supplement the direct absorption route. In Section 3.2.1, we pointed out that photodissociation often yields one or more electronically excited fragments. Dissociation of water vapour (for example, by Lyman-α radiation) yields some $OH(A^2\Sigma^+)$

$$H_2O + h\nu \rightarrow OH(A^2\Sigma^+) + H \tag{7.3}$$

that contributes a small intensity to the OH UV dayglow. One of the most important excitation reactions of this kind involves ozone photolysis. At $\lambda <$ 310 nm *both* the atomic O *and* the molecular O_2 fragments are excited (see Section 3.2.1), and contribute to the oxygen airglow, as will be discussed in Section 7.4.1. Photoionization is a particular case of photofragmentation and can lead to excitation in the positive ion. The process is probably responsible for exciting a variety of N_2^+ states such as $A^2\Pi_u$, and $B^2\Sigma_u^+$ that make important contributions to the dayglow throughout the visible and near-infrared regions

$$N_2(X^1\Sigma_g^+) + h\nu\,(\lambda \leq 79.6\text{ nm}) \rightarrow N_2^+(B^2\Sigma_u^+, A^2\Pi_u, X^2\Sigma_g^+) + e. \tag{7.4}$$

Photoelectrons produced in ionization processes such as (7.4) may be created with excess energy, since solar photon energies can be appreciably greater than ionization potentials. At high altitudes, the electrons have a mean kinetic energy of 10 eV. Inelastic collisions of the electrons with molecules generally lead to the dissipation of the excess energy as vibration and rotation. Sometimes, however, low-lying excited electronic states may be populated by the electron impact. Several excited states of neutral N_2 seem to be formed by this mechanism

$$N_2(X^1\Sigma_g^+) + e(\text{`hot'}) \rightarrow N_2(C^3\Pi_u, B^3\Pi_g, A^3\Sigma_u^+) + e, \tag{7.5}$$

and transitions of systems such as C → B ('Second positive'), B → A ('First positive'), and A → X ('Vegard-Kaplan') are significant components of the dayglow. Note that the C → B and B → A radiative transitions populate the lower—but still excited—electronic levels by cascading from the higher ones. Radiationless transitions can also populate, in an intramolecular energy-transfer step, excited states distinct from those initially excited. Rather complicated processes of this kind occur in the N_2 system: not only does the upper B state populate the lower A in a radiationless crossing, but high vibrational levels of the A state can cross back to lower vibrational levels of the B. We shall meet cases of *inter*molecular energy transfer, in which a separate and chemically distinct species is excited, in Section 7.4.1. Reference back to Fig. 3.1, which depicts the fates of electronically excited species, will show that pathways (i) to (v) thus all make their contribution to some feature or other of the dayglow.

At night, the source of photochemical excitation is removed, yet the existence of the nightglow shows that excited species persist. In general, this result must mean that solar energy has been stored during the day, and that reactions are releasing the energy at night. Neutral atoms (especially oxygen) are a very important energy reservoir at altitudes below ~ 100 km, while ions are very important at higher levels. We shall see in Section 7.4.1 how several atomic and molecular transitions in the nightglow owe their existence to atomic oxygen. Reaction exothermicity is used to populate preferentially certain states of the species, which are thus out of thermal equilibrium with their surroundings. Emission of *chemiluminescence* then contributes to the nightglow. A typical example from the laboratory is the reaction

$$NO + O_3 \rightarrow NO_2^* + O_2 \tag{7.6}$$

in which emission is seen from NO_2^* populated virtually up to the exothermicity (205 kJ mol^{-1}) of the reaction. For many years, it was not clear if reaction (7.6) is of significance in the airglow, but an airglow layer in the stratosphere (altitude 40–60 km) thought to arise from the process has now been observed by the WINDII instrument on UARS (p. 526). NO_2 emission is also implicated in a weak background continuum of the nightglow, and it may be produced in the *two-body*, radiatively stabilized, combination of O with NO

$$NO + O \rightarrow NO_2 + h\nu. \tag{7.7}$$

Dissociative recombination of positive ions with electrons (Section 6.2, pp. 490–1) liberates the molecular ionization energy in the fragments, and electronic excitation may result. Excited O-atoms are formed by dissociative recombination of O_2^+ (cf p. 496), and the first electronically excited state of N, the ^2D, is probably populated in the ionosphere by the analogous reaction with NO^+

$$NO^+ + e \rightarrow N(^2D) + O. \tag{7.8}$$

Nightglow excitation mechanisms can obviously operate during the day, but they are *usually* dominated while the Sun is present by processes more directly utilizing solar energy. Some exceptions do exist, a notable one being the chemical excitation of *vibrationally* excited OH radicals described in Section 7.4.3.

7.3 Airglow intensities and altitude profiles

Emission from an excited species A* is described by a rate law

$$I = k_r[A^*] \tag{7.9}$$

where I is here a total emission rate in quanta (or photons) per unit volume per unit time. The rate coefficient, k_r, is equivalent to the transition probability, or

Einstein 'A' coefficient for spontaneous emission; a radiative lifetime, τ, is often defined by the relation $\tau = \ln 2/k_r$. To a first approximation, k_r is independent of concentration or pressure (although for certain forbidden transitions, collision-induced processes do endow k_r with some pressure dependence).

Radiation is an isotropic process, so that each volume element of the airglow emits equally in all directions. A human observer or an instrument does not integrate all this radiation, but rather perceives a 'brightness' that depends on the flux of photons per unit area per unit time. For this reason, airglow brightnesses are usually measured in units of the *Rayleigh* (R). One Rayleigh is the brightness of a source emitting 10^6 photons cm^{-2} s^{-1} in all directions.[a] The convenience of the unit lies in the direct relationship between it and the emission rate of eqn (7.9), since (for an optically thin medium) I need only be multiplied by the depth of the emitting layer to give the Rayleigh brightness. For correct scaling of the 10^6 factor in the definition, the depth has to be measured in units of 10^6cm (10 km), which is, in fact, a typical thickness for airglow emission layers. The name of the unit honours Robert John Strutt, Fourth Lord Rayleigh, who performed much pioneering work on the airglow. He is sometimes called 'the airglow Rayleigh' to distinguish him from his father, the Third Lord Rayleigh, or 'scattering Rayleigh' (cf. p. 85).

Intensities of airglow features are related to atmospheric concentrations of excited species through the transition probability, k_r, of eqn (7.9). Large intensities of strong emitters may not necessarily correspond to higher concentrations than lower intensities of weak emitters. For example, the dayglow intensity of the N(^{2}D–^{4}S) lines at $\lambda \sim 520$ nm is only 90 R, but since k_r for this highly forbidden transition is ~7.4×10^{-6} s^{-1} (lifetime 26 hours) the concentration of N(^{2}D) in our hypothetical layer of 10 km thickness would be 10^7 atom cm^{-3}. Emission from sodium in the dayglow, of the allowed Na(^{2}P–^{2}S) resonance transition at $\lambda \sim 589$ nm, is more than 300 times as intense (30 kR). However, k_r is $\sim 6.3 \times 10^7$ s^{-1} (lifetime 12 nanoseconds) which suggests a concentration of excited sodium atoms of a few hundred per cubic metre (4.8×10^{-4} atom cm^{-3}) for a 10 km-thick emitting layer.

Radiative decay inevitably competes with production of emitting excited species. Non-radiative intra- and intermolecular processes (cf. Fig. 3.1) provide additional loss mechanisms. For many atmospherically significant airglow emitters, physical quenching [pathway (vi) in the figure] is the most important of these processes. If we write the quencher species as M and the rate coefficient as k_q, then the generalized excitation–de-excitation mechanism becomes

[a]The light is radiated into a solid angle of 4π steradians, so that the source is emitting $10^6/4\pi$ photons cm^{-2} steradian^{-1} s^{-1}, a quantity called the *surface brightness* (sometimes 'intensity') in optical measurements.

$$\text{Source} \rightarrow \text{A*} \qquad \text{Excitation; Rate} = P \qquad (7.10)$$

$$\text{A*} \rightarrow \text{A} + h\nu \qquad \text{Emission; Rate} = k_r[\text{A*}] \qquad (7.11)$$

$$\text{A*} + \text{M} \rightarrow \text{A} + \text{M} \qquad \text{Quenching; Rate} = k_q[\text{A*}][\text{M}]. \qquad (7.12)$$

So long as the excitation rate, P, does not change rapidly, then a steady state for [A*] may be set up, and

$$I = k_r[\text{A*}] = k_r P/(k_r + k_q[\text{M}])$$
$$= P(1 + k_q[\text{M}]/k_r)^{-1}. \qquad (7.13)$$

This equation should be compared with eqns (3.57) to (3.59) in Section 3.5, where the quantitative justification of steady-state treatments is to be found.

In the discussion that follows we ignore physical transport in order to highlight the chemical kinetic controls on airglow. Physical quenching determines the intensity of the airglow for a given excitation rate according to the relative magnitudes of k_r and k_q[M]. For a species such as N(^{2}D), with k_r less than 10^{-5} s^{-1}, quenching dominates at all emitting altitudes. Reported values of k_q for quenching of N(^{2}D) by atmospheric gases are of the order of 10^{-11} cm^3 molecule-^{1} s^{-1}, so that deactivation is dominated by radiation only for [M] $\leq 10^6$ molecule cm^{-3}. Total particle concentrations of this magnitude are found at altitudes above ~ 500 km in the Earth's atmosphere. The peak of N(^{2}D) emission in the dayglow appears to originate at ~ 200 km, where [M] ~10^{10} particle cm^{-3}, so that only one in 10^4 of the atoms excited can actually emit. Airglow emission profiles are obviously dependent both on the variation of the excitation rate, P, with altitude, and on the [M]-dependence of the function $(1 + k_q[\text{M}]/k_r)^{-1}$ found in eqn (7.13). The arguments that we have presented in earlier chapters about the compromise between atmospheric penetration of energetic solar ultraviolet and the concentration of reactive species apply equally to airglow excitation, and layer-like airglow regions can be expected. Species requiring high energy for their production and excitation are generally found at elevated altitudes. As an additional feature in the airglow, however, the lower boundary of the emitting layer is often sharpened by increased rates of loss by quenching acting together with decreased rates of excitation.

In circumstances where P changes rapidly with time, the steady state may not be maintained. Such a situation may arise at twilight or dawn, or during an eclipse, if the excitation process is dependent on the presence of sunlight. For the extreme case, where P goes to zero instantaneously, airglow intensity will decay with a first-order rate determined by the composite coefficient $(k_r + k_q[\text{M}])$. More realistic cases can be treated numerically, and allowance made for physical transport. Observations of the time dependence of the airglow intensity during periods of change can clearly provide information about [M], and hence emission altitude, if k_q and k_r are known, or conversely about the rate constants if the emission altitude can be estimated.

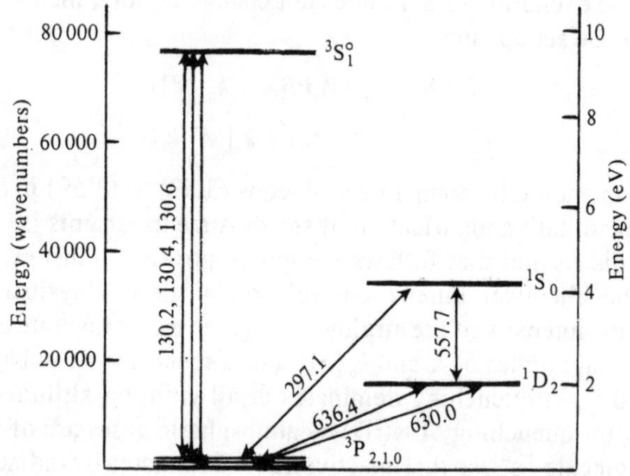

Fig. 7.3. Some low-lying energy levels of atomic oxygen. Atmospherically important optical transitions are shown by the arrows, and the wavelengths of the lines are given in nanometres. Data from Moore, C.E., *Atomic energy levels*, Vol. 1, NSRDS–NBS35, Washington DC, 1971.

7.4 Specific emission sources

7.4.1 Atomic and molecular oxygen

Excited states of O and O_2 make an extremely important series of contributions to the airglow of Earth and other planets. Figures 7.3 and 7.4 introduce this topic by showing some transitions in O and O_2 that can be observed at ground level. All the relatively long wavelength ($\lambda > 300$ nm) transitions are forbidden by the electric-dipole rules, and the excited states are therefore metastable.

The O($^1S \rightarrow {}^1D$) transition at $\lambda = 557.7$ nm was the first component of the airglow to be identified, by high-resolution spectroscopy, with a specific atomic or molecular event. Because the radiation was already known in the aurora, this well-known line is often called the *auroral green line*, even when it originates in the airglow. Both the green line, and the red doublet at $\lambda =$ 630.0 and 636.4 nm, due to the O($^1D \rightarrow {}^3P_{2,1}$) transition, seem to arise from two different altitude regions of the atmosphere. The high-altitude excitation processes include electron impact and dissociative recombination of O_2^+, which can yield both O(^1S) and O(^1D)

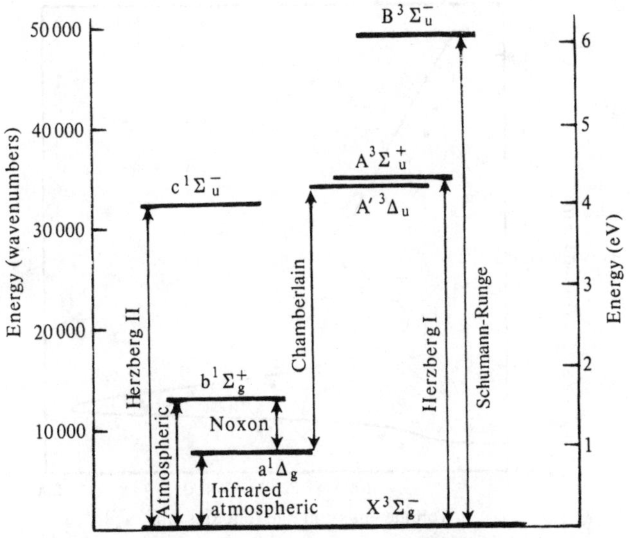

Fig. 7.4. Nomenclature of some optical transitions in molecular oxygen. With the exception of the B ↔ X (Schumann–Runge) system, all these transitions are forbidden by electric-dipole selection rules. Data from Krupenie, P. H. *J. phys. Chem. Ref. Data* **1**, 423 (1972).

$$O_2^+ + e \rightarrow O(^1S) + O(^3P) + 2.78 \text{ eV} \qquad (7.14)$$

$$O_2^+ + e \rightarrow O(^1D) + O(^3P) + 4.99 \text{ eV}. \qquad (7.15)$$

This two-channel dissociative recombination process is, of course, one of the two most important charge neutralization steps in the F region, the other being reaction (7.8) (cf. Sections 6.2 and 6.4.1). Reactions (7.14) and (7.15), with ions produced by energetic solar particles (see Section 6.3), are likely to be significant auroral sources of $O(^1S)$ and $O(^1D)$. The airglow observations require the yields of $O(^1S)$ from reaction (7.14) to reach values as high as 0.23. Theoretical calculations suggest yields in the range 0.0012–0.0016 for direct dissociative recombination of $O_2^+(v = 0)$; an indirect mechanism, involving initial capture into a repulsive state of O_2 followed by dissociation of another state of different symmetry, is required to explain the higher yields demanded by the observations.

As described on pp. 496–7, deliberate chemical modification of the night-time ionosphere was achieved in the RED AIR experiment. According to the changes of intensity of emission from $O(^1D)$ following injection of CO_2 above and below the peak of the F region, the quantum yield of reaction (7.15) has a mild altitude dependence, decreasing by 16 per cent from 275 to 350 km. In turn, this result suggests that there is some dependence of the yield on the

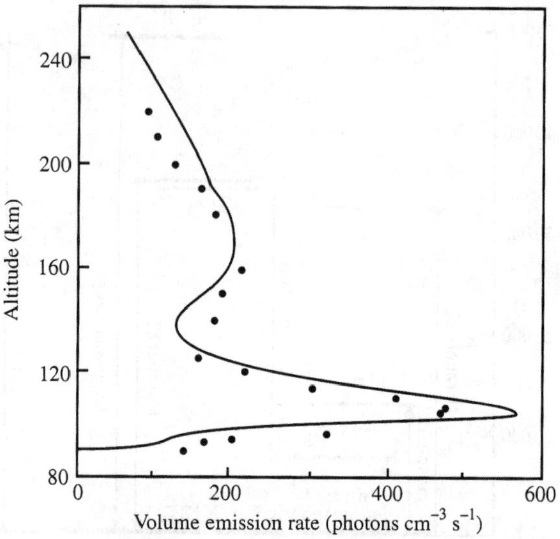

Fig. 7.5. Measured (WINDII) and modelled O(^{1}S) green-line dayglow emission profiles for 53°S, 126°E. From Tyagi, S. and Singh, V. *Ann. Geophysicae* **16**, 1599 (1998).

degree of vibrational excitation in the O_2^+ ion. However, the yield of O(^{1}S) from reaction (7.14) seems even more energy-dependent, and it is probably the vibrational energy dependence of the [O(^{1}S)]/[O(^{1}D)] ratio in the recombination step that accounts for the variability of enhancement of the oxygen lines in auroras.

Reactions of neutral species are responsible for the lower altitude (90–100 km) O-atom airglow. The main emitting layer of O(^{1}S) is in this region, and the excited atoms are formed by day and by night in a process that we shall describe in detail later. Figure 7.5 shows measurements of the O(^{1}S) green-line emission made with the WINDII satellite instrument (p. 526). The two emitting layers are clearly evident in the observations. The solid line shows the predictions of one model in which several excitation processes are included. For the particular location and time to which the figure refers, the agreement between model and observation is quite good throughout the two emitting regions, although there is a tendency (more pronounced in results for other times and places) for the model to overestimate O(^{1}S) concentrations in the lower layer. A nightglow profile will be included later in Fig. 7.11.

A little O(^{1}D) is populated radiatively in the (^{1}S → ^{1}D) emission of the green line. By day, however, two familiar photodissociation steps can generate O(^{1}D). Optical dissociation of O_2 in the Schumann–Runge continuum ($\lambda \sim 175$ nm; see Section 3.2.1) yields one ^{1}D and one ^{3}P atom

$$O_2 + h\nu(\lambda \lesssim 175\text{ nm}) \rightarrow O(^1D) + O(^3P). \qquad (7.16)$$

Maximum solar energy absorption, and thus the maximum $O(^1D)$ production rate, in the Schumann–Runge system occurs at altitudes of 80–110 km, corresponding quite well with the maximum in $O(^1D)$ daytime emission. At longer wavelengths, and thus potentially at lower altitudes, ozone photolysis is a source of $O(^1D)$, as discussed at length in Section 3.2.1

$$O_3 + h\nu(\lambda < 310\text{ nm}) \rightarrow O(^1D) + O_2(^1\Delta_g). \qquad (7.17)$$

The maximum in $O(^1D)$ production rate from this reaction arises at an altitude of ~ 40 km. However, quenching of $O(^1D)$ by N_2 and O_2 is efficient, and at 40 km the collisional lifetime is less than 1 ps, so that radiation (lifetime 110 s) is very improbable.

The $^1\Delta_g$ molecular fragment produced in reaction (7.17) is, by contrast with $O(^1D)$, only very weakly quenched by N_2 or O_2. In the upper stratosphere, the collisional lifetime is of the order of tens to hundreds of seconds. Although the radiative lifetime is also very long (44 min) for the highly forbidden $O_2(^1\Delta_g \rightarrow {}^3\Sigma_g^-)$ Infrared Atmospheric Band at λ = 1270 nm (see p. 526 and Fig. 7.4), a sensible fraction of $O_2(^1\Delta_g)$ can emit because of the inefficiency of physical deactivation. In fact, the Infrared Atmospheric Band is the most intense feature of the dayglow, the (0, 0) transition (observed at altitudes high enough to avoid self-absorption, see p. 526) having an intensity of 20 MR. Consideration of the energy balance alone points to ozone photolysis as the source of $O_2(^1\Delta_g)$, since no species other than ozone absorbs enough sunlight at the altitudes where the dayglow emission originates. Peak concentrations of $O_2(^1\Delta_g)$, determined by rocket photometry, are typically 2×10^{10} molecule cm^{-3} at 50–60 km altitude. Concentrations of this magnitude mean that the metastable excited molecules are present at the parts per million level. Laboratory measurements of the quantum yield for singlet oxygen production in reaction (7.17), and of rate coefficients for quenching by atmospheric gases, may be put together with experimentally determined atmospheric ozone concentrations and solar irradiances to predict an $[O_2(^1\Delta_g)]$–altitude profile. The results of such a calculation are displayed as the solid line in Fig. 7.6, while some rocket measurements of $[O_2(^1\Delta_g)]$ are shown as circles on the figure. Agreement of this kind may be taken as strong confirmation of the mechanism suggested for daytime excitation of $O_2(^1\Delta_g)$. After the Sun falls below the horizon, or is eclipsed, concentrations of $O_2(^1\Delta_g)$ fall rapidly. An analysis of the intensity–time dependence in eclipses has been used to derive a composite quenching rate coefficient by atmospheric gases of 3×10^{-19} cm^3 molecule^{-1} s^{-1}. This value is almost exactly identical to that calculated for air from the laboratory quenching data on the individual atmospheric gases. Confidence in the model is thus so great that dayglow measurements of the Infrared Atmospheric Band can be used to derive atmospheric ozone concentrations and profiles. For example, a secondary peak in emission

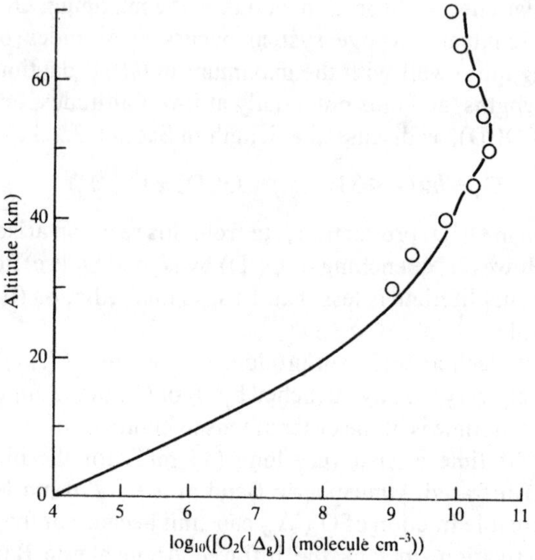

Fig. 7.6. Atmospheric concentrations of $O_2(^1\Delta_g)$ during the day. Experimental values (circles) are derived from rocket measurements of the Infrared Atmospheric Band intensity in the dayglow, while the calculations (solid line) are based on laboratory kinetic and spectroscopic data. From Crutzen, P.J., Jones, I.T.N., and Wayne, R.P. *J. geophys. Res.* **76**, 1490 (1971).

intensity found at a height of ~ 90 km is probably to be associated with a secondary maximum in $[O_3]$ at this altitude. One of the instruments on the Solar Mesosphere Explorer (SME) satellite routinely inverts the airglow intensities at $\lambda = 1270$ nm, taking into account quenching rates, to derive ozone concentrations. Another example of the use of airglow intensities to derive upper-atmosphere ozone concentrations is shown in Fig. 7.7. In this case, rocket measurements of the $O_2\,(^1\Delta_g)$ airglow were made during the evening twilight near the equator. Production of $O_2\,(^1\Delta_g)$ in reaction (7.17) was the most important process, but several minor routes were taken into account in inverting the airglow intensities to generate the ozone profile.

The Infrared Atmospheric Band at $\lambda = 1270$ nm is present in the airglow of both Mars and Venus. As discussed in Section 7.1, Doppler-shifted lines can be observed by telescope. Intensities on Mars are as high as 26 MR in the winter north-polar region. Ozone photolysis, as on Earth, appears to be the only reasonable source of the singlet oxygen. Martian atmospheric ozone concentrations are quite small (see Section 8.3); quenching of $O_2(^1\Delta_g)$ by CO_2, the major component of the atmosphere, is very slow, so that high intensities of the Infrared Atmospheric Band are still possible. Ozone photolysis cannot

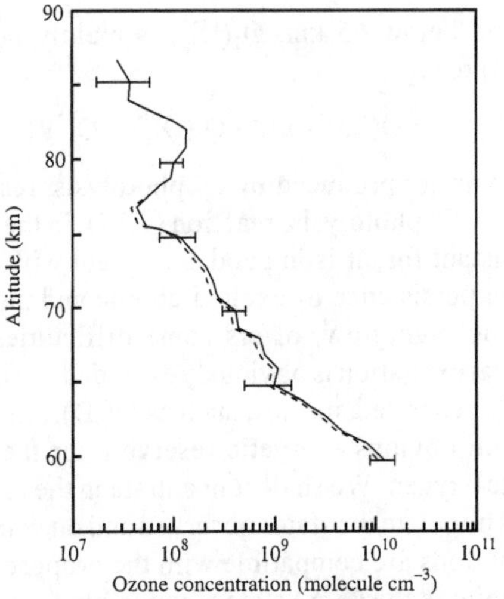

Fig. 7.7. Ozone height profile in the mesosphere derived from measurements of the twilight airglow of $O_2(^1\Delta_g)$. From Batista, P.P., Takahasi, H., Clemestra, B.R., and Llewellyn, E.J. *J. geophys. Res.* **101**, 7917 (1996).

be the source of $O_2(^1\Delta_g)$ on Venus, because the column emission rate on the day side (1.8 MR) is close to the rate at which O atoms are produced by CO_2 photolysis. Photodissociation of ozone can account for only 130 kR of the emission. Favoured alternative excitation reactions have included the exothermic processes

$$Cl + O_3 \rightarrow ClO + O_2 + 161 \text{ kJ mol}^{-1} \qquad (7.18)$$

$$ClO + O \rightarrow Cl + O_2 + 236 \text{ kJ mol}^{-1}. \qquad (7.19)$$

Each reaction is sufficiently exothermic to populate either $O_2(^1\Delta_g)$ or $O_2(^1\Sigma_g^+)$. Together, they represent the Cl_x catalytic chain (see Section 4.4.1), but the chain needs to be fed by O_3 as well as by O. Laboratory evidence for the excitation of $O_2(^1\Delta_g)$ in either reaction is very meagre, and the presence of chlorine-containing species on Venus is currently disputed! Interpretation of $O_2(^1\Delta_g)$ airglow intensities on Venus thus remains uncertain.

The second excited singlet of O_2, the $b^1\Sigma_g^+$ state (Fig. 7.4), also produces a strong emission system in the dayglow. Just beyond the visible region, at λ = 762 nm, the forbidden transition $O_2(^1\Sigma_g^+ \rightarrow {}^3\Sigma_g^-)$ is readily observed by rocket photometry. Radiation of this system is called the *Atmospheric Band*. Between 65 and 100 km, the dominant excitation mechanism is resonance absorption of solar radiation

$$O_2 + h\nu(\lambda = 762\ \text{nm}) \rightarrow O_2(^1\Sigma_g^+). \tag{7.20}$$

Above 100 km, and below 65 km, $O_2(^1\Sigma_g^+)$ is mainly populated by energy transfer from $O(^1D)$ to O_2

$$O(^1D) + O_2 \rightarrow O_2(^1\Sigma_g^+) + O(^3P). \tag{7.21}$$

Excited oxygen atoms are produced by O_2 photolysis, reaction (7.16), in the higher region, and by O_3 photolysis, reaction (7.17), in the lower. The derived quenching rate constant for air is in good agreement with laboratory data.

The night-time persistence of excited atomic and molecular oxygen, as demonstrated by the nightglow, offers some difficulties in interpretation. Direct photochemical excitation is obviously excluded. So is indirect excitation by photochemically generated species, such as $O(^1D)$, whose lifetime against reaction is short. An obvious energetic reservoir for the oxygen species is ground-state atomic oxygen. We shall concentrate in the rest of this section on how O atoms could populate the states observed and on whether the intensities and altitude distributions are compatible with the proposed mechanisms.

Our starting point is the $O(^1S)$ state responsible for the green line of the nightglow. For years, controversy existed over the relative merits of a one-step termolecular reaction

$$O + O + O \rightarrow O(^1S) + O_2, \tag{7.22}$$

suggested initially by Chapman, and a two-step scheme,

$$O + O + M \rightarrow O_2^* + M \tag{7.23}$$

$$O + O_2^* \rightarrow O(^1S) + O_2, \tag{7.24}$$

due to Barth; O_2^* is some sufficiently energy-rich state of O_2 whose identity we shall shortly discuss. Both schemes predict a production rate of $O(^1S)$ that is proportional to $[O]^3$. Anomalies in the laboratory data have now been resolved, and theoretical and experimental considerations all favour the two-step 'Barth' mechanism.

Recombination of ground-state O atoms in reaction (7.23) could, in principle, populate any or all of the singlet, triplet (and quintet) molecular states that correlate with $O(^3P)$. The potential energy diagram of Fig. 3.2 shows some of these states. Excitation of $O(^1S)$ in reaction (7.24) requires 4.19 eV. The $v' = 0$ levels of $c^1\Sigma_u^-$ (4.05 eV), $A'^3\Delta_u$ (4.24 eV), and $A^3\Sigma_u^+$ (4.34 eV) all lie near this energy, although for $c^1\Sigma_u^-(v' = 0)$ energy transfer must be slightly endothermic. Recent experimental evidence has, in fact, drawn particular attention to the $c^1\Sigma_u^-$ state. Nightglow emission ($\lambda \sim 400$–650 nm) from Venus that was detected by the Soviet Venera 9 spacecraft was subsequently identified as resulting from a $c^1\Sigma_u^- \rightarrow X^3\Sigma_g^-$ ('Herzberg II': see Fig. 7.4) progression in O_2 overlapped by weaker bands of the $A'^3\Delta_u \rightarrow a^1\Delta_g$ ('Chamberlain') system. Re-examination of terrestrial airglow spectra in the light of these findings revealed that the Herzberg II bands, due to $O_2(c^1\Sigma_u^-)$, are also present in the Earth's nightglow. Intensities are, however, much weaker on Earth (~100 R) than on Venus (2700 R), and the vibrational distribution is totally different, with the terrestrial distribution peaking for $v' = 7$ and the Venus spectrum showing emission only from $v' = 0$. On Mars, the Herzberg II system is even weaker, the intensity being ≤ 30 R. Identification of the state of O_2^* that excites $O(^1S)$ is helped by the Venusian nightglow studies. Green-line emission is very weak (< 5 R) on Venus, so that the precursor state present in the Earth's atmosphere is virtually absent in the upper atmosphere of Venus. But $v' = 0$ of $c^1\Sigma_u^-$, $A'^3\Delta_u$, and $A^3\Sigma_u^+$ are all known to be formed on Venus. The difference between the species is that the $c^1\Sigma_u^-$ state lies 0.14 eV below the $O(^1S)$ excitation energy, while the A′ and A states lie above. An activation energy of the order of the endothermicity would make reaction (7.24) slow and the emission weak, since the temperature is only ~ 188 K at those altitudes in the Venusian atmosphere where [O] is highest. Vibrational relaxation of $O_2(c^1\Sigma_u^-)$ is much less pronounced in the Earth's atmosphere than on Venus, with peak intensities arising from $v' = 7$. Reaction (7.24) becomes exothermic for $v' > 2$ or 3, so that the process can be an efficient source of $O(^1S)$ on Earth.

Laboratory studies have demonstrated explicitly that $O_2(c^1\Sigma_u^-)$ is, indeed, formed in the recombination of $O(^3P)$, so that reaction (7.23) can be written out in full for production of the $c^1\Sigma_u^-$ state

$$O + O + M \rightarrow O_2(c^1\Sigma_u^-) + M. \tag{7.25}$$

Emission of the c → X, Herzberg II, bands is greatest when carbon dioxide is used as the third-body M, but CO_2 is not essential. Carbon dioxide has a peculiar influence either because it is a particularly effective third-body, or because it encourages vibrational relaxation in the recombining O atoms to funnel most of the excited population to $v' = 0$ of $O_2(c^1\Sigma_u^-)$. Venus, with a CO_2 atmosphere, therefore has a relatively large concentration of the excited molecules, but with $v' = 0$; Earth, on the other hand, has a lower concentration, but with v' up to 7, because M is predominantly N_2 and O_2. Emission of the Herzberg II system from Mars is very weak indeed.

At first sight, recombination of $O(^3P)$ in reaction (7.23) is a plausible night-time source not only of $O_2(c^1\Sigma_u^-)$, but also of the $a^1\Delta_g$ and $b^1\Sigma_g^+$ states that give rise to the Infrared Atmospheric and Atmospheric Bands of the nightglow. Channelling into the available molecular electronic states might be expected to be determined statistically, according to the electronic degeneracies. Much depends, however, on the energy barriers experienced by the atoms at large internuclear separations as they approach each other on the various potential curves. Nightglow concentrations of $O_2(^1\Delta_g)$ such as those derived for Fig. 7.2, would require 25 per cent of O atom recombination to follow the $O_2(^1\Delta_g)$ curve, which seems an intolerably large proportion. Theoretical considerations, in fact, favour $^5\Pi_g$ as the dominant initial state, with the lower electronic levels being filled by radiationless and collisional processes. Quite apart from the problems with absolute rates of $O_2(^1\Delta_g)$ and $O_2(^1\Sigma_g^+)$ production, the simple recombination mechanism also fails to predict correctly the altitudes and thicknesses of the emitting layers of the various bands, and to reconcile them with the altitude of [O] maximum.

An alternative mechanism for nightglow excitation calls again on the $c^1\Sigma_u^-$ state of O_2 as a precursor species. In this scheme, $O_2(c^1\Sigma_u^-)$ is formed vibrationally excited in the O-atom recombination step (perhaps with the $^5\Pi_g$ as an intermediate state). Quenching of $O_2(c^1\Sigma_u^-)$ by atomic oxygen possesses a large rate coefficient, but the products depend on the degree of vibrational excitation. Higher levels ($v' > 2$ or 3) can excite $O(^1S)$ in reaction (7.24); lower levels still have enough energy to produce $O_2(a^1\Delta_g)$ as a quenching product. Molecular oxygen quenches $O_2(c^1\Sigma_u^-)$ to $O_2(b^1\Sigma_g^+)$ in this hypothesis. The set of quenching reactions is thus

$$O_2(c^1\Sigma_u^-, v' > 2\text{–}3) + O \rightarrow O(^1S) + O_2(X^3\Sigma_g^-) \qquad (7.26)$$

$$O_2(c^1\Sigma_u^-, v' < 2\text{–}3) + O \rightarrow O_2(a^1\Delta_g) + O \qquad (7.27)$$

$$O_2(c^1\Sigma_u^-, v') + O_2 \rightarrow O_2(b^1\Sigma_g^+) + O_2. \qquad (7.28)$$

For the higher v' levels at least, quenching of $O_2(c^1\Sigma_u^-)$ by the major atmospheric constituents (N_2, O_2, and CO_2) is very rapid, so that $O(^1S)$ formation in reaction (7.26) can only represent a minor fate of the $O_2(c^1\Sigma_u^-$, $v'>2\text{–}3)$ molecules. However, the terrestrial airglow observations require only a relatively small fractional yield of $O(^1S)$ from the recombination of $O(^3P)$. The intensity of the upper emission layer[a] of the Infrared Atmospheric Band (Fig. 7.2) is expected to roughly follow $[O(^3P)]$ but is possibly shifted to slightly lower altitudes than the peak at ~ 98 km of the green line due to $O(^1S)$, since the increased vibrational relaxation at higher densities favours reaction (7.27) over (7.26). Maximum production rates for $O_2(b^1\Sigma_g^+)$ are determined by the product $[O_2]$ [M] as well as $[O]^2$, so that the peak Atmospheric Band intensity should come from yet lower altitudes (~ 94 km), in agreement with the observations. Thus the mechanism, although hypothetical, does consistently explain the nightglow emissions of the three singlet states of O_2.

7.4.2 Atomic sodium

Airglow emissions from atomic sodium and other metals are of interest because the *allowed*, resonance, transitions fall within the visible region and are therefore observable from the Earth's surface. Metallic ions have been observed consistently at D- and E-region heights by rocket-borne mass spectrometers, and, as pointed out in Section 6.4.2, ablation of meteors is the

[a]An entirely different mechanism, perhaps involving vibrationally excited OH (Section 7.4.3), must be responsible for the lower peak.

major source of the metals. The mesospheric layer of neutral sodium atoms lies roughly between 82 and 105 km, with peak concentrations being found typically at ~ 91 km. Normal peak concentrations are ~ 5×10^3 atom cm^{-3}. Increases in concentration observed during meteor showers support the idea of a meteor ablation source for the atoms (see p. 526). Dayglow emission of the sodium yellow $Na(^2P_{1/2,\ 3/2} \rightarrow {}^2S)$ resonance doublet (λ = 589.6, 589.0 nm) is excited by direct absorption of solar radiation. Intensities are around 30 kR. As pointed out in Section 7.3, this intensity corresponds to an excited state concentration of only ~ 5×10^{-4} atom cm^{-3}, so that roughly one atom in 10^7 is excited. Artificial excitation of the resonance emission provides a method for the measurement of concentration–altitude profiles of sodium from the ground. A dye laser is tuned to one of the resonance lines, and a short pulse of light directed into the atmosphere. Sodium atoms present in the atmosphere can resonantly absorb and re-emit the radiation. Examination of the intensity of the scattered light, and the time delay between outgoing and returning pulses, allows the profile to be derived. By analogy with radio-frequency radar sounding, the laser technique is called *lidar*. High-resolution lidar measurements also allow wind and temperatures to be calculated for the mesosphere, in the same way that airglow emission spectra are used (see p. 526).

Nightglow intensities of the yellow sodium lines are much weaker (~ 100 R) than dayglow ones, but that they are present at all means that some reaction must be capable of exciting $Na(^2P)$. Figure 7.8 illustrates an emission-rate profile measured by rocket: in addition, the figure shows a profile of ground-state Na measured at the launch site by lidar. Interferometric measurements on the spectral line shapes show that the $Na(^2P)$ atoms possess a kinetic energy too small to arise from excitation in ionic processes such as dissociative recombination. Chemical reaction, however, implies consumption of atomic sodium, and the elemental sodium in the layer would soon be depleted unless a regenerative step also occurs. An oxidation–reduction cycle, involving the species NaO, is the only plausible scheme. The cycle starts with the oxidation of sodium by O_3

$$Na + O_3 \rightarrow NaO + O_2, \tag{7.29}$$

and this reaction is followed by reduction of NaO by atomic oxygen to yield either excited, ^2P, or ground-state, ^2S, sodium atoms

$$NaO + O \xrightarrow{\alpha} Na(^2P) + O_2, \tag{7.30}$$

$$NaO + O \xrightarrow{1-\alpha} Na(^2S) + O_2. \tag{7.31}$$

Involvement of ozone is inevitable in view of the emission intensity. Given that [Na] ~ 5×10^3 atom cm^{-3} at its peak, and with 100 ^2P atoms excited per second (100 R in a 10 km layer), the product of rate coefficient and co-reactant must

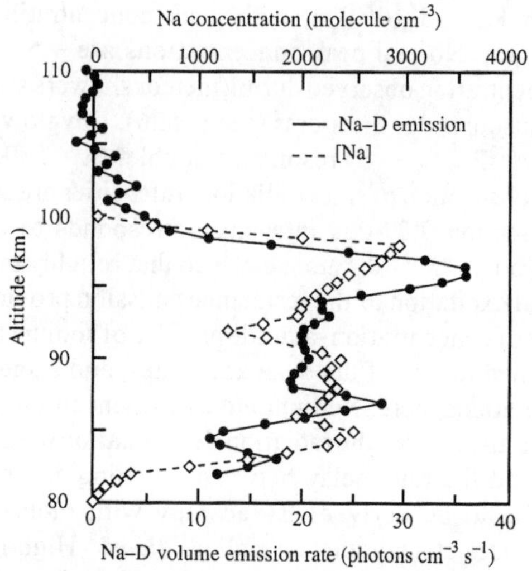

Fig. 7.8. Volume emission rate from $Na(^2P)$ measured by rocket, and the ground-state Na concentration measured from the ground by lidar. From Takahasi, H., Melo, S.M.L., Clemesha, B.R., Simonich, D.M., Stegman, J. ,and Witt G. *J. geophys. Res.* **101**, 4033 (1996).

be at least $100/(5 \times 10^3) \sim 2 \times 10^{-2}$ s^{-1}, even if all the product NaO species lead to excitation of $Na(^2D)$ in reaction (7.30). Laboratory measurements have shown that all three reactions possess very large rate coefficients, but since [O] is orders of magnitude larger than $[O_3]$ at altitudes of around 90 km, as demonstrated later (cf. Fig. 7.10), reaction (7.29) is the rate-determining step. The rate coefficient as determined in the laboratory for this latter process is about 6×10^{-10} cm^3 molecule^{-1} s^{-1} at mesospheric temperatures (see end of paragraph). The *minimum* required concentration of co-reactant is thus (2 × f06 × 10^{-10}) ~ 3×10^7 molecule cm^{-3}. Ozone is almost certainly the only reactive species in sufficient abundance at 90 km that could give an energy-rich product (NaO). Regeneration of atomic sodium requires a reduction process that is at least as fast as the oxidation reaction in the mesosphere, and that is sufficiently exothermic to excite $Na(^2P)$. Nightglow measurements show that the fraction, α, of reactions of O with NaO that lead to $Na(^2P)$ is about 0.1. Theory suggests that these conditions can be met if the reactions are of the electron-jump type; the participation of the electron-jump mechanism in reaction (7.29) as well can explain the very large rate coefficient for the process, which is about twice the limit allowed by hard-sphere collision theory.

The appearance in the atmosphere of a highly reactive species such as atomic Na is rather surprising, and since the first observations of the layer in the 1950s has posed a considerable puzzle. Cyclic regeneration of free elemental sodium is, of course, demanded if sodium undergoes any reaction,

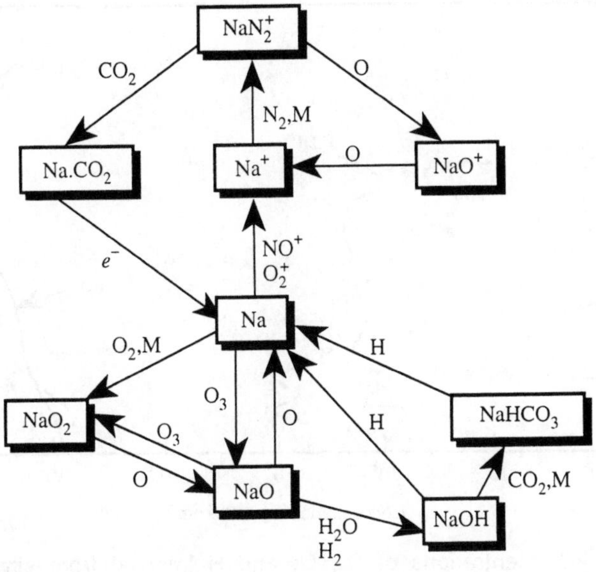

Fig. 7.9. Flow diagram of major processes in the mesospheric chemistry of neutral and ionized atomic sodium. Diagram provided by J.M.C. Plane, 1998.

regardless of the emission of light. Reaction (7.30), followed by radiation and by reaction (7.29), provides the necessary sequence so far as NaO is concerned. A combination of laboratory kinetic studies and atmospheric modelling has now shown that sodium undergoes a quite complex series of regenerative cycles involving both neutral and ionic species. Figure 7.9 shows the major reactions in flow-diagram form. Species such as NaO_2, formed in the three-body reaction

$$Na + O_2 + M \rightarrow NaO_2 + M, \quad (7.32)$$

for example, participate in the cycling process. The major neutral sink is sodium bicarbonate, $NaHCO_3$, which is formed by a sequence of reactions involving O_3, H_2O, and CO_2. However, $NaHCO_3$, as well as NaO_2 and NaOH, are cycled back to atomic Na by reactions involving O and H. Note that these reactions represent *reductions*: although most of the atmosphere is an oxidizing medium, at altitudes between 80 and 100 km the atmosphere becomes strongly reducing because of a significant increase in atomic O and H relative to O_3 and O_2. At 90 km, almost all of the sodium is in the form of free atoms, while above 100 km, photoionization and charge-transfer from ambient ions such as NO^+ (see Section 6.4.2) produces Na^+, which is cycled back to Na by forming clusters that then undergo dissociative recombination in reaction with electrons.

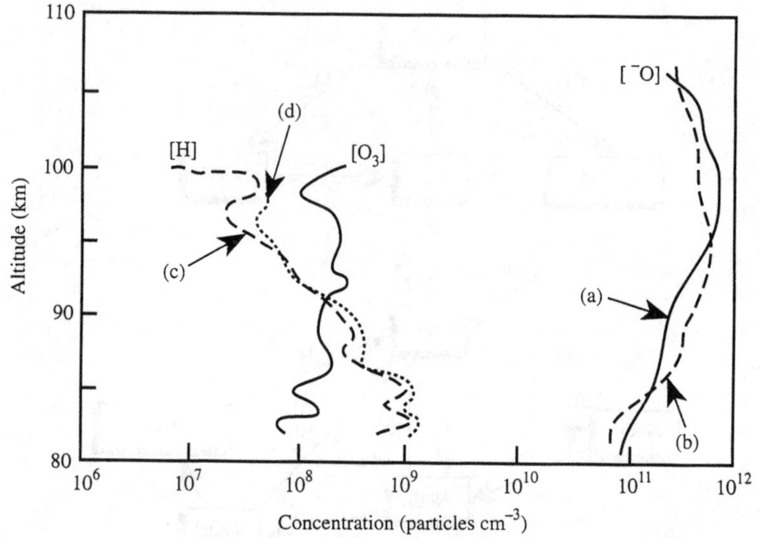

Fig. 7.10. Concentrations of O_3, O, and H inferred from simultaneous measurements of Na(^{2}P) (shown in Fig. 7.8), $O_2(^1\Sigma_g^+)$ Atmospheric Band, and OH (8,3) Meinel band intensities. Curve (a) is the O-atom profile derived from the intensity of the Atmospheric Band, while curve (b) is that predicted by a model. Curve (c) is the H-atom profile obtained from the intensity of the Atmospheric Band, and curve (d) is that derived from the OH band. All calculations assume that $\alpha = 0.1$ (see text). Source as for Fig. 7.8.

Figure 7.10 shows how a state-of-the-art rocket airglow measurement may be used to establish concentrations of minor constituents in the mesosphere. In the experiments summarized in this figure and Fig. 7.8, a total of ten airglow photometers (six forward-looking, and four side-looking) were carried on the same rocket. They observed the OI 557.5nm line, O_2 Herzberg I bands, O_2 Atmospheric Band, Na D lines, OH Meinel (8,3) band, and airglow continuum. A lidar system operating at the ground at the launch site was used to measure Na-atom concentrations at the time of the rocket experiment. Figure 7.10 shows the concentrations of O, H, and O_3 inferred from the measurements. The model calculations on which the displayed profiles are based all use a value of $\alpha = 0.1$, in accordance with the value suggested earlier. Note that these determinations clearly support the view that [O] » [O_3], which has been a feature of several points of our discussion of the mesosphere.

An interesting phenomenon associated with meteor train luminosity may have one explanation in the cyclic chain processes (7.29) and (7.30). Most meteors that produce a visible train on entering the Earth's atmosphere show a very short-lived luminosity, following the impact of the body with the atmosphere. In about one in 125 000 visual meteor events, however, the train lasts for more than ten minutes; enduring luminosity for over an hour has been recorded. Emissions of sodium and magnesium (as well as, perhaps, of the O_2 Atmospheric Band) have been identified in these persistent trains. The afterglows are most strongly emitted from altitudes of ~ 90 km. Oxidation and reduction in a cycle, as proposed for sodium in the airglow, seems an attractive way of explaining long-lasting excitation of additional sodium ablated off a meteor.

7.4.3 Hydroxyl radicals

Vibrational emission phenomena are ordinarily thought of in connection with thermally equilibrated radiative transfer in the atmosphere (Sections 2.2.2 to 2.2.4). However, vibrationally *dis*-equilibrated OH radicals make a strong contribution to the longwave visible and near infrared airglow, and offer a rare example of vibrational, rather than electronic, airglow emission. The OH airglow has additional interest because, as we shall see later, it provided the first hint of the catalytic chains that are now known to be so important in atmospheric ozone chemistry (Section 4.4.1). Vibrational–rotational transitions among the nine lowest levels, v'', in the ground electronic state of OH give rise to the observed emission system, known as the *Meinel Bands* of OH. Transitions are observed corresponding not only to the allowed $\Delta v'' = 1$ fundamental, but also to the forbidden $\Delta v'' > 1$ overtones. Fundamentals lie in the wavelength range ~4500 nm ($v'' = 9 \rightarrow 8$) to ~2800 nm ($v'' = 1 \rightarrow 0$), and are intrinsically the strongest features, but are relatively difficult to detect because they lie well into the infrared. Overtone transitions of the $\Delta v'' = 4, 5...$ series lie at $\lambda < 1000$ nm, and, although much weaker because of their forbidden nature, they are more readily observed without interference from the thermal background of atmosphere or detectors than the fundamentals. Even $\Delta v'' = 9$, for $v'' = 9 \rightarrow 0$, is detectable just in the ultraviolet ($\lambda \sim 382$ nm), although the intensity of the band is nearly eight orders of magnitude weaker than the bands of the $\Delta v'' = 1$ series. Figure 7.11 shows sample photometric measurements of one of the Meinel bands, OH (9, 4), obtained with the HRDI instrument on UARS (p. 181); simultaneous measurements of the (0, 0) transition of the Atmospheric Band, $O_2(^1\Sigma_g^+) \rightarrow O_2(^3\Sigma_g^-)$, and of the $O(^1S)$ green line are also displayed in the figure.

Evaluation of possible excitation mechanisms starts from the apparent sharp cut-off in excitation at $v'' = 9$, corresponding to ~ 312 kJ mol^{-1}, with no emission from $v'' = 10$ (~ 337 kJ mol^{-1}) being seen. A reaction is needed for excitation whose exothermicity lies between 312 and 337 kJ mol^{-1}. Atomic hydrogen reacts with ozone to form OH in exactly this way

$$H + O_3 \rightarrow OH + O_2 + 322 \text{ kJ mol}^{-1}. \qquad (7.33)$$

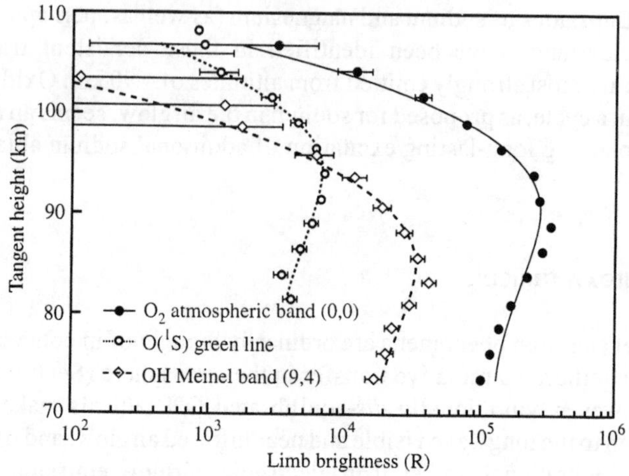

Fig. 7.11. HRDI altitude-profile measurements of some airglow features, including the Meinel (9, 4) band of OH. From Yee, J.-H., Crowley, G., Roble, R.G., Skinner, W.R., Burrage, M.D., and Hays, P.B. *J. geophys. Res.* **102**, 19949 (1997).

Laboratory investigations show that this reaction is indeed strongly chemiluminescent, with OH ($v'' < 9$) being the emitter. The immediate question for atmospheric excitation is why reaction (7.33) does not consume all available H atoms if it is fast enough to give the observed intensities. An explanation came from laboratory studies of the rate of the reaction

$$O + OH \rightarrow O_2 + H, \tag{7.34}$$

which was shown to be fast enough to regenerate H atoms in the atmosphere. Cyclic regeneration of H also explains a further result of the laboratory studies that was initially disturbing. Addition of ozone to atomic oxygen led to weak emission of the OH Meinel bands *even though hydrogen was absent and the reactants supposedly dry*! Realization that traces of H or H_2O (source of OH) could participate in a catalytic cycle made the result less surprising, and emphasized the need for purity and dryness in reactants. The same realization afforded an explanation of why the apparent rate constant for the reaction

$$O + O_3 \rightarrow 2O_2 \tag{7.35}$$

was much larger when measured in experiments where [O] was high than it was when the atoms were very dilute. Because sufficient precautions had not been taken, traces of H or OH catalysed O_3 decomposition in the cycle (7.33) + (7.34) in the high [O] experiments. From this interpretation of the laboratory results, it is a short step to the postulation of HO_x catalytic cycles in the stratospheric ozone destruction by atomic oxygen (Section 4.4.1).

Further support for the H–O_3 mechanism in exciting the OH Meinel Bands of the atmosphere comes from the diurnal intensity variations. Dayglow intensities totalled over all bands are typically 4.5 MR. Reaction (7.33) can give the required excitation rate, so long as cascading between different v'' levels can give several photons for each excitation event. Nightglow intensities are not much less, but a curious phenomenon is observed in both morning dawn and evening twilight periods. At both periods, the intensity decreases rapidly and then recovers nearly to the earlier value. Excited OH production is suppressed above ~ 85 km by day, when $[O_3]$ is depleted, but enhanced at lower altitudes, where the supply of O atoms is rate-determining. During the twilight periods, concentrations of O, O_3, and H vary in a complicated manner, but models predict a dip in emission intensity consistent with that seen.

This concluding section of the chapter illustrates how airglow studies give an understanding of atmospheric processes not easily available from other types of investigation. Complemented by reliable laboratory experiments on excited species, airglow is thus a prime source of information about the atmosphere of Earth and the other planets.

Bibliography

General introductions to the study of airglow

The solar-terrestrial environment. Hargreaves, J.K. (Cambridge University Press, Cambridge, 1992).

The radiating atmosphere. McCormack, B.M. (ed.) (D. Reidel, Dordrecht,1971).

Spectroscopic emissions. Chapter 7 in *Physics and chemistry of the upper atmosphere*. Rees, M.H. (Cambridge University Press, Cambridge, 1989).

A review of the photochemistry of selected nightglow emissions from the mesopause. Meriwether, J.W., Jr. *J. geophys. Res.* **94**, 14629 (1989).

The EUV dayglow at high spectral resolution. Morrison, M.D., Bowers, C.W., Feldman, P.D., and Meier, R.R. *J. geophys. Res.* **95**, 4113 (1990).

Upper atmosphere models and research. Ryecroft, M.J., Kasting, G.M., and Rees, D. (eds.) *Adv. Space Res.* **10**, No. 6 (1990).

Models for aurora and airglow emissions from other planetary atmospheres. Fox, J.L. *Canad. J. Phys.* **64**, 1631 (1986).

The visible airglow experiment—a review. Hays, P.B., Abreu, V.J., Solomon, S.C., and Yee, J. *Planet. Space Sci.* **36**, 21 (1988).

The airglow. McEwan, M.J. and Phillips, L.F., Chapter 5 of *Chemistry of the atmosphere*. (Edward Arnold, London, 1975).

Physics of the aurora and airglow. Chamberlain, J.W. (Academic Press, New York, 1961).

The light of the night sky, Roach, F.E. and Gordon, J.L. (D. Reidel, Dordrecht, 1973).

These publications provide a background to the chemistry of the regions of the atmosphere where many of the airglow emissions originate.

A three-dimensional dynamic model of the minor constituents of the mesosphere. Sonnemann, G., Kremp, C., Ebel, A., and Berger, U. *Atmos. Environ.* **32**, 3157 (1998).

Microwave observations and modeling of O_3, H_2O and HO_2 in the mesosphere. Clancy, R.T., Sandor, B.J., Rusch, D.W., and Muhleman, D.O. *J. geophys. Res.* **99**, 5465 (1994).

Neutral and ion chemistry of the upper atmosphere. Thomas, L., in *Handbuch der Physik* Rawer, K. (ed.) XLIX/6, 7 Part VI. (Springer-Verlag, Berlin, 1982).

Sections 7.1 and 7.2

Excited states and emission features

The excited state in atmospheric chemistry. Marston, G. *Chem. Soc. Revs.* **25**, 33 (1996).

The role of metastable species in the thermosphere. Torr, M.R. and Torr, D.G. *Rev. Geophys. & Space Phys.* **20**, 91 (1982).

Atomic and molecular emissions in the middle ultraviolet dayglow. Bucsela, E.J., Cleary, D.D., Dymond, K.F., and McCoy, R.P. *J. geophys. Res.* **103**, 29215 (1998).

EURD observations of EUV nightime airglow lines. LopezMoreno, J.J., Morales, C., Gomez, J.F., Trapero, J., Bowyer, S., Edelstein, J., Lampton, M., and Korpela, E.J. *Geophys. Res. Letts.* **25**, 2937 (1998).

Ground based spectroscopic studies of sunlit airglow and aurora. Chakrabarti, S. *J. atmos. solar-terrest. Phys.* **60**, 1403 (1998).

A new airglow layer in the stratosphere. Evans, W.F.J., and Shepherd, G.G. *Geophys. Res. Letts.* **23**, 3623 (1996).

Airglow measurements of possible changes in the ionosphere and middle atmosphere. Shepherd, G.G., Siddiqi, N.J., Wiens, R.H., and Zhang, S.P. *Advan. Space Res.* **20**, 2127 (1997).

Two-dimensional spectral analysis of mesospheric airglow image data. Garcia, F.J., Taylor, M.J., and Kelley, M.C. *Appl. Optics* **36**, 7374 (1997).

Transient decreases of Earth's far-ultraviolet dayglow. Frank, L.A. and Sigwarth, J.B. *Geophys. Res. Letts.* **24**, 2423 (1997).

Global change in the mesosphere lower thermosphere region: Has it already arrived? Thomas, G.E. *J. atmos. terrest. Phys.* **58**, 1629 (1996).

Detection of polar dayglow during rocket research of atmosphere emission. Khokhlov, V.N. *Geomag. Aeronom.* **36**, 216 (1996).

Ultraviolet sky as observed by the shuttle-borne ultraviolet imaging telescope. Waller, W.H., Marsh, M., Bohlin, R.C., Cornett, R.H., Dixon, W.V., Isensee, J.E., Murthy, J., O'Connell, R.W., Roberts, M.S., Smith, A.M., and Stecher, T.P. *Astronom. J.* **110**, 1255 (1995).

The middle ultraviolet dayglow spectrum. Cleary, D.D., Gnanalingam, S., McCoy, R.P., Dymond, K.F., and Eparvier, F.G. *J. geophys. Res.* **100**, 9729 (1995).

Determination of atmospheric composition and temperature from UV airglow. Meier, R. R. and Anderson, D.E., Jr. *Planet. Space Sci.* **31**, 967 (1983).

Low latitude airglow. Meier, R.R. *Rev. Geophys. & Space Phys.* **17**, 485 (1979).

Photographic observations of the Earth's airglow from space. Mende, S.R., Banks, P.M., Nobles, R., Garriot, O.K., and Hoffman, J. *Geophys. Res. Letts.* **10**, 1108 (1983).

Atomic oxygen and ozone in the mesosphere and thermosphere

Springtime transition in lower thermospheric atomic oxygen. Shepherd, G.G., Stegman, J., Espy, P., McLandress, C., Thuillier, G., and Wiens, R.H. *J. geophys. Res.* **104,** 213 (1999).

Estimation of atomic oxygen concentrations from measured intensities of infrared nitric oxide radiation. Kirillov, A.S., and Aladjev, G.A. *Ann. Geophys.* **16,** 847 (1998).

Odd oxygen measurements during the Noctilucent Cloud 93 rocket campaign. Gumbel, J., Murtagh, D.P., Espy, P.J., Witt, G., and Schmidlin, F.J. *J. geophys. Res.* **103,** 23399 (1998).

Long-term changes of ozone and atom oxygen high-latitude distributions in lower thermosphere. Semenov, A.I. *Geomag. Aeronom.* **37,** 132 (1997).

Laboratory measurements required for upper atmospheric remote sensing of atomic oxygen. McDade, I.C. *Advan. Space Res.* **19,** 653 (1997).

Monte Carlo studies of the resonance fluorescence technique for atmospheric atomic oxygen measurements. Gumbel, J. and Witt, G. *J. Quant. Spect. Radiat. Transfer* **58,** 1 (1997).

Rocket observation of atomic oxygen and night airglow: Measurement of concentration with an improved resonance fluorescence technique. Kita, K., Imamura, T., Iwagami, N., Morrow, W.H., and Ogawa, T. *Ann. Geophys.* **14,** 227 (1996).

Atomic hydrogen and ozone concentrations derived from simultaneous lidar and rocket airglow measurements in the equatorial region. Takahashi, H., Melo, S.M.L., Clemesha, B.R., Simonich, D.M., Stegman, J., and Witt, G. *J. geophys. Res.* **101,** 4033 (1996).

A reference model for atomic oxygen in the terrestrial atmosphere. Llewellyn, E.J. and McDade, I.C. *Advan. Space Res.* **18,** 209 (1996).

Specific emission systems. See also Sections 7.4.1–7.4.3.

Electronic structure of excited states of selected atmospheric systems. Michels, H.H., in *The excited state in chemical physics*, Vol. 2 McGowan, J.W. (ed.) (John Wiley, New York, 1979).

Infrared emission spectra of nitric oxide following the radiative association of nitrogen atoms and oxygen atoms. Sun, Y., and Dalgarno, A. *J. Quant. Spect. Radiat. Transfer* **55,** 245 (1996).

N_2 triplet band systems and atomic oxygen in the dayglow. Broadfoot, A.L., Hatfield, D.B., Anderson, E.R., Stone, T.C., Sandel, B.R., Gardner, J.A., Murad, E., Knecht, D.J., Pike, C.P., and Viereck, R.A. *J. geophys. Res.* **102,** 11567 (1997).

NI 8680 and 8629 Å multiplets in the dayglow. Bucsela, E.J. and Sharp, W.E. *J. geophys. Res.* **102,** 2457 (1997).

First optical detection of atomic deuterium in the upper atmosphere from Spacelab I Bertaux, J.L., Goutail, F., Dimarellis, E., Kockarts, G., and Van Ransbeek, E. *Nature, Lond.* **309**, 771 (1984).

The wind-imaging interferometer (WINDII)

Characterization of the wind imaging interferometer. Hersom, C.H. and Shepherd, G.G. *Appl. Optics* **34,** 2871 (1995).

WINDII on UARS: A view of upper mesosphere and lower thermosphere dynamics. Shepherd, G.G. *J. Geomag. Geoelectr.* **48,** 125 (1996).

Use of airglow to probe winds, atmospheric waves, and atmospheric dynamics in general

Large airglow enhancements produced via wave-plasma interactions in sporadic E. Djuth, F.T., Bernhardt, P.A., Tepley, C.A., Gardner, J.A., Kelley, M.C., Broadfoot, A.L., Kagan, L.M., Sulzer, M.P., Elder, J.H., Selcher, C., Isham, B., Brown, C., and Carlson, H.C. *Geophys. Res. Letts.* **26,** 1557 (1999).

Measuring gravity wave momentum fluxes with airglow imagers. Gardner, C.S., Gulati, K., Zhao, Y.C., and Swenson, G. *J. geophys. Res.* **104,** 11903 (1999).

Simultaneous measurements of airglow OH emission and meteor wind by a scanning photometer and the MU radar. Takahashi, H., Batista, P.P., Buriti, R.A., Gobbi, D., Nakamura, T., Tsuda, T., and Fukao, S. *J. atmos. solar-terrest. Phys.* **60,** 1649 (1998).

Seasonal and diurnal variations of wind and temperature near the mesopause from Fabry-Perot interferometer observations of OH Meinel emissions. Choi, G.H., Monson, I.K., Wickwar, V.B., and Rees, D. *Advan. Space Res.* **21,** 847 (1998).

WINDII observations of the 558 nm emission in the lower thermosphere: The influence of dynamics on composition. Shepherd, G.G., Roble, R.G., McLandress, C., and Ward, W.E. *J. atmos. solar-terrest. Phys.* **59,** 655 (1997).

A fundamental theorem of airglow fluctuations induced by gravity waves. Hines, C.O. *J. atmos. solar-terrest. Phys.* **59,** 319 (1997).

Mean vertical wind in the mesosphere lower thermosphere region (80–120 km) deduced from the WINDII observations on board UARS. Fauliot, V., Thuillier, G., and Vial, F. *Ann. Geophys.* **15,** 1221 (1997).

Investigations of thermospheric ionospheric dynamics with 6300 Å images from the Arecibo Observatory. Mendillo, M., Baumgardner, J., Nottingham, D., Aarons, J., Reinisch, B., Scali, J., and Kelley, M. *J. geophys. Res.* **102,** 7331 (1997).

An analysis of wind imaging interferometer observations of $O(^1S)$ equatorial emission rates using the thermosphere–ionosphere–mesosphere–electrodynamics general circulation model. Roble, R.G. and Shepherd, G.G. *J. geophys. Res.* **102,** 2467 (1997).

Statistics of gravity waves seen in O_2 nightglow over Bear Lake Observatory. Wiens, R.H., Wang, D.Y., Peterson, R.N., and Shepherd, G.G. *J. geophys. Res.* **102,** 7319 (1997).

Two day wave induced variations in the oxygen green line volume emission rate: WINDII observations. Ward, W.E., Solheim, B.H., and Shepherd, G.G. *Geophys. Res. Letts.* **24,** 1127 (1997).

Full-wave modeling of small-scale gravity waves using airborne lidar and observations of the Hawaiian airglow (ALOHA-93) $O(^1S)$ images and coincident Na wind/temperature lidar measurements. Hickey, M.P., Taylor, M.J., Gardner, C.S., and Gibbons, C.R. *J. geophys. Res.* **103,** 6439 (1998).

Characteristics of atmospheric waves in the tidal period range derived from zenith observations of $O_2(0–1)$ Atmospheric and OH(6–2) airglow at lower midlatitudes. Reisin, E.R. and Scheer, J. *J. geophys. Res.* **101,** 21223 (1996).

All-sky measurements of short-period waves imaged in the OI(557.7 nm), Na(589.2 nm) and near-infrared OH and $O_2(0, 1)$ nightglow emissions during the ALOHA-93 campaign. Taylor, M.J. and Bishop, M.B. *Geophys. Res. Letts.* **22,** 2833 (1995).

Photochemical–dynamical modeling of the measured response of airglow to gravity-waves. 1. Basic model for OH airglow. Makhlouf, U.B., Picard, R.H., and Winick, J.R. *J. geophys. Res.* **100,** 11289 (1995).

Temperatures as deduced from airglow observations

Seasonal temperature variations in the mesopause region at mid-latitude: comparison of lidar and hydroxyl rotational temperatures using WINDII/UARS OH Height profiles. She, C.Y., and Lowe, R.P. *J. atmos. solar-terrest. Phys.* **60,** 1573 (1998).

Mesospheric rotational temperatures determined from the OH(6–2) emission above Adelaide, Australia. Hobbs, B.G., Reid, I.M., and Greet, P.A. *J. atmos. terrest. Phys.* **58,** 1337 (1996).

Tides in emission rate and temperature from the O_2 nightglow over bear lake observatory. Wiens, R.H., Zhang, S.P., Peterson, R.N., and Shepherd, G.G. *Geophys. Res. Letts.* **22,** 2637 (1995).

Section 7.3

Application of tomographic inversion in studying airglow in the mesopause region. Nygren, T., Taylor, M.J., Lehtinen, M.S., and Markkanen, M. *Ann. Geophys.* **16,** 1180 (1998).

Annual variation of airglow heights derived from wind measurements. Plagmann, M., Marsh, S.H., Baggaley, W.J., Bennett, R.G.T., Deutsch, K.A., Fraser, G.J., Hernandez, G., Lawrence, B.N., Plank, G.E., and Smith, R.W. *Geophys. Res. Letts.* **25,** 4457 (1998).

Rocket measurements of the equatorial airglow: MULTIFOT 92 database. Takahasi, H., Clemesha, B.R., Simonich, D.M., Melo, S.M.L., Teixeira, N.R., Stegman, J., and Witt, G. *J. atmos. terr. Phys.* **58,** 1943 (1996).

Atomic oxygen concentrations from rocket airglow observations in the equatorial region. Melo, S.M.L., Takahashi, H., Clemesha, B.R., Batista, P.P., and Simonich, D.M. *J. atmos. terrest. Phys.* **58,** 1935 (1996).

Variability of oxygen-atom concentration in the lower thermosphere observed by rocket experiments. Kita, K., Imamura, T., Iwagami, N., Morrow, W.H., and Ogawa, T. *Advan. Space Res.* **17,** 85 (1995).

Section 7.4.1

The photochemistry of the MLT oxygen airglow emissions and the expected influences of tidal perturbations. McDade, I.C. *Advan. Space Res.* **21,** 787 (1998).

Atomic oxygen contributions to the airglow

OI emissions. Noxon, J.F., in *Physics and chemistry of upper atmospheres*, pp. 213–18 (D. Reidel, Dordrecht, 1973).

The OI (6300 Å) airglow. Hays, P.B., Rusch, D.W., Roble, R.G., and Walker, J.C.G. *Rev. Geophys. & Space Phys.* **16,** 225 (1978).

High resolution 2-D maps of OI 630.0 nm thermospheric dayglow from equatorial latitudes. Raju, D.P. and Sridharan, R. *Ann. Geophys.* **16,** 997 (1998).

Nightglow zenith emission rate variations in O(^{1}S) at low latitudes from wind imaging interferometer (WINDII) observations. Zhang, S.P., Wiens, R.H., Solheim, B.H., and Shepherd, G.G. *J. geophys. Res.* **103,** 6251 (1998).

A model for the response of the atomic oxygen 557.7nm and the OH Meinel airglow to atmospheric gravity waves in a realistic atmosphere. Makhlouf, U.B., Picard, R.H., Winick, J.R., and Tuan, T.F. *J. geophys. Res.* **103,** 6261 (1998).

N_2 triplet band systems and atomic oxygen in the dayglow. Broadfoot, A.L., Hatfield, D.B., Anderson, E.R., Stone, T.C., Sandel, B.R., Gardner, J.A., Murad, E., Knecht, D.J., Pike, C.P., and Viereck, R.A. *J. geophys. Res.* **102,** 11567 (1997).

Post-sunset wintertime 630.0 nm airglow perturbations associated with gravity waves at low latitudes in the South American sector. Sobral, J.H.A., Borba, G.L., Abdu, M.A., Batista, I.S., Sawant, H., Zamlutti, C.J., Takahashi, H., and Nakamura, Y. *J. atmos. solar-terrest. Phys.* **59,** 1611 (1997).

Dynamical influences on atomic oxygen and 5577 Å emission rates in the lower thermosphere. Coll, M.A.I. and Forbes, J.M. *Geophys. Res. Letts.* **25,** 461 (1998).

A model for 5577 Å and 8446 Å atomic oxygen dayglow emission at midaltitudes. Singh, V. and Tyagi, S. *Advan. Space Res.* **20,** 1145 (1997).

Horizontal gradients of the nocturnal OI 557.7nm and OI 630nm photoemission rates in the equatorial ionosphere based on rocket electron density data. Sobral, J.H.A., Abdu, M.A., Muralikrishna, P., Takahashi, H., Sawant, H.S., Zamlutti, C.J., and de Aquino, M.G. *Advan. Space Res.* **20,** 1317 (1997).

The O(^{1}D) dayglow emission as observed by the wind imaging interferometer on UARS. Singh, V., McDade, I.C., Shepherd, G.G., Solheim, B.H., and Ward, W.E. *Advan. Space Res.* **17,** 11 (1995).

The 'auroral green line'

The morphology of oxygen greenline dayglow emission. Tyagi, S., and Singh, V. *Ann. Geophys.* **16,** 1599 (1998).

The source of green light emission determined from a heavy-ion storage ring experiment. Kella, D., Vejby Christensen, L., Johnson, P.J., Pedersen, H.B., and Anderson, L.H. *Science* **276,** 1530 (1997).

The O(^{1}S) dayglow emission as observed by the WIND imaging interferometer on the UARS. Singh, V., McDade, I.C., Shepherd, G.G., Solheim, B.H., and Ward, W.E. *Ann. Geophys.* **14,** 637 (1996).

Rocket photometry and the lower-thermospheric oxygen nightglow. Greer, R.G.H., Murtagh, D.P., McDade, I.C., Llewellyn, E.J., and Witt, G. *Phil. Trans. R. Soc.* **A323,** 579 (1987).

Excitation of 557.7-nm OI line in nightglow. Bates, D.R. *Planet. Space Sci.* **36,** 883 (1988).

The excitation of O(^{1}S) and O_2 bands in the nightglow: a brief review and preview. McDade, I.C. and Llewellyn, E.J. *Canad. J. Phys.* **64**, 1626 (1986).
Excitation of O(^{1}S) and emission of 5577Å radiation in aurora. Rees, M.H. *Planet. Space Sci.* **32**, 373 (1984).
The green light of the night sky. Bates, D.R. *Planet. Space Sci.* **29**, 1061 (1981).

On the mechanism and yields of O(^{1}S) and O(^{1}D) formation in the dissociative recombination of O_2

Oxygen green and red line emission and O_2^+ dissociative recombination. Bates, D.R., *Planet. Space Sci.* **38**, 889 (1990).
O(^{1}S) from dissociative recombination of O_2^+: nonthermal line profile measurements from dynamics explorer. Killeen, T.L. and Hays, P.B. *J. geophys. Res.* **88**, 10163 (1983).
The dissociative recombination of O_2^+: the quantum yield of O(^{1}S) and O(^{1} D). Abreu, V.J., Solomon, S.C., Sharp, W.E., and Hays, P.B. *J. geophys. Res.* **88**, 4140(1983).

Molecular oxygen

The review listed first includes a section on airglow emission.

Reactions of singlet molecular oxygen in the gas phase. Wayne, R.P. in *Singlet oxygen*, Vol. 1 Frimer, A.A. (ed.) (CRC Press, Florida, 1985).
Singlet oxygen airglow. Wayne, R.P. *J. Photochem.* **25**, 345 (1984).
Tidal influences on O_2 atmospheric band dayglow: HRDI observations vs. model simulations. Marsh, D.R., Skinner, W.R., and Yudin, V.A. *Geophys. Res. Letts.* **26,** 1369 (1999).
Extreme intensity variations of O_2(b) airglow induced by tidal oscillations. Scheer, J. and Reisen, E.R. *Advan. Space Res.* **21,** 827 (1998).
Tidal mechanisms of dynamical influence on oxygen recombination airglow in the mesosphere and lower thermosphere. Ward, W.E. *Advan. Space Res.* **21,** 795 (1998).
Global simulations and observations of O(^{1}S), O_2($^1\Sigma$) and OH mesospheric nightglow emissions. Yee, J.H., Crowley, G., Roble, R.G., Skinner, W.R., Burrage, M.D., and Hays, P.B. *J. geophys. Res.* **102,** 19949 (1997).
The isotopic oxygen nightglow as viewed from Mauna Kea. Slanger, T.G., Huestis, D.L., Osterbrock, D.E., and Fulbright, J.P. *Science* **277,** 1485 (1997).
The O_2 Herzberg I bands in the equatorial nightglow. Melo, S.M.L., Takahashi, H., Clemesha, B.R., and Stegman, J. *J. atmos. solar-terrest. Phys.* **59,** 295 (1997).
Decay of $O_2(a^1\Delta_g)$ in the evening twilight airglow: Implications for the radiative lifetime. Pendleton, W.R., Baker, D.J., Reese, R.J., and O'Neil, R.R. *Geophys. Res. Letts.* **23,** 1013 (1996).
Spectral imaging of O_2 infrared atmospheric airglow with an InGaAs array detector. Doushkina, V.V., Wiens, R.H., Thomas, P.J., Peterson, R.N., and Shepherd, G.G. *Appl. Optics* **35,** 6115 (1996).
Metastable oxygen emission bands. Slanger, T.G. *Science* **202,** 751 (1978).
The oxygen nightglow. Bates, D.R. in *Progress in atmospheric physics* Rodrigo, R., Lopez-Moreno, J.J., Lopez-Puertas, M., and Molina, A. (D. Reidel, Dordrecht, 1988).

Use of molecular oxygen emission to determine O and O_3 concentrations

Mesospheric ozone concentration at an equatorial location from the 1.27-μm O_2 airglow emission. Batista, P.P., Takahasi, H., Clemesha, B.R., and Llewellyn, E.J. *J. geophys. Res.* **101**, 7917 (1996).

On the utility of the molecular-oxygen dayglow emissions as proxies for middle atmospheric ozone. Mlynczak, M.G. and Olander, D.S. *Geophys. Res. Letts.* **22**, 1377 (1995).

SME observations of $O_2(^1\Delta_g)$ nightglow: an assessment of the chemical production mechanisms. Howell, C.D., Michelangeli, D.V., Allen, M., Yung, Y.L., and Thomas, R.J. *Planet. Space Sci.* **38**, 529 (1990).

Molecular oxygen airglow from Venus. See also Sections 8.2 and 8.3.

Ground-based near-infrared observations of the Venus nightside: 1.27μm $O_2(a^1\Delta_g)$ airglow from the upper atmosphere. Crisp, D., Meadows, V.S., Bezard, B., deBergh, C., Maillard, J.P., and Mills, F.P. *J. geophys. Res.* **101**, 4577 (1996).

The Infrared Atmospheric Band

ETON 6: A rocket measurement of the O_2 infrared atmospheric (0,0) band in the nightglow. McDade, I.C., Llewellyn, E.J., Greer, R.G.H., and Murtagh, D.P. *Planet. Space Sci.* **35**, 1541 (1987).

A global study of $^1\Delta$ airglow: day and twilight. Noxon, J.F. *Planet. Space Sci.* **30**, 545 (1982).

$O_2(^1\Delta_g)$ in the atmosphere. Llewellyn, E.J., Evans, W.F.J., and Wood, H.C., in *Physics and chemistry of upper atmospheres*, pp. 193–202. (D. Reidel, Dordrecht, 1973).

Measurement of $O_2(a^1\Delta_g)$ emission in a total solar eclipse. Bantle, M., Llewellyn, E.J., and Solheim, B.H. *J. atmos. terr. Phys.* **46**, 265 (1984).

Rocket-borne photometric measurements of $O_2(^1\Delta_g)$, green line and OH Meinel bands in the nightglow. Lopez-Moreno, J.J., Vidal, S., Rodrigo, R., and Llewellyn, E.J. *Ann. Geophysicae* 2, 61 (1984).

A simultaneous measurement of the height profiles of the night airglow OI 5577 A, O_2 Herzberg and atmospheric bands. Ogawa, T., Iwagami, N., Nakamura, M., Takano, M., Tanabe, A., Takechi, A., Miyashita, A., and Suzuki, K.J. *Geomagn. Geoelectr.* **39**, 211 (1987).

The $c^1\Sigma_u^-$ state of O_2 and the mechanism of excitation of oxygen airglow. The first paper describes the identification of emission from $O_2(c^1\Sigma_u^-)$ in the Earth's nightglow. Subsequent papers discuss the possible involvement of O-atom recombination in excitation of various oxygen airglow features, and consider whether an excited state of O_2—conceivably $O_2(c^1\Sigma_u^-)$— might be a precursor in a two step mechanism.

$O_2(c^1\Sigma_u^- \rightarrow X^3\Sigma_g^-)$ emission in the terrestrial nightglow. Slanger, T.G. and Huestis, D.L. *J. geophys. Res.* **86**, 3551 (1981).

Association of atomic oxygen and airglow excitation mechanisms. Wraight, P.C. *Planet. Space Sci.* **30**, 251 (1982).

Auroral population of the $O_2(b^1\Sigma_g^+, v')$ levels. Henriksen, K. and Sivjee, G.G. *Planet. Space Sci.* **38**, 835 (1990).
The excitation of $O_2(b^1\Sigma_g^+)$ in the nightglow. Greer, R.G.H., Llewellyn, E.J., Solheim, B.H., and Witt, G. *Planet. Space Sci.* **29**, 383 (1981).
An assessment of the proposed $O(^1S)$ and $O_2(b^1\Sigma_g^+)$ nightglow excitation parameters. Murtagh, D.P., Witt, G., Stegman, J., McDade, I.C., Llewellyn, E.J., Harris, F., and Greer, R.G.H. *Planet. Space Sci.* **38**, 43 (1990).
Excitation of oxygen emissions in the night airglow of the terrestrial planets. Krasnopolsky, V.A. *Planet. Space Sci.* **29**, 925 (1981).
On the excitation of oxygen emissions in the airglow of the terrestrial planets. Llewellyn, E.J., Solheim, B.H., Witt, G., Stegman, J., and Greer, R.G.H. *J. Photochem.* **12**, 179 (1980).
O_2-triplet emissions in the nightglow. Murtagh, D.P., Witt, G., and Stegman, J. *Canad. J. Phys.* **64**, 1587 (1986).
Excitation of Herzberg I and II bands in the atmospheres of Earth and Venus. Parisot, J.-P. *Ann. Geophysicae* 86, 481 (1986).

Laboratory studies directed at aspects of molecular oxygen airglow.

Temperature dependence of the collisional removal of $O_2(b^1\Sigma_g^+, v = 1$ and $2)$ at 110–260 K, and atmospheric applications. Hwang, E.S., Bergman, A., Copeland, R.A., and Slanger, T.G. *J. Chem. Phys.* **110,** 18 (1999).
Integrated absorption intensity and Einstein coefficients for the O_2 $a^1\Delta_g - X^3\Sigma_g^-(0,0)$ transition: A comparison of cavity ringdown and high resolution Fourier transform spectroscopy with a long-path absorption cell. Newman, S.M., Lane, I.C., Orr-Ewing, A.J., Newnham, D.A., and Ballard, J. *J. Chem. Phys.* **110,** 10749 (1999).
Collisional removal of $O_2(b^1\Sigma_g^+, v = 1,2)$ by O_2, N_2, and CO_2. Bloemink, H.I., Copeland, R.A., and Slanger, T.G. *J. Chem. Phys.* **109,** 4237 (1998).
Vibrational-level-dependent yields of $O_2(b^1\Sigma_g^+, v = 0)$ following collisional removal of $O_2(A^3\Sigma_u^+, v)$. Shiau, T.P., Hwang, E.S., Buijsse, B., and Copeland, R.A. *Chem. Phys. Letts.* **282,** 369 (1998).
Rotational line strengths and self-pressure-broadening coefficients for the 1.27μm, $a^1\Delta_g - X^3\Sigma_g^-$ $v = 0$–0 band of O_2. Lafferty, W.J., Solodov, A.M., Lugez, C.L., and Fraser, G.T. *Appl. Optics* **37,** 2264 (1998).
Collisional removal of $O_2(c^1\Sigma_u^-, v = 9)$ by O_2, N_2, and He. Copeland, R.A., Knutsen, K., Onishi, M.E., and Yalcin, T. *J. Chem. Phys.* **105,** 10349 (1996).
Temperature dependence of the collisional removal of $O_2(A^3\Sigma_u^+, v = 9)$ with O_2 and N_2. Hwang, E.S., and Copeland, R.A. *Geophys. Res. Letts.* **24,** 643 (1997).
Oxygen band system transition arrays. Bates, D.R. *Planet. Space Sci.* **37**, 881 (1989).

Section 7.4.2

Meteoric metals in the atmosphere. See also Section 6.4.2.

Laboratory studies of the chemistry of meteoric metals. Plane, J.M.C. and Helmer, M. in *Research in chemical kinetics*. Hancock, G. and Compton, R.G. (eds.) (Elsevier, Amsterdam, 1994).

Characteristics of the sporadic Na layers observed during the airborne lidar and observations of Hawaiian airglow airborne noctilucent cloud (ALOHA/ANLC-93) campaigns. Qian, J., Gu, Y.Y., and Gardner, C.S. *J. geophys. Res.* **103,** 6333 (1998).
An ion–molecule mechanism for the formation of neutral sporadic Na layers. Cox, R.M., and Plane, J.M.C. *J. geophys. Res.* **103,** 6349 (1998).
Dynamical and chemical aspects of the mesospheric Na 'wall' event on October 9, 1993 during the Airborne Lidar and Observations of Hawaiian Airglow (ALOHA) campaign. Swenson, G.R., Qian, J., Plane, J.M.C., Espy, P.J., Taylor, M.J., Turnbull, D.N., and Lowe, R.P. *J. geophys. Res.* **103,** 6361 (1998).
Lidar observations of atmospheric sodium at an equatorial location. Clemesha, B.R., Simonich, D.M., Batista, P.P., and Batista, I.S. *J. atmos. solar-terrest. Phys.* **60,** 1773 (1998).
Comprehensive model for the atmospheric sodium layer. McNeil, W.J., Murad, E., and Lai, S.T. *J. geophys. Res.* **100,** 16847 (1995).
Mg^+ and other metallic emissions observed in the thermosphere. Viereck, R.A., Murad, E., Lai, S.T., Knecht, D.J., Pike, C.P., Gardner, J.A., Broadfoot, A.L., Anderson, E.R., and McNeil, W.J. *Advan. Space Res.* **18,** 61 (1995).
The chemistry of meteoric metals in the upper atmosphere. Plane, J.M.C. *Int. Rev. Phys. Chem.* **10,** 55 (1991).
A meteor ablation model of the sodium and potassium layers. Hunten, D.M. *Geophys. Res. Letts.* **8,** 369 (1981).

Sodium airglow

The first paper describes a concerted study in which ten airglow photometers were launched on a rocket to measure sodium, hydroxyl, and atomic and molecular oxygen airglows, and a ground-based sodium lidar operated at the launch site provided simultaneous vertical profiles of atmospheric sodium density.

Experimental evidence for photochemical control of the atmospheric sodium layer. Clemesha, B.R., Simonich, D.M., Takahashi, H., Melo, S.M.L. and Plane, J.M.C., *J. geophys. Res.* **100,** 18909 (1995).
Mesospheric Na layer at 40 degrees N: Modeling and observations. Plane, J.M.C., Gardner, C.S., Yu, J.R., She, C.Y., Garcia, R.R., and Pumphrey, H.C. *J. geophys. Res.* **104,** 3773 (1999).
Excitation mechanism of the mesospheric sodium nightglow. Herschbach, D.R., Kolb, C.E., Worsnop, D.R., and Shi, X. *Nature, Lond.* **356,** 414 (1992)
Night-time Na D emission observed from a polar orbit DMSP satellite. Newman, A.L. *J. geophys. Res.* **93,** 4067 (1988).
Excitation of the Na D-doublet of the airglow. Bates, D.R. and Ojha, P.C. *Nature, Lond.* **286,** 790 (1980).
Laser sounding of atmospheric sodium: interpretation in terms of global atmospheric parameters. Mégie, G. and Blamont, J.E. *Planet. Space Sci.* **25,** 1093 (1977).
Sodium nightglow and gravity waves. Molina, A. *J. atmos. Sci.* **40,** 2444 (1983).

Chemistry of sodium and other meteoric metals in the mesosphere.

These papers describe laboratory studies that provide the background for the various interpretations put forward. The last citation in the list is a seminal contribution that presents suggestions and experimental measurements for kinetic parameters based on an 'electron jump' model.

Laboratory studies of the chemistry of meteoric metals. Plane, J.M.C. and Helmer, M. in *Research in chemical kinetics*, Hancock, G. and Compton, R.G., (eds.) (Elsevier, Amsterdam, 1994).

A study of the reaction of Fe^+ with O_3, O_2 and N_2. Rollason, R.J. and Plane, J.M.C. *J. Chem. Soc. Faraday Trans.* **94**, 3067 (1998).

The heat of formation of $NaO^+(X^3\Sigma^-)$ and $NaO(X^2\Pi)$. Lee, E.P.F., Soldan, P., and Wright, T.G. *Chem. Phys. Letts.* **295,** 354 (1998).

An experimental and theoretical study of the clustering reactions between Na^+ ions and N_2, O_2 and CO_2. Cox, R.M. and Plane, J.M.C. *J. Chem. Soc. Faraday Trans.* **93,** 2619 (1997).

Excitation mechanism of the mesospheric sodium airglow. Herschbach, D.R., Kolb, C.E., Worsnop, D.R. and Shi, X. *Nature, Lond.* **356,** 414 (1992)

A kinetic investigation of the reactions $Na + O_3$ and $NaO + O_3$ over the temperature range 207–307 K. Plane, J.M.C., Nien, C.-F., Allen, M.R., and Helmer, M. *J. phys. Chem.* **97,** 4459 (1993).

Gas phase chemical kinetics of sodium in the upper atmosphere. Kolb, C.W. and Elgin J.B. *Nature, Lond.* **263,** 488 (1976).

Meteor trails

The source of enduring meteor train luminosity. Baggaley, W. J. *Nature, Lond.* **289,** 530 (1981).

The excitation of spectral lines in faint meteor trains. Poole, L.M.G. *J. atmos. terr. Phys.* **41,** 53 (1979).

Section 7.4.3

Satellite measurements of hydroxyl in the mesosphere. Conway, R.R., Stevens, M.H., Cardon, J.G., Zasadil, S.E., Brown, C.M., Morrill, J.S., and Mount, G.H. *Geophys. Res. Letts.* **23**, 2093 (1996).

The detection of the hydroxyl nightglow layer in the mesosphere by ISAMS/UARS. Zaragoza, G., Lopez Puertas, M., Lopez Valverde, M.A., and Taylor, F.W. *Geophys. Res. Letts.* **25,** 2417 (1998).

Analytical models for the responses of the mesospheric OH* and Na layers to atmospheric gravity waves. Swenson, G.R. and Gardner, C.S. *J. geophys. Res.* **103,** 6271 (1998).

OH(6–2) spectra and rotational temperature measurements at Davis, Antarctica. Greet, P.A., French, W.J.R., Burns, G.B., Williams, P.F.B., Lowe, R.P., and Finlayson, K. *Ann. Geophys.* **16,** 77 (1998).

Variations of the distribution of population of hydroxyl molecule vibrating levels of menopause. Shefov, N.N., Semenov, A.I., Yurchenko, O.T., Tikhonova, V.V., and Novikov, N.N. *Geomag. Aeronom.* **38,** 187 (1998).

High resolution spectrography of hydroxyl in the night airglow. Elsworth, Y., Lopez-Moreno, J.J., and James, J.F. *J. atmos. solar-terrest. Phys.* **59,** 117 (1997).

Determination of the concentration of mesopause atomic oxygen by hydroxylic emission. Perminov, V.I., and Yarov, V.N. *Geomag. Aeronom.* **36,** 167 (1996).

Mesospheric oxygen densities inferred from night-time OH Meinel band emissions. McDade, I.C. and Llewellyn, E.J. *Planet. Space Sci.* **36**, 897 (1988).
An experimental study of the nightglow OH(8–3) band emission process in the equatorial mesosphere. Melo, S.M.L., Takahashi, H., Clemesha, B.R., and Simonich, D.M. *J. atmos. solar-terrest. Phys.* **59**, 479 (1997).
WINDII/UARS observation of twilight behaviour of the hydroxyl airglow, at mid-latitude equinox. Lowe, R.P., LeBlanc, L.M., and Gilbert, K.L. *J. atmos. terrest. Phys.* **58**, 1863 (1996).
Empirical model of hydroxyl emission variations. Semenov, A.I., and Shefov, N.N. *Geomag. Aeronom.* **36**, 68 (1996).
A model study of the temporal behaviour of the emission intensity and rotational temperature of the OH Meinel bands for high-latitude summer conditions. Lopez Gonzalez, M.J., Murtagh, D.P., Espy, P.J., Lopez Moreno, J.J., Rodrigo, R., and Witt, G. *Ann. Geophys.* **14**, 59 (1996).
Rocket-borne measurements of horizontal structure in the OH(8,3) and Na-D airglow emissions. Clemesha, B.R., and Takahashi, H. *Advan. Space Res.* **17**, 81 (1995).
Rocket measurements of the altitude distributions of the hydroxyl airglow. Baker D.J. and Stair, A.T., Jr. *Phys. Scri.* **37#**, 611 (1988).
Seasonal variability of the OH Meinel bands. Le Texier, H., Solomon, S., and Garcia, R.R. *Planet. Space Sci.* **35**, 977 (1987).
Deactivation of vibrational states of hydroxyl molecules with atomic and molecular oxygen in the mesopause area. Perminov, V.I., Semenov, A.I., and Shefov, N.N. *Geomag. Aeronom.* **38**, 100 (1998).
Energy transfer in the ground state of OH: Measurements of OH(v = 8,10,11) removal. Dyer, M.J., Knutsen, K., and Copeland, R.A. *J. Chem. Phys.* **107**, 7809 (1997).
Kinetic parameters for OH nightglow modeling consistent with recent laboratory measurements. Adler-Golden, S. *J. geophys. Res.* **102**, 19969 (1997).
Rotational relaxation of high-N states of OH ($X^2\Pi$, v = 1–3) by O_2. Holtzclaw, K.W., Upschulte, B.L., Caledonia, G.E., Cronin, J.F., Green, B.D., Lipson, S.J., Blumberg, W.A.M., and Dodd, J.A. *J. geophys. Res.* **102**, 4521 (1997).
Kinetics of excitation of mesopause hydroxyl emission and the role of oscillating relaxation upper vibrational levels. Grigoreva, V.M., Gershenzon, Y.M., Semenov, A.I., Umanskii, S.Y., Shalashilin, D.V., and Shefov, N.N. *Geomag. Aeronom.* **37**, 136 (1997).
Collisional removal of OH($X^2\Pi$, v = 7) by O_2, N_2, CO_2, and N_2O. Knutsen, K., Dyer, M.J., and Copeland, R.A. *J. chem. Phys.* **104**, 5798 (1996).
Kinetic parameters related to sources and sinks of vibrationally excited OH in the nightglow. McDade, I.C. and Llewellyn, E.J. *J. geophys. Res.* **92**, 7643 (1987).

8 Extraterrestrial atmospheres

8.1 Introduction

Our knowledge about the bodies in the solar system and their atmospheres has expanded phenomenally in the last few decades. Increasing sophistication of Earth-based investigations has allowed detailed examination of atmospheres from afar. But the coming of the 'space age' has provided a wealth of information that cannot be obtained at all from Earth's surface. Instruments can be borne aloft beyond our own atmosphere; 'fly-bys', orbiters, and landers have reached the atmospheres of distant planets to give the result of *in-situ* measurements. Table 8.1 shows some of the more important space missions that have included atmospheric observations. This chapter presents some features of atmospheric chemistry as it now appears in the light of recent research. The solar system bodies under consideration include Saturn's satellite Titan, which has a massive atmosphere, as well as the planets. The atmospheres of these bodies seem to divide into two classes: those of the Inner

Table 8.1 Some important earlier and proposed missions

Date		Name	Objective
July–Sept	1976	Viking 1,2	Mars entry science and lander
Dec	1978	Pioneer Venus	Orbiter and 'Bus' carrying entry probes to Venus
Dec	1978	Venera 11, 12	Soviet probes to Venus
March	1979	Voyager 1	Closest approach to Jupiter
July	1979	Voyager 2	Closest approach to Jupiter
Sept	1979	Pioneer 11	First ever fly-by of Saturn
Nov	1980	Voyager 1	Closest approach to Saturn
Aug	1981	Voyager 2	Closest approach to Saturn
March	1982	Venera 13, 14	Soviet sounders to Venus
July	1985	'VEGA'	Venus fly-by and on to
March	1986	Venera 15, 16	Comet Halley (USSR/France)
March	1986	Giotto	Comet Halley mission (ESA)
Jan	1986	Voyager 2	First ever fly-by of Uranus
Aug	1989	Voyager 2	First ever fly-by of Neptune and Triton
–	1991	Magellan	Venus mapping
Dec	1995	Galileo	Jupiter, Io long-term study
–	1997	Pathfinder	Mars orbiter, 'Sojourner' surface rover
Dec	1997	Galileo Europa	Europa 'ice' and ionosphere
–	1999	NEAR	Near Earth Asteroid (433 Eros) Rendezvous
Sept	1999	Mars Climate	Volatiles and climate history (lost on Mars entry)
–	2003	Rosetta	Comet orbiter and lander (ESA)
July	2004	Cassini	Saturn orbiter, Titan probe 'Huygens' (ESA)
–	2004	Stardust	Cometary dust and volatiles
April	2006	Champollion	Comet rendezvous and lander

Table 8.2 Composition of the atmosphere of Venus[a]

Species	Mole fraction	Remarks
CO_2	96.5	
N_2	3.5	
H_2O	2×10^{-5}	surface
	6×10^{-5}	22 km
	1.5×10^{-4}	42 km
Ar	7×10^{-5}	
CO	1.7×10^{-6}	surface
	3.0×10^{-5}	42 km
	4.5×10^{-5}	cloud top
	1×10^{-3}	100 km
O_2	$<3 \times 10^{-7}$	cloud top
He	12×10^{-6}	
Ne	7×10^{-6}	
HCl	6×10^{-7}	cloud top
COS	3×10^{-7}	
SO_2	$5\text{–}100 \times 10^{-9}$	
Kr	2.5×10^{-8}	
SO	2×10^{-8}	
HF	5×10^{-9}	
Xe	1.9×10^{-9}	

[a]From Yung, Y.L. and DeMore, W.B. *Photochemistry of planetary atmospheres*, Oxford University Press, Oxford, 1999.

and the Outer planets. Any gases left over when the inner planets (Venus, Earth, Mars) were formed were lost, and their present atmospheres have been acquired by outgassing from the solid planet followed by chemical modification and evolution (see Chapter 9). Carbon dioxide characterizes the atmospheres of these planets, although Earth is a special case because life has converted CO_2 to O_2 (see Chapters 1 and 9). Outer planets (Jupiter, Saturn, Uranus, Neptune) have retained an atmosphere similar to that with which they were formed because the large escape velocities (cf. Section 2.3.2 and Table 2.1) inhibit loss. Titan's peculiar atmosphere (Section 8.5.1) is probably a consequence of the low escape velocity (Table 2.1) allowing the atmosphere to have evolved more than that of the parent planet, Saturn.

8.2 Venus

8.2.1 Atmospheric composition

Venus is the nearest planet to the Sun to possess an atmosphere, which is about a hundred times as massive as that of the Earth (cf. Table 1.1). Table 8.2 presents data for the composition of the Venusian atmosphere in more detail than that provided in Table 1.1. Carbon dioxide is the major constituent (96.5 per cent) with ~3.5 per cent of N_2 as the next most abundant species. Incoming solar energies (~2600 W m^{-2}) are nearly twice as great as on Earth, and radiation trapping by the CO_2 leads to high surface temperatures (~730 K: cf. Section 2.2). Optical observations of Venus from a distance are hindered by an unbroken layer of cloud, whose top extends to a height of about 70 km. Our knowledge of the planet has been greatly extended by the Mariner (USA: 1962–74) and Venera 4–16 (USSR: 1967–1986) series of space probes. Our perception of Venus underwent a revolution in December 1978 when several spacecraft reached the planet. Five probes belonged to the Pioneer-Venus (USA) multiprobe mission. Four of the craft reached the surface, two on the night-side and two on the day-side, while the fifth craft was the orbiter, which continues to circle the planet. In the same month, two Soviet spacecraft (Venera 11, 12) reached Venus, and each landed a probe on the surface. The instrumentation included mass spectrometers (for ionic and for neutral species), gas chromatographs, particle-size spectrometers, UV spectrometers, nephelometers (cloud-density investigations), X-ray spectrometers, thunderstorm detectors, and so on. Seven gas analysers encountered the atmosphere and obtained data on the chemical composition from about 700 km altitude down to the surface. Studies of Venus were continued with the Soviet Venera 13 and 14 lander probes and the Venera 15 and 16 orbiters. Some problems of identification remain. Relatively low-resolution mass spectral peaks could have more than one origin: for example, $m/e = 64$ could be due to SO_2 or S_2. Retention times on chromatographic columns are also subject to misinterpretation. Contamination of mass spectral data, especially by H_2O, is always a difficulty, and one Pioneer-Venus analyser had the problem compounded because the inlet used for sampling became blocked by a cloud droplet between 50 and 28 km during the probe descent. Nevertheless, the results of the different experiments are reasonably consistent, and form the basis of the Venus entries in Tables 1.1 and 8.2 (to which reference should be made in subsequent discussion of atmospheric concentrations). The existence of COS (carbonyl sulphide), originally reported to be a major sulphur compound in the lowest part of the atmosphere, is doubtful there, and its mixing ratio at higher altitudes is a fraction of a part per million. A serious discrepancy in the data exists for the water vapour abundance, with mixing ratios from 2×10^{-5} (optical spectroscopy) to 5×10^{-3} (gas chromatography) being suggested by different experiments. One of the most significant findings from the Pioneer-Venus and Venera missions concerns the absolute and relative isotopic abundances of the inert gases; we shall discuss this topic in connection with the evolution of planetary atmospheres (Section 9.2.1).

Chlorine-containing compounds are of potential importance in Venusian photochemistry, as will appear later. Hydrogen chloride has been identified (as has HF) by Earth-based infrared spectrometry. These studies only give concentrations above the Venusian cloud tops (> 70 km), of course; for HCl, the mixing ratio is $4.2 \pm 0.7 \times 10^{-7}$. Claims have been made for the identification of molecular chlorine in the Venera 11, 12 optical spectroscopy results (mixing ratio $\sim 10^{-7}$). Pioneer-Venus data also support the presence of Cl_2. Ultraviolet absorption was generally ascribed to sulphur species prior to the Pioneer-Venus mission. Models of transparent H_2SO_4 clouds on top of the lower S_8 sulphur clouds appeared consistent with UV reflectance measurements. However, constraints placed on cloud models by the Pioneer-Venus observations seem to rule out sulphur as the UV absorber. Instead, SO_2 appears to be the major source of opacity at short wavelengths, but at $\lambda > 320$ nm a second absorber is required. A tenable candidate is Cl_2 if it is present at concentrations of about 1 p.p.m.

8.2.2 Clouds

The clouds provide, from the chemist's point of view, a convenient division between 'lower' and 'upper' atmospheres of Venus. As we discussed in Sections 2.3.1 and 2.3.2, the thermal structure of the Venusian atmosphere allows formal identification of troposphere, stratosphere, mesosphere, and thermosphere. Very roughly, the region below the clouds corresponds to the troposphere, while the stratosphere extends from the cloud tops to ~110 km. The clouds also absorb or reflect almost all solar ultraviolet radiation. Photochemical change is thus confined to that part of the atmosphere above the clouds, while below the clouds virtually all chemical change is thermal in nature.

Cloud structure on Venus is quite complex, with three major layers, centred on 51, 54, and 62 km, being identified. In addition, there appear to be lower pre-cloud and haze regions. Three kinds of cloud particles are present. 'Mode 1' are small, ubiquitous aerosols, which extend below the main layers to within 31 km of the surface. 'Mode 2' particles are ~2.0 μm in diameter, and are seen in all three cloud layers. The refractive index of these particles is 1.44, and they are almost certainly concentrated sulphuric acid (75 per cent H_2SO_4 + 25 per cent H_2O gives the required refractive index). 'Mode 3' particles are larger solid crystals of needle, plate, or dendritic shape, restricted to the middle and lower cloud layers. Sulphuric acid is thus not the only constituent of the Venusian clouds, and, in fact, solid mode 3 particles are about ten times more abundant (mass loading) than mode 2 in the lower two layers.

X-ray fluorescence experiments on Venera 15 and 16 ('Vega': see Section 8.7) suggest that metal chlorides are important components of the clouds, with Fe_2Cl_6 constituting one per cent of the column mass loading above 47 km and Al_2Cl_6 being the dominant chlorine-bearing species in the middle cloud layer.

Phosphorus compounds also seem to be present in the cloud layers (Venera 15, 16), with P_4O_6 as the main phosphorus-bearing gas. Reaction of P_4O_6 with H_2SO_4 droplets converts the latter into phosphoric acid (H_3PO_4)

droplets with release of sulphur dioxide. The H_3PO_4 droplets lose water when they descend to about 25 km, and evaporation of P_4O_{10} should occur. One interpretation of the Venera data thus now suggests a cloud composition of H_2SO_4 (altitude 52–62 km, mass loading 5 mg m^{-3}), Fe_2Cl_6 (47–52 km, 0.2 mg m^{-3}), H_3PO_4 ($\leq$ 52 km, 5 mg m^{-3}), Al_2Cl_6 (53–58 km, 3 mg m^{-3}), and S_8 (52–62 km, total sulphur polymers ca. 0.1 of H_2SO_4 by mass). Further analysis of the data is needed to see how far the X-ray fluorescence results are supported by the mass-spectrometric and gas-chromatographic data obtained in the same mission.

8.2.3 Lightning

One of the more unusual experiments on the Venera 11, 12 probes was the 'Groza' ('Thunderstorm') search for low-frequency electromagnetic radiation pulses (*sferics*) which must be emitted by lightning discharges if they are present in the atmosphere of Venus. On both probes, groups of pulses were discovered whose character was like terrestrial thunderstorm radiation. Additional support for Venusian thunderstorms comes from a plasma wave detector on the Pioneer-Venus orbiter. Impulsive signals, of a kind not normally encountered, are occasionally seen when the orbiter penetrates the night-time ionosphere, and these impulses are consistent with an origin from lightning. Re-examination of optical spectrometer data from the earlier Venera 9, 10 missions showed that lightning had been seen, and that an extensive thunderstorm had been encountered with a flashing frequency of 100 flashes per second. Ultraviolet and blue emissions were observed, pointing to the clouds rather than the surface as the location of the lightning, because the observed radiation could not have penetrated the lower atmosphere. Estimates for the mean rate of energy dissipation by lightning in the Venusian atmosphere range from 4×10^{-8} J cm^{-2} s^{-1} to as much as fifty times this value.

Lightning discharges may affect cloud chemistry by pyrolysing CO_2 to provide a local source of CO, far removed from the higher altitude CO_2 photolysis region (Section 8.2.5). Possibly more important, NO may be produced. Laboratory investigations of simulated lightning discharges in an oxygen-deficient mixture of CO_2 (95 per cent) and N_2 (5 per cent), representing the composition of the Venusian atmosphere, show that around 4×10^{15} molecule J^{-1} of NO are formed. Nearly 10^{12} molecule cm^{-2} s^{-1} of NO could thus be generated in the Venusian clouds.

Oxides of nitrogen may play an interesting role in the oxidation of SO_2 to SO_3 or H_2SO_4 present in the clouds, and may even be incorporated into the 'mode 3' solid aerosol. Commercial production of sulphuric acid was, until the early years of this century, carried out by the 'Lead Chamber' process in which nitrogen oxides were carrier catalysts for the oxidation of SO_2. The process may be summarized by the reaction sequence

$$NO + NO_2 + 2H_2SO_4 \rightarrow 2NOHSO_4 + H_2O \quad (8.1)$$

$$2NOHSO_4 + SO_2 + 2H_2O \rightarrow 3H_2SO_4 + 2NO \quad (8.2)$$

$$\text{Net} \quad NO_2 + SO_2 + H_2O \rightarrow H_2SO_4 + NO$$

Nitrosylsulphuric acid ($NOHSO_4$) is an essential part of the scheme (although the detailed mechanism may involve formation of HNO_2 and oxidation of H_2SO_3 in the aqueous phase). Venusian oxidation of SO_2 could follow the same route if sufficient NO_x is present. A further observation from Lead Chamber operation is especially interesting in this context. If not enough H_2O is used, the process can be interrupted before reaction (8.2) occurs, and solid $NOHSO_4$ occasionally forms midair in the shape of feathery crystals. As a conjecture, then, the mode 3 cloud particles may be $NOHSO_4$. Solid $NOHSO_4$ melts at 347 K, the temperature near the border of the middle and lower cloud levels, where the number of cloud particles rapidly decreases. A dilute solution of $NOHSO_4$ (1.4 per cent) in H_2SO_4 (85 per cent) shows visible and UV absorption compatible with the Venusian spectrophotometric observations (cf. Section 8.2.1). Whether or not $NOHSO_4$ is present on Venus, or plays a part in SO_2 oxidation, will depend on NO_x concentrations. Mixing ratios in the parts per million range seem needed, which demand mean NO_x lifetimes of at least thousands of years if lightning is the source of NO_x. Unfortunately, neither gas-chromatographic nor mass-spectrometric data from the recent *in-situ* measurements provide evidence to resolve the question.

8.2.4 Sub-cloud chemistry

Chemistry in the region below the clouds seems to be dominated by the thermal reactions of sulphur- and carbon-containing species. In view of the high temperature and pressure, a starting point in geochemical concepts of the troposphere has been the assumption of complete chemical equilibrium. The surface is often considered to promote rapid equilibration by catalytic acceleration of chemical processes. The equilibrium treatment has to be adopted as a first approximation, because kinetic data are not available for the individual elementary reactions. Rates may, however, be low enough that vertical transport and mixing dominate over chemical reactions in the whole lower atmosphere. Atmospheric composition should then be approximately constant throughout the sub-cloud region, and can be determined from thermochemical equilibrium calculations for conditions near the surface.

The *in-situ* measurements cast some doubt on both the constancy of composition and the equilibrium assumption itself. Mixing ratios of COS calculated from the equilibrium constant for the reaction

$$3CO + SO_2 \rightleftharpoons 2CO_2 + COS \quad (8.3)$$

are ~10^{-4} near the planet's surface, and reach 10^{-2} by an altitude of 10 km, taking the values of SO_2 and CO concentration given in Table 8.2. Pioneer-Venus measurements set the limit of detectability at a mixing ratio of 10^{-5} and even Venera 13, 14 data do not exceed 4×10^{-5}; the current best estimate is as low as 3×10^{-7}. Thus chemical equilibrium does not seem to be attained for COS. For S_2 formation, the overall equilibrium can be written

$$4CO + 2SO_2 \rightleftharpoons 4CO_2 + S_2. \tag{8.4}$$

Thermochemical data suggest an S_2 mixing ratio of 1.3×10^{-6} using the concentrations of Table 8.2, but, because of the $[CO]^4$ dependence, a value of 1.8×10^{-7} is obtained for a slightly smaller CO mixing ratio of 1.7×10^{-5} (measured by Venera 12 at 12 km). Venera orbiter spectrometer results suggest a surface mixing ratio for S_2 of 1.6×10^{-7}, which would be consistent with the lower [CO] value. Thus S_2 concentrations may support the concept of tropospheric chemical equilibrium at the surface. Above the surface, it seems that kinetic constraints must exist. Ascending and descending gas fluxes in the vertical atmospheric motion may possibly have differing compositions, with the ascending gases containing components of crustal origin (e.g. COS, S_2, H_2S, HCl, etc.) and the descending gases containing products of photolysis at the cloud tops (SO_3, Cl_2, O_2, etc.). In this case, the atmospheric composition should be regarded as reflecting a steady-state condition rather than a thermodynamic equilibrium one. At the surface, chemical equilibrium may be achieved among all gaseous species except CO (for which the measured mixing ratio is nearly constant with altitude). This view then allows CO and O_2 to be out of equilibrium with each other and with the COS/SO_2 system in the free troposphere. Surface equilibration demands a substantial amount of sulphur trioxide (SO_3) to be present. The data of Table 8.2 require there to be almost ten times as much SO_3 as SO_2, but none of the *in-situ* experiments were sensitive to SO_3, so that its presence remains hypothetical.

Elementary reaction steps can be written that explain the interconversions between the different species. Most schemes start by production of SO and thence atomic sulphur

$$SO_2 + CO \rightarrow SO + CO_2 \tag{8.5}$$

$$SO + CO \rightarrow S + CO_2 \tag{8.6}$$

$$SO + SO_2 \rightarrow S + SO_3. \tag{8.7}$$

Gas-phase S_2 may be formed *via* COS as an intermediate

$$CO + S + M \rightarrow COS + M \tag{8.8}$$

$$S + COS \rightarrow S_2 + CO \tag{8.9}$$

$$SO + COS \rightarrow S_2 + CO_2. \tag{8.10}$$

Concentrations of species such as COS and CO are determined by the competition between production and loss processes. Observational data for COS from the Pioneer-Venus mission were confined to an upper limit of a few p.p.m. Preliminary analysis of Venera 13, 14 (1982) results has suggested slightly higher concentrations (40 ± 20 p.p.m.) together with some H_2S (80 ± 40 p.p.m.) in the altitude range 27 to 37 km, but these data are still subject to considerable refinement. It is, however, clear that the reduced gases COS and H_2S are about an order of magnitude less abundant than SO_2. Low concentrations of CO also suggest that reactions consuming the gas are efficient relative to those forming it below the clouds. Spectroscopic data from the Venera probes give strong circumstantial evidence for the presence of S_2, S_3, and higher sulphur oligomers, S_x. Reactions involving the transfer of S from COS, such as

$$S_2 + COS \rightarrow CO + S_3, \tag{8.11}$$

and similar processes involving $S_3 \ldots S_x$, could conceivably have quite large rate coefficients, but they are unlikely to be important in the Venusian troposphere because of the small concentrations of COS. Sequential addition of the type

$$S_n + S_m (+ M) \rightarrow S_{n+m} (+ M) \tag{8.12}$$

is an alternative source of S_x ($x = n + m$), with the stable S_8 probably being the terminal species. Photodissociation of polysulphur may be the process that drives Venusian sulphur chemistry out of thermal equilibrium, and that initiates the reversal of the reducing steps (8.5) to (8.10). Because the $S_{(x-1)}$–S bond strength is relatively weak for $x > 3$, photodissociation thresholds for the steps

$$S_x + h\nu \rightarrow S(^*) + S_{(x-1)} \tag{8.13}$$

can be displaced to longer wavelengths compared to the S_2 limit (285 nm) and into the spectral region where some radiation penetrates the clouds. Excited S atoms may be a product of the dissociation, so that hydrogen-abstraction reactions such as

$$S^* + H_2O \rightarrow SH + OH \tag{8.14}$$

$$S^* + HCl \rightarrow SH + Cl \tag{8.15}$$

could become thermochemically possible. Radicals like SH and OH initiate further chemistry, with SH playing a part in H_2S and sulphane (H_yS_z) formation. Several cyclic processes can be written that utilize the photolysis of S_3 below the cloud tops. A typical sequence is

Cycle 1:

$$S_3 + h\nu \rightarrow S_2 + S^* \tag{8.16}$$

$$S^* + H_2O \rightarrow SH + OH \tag{8.14}$$

$$OH + CO \rightarrow CO_2 + H \tag{8.17}$$

$$H + SH \rightarrow H_2 + S \tag{8.18}$$

$$S + S_2 + M \rightarrow S_3 + M \tag{8.19}$$

Net $\quad CO + H_2O + h\nu \rightarrow CO_2 + H_2$

Such a reaction sequence may be a source of H_2 and a sink of H_2O in the lower atmosphere that could balance the reverse net reaction occurring above the clouds. Since the concentrations of the S_x precursors of S* are temperature-dependent, the role of these molecules, especially S_3 ('thiozone'), may bear a remarkable analogy to the role of ozone in the Earth's upper atmosphere.

8.2.5 Stratospheric chemistry

Three inter-related questions about the chemistry of the Venusian stratosphere deserve particular attention. First, what controls the extent of CO_2 photolysis, and the abundances of CO and O_2? Secondly, what part does the SO_2 present above the cloud tops play in stratospheric chemistry? And, thirdly, what is the abundance of H_2 in the bulk atmosphere?

Photolysis of CO_2 occurs readily above the cloud tops

$$CO_2 + h\nu\,(\lambda \le 204\ \text{nm}) \rightarrow CO + O, \tag{8.20}$$

but the recombination reaction

$$O + CO + M \rightarrow CO_2 + M. \tag{8.21}$$

is spin forbidden and slow. Atomic oxygen is lost mainly in the direct and indirect (see later) formation of O_2, which we represent by the equation

$$O + O + M \rightarrow O_2 + M \tag{8.22}$$

Measurements of the Infrared Atmospheric Band airglow indicate that a substantial fraction of CO_2 photolysis leads to O_2 production, as discussed in Section 7.4.1. Without some oxidation step for CO, the present-day abundance (~45 p.p.m. at the cloud tops) could be produced photochemically in about 200 years, and the stratospheric O_2 concentration (~1 p.p.m.) would be formed in a few years. The stoicheiometry implied by reactions (8.20) and (8.22) would predict $[CO]/[O_2] = 2$, whereas the measured ratio is at least 45. Some additional reactions are therefore demanded that will oxidize CO back to CO_2

and that will break O–O bonds. Catalytic cycles, such as those important in terrestrial chemistry (Sections 4.4.1 and 5.3) are likely to be involved.

The simplest possible oxidation of CO starts with the reaction (Section 5.3.5) with OH

$$CO + OH \rightarrow CO_2 + H. \tag{8.17}$$

We shall see (Section 8.3.2) that a cycle based on this reaction is very important on Mars. However, the cycle cannot account for the efficient removal of O_2 on Venus, because there is insufficient H_2O to yield the necessary concentrations of HO_x radicals.

Considerable interest has been focused on the possible roles of ClO_x and SO_x species in catalytic cycles since the discovery of HCl and SO_2 in the upper atmosphere, and chlorine-containing compounds in the lower atmosphere (pp. 563–4). In the Earth's atmosphere, ClO_x only catalyses the conversion of odd oxygen to O_2 (Section 4.4.1). On Venus, however, a dual role is possible, with two sequences leading to recombination of O with CO. Three-body formation of chloroformyl radicals (ClCO) and then of peroxychloroformyl radicals ($ClCO_3$) is involved in both cycles

Cycle 2:

$$Cl + CO + M \rightarrow ClCO + M \tag{8.23}$$

$$ClCO + O_2 + M \rightarrow ClCO_3 + M \tag{8.24}$$

$$ClCO_3 + O \rightarrow Cl + CO_2 + O_2 \tag{8.25}$$

Net $$CO + O \rightarrow CO_2$$

Cycle 3:

$$Cl + CO + M \rightarrow ClCO + M \tag{8.23}$$

$$ClCO + O_2 + M \rightarrow ClCO_3 + M \tag{8.24}$$

$$ClCO_3 + Cl \rightarrow Cl + CO_2 + ClO \tag{8.26}$$

$$O + ClO \rightarrow O_2 + Cl \tag{8.27}$$

Net $$CO + O \rightarrow CO_2.$$

We note that these cycles convert CO back to CO_2, and consume atomic oxygen, but they do not break the O–O bond and do not, therefore, lead to the destruction of O_2. Sulphur-containing species seem capable of catalysing the (photo)dissociation of O_2. Here, the initial step is photolysis of SO_2

$$SO_2 + h\nu\,(\lambda \leq 219\text{ nm}) \rightarrow SO + O, \tag{8.28}$$

followed by atomic-sulphur formation from SO by a photochemical and a thermal step

$$SO + h\nu(\lambda \leq 235 \text{ nm}) \rightarrow S + O \tag{8.29}$$

$$SO + SO \rightarrow S + SO_2. \tag{8.30}$$

Atomic sulphur is then responsible for the fission of the O–O bond

$$S + O_2 \rightarrow SO + O. \tag{8.31}$$

An important cycle has recently been identified following the discovery of a reaction between ClO and SO. The cycle thus links ClO_x and SO_x species

Cycle 4:

$$SO_2 + h\nu(\lambda \leq 219 \text{ nm}) \rightarrow SO + O \tag{8.28}$$

$$Cl + CO + M \rightarrow ClCO + M \tag{8.23}$$

$$ClCO + O_2 + M \rightarrow ClCO_3 + M \tag{8.24}$$

$$ClCO_3 + Cl \rightarrow Cl + CO_2 + ClO \tag{8.26}$$

$$ClO + SO \rightarrow Cl + SO_2 \tag{8.32}$$

Net $\quad CO + O_2 + h\nu \rightarrow CO_2 + O$

A key feature of the scheme is the coupling of reactions (8.28) and (8.32), which has the effect of using SO_2 photosensitization to break a Cl–O bond. Photolysis of SO_2 thus synergizes the ClO_x catalysis. The mechanism bears an interesting analogy with the tropospheric oxidation cycle on Earth (Section 5.3)

Cycle 5:

$$CO + OH \rightarrow CO_2 + H \tag{8.17}$$

$$H + O_2 + M \rightarrow HO_2 + M \tag{8.33}$$

$$HO_2 + NO \rightarrow NO_2 + OH \tag{8.34}$$

$$NO_2 + h\nu \rightarrow NO + O \tag{8.35}$$

Net $\quad CO + O_2 + h\nu \rightarrow CO_2 + O$

with SO_2 playing the part of NO_2, and ClO_x playing the part of HO_x.

Cycle 4 may, indeed, be supplemented by cycles such as 5 on Venus if sufficient NO_x is present in the atmosphere. Although some NO is undoubtedly present above the clouds, because NO emission bands are seen in the ultraviolet airglow, it is not clear whether the lightning source (Section 8.2.3) in reality supplies enough NO to make cycle 5 important. Similar arguments apply to HO_x catalytic schemes that demand high concentrations of H_2. Mass spectrometric measurements of H_2 mixing ratios are the subject of some controversy, since the $m/e = 2$ ion may be either H_2^+ or D^+ (and the $m/e = 19$ ion can likewise be either H_3O^+ or HDO^+). Indeed, low H_2 favours the role of ClO_x chemistry on Venus. The terrestrial sink reaction (Sections 4.4.2 and 4.4.3),

$$Cl + CH_4 \rightarrow HCl + CH_3, \tag{8.36}$$

is replaced on Venus by the analogous process

$$Cl + H_2 \rightarrow HCl + H, \tag{8.37}$$

and excess H_2 will suppress catalytic cycles involving ClO_x. Important processes reverse the formation of HCl, of which the dominant one is photochemical

$$HCl + h\nu \rightarrow H + Cl. \tag{8.38}$$

In addition, thermal processes destroy HCl

$$H + HCl \rightarrow H_2 + Cl \tag{8.39}$$

$$O + HCl \rightarrow OH + Cl \tag{8.40}$$

$$OH + HCl \rightarrow H_2O + Cl, \tag{8.41}$$

with the consequent formation of H_2 and H_2O. Molecular and atomic hydrogen are transferred to the Venusian exosphere, so that the rate of escape from the atmosphere is determined by the rate of HCl photolysis. Water formation in reaction (8.41) is a permanent sink in the Venusian stratosphere, since H_2O photolysis is prevented by the absorption of short-wave radiation, and H_2O is scavenged by SO_3 to form H_2SO_4 (see p. 574). The net chemical change in the stratosphere is thus photochemical conversion of HCl and CO_2 to form H_2O, CO, and Cl_2, and the build-up and downward flux of free chlorine is the only way to conserve the atmospheric oxidation state. Reversal of photochemical change by thermochemical equilibrium chemistry must complete the cycle in the lower atmosphere. Imperfect balance in the cycle leads to a permanent and irreversible leakage of hydrogen by escape (cf. Section 9.3).

Figures 8.1 and 8.2 summarize the results of a one-dimensional model calculation based on the scheme of cycle 4, with no source of H_2 from below the clouds. Note that the mixing ratio for H_2 reaches 10^{-7} by 100 km, even though it is 10^{-13} at the lower boundary. As can be seen from Fig. 8.1, the model matches the measured concentration and apparent scale height for SO_2 up to ~70 km, and the CO measurements up to ~100 km. The limit for O_2 concentration (derived from Earth-based optical measurements) is also consistent with the model for heights less than ~70 km. Figure 8.2 shows that by far the most abundant intermediate species in the lower stratosphere is Cl, with O dominating only above about 90 km. One particularly significant conclusion of the model is that rates of production and loss of CO are about

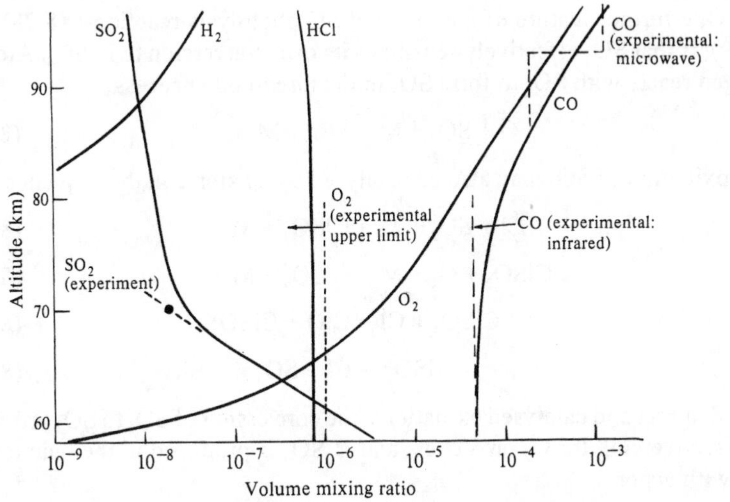

Fig. 8.1. Venusian abundances of bulk gases predicted by a model based on extended chlorine chemistry. Some experimental data are provided for comparison: the upper limit for O_2 is derived from measurements from Earth. Data of Yung, Y.L. and DeMore, W.B. *Icarus* **51**, 199 (1982).

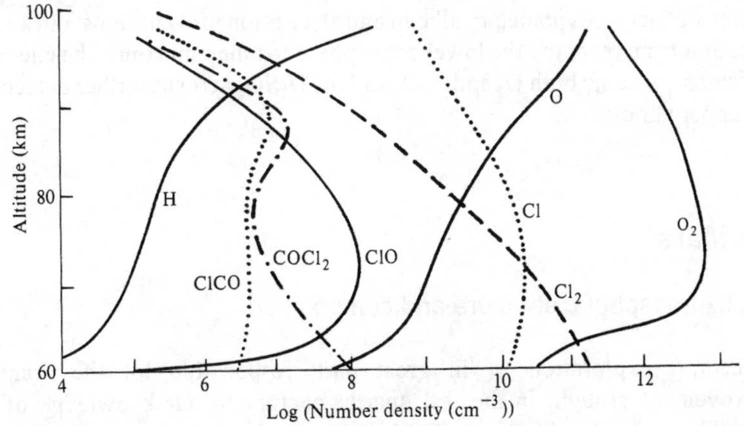

Fig. 8.2. Concentrations of minor and intermediate species derived by the model calculations of the Venusian atmosphere. Data from same source as used for Fig. 8.1.

equal at ~80 km (i.e. there is a photochemical stationary state there). Without such a steady state, it is difficult to explain observations that mesospheric CO concentrations drop at night. In other models, that adopt 'high' H_2 and NO concentrations, the bulk of CO is destroyed around 70 km, and there is no significant chemical sink in the mesosphere between 80 and 90 km; there should then be no diurnal variation in CO.

One further feature of the SO_2 and SO photolysis reactions, (8.28) and (8.29), is that SO_2 effectively sensitizes its own conversion to H_2SO_4. Atomic oxygen reacts with SO_2 to form SO_3 in the three-body process

$$O + SO_2 + M \rightarrow SO_3 + M. \tag{8.42}$$

The oxidation of SO_2 may also be catalysed by Cl atoms in the sequence

$$Cl + SO_2 + M \rightarrow ClSO_2 + M \tag{8.43}$$

$$ClSO_2 + O_2 + M \rightarrow ClSO_4 + M \tag{8.44}$$

$$ClSO_4 + Cl \rightarrow ClO + ClSO_3^* \tag{8.45}$$

$$ClSO_3^* \rightarrow Cl + SO_3. \tag{8.46}$$

In both direct and catalysed oxidations, the conversion of SO_2 to SO_3 provides an effective sink for O_2 on Venus, and H_2SO_4 is produced in the reaction of SO_3 with water

$$SO_3 + H_2O + M \rightarrow H_2SO_4 + M. \tag{8.47}$$

The photochemical aerosols produced by reactions (8.42) to (8.47) constitute the cloud layer that separates troposphere from stratosphere in the Venusian atmosphere. The consequences of cloud formation are profound, because scattering increases planetary albedo and absorption prevents most ultraviolet radiation from reaching the lower atmosphere. At the same time, the chemical reactions consume both O_2 and H_2O, and the H_2SO_4 aerosol further desiccates the upper atmosphere.

8.3 Mars

8.3.1 Atmospheric structure and composition

Spacecraft exploration is, in great part, responsible for the dramatic improvement brought in the last quarter century to our knowledge of the atmosphere of Mars. Starting with Mariners 4 to 9 (1965–72), and Mars 2 to 6 (1971–4), the programme culminated with the spectacularly successful Viking landers and orbiters, which reached Mars in the summer of 1976. Detailed information on the composition and meteorology of the Martian atmosphere came from *in-situ* and remote-sensing measurements taken over more than a Martian year. In contrast to the atmosphere of Venus, the Martian atmosphere is characterized by the variability of its physical parameters with the season, time of day, latitude and longitude, the time in the solar cycle, and with meteorological events such as dust storms and clouds, and other dynamical phenomena.

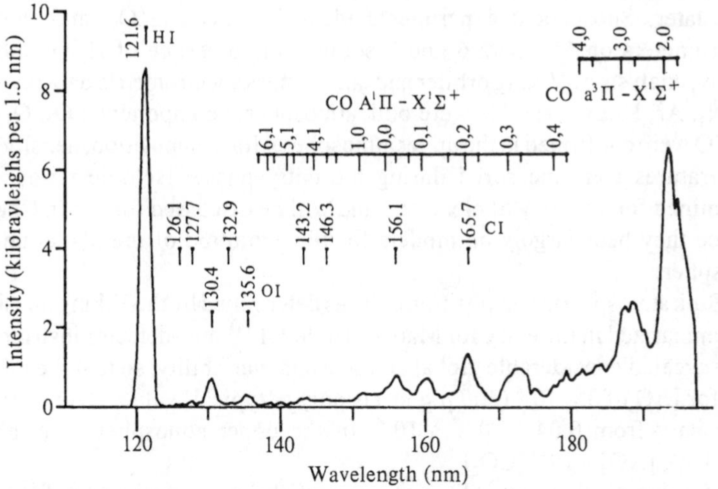

Fig. 8.3. Mars airglow spectrum, 110–190 nm. To obtain this spectrum, 120 individual limb observations, at 1.5 nm resolution, were averaged. Mariner 9 data, from Barth, C.A., Stewart, A.J., Hord, C.W., and Lane, A.L. *Icarus* **17**, 457 (1972).

Mars has a rather tenuous atmosphere composed mainly of CO_2. Exerting a surface pressure of only 6–8 mbar (4.5–6.0 Torr), the atmosphere's total mass is less than one per cent of that of the Earth. Surface temperatures calculated from radiative transfer models (Section 2.2, Table 2.1) are ~217K, in reasonable agreement with direct measurement. During the winter months, a thin sheet of cloud (the *polar hood*, probably composed of ice crystals) gradually spreads from the polar regions to middle latitudes. During the rest of the year, the atmosphere is generally cloud-free, but the surface is sometimes obscured for several weeks by dust raised by winds from the surface. Layered deposits cover the poles, and the area of the polar caps expands greatly during the local winter. Surface temperatures fall as low as 125 K at these periods: condensation from a CO_2 vapour pressure of 6.8 mbar becomes possible at 148 K, and the winter caps are probably a mixture of water-ice and solid CO_2. Residual summer polar caps must consist largely of water-ice, since the cap temperatures are then 200–205 K; the thickness is estimated as between 1 and 1000 m. A permafrost layer covers most of the planet in winter, but, by recession towards the poles in spring and summer, this reservoir releases substantial quantities of water vapour to the atmosphere. Some further aspects of the effects of CO_2 condensation and the small heat capacity of the atmosphere on pressure variability, temperature structure, and circulation are discussed at appropriate points in Chapter 2.

Carbon dioxide was first identified in 1947 as a constituent of the Martian atmosphere by optical spectroscopy from Earth, and H_2O was discovered 16 years later. Subsequent experiments identified CO and O_2, and the UV spectrometers on Mariners 6 and 7 showed the presence of H, O, and O_3. Finally, analysis of Viking orbiter and lander mass spectrometric data revealed that N_2, Ar, Kr, Xe, and Ne were bulk atmospheric components; O_2, O, NO, and CO were confirmed in the upper atmosphere. Ion composition, density, and temperatures were measured during the entry phase. Isotopic ratios were determined for a variety of elements, and will be discussed further in Chapter 9 since they bear largely on models for the evolution of the planet and its atmosphere.

Bulk atmospheric composition data, as determined by the Viking missions, are summarized in the entry for Mars in Table 1.1. Water-detector instruments have revealed considerable global and seasonal variability, so that the mixing ratio for H_2O of 3×10^{-4} (300 p.p.m.) is only a 'typical' value. Ozone mixing ratios vary from 0.04 to 0.2×10^{-6}. In the upper atmosphere (i.e. above ~100 km), [NO] ~ 10^{-4} [CO_2].

Mariner airglow data illustrate one of the most interesting facets of Martian atmospheric chemistry. Part of the ultraviolet spectrum obtained from Mariner 9 is reproduced in Fig. 8.3. Amongst the transitions seen are the resonance lines of H (λ = 121.6 nm), O (λ ~ 130.4 nm), and molecular band systems of CO, while, at longer wavelengths, systems due to CO^+ and CO_2^+ also appear. Apparent emission rates may be plotted as a function of altitude (Fig. 8.4), and the results show large differences in behaviour for the several airglow features. With the exception of H and O lines (and a portion of the CO_2^+ $A^2\Pi \rightarrow X^2\Pi$ transition, not included in our figure), the scale heights are small (~18 km) and identical. With the same exceptions, which are excited by resonance scattering, the emissions are probably excited by the action of solar photons and photoelectrons on CO_2, and thus have the same scale height as CO_2 itself. These two conclusions mean that the upper atmosphere is essentially undissociated and cold. Since atomic oxygen emissions are observed above 250 km altitude, there must be *some* dissociation of CO_2. Part of the CO_2^+ (A → X) emission arises from fluorescent scattering, so that CO_2^+ is a constituent (but not necessarily a major one) of the Martian ionosphere.

The large scale heights, in the upper atmosphere, of oxygen (at mixing ratios of 5 to 10×10^{-3}) and hydrogen (at an almost *constant* concentration of 3×10^4 atom cm^{-3}), indicate clearly that these atomic species are escaping the gravitational field of the planet (Section 2.3.2). The base of the exosphere (*exobase*) lies at ~230 km on Mars. Although temperatures are relatively low (~320 K), thermal escape of hydrogen is possible because of the small value of *g*, and hence escape velocity, on Mars (~5 km s^{-1}: Table 2.1). Escape fluxes of about 1.2×10^8 atom cm^{-2} s^{-1} of hydrogen are predicted for the measured concentration of 3×10^4 atom cm^{-3}. That any hydrogen remains in the exosphere therefore means that there is an equivalent source, presumably

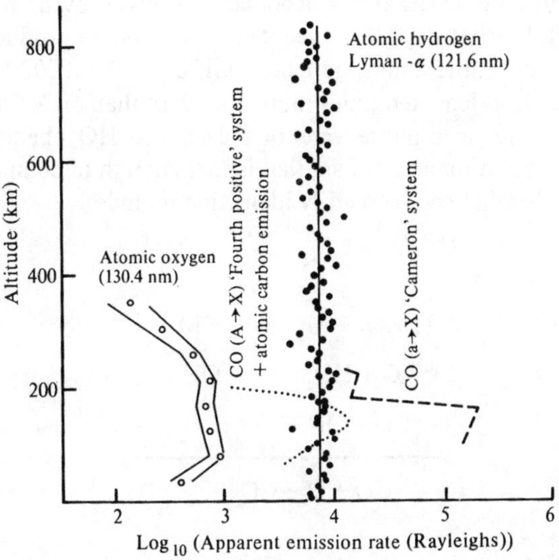

Fig. 8.4. Altitude profiles for dayglow emission features determined from the Mariner 9 data. The line through the HI (Lyman-α) data is a least-squares fit, and the envelope on the OI points indicates error limits. Source as for Fig. 8.3; and McEwan, M.J. and Phillips, L.F. *Chemistry of the atmosphere*, Edward Arnold, London, 1975.

dissociation of a hydrogen-bearing molecule such as H_2O, which is currently operating in the Martian atmosphere. We shall see in Section 8.3.3 that the four H atoms are liberated through the intermediacy of ionic processes. Furthermore, O-atom escape also involves ions, and the rates of oxygen and hydrogen escape processes are self-regulating to be equivalent to loss of H_2O. Before we turn to a consideration of the Martian ionosphere, however, we examine those aspects of the 'neutral' photochemistry that govern the concentrations of minor constituents such as CO and H- and O-bearing molecules.

8.3.2 Carbon dioxide photochemistry

Mars, like Venus, has an atmosphere whose bulk chemistry is dominated by carbon dioxide photolysis

$$CO_2 + h\nu(\lambda \leq 204 \text{ nm}) \rightarrow CO + O. \quad (8.20)$$

As on Venus, it is again necessary to find a catalytic route for the recombination of O and CO, since the direct process

$$O + CO + M \rightarrow CO_2 + M \quad (8.21)$$

cannot possibly maintain CO and O_2 at concentrations as low as those observed (Section 8.2.5). Present-day CO concentrations could be produced in two to three years, and the entire atmosphere be modified in about 2000 years. Water vapour on Mars is at least ten times more abundant than on Venus, and there is evidence for condensed-phase water on the surface. HO_x chemistry can thus provide a recombination mechanism that is fast enough to compete with CO_2 photolysis. Cycles that can be readily identified include

Cycle 6:

$$H + O_2 + M \rightarrow HO_2 + M \tag{8.33}$$

$$O + HO_2 \rightarrow O_2 + OH \tag{8.48}$$

$$CO + OH \rightarrow CO_2 + H \tag{8.17}$$

Net $\quad CO + O \rightarrow CO_2$

Cycle 7:

$$O + O_2 + M \rightarrow O_3 + M \tag{8.49}$$

$$H + O_3 \rightarrow O_2 + OH \tag{8.50}$$

$$CO + OH \rightarrow CO_2 + H \tag{8.17}$$

Net $\quad CO + O \rightarrow CO_2.$

Odd-hydrogen compounds are supplied by photochemical decomposition of H_2O either directly in photolysis,

$$H_2O + h\nu \rightarrow OH + H, \tag{8.51}$$

or by reaction with $O(^1D)$ derived from CO_2, O_2, and O_3 photolysis (cf. Sections 3.2.1, 4.4.3, and 5.3.1)

$$O(^1D) + H_2O \rightarrow OH + OH. \tag{8.52}$$

Water may be re-formed by the reaction

$$OH + HO_2 \rightarrow H_2O + O_2, \tag{8.53}$$

but there is a small net sink for H_2O at low altitudes associated with the formation of molecular hydrogen

$$H + HO_2 \rightarrow H_2 + O_2. \tag{8.54}$$

It is the supply of H_2 from this process that ultimately limits the rate at which hydrogen escapes from the exosphere. Since the reaction is fed by the leak of

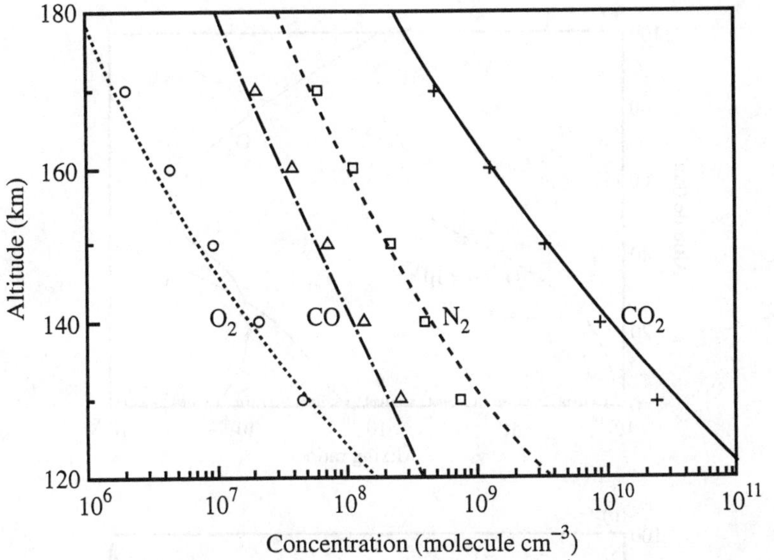

Fig. 8.5. Time-averaged model (lines) and Viking 1 measurements (symbols) of the bulk constituents of the Martian atmosphere. From Yung, Y.L. and DeMore, W.B. *Photochemistry of planetary atmospheres*, Oxford University Press, Oxford, 1999.

H_2O that is permanently destroyed, escape of two H atoms is necessarily accompanied by the escape of one O atom, a point to which we shall return in the next section.

Molecular oxygen is formed mainly in the reaction

$$O + OH \rightarrow O_2 + H, \tag{8.55}$$

the catalytic cycles having no net effect on O_2 concentrations. Ozone is created largely by the O + O_2 combination reaction of cycle 7 (reaction 8.49)), but its destruction is dominated by photolysis, rather than by removal in reaction (8.50). Concentrations of O_3 therefore reflect in a direct fashion the concentrations of O, which are themselves highly dependent on the odd-hydrogen reactions of cycles 6 and 7. Spatial variability of O_3 is very large on Mars (it approaches a factor of 30), and it seems reasonable that this variability should reflect large inhomogeneities in the spatial distribution of odd hydrogen, with [O_3] large where [HO_x] is small. Part of the variation can be ascribed to condensation of H_2O in low-temperature regions, but additional factors such as changing vertical and horizontal transport of O and O_3 are probably also involved.

Atmospheric concentrations of the gases such as CO_2, CO, N_2, and O_2, and reaction intermediates (e.g. H, OH, HO_2, O_3, and O) can be calculated using

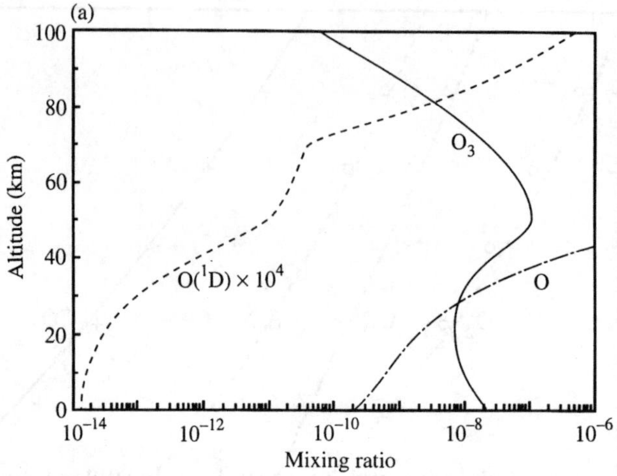

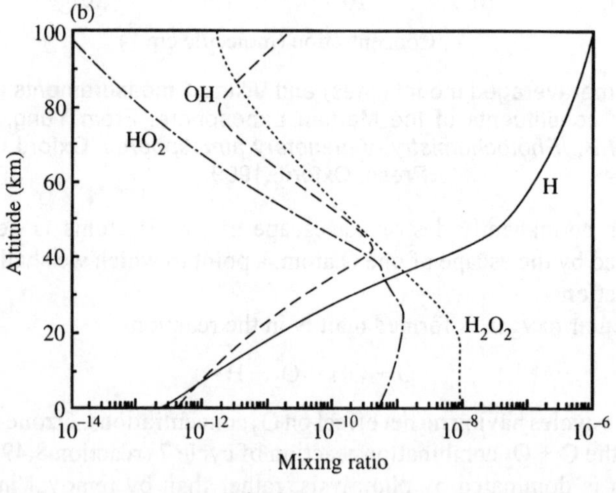

Fig. 8.6. Martian concentration profiles for (a) $O(^3P)$, $O(^1D)$ and O_3; (b) H, OH, HO_2, and H_2O_2 predicted by a one-dimensional model. The averaged model atmosphere used in these calculations is described by Fig. 8.5; the source is the same as for that figure.

appropriate models. Figures 8.5 and 8.6 show the predictions of a one-dimensional model in which over 100 chemical reactions are incorporated. Figure 8.5 shows results for the gases that were measured in the Viking 1 entry, and it is evident that the agreement between experiment and model prediction is excellent. Considerable confidence must thus attach to the predictions for other species represented in Fig. 8.6.

Progress in heterogeneous chemistry associated with interpretations of the Earth's Antarctic ozone hole (Section 4.7.3) has prompted suggestions that heterogeneous processes might also play a part in the chemistry of the Martian atmosphere. The widespread presence of dust means that a molecule in the Martian atmosphere collides, on average, with a silicate dust particle every thousand seconds. Some conversion of CO to CO_2 could thus plausibly occur on the surfaces of dust grains. Similarly, water-ice particles could denitrify the Martian atmosphere in the same way that they denitrify the Earth's polar stratosphere. In that case, large amounts of the oxides of nitrogen might be sequestered on the Martian surface. For the time being, such ideas remain purely speculative, but they do illustrate the growing awareness of the chemical, as well as physical, effects of suspended particles in atmospheres (cf. Sections 1.3, 4.7.4, 4.7.7, and 5.8).

8.3.3 Ionospheric chemistry

The ionosphere of Mars merits special attention because of the part that it may play in determining the bulk composition of the atmosphere. Ionic processes can release sufficient kinetic energy for product translational velocities to exceed the rather small Martian escape velocity ($\sim 5\ \mathrm{km\,s^{-1}}$). It is therefore important to understand what factors control the ion composition and the rates of ionization in the Martian atmosphere.

Figure 8.7 is a concentration–altitude profile obtained by the Viking 1 lander for ions in the Martian atmosphere: it represents the first *in-situ* measurement of ions from the ionosphere of a planet other than our own. Two interesting features of the profile are immediately apparent. First, the dominant ion throughout the atmosphere is O_2^+, and CO_2^+ follows a similar-shaped profile, but is about ten times less abundant. Secondly, the total ion density shows a single peak in concentration at an altitude of about 130 km, in contrast with the multi-layered structure of the Earth's ionosphere (Section 6.1.3). Explanations of both features turn out to be related.

Primary ion formation in the atmosphere of Mars (as in that of Venus) is achieved by photoionization of the main atmospheric component, CO_2

$$CO_2 + h\nu(\lambda \lesssim 93\ \mathrm{nm}) \rightarrow CO_2^+ + e. \tag{8.56}$$

However, the CO_2^+ ion reacts very rapidly with neutral atomic oxygen to yield O and O_2

$$CO_2^+ + O \rightarrow O^+ + CO_2 \tag{8.57}$$

$$CO_2^+ + O \rightarrow O_2^+ + CO \tag{8.58}$$

in comparable amounts. An even faster reaction between O^+ and CO_2

$$O^+ + CO_2 \rightarrow O_2^+ + CO, \tag{8.59}$$

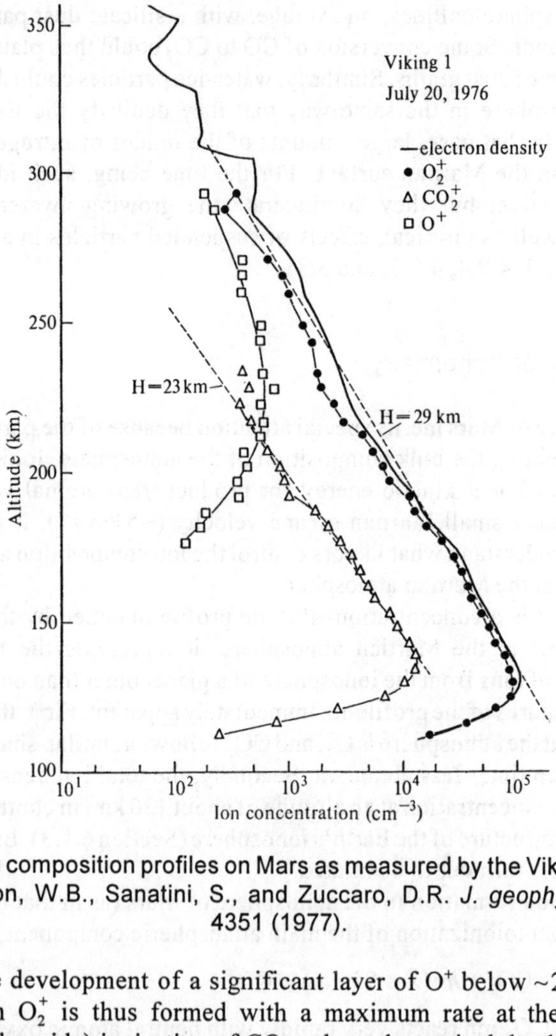

Fig. 8.7. Ion composition profiles on Mars as measured by the Viking 1 lander. From Hanson, W.B., Sanatini, S., and Zuccaro, D.R. *J. geophys. Res.* **82**, 4351 (1977).

prevents the development of a significant layer of O^+ below ~200 km. The terminal ion O_2^+ is thus formed with a maximum rate at the altitude of maximum (day-side) ionization of CO_2, where the optical depth is unity for solar ultraviolet radiation in the approximate wavelength range 20–90 nm. Such behaviour is the analogue of an F_1 mechanism in terms of the Earth's ionosphere (Section 6.4.1), and there is no Martian F_2 peak comparable to Earth's. It will be recalled that the F_2 peak arises because of the bottle-neck in ion–electron recombination caused by the step that forms molecular ions

$$O^+ + N_2 \rightarrow NO^+ + N. \tag{8.60}$$

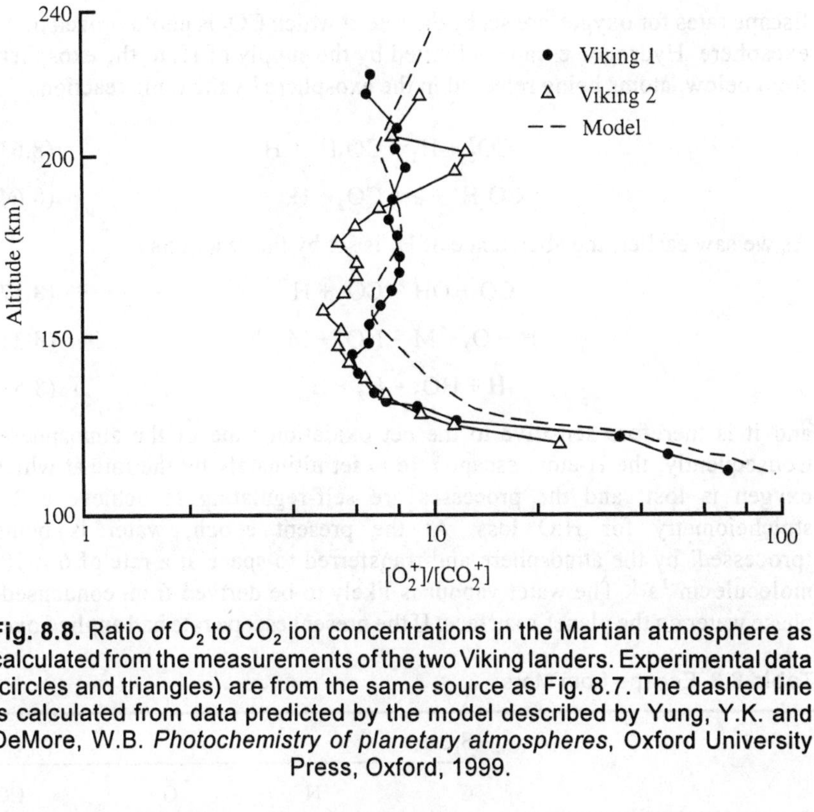

Fig. 8.8. Ratio of O_2 to CO_2 ion concentrations in the Martian atmosphere as calculated from the measurements of the two Viking landers. Experimental data (circles and triangles) are from the same source as Fig. 8.7. The dashed line is calculated from data predicted by the model described by Yung, Y.K. and DeMore, W.B. *Photochemistry of planetary atmospheres*, Oxford University Press, Oxford, 1999.

Curiously, the rate coefficient for reaction (8.59) with CO_2 is 1000 times larger than that for reaction (8.60) with N_2, so that conversion of atomic to molecular ions is not a rate-determining step in the Martian ionosphere at the altitudes of maximum ionization. Calculations of [CO_2^+], [O_2^+] (and [O^+]) profiles match the experiments well, and the ratio [O_2^+]/[CO_2^+] is correctly predicted, as can be seen in Fig. 8.8.

Ion–electron recombination

$$AB^+ + e \rightarrow A + B \tag{8.61}$$

can release, as translation apart of fragments A and B, an energy up to the ionization potential of AB. Table 8.3 shows the energy that can be released in the recombination products of several possible Martian ions, together with the energy required for the fragments to escape the gravitational field of the planet (cf. Section 2.3.2). Products for which energies are listed in bold-face type can therefore escape. Most important of these in the present-day atmosphere must be atomic oxygen from recombination of O_2^+, the dominant ion,

$$O_2^+ + e \rightarrow O^*(^1D, ^1S) + O. \qquad (8.62)$$

Escape rates for oxygen are set by the rate at which CO_2 is photoionized in the exosphere. Hydrogen escape is limited by the supply of H_2 to the exosphere from below, atoms being released in the exosphere by the ionic reactions

$$CO_2^+ + H_2 \rightarrow CO_2H^+ + H \qquad (8.63)$$

$$CO_2H^+ + e \rightarrow CO_2 + H. \qquad (8.64)$$

As we saw earlier, the abundance of H_2 is set by the reactions

$$CO + OH \rightarrow CO_2 + H \qquad (8.17)$$

$$H + O_2 + M \rightarrow HO_2 + M \qquad (8.33)$$

$$H + HO_2 \rightarrow H_2 + O_2, \qquad (8.54)$$

and it is therefore sensitive to the net oxidation state of the atmosphere. Consequently, the H-atom escape rate is set ultimately by the rate at which oxygen is lost, and the processes are self-regulating to achieve a 2:1 stoicheiometry for H:O loss. At the present epoch, water is being 'processed' by the atmosphere and transferred to space at a rate of 6×10^7 molecule cm^{-2} s^{-1}. The water vapour is likely to be derived from condensed-phase water on the planet's surface. If the present escape rate had applied over

Table 8.3 Escape from Mars

	Species			
	C	N	O	CO
Escape energy (kJ mol^{-1})	144	168	192	336
Reaction exothermicity in each fragment (kJ mol^{-1})				
$CO_2^+ + e$	–	–	**509**	291
$O_2^+ + e$	–	–	**239**	–
$N_2^+ + e$	–	**279**	–	–
$CO^+ + e$	**159**	–	119	–
$NO^+ + e$	–	138	121	–

Note: Reaction exothermicities are calculated for electronic ground-state products *except* for O_2^+, where the lowest energy channel is assumed to be $O(^3P) + O(^1D)$. If excited products are formed in other recombinations, less energy will be available as translation: e.g. if N_2^+ yields $N(^4S) + N(^2D)$, the reaction exothermicity is 164 kJ mol^{-1} for each atom.

the life of the solar system (~4.6 × 10^9 yr), the water lost would have coated the surface of Mars with ice to an average depth of 2.5 m. In reality, solar ultraviolet fluxes are likely to have been greater in the early history of the solar system (cf. p. 687, Section 9.4), and the greater resultant escape rates increase the estimates of surface depth lost by a factor of up to eight (i.e. to 20 m). Some geological evidence can be interpreted to suggest much greater initial inventories of water (up to 500 m), so that additional mechanisms for hydrogen loss may have operated in earlier periods of the planetary history. An apparent substantial D/H enrichment (Section 9.3) requires such enhanced loss rates in the past.

Escape of species for which there is not a surface source raises the interesting question of whether the atmosphere is undergoing continual change in its composition. Viking isotopic measurements of the $^{15}N/^{14}N$ ratio show an enrichment of about 1.6 compared with the terrestrial abundance (the ratio is 3.66 × 10^{-3} on Earth, and 5.94 × 10^{-3} on Mars). Such enrichment could be brought about on Mars, from N_2 that initially possessed the terrestrial composition, by preferential escape of the lighter ^{14}N isotope over geological time. Translationally 'hot' nitrogen, able to escape, can be formed by recombination of N_2^+ (Table 8.3), or by impact dissociation of N_2 by energetic photoelectrons. The preference for loss of ^{14}N over ^{15}N is only small, so that large total amounts of nitrogen must have been lost during the planet's history. Initial partial pressures of N_2 must have been at least 1.3 mbar (they are now 0.17 mbar). Some calculations that allow for sinks of nitrogen such as fixation by lightning suggest that the early Martian atmosphere may have contained as much as 100 mbar of N_2 and 10 bar of CO_2. On the assumption that the Martian H/N ratio was the same as on Earth, these amounts of N_2 imply a surface layer of H_2O as much as 500 m deep, in accordance with the estimates based on geological evidence. Mars therefore provides an example of an evolving atmosphere. Loss of nitrogen and other constituents of the Martian atmosphere is evidently a result of the small gravitational acceleration of this particular planet. A more general discussion of atmospheric evolution is presented in Chapter 9.

8.4 Jupiter and Saturn

Beyond the asteroid belt lie the giant planets Jupiter, Saturn, Uranus, and Neptune, with their extensive satellite systems, and Pluto, which is more like a satellite than any of its giant companions. Jupiter and Saturn have low mean densities, about one-quarter and one-eighth that of Earth, respectively, suggesting that they are composed almost entirely of light elements. Our knowledge of the Jovian and Saturnian systems has been advanced enormously first by the Pioneer 10 and 11, and then by the Voyager 1 and 2 encounters. Voyager 2 has provided data of a similar quality about Uranus, Neptune, and

Table 8.4 Composition of the atmospheres of Jupiter and Saturn.

Constituent	Volume mixing ratio	
	Jupiter	Saturn
H_2	0.898	0.963
He	0.102 (0.136[a])	0.0325
CH_4	3×10^{-3}	4.5×10^{-3}
NH_3	2.6×10^{-3} [b]	$0.5–2.0 \times 10^{-4}$
H_2O	4.0×10^{-6} [b]	$< 2.0 \times 10^{-8}$
C_2H_6	5.8×10^{-6}	7.0×10^{-6}
PH_3	7.0×10^{-7}	1.4×10^{-6}
CH_3D	2.0×10^{-7}	3.9×10^{-7}
C_2H_2	1.1×10^{-7}	3.0×10^{-7}
CO	1.6×10^{-9}	1.0×10^{-9}
HCN	2×10^{-9} [c]	$< 4.0 \times 10^{-9}$
GeH_4	7×10^{-10}	0.4×10^{-9}
C_2H_4	7×10^{-9} [d]	–
CH_3C_2H	2.5×10^{-9} [d]	no estimate[d]
C_3H_3	–	no estimate[d]
C_6H_6	2×10^{-9} [d]	–
H_2S	$< 3.3 \times 10^{-8}$	$< 2.0 \times 10^{-7}$

[a] Result from Galileo helium abundance detector.
[b] Value at 1–4 bar.
[c] Tropopause and below.
[d] Tentative identification.

Data summarized by Strobel, D.F. *Int. Rev. phys. Chem.* **3**, 145 (1983); Atreya, S.K. *Atmospheres and ionospheres of the outer planets and their satellites*, Springer-Verlag, Heidelberg, 1986; and Yung, Y.K. and DeMore, W.B. *Photochemistry of planetary atmospheres*, Oxford University Press, Oxford, 1999, who give references to the original publications.

some of their satellites as well. It has been said that in the few hours of the Voyager encounters, more was learnt about the planets and satellites than had been found out in the rest of human history!

More recently, the Galileo mission to Jupiter and the satellites Europa, Io, and Callisto has made a further huge contribution to knowledge. Following the launch of Galileo from the space shuttle Atlantis in 1989, the spacecraft used the gravity of Venus once and that of Earth twice to gain enough energy to reach Jupiter. On the voyage, Galileo took the first close-up images of an asteroid (Gaspra), and discovered the first known moon of an asteroid, Dactyl at Ida. Galileo also had the unique opportunity of being able to obtain images of the crash of comet Shoemaker–Levy 9 into Jupiter's atmosphere. An atmospheric probe was released in July 1995, and reached the planet in December. Data were returned during the probe's survival period of 59 minutes on atmospheric temperature, pressure, composition, winds, and lightning. Meanwhile the orbiter started a tour of the Jovian system that was

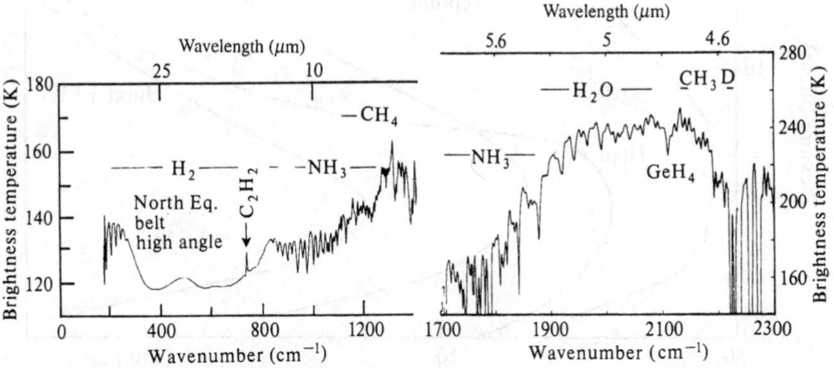

Fig. 8.9. Thermal emission spectrum from Jupiter recorded by the Infrared Interferometric Spectrometer (IRIS) carried by Voyager 1. Strong spectral features are seen for the gases H_2, C_2H_2, NH_3, CH_4, H_2O, GeH_4, and CH_3D, while between 1100 and 1200 cm^{-1} there is significant absorption by PH_3 (Q branch at 1122 cm^{-1}). From Hanel, R., Conrath, B., Flasar, M., Kunde, V., Lowman, P., Maguire, W., Pearl, J., Pirraglia, J., Samuelson, R., Gautier, D., Gierasch, P., Kumar, S., and Ponnamperuma, C. *Science* **204**, 972 (1979).

extended until the end of 1999 as the Galileo-Europa mission, with focused objectives on 'ice, water, and fire': the ice on Europa, the water in Jupiter's atmosphere, and the fire of Io's volcanoes, along with investigations of Io's torus (see later) and of the properties of Callisto.

The turn of the new millennium is an exciting time, too, for studies of Saturn and its satellite Titan. The Cassini mission is, perhaps, the most ambitious effort in planetary space exploration ever mounted. A joint endeavour of NASA, the European Space Agency (ESA) and the Agenzia Spaziale Italiana (ASI), Cassini will study the Saturnian system over a four-year period. Cassini was launched in 1997, and will reach Saturn in 2004, after using gravity-assist manoeuvres at Venus, Earth, and Jupiter to increase the speed of the spacecraft. The twelve scientific instruments will conduct studies of the planet, its moons, rings, and magnetic environment. Cassini carries the probe Huygens that will be released in November 2004 from the main spacecraft to parachute for 2.5 hours through the atmosphere to the surface of Saturn's moon Titan. Instruments on board will measure temperature, pressure, density, and energy balance in the atmosphere and, as the probe breaks through the cloud deck, a camera will capture pictures of Titan's clouds and surface. If the probe survives the landing, it can possibly return data from Titan's

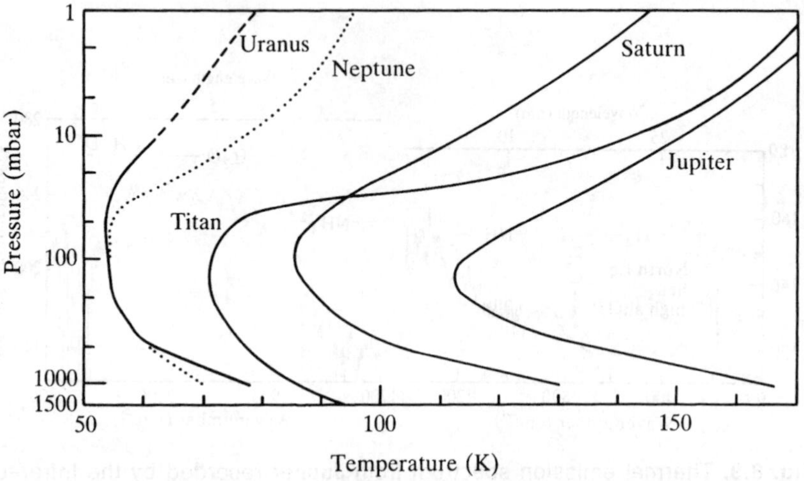

Fig. 8.10. Temperature–pressure structure of the Jovian planets, and of Saturn's satellite, Titan. Data from Strobel, D.F. *Int. Rev. Phys. Chem.* **3**, 145 (1983); Hanel, R., *et al. Science* **233**, 70 (1986); and Conrath, B., *et al. Science* **246**, 1454 (1989).

surface. The touchdown could be on solid ground, ice, or even in a lake of liquid methane or ethane. One instrument will detect if Huygens is bobbing in such liquid, and other instruments can determine the chemical composition of that liquid.

There is obviously much science to be learned in the future, but now we return to what has already been discovered. Atmospheric composition was investigated by several instruments on the Voyager spacecraft. Experiments studying infrared radiation (IRIS), ultraviolet spectroscopy (UVS), photo-polarimetry (PPS: aerosols), and radioscience (RSS: ions) were of particular value. Figure 8.9 indicates the type of spectrum obtained in the IRIS experiments (Voyager 1 at Jupiter): from the data, the identity and concentration of infrared emitters can be deduced. Table 8.4 describes the chemical compositions of the Jovian and Saturnian atmospheres as they now appear after the Voyager missions. As expected, the bulk of the atmosphere is made up of hydrogen and helium. The fractional abundance of He is, however, markedly smaller than the solar ratio (molar ratio 0.16), especially in the atmosphere of Saturn. This result suggests that helium has undergone gravitational separation from hydrogen within the interiors of the planets. The CH_4/H_2 ratio in Jupiter's atmosphere is roughly 2.1 times the solar equivalent value (although that may not be the case for the bulk planetary composition).

Analysis of NH_3 data also leads to the conclusion that the nitrogen/hydrogen ratio is enhanced by a factor of two over the solar ratio, while oxygen and germanium seem to be *depleted* by factors of ~50–150 and ~10, respectively. Taken at face value, these elemental ratios seem to argue against the concept of formation of the planetary system from a primitive homogeneous nebula in favour of an accretion hypothesis (see p. 657).

Deep in the atmosphere, thermal chemistry yields compounds of the elements consistent with thermochemical equilibrium at the temperatures and pressures encountered (although the detection of PH_3 and GeH_4 implies the occurrence of disequilibrating processes). Photochemistry in the atmospheres of Jupiter and Saturn can convert CH_4 to heavier hydrocarbons, and NH_3 to N_2H_4, as we shall see shortly. Our current understanding is that the photochemical products are transported downwards to the hot, dense interiors, where thermal decomposition and subsequent reaction with H_2 recycle CH_4 and NH_3. Some of the possible photochemically generated species are condensable at the low temperatures encountered in the atmospheres, so that a knowledge of the vertical temperature structure of the atmosphere is needed for proper interpretation of the photochemistry. Figure 8.10 gives temperature–altitude profiles for the outer planets and for Titan, over the regions of photochemical interest. On Jupiter and Saturn, the temperature profile starts at the transition level from an adiabatic lapse rate (convectively controlled) to an increasing temperature above (radiatively controlled). The temperature minima act as 'cold traps' which limit mixing ratios of condensable gases in the upper atmospheres. Thus NH_3 is limited to a mixing ratio of less than 10^{-7} above the Jovian minimum of ~110 K, while the cooler 'tropopause' of Saturn (~85 K) prevents detectable concentrations of NH_3 reaching the upper atmosphere. An atmospheric 'parcel' ascending from great depths will lose constituents as they condense out as various liquid or solid phases. Condensates remain as aerosols at the appropriate levels. For example, on Jupiter, dense water clouds form at ~270 K, while near the 200 K level, H_2S is thought to react with NH_3 to form a cloud of solid NH_4SH particles. White crystals of ammonia precipitate out at ~154 K to produce the visible upper cloud layer.[a]

Above the clouds, photochemical transformations can take place. Although the chemistry of the Jovian and Saturnian atmospheres is very different from anything we have encountered so far, one well-established principle still applies, and that is that the processes requiring the highest energy occur highest in the atmosphere, because short-wavelength radiation is generally attenuated more rapidly than longer wavelength radiation. Figure 8.11 provides a pictorial summary of some of the most important chemical

[a]This picture of cloud formation has been generally accepted for many years. However, data from the nephelometer carried on the Galileo probe to Jupiter in 1995 seem at variance with the idea. None of the expected thick, dense clouds were found, and concentrations of particles in the immediate vicinity of the probe were minimal. Only one distinct cloud structure was detected, possibly that of the NH_4SH layer. Yet observations from Earth and Voyager suggest strongly that Jupiter is enshrouded with clouds. It may be, of course, that the Galileo probe site was atypical.

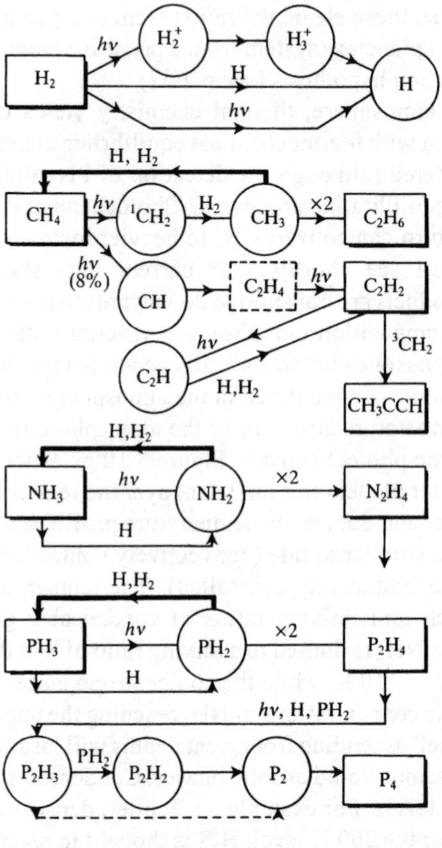

Fig. 8.11. Most important steps in the chemistry of the atmospheres of Jupiter and Saturn.

steps. Atomic hydrogen is formed photochemically from molecular hydrogen, and the chemistry of the atmospheres of Jupiter and Saturn is greatly influenced by the reactions of other species with both H and H_2. Hydrides such as CH_4, NH_3, and PH_3 also undergo photolysis to produce the intermediates CH_2, CH, NH_2, and PH_2 that participate in further reactions to yield some of the compounds observed experimentally. Note that the back reactions of CH_3 (derived from CH_2), NH_2, and PH_2 with both H and H_2 regenerate the starting reactants, and dominate over the formation of the more complex products.

We shall now examine the individual steps in more detail. Hydrogen is the dominant constituent of the atmospheres under discussion. It absorbs to dissociate only at $\lambda < 100$ nm, with a dissociation continuum having its onset

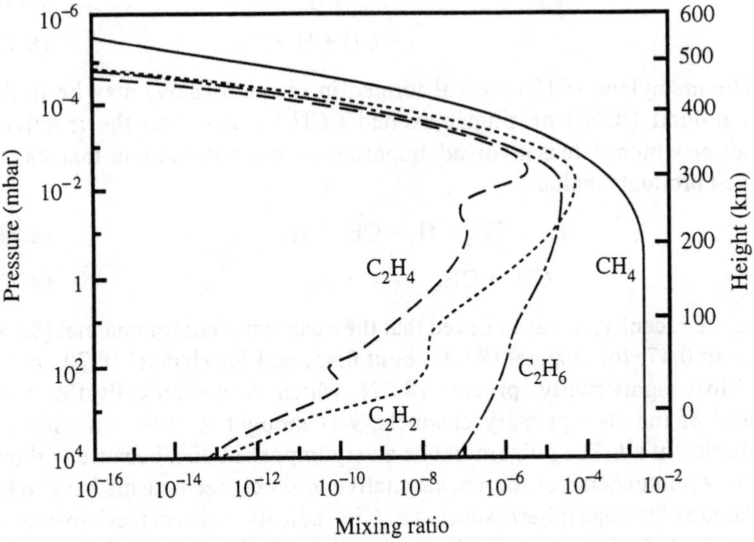

Fig. 8.12. Model calculation of altitude profiles for some hydrocarbons in the Jovian atmosphere. From Gladstone, G.R., Allen, M., and Yung, Y.L. *Icarus* **119**, 1 (1996).

at $\lambda = 84.5$ nm, and an ionization continuum starting at $\lambda = 80.4$ nm. Ionization of H_2 leads to the formation of at least two H atoms *via* the reactions

$$H_2^+ + H_2 \rightarrow H_3^+ + H \quad (8.65)$$

$$H_3^+ + e \rightarrow H_2 + H \text{ (or 3H)} \quad (8.66)$$

so that the effect for neutral chemistry of either ionization or dissociation is H-atom formation. Since the three-body recombination of H atoms is exceedingly slow at ionospheric pressures, there is a net downward flux of H atoms from the ionosphere to lower altitudes, where methane and ammonia photochemistry are important. Methane photolysis requires photons with $\lambda \lesssim 145$ nm, and occurs preferentially high in the stratosphere. Ammonia photolysis is driven by photons with $\lambda \lesssim 160$ nm, and occurs primarily near the tropopause, where condensation and photolysis have not severely depleted the species.

One of the most interesting aspects of the chemistry of the outer solar system is the synthesis of organic compounds. Since methane is a relatively abundant source molecule, it is necessary to investigate how it might be converted to more complex molecules. Intense solar Lyman-α radiation ($\lambda =$ 121.6 nm) dissociates CH_4 *via* several primary paths

$$CH_4 + h\nu \rightarrow CH_3 + H \quad (8.67)$$

$$\rightarrow {}^1CH_2 + H_2 \quad (8.68)$$

$$\rightarrow {}^{1,3}CH_2 + 2H \quad (8.69)$$

$$\rightarrow CH + H + H_2. \quad (8.70)$$

The methylene (CH_2) radical formed in reaction (8.69) may be in the triplet ground (${}^3CH_2$) or singlet excited (${}^1CH_2$) states, and the reactivity depends on which state is involved. Important secondary reactions that lead to observed products include

$$CH_2 + H_2 \rightarrow CH_3 + H, \quad (8.71)$$

$$CH_3 + CH_3 \xrightarrow{(M)} C_2H_6. \quad (8.72)$$

Until quite recently, it was believed that the quantum yield for channel (8.68) was about 0.47, for channel (8.69) about 0.45, and for channel (8.70) about 0.08. Most significantly, process (8.67), which is energetically the most favoured of the four primary channels, was thought to make virtually no contribution at all. The point might be a very important one, because without this direct photochemical source, alternative indirect mechanisms have to be postulated as the atmospheric sources of CH_3 radicals. Indirect mechanisms of this kind include formation of CH_2 in processes (8.68) or (8.69) followed by reaction in (8.71), or hydrogen abstraction from CH_4 by other radicals. However, experiments on the dynamics of photodissociation of CH_4 appear to contradict the earlier results for the quantum efficiencies of the different channels. At $\lambda = 121.6$ nm, at least, roughly 50 per cent of the dissociation events yield an H atom together with a CH_3 radical possessing high levels of vibrational and rotational excitation. Such a yield of CH_3 might, indeed, explain the otherwise surprisingly high concentrations of C_2H_6 found in the atmospheres of Jupiter and Saturn (and Titan: see Section 8.5.1). Most models have not, however, yet used the newer data, and older ones have tended to predict much lower values (near unity) for $[C_2H_6]$ relative to $[C_2H_2]$ compared to the ratio actually found (from 40 to 200).

Ethene (ethylene) is formed by the reaction

$$CH + CH_4 \rightarrow C_2H_4 + H, \quad (8.73)$$

but the C_2H_4 is rapidly photolysed to ethyne (acetylene)

$$C_2H_4 + h\nu \rightarrow C_2H_2 + H_2, \quad (8.74)$$

which is itself photochemically rather stable, because its dissociation products (C_2H and C_2) react with H_2 to regenerate C_2H_2. As a result, concentrations of C_2H_2 in the upper atmospheres are much greater than those of C_2H_4. Higher hydrocarbons, and possibly even polymeric materials, can be formed by reactions of C_2H_2, as for example with ${}^3CH_2$

$$^3CH_2 + C_2H_2 \xrightarrow{(M)} CH_3C_2H \tag{8.75}$$

to yield one of the observed products, propyne (methylacetylene).

Modelling is beginning to prove successful for the atmospheric chemistry of Jupiter. Figure 8.12 shows predicted concentrations for one such model that employs nearly 200 photochemical and thermal reactions in its chemistry. The lower boundary is set at the level where the pressure is 6 bar, well below the tropopause region (0.1 bar). At this lower boundary, the mixing ratios of H_2, He, and CH_4 are taken more or less to be those given in Table 8.4. Concentrations of all other species are calculated in the model. One of the problems with this kind of study is that, for many of the chemical processes (and especially for the 78 photochemical ones included), rate coefficients, activation energies, and quantum yields are a matter of estimate or conjecture since they have not been studied in the laboratory. Nevertheless, the results look encouraging, and both relative and absolute concentrations of C_2 hydrocarbons seem quite well captured. The authors of this model did not use the newer data for the quantum yield of reaction (8.67), but they suggest that the indirect route involving CH_2 and reaction (8.71) can replace the direct formation of CH_3 with comparable efficiency. In any case, it is claimed that 80 per cent of the CH_3 in the atmosphere is a consequence of hydrogen abstraction from CH_4 by other radicals, such as C_2H (see Section 8.5, and reactions (8.90) and (8.91) for a discussion of this type of process on Titan). The model predicts peak mixing ratios of the C_2 hydrocarbons around altitudes corresponding to 10^{-6} bar (which is where the maximum rate of destruction of CH_4 occurs). Above and below this altitude, the concentrations of the C_2-hydrocarbons decrease because of their destruction and removal. At the peak, corresponding to $5–10 \times 10^{-6}$ bar, C_2H_2 has a mixing ratio of 40 p.p.m., and is the most abundant of all disequilibrium species in the model.

Although the neutral mass spectrometer on the Galileo probe detected elements such as carbon, nitrogen, and sulphur, few complex organic compounds were found. According to the lightning and radio emission detector, lightning is three to ten times less common on Jupiter than on Earth, for any given surface area (but those storms that do occur seem much stronger than on Earth). The dearth of complex organic molecules, which are often produced by lightning, seems consistent with the relative lack of storm activity.

Ammonia photochemistry in the Jovian and Saturnian atmospheres is rather simple, since the primary step for the available wavelengths (and in the presence of H_2) can be written

$$NH_3 + h\nu \rightarrow NH_2 + H. \tag{8.76}$$

Hydrazine formation or regeneration of ammonia follow

$$NH_2 + NH_2 \xrightarrow{(M)} N_2H_4, \tag{8.77}$$

$$NH_2 + H \xrightarrow{(M)} NH_3, \tag{8.78}$$

with condensation of the N_2H_4 as a haze. Entirely analogous reactions describe the main part of phosphine photochemistry

$$PH_3 + h\nu \rightarrow PH_2 + H \tag{8.79}$$

$$PH_2 + PH_2 \xrightarrow{(M)} P_2H_4 \tag{8.80}$$

$$PH_2 + H \xrightarrow{(M)} PH_3 \tag{8.81}$$

and solid P_2H_4 is probably formed as a condensation product. Coupling between NH_3 and PH_3 chemistry is brought about by the reaction

$$H + PH_3 \rightarrow PH_2 + H_2, \tag{8.82}$$

since the atomic hydrogen liberated by NH_3 photolysis in reaction (8.76) accelerates the conversion of PH_3 to P_2H_4. Both PH_3 and NH_3 concentrations decrease rapidly above the tropopause, and measurements utilizing ultraviolet absorption therefore detect NH_3 at concentrations orders of magnitude smaller than those using infrared techniques that penetrate deeper into the atmosphere.

Both NH_3 and PH_3 photochemistry have minor pathways that we shall exemplify in the case of PH_3. P_2H_4 is susceptible to photolysis, or attack by H and PH_2, to yield P_2H_3 and then P_2H_2

$$P_2H_4 + h\nu\,(\text{or H, } PH_2) \rightarrow P_2H_3 + H\,(\text{or } H_2,\ PH_3) \tag{8.83}$$

$$P_2H_3 + PH_2\,(\text{or H, } P_2H_3) \rightarrow P_2H_2 + PH_3\,(\text{or } H_2,\ P_2H_4). \tag{8.84}$$

Elemental phosphorus is an end-product

$$P_2H_2 \rightarrow P_2 + H_2 \tag{8.85}$$

$$2P_2H_3 \rightarrow P_4H_6 \rightarrow 2PH_3 + P_2 \tag{8.86}$$

$$P_2 + P_2 \xrightarrow{(M)} P_4. \tag{8.87}$$

Likewise, N_2 is the end-product of the N_2H_4 sequence. Were it not for the condensation of N_2H_4, and thus its transport to the troposphere and protection from chemical and photon attack, N_2 would be the ultimate product of the photolysis of NH_3.

Formation of P_4 in reaction (8.87) brings us inevitably to a discussion of the red and yellow colourings seen on Jupiter and Saturn. Colours of a wide variety are found on the bodies of the outer solar system, but their sources are not known. Red phosphorus is one obvious candidate, but there has been some controversy over the actual colour of the phosphorus formed during photolysis of PH_3. Spectroscopy authenticates the allotrope as 'red', but visually the colours range from yellow to violet. Sulphur species offer another plausible inorganic contributor to the observed colourings. Coloration by complex organic molecules is another possibility that carries additional interest because of the implications for the origin of life in reducing atmospheres. Simple hydrocarbons such as C_2H_6 and C_2H_2 do not accumulate in sufficient concentrations on Jupiter and Saturn to condense, nor do they absorb in the visible region. What is needed is complex molecules with conjugated bonds (e.g. polyacetylenes) and nitriles. In the context of complex molecule formation, the observed presence of HCN may be significant. Convection from the hot, dense interior atmosphere cannot be the source of the molecule, nor can local heating due to lightning discharges produce as much as is found. One promising pathway to HCN generation seems to be the ultraviolet photolysis of the cyclic isomer of C_2H_5N, aziridine (ethyleneimine), itself formed indirectly from NH_2 radicals and C_2H_2. Laboratory studies show that aziridine is photolysed by vacuum ultraviolet light to yield HCN so that the pathway is plausible, although not substantiated, for the planetary atmospheres.

Although the atmospheres of Jupiter and Saturn are usually thought of as totally reducing, the oxygen abundance found by the Voyager mission would require substantial amounts of H_2O and other oxygen species. However, the Galileo probe told a rather different story. The atmosphere was much drier than anticipated. And, rather than finding an oxygen abundance twice or more that of the Sun (based on the Jovian H_2O abundance), the apparent abundance on Jupiter was one-fifth, or even less, that of the Sun. This high depletion of O_2 relative to the Sun will demand new ways of thinking about Jupiter's formation and evolution. Most H_2O condenses out below the 'photochemical' regions, but the discovery of CO on Jupiter proves that photochemical processes involving oxygen species do occur. At one time, it was thought that an extraplanetary source of H_2O might be needed to explain the CO. However, measurements of the linewidths of the CO spectra have now demonstrated that the CO is formed by the reaction of methane with water in the planetary interior. Oxygen in any form is virtually bound to be converted ultimately to CO. Once formed, CO cannot be photolysed, and it is not attacked by OH in the lower stratosphere because condensation of water removes the OH source that would oxidize it to CO_2, while CO_2 photolysis itself rapidly re-forms carbon monoxide. Nevertheless, current models suggest that the mixing ratio of CO would be negligible near the part of the planet probed by spectroscopy (above a pressure of about 1 bar) if concentrations were determined by stationary-state chemistry alone. It seems likely that vigorous mixing between the deep atmosphere and the troposphere transports CO-rich air to an altitude where it is observable.

8.5 Titan, Io, Europa, and Callisto

8.5.1 Titan

Saturn's largest satellite, Titan, is the only satellite in the solar system to possess a massive atmosphere. As early as 1908, visual observations had suggested the existence of an atmosphere, and by 1944 absorption bands of

Table 8.5 Composition of Titan's atmosphere

Constituent	Volume mixing ratio		
	Near surface	Stratosphere (40–100 km)	Mesosphere and thermosphere
N_2	>0.97	>0.97	>0.97 (3900 km)
CH_4	$<3 \times 10^{-2}$	$1\text{–}3 \times 10^{-2}$	0.08 (1140 km)
CH_3D	–	$6.4 \times 10^{-4} \times [CH_4]$	–
H_2	$2 \pm 1 \times 10^{-3}$	2.0×10^{-3}	–
CO	$10 \pm 5 \times 10^{-5}$	6×10^{-5}	–
CO_2	–	1.5×10^{-9}	–
H_2O	–	$<1 \times 10^{-9}$	–
Ar	<0.13	–	$<6 \times 10^{-2}$
Ne	$<2 \times 10^{-3}$	–	$<1 \times 10^{-2}$
C_2H_6	–	2×10^{-5}	–
C_2H_4	–	4×10^{-7}	–
C_2H_2	–	2×10^{-6}	$1\text{–}2 \times 10^{-2}$ (840 km)
C_3H_8	–	$2\text{–}4 \times 10^{-6}$	–
CH_3CCH	–	3×10^{-8}	–
CHCCCH	–	$1\text{–}10 \times 10^{-8}$	–
HCN	–	2×10^{-7}	$<5 \times 10^{-4}$
C_2N_2	–	$1\text{–}10 \times 10^{-8}$	–
HCCCN	–	$1\text{–}10 \times 10^{-8}$	–

Data summarized by Strobel, D.F. *Int. Rev. Phys. Chem.* **3**, 145 (1983); Yung, Y.K., Allen, M., and Pinto, J.P. *Astrophys. J. Supp. Ser.* **55**, 465 (1984); and Yung, Y.K. and DeMore, W.B. *Photochemistry of planetary atmospheres*, Oxford University Press, Oxford, 1999, who give references to the original publications.

methane had been discovered. With the Voyager fly-bys have come our first definitive data about the atmospheric composition and structure. Ultraviolet spectroscopy (UVS) experiments detected emission lines from molecular and atomic nitrogen in the upper atmosphere (Fig. 8.13), characteristic of electron-excited N_2. Together with measurements of temperature and scale height, the UVS results show that the atmosphere is predominantly N_2, with methane a minor constituent. Temperature and pressure at the surface were found to be 94.5 ± 0.4 K and 1.5 bar, so that the pressure is roughly 50 per cent greater than that on Earth. Titan is the only body in the solar system besides Earth that has an atmosphere composed largely of nitrogen. Voyager's infrared (IRIS) experiments detected a suite of hydrocarbons and nitrogen compounds in addition to CH_4, and our present knowledge of Titan's atmospheric composition is given in Table 8.5. Television pictures returned by the Voyager cameras showed that the satellite is covered by coloured clouds. The clouds must be aerosols derived from the gaseous organic compounds, and are thus Titan's equivalent of photochemical smog (cf. Section 5.10.7)! Above the

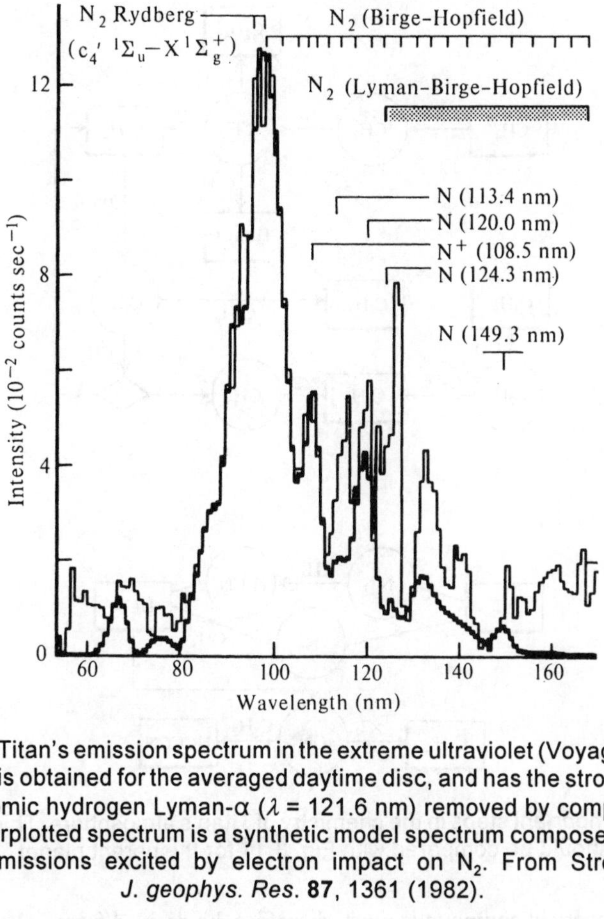

Fig. 8.13. Titan's emission spectrum in the extreme ultraviolet (Voyager 1). The spectrum is obtained for the averaged daytime disc, and has the strong feature due to atomic hydrogen Lyman-α (λ = 121.6 nm) removed by computer. The heavy overplotted spectrum is a synthetic model spectrum composed of N_2, N, and N^+ emissions excited by electron impact on N_2. From Strobel, D.F. *J. geophys. Res.* **87**, 1361 (1982).

coloured cloud layers, which extend from the surface to an altitude of ~ 200 km, lies a thinner haze layer of aerosol particles. Absorption of solar radiation by the various aerosols leads to a net heating above an altitude of ~ 40 km, and thus to a temperature inversion in Titan's 'stratosphere'. The curve for Titan in Fig. 8.10 shows that the minimum temperature, at the tropopause, is about 70 K. Titan's atmosphere is reducing in bulk chemical composition, the oxidation state lying between that of Jupiter and Saturn on the one hand, and the terrestrial planets on the other. For many of the organic compounds listed in Table 8.5, the measured abundances exceed the saturated vapour pressures at the tropopause, so that the lower atmosphere cannot be the source of complex molecules found in the stratosphere, even though certain species (e.g. HCN) might be abundant on the surface and in the interior of the satellite.

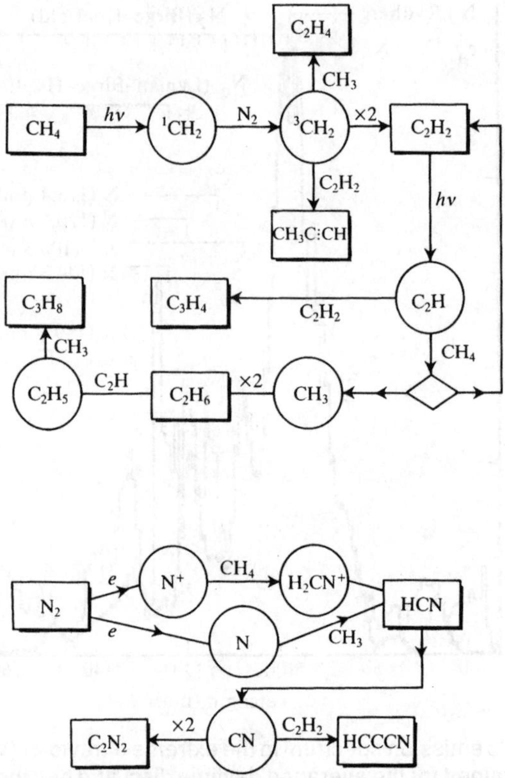

Fig. 8.14. Important steps in the chemistry of Titan's atmosphere. This diagram should be compared with Fig. 8.11 for the parent planet.

The observed compounds must, therefore, be derived from volatile parent molecules. Nitrogen does not condense at the temperatures of the tropopause, although CH_4 clouds may exist. The surface temperature is probably just above the triple-point of CH_4 (90.7 K), so that liquid methane pools or oceans may cover the satellite's surface. Negligible amounts of NH_3 are present in the stratosphere because of the 70 K 'cold-trap' temperature. Any N_2H_4 formed by ammonia photolysis [reactions (8.76) and (8.77)] near the surface will also be trapped out. We shall see in Section 9.2.2 that ammonia photochemistry is one possible source of Titan's present-day nitrogen. In that case, trapping of NH_3 must not have been significant, and the atmosphere must have been at least 50 K warmer than it is now, during about 4 per cent of its evolutionary history.

Escape of atomic and molecular hydrogen from Titan's atmosphere is ensured by the combination of very low escape velocity (2.1 km s^{-1}: Table 2.1), extended atmosphere, and 'warm' thermosphere (~ 186 K). Thermal escape of H and H_2 can cope with the rate of hydrogen formation in direct and catalysed photolysis of hydrocarbons, principally CH_4. Build-up of heavier hydrocarbons, such as C_2H_6 and C_3H_8, at the expense of CH_4, therefore takes place, and H_2 is a minor constituent of Titan's atmosphere. Recycling back to CH_4 is impossible, so that there is a one-way evolution towards more complex organic species that condense at the tropopause and are ultimately deposited on the surface. Hydrogen lost from the satellite has formed a doughnut-shaped torus around Saturn, through which Titan is continually sweeping.

The continuing production of complex species leads to the formation of the photochemical aerosol that in turn absorbs sunlight to give a thermal inversion and yield dynamical stability of the atmosphere. There are obvious parallels with the ozone layer in Earth's atmosphere.

Absence of H_2, and the presence of abundant N_2, modify Titan's atmospheric photochemistry considerably from that of Jupiter or Saturn. On the planets, radicals such as CH_3 or NH_2 derived from CH_4 or NH_3 react most frequently by abstraction of hydrogen from H_2 or by three-body association with H-atoms to return to the starting compound in a 'do nothing' cycle. Such reactions cannot occur on Titan, so that the less hydrogen-rich hydrocarbons are favoured. Mixing ratios of C_2H_6 are four times greater on Titan than on Saturn, those of C_2H_2 are 27 times larger, and C_2H_4 is detected on the satellite but not on the planet. The extensive airglow of N_2, N, and N^+ emission features means that N_2 is excited, dissociated, and ionized by electron impact processes in the thermosphere. Reactions of the energetic nitrogen species must therefore be included in any consideration of chemistry in Titan's atmosphere. Figure 8.14 provides an overview of some of the most important steps leading to identified products, and should be compared with Fig. 8.11 for Jovian chemistry. The three key differences are the absence of back-reactions involving H and H_2, the presence of processes involving N and N^+, and the quenching of $^1CH_2$ to $^3CH_2$ by N_2 and the consequent formation of C_2H_4, C_2H_2, and C_3H_4 from the triplet.

Because N_2 is the bulk constituent of Titan's atmosphere, this latter quenching means that $^3CH_2$ is the net product of CH_4 photolysis on Titan (with only minor pathways to CH and possibly CH_3 [cf. reactions (8.67)–(8.70), and the discussion of CH_3 formation on p. 592]). These triplet radicals react primarily to form C_2H_2, so that the first stages in Titan's hydrocarbon photochemistry are

$$CH_4 + h\nu \xrightarrow{N_2} {}^3CH_2 + H_2 \text{ (or 2H)} \tag{8.88}$$

$$^3CH_2 + {}^3CH_2 \rightarrow C_2H_2 + H_2 \text{ (or 2H)}, \tag{8.89}$$

followed by photolysis of C_2H_2 to yield C_2H radicals, which catalyse dissociation of CH_4

$$C_2H_2 + h\nu \rightarrow C_2H + H \quad (8.90)$$

$$C_2H + CH_4 \rightarrow C_2H_2 + CH_3 \quad (8.91)$$

$$\text{Net} \quad CH_4 \xrightarrow{h\nu} CH_3 + H$$

Reactions of the methyl (CH_3), methylene (CH_2), and ethynyl (C_2H) radicals can then yield many of the compounds observed in the atmosphere

$$^3CH_2 + CH_3 \rightarrow C_2H_4 + H \quad (8.92)$$

$$CH_3 + CH_3 \xrightarrow{(M)} C_2H_6 \quad (8.72)$$

$$C_2H + C_2H_6 \rightarrow C_2H_2 + C_2H_5 \quad (8.93)$$

$$C_2H_5 + CH_3 \xrightarrow{(M)} C_3H_8 \quad (8.94)$$

$$C_2H + C_2H_2 \rightarrow H + HC{\vdots}CC{\vdots}CH \text{ (diacetylene)} \quad (8.95)$$

$$^3CH_2 + C_2H_2 \xrightarrow{(M)} CH_2{:}C{:}CH_2 \text{ (allene)} \quad (8.96)$$

$$CH_2{:}C{:}CH_2 \rightarrow CH_3C{\vdots}CH \text{ (methylacetylene)}. \quad (8.97)$$

Allene ($CH_2{:}C{:}CH_2$) is the isomer of C_3H_4 favoured in reaction (8.96), but is not found on Titan: isomerization can yield the methylacetylene ($CH_3C{\vdots}CH$) observed. Polyacetylenes can be formed by successive reactions analogous to eqn (8.95), but their growth may be inhibited by H atoms even at the low [H] found on Titan. It is thus not clear at present whether such molecules are responsible for aerosol formation.

Chemical models have been developed to describe the atmosphere of Titan that appear to work quite well for the hydrocarbon species. Figure 8.15 shows the results of one such model. The eddy diffusion coefficient used to describe vertical transport in the model was constrained by an altitude profile for HCN (see next paragraph) obtained by ground-based millimetre-wave spectroscopy. The overall chemical scheme contains 62 species in 249 reactions! Of special interest is the inclusion of direct formation of CH_3 in the reaction

$$CH_4 + h\nu \rightarrow CH_3 + H. \quad (8.67)$$

Different variants of the photolysis scheme give only slightly differing results so long as reaction (8.67) contributes roughly 50 per cent to the photolytic loss of CH_4. At the lower boundary in the model, CH_4 is set at 4.4 per cent, and at 1140 km the calculated mole fraction of CH_4 is 10 per cent, quite close to the 8 ± 3 per cent found by the UV spectrometer on Voyager at this altitude (see Table 8.5). For the other hydrocarbons, agreement between model and observation is good or acceptable for C_2H_2, C_2H_6, and C_3H_8: the boxes in the figure show the error ranges for altitude and concentration of the Voyager measurements. On the other hand, C_2H_4 is underestimated in the model by a

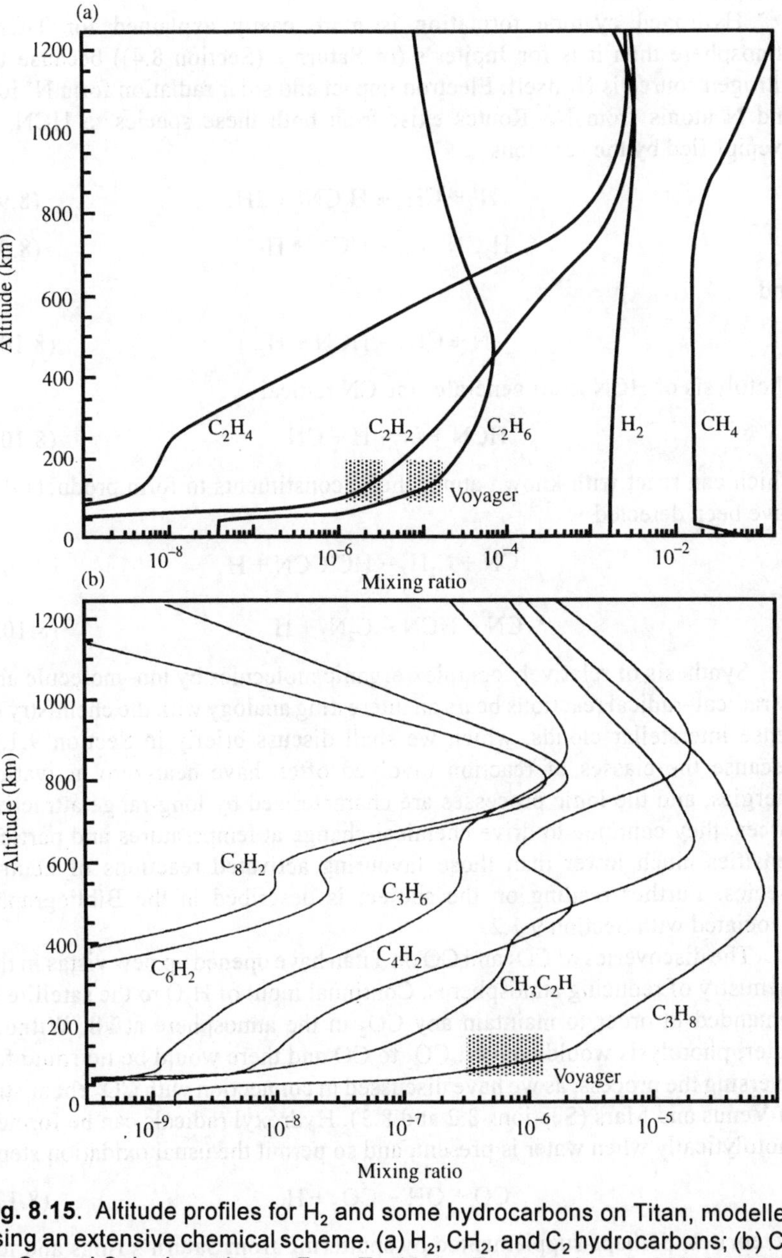

Fig. 8.15. Altitude profiles for H_2 and some hydrocarbons on Titan, modelled using an extensive chemical scheme. (a) H_2, CH_2, and C_2 hydrocarbons; (b) C_3, C_4, C_6, and C_8 hydrocarbons. From Toublanc, D., *et al.* *Icarus* **113**, 2 (1995).

factor of almost 30, while both C_3H_4 (propyne) and C_4H_2 (diacetylene) are overestimated by a factor of 3–5. Some details of the chemistry evidently still need refinement if the Voyager observations have themselves been correctly interpreted.

Hydrogen cyanide formation is more easily explained for Titan's atmosphere than it is for Jupiter's (or Saturn's (Section 8.4)) because the nitrogen source is N_2 itself. Electron impact and solar radiation form N^+ ions and N atoms from N_2. Routes exist from both these species to HCN, as exemplified by the reactions

$$N^+ + CH_4 \rightarrow H_2CN^+ + 2H \tag{8.98}$$

$$H_2CN^+ + e \rightarrow HCN + H \tag{8.99}$$

and

$$N + CH_3 \rightarrow HCN + H_2. \tag{8.100}$$

Photolysis of HCN itself generates the CN radical

$$HCN + h\nu \rightarrow H + CN, \tag{8.101}$$

which can react with known atmospheric constituents to form products that have been detected

$$CN + C_2H_2 \rightarrow HC{:}CCN + H \tag{8.102}$$

$$CN + HCN \rightarrow C_2N_2 + H. \tag{8.103}$$

Synthesis of relatively complex organic molecules by ion–molecule and by radical–radical reactions bears an interesting analogy with the chemistry of dense interstellar clouds, which we shall discuss briefly in Section 9.1.2. Because the classes of reaction involved often have near-zero activation energies, and the ionic processes are characterized by long-range attractive forces, they continue to drive chemical change at temperatures and particle densities much lower than those favouring activated reactions of neutral species. Further reading on the subject is described in the Bibliography associated with Section 9.1.2.

The discoveries of CO_2 and CO on Titan have opened up new vistas in the chemistry of reducing atmospheres. Continual input of H_2O to the satellite is demanded in order to maintain any CO_2 in the atmosphere at all. Without water, photolysis would convert CO_2 to CO and there would be no route for reversing the process, as we have discussed in connection with CO_2 chemistry on Venus and Mars (Sections 8.2 and 8.3). Hydroxyl radicals can be formed photolytically when water is present, and so permit the usual oxidation step

$$CO + OH \rightarrow CO_2 + H. \tag{8.17}$$

Meteorites, perhaps supplemented by sputtering from Saturn's rings and icy satellites, can account for the requisite water supply. Input of H_2O can also be responsible, *via* OH radicals, for the partial oxidation of CH_4 to CO itself. However, the evidence so far available does not exclude the possibility of Titan having been formed with an atmosphere containing CO (and, in fact, N_2: see Section 9.2.2). Indeed, this argument derives some support from observations of CO-ice on the surfaces of Triton and Pluto (Section 8.6). Titan's CO_2, on the other hand, is certainly photochemical in origin. Carbon monoxide in the contemporary atmosphere of Titan appears to be roughly in a steady state between formation and oxidation to CO_2. Carbon dioxide is removed by photolysis (and perhaps by reaction with CH_2), but at a rate insufficient to overcome its production. Its vapour pressure therefore builds up to reach saturation near the tropopause. Solid CO_2 condenses and is precipitated to the surface. The measured CO_2 abundance is consistent with saturation at 75 K and a total pressure of ~ 100 mbar, so that CO_2 seems to be in a steady state, controlled by condensation and sublimation. If the large influx of water postulated for Titan is borne out by further observation, H_2O vapour would have to be considered as a potentially important species on Saturn as well. Spectroscopy shows that CO on Jupiter is present predominantly in the troposphere (p. 595): it is formed by reactions deep in the interior. For Saturn, the altitude distribution of CO is not known, but an influx of water would make a stratospheric source of CO likely. The chemistry of H_2O, CO, and CO_2 may ultimately prove to be as significant to reducing atmospheres as that of the 'trace constituent' reduced compounds (e.g. CH_4, H_2) is to oxidizing ones.

8.5.2 Io, Europa, and Callisto

A massive atmosphere like Titan's is not found on any of the other satellites of the gas giant planets. However, Jupiter's satellite Io is able to retain a tenuous atmosphere which is of interest because its composition is quite unlike that of any other body in the solar system. One of the first indications that Io might have an atmosphere was provided by a brightening of the satellite in the period immediately after its emergence from an eclipse, suggesting that condensation of an atmospheric gas had occurred during the period of darkness. Experiments on Pioneer 10 established the existence of an ionosphere with relatively high electron densities, and thus demonstrated that the atmosphere was present, although surface pressures were perhaps only 10^{-7} of those on Earth. Later experiments set the pressure even lower, as explained later. Atomic emission lines from sodium (and weaker ones from potassium) show that alkali metals are present in a torus of atoms that have escaped from Io, but are still orbiting Jupiter. The highlight of the visits of the Voyager craft to Io was the discovery of active volcanoes on the satellite. The infrared absorption experiment on Voyager identified SO_2 in the atmosphere above one of the volcanic hot spots, and the fly-bys showed that sulphur and oxygen (as ions) made up a giant plasma torus encircling Jupiter at Io's orbital distance (see later). The International Ultraviolet Explorer (IUE) has also detected optical emissions from *neutral* sulphur and oxygen atoms near Io. These observations, coupled with a further inference that SO_2 and S are present in condensed phases at the satellite's surface, suggest an atmosphere in which SO_2 is a major constituent, with minor amounts of the photochemically derived products S and O, and some Na and K. The Galileo orbiter has subsequently established that Io is probably the Solar system's volcanically most active body. The energy that drives the volcanoes derives from tidal heating associated with a resonance between the orbits of Io and Europa. The process releases an order of magnitude more power for a given surface area than the geothermal flux on Earth. Temperatures of the volcanoes are sometimes as high as 1800 °C, substantially higher than those of Earth's volcanoes. It is surmised that lava made of a silicate material rich in magnesium erupts from below Io's surface. Further understanding of Io's volcanism should follow the sampling of a volcanic plume when the Galileo orbiter flies 500 km above the active volcano Pillan Patera in late 1999.

Sulphur dioxide has a vapour pressure of 10^{-9} bar or less at the temperatures (< 90 K) in polar regions and on the night-side, but, at the subsolar point, temperatures, and thus vapour pressures, are much higher (~ 130 K and > 10^{-7} bar). The simplest view of the atmosphere of Io is thus one in which a relatively dense atmosphere exists near the volcanic plumes and under the Sun, but which becomes increasingly thin towards the poles and the night-side. Microwave observations are consistent with this hypothesis, since they show 4–35 × 10^{-9} bar of SO_2 covering 3–15 per cent of the surface, similar to that expected if the gas-phase SO_2 were in equilibrium with the surface frost. Circumstantial evidence concerning flow dynamics suggests the presence in addition of a non-condensable gas, such as O_2, at pressures of about 20 × 10^{-9} bar, but there is no direct observational support as yet for this hypothesis.

The primary pathways for photodissociation of SO_2 are

$$SO_2 + h\nu(\lambda < 221 \text{ nm}) \rightarrow SO + O \tag{8.104}$$

$$SO_2 + h\nu(\lambda < 207 \text{ nm}) \rightarrow S + O_2, \tag{8.105}$$

Two important secondary reactions can also act as sources of S and O atoms

$$SO + SO \rightarrow SO_2 + S \tag{8.106}$$

$$S + O_2 \rightarrow SO + O. \tag{8.107}$$

Atomic oxygen can readily escape from Io, but the relatively slower thermal escape of atomic sulphur may lead to net accumulations of sulphur in the atmosphere and at the surface.

Ionospheric species are produced from photoionization and electron-impact ionization of SO_2, S, and O, with a balancing charge neutralization *via* dissociative recombination (cf. reaction (6.8) and the associated discussion in Section 6.2) of the molecular ion

$$SO_2^+ + e \rightarrow SO + O. \qquad (8.108)$$

Attempts have been made at modelling the neutral and ionized atmospheres of Io, but more data are needed for resolution of several problems. The Galileo spacecraft is scheduled to make a fly-by of Io at an altitude of 1000 km (as well as remaining in the Jovian system for several years), and the experiments it carries should complement ground-based observations in improving our understanding of the atmosphere of this interesting satellite.

Galileo has detected three *visible* airglow emissions from Io while the satellite was eclipsed by Jupiter. Bright blue glows emanate from volcanic plumes, and probably result from electron impact on SO_2; weaker red and green emissions from other regions may be produced by O and Na. The torus surrounding Io was discovered even before the Voyager encounters, but Voyager found a spectacular ultraviolet emission display from doubly and triply charged ions of O and S. Escape velocities from Io are small; those from Jupiter are at least ten times larger. Atoms can thus leave the moon to orbit around Jupiter, but not escape from Jupiter itself, and they form the torus. In the immediate vicinity of Io, the cloud contains neutral species, especially Na, but also K, O, and S. The lifetimes of these atoms against ionization is only a few hours, and beyond Io the torus contains virtually only ions. Terrestrial telescope observations have also identified emissions from Cl ions in the torus. Chemical reactions may thus actually produce NaCl in the atmosphere. Speculations about the origin of chlorine centre on emission from volcanoes or break-up of NaCl on Io's surface following impact from charged particles. The width of the torus is roughly the same as the radius of Jupiter, and the maximum ion density (a few thousand per cm^3) is found at just under six Jupiter radii from the planet.

Europa, like Io, is also providing surprises as the Galileo Europa orbiter mission unfolds in the late 1990s. The moon possesses more water than the total amount found on Earth, and appears to have had, in recent geological history, a salty ocean under its icy cracked and frozen surface. Internal tidal friction may have been sufficient to melt the ice. An ionosphere has been detected on Europa by radio occultation experiments performed during the Galileo encounters. By inference, there is also an atmosphere. Ion densities are only 10^4 ion cm^{-3} at most (up to 25 times less than found in the atmosphere of Jupiter). It seems likely that charged particles trapped in Jupiter's magnetosphere hit the icy surface of Europa, and eject and ionize water molecules. If that is the case, the most probable constituent of the tenuous neutral atmosphere would be oxygen. Yet more surprising, the near-infrared mapping spectrometer on the Galileo orbiter has detected hydrogen peroxide, H_2O_2, on the surface. Again, the source seems likely to be radiolysis of H_2O by energetic particles from Jupiter. H_2O_2 is, of course, photochemically labile, and can produce OH radicals that in turn could be a source of H_2 and O_2 in the atmosphere of the satellite. Other species detected on Europa's surface by the Galileo orbiter include SO_2 and CO_2. Since these compounds have a detectable

Table 8.6 Composition of the atmospheres of Uranus and Neptune

Species	Uranus	Neptune	Comment
H_2	0.825 ± 0.033	0.80 ± 0.03	
He	0.152 ± 0.033	0.19 ± 0.032	
CH_4	2.3×10^{-2}	$1.0–2.0 \times 10^{-2}$	troposphere
	2.0×10^{-5}	$6.0–50.0 \times 10^{-4}$	stratosphere
C_2H_6	$1–20 \times 10^{-9}$	$1.0–4.0 \times 10^{-6}$	
C_2H_2	1.0×10^{-8}	$2.0–7.4 \times 10^{-8}$	
CO	$< 3.0 \times 10^{-8}$	1.2×10^{-6}	
HCN	$< 1.0 \times 10^{-10}$	1.0×10^{-9}	
HD	1.48×10^{-4}	1.92×10^{-4}	$(D/H)_{Uranus} = 9.0 \times 10^{-5}$
CH_3D	8.3×10^{-6}	1.2×10^{-5}	$(D/H)_{Neptune} = 1.2 \times 10^{-4}$
H_2S	$< 8.0 \times 10^{-7}$	$< 3.0 \times 10^{-6}$	
NH_3	$< 1.0 \times 10^{-7}$	$< 6.0 \times 10^{-7}$	

Based on data presented by Yung, Y.K. and DeMore, W.B. *Photochemistry of planetary atmospheres*, Oxford University Press, Oxford, 1999.

vapour pressure at the temperature of Europa's surface, it follows that they must make some contribution to the atmosphere.

Even Callisto looks as though it might have a salty ocean deep underneath the heavily cratered surface of the moon. If that is the case, processes similar to those occurring on Europa could generate an atmosphere and ionosphere on Callisto. Data from the remaining flybys of Callisto during the Galileo mission during 1999 will be examined most carefully to look for signs of such an atmosphere.

8.6 Uranus, Neptune, Triton, and Pluto

Beyond Saturn lies a cold and dim part of the solar system hardly conducive to an active chemistry. However, some of the bodies at these great distances possess atmospheres with interesting features that are only now slowly becoming apparent. The successful encounters of Voyager 2 with the Uranian and Neptunian systems in 1986 and 1989 have brought a wealth of observational data that will take years for detailed analysis. Even the preliminary results have vastly enhanced the knowledge that has been built up from ground-based observations. Table 8.6 shows current estimates of the concentrations of various species in the atmospheres of the two planets. Uranus and Neptune are much smaller than Jupiter and Saturn, with masses only a few per cent of the mass of Jupiter, but they are denser, suggesting a rocky or icy core and an overall departure from solar composition. However, the composition of the atmospheres of Uranus and Neptune is closer to that presumed for the Sun than that of Jupiter or Saturn. The atmospheres are mainly hydrogen, and contain helium with a mole fraction abundance of about 0.15, almost exactly that of the Sun. Table 8.4 shows that, according to the Voyager measurements, helium is depleted with respect to this abundance on Jupiter, and, especially, on Saturn (although the Galileo measurements of helium at Jupiter seem to show a value nearer the solar abundance). Gravitational separation of helium towards the centre of the planet, which probably accounts for depletions higher up on the gas giants, cannot, of course, arise on planets with solid cores. There is certainly plenty of dynamic activity within the atmospheres. Neptune must be one of the windiest places in the solar system, with differential wind speeds approaching the speed of sound!

Quadrupole and pressure-induced bands of H_2 measured from Earth had demonstrated the presence of H_2 in the atmospheres before the Voyager encounters. Airglow and auroral emissions from atomic and molecular hydrogen in the vacuum ultraviolet region confirm that H and H_2 are present up to great altitudes. Excitation is effected either by impact of low-energy electrons or by Rayleigh and resonance scattering of sunlight (cf. Sections 2.6 and 7.2). Spectra of the planets are crossed by strong absorption bands of CH_4 in their atmospheres, and the visible component of the absorption lends a greenish colour to the planets. The Voyager instruments detected clouds of methane in the atmospheres of both Uranus and Neptune. Below the cloud decks, CH_4 could possess abundances of two per cent (Uranus) and not less than one per cent (Neptune). These abundances are 10 to 20 times the solar values, as might be expected if the planets accreted from ice-rich material. Above the clouds, there seems to be a significant difference in methane loading between the two planets. On Uranus, the mixing ratio is 2×10^{-5}, while on Neptune the estimates of the mixing ratio range from 6 to 50×10^{-4}, 30 to 250 times more abundant. This result may reflect the greater mixing of material from below the clouds in the case of Neptune, although it could also show a real difference in atmospheric composition. Acetylene, C_2H_2, is found on both planets.

The production of C_2H_2 and C_2H_6 in the atmospheres of Uranus and Neptune is likely to be a result of photochemical conversions of CH_4 similar to those that operate in the atmospheres of Jupiter and Saturn [cf. Fig. 8.11 and reactions (8.67) to (8.74)]. Photolysis of C_2H_6, and any C_2H_4 formed, will provide a further source of C_2H_2. Synthesis of higher hydrocarbons is likely, and aerosols may be formed. Changes in the brightness of Neptune are in antiphase with changes in solar activity (as they are on Titan), suggesting that sunlight is connected with the formation of aerosol particles that decrease the planetary albedo. Hazes and clouds are probably present in both atmospheres, although the densities and height distributions are very different. Uranus appears much clearer to great depths than does Neptune. Indeed, Neptune's atmosphere is so laden with aerosol particles that heating occurs in the atmosphere and a temperature inversion arises as in Titan's atmosphere. Figure 8.10 shows the temperature–pressure structure of the atmospheres of these planets; the 'tropopause' is evident for Neptune. According to a comparison of chemical models of the atmospheres of Jupiter and Neptune, one main difference resides in the lower temperature of Neptune. Condensation of C_2H_6 and C_2H_2 plays a major role for Neptune, and accounts for the lower concentrations of these hydrocarbons there compared with those on Jupiter.

Condensation also removes ammonia on Uranus and Neptune, and explains why NH_3 has not been detected in the atmospheres of those planets, but is found at similar abundances on Jupiter and Saturn. HCN has been detected in the atmosphere of Neptune, but not in the atmosphere of any other planet except Earth. It is not yet clear if the HCN has its origin in the interior of the planet or in photochemical production in the stratosphere. There are problems with both mechanisms. Transport of HCN from the interior requires passage through the cold tropopause, which would be expected to trap out almost all the gas. Photochemical production, on the other hand, requires a mixing ratio of N_2 of the order of 10^{-3}, making N_2 the dominant form of nitrogen in Neptune's atmosphere, and much greater than expected from equilibrium calculations. High N_2 concentrations could explain He abundances in excess of the value for the Sun (cf. Table 8.6), but there are other ways of interpreting this observation. A better understanding of HCN in the atmosphere will doubtless appear as further experimental data become available.

Carbon monoxide is another molecule whose presence in the atmosphere of Neptune is something of a puzzle. The observed mixing ratio (Table 8.6) is about three orders of magnitude larger than that found on Jupiter or Saturn (Table 8.4). Yet CO concentrations on Uranus are at least 40 times smaller. The large abundance on Neptune rules out an extraplanetary source, and the internal heat source on Neptune is likely to provide the dynamical activity that allows transport of CO upwards. Uranus does not possess such a heat source, and that is doubtless one of the main factors leading to the difference in concentrations of CO on the two planets. Radiotelescope microwave emission and absorption measurements, and UV observations from the Hubble Space Telescope, have confirmed the virtual absence of CO on Uranus, and its presence with a nearly constant mixing ratio throughout the stratosphere and the troposphere of Neptune. This uniformity of the mole fraction of CO in the wide pressure range 10^{-5} bar to 4 bar confirms an internal origin for CO, although the mechanism for its transport from the deep interior to the observable atmosphere still needs to be worked out in terms of atmospheric dynamics.

The outer solar system contains numerous smaller solid bodies in addition to the two large planets. All the bodies appear to be composed mainly of ices and other compounds, and all are probably aggregates of the solids that condensed from the solar nebula far from the heat of the proto-Sun. Pluto is the most familiar of these bodies. Often regarded as a planet, it has an unusual orbit, and may once have been the satellite of one of the outer planets. It has its own satellite, Charon, that may or may not have been captured. Triton is another peculiar body that may have been captured. It is the largest satellite of

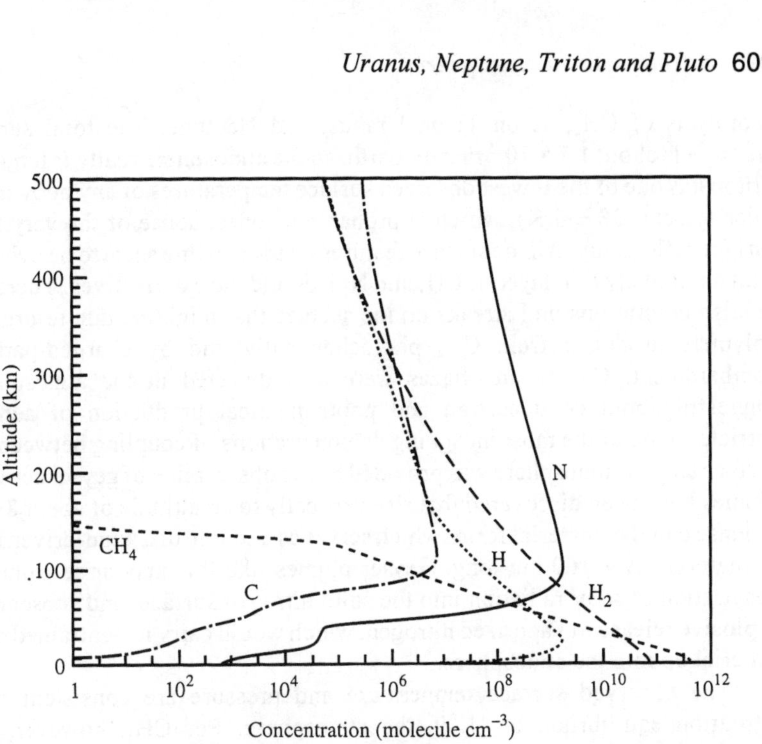

Fig. 8.16. Altitude profiles for some important minor constituents of the atmosphere of Triton. From Strobel, D.F. and Summers, M.E. *Triton's upper atmosphere and ionosphere* in Cruikshank, D.P. (ed.) *Neptune and Triton*, University of Arizona Press, Tucson, 1995.

what was the most distant planet at the time of the Voyager encounter, Neptune (Pluto then being closer to the Sun, although the two planets have now reversed relative positions). Within 10–100 million years it will be ruptured as it approaches its parent.

Both Triton and Pluto, at least, have tenuous atmospheres according to ground-based observations. Before the Voyager fly-by of Triton, the nature of this interesting satellite's atmosphere could only be inferred from the presence of CH_4 and N_2 surface frosts. Ultraviolet spectroscopy experiments on Voyager 2 show strong airglow features from neutral molecular nitrogen and ionized atomic nitrogen, similar in some ways to the airglow of Titan (cf. Section 8.5.1 and Fig. 8.13). These results show that the atmosphere is predominantly made up of N_2, but that CH_4 is also present in the lower atmosphere with a mixing ratio of about 10^{-4}. Post-Voyager observations showed the presence of CH_4, CO, and CO_2 on the surface of Triton. The first two of these compounds, but not CO_2, have enough vapour pressure at $T = 38$ K to make some detectable contribution to the atmosphere; so do Ar and N_2. A very intense emission feature from H Lyman-α in the airglow spectrum shows that atomic hydrogen is present in the atmosphere, and it must be derived from photolysis of CH_4, as on Titan, Uranus, and Neptune. The total surface pressure is about 1.7×10^{-5} that on Earth, so the atmosphere really *is* tenuous! Triton has one of the lowest observed surface temperatures of any body in the solar system (38 ± 4 K), which is probably a consequence of the very high surface reflectivity. All the surface features on the satellite seem to be overlain with a relatively thin layer of CH_4 and N_2 ices and their derivatives. There are reddish colorations and streaks on the surface that might be due to organic polymers produced from CH_4 photochemically and by charged-particle bombardment. Clouds and hazes were also detected in the atmosphere, suggesting both condensation and photochemical production of aerosol particles. One of the most intriguing demonstrations of coupling between the surface and the atmosphere was provided by the observation of geyser activity. Plumes have been discovered that rise vertically to an altitude of about 8 km. A dense cloud of material forms which serves as a source of a wind-driven trail of material over 100 km long. Geyser plumes like this may arise from the penetration of solar radiation into the solid nitrogen surface, and subsequent explosive release of vaporized nitrogen, which would carry ice-entrained dark material up into the atmosphere.

The observed surface temperature and pressure are consistent with saturation equilibrium of N_2 in the atmosphere. For CH_4, however, the abundance seems to be far below saturation, and at 40 km the mixing ratio is only 2.5×10^{-6}, some 30 times smaller than expected if the concentration were controlled by temperatures at the surface and in the lowest part of the atmosphere (say up to 10 km). The mixing ratios for CH_4 imply a decrease in concentration with altitude much more rapid than predicted by a hydrostatic distribution (Section 2.1). A possible explanation is that methane is removed photochemically, and that vertical mixing is weak enough that the concentration gradients can be maintained. If this interpretation is correct, then photochemistry will occur predominantly in the first 20–30 km of the atmosphere. Photolysis of CH_4 by solar and interstellar Lyman-α ($\lambda = 121.6$ nm) radiation yields CH_3, CH_2, CH, H_2, and H [cf. reactions (8.67) to (8.70)]. Because nitrogen is present, Titan's atmosphere may provide a better analogy with Triton's than does Jupiter's. As illustrated in Fig. 8.14, C_2 hydrocarbons such as C_2H_6, C_2H_4, and C_2H_2 may be formed initially, but because the atmosphere of Triton is so cold, the heavier hydrocarbons condense without undergoing photolysis or further reaction, and hydrogen escapes from the satellite. According to one calculation, the flux of C_2 hydrocarbons to the surface would have been sufficient to create a layer on Triton 6 m thick over the life-span of the solar system. The atomic and molecular hydrogen produced ultimately escapes from Triton to enter Neptune's magnetosphere.

One surprising difference between Triton and Titan that was discovered in the Voyager encounter was the presence of a substantial ionosphere on Triton possessing an unusually high electron density. An obvious candidate for the major ion would be N^+, formed by photoionization and electron impact in the processes

$$N_2 + h\nu \rightarrow N^+ + N + e \tag{8.109}$$

$$N_2 + e \rightarrow N^+ + N + 2e. \tag{8.110}$$

However, N^+ ions are likely to undergo very rapid losses, and the rates of processes (8.109) and (8.110) with the available fluxes of solar EUV and precipitated magnetospheric electrons may not be adequate to maintain the electron concentrations detected in the Voyager experiments. A suggested alternative is that C^+ ions might be the major reservoir of positive charge. Neutral atomic carbon can be formed from CO by the pair of reactions

$$CO + h\nu \rightarrow CO^+ + e \tag{8.111}$$

$$CO^+ + e \rightarrow C + O. \tag{8.112}$$

Since C has the low ionization potential of 11.26 eV, most other ions will transfer charge to this atom in processes such as

$$N_2^+ + C \rightarrow C^+ + N_2 \tag{8.113}$$

$$N^+ + C \rightarrow C^+ + N. \tag{8.114}$$

Unlike N^+, the possible loss processes for C^+ are all slow, so that C^+ concentrations (and hence electron densities) can build up to much higher values than if N^+ were the only atomic ion.

Despite the daunting potential difficulties, attempts have been made to model the minor constituents in the atmosphere of Triton. Figure 8.16 shows the predicted altitude profiles of one such model. The parent hydrocarbon CH_4 is photolysed most rapidly near the surface, and its concentration decreases rapidly with altitude because of the photodissociation. The most abundant products of this photochemistry are H, H_2, C_2H_4, and C_2H_2, although concentrations of H and H_2 are limited by their escape from the atmosphere. Higher hydrocarbons are removed by condensation at the surface, and C_2H_6 is absent from this model because the direct production of CH_3 in reaction (8.67) has not been included and other sources of CH_3 are weak in Triton's atmosphere. Of the atomic species, C, N, and O are the most important, and the presence of N and O allows the formation of HCN and CO_2.

For Pluto, the experimental information comes mainly from ground-based observations and the Infrared Astronomical Satellite (IRAS). The near-infrared spectrum of Pluto is dominated by bands similar to those produced by solid (i.e. surface) methane, although there are also features that can be ascribed to gas-phase absorption. There are variations of absorption with the planet's orbital phase that imply a system with surface frost as well as atmospheric absorption. Comparison of the infrared reflectance spectrum with a model spectrum suggests that the surface ice is largely composed of solid N_2 (98 per cent), with traces of CH_4 (1.5 per cent) and CO (0.5 per cent). CO_2, if present, makes up less than 0.07 per cent of the surface (in distinction to Triton, where it contributes 0.2 per cent). Note that H_2O ice is seen neither on Pluto nor on Triton, although its presence has been demonstrated on the surface of Charon, Pluto's satellite. The presence of the ices implies an atmosphere that at least corresponds to the saturated vapour pressures of the constituents of the ice. However, the temperature of Pluto's surface is far from certain, and the vapour pressure of N_2, CH_4, and CO are all sensitive functions of temperature. Current best measurements put the surface temperature in the range 30–44 K, with the middle of the range thought most probable. The vapour pressure of N_2 increases from 0.1 to more than 500 μbar over the full temperature range. A star-occultation observation has demonstrated clearly the existence of a true atmosphere around Pluto, because the light of the star dimmed slowly rather than abruptly. These measurements imply an atmosphere with a pressure of 0.78 μbar and a temperature of 117 K at the altitude where half the starlight was attenuated (ca. 1214 km). There must obviously be a significant temperature gradient somewhere between the surface and the half-light level, and different models interpret the behaviour in different ways. One model, which sets the temperature drop within the first 100 km of the surface, predicts a surface pressure of 3 μbar.

Pluto is small (radius ca. 1500 km), and the escape velocity is only 0.95 km s^{-1}, less than that of Titan. It has been suggested that the rate of escape of methane from Pluto could be so great that the whole mass of the planet could have been lost over its life unless some heavier gas (e.g. nitrogen or argon) were present to limit the escape rate by providing a diffusional barrier. Allowing for energy transport in the atmosphere, however, greatly reduces the predicted escape rate, and it seems that the loss of CH_4 over the age of the solar system could be as little as 0.5 per cent of the planetary mass. There is thus no need to invoke a heavier secondary gas to maintain the CH_4 on Pluto. On Pluto's smaller-mass companion, Charon, methane is absent because of hydrodynamic escape.

It seems likely that photochemistry similar to that on Titan or Triton will occur. Solar fluxes at both Pluto and Triton are roughly ten times lower than at Titan, so that solar photochemistry is correspondingly slower. In these circumstances, chemistry initiated by charged particles and cosmic rays is likely to become relatively more important, but the secondary steps are likely to be similar to those already encountered. Pluto has a red tint to its surface, suggesting that methane is being converted to higher hydrocarbons as it is in Triton's atmosphere. Abundances of CH_4 and CO on Pluto are an order of magnitude higher on Pluto than on Triton. As a consequence, photochemistry can occur at higher altitudes, and the products thus avoid immediate condensation on the cold surface. In addition, the extended atmosphere of Pluto means that, although the planet is further from the Sun than Triton, a larger area of atmosphere is exposed to the Sun, and the total dose of EUV radiation received by the two bodies may be comparable.

Voyager 2 took advantage of an opportunity that arises only once in 176 years when a unique alignment of the planets made feasible a visit to them. The spacecraft and its instrumentation made a remarkable journey that lasted more than 12 years from the launch and successfully returned information, much dramatic and unexpected, about the planets and their atmospheres. The Voyager and Pioneer spacecraft have now passed on beyond Neptune and Triton to become the most distant identifiable objects in the solar system with the exception of the comets, and it is to these latter remarkable bodies that we turn finally in this chapter.

8.7 Comets

Comets are accompanied by extended tails of gases that might be thought of as atmospheres. In the classical model, a comet nucleus consists of dust and ice in the manner of a very loosely packed snowball. When a comet's orbit brings it near enough to the Sun, heating of the outer layers vaporizes material that is released to space. The familiar coma and tail of a comet are formed by the sublimation of ices; the vaporized gases entrain dust and ice particles. Photochemical reactions can alter the composition of the gases just as in any other atmosphere, but almost all of the gases released from the nucleus are lost to space. Cometary science changed dramatically in the mid-1980s with spacecraft encounters with cometary bodies, especially with comet Halley in March 1986. Six spacecraft from four agencies approached the comet as a 'Halley Armada'. Giotto was the mission of the European Space Agency, and two craft of the Venera series (cf. Section 8.2) were sent *via* Venus as a Franco–Soviet venture. 'Venera' was renamed 'Vega' for the encounters, which is a pun: the word is a contraction of Venera–Halley, but Russian has no 'H', and 'G' has to be used instead, turning the name into a bright star! Giotto and Vega carried mass spectrometers for both neutral and ionized species, infrared spectrometers to study the gases, and particulate impact analysers to investigate the chemical and isotopic composition of the solid dust-particles. The *in-situ* observations were backed up by a full-scale remote ground- and satellite-based campaign to provide the most intensive study of a comet ever attempted.

Molecules found in the coma of comet Halley that are likely to be parent species from the nucleus include CO, CO_2, CH_4, NH_3, N_2, HCHO, HCN, and saturated and unsaturated hydrocarbons, in addition to water vapour. Carbon monoxide is the most abundant species after water, perhaps making up 20 per cent of the gases leaving the nucleus. The parent molecules are then the source of daughter atoms, radicals, and ions that have been known for some time from their optical emission spectra. Estimated rates of gas sublimation suggest that the comet loses about one metre of its thickness every time it orbits the Sun, so that the life of the comet is limited to less than 10 000 revolutions. The dust component is weighted towards large abundances of very small particles, some of which have a mass of as little as 10^{-17} g (which might correspond to 100 atoms). Several types of dust particle were discovered. They include particles rich in the lighter elements, C, H, O, and N (and named *CHON particles*), and others that are rich in heavier elements such as Mg (and are probably dominated by silicates). The weight ratio of gas:dust is about 0.9, so that, for every 100 atoms of Mg in the dust, there are 312 atoms of carbon in the gas. A wide variety of organic molecules has been inferred to be present in the CHON particles, including saturated and unsaturated hydrocarbons, compounds possessing –CN and –NH– groups, and ring compounds such as hydrocarbons, pyrrole, pyridine, purine, adenine, and xanthine. The dust grains serve as an extended source of gas in the inner coma. For example, a large source of CO extends up to about 15 000 km from the nucleus, and constitutes more than half the total CO flux. Strange jets contain CN radicals up to 50 000 km from the nucleus, and may be related to a source involving CHON particles. The CHON particles have an abundance of carbon almost identical to the cosmic abundance, and this result, and some other observations of atomic and isotopic abundances, suggests that dust from Halley's comet is the most unaltered material from the early solar system yet analysed in man's experiments.

The nucleus of comet Halley has an extremely low albedo (ca. 0.04), making it one of the darkest objects in the solar system. Bright dust jets come from relatively small active sites in the sunlit hemisphere. The surface darkness may be a consequence of residual complex and non-volatile organic species left near the surface. Measurements by the ion mass spectrometers show the presence of the dissociation products of methane. If a low-temperature condensate such as methane is present in the near-surface layers of the nucleus, then those layers must have remained unaltered until they were exposed to the Sun. Other compounds besides CH_4, such as CO and CO_2, are also present near the surface of comet Halley, and these volatile materials cause pronounced activity of the comet when it is still up to five times more distant from the Sun than the Earth is. This evidence, and the measured atomic abundances in the dusts and gases, make it likely that the comet is made up of pristine matter from the solar nebula, and that it must have formed in the outermost reaches of the solar system. The comets are thus among the most primitive objects that remain in the solar system, and their study provides one line of evidence about the nature of the system soon after its formation. Data from Giotto and Vega are still being analysed and providing more detailed information. Various plans have been made for spacecraft missions to rendezvous with comets, perform *in-situ* measurements, and perhaps even return samples to Earth. Although these projects have been subject to cancellation or modification as a result of budgetary constraints, at the time of writing several seem viable. Included in the missions are NASA's Stardust and ESA's Rosetta. Stardust, launched in February 1999, will fly within 100 km of comet P/Wild-2 in early 2004, and collect cometary dust and volatile materials. Real-time compositional analysis will be carried out throughout the mission, and the spacecraft will return to Earth in January 2006 to drop off the samples using a streamlined, low-cost reentry capsule. Rosetta includes a neutral and ion mass spectrometer and a gas chromatograph on the orbiter, and two gas chromatograph–mass spectrometer systems on the lander to study composition, isotopic abundances, and complex organic molecules in cometary material. The current mission profile calls for a launch in January 2003, flybys of Mars and Earth in 2005, passage through the asteroid belt in 2007–2008, and the encounter with comet P/Wirtanen in late 2011. These are grand and ambitious plans indeed! However, whatever the outcome of such future missions, knowledge already gained over the last twenty-five years from the spacecraft investigations of the planets, their satellites, and the comets has enabled us to construct informed views about the past and probable future of Earth and other planets and their atmospheres. These extrapolations are the subject of our next chapter.

Bibliography

The first book listed describes the Solar System as it appears after several decades of planetary exploration. Atmospheres are discussed in this context, and the book makes an excellent introduction to the study of the planets and their atmospheres. Yung and DeMore focus specifically on atmospheric chemistry, and their book is an outstandingly useful resource. A more popularized account follows, but it is an extremely useful source of information. The fourth publication is a book prepared to accompany a BBC Television series, and it is also a highly readable introduction to the solar system, the planets, and their satellites.

Physics and chemistry of the solar system. Lewis, J.S., 2nd edn. (Academic Press, San Diego, 1997).

Photochemistry of planetary atmospheres. Yung, Y.K. and DeMore, W.B. (Oxford University Press, Oxford, 1999).

The new solar system. Beatty, J.K., Petersen, C.C., and Chaikin, A. (eds.) 4th edn. (Cambridge University Press, Cambridge, 1999).

The planets. McNab, D. and Younger, J. (BBC Worldwide, London, 1999).

The photochemistry of atmospheres. Levine, J.S. (ed.) (Academic Press, Orlando, 1985).

Atmospheric compositions: key similarities and differences. Pepin, R.O., in *Origin and evolution of planetary and satellite atmospheres*. Atreya, S.K., Pollack, J.B., and Matthews, M.S. (eds.) (University of Arizona Press, Tucson, 1989).

Planetary atmospheres. Hunten, D.M., Pepin, R.O., and Owen, T. in *Meteorites and the early solar system*. Kerridge, J.F., and Matthews, M.S. (eds.) (University of Arizona Press, Tucson, 1989).

Chemical processes in the solar system: a kinetic perspective. McElroy, M.B., *Int. Rev. Sci., Phys. Chem. Ser. 2*, Vol. 9 (*Chemical kinetics*), 127 (1975).

Section 8.2

Venus

These references cover most aspects of composition, structure, meteorology, and chemical transformation in the Venusian atmosphere as seen after the Pioneer–Venus and Venera missions.

The atmospheres of Venus, Earth, and Mars: A critical comparison. Prinn, R.G. and Fegley, B., Jr. *Ann. Rev. planet. Space Sci.* **15**, 171 (1987).

Photochemistry of the atmospheres of Mars and Venus. Krasnopolsky, V.A. (Springer Verlag, Berlin, 1986).

Volcanism and tectonics on Venus. Nimmo, F. and McKenzie, D. *Annu. Rev. Earth planet. Sci.* **26,** 23 (1998).

The volcanoes and clouds of Venus. Prinn, R.G. *Scient. Am.* **252** (3), 36 (March 1985).

Remote sensing of planetary atmospheres: Venus. Taylor, F.W. *Advan. Space Res.* **21,** 409 (1998).

The atmosphere of Venus. Schubert, G. and Covey, C. *Scient. Am.* **245** (7), 44 (July 1981).

Magellan – a new view of Venus geology and geophysics. Bindschadler, D.L. *Rev. Geophys.* **33,** 459 (1995).

Magellan arrives at Venus. Grimm, R.E., guest editor. Collected papers in special issue (August 1990) of *Geophys. Res. Letts.* **17** (1990).

The Venus atmosphere. Keating, G.M. (ed.) *Adv. Space Res.* **10**, No. 5 (1990).

Venus. Hunten, D.M., Colin, L., Donahue, T., and Moroz, V.I. (eds.) (University of Arizona Press, Tucson, 1983).

The atmosphere of Venus. Moroz, V.I. *Space Sci. Rev.* **29**, 3 (1981).

Pioneer Venus (special issues). *J. geophys. Res.* **85**, No. A13 (1980); *Science* **205**, 41 et seq. (1979); *Science* **203**, 743 et seq. (1979).

The deep atmosphere of Venus revealed by high-resolution night-side spectra. Bezard, B., de Bergh, C., Crisp, D., and Maillard, J.-P. *Nature, Lond.* **345**, 508 (1990).

Section 8.2.1

An exospheric perspective of isotopic fractionation of hydrogen on Venus. Hedges, R.R. *J. geophys. Res.* **104,** 8463 (1999).

Nonthermal escape of the atmospheres of Venus, Earth, and Mars. Shizgal, B.D. and Arkos, G.G. *Rev. Geophys.* **34,** 483 (1996).

Hydrodynamic escape of oxygen from primitive atmospheres: Applications to the cases of Venus and Mars. Chassefiere, E. *Icarus* **124,** 537 (1996).

Impact erosion of planetary atmospheres: Some surprising results. Newman, W.I., Symbalisty, E.M.D., Ahrens, T.J., and Jones, E.M. *Icarus* **138,** 224 (1999).

Wind effects on the entry of meteorites into the atmosphere of Venus. Cordero, G. and Mendoza, B. *Planet. Space Sci.* **47,** 345 (1999).

Hydrous silicates and water on Venus. Zolotov, M.Y., Fegley, B., and Lodders, K. *Icarus* **130,** 475 (1997).

Volcanic degassing of argon and helium and the history of crustal production on Venus. Namiki, N. and Solomon, S.C. *J. geophys. Res.* **103,** 3655 (1998).

Water vapour in the lower atmosphere of Venus: A new analysis of optical spectra measured by entry probes. Ignatiev, N.I., Moroz, V.I., Moshkin, B.E., Ekonomov, A.P., Gnedykh, V.I., Grigoriev, A.V., and Khatuntsev, I.V. *Planet. Space Sci.* **45,** 427 (1997).

CO_2 amount on Venus constrained by a criterion of topographic greenhouse instability. Hashimoto, G.L., Abe, Y., and Sasaki, S. *Geophys. Res. Letts.* **24,** 289 (1997).

Temporal and spatial variations in the Venus mesosphere retrieved from Pioneer Venus OIR. Irwin, P.G.J. *Advan. Space Res.* **19,** 1169 (1997).

The 825–1110 angstrom EUV spectrum of Venus. Stern, S.A., Slater, D.C., Gladstone, G.R., Wilkenson, E., Cash, W.C., Green, J.C., Hunten, D.M., Owen, T.C., and Paxton, L. *Icarus* **122,** 200 (1996).

Observations of the CO bulge on Venus and implications for mesospheric winds. Gurwell, M.A., Muhleman, D.O., Shah, K.P., Berge, G.L., Rudy, D.J., and Grossman, A.W. *Icarus* **115,** 141 (1995).

Measurements of the Venus lower atmosphere composition: a critical comparison. Hoffman, J.H., Oyama, V.I., and von Zahn, U. *J. geophys. Res.* **85,** 7871 (1980).

The oxidation state of the lower atmosphere and surface of Venus. Fegley, B., Zolotov, M.Y., and Lodders, K. *Icarus* **125,** 416 (1997).

A model of thermochemical equilibrium in the near-surface atmosphere of Venus. Zolotov, M.Y. *Geokhimiya* **100,** 26327 (1995).

Atmosphere–surface interactions on Venus and implications for atmospheric evolution. Khodakovsky, I.L. *Planet. Space Sci.* **30,** 803 (1982).

Section 8.2.2

Clouds are discussed at length in the more general references given above. These papers deal with additional points, including high altitude hazes, and clouds on the night side of the planet.

Venus cloud formation in the meridional circulation. Imamura, T. and Hashimoto, G.L. *J. geophys. Res.* **103,** 31349 (1998).

A numerical microphysical model of the condensational Venus cloud. James, E.P., Toon, O.B., and Schubert, G. *Icarus* **129,** 147 (1997).

Vega mission results and the chemical composition of Venusian clouds. Krasnopolsky, A. *Icarus* **80,** 210 (1989).

Cloud structure on the dark side of Venus. Allen, D.A. and Crawford, J.W. *Nature, Lond.* **307,** 222 (1984).

A re-examination of the evidence for large, solid particles in the clouds of Venus. Knollenberg, R.G. *Icarus* 57, **161** (1984).

Large, solid particles in the clouds of Venus: do they exist? Toon, O.B., Ragent, B., Colburn, D., Blamont, J., and Cot, C. *Icarus* **57,** 143 (1984).

Venus: mesospheric hazes of ice, dust, and acid aerosols. Turco, R.P., Toon, O.B., Whitten, R.C., and Keesee, R.G. *Icarus* **53,** 18 (1983).

Section 8.2.3

Lightning may be of importance in Venusian atmospheric chemistry. The first reference here discusses the general phenomenon of lightning, and the others explore aspects of lightning relevant to Venusian observations.

Optical detection of lightning on Venus. Hansell, S.A., Wells, W.K., and Hunten, D.M. *Icarus* **117**, 345 (1995).
Lightning generation in planetary atmospheres. Levin, Z., Borucki, W.J., and Toon, O.B. *Icarus* **56**, 80 (1983).
Lightning measurements from the Pioneer Venus orbiter. Scarf, F.L. and Russell, C.T. *Geophys. Res. Letts.* **10**, 1192 (1983).
Comparison of Venusian lightning observations. Borucki, W.J. *Icarus* **52**, 354 (1982).
Lightnings and nitric oxide on Venus. Krasnopolsky, V.A. *Planet. Space Sci.* **31**, 1363 (1983).
Production of nitric oxide by lightning on Venus. Levine, J.S., Gregory, G.L., Harvey, G.A., Howell, W.E., Borucki, W.J., and Orville, R.E. *Geophys. Res. Letts.* **9**, 893 (1982).

Section 8.2.4

Sulphur chemistry

Is disulfur monoxide a second absorber on Venus? Na, C.Y. and Esposito, L.W. *Icarus* **125,** 364 (1997).
VEGA 1 and VEGA 2 entry probes: An investigation of local UV absorption (220–400 nm) in the atmosphere of Venus (SO_2, aerosols, cloud structure). Bertaux, J.L., Widemann, T., Hauchecorne, A., Moroz, V.I., and Ekonomov, A.P. *J. geophys. Res.* **101,** 12709 (1996).
Laboratory measurement of the temperature dependence of gaseous sulfur dioxide (SO_2) microwave absorption with application to the Venus atmosphere. Suleiman, S.H., Kolodner, M.A., and Steffes, P.G. *J. geophys. Res.* **101,** 4623 (1996).
Formation of carbonyl sulfide (OCS) from carbon monoxide and sulfur vapor and applications to Venus. Hong, Y., and Fegley, B. *Icarus* **130,** 495 (1997).
The formation of carbonyl sulfide (OCS) in Venus' lower atmosphere. Hong, Y. and Fegley, B. *Meteoritics & planet. Sci.* **32,** A61 (1997).
The microwave absorption and abundance of sulfuric acid vapor in the Venus atmosphere. Kolodner, M.A., and Steffes, P.G. *Icarus* **132,** 151 (1998).
The sulfur vapor pressure over pyrite on the surface of Venus. Hong, Y. and Fegley, B. *Planet. Space Sci.* **46,** 683 (1998).
Why pyrite is unstable on the surface of Venus. Fegley, B. *Icarus* **128,** 474 (1997).
H_2O–H_2SO_4 system in Venus' clouds and OCS, CO and H_2SO_4 profiles in Venus' troposphere. Krasnopolsky, V.A. and Pollack, J.B. *Icarus* **109**, 58 (1994).
International Ultraviolet Explorer observation of Venus SO_2 and SO. Na, C.Y., Esposito, L.W., and Skinner, T.E. *J. geophys. Res.* **95**, 7485 (1990).
Sulfur dioxide: episodic injection shows evidence for active Venus volcanism. Esposito, L.W. *Science* **223**, 1072 (1984).

Sulfur trioxide in the lower atmosphere of Venus? Craig, R.A., Reynolds, R.T., Ragent, B., Carle, G.C., Woeller, F., and Pollack, J.B. *Icarus* **53**, 1 (1983).

The clouds of Venus: sulfuric acid by the lead chamber process? Sill, G.T. *Icarus* **53**, 10 (1983).

On the possible roles of gaseous sulfur and sulfanes in the atmosphere of Venus. Prinn, R.G. *Geophys. Res. Letts.* **6**, 807 (1979).

Section 8.2.5

Catalytic processes in the atmospheres of Earth and Venus. DeMore, W.B. and Yung, Y.L. *Science* **217**, 1209 (1982).

Photochemistry of the stratosphere of Venus: implications for atmospheric evolution. Yung, Y.L. and DeMore, W.B. *Icarus* **51**, 199 (1982).

Venusian airglow and oxygen spectroscopy

The impact of gravity waves on the Venus thermosphere and O_2 IR nightglow. Zhang, S., Bougher, S.W., and Alexander, M.J. *J. geophys. Res.* **101**, 23195 (1996).

The 825–1110 Å EUV spectrum of Venus. Stern, S.A., Slater, D.C., Gladstone, G.R., Wilkenson, E., Cash, W.C., Green, J.C., Hunten, D.M., Owen, T.C., and Paxton, L. *Icarus* **122**, 200 (1996).

Ground-based near-infrared observations of the Venus nightside: The thermal structure and water abundance near the surface. Meadows, V.S. and Crisp, D. *J. geophys. Res.* **101**, 4595 (1996).

Ground-based near-infrared observations of the Venus nightside: 1.27μm $O_2(a^1\Delta_g)$ airglow from the upper atmosphere. Crisp, D., Meadows, V.S., Bezard, B., de Bergh, C., Maillard, J.P., and Mills, F.P. *J. geophys. Res.* **101**, 4577 (1996).

Hot hydrogen and oxygen atoms in the upper atmosphere of Venus and Mars. Nagy, A.F., Kim, J., and Cravens, T.E. *Ann. Geophys.* **8**, 251 (1990).

Identification of the UV nightglow from Venus. Feldman, P.D., Moos, H.W., Clarke, J.T. and Lane, A.L. *Nature, Lond.* **279**, 221 (1979).

Ultraviolet night airglow of Venus. Stewart, A.I. and Barth, C.A. *Science* **205**, 59 (1979).

Spectroscopy of molecular oxygen in the atmospheres of Venus and Mars. Trauger, J.T. and Lunine, J.I. *Icarus* **55**, 272 (1983).

The upper atmosphere and ionosphere of Venus

Pioneer Venus orbiter contributions to a revised Venus reference ionosphere. Brace, L.H., Grebowsky, J.M., and Kliore, A.J. *Advan. Space Res.* **19**, 1203 (1997).

The dayside Venus ionosphere. 1. Pioneer–Venus retarding potential analyser experimental observations. Miller, K.L., Knudsen, W.C., and Spenner, K. *Icarus* **57**, 386 (1984).

A two-dimensional model of the ionosphere of Venus. Cravens, T.E., Crawford, S.L., Nagy, A.F., and Gombosi, T.I. *J. geophys. Res.* **88**, 5595 (1983).

The chemistry of metastable species in the Venusian ionosphere. Fox, J.L. *Icarus* 51, **248** (1982).

Variations in ion and neutral composition of Venus: Evidence of solar control of the formation of predawn bulges in H^+ and He. Taylor, H.A., Jr., Mayr, H., Brinton, H., Niemann, H., Hartle, R., and Daniell, R.E., Jr. *Icarus* **52**, 211 (1982).
Day and night models of the Venus thermosphere. Massie, S.T., Hunten, D.M., and Sowell, D.L. *J. geophys. Res.* **88**, 3955 (1983).
Global empirical model of the Venus thermosphere. Hedin, A.E., Niemann, H.B., Krasprzak, W.T., and Seiff, A. *J. geophys. Res.* **88**, 73 (1983).
A two-dimensional model of the nightside ionosphere of Venus. Bougher, S.W. and Cravens, T.E. *J. geophys. Res.* **89**, 3837 (1984).

Section 8.3

Mars. Kieffer, H.H., Jakosky, B.M., Snyder, C.W., and Matthews, M.S. (eds.) (University of Arizona Press, Tucson, 1992).
Chemical composition of the atmosphere of Mars. Moroz, V.I. *Advan. Space Res.* **22,** 449 (1998).
Mars. Spohn, T., Sohl, F., and Breuer, D. *Astron. Astrophys. Rev.* **8,** 181 (1998).
Photochemistry of the atmospheres of Mars and Venus. Krasnopolsky, V.A. (Springer-Verlag, Berlin, 1986).
Mars' atmosphere: A study with microstation network. Marov, M.Y. and Walberg, G.D. *Advan. Space Res.* **22,** 459 (1998).
Atmosphere–surface interactions on Mars: $\Delta\,^{17}O$ measurements of carbonate from ALH 84001. Farquhar, J., Thiemens, M.H., and Jackson, T. *Science* **280,** 1580 (1998).
Photochemical mapping of Mars. Krasnopolsky, V.A. *J. geophys. Res.* **102,** 13313 (1997).
New composite spectra of Mars, 0.4–5.7 μm. Erard, S. and Calvin, W. *Icarus* **130,** 449 (1997).
Stability of the Martian atmosphere: possible role of heterogeneous chemistry. Atreya, S.K. and Blamont, J.E. *Geophys. Res. Letts.* **17**, 287 (1990).
The composition of the Martian atmosphere. Owen, T. *Adv. Space Res.* **2**, 75 (1982).

Some missions to Mars that have provided data about the atmosphere

Results of the Mars Pathfinder atmospheric structure investigation. Magalhaes, J.A., Schofield, J.T., and Seiff, A. *J. geophys. Res.* **104,** 8943 (1999).
Overview of the Mars Pathfinder Mission: Launch through landing, surface operations, data sets, and science results. Golombek, M.P. and 54 others. *J. geophys. Res.* **104,** 8523 (1999).
The Planet-B neutral gas mass spectrometer. Niemann, H.B. and 17 others. *Earth Planets & Space* **50,** 785 (1998).
The Ion Mass Imager on the Planet-B spacecraft. Norberg, O., Yamauchi, M., Lundin, R., Olsen, S., Borg, H., Barabash, S., Hirahara, M., Mukai, T., and Hayakawa, H. *Earth Planets & Space* **50,** 199 (1998).
The Mars Oxidant experiment (MOx) for Mars '96. McKay, C.P., Grunthaner, F.J., Lane, A.L., Herring, M., Bartman, R.K., Ksendzov, A., Manning, C.M., Lamb, J.L., Williams, R.M., Ricco, A.J., Butler, M.A., Murray, B.C., Quinn, R.C., Zent, A.P., Klein, H.P., and Levin, G.V. *Planet. Space Sci.* **46,** 769 (1998).
Vikings at Mars: (special reports), *J. geophys. Res.* **82**, No. 28 (30 Sept. 1977).

Evolution of the atmosphere

The rôle of ionic processes is explained in the text at the end of Section 8.3.3.

Early Mars climate models. Haberle, R.M. *J. geophys. Res.* **103,** 28467 (1998).
Upper limits to the outflow of ions at Mars: Implications for atmospheric evolution. Fox, J.L. *Geophys. Res. Letts.* **24,** 2901 (1997).
On the outflow of O_2^+ ions at Mars. Kar, J., Mahajan, K.K., and Kohli, R. *J. geophys. Res.* **101,** 12747 (1996).
The N-15/N-14 isotope fractionation in dissociative recombination of N_2^+. Fox, J.L. and Hac, A. *J. geophys. Res.* **102,** 9191 (1997).
Production and escape of nitrogen from Mars. Fox, J.L. *J. geophys. Res.* **98,** 3297 (1993).
Nonthermal escape of the atmospheres of Venus, Earth, and Mars. Shizgal, B.D., and Arkos, G.G. *Rev. Geophys.* **34,** 483 (1996).
Hydrodynamic escape of oxygen from primitive atmospheres: Applications to the cases of Venus and Mars. Chassefiere, E. *Icarus* **124,** 537 (1996).
Atmospheric loss since the onset of the Martian geologic record: Combined role of impact erosion and sputtering. Brain, D.A. and Jakosky, B.M. *J. geophys. Res.* **103,** 22689 (1998).
Loss of atmosphere from Mars due to solar wind induced sputtering. *Science* **268,** 697 (1995).
Loss of H and O from Mars: Implications for the planetary water inventory. Lammer, H., Stumptner, W., and Bauer, S.J. *Geophys. Res. Letts.* **23,** 3353 (1996).
The history of Martian volatiles. Jakosky, B.M. and Jones, J.H. *Rev. Geophys.* **35,** 1 (1997).
Photochemistry and evolution of Mars' atmosphere, a Viking perspective. McElroy, M.B., Kong, T.Y., and Yung, Y.L. *J. geophys. Res.*, **82,** 4379 (1977).

Prospects for existence of life on Mars

Life on Mars: chemical arguments and clues from Martian meteorites. Brack, A., and Pillinger, C.T. *Extremophiles* **2,** 313 (1998).
Prospects for the evolution of life on Mars —Viking 20 years later. Mancinelli, R.L. *Advan. Space Res.* **22,** 471 (1998).
The search for life on Mars. McKay, C.P. *Origins Life Evol. Biosphere* **27,** 263 (1997).
Possible complex organic compounds on Mars. Kobayashi, K., Sato, T., Kajishima, S., Kaneko, T., Ishikawa, Y., and Saito, T. *Advan. Space Res.* **19,** 1067 (1997).
Organic degradation under simulated Martian conditions. Stoker, C.R. and Bullock, M.A. *J. geophys. Res.* **102,** 10881 (1997).
The search for nitrogen compounds on the surface of Mars. Mancinelli, R.L. *Advan. Space Res.* **18,** 241 (1996).

Ozone on Mars

Minimal aerosol loading and global increases in atmospheric ozone during the 1996–1997 Martian northern spring season. Clancy, R.T., Wolff, M.J., and James, P.B. *Icarus* **138,** 49 (1999).

Mars ozone measurements near the 1995 aphelion: Hubble space telescope ultraviolet spectroscopy with the faint object spectrograph. Clancy, R.T., Wolff, M.J., James, P.B., Smith, E., Billawala, Y.N., Lee, S.W., and Callan, M. *J. geophys. Res.* **101,** 12777 (1996).

Mariner 9 Ultraviolet Spectrometer Experiment: Seasonal variation of ozone on Mars. Barth, C.A., Hord, C.W., Stewart, A.I., Lane, A.L., Dick, M.L., and Anderson, G.P. *Science* **179**, 795 (1973).

The global distribution of O_3 on Mars. Kong, T.Y. and McElroy, M.B. *Planet. Space Sci.* **25**, 839 (1977).

Water vapour

Measurements of the atmospheric water vapor on Mars by the Imager for Mars Pathfinder. Titov, D.V., Markiewicz, W.J., Thomas, N., Keller, H.U., Sablotny, R.M., Tomasko, M.G., Lemmon, M.T., and Smith, P.H. *J. geophys. Res.* **104,** 9019 (1999).

Mars water abundance: An estimate from D/H ratio in SNC meteorites. Ghosh, S. and Mahajan, K.K. *J. geophys. Res.* **103,** 5919 (1998).

The Mars water cycle: Determining the role of exchange with the regolith. Jakosky, B.M., Zent, A.P., and Zurek, R.W. *Icarus* **130,** 87 (1997).

Clouds

CO_2 ice clouds in the upper atmosphere of Mars. Clancy, R.T. and Sandor, B.J. *Geophys. Res. Letts.* **25,** 489 (1998).

Infrared spectral imaging of Martian clouds and ices. Klassen, D.R., Bell, J.F., Howell, R.R., Johnson, P.E., Golisch, W., Kaminski, C.D., and Griep, D. *Icarus* **138,** 36 (1999).

Cloud formation under Mars Pathfinder conditions. Colaprete, A., Toon, O.B., and Magalhaes, J.A. *J. geophys. Res.* **104,** 9043 (1999).

Noctilucent clouds on Mars. Lazarev, A.I. *J. opt. Technol.* **66,** 313 (1999).

Variations in Martian surface composition and cloud occurrence determined from thermal infrared spectroscopy: Analysis of Viking and Mariner 9 data. Christensen, P.R. *J. geophys. Res.* **103,** 1733 (1998).

Detection and monitoring of H_2O and CO_2 ice clouds on Mars. Bell, J.F., Calvin, W.M., OckertBell, M.E., Crisp, D., Pollack, J.B., and Spencer, J. *J. geophys. Res.* **101,** 9227 (1996).

Ice hazes and clouds in the Martian atmosphere as derived from the Phobos/KRFM data. Petrova, E., Keller, H.U., Markiewicz, W.J., Thomas, N., and Wuttke, M.W. *Planet. Space Sci.* **44,** 1163 (1996).

Dust and aerosols

Properties of dust in the Martian atmosphere from the Imager on Mars Pathfinder. Tomasko, M.G., Doose, L.R., Lemmon, M., Smith, P.H., and Wegryn, E. *J. geophys. Res.* **104,** 8987 (1999).

North polar dust storms in early spring on Mars. James, P.B., Hollingsworth, J.L., Wolff, M.J., and Lee, S.W. *Icarus* **138,** 64 (1999).

Electrostatic discharge in Martian dust storms. Melnik, O. and Parrot, M. *J. geophys. Res.* **103,** 29107 (1998).
Dust storm impacts on the Mars upper atmosphere. Bougher, S.W., Murphy, J., and Haberle, R.M. *Advan. Space Res.* **19,** 1255 (1997).
Aerosol component of the Martian atmosphere and its variability from the results of infrared radiometry in the Termoscan/Phobos-2 experiment. Titov, D.V., Moroz, V.I., and Gektin, Y.M. *Planet. Space Sci.* **45,** 637 (1997).
Observations of Martian aerosols with the imager for Mars pathfinder. Thomas, N., Keller, H.U., Markiewicz, W.J., and Smith, P.H. *Advan. Space Res.* **19,** 1271 (1997).
Martian dust storms: A review. Fernandez, W. *Earth Moon & Planets* **77,** 19 (1997).

Helium

Helium on Mars: EUVE and PHOBOS data and implications for Mars' evolution. Krasnopolsky, V.A. and Gladstone, G.R. *J. geophys. Res.* **101,** 15765 (1996).
Helium observation in the Martian ionosphere by an X-ray ultraviolet scanner on Mars orbiter NOZOMI. Nakamura, M., Yamashita, K., Yoshikawa, I., Shiomi, K., Yamazaki, A., Sasaki, S., Takizawa, Y., Hirahara, M., Miyake, W., Saito, Y., and Chakrabarti, S. *Earth Planets & Space* **51,** 61 (1999).

Carbon dioxide and carbon monoxide

The Mars seasonal CO_2 cycle and the time variation of the gravity field: A general circulation model simulation. Smith, D.E., Zuber, M.T., Haberle, R.M., Rowlands, D.D., and Murphy, J.R. *J. geophys. Res.* **104,** 1885 (1999).
CO_2 snowfall on Mars: Simulation with a general circulation model. Forget, F., Hourdin, F., and Talagrand, O. *Icarus* **131,** 302 (1998).
Carbonic acid in the gas phase and its astrophysical relevance. Hage, W., Liedl, K.R., Hallbrucker, A., and Mayer, E. *Science* **279,** 1332 (1998).
Carbonic acid on Mars? Strazzulla, G., Brucato, J.R., Cimino, G., and Palumbo, M.E. *Planet. Space Sci.* **44,** 1447 (1996).
Observations of CO in the atmosphere of Mars in the (2–0) vibrational band at 2.35 microns. Billebaud, F., Rosenqvist, J., Lellouch, E., Maillard, J.P., Encrenaz, T., and Hourdin, F. *Astron. Astrophys.* **333,** 1092 (1998).
Kinetics of the heterogeneous reaction $CO + O \rightarrow CO_2$ on inorganic oxide and water ice surfaces: Implications for the Martian atmosphere. Choi, W.Y. and Leu, M.T. *Geophys. Res. Letts.* **24,** 2957 (1997).
Uniqueness of a solution of a steady-state photochemical problem—applications to Mars. Krasnopolsky, V.A. *J. geophys. Res.* **100,** 3263 (1995).
A photochemical model of the Martian atmosphere. Nair, H., Allen, M., Anbar, A.D., Yung, Y.L., and Clancy, R.T. *Icarus* **111,** 124 (1994).

Airglow and ultraviolet spectroscopy; hot atoms

Mariner 9 Ultraviolet Spectrometer Experiment: Mars airglow spectroscopy and variations in Lyman alpha. Barth, C.A., Stewart, A.I., Hord, C.W., and Lane, A.L. *Icarus* **17,** 457 (1972).

Ultraviolet spectroscopy of the inner solar system from Mariner 10. Broadfoot, A.L. *Rev. Geophys. & Space Phys.* **14**, 625 (1976).

Solar cycle variability of hot oxygen atoms at Mars. Kim, J., Nagy, A.F., Fox, J.L., and Cravens, T.E. *J. geophys. Res.* **103,** 29339 (1998).

Spectrum of hot O at the exobases of the terrestrial planets. Fox, J.L. *J. geophys. Res.* **102,** 24005 (1997).

Hot hydrogen and oxygen atoms in the upper atmosphere of Venus and Mars. Nagy, A.F., Kim, J., and Cravens, T.E. *Ann. Geophys.* **8,** 251 (1990).

The ionosphere

Chemistry of the nightside ionosphere of Mars. Haider, S.A. *J. geophys. Res.* **102,** 407 (1997).

Charge exchange near Mars: The solar wind absorption and energetic neutral atom production. Kallio, E., Luhmann, J.G., and Barabash, S. *J. geophys. Res.* **102,** 22183 (1997).

Conditions in the Martian ionosphere/atmosphere from a comparison of a thermospheric model with radio occultation data. Breus, T.K., Pimenov, K.Y., Izakov, M.N., Krymskii, A.M., Luhmann, J.G., and Kliore, A.J. *Planet. Space Sci.* **46,** 367 (1998).

Numerical models of the Martian coupled thermosphere and ionosphere. Winchester, C. and Rees, D. *Advan. Space Res.* **15,** 51 (1995).

Solar wind proton deposition into the Martian atmosphere. Brecht, S.H. *J. geophys. Res.* **102,** 11287 (1997).

The depletion of the solar wind near Mars. Lichtenegger, H., Dubinin, E., and Ip, W.H. *Advan. Space Res.* **20,** 143 (1997).

This reference is also relevant to the Venusian ionosphere

Ionospheres of the terrestrial planets. Schunk, R.W. and Nagy, A.F. *Rev. Geophys. & Space Phys.* **18,** 813 (1980).

As explained in the text, ionospheric chemistry may play an important part in the escape of species from Mars, and thus in the evolution of the atmosphere. Appropriate references are given earlier under the heading 'Evolution of the atmosphere'.

Sections 8.4 to 8.6

The outer solar system

The composition of outer planet atmospheres. Gautier, D. and Owen, T. in *Origin and evolution of planetary and satellite atmospheres*. Atreya, S.K., Pollack, J.B., and Matthews, M.S. (eds.) (University of Arizona Press, Tucson, 1989).

Atmospheres and ionospheres of the outer planets and their satellites. Atreya, S.K. (Springer-Verlag, Berlin, 1986).

Saturn. Gehrels, T., and Matthews, M.S. (eds.) (University of Arizona Press, Tucson, 1984).

The atmospheres of the outer planets. Hunt, G.E., *Annu. Rev. Earth & Planet. Sci.* **11**, 415 (1983).
The atmospheres of the outer planets and satellites. Trafton, L. *Rev. Geophys. & Space Phys.* **19**, 43 (1981).
Jupiter and Saturn. Ingersoll, A.P. *Scient. Am.* **254**, 66 (Dec 1981).

Reports on the Galileo, Voyager, and Pioneer 11 mission findings

Galileo mission to Jupiter (special report). *Science* **274**, 377–413 (1996).
The Galileo probe mission to Jupiter: Science overview. Young, R.E. *J. geophys. Res.* **103,** 22775 (1998).
Chemical composition measurements of the atmosphere of Jupiter with the Galileo Probe mass spectrometer. Niemann, H.B., Atreya, S.K., Carignan, G.R., Donahue, T.M., Haberman, J.A., Harpold, D.N., Hartle, R.E., Hunten, D.M., Kasprzak, W.T., Mahaffy, P.R., Owen, T.C., and Spencer, N.W. *Advan. Space Res.* **21,** 1455 (1998).
The composition of the Jovian atmosphere as determined by the Galileo probe mass spectrometer. Niemann, H.B., Atreya, S.K., Carignan, G.R., Donahue, T.M., Haberman, J.A., Harpold, D.N., Hartle, R.E., Hunten, D.M., Kasprzak, W.T., Mahaffy, P.R., Owen, T.C., and Way, S.H. *J. geophys. Res.* **103,** 22831 (1998).
The Galileo probe mass spectrometer: Composition of Jupiter's atmosphere. Niemann, H.B., Atreya, S.K., Carignan, G.R., Donahue, T.M., Haberman, J.A., Harpold, D.N., Hartle, R.E., Hunten, D.M., Kasprzak, W.T., Mahaffy, P.R., Owen, T.C., Spencer, N.W., and Way, S.H. *Science* **272,** 846 (1996).
Further evaluation of waves and turbulence encountered by the Galileo Probe during descent in Jupiter's atmosphere. Seiff, A., Kirk, D.B., Mihalov, J., and Knight, T.C.D. *Geophys. Res. Letts.* **26,** 1199 (1999).
Galileo probe measurements of thermal and solar radiation fluxes in the Jovian atmosphere. Sromovsky, L.A., Collard, A.D., Fry, P.M., Orton, G.S., Lemmon, M.T., Tomasko, M.G., and Freedman, R.S. *J. geophys. Res.* **103,** 22929 (1998).
New constraints on the composition of Jupiter from Galileo measurements and interior models. Guillot, T., Gautier, D., and Hubbard, W.B. *Icarus* **130,** 534 (1997).
Vertical structure of Jupiter's atmosphere at the Galileo probe entry latitude. Chanover, N.J., Kuehn, D.M., and Beebe, R.F. *Icarus* **128,** 294 (1997).
Galileo probe: *in situ* observations of Jupiter's atmosphere. Young, R.E., Smith, M.A., and Sobeck, C.K. *Science* **272,** 837 (1996).
Structure of Jupiter's upper atmosphere: Predictions for Galileo. Yelle, R.V., Young, L.A., Vervack, R.J., Young, R., Pfister, L., and Sandel, B.R. *J. geophys. Res.* **101,** 2149 (1996).
Voyagers at Jupiter and Saturn (special reports): Jupiter: Voyager 1, *Science* **204**, 945 *et seq.* (1979): Voyager 2, *Science* **206**, 925 *et seq.* (1979); Saturn: Voyager 1, *Science* **212**, 159 *et seq.* (1981): Voyager 2, *Science* **215**, 499 *et seq.* (1981); Voyagers 1, 2, *J. geophys. Res.* **88**, 8625 *et seq.* (1983).
Pioneer 11 at Saturn (special reports). *Science* **207**,400 (1980); *J. geophys. Res.* **85**, 5651 (1980).

Section 8.4

The Jovian system: the last outpost for life? Hiscox, J. *Astron. Geophys.* **40,** 22 (1999).

Jupiter's atmosphere. Drossart, P. *Recherche* **472,** 903 (1996).

Photochemistry and thermal chemistry in the atmospheres of the outer solar system

Hydrocarbon photochemistry in the upper atmosphere of Jupiter. Gladstone, G.R., Allen, M., and Yung, Y.L. *Icarus* **119,** 1 (1996).
A box model of the photolysis of methane at 123.6 and 147 nm—comparison between model and experiment. Smith, N.S. and Raulin, F. *J. Photochem. Phobiol.* **A124,** 101 (1999).
Chemical equilibrium abundances in brown dwarf and extrasolar giant planet atmospheres. Burrows, A. and Sharp, C.M. *Astrophys. J.* **512,** 843 (1999).
Nonadiabatic models of Jupiter and Saturn. Guillot, T., Chabrier, G., Morel, P., and Gautier, D. *Icarus* **112,** 354 (1994).
Photochemistry and clouds of Jupiter, Saturn, and Uranus. Atreya, S.K., and Romani, P.N. in *Recent advances in planetary meteorology*. Hunt, G.E. (ed.) (Cambridge University Press, Cambridge, 1985).
The photochemistry of the atmospheres of the outer planets and their satellites. Strobel, D.F. in *The photochemistry of atmospheres*. Levine, J.S. (ed.) (Academic Press, Orlando, 1985).
Composition and chemistry of Saturn's atmosphere. Prinn, R.G., Larson, H.P., Caldwell, J.J., and Gautier, D. in *Saturn*, Gehrels, T. and Matthews, M.S. (eds.) (University of Arizona Press, Tucson, 1984).
Aeronomy of the major planets: photochemistry of ammonia and hydrocarbons. Strobel, D.F. *Rev. Geophys. & Space Phys.* **13**, 372 (1975).
Photochemistry of the reducing atmospheres of Jupiter, Saturn and Titan. Strobel, D.F. *Int. Rev. phys. Chem.* **3**, 145 (1983).
Jovian H_2 dayglow emission. McGrath, M.A., Ballester, G.E., and Moos, L.W. *J. geophys. Res.* **95**, 10365 (1990).

Aerosols and clouds in Jupiter's atmosphere.

A comparison of the structure of the aerosol layers in the great red spot of Jupiter and its surroundings before and after the 1993 SEB disturbance. Munoz, O., Moreno, F., Molina, A., and Ortiz, J.L. *Astron. Astrophys.* **344,** 355 (1999).
Constraints on the tropospheric cloud structure of Jupiter from spectroscopy in the 5μm region: A comparison between Voyager/IRIS, Galileo/NIMS, and ISO/SWS spectra. Roos-Serote, M., Drossart, P., Encrenaz, T., Carlson, R.W., and Leader, F. *Icarus* **137,** 315 (1999).
Reflected spectra and albedos of extrasolar giant planets. I. Clear and cloudy atmospheres. Marley, M.S., Gelino, C., Stephens, D., Lunine, J.I., and Freedman, R. *Astrophys. J.* **513,** 879 (1999).
Cloud structure and atmospheric composition of Jupiter retrieved from Galileo near-infrared mapping spectrometer real-time spectra. Irwin, P.G.J., Weir, A.L., Smith, S.E., Taylor, F.W., Lambert, A.L., Calcutt, S.B., Cameron-Smith, P.J., Carlson, R.W., Baines, K., Orton, G.S., Drossart, P., Encrenaz, T., and Roos-Serote, M. *J. geophys. Res.* **103,** 23001 (1998).
The structure of the stratospheric aerosol layer in the equatorial and south polar regions of Jupiter. Moreno, F. *Icarus* **124,** 632 (1996).

The clouds of Jupiter: Results of the Galileo Jupiter mission probe nephelometer experiment. Ragent, B., Colburn, D.S., Rages, K.A., Knight, T.C.D., Avrin, P., Orton, G.S., Yanamandra-Fisher, P.A., and Grams, G.W. *J. geophys. Res.* **103,** 22891 (1998).

Phosphine photochemistry, and the possibility that there is an interaction with ammonia photochemistry; organophosphorus compounds; arsine chemistry.

High-resolution 10μm spectroscopy of ammonia and phosphine lines on Jupiter. Lara, L.M., Bezard, B., Griffith, C.A., Lacy, J.H., and Owen, T. *Icarus* **131,** 317 (1998).

On the latitude variation of ammonia, acetylene, and phosphine altitude profiles on Jupiter from HST Faint Object Spectrograph observations. Edgington, S.G., Atreya, S.K., Trafton, L.M., Caldwell, J.J., Beebe, R.F., Simon, A.A., West, R.A., and Barker, C. *Icarus* **133,** 192 (1998).

Ammonia abundance in Jupiter's atmosphere derived from the attenuation of the Galileo probe's radio signal. Folkner, W.M., Woo, R., and Nandi, S. *J. geophys. Res.* **103,** 22847 (1998).

Regioselectivity of the photochemical addition of phosphine to unsaturated hydrocarbons in the atmospheres of Jupiter and Saturn. Guillemin, J.C., LeSerre, S., and Lassalle, L. *Advan. Space Res.* **19,** 1093 (1997).

The tropospheric abundances of NH_3 and PH_3 in Jupiter great red spot, from Voyager IRIS observations. Griffith, C.A., Bezard, B., Owen, T., and Gautier, D. *Icarus* **98,** 82 (1992).

Photochemistry of phosphine and Jupiter's great red spot. Noy, N., Pudolak, M., and Bar-Nun, A. *J. geophys. Res.* **86,** 11985 (1981).

Phosphine photochemistry in the atmosphere of Saturn. Kaye, J.A., and Strobel, D.F. *Icarus* **59,** 314 (1984).

Gas phase synthesis of organophosphorus compounds and the atmospheres of the giant planets. Bossard, A.R., Kagama, R., and Rawlin, F. *Icarus* **67,** 305 (1986).

The chemistry of arsine (AsH_3) in the deep atmospheres of Saturn and Jupiter. Fegley, M.B. *Bull. Amer. Astron. Soc.* **20,** 879 (1988).

The abundance of AsH_3 in Jupiter. Noll, K.S., Larson, H.P., and Geballe, T.R. *Icarus* **83,** 494 (1990).

Some minor components of the Jovian atmosphere

Methane, ammonia, and temperature measurements of the Jovian planets and Titan from CCD spectrophotometry. Karkoschka, E. *Icarus* **133,** 134 (1998).

Observations of CH_4, C_2H_6, and C_2H_2 in the stratosphere of Jupiter. Sada, P.V., Bjoraker, G.L., Jennings, D.E., McCabe, G.H., and Romani, P.N. *Icarus* **136,** 192 (1998).

The abundances of ethane and acetylene in the atmospheres of Jupiter and Saturn. Noll, K.S., Knacke, R.F., Tokunaga, A.T., Lacy, J.H., Beck, S., and Serabyn, E. *Icarus* **65,** 257 (1986).

Halogens in the giant planets: Upper limits to HBr in Saturn and Jupiter. Noll, K.S. *Icarus* **124,** 608 (1996).

The presence of hydrogen cyanide is controversial; if present, it is difficult to find a plausible route to its formation.

Broadband submillimeter spectroscopy of HCN, NH_3, and PH_3, in the troposphere of Jupiter. Davis, G.R., Naylor, D.A., Griffin, M.J., Clark, T.A., and Holland, W.S. *Icarus* **130,** 387 (1997).
Nondetection of hydrogen cyanide on Jupiter. Bezard, B., Griffith, C., Lacy, J., and Owen, T. *Icarus* **118**, 384 (1995).
HCN formation on Jupiter: The coupled photochemistry of ammonia and acetylene. Kaye, J.A. and Strobel, D.F. *Icarus* **54**, 417 (1983).

The discovery of CO prompted speculations that the source of oxygen might be extraplanetary, although that idea is now discounted. The amounts of oxygen-containing species in the atmospheres are important in determining the existence of oxidized carbon compounds; the references to CO and CO_2 on Titan (see Section 8.5.1 below) are also relevant to a discussion of this point.

Detection of H_2O in the splash phase of G- and R-impacts from NIMS–Galileo. Encrenaz, T., Drossart, P., Carlson, R.W., and Bjoraker, G. *Planet. Space Sci.* **45,** 1189 (1997).
CO in the solar system. Encrenaz, T. *IAU Symposia* **125,** 245 (1997).
External supply of oxygen to the atmospheres of the giant planets. Feuchtgruber, H., Lellouch, E., de Graauw, T., Bezard, B., Encrenaz, T., and Griffin, M. *Nature, Lond.* **389,** 159 (1997).
X-ray and EUV emission spectra of oxygen ions precipitating into the Jovian atmosphere. Kharchenko, V., Liu, W.H., and Dalgarno, A. *J. geophys. Res.* **103,** 26687 (1998).
The origin and vertical distribution of carbon monoxide on Jupiter. Noll, K.S., Knacke, R.F., Geballe, T.R., and Tokunaga, A.T. *Astrophys. J.* **324**, 1210 (1988).
Chemical constraints on the water and total oxygen abundances in the deep atmosphere of Jupiter. Fegley, M.B., Jr. and Prinn, R. *Astrophys. J.* **324**, 621 (1988).
Detection of carbon monoxide in Saturn. Noll, K.S., Knacke, R.F., Geballe, T.R., and Tokunaga, A.T. *Astrophys. J.* **309**, L91 (1986).

Lightning probably occurs in the atmospheres of the outer planets. See also the second reference for Section 8.2.3.

Measurements of radio frequency signals from lightning in Jupiter's atmosphere. Rinnert, K., Lanzerotti, L.J., Uman, M.A., Dehmel, G., Gliem, F.O., Krider, E.P., and Bach, J. *J. geophys. Res.* **103,** 22979 (1998).
Moist convection and the abundances of lightning-produced CO, C_2H_2, and HCN on Jupiter. Podolak, M. and Bar-Nun, A. *Icarus* **75**, 566 (1988).
Lightning activity on Jupiter. Borucki, W.J., Bar-Nun, A., Scarf, F.L., Cook, A.F., and Hunt, G.E. *Icarus* **52**, 492 (1982).

The heat of conversion from p-H_2 to o-H_2 could drive high zonal flows in the atmospheric 'wind' systems. Voyager infrared measurements appear to support the idea of o–p disequilibrium.

Thermal structure and para hydrogen fraction on the outer planets from Voyager IRIS measurements. Conrath, B.J., Gierasch, P.J., and Ustinov, E.A. *Icarus* **135,** 501 (1998).

Global variation of the para-hydrogen fraction in Jupiter's atmosphere and implications for dynamics on the outer planets. Conrath, B.J. and Gierasch, P.J. *Icarus* **57**, 184 (1984).

Dynamical consequences of orthohydrogen–parahydrogen disequilibrium on Jupiter and Saturn. Gierasch, P.J. *Science* **219**, 847 (1983).

Evidence for disequilibrium of ortho and para hydrogen on Jupiter from Voyager IRIS measurements. Conrath, B.J. and Gierasch, P.J. *Nature, Lond.* **306**, 571 (1983).

Jovian aurora and airglow

Jovian aurorae. Kim, S.J., Lee, D.H., and Kim, Y.H. *Rep. Prog. Phys.* **61,** 525 (1998).

Auroral Lyman alpha and H_2 bands from the giant planets. 3. Lyman alpha spectral profile including charge exchange and radiative transfer effects and H_2 color ratios. Rego, D., Prange, R., and Ben-Jaffel, L. *J. geophys. Res.* **104,** 5939 (1999).

HST spectra of the Jovian ultraviolet aurora: Search for heavy ion precipitation. Trafton, L.M., Dols, V., Gerard, J.C., Waite, J.H., Gladstone, G.R., and Munhoven, G. *Astrophys. J.* **507,** 955 (1998).

Galileo orbiter ultraviolet observations of Jupiter aurora. Ajello, J., Shemansky, D., Pryor, W., Tobiska, K., Hord, C., Stephens, S., Stewart, I., Clarke, J., Simmons, K., McClintock, W., Barth, C., Gebben, J., Miller, D., and Sandel, B. *J. geophys. Res.* **103,** 20125 (1998).

Self-absorption by vibrationally excited H_2 in the Astro-2 Hopkins ultraviolet telescope spectrum of the Jovian aurora. Wolven, B.C. and Feldman, P.D. *Geophys. Res. Letts.* **25,** 1537 (1998).

Imaging Jupiter's aurora at visible wavelengths. Ingersoll, A.P., Vasavada, A.R., Little, B., Anger, C.D., Bolton, S.J., Alexander, C., and Tobiska, W.K. *Icarus* **135,** 251 (1998).

Detection of self-reversed Ly alpha lines from the Jovian aurorae with the Hubble Space Telescope. Prange, R., Rego, D., Pallier, L., Ben-Jaffel, L., Emerich, C., Ajello, J., Clarke, J.T., and Ballester, G.E. *Astrophys. J.* **484,** L169 (1997).

Mid-to-low latitude H_3^+ emission from Jupiter. Miller, S., Achilleos, N., Ballester, G.E., Lam, H.A., Tennyson, J., Geballe, T.R., and Trafton, L.M. *Icarus* **130,** 57 (1997).

Temperatures and altitudes of Jupiter's ultraviolet aurora inferred from GHRS observations with the Hubble Space Telescope. Kim, Y.H., Fox, J.L., and Caldwell, J.J. *Icarus* **128,** 189 (1997).

Analysis of Jovian auroral H Ly-α emission (1981–1991). Harris, W., Clarke, J.T., McGrath, M.A., and Ballester, G.E. *Icarus* **123,** 350 (1996).

The ultraviolet spectrum of the Jovian dayglow. Liu, W.H. and Dalgarno, A. *Astrophys. J.* **462,** 502 (1996).

Jupiter He 584 angstrom dayglow — new results. Vervack, R.J., Sandel, B.R., Gladstone, G.R., McConnell, J.C., and Parkinson, C.D. *Icarus* **114,** 163 (1995).

4 micron high-resolution spectra of Jupiter in the north equatorial belt — H_3^+ emissions and the C^{12}:C^{13} ratio. Marten, A., DeBergh, C., Owen, T., Gautier, D., Maillard, J.P., Drossart, P., Lutz, B.L., and Orton, G.S. *Planet. Space Sci.* **42,** 391 (1994).

Helium and other noble gases

Helium in Jupiter's atmosphere: Results from the Galileo probe Helium Interferometer Experiment. Von Zahn, U., Hunten, D.M., and Lehmacher, G. *J. geophys. Res.* **103,** 22815 (1998).

The helium mass fraction in Jupiter's atmosphere. Von Zahn, U. and Hunten, D.M. *Science* **272,** 849 (1996).

Strange xenon in Jupiter. Manuel, O., Windler, K., Nolte, A., Johannes, L., Zirbel, J., and Ragland, D. *J. radioanal. nuc. Chem.* **238,** 119 (1998).

Isotope fractionation

Galileo probe measurements of D/H and He^3/He^4 in Jupiter's atmosphere. Mahaffy, P.R., Donahue, T.M., Atreya, S.K., Owen, T.C., and Niemann, H.B. *Space Sci. Rev.* **84,** 251 (1998).

Deuterium enrichment in giant planets. Lecluse, C., Robert, F., Gautier, D., and Guiraud, M. *Planet. Space Sci.* **44,** 1579 (1996).

The Jovian ionosphere

Jupiter's ionosphere: New results from Voyager 2 radio occultation measurements. Hinson, D.P., Twicken, J.D., and Karayel, E.T. *J. geophys. Res.* **103,** 9505 (1998).

Jupiter's ionosphere: Results from the first Galileo radio occultation experiment. Hinson, D.P., Flasar, F.M., Kliore, A.J., Schinder, P.J., Twicken, J.D., and Herrera, R.G. *Geophys. Res. Letts.* **24,** 2107 (1997).

JIM: A time-dependent, three-dimensional model of Jupiter's thermosphere and ionosphere. Achilleos, N., Miller, S., Tennyson, J., Aylward, A.D., Mueller-Wodarg, I., and Rees, D. *J. geophys. Res.* **103,** 20089 (1998).

The chemistry of hydrocarbon ions in the Jovian ionosphere. Kim, Y.H. and Fox, J.L. *Icarus* **112,** 310 (1994).

Representative laboratory studies of relevance to the Jovian atmosphere

Low-temperature rate coefficients for reactions of the ethynyl radical (C_2H) with C_3H_4 isomers methylacetylene and allene. Hoobler, R.J. and Leone, S.R. *J. phys. Chem.* A **103,** 1342 (1999).

Recent rate constant and product measurements of the reactions $C_2H_3 + H_2$ and $C_2H_3 +$ H — Importance for photochemical modeling of hydrocarbons on Jupiter. Romani, P.N. *Icarus* **122,** 233 (1996).

Line profile of H Lyman alpha from dissociative excitation of H_2 with application to Jupiter. Ajello, J.M., Kanik, I., Ahmed, S.M., and Clarke, J.T. *J. geophys. Res.* **100,** 26411 (1995).

An extraordinary opportunity to observe perturbations in the Jovian atmosphere was afforded when the comet Shoemaker–Levy 9 collided with the planet.

Ejecta pattern of the impact of Comet Shoemaker–Levy 9. Pankine, A.A. and Ingersoll, A.P. *Icarus* **138,** 157 (1999).

The evolution of debris from Comet D Shoemaker–Levy 9 on Jupiter. Little, J.E., Fitzsimmons, A., Andrews, P.J., Catchpole, R., Walton, N., and Williams, I.P. *Icarus* **131,** 334 (1998).

Collision of comet Shoemaker–Levy 9 with Jupiter. Encrenaz, T. *Comptes Rendus Ser. II, Fasc. B* **324,** 591 (1997).

Infrared spectroscopy of Jupiter's atmosphere after the A and E impacts of Comet Shoemaker–Levy 9. Kim, S.J., Orton, G.S., Dumas, C., and Kim, Y.H. *Icarus* **120,** 326 (1996).

Atomic emission lines in the spectrum observed after the impact between the L-fragment of comet D Shoemaker–Levy 9 and Jupiter. Costa, R.D.D., Pacheco, J.A.D., Singh, P.D., de Almeida, A.A., and Codina-Landaberry, S.J. *Astrophys. J.* **485,** 380 (1997).

Ly-α induced fluorescence of H_2 and CO in Hubble Space Telescope spectra of a comet Shoemaker–Levy 9 impact site on Jupiter. Wolven, B.C., Feldman, P.D., Strobel, D.F., and McGrath, M.A. *Astrophys. J.* **475,** 835 (1997).

Far-uv emissions from the SL9 impacts with Jupiter. Ballester, G.E., Harris, W.M., Gladstone, G.R., Clarke, J.T., Prange, R., Feldman, P.D., Combi, M.R., Emerich, C., Strobel, D.F., Talavera, A., Budzien, S.A., Vincent, M.B., Livengood, T.A., Jessup, K.L., McGrath, M.A., Hall, D.T., Ajello, J.M., Ben-Jaffel, L., Rego, D., Fireman, G., Woodney, L., Miller, S., and Liu, X. *Geophys. Res. Letts.* **22,** 2425 (1995).

EUVE observations of Jupiter during the impact of comet Shoemaker–Levy 9. Gladstone, G.R., Hall, D.T., and Waite, J.H. *Science* **268,** 1595 (1995).

Nature and source of organic matter in the Shoemaker–Levy 9 Jovian impact blemishes. Wilson, P.D. and Sagan, C. *Icarus* **129,** 207 (1997).

High temperature chemistry in the fireballs formed by the impacts of comet P/Shoemaker–Levy 9 in Jupiter. Borunov, S., Drossart, P., Encrenaz, T., and Dorofeeva, V. *Icarus* **125,** 121 (1997).

The formation of HCS and HCSH molecules and their role in the collision of comet Shoemaker–Levy 9 with Jupiter. Kaiser, R.I., Ochsenfeld, C., Head-Gordon, M., and Lee, Y.T. *Science* **279,** 1181 (1998).

Carbon monoxide in Jupiter after Comet Shoemaker–Levy 9. Noll, K.S., Gilmore, D., Knacke, R.F., Womack, M., Griffith, C.A., and Orton, G. *Icarus* **126,** 324 (1997).

Carbon monoxide in Jupiter after the impact of comet Shoemaker–Levy 9. Lellouch, E., Bezard, B., Moreno, R., Bockelee-Morvan, D., Colom, P., Crovisier, J., Festou, M., Gautier, D., Marten, A., and Paubert, G. *Planet. Space Sci.* **45,** 1203 (1997).

Hydrogen cyanide polymers from the impact of comet P/Shoemaker–Levy 9 on Jupiter. Matthews, C.N. *Advan. Space Res.* **19,** 1087 (1997).

The collision of comet Shoemaker–Levy 9 with Jupiter — detection and evolution of HCN in the stratosphere of the planet. Marten, A., Gautier, D., Griffin, M.J., Matthews, H.E., Naylor, D.A., Davis, G.R., Owen, T., Orton, G., Bockelee-Morvan, D., Colom, P., Crovisier, J., Lellouch, E., Depater, I., Atreya, S., Strobel, D., Han, B., and Sanders, D.B. *Geophys. Res. Letts.* **22,** 1589 (1995).

Chemical and thermal response of Jupiter atmosphere following the impact of comet Shoemaker–Levy 9. Lellouch, E., Paubert, G., Moreno, R., Festou, M.C., Bezard, B., Bockelee-Morvan, D., Colom, P., Crovisier, J., Encrenaz, T., Gautier, D., Marten, A., Despois, D., Strobel, D.F., and Sievers, A. *Nature, Lond.* **373,** 592 (1995).

The impact of comet Shoemaker–Levy 9 on the Jovian ionosphere and aurorae. Miller, S., Achilleos, N., Lam, H.A., Dinelli, B.M., and Prange, R. *Planet. Space Sci.* **45,** 1237 (1997).

Saturn

Investigation of Saturn's atmosphere by Cassini. Taylor, F.W., Calcutt, S.B., Irwin, P.G.J., Nixon, C.A., Read, P.L., Smith, P.J.C., and Vellacott, T.J. *Planet. Space Sci.* **46,** 1315 (1998).

Cassini UVIS observations of Saturn's rings. Esposito, L.W., Colwell, J.E., and McClintock, W.E. *Planet. Space Sci.* **46,** 1221 (1998).

Imaging spectroscopy of Saturn and its satellites: VIMS–V onboard Cassini. Capaccioni, F., Coradini, A., Cerroni, P., and Amici, S. *Planet. Space Sci.* **46,** 1263 (1998).

Rings around planets, atmospheric superrotation, and their great spots. Kundt, W. and Luttgens, G. *Astrophys. Space Sci.* **257,** 33 (1998).

Fluorescent hydroxyl emissions from Saturn's ring atmosphere. Hall, D.T., Feldman, P.D., Holberg, J.B., and McGrath, M.A. *Science* **272,** 516 (1996).

He 584 angstrom dayglow at Saturn: A reassessment. Parkinson, C.D., Griffioen, E., McConnell, J.C., and Sandel, B.R. *Icarus* **133,** 210 (1998).

New analysis of the Voyager UVS H Lyman-α emission of Saturn. Ben-Jaffel, L., Prange, R., Sandel, B.R., Yelle, R.V., Emerich, C., Feng, D., and Hall, D.T. *Icarus* **113,** 91 (1995).

Saturn's hydrogen aurora: Wide field and planetary camera 2 imaging from the Hubble Space Telescope. Trauger, J.T., Clarke, J.T., Ballester, G.E., Evans, R.W., Burrows, C.J., Crisp, D., Gallagher, J.S., Griffiths, R.E., Hester, J.J., Hoessel, J.G., Holtzman, J.A., Krist, J.E., Mould, J.R., Sahai, R., Scowen, P.A., Stapelfeldt, K.R., and Watson, A. *J. geophys. Res.* **103,** 20237 (1998).

Detection of methyl radicals (CH_3) on Saturn. Bezard, B., Feuchtgruber, H., Moses, J.I., and Encrenaz, T. *Astron. Astrophys.* **334,** L41 (1998).

First results of ISO–SWS observations of Saturn: Detection of CO_2, CH_3C_2H, C_4H_2 and tropospheric H_2O. De Graauw, T., Feuchtgruber, H., Bezard, B., Drossart, P., Encrenaz, T., Beintema, D.A., Griffin, M., Heras, A., Kessler, M., Leech, K., Lellouch, E., Morris, P., Roelfsema, P.R., Roos Serote, M., Salama, A., Van den bussche, B., Valentijn, E.A., Davis, G.R., and Naylor, D.A. *Astron. Astrophys.* **321,** L13 (1997).

Hydrogen dimer features in the 2μm spectra of Saturn and Neptune. Trafton, L.M., Kim, S.J., Geballe, T.R., and Miller, S. *Icarus* **130,** 544 (1997).

First detection of the 56μm rotational line of HD in Saturn's atmosphere. Griffin, M.J., Naylor, D.A., Davis, G.R., Ade, P.A.R., Oldham, P.G., Swinyard, B.M., Gautier, D., Lellouch, E., Orton, G.S., Encrenaz, T., de Graauw, T., Furniss, I., Smith, H., Armand, C., Burgdorf, M., Di Giorgio, A., Eward, D., Gry, C., King, K.J., Lim, T., Molinari, S., Price, M., Sidher, S., Smith, A., Texier, D., Trams, N., Unger, S.J., and Salama, A. *Astron. Astrophys.* **315,** L389 (1996).

External supply of oxygen to the atmospheres of the giant planets. Feuchtgruber, H., Lellouch, E., de Graauw, T., Bezard, B., Encrenaz, T., and Griffin, M. *Nature, Lond.* **389,** 159 (1997).

Thermal plasma and neutral gas in Saturn's magnetosphere. Richardson, J.D. *Rev. Geophys.* **36,** 501 (1998).

Evidence of a source of energetic ions at Saturn. Paranicas, C., Cheng, A.F., Mauk, B.H., Keath, E.P., and Krimigis, S.M. *J. geophys. Res.* **102,** 17459 (1997).

Satellites of Jupiter and Saturn

These satellites are an interesting collection of bodies, and several may possess a tenuous atmosphere, perhaps sweeping through a torus around the parent planet. Saturn's satellites Rhea and Dione may even possess some ozone!

Outer planets and their major satellites — atmospheric and surface properties. Lunine, J.I. *Rev. Geophys.* **33**, 509 (1995).

Origin and bulk chemical composition of the Galilean satellites and the primitive atmosphere of Jupiter: A pre-Galileo analysis. Prentice, A.J.R. *Earth Moon and Planets* **73,** 237 (1996).

The satellites of Jupiter and Saturn. Morrison, D. *Annu. Rev. Astron. Astrophys.* **20,** 469 (1982).

Detection of ozone on Saturn's satellites Rhea and Dione. Noll, K.S., Roush, T.L., Cruikshank, D.P., Johnson, R.E., and Pendleton, Y.J. *Nature, Lond.* **388,** 45 (1997).

Section 8.5.1

Titan: the Earth-like moon. Coustenis, A. and Taylor, F. (World Scientific Publishing, 1999).

Present state and chemical evolution of the atmospheres of Titan, Triton, and Pluto. Lunine, J.I., Atreya, S.K., and Pollack, J.B. in *Origin and evolution of planetary and satellite atmospheres*. Atreya, S.K., Pollack, J.B., and Matthews, M.S. (eds.) (University of Arizona Press, Tucson, 1989).

Huygens: science, payload and mission. Lebreton, J.-P. (coordinator) (ESA Publications Division, Noordwijk, 1997).

Titan in the solar system. Taylor, F.W. and Coustenis, A. *Planet. Space Sci.* **46,** 1085 (1998).

Voyagers at Titan (special reports). *J. geophys. Res.* **87**, 1351 *et seq.* (1982); *J. geophys. Res.* **88**, 8625 *et seq*. (1983).

Titan's atmosphere from Voyager infrared observations. 1. The gas composition of Titan's equatorial region. Coustenis, A., Bezard, B., and Gautier, D. *Icarus* **80**, 54 (1989).

The composition and origin of Titan's atmosphere. Owen, T. *Planet. Space Sci.* **30,** 833 (1982).

Physical and meteorological properties of Titan's atmosphere

The composition of Titan's atmosphere: a meteorological perspective. Flasar, F.M. *Planet. Space Sci.* **46,** 1109 (1998).

Seasonal variation of Titan's atmospheric structure simulated by a general circulation model. Tokano, T., Neubauer, F.M., Laube, M., and McKay, C.P. *Planet. Space Sci.* **47,** 493 (1999).

The ultraviolet albedo of Titan. McGrath, M.A., Courtin, R., Smith, T.E., Feldman, P.D., and Strobel, D.F. *Icarus* **131,** 382 (1998).

The dynamic meteorology of Titan. Flasar, F.M. *Planet. Space Sci.* **46,** 1125 (1998).

An exobiological view of Titan and the Cassini–Huygens mission. Raulin, F., Coll, P., Coscia, D., Gazeau, M.C., Sternberg, R., Bruston, P., Israel, G., and Gautier, D. *Advan. Space Res.* **22,** 353 (1998).

Titan's thermal emission spectrum—reanalysis of the Voyager infrared measurements. Courtin, R. and Gautier, D. *Icarus* **114,** 144 (1995).

Temperature lapse rate and methane in Titan's troposphere. McKay, C.P., Martin, S.C., Griffith, C.A., and Keller, R.M. *Icarus* **129,** 498 (1997).

Importance of phase changes in Titan's lower atmosphere. Tools for the study of nucleation. Guez, L., Bruston, P., Raulin, F., and Regnaut, C. *Planet. Space Sci.* **45,** 611 (1997).

General aspects of the composition and chemistry of the atmosphere

Vertical distribution of Titan's atmospheric neutral constituents. Lara, L.M., Lellouch, E., Lopez-Moreno, J.J., and Rodrigo, R. *J. geophys. Res.* **101,** 23261 (1996).

Photochemical modeling of Titan's atmosphere. Toublanc, D., Parisot, J.P., Brillet, J., Gautier, D., Raulin, F., and McKay, C.P. *Icarus* **113,** 2 (1995).

Modeling of methane photolysis in the reducing atmospheres of the outer solar system. Smith, N.S. and Raulin, F. *J. geophys. Res.* **104,** 1873 (1999).

Titan's atmosphere composition — certainties and speculations. Gautier, D. *Advan. Space Res.* **15,** 295 (1994).

Titan's atmosphere from Voyager infrared observations. 3. Vertical distributions of hydrocarbons and nitriles near Titan's north-pole. Coustenis, A., Bezard, B., Gautier, D., Marten, A., and Samuelson, R. *Icarus* **89,** 152 (1991).

Titan's atmosphere from Voyager infrared observations. 2. The CH_3D abundance and D/H ratio from the 900–1200cm^{-1} spectral region. Coustenis, A., Bezard, B., and Gautier, D. *Icarus* **82,** 67 (1989).

Titan's atmosphere from Voyager infrared observations. 1. The gas-composition of Titan's equatorial region. Coustenis, A., Bezard, B., and Gautier, D. *Icarus* **80,** 54 (1989).

Hydrocarbons and more complex organic molecules. Species such as ethane may condense to form oceans on the surface of the satellite.

Evidence from scanning electron microscopy of experimental influences on the morphology of Triton and Titan tholins. De Vanssay, E., McDonald, G.D., and Khare, B.N. *Planet. Space Sci.* **47,** 433 (1999).

Methane, ammonia, and temperature measurements of the Jovian planets and Titan from CCD-spectrophotometry. Karkoschka, E. *Icarus* **133,** 134 (1998).

Vertical distribution of Titan's atmospheric neutral constituents. Lara, L.R., Lellouch, E., Lopez-Moreno, J.J., and Rodrigo, R. *J. geophys. Res.* **103,** 25775 (1998).

Steady-state model for methane condensation in Titan's troposphere. Samuelson, R.E., and Mayo, L.A. *Planet. Space Sci.* **45,** 949 (1997).

Photochemical sources of non-thermal neutrals for the exosphere of Titan. Cravens, T.E., Keller, C.N., and Ray, B. *Planet. Space Sci.* **45,** 889 (1997).

Properties of the main high molecular weight hydrocarbons in Titan's atmosphere. Dimitrov, V. and Bar Nun, A. *Prog. React. Kinet.* **22,** 67 (1997).

Metastable diacetylene reactions as routes to large hydrocarbons in Titan's atmosphere. Zwier, T.S. and Allen, M. *Icarus* **123,** 578 (1996).

Photochemical growing of complex organics in planetary atmospheres. Raulin, F. and Bruston, P. *Advan. Space Res.* **18,** 41 (1996).

Gaseous abundances and methane supersaturation in Titan's troposphere. Samuelson, R.E., Nath, N.R., and Borysow, A. *Planet. Space Sci.* **45,** 959 (1997).

Ethane abundance on Titan. Kostiuk, T., Fast, K., Livengood, T.A., Goldstein, J., Hewagama, T., Buhl, D., Espenak, F., and Ro, K.H. *Planet. Space Sci.* **45,** 931 (1997).

Hydrogen cyanide, cyanoacetylene, and other nitrogen-containing molecules

Millimeter and submillimeter heterodyne observations of Titan, retrieval of the vertical profile of HCN and the C-12/C-13 ratio. Hidayat, T., Marten, A., Bezard, B., Gautier, D., Owen, T., Matthews, H.E., and Paubert, G. *Icarus* **126,** 170 (1997).

Stratospheric profile of HCN on Titan from millimeter observations. Tanguy, I., Bezard, B., Marten, A., Gautier, D., Gerard, E., Paubert, G., and Lecacheux, A. *Icarus* **85,** 43 (1990).

Mechanism of cyanoacetylene photochemistry at 185 and 254 nm. Clarke, D.W. and Ferris, J.P. *J. geophys. Res.* **101,** 7575 (1996).

Photochemistry of cyanoacetylene at 193.3 nm. Seki, K., He, M.Q., Liu, R.Z., and Okabe, H. *J. phys. Chem.* **100,** 5349 (1996).

C_4N_2 ice in Titan's north polar stratosphere. Samuelson, R.E., Mayo, L.A., Knuckles, M.A., and Khanna, R.J. *Planet. Space Sci.* **45,** 941 (1997).

An update of nitrile chemistry on Titan. Yung, Y.L. *Icarus* **72,** 468 (1987).

Electron impact excitation, and possible N-containing ion chemistry, in Titan's atmosphere.

EUV emission from Titan's upper atmosphere: Voyager 1 encounter. Strobel, D.F. and Shemansky, D.E. *J. geophys. Res.* **87**, 1361 (1982).

Galactic cosmic rays and N_2 dissociation on Titan. Capone, L.A., Dubach, J., Prasad, S.S., and Whitten, R.C. *Icarus* **55,** 73 (1983).

The discovery of the oxidized species CO and CO_2 in the reducing atmosphere of Titan has opened the way to a greater understanding of the detailed chemistry. Continual input is required from a source of oxygen, probably H_2O.

CO_2 on Titan. Samuelson, R.E., Maguire, W.C., Hanel, R.A., Kunde, V.G., Jennings, D.E., Yung, Y.L., and Aikin, A.C. *J. geophys. Res.* **88**, 8709 (1983).

Titan's 5μm spectral window: Carbon monoxide and the albedo of the surface. Noll, K.S., Geballe, T.R., Knacke, R.F., and Pendleton, Y.J. *Icarus* **124,** 625 (1996).

Millimeter and submillimeter heterodyne observations of Titan, the vertical profile of carbon monoxide in its stratosphere. Hidayat, T., Marten, A., Bezard, B., Gautier, D., Owen, T., Matthews, H.E., and Paubert, G. *Icarus* **133,** 109 (1998).

Evidence for water vapor in Titan's atmosphere from ISO/SWS data. Coustenis, A., Salama, A., Lellouch, E., Encrenaz, T., Bjoraker, G.L., Samuelson, R.E., de Graauw, T., Feuchtgruber, H., and Kessler, M.F. *Astron. Astrophys.* **336,** L85 (1998).

Abundance of carbon monoxide in the stratosphere of Titan from millimeter heterodyne observations. Marten, A., Gautier, D., Tanguy, L., Lecacheux, A., Rekolen, C., and Paubert, G. *Icarus* **76,** 558 (1988).

Titan: discovery of carbon monoxide in its atmosphere. Lutz, B.L., deBergh, C., and Owen, T. *Science* **220,** 1374 (1983).

The upper atmosphere, ionosphere, and torus of Titan.

Modeling the production and the imaging of energetic neutral atoms from Titan's exosphere. Amsif, A., Dandouras, J., and Roelof, E.C. *J. geophys. Res.* **102,** 22169 (1997).

Ion–molecule chemistry in Titan's ionosphere. Anicich, V.G. and McEwan, M.J. *Planet. Space Sci.* **45,** 897 (1997).

Titan's ion exosphere wake: A natural ion mass spectrometer? Luhmann, J.G. *J. geophys. Res.* **101,** 29387 (1996).

Corona discharge of Titan's troposphere. Navarro-Gonzalez, R. and Ramirez, S.I. *Advan. Space Res.* **19,** 1121 (1997).

Titan's atmosphere: possible clustering reactions of $HCNH^+$, $HCNH^+ (N_2)$ and $HCNH^+ (CH_4)$ ions with acetylene. Vacher, J.R., Le Duc, E., and Fitaire, M. *Planet. Space Sci.* **47,** 151 (1999).

Interaction of solar flare X-rays with the atmosphere of Titan. Banaszkiewicz, M., and Zarnecki, J.C. *Planet. Space Sci.* **47,** 35 (1999).

Ion–molecule reactions relevant to Titan's ionosphere. McEwan, M.J., Scott, G.B.I., and Anicich, V.G. *Int J. mass Spect.* **172,** 209 (1998).

Dynamic escape of H from Titan as consequence of sputtering induced heating. Lammer, H., Stumptner, W., and Bauer, S.J. *Planet. Space Sci.* **46,** 1207 (1998).

Model of Titan's ionosphere with detailed hydrocarbon ion chemistry. Keller, C.N., Anicich, V.G., and Cravens, T.E. *Planet. Space Sci.* **46,** 1157 (1998).

Titan's ionosphere: A review. Nagy, A.F. and Cravens, T.E. *Planet. Space Sci.* **46,** 1149 (1998).

Model of Titan's ionosphere with detailed hydrocarbon ion chemistry. Keller, C.N., Anicich, V.G., and Cravens, T.E. *Planet. Space Sci.* **46,** 1157 (1998).

Detection of Titan's ionosphere from Voyager 1 radio occultation observations. Bird, M.K., Dutta Roy, R., Asmar, S.W., and Rebold, T.A. *Icarus* **130,** 426 (1997).

Hydrocarbon ions in the ionosphere of Titan. Fox, J.L., and Yelle, R.V. *Geophys. Res. Letts.* **24,** 2179 (1997).

Ion–molecule clustering reactions of $HCNH^+$ ion with N_2 and/or CH_4 in N_2–CH_4 mixtures. Vacher, J.R., Le Duc, E., and Fitaire, M. *Planet. Space Sci.* **45,** 1407 (1997).

Recent advances in planetary ionospheres. Kar, J. *Space Sci. Rev.* **77,** 193 (1996).

A low-pressure study of C_2N_2 ion chemistry. McEwan, M.J. and Anicich, V.G. *J. phys. Chem.* **99,** 12204 (1995).

A model of the ionosphere of Titan. Keller, C.N. et al. *J. geophys. Res.* **97,** 12117 (1992).

Triton's ionospheric source: Electron precipitation or photoionization. Sittler, E.C. and Hartle, R.E. *J. geophys. Res.* **101,** 10863 (1996).
Distribution of molecular hydrogen in atmosphere of Titan. Bertaux, J.L. and Kockarts, G. *J. geophys. Res.* **88,** 8716 (1983).
Titan: Upper atmosphere and torus. Strobel, D.F., in *Saturn*. Gehrels, T. (ed.) Chapter IV. (University of Arizona Press, Tucson, 1983).
Titan's gas and plasma torus. Eviatur, A. and Podolak, M. *J. geophys. Res.* **88,** 833 (1983).
An estimate of the H_2 density in the atomic hydrogen cloud of Titan. Ip, W.-H. *J. geophys. Res.* **89,** 2377 (1984).

Hazes and aerosols in Titan's atmosphere may be the analogue of photochemical smog on Earth.

Elemental composition, solubility, and optical properties of Titan's organic haze. McKay, C.P. *Planet. Space Sci.* **44,** 741 (1996).
Titan's atmospheric haze: the case for HCN incorporation. Lara, L.M., Lellouch, E., and Shematovich, V. *Astron. Astrophys.* **341,** 312 (1999).
A model of energy-dependent agglomeration of hydrocarbon aerosol particles and implication to Titan's aerosols. Dimitrov, V., and Bar Nun, A. *J. Aerosol Sci.* **30,** 35 (1999).
Review and latest results of laboratory investigations of Titan's aerosols. Coll, P., Coscia, D., Gazeau, M.C., Guez, L., and Raulin, F. *Origins Life Evol. Biosphere* **28,** 195 (1998).
Transient clouds in Titan's lower atmosphere. Griffith, C.A., Owen, T., Miller, G.A., and Geballe, T. *Nature, Lond.* **395,** 575 (1998).
Titan haze: Structure and properties of cyanoacetylene and cyanoacetylene–acetylene photopolymers. Clarke, D.W. and Ferris, J.P. *Icarus* **127,** 158 (1997).
New planetary atmosphere simulations: Application to the organic aerosols of Titan. Coll, P., Cosia, D., Gazeau, M.C., and Raulin, F. *Advan. Space Res.* **19,** 1113 (1997).
A mechanism for the formation of aerosol concentrations in the atmosphere of Titan. Elperin, T., Kleeorin, N., Podolak, M., and Rogachevskii, I. *Planet. Space Sci.* **45,** 923 (1997).
An adequate kinetic model of photochemical aerosol formation in Titan's atmosphere. Dimitrov, V. and Bar Nun, A. *Advan. Space Res.* **19,** 1103 (1997).
An adequate kinetic model of the photochemical formation of hydrocarbon aerosols in Titan's atmosphere. Dimitrov, V. and Bar Nun, A. *Prog. React. Kinet.* **22,** 3 (1997).
The Titan haze revisited—magnetospheric energy sources and quantitative tholin yields. Thompson, W.R., McDonald, G.D., and Sagan, C. *Icarus* **112,** 376 (1994).
Vertical distribution of scattering hazes in Titan's upper atmosphere. Rages, K. and Pollack, J.B. *Icarus* **55,** 50 (1983).
Size estimates of Titan's aerosols based on Voyager high-phase-angle images. Rages, K., Pollack, J.B., and Smith, P.H. *J. geophys. Res.* **88,** 8721 (1983).
Inhomogeneous models of Titan's aerosol distribution. Podolak, M., Bar-Nun, A., Noy, N., and Giver, L.P. *Icarus* **57,** 72 (1984).
Titan: far infrared and microwave remote sensing of methane clouds and organic hazes. Thompson, W.R. and Sagan, C. *Icarus* **60,** 236 (1984).
Production and condensation of organic gases in the atmosphere of Titan. Sagan, C. and Thompson, W.R. *Icarus* **59,** 133 (1984).

There has been speculation about the possibility of atmospheric chemistry maintaining enough C_2H_6 to supply a condensed 'ocean' on the satellite. The first report shows that CH_4 does not form oceans or global cloud, although there may be a methane ice haze high in the troposphere.

Is Titan wet or dry? Eshelman, V.R., Lindel, G.F., and Tyler, G.L. *Science* **221**, 53 (1983).
Hiding Titan's ocean: Densification and hydrocarbon storage in an icy regolith. Kossacki, K.J. and Lorenz, R.D. *Planet. Space Sci.* **44,** 1029 (1996).
Does Titan have an ocean? A review of current understanding of Titan's surface. Lunine, J.I. *Rev. Geophys.* **31**, 133 (1993).
Titan hypothesized ocean properties — the influence of surface-temperature and atmospheric composition uncertainties. Dubouloz, N., Raulin, F., Lellouch, E., and Gautier, D. *Icarus* **82,** 81 (1989).
Titan's atmosphere and hypothesized ocean — a reanalysis of the Voyager-1 radio-occultation and IRIS 7.7μm data. Lellouch, E., Coustenis, A., Gautier, D., Raulin, F., Dubouloz, N., and Frere, C. *Icarus* **79,** 328 (1989).
Photochemically driven collapse of Titan's atmosphere. Lorenz, R.D., McKay, C.P. and Lunine, J.I. *Science* **275**, 642 (1997).

The atmosphere of Titan has clearly evolved over time. The evolutionary behaviour may possibly be a model for the development of the Earth's early atmosphere.

Some speculations on Titan's past, present and future. Lunine, J.I., Lorenz, R.D., and Hartmann, W.K. *Planet. Space Sci.* **46,** 1099 (1998).
Titan and the origin of life on Earth. Owen, T., Raulin, F., McKay, C.P., Lunine, J.I., Lebreton, J.P., and Matson, D.L. *ESA Bulletin* **24,** 2905 (1997).
Chemical evolution on Titan: Comparisons to the prebiotic earth. Clarke, D.W. and Ferris, J.P. *Origins Life Evol. Biosphere* **27,** 225 (1997).
A massive early atmosphere on Titan. Lunine, J.I. and Nolan, M.C. *Icarus* **100**, 221 (1992).
Chemistry and evolution of Titan's atmosphere. Strobel, D.F. *Planet. Space Sci.* **30,** 839 (1982).

Section 8.5.2

Production of O_2 on icy satellites by electronic excitation of low-temperature water ice. Sieger, M.T., Simpson, W.C., and Orlando, T.M. *Nature, Lond.* **394,** 554 (1998).
Sputtering of water ice surfaces and the production of extended neutral atmospheres. Shi, M., Baragiola, R.A., Grosjean, D.E., Johnson, R.E., Jurac, S., and Schou, J. *J. geophys. Res.* **100,** 26387 (1995).

Io

Io's tenuous atmosphere. Johnson, T.V. and Matson, D.L. in *Origin and evolution of planetary and satellite atmospheres*. Atreya, S.K., Pollack, J.B., and Matthews, M.S. (eds.) (University of Arizona Press, Tucson, 1989).

Io. Johnson, T.V. and Soderblum, L.A. *Scient. Am.* **249**, 60 (Dec. 1983).
A search for new species in Io's extended atmosphere. Na, C.Y., Trafton, L.M., Barker, E.S., and Stern, S.A. *Icarus* **131,** 449 (1998).
Io's atmosphere: Not yet understood. Lellouch, E. *Icarus* **124,** 1 (1996).
Photochemistry and vertical transport in Io's atmosphere and ionosphere. Summers, M.E., and Strobel, D.F. *Icarus* **120,** 290 (1996).
Io meteorology: how atmospheric pressure is controlled locally by volcanoes and surface frosts. Ingersoll, A.P. *Icarus* **81**, 298 (1989).
Io's patchy SO_2 atmosphere as measured by the Galileo ultraviolet spectrometer. Hendrix, A.R., Barth, C.A., and Hord, C.W. *J. geophys. Res.* **104,** 11817 (1999).
The distribution of sulfur dioxide and other infrared absorbers on the surface of Io. Carlson, R.W., Smythe, W.D., Lopes-Gautier, R.M.C., Davies, A.G., Kamp, L.W., Mosher, J.A., Soderblom, L.A., Leader, F.E., Mehlman, R., Clark, R.N., and Fanale, F.P. *Geophys. Res. Letts.* **24,** 2479 (1997).
On the excitation of Io's atmosphere by the photoelectrons: Application of the analytical yield spectral model of SO_2. Bhardwaj, A. and Michael, M. *Geophys. Res. Letts.* **26,** 393 (1999).
Volcanic origin of disulfur monoxide (S_2O) on Io. Zolotov, M.Y. and Fegley, B. *Icarus* **133,** 293 (1998).
Volcanic production of sulfur monoxide (SO) on Io. Zolotov, M.Y. and Fegley, B. *Icarus* **132,** 431 (1998).
Sputtering products of sodium sulfate: Implications for Io's surface and for sodium-bearing molecules in the Io torus. Wiens, R.C., Burnett, D.S., Calaway, W.F., Hansen, C.S., Lykke, K.R., and Pellin, M.J. *Icarus* **128,** 386 (1997).
Galileo search for SO_2-frost condensation on Io's nightside. Simonelli, D.P., Veverka, J., Senske, D.A., Fanale, F.P., Schubert, G., and Belton, M.J.S. *Icarus* **135,** 166 (1998).
First observational evidence for condensation of Io's SO_2 atmosphere on the nightside. Buratti, B.J., Mosher, J.A., and Terrile, R.J. *Icarus* **118,** 418 (1995).
SO_2–rock interaction on Io. 2. Interaction with pure SO_2. Burnett, D.S., Goreva, J., Epstein, S., Haldemann, S.L., Johnson, M.L., and Rice, A. *J. geophys. Res.* **102,** 19371 (1997).
On the dissociative ionization of SO_2 in Io's atmosphere. Michael, M. and Bhardwaj, A. *Geophys. Res. Letts.* **24,** 1971 (1997).
SO_2 distributions on Io. Sartoretti, P., Belton, M.J.S., and McGrath, M.A. *Icarus* **122,** 273 (1996).
Detection of sulfur monoxide in Io's atmosphere. Lellouch, E., Strobel, D.F., Belton, M.J.S., Summers, M.E., Paubert, G., and Moreno, R. *Astrophys. J.* **459,** L107 (1996).
The state of SO_2 on Io's surface. Kerton, C.R., Fanale, F.P., and Salvail, J.R. *J. geophys. Res.* **101,** 7555 (1996).
The gaseous sulfur dioxide abundance over Io's leading and trailing hemispheres.Trafton, L.M., Caldwell, J.J., Barnet, C., and Cunningham, C.C. *Astrophys. J.* **456,** 384 (1996).
The global distribution, abundance, and stability of SO_2 on Io. McEwen, A.S., Johnson, T.V., Matson, D.L., and Soderblom, L.A. *Icarus* **75**, 450 (1988).
Sulfur dioxide on Io: spatial distribution and physical state. Howell, H.R., Cruikshank D.P., and Fanale, F.P. *Icarus* **57**, 83 (1984).

The 2.5–5.0 micrometre spectra of Io: evidence for H_2S and H_2O frozen into SO_2. Salama, F., Allamandola, L.J., Witterborn, F.C., Cruikshank, D.P., Sandford, S.A., and Bregman, J.D. *Icarus* **83**, 66 (1990).

Io's atmosphere from microwave detection of SO_2. Lellouch, E., Belton, M., de Pater, I., Gulkis, S., and Encrenaz, T. *Nature, Lond.* **346**, 639 (1990).

Far-ultraviolet imaging spectroscopy of Io's atmosphere with HST/STIS. Roesler, F.L., Moos, H.W., Oliversen, R.J., Woodward, R.C., Retherford, K.D., Scherb, F., McGrath, M.A., Smyth, W.H., Feldman, P.D., and Strobel, D.F. *Science* **283,** 353 (1999).

Determination of the neutral number density in the Io torus from Galileo–EPD measurements. Lagg, A., Krupp, N., Woch, J., Livi, S., Wilken, B., and Williams, D.J. *Geophys. Res. Letts.* **25,** 4039 (1998).

Ground-based observations of Io. Matson, D.L., Johnson, T.V., Blaney, D.L., and Veeder, G.J. *Rev. Geophys.* **33**, 505 (1995).

Active volcanism on Io as seen by Galileo SSI. McEwen, A.S., Keszthelyi, L., Geissler, P., Simonelli, D.P., Carr, M.H., Johnson, T.V., Klaasen, K.P., Breneman, H.H., Jones, T.J., Kaufman, J.M., Magee, K.P., Senske, D.A., Belton, M.J.S., and Schubert, G. *Icarus* **135,** 181 (1998).

The Pele plume (Io): Observations with the Hubble Space Telescope. Spencer, J.R., Sartoretti, P., Ballester, G.E., McEwen, A.S., Clarke, J.T., and McGrath, M.A. *Geophys. Res. Letts.* **24,** 2471 (1997).

Io's sodium corona and spatially extended cloud: A consistent flux speed distribution. Smyth, W.H. and Combi, M.R. *Icarus* **126,** 58 (1997).

High resolution spectra of Io's neutral potassium and oxygen clouds. Thomas, N. *Astron. Astrophys.* **313,** 306 (1996).

Production of hydrogen ions at Io. Frank, L.A. and Paterson, W.R. *J. geophys. Res.* **104,** 10345 (1999).

Coupling the plasma interaction at Io to Jupiter. Crary, F.J. and Bagenal, F. *Geophys. Res. Letts.* **24,** 2135 (1997).

Plasma injection near Io. Hill, T.W. and Pontius, D.H. *J. geophys. Res.* **103,** 19879 (1998).

Galileo radio occultation measurements of Io's ionosphere and plasma wake. Hinson, D.P., Kliore, A.J., Flasar, F.M., Twicken, J.D., Schinder, P.J., and Herrera, R.G. *J. geophys. Res.* **103,** 29343 (1998).

Io's plasma environment during the Galileo flyby: Global three-dimensional MHD modeling with adaptive mesh refinement. Combi, M.R., Kabin, K., Gombosi, T.I., De Zeeuw, D.L., and Powell, K.G. *J. geophys. Res.* **103,** 9071 (1998).

Energy escape rate of neutrals from Io and the implications for local magnetospheric interactions. Smyth, W.H. *J. geophys. Res.* **103,** 11941 (1998).

A three-dimensional azimuthally symmetric model atmosphere for Io. 2. Plasma effect on the surface. Wong, M.C. and Johnson, R.E. *J. geophys. Res.* **101,** 23255 (1996).

Ejection of sodium from sodium sulfide by the sputtering of the surface of Io. Chrisey, D.B., Johnson, R.E., Boring, J.W., and Phipps, J.A. *Icarus* **75**, 233 (1988).

Escape of sulfur and oxygen from Io. Cheng, A.F. *J. geophys. Res.* 89, 3939 (1984).

The SO_2 atmosphere of Io: ion chemistry, atmospheric escape, and models corresponding to the Pioneer 10 radio occultation measurements. Kumar, S. *Icarus* **61**, 101 (1985).

Europa

Europa's surface composition and sputter-produced ionosphere. Johnson, R.E., Killen, R.M., Waite, J.H., and Lewis, W.S. *Geophys. Res. Letts.* **25,** 3257 (1998).

Hot oxygen corona at Europa. Nagy, A.F., Kim, J., Cravens, T.E., and Kliore, A.J. *Geophys. Res. Letts.* **25,** 4153 (1998).

The far-ultraviolet oxygen airglow of Europa and Ganymede. Hall, D.T., Feldman, P.D., McGrath, M.A., and Strobel, D.F. *Astrophys. J.* **499,** 475 (1998).

Europa's oxygen exosphere and its magnetospheric interaction. Ip, W.H. *Icarus* **120,** 317 (1996).

Hydrogen peroxide on the surface of Europa. Carlson, R.W., Anderson, M.S., Johnson, R.E., Smythe, W.D., Hendrix, A.R., Barth, C.A., Soderblom, L.A., Hansen, G.B., McCord, T.B., Dalton, J.B., Clark, R.N., Shirley, J.H., Ocampo, A.C., and Matson, D.L. *Science* **283,** 2062 (1999).

Interaction of the Jovian magnetosphere with Europa: Constraints on the neutral atmosphere. Saur, J., Strobel, D.F., and Neubauer, F.M. *J. geophys. Res.* **103,** 19947 (1998).

The ionosphere of Europa from Galileo radio occultations. Kliore, A.J., Hinson, D.P., Flasar, F.M., Nagy, A.F., and Cravens, T.E. *Science* **277,** 355 (1997).

Discovery of an extended sodium atmosphere around Europa. Brown, M.E. and Hill, R.E. *Nature, Lond.* **380,** 229 (1996).

Callisto

A tenuous carbon dioxide atmosphere on Jupiter's moon Callisto. Carlson, R.W. *Science* **283,** 820 (1999).

Ganymede

O_2/O_3 microatmospheres in the surface of Ganymede. Johnson, R.E. and Jesser, W.A. *Astrophys. J.* **480,** L79 (1997).

Oxygen on Ganymede: Laboratory studies. Vidal, R.A., Bahr, D., Baragiola, R.A., and Peters, M. *Science* **276,** 1839 (1997).

Hubble space telescope detects thin oxygen atmosphere on Ganymede. Wong, M.C. and Johnson, R.E. *J. geophys. Res.* **101,** 23243 (1996).

Detection of ozone on Ganymede. Noll, K.S., Johnson, R.E., Lane, A.L., Domingue, D.L., and Weaver, H.A. *Science* **273,** 341 (1996).

A search for a sodium atmosphere around Ganymede. Brown, M.E. *Icarus* **126,** 236 (1997).

Polar "caps" on Ganymede and Io revisited. Johnson, R.E. *Icarus* **128,** 469 (1997).

Energetic ion sputtering effects at Ganymede. Ip, W.H., Williams, D.J., McEntire, R.W., and Mauk, B. *Geophys. Res. Letts.* **24,** 2631 (1997).

Galileo ultraviolet spectrometer observations of atomic hydrogen in the atmosphere of Ganymede. Barth, C.A., Hord, C.W., Stewart, A.I.F., Pryor, W.R., Simmons, K.E., McClintock, W.E., Aiello, J.M., Naviaux, K.L., and Aiello, J.J. *Geophys. Res. Letts.* **24,** 2147 (1997).

Section 8.6

Voyagers at Uranus and Neptune (special reports): Uranus: *Science* **233**, 39–109 (1986); *J. geophys. Res.* **92**, 14873–15375 (1987); Neptune: *Science* **246**, 1417–1501 (1989); *Geophys. Res. Letts.* **17**, 1643–1772 (1991); *J. geophys. Res.* **96**, 18903–19268 (1991).

Uranus

Uranus. Miner, E.D. (Wiley, Chichester, 1998).

Uranus. Bergstrahl, J.T., Miner, E.D., and Matthews, M.S. (eds.) (University of Arizona Press, Tucson, AZ, 1991).

Uranus. Ingersoll, A.P. *Scient. Am.* **256**(1), 20 (Jan 1987).

Comparative models of Uranus and Neptune. Podolak, M., Weizman, A., and Marley, M. *Planet. Space Sci.* **43,** 1517 (1995).

Uranus deep atmosphere revealed. De Pater, I., Romani, P.N., and Atreya, S.K. *Icarus* **82**, 288 (1989).

Photochemistry of the atmosphere of Uranus. Summers, M.E. and Strobel, D.F. *Astrophys. J.* **346,** 495 (1989).

Detection of HD in the atmospheres of Uranus and Neptune: a new determination of the D/H ratio. Feuchtgruber, H., Lellouch, E., Bezard, B., Encrenaz, T., de Graauw, T., and Davis, G.R. *Astron. Astrophys.* **341,** L17 (1999).

Thermal structure of Uranus' atmosphere. Marley, M.S. and McKay, C.P. *Icarus* **138,** 268 (1999).

The Uranian geometric albedo: An analysis of atmospheric scatterers in the near-infrared. Walter, C.M. and Marley, M.S. *Icarus* **132,** 285 (1998).

Thermal structure and para hydrogen fraction on the outer planets from Voyager IRIS measurements. Conrath, B.J., Gierasch, P.J., and Ustinov, E.A. *Icarus* **135,** 501 (1998).

ISO observations of Uranus: The stratospheric distribution of C_2H_2 and the eddy diffusion coefficient. Encrenaz, T., Feuchtgruber, H., Atreya, S.K., Bezard, B., Lellouch, E., Bishop, J., Edgington, S., de Graauw, T., Griffin, M., and Kessler, M.F. *Astron. Astrophys.* **333,** L43 (1998).

Determination of the composition and state of icy surfaces in the outer solar system. *Annu. Rev. Earth planet. Sci* **25,** 243 (1997).

Outer planet ionospheres: A review of recent research and a look toward the future. Waite, J.H., Lewis, W.S., Gladstone, G.R., Cravens, T.E., Maurellis, A.N., Drossart, P., Connerney, J.E.P., Miller, S., and Lam, H.A. *Advan. Space Res.* **20,** 243 (1997).

Variation in the H_3^+ emission of Uranus. Lam, H.A., Miller, S., Joseph, R.D., Geballe, T.R., Trafton, L.M., Tennyson, J., and Ballester, G.E. *Astrophys. J.* **474,** L73 (1997).

Atomic hydrogen corona of Uranus. Herbert, F. and Hall, D.T. *J. geophys. Res.* **101,** 10877 (1996).

Broad-band spectroscopic detection of the CO J = 3–2 tropospheric absorption in the atmosphere of Neptune. Naylor, D.A., Davis, G.R., Griffin, M.J., Clark, T.A., Gautier, D., and Marten, A. *Astron. Astrophys.* **291,** 151 (1994).

Chemical models of the deep atmosphere of Uranus. Fegley, M.B., Jr. and Prinn, R. *Astrophys. J.* **307**, 852 (1986).

First observations of CO and HCN on Neptune and Uranus at millimeter wavelengths and their implications for atmospheric chemistry. Marten, A., Gautier, D., Owen, T., Sanders, D.B., Matthews, H.E., Atreya, S.K., Tilanus, R.P.J., and Deane, J.R. *Astrophys. J.* **406,** 285 (1993).

An analysis of the Voyager 2 ultraviolet spectrometer occultation data at Uranus—inferring heat sources and model atmospheres. Stevens, M.H., Strobel, D.F., and Herbert, F. *Icarus* **101**, 45 (1993).

Nature of the stratospheric haze on Uranus: evidence for condensed hydrocarbons. Pollack, J.B., Rages, K., Pope, S.K., Tomasko, M.G.. Romani, P.N., and Atreya, S.K. *J. geophys. Res.* **92**, 15037 (1987).

Methane abundance in the atmosphere of Uranus. Teifel, V.G. *Icarus* **53**, 389 (1983).

Neptune

Neptune and Triton. Cruikshank, D.P. (ed.) (University of Arizona Press, Tucson, AZ, 1995).

Neptune's atmosphere revealed. Romani, P.N., de Pater, I., and Atreya, S.K. *Geophys. Res. Letts.* **16**, 933 (1989).

Detection of the methyl radical on Neptune. Bezard, B., Romani, P.N., Feuchtgruber, H., and Encrenaz, T. *Astrophys. J.* **515,** 868 (1999).

Effect of chemical kinetics uncertainties on hydrocarbon production in the stratosphere of Neptune. Dobrijevic, M., and Parisot, J.P. *Planet. Space Sci.* **46,** 491 (1998).

The plasmasphere of Neptune. Huang, T.S., Ho, C.W., and Alexander, C.J. *J. geophys. Res.* **103,** 20267 (1998).

Neptune's stratospheric winds from three central flash occultations. French, R.G., McGhee, C.A., and Sicardy, B. *Icarus* **136,** 27 (1998).

EPIC simulations of time-dependent, three-dimensional vortices with application to Neptune's Great Dark Spot. Le Beau, R.P. and Dowling, T.E. *Icarus* **132,** 239 (1998).

Voyager 2 ultraviolet spectrometer solar occultations at Neptune: photochemical modelling of the 125–165 nm lightcurves. Bishop, J., Romani, P.N., and Atreya, S.K. *Planet. Space Sci.* **46,** 1 (1998).

HCN formation under electron impact: Experimental studies and application to Neptune's atmosphere. Gazeau, M.C., Cottin, H., Guez, L., Bruston, P., and Raulin, F. *Advan. Space Res.* **19,** 1135 (1997).

Hydrogen dimer features in the 2μm spectra of Saturn and Neptune. Trafton, L.M., Kim, S.J., Geballe, T.R., and Miller, S. *Icarus* **130,** 544 (1997).

Atmospheric structure of Neptune in 1994, 1995, and 1996: HST imaging at multiple wavelengths. Hammel, H.B. and Lockwood, G.W. *Icarus* **129,** 466 (1997).

Metal ions in the atmosphere of Neptune. Lyons, J.R. *Science* **267**, 648 (1995).

Constraints on N_2 in Neptune atmosphere from Voyager measurements. Conrath, B.J., Gautier, D., Owen, T.C., and Samuelson, R.E. *Icarus* **101,** 168 (1993).

The CO abundance on Neptune from HST observations. Courtin, R., Gautier, D., and Strobel, D. *Icarus* **123,** 37 (1996).

The origin of carbon monoxide in Neptune's atmosphere. Lodders, K. and Fegley, B. *Icarus* **112**, 368 (1994).

CO in the troposphere of Neptune—detection of the J = 1–0 line in absorption. Guilloteau, S., Dutrey, A., Marten, A., and Gautier, D. *Astron. Astrophys.* **279,** 661 (1993).

Methane photochemistry on Neptune: ethane and acetylene mixing ratios and haze production. Romani, P.N. et al. *Icarus* **106**, 442 (1993).

Methane photochemistry and haze production on Neptune. Romani, P.N. and Atreya, S.K. *Icarus* **74**, 424 (1988).

Stratospheric hazes from CH_4 photochemistry on Neptune. Romani, P.N. and Atreya, S.K. *Geophys. Res.* Letts. **16**, 941 (1989).

The helium abundance of Neptune from Voyager measurements. Conrath, B.J., Gautier, D., Lindal, G.F., Samuelson, R.E., and Shaffer, W.A. *J. geophys. Res.* **96,** 18907 (1991).

Infrared observations of the Neptunian system. Conrath, B., Flasar, F.M., Hanel, R., Kunde, V., Maguire, W., Pearl, J., Pirraglia, J., Samuelson, R., Gierasch, P., Weir, A., Bezard, B., Gautier, D., Cruikshank, D., Horn, L., Springer, R., and Shaffer, W. *Science* **246,** 1454 (1989).

Even in the outermost reaches of the Solar System, bodies such as Triton and Pluto may possess atmospheres.

Triton

Formation of the satellites of the outer solar system: sources of their atmospheres. Coradini, A., Cerroni, P., Magni, G., and Federico, C. in *Origin and evolution of planetary and satellite atmospheres*. Atreya, S.K., Pollack, J.B., and Matthews, M.S. (eds.) (University of Arizona Press, Tucson, 1989).

Triton: A satellite with an atmosphere. Cruickshank, D.P. and Silvaggio, P.M. *Astrophys. J.* **233**, 1016 (1979).

Photochemistry of Triton's atmosphere and ionosphere. Krasnopolsky, V.A. and Cruikshank, D.P. *J. geophys. Res.* **100,** 21271 (1995).

Evidence from scanning electron microscopy of experimental influences on the morphology of Triton and Titan tholins. De Vanssay, E., McDonald, G.D., and Khare, B.N. *Planet. Space Sci.* **47,** 433 (1999).

Global warming on Triton. Elliot, J.L., Hammel, H.B., Wasserman, L.H., Frenz, O.G., McDonald, S.W., Person, M.J., Olkin, C.B., Spencer, J.R., Stansberry, J.A., Buie, M.W., Pasachoff, J.M., Babcock, B.A., and McConnochie, T.H. *Nature, Lond.* **393,** 765 (1998).

The thermal structure of Triton's atmosphere: Results from the 1993 and 1995 occultations. Olkin, C.B., Elliot, J.L., Hammel, H.B., Cooray, A.R., McDonald, S.W., Foust, J.A., Bosh, A.S., Buie, M.W., Millis, R.L., Wasserman, L.H., Dunham, E.W., Young, L.A., Howell, R.R., Hubbard, W.B., Hill, R., Marcialis, R.L., McDonald, J.S., Rank, D.M., Holbrook, J.C., and Reitsema, H.J. *Icarus* **129,** 178 (1997).

Triton's distorted atmosphere. Elliot, J.L., Stansberry, J.A., Olkin, C.B., Agner, M.A., and Davies, M.E. *Science* **278,** 436 (1997).

The role of an internal heat source for the eruptive plumes on Triton. Duxbury, N.S., and Brown, R.H. *Icarus* **125,** 83 (1997).

Triton's ionospheric source: Electron precipitation or photoionization. Sittler, E.C. and Hartle, R.E. *J. geophys. Res.* **101,** 10863 (1996).

The Triton Neptune plasma interaction. Hoogeveen, G.W. and Cloutier, P.A. *J. geophys. Res.* **101,** 19 (1996).

The emissivity of volatile ices on Triton and Pluto. Stansberry, J.A., Pisano, D.J., and Yelle, R.V. *Planet. Space Sci.* **44,** 945 (1996).
A model for the overabundance of methane in the atmospheres of Pluto and Triton. Stansberry, J.A., Spencer, J.R., Schmitt, B., Benchkoura, A.I., Yelle, R.V., and Lunine, J.I. *Planet. Space Sci.* **44,** 1051 (1996).
On the haze model for Triton. Krasnopolsky, V.A. et al. *J. geophys. Res.* **98,** 17123 (1993).
Solar control of the upper atmosphere of Triton. Lyons, J.R. et al. *Science* **256,** 204 (1992).
Measurement of the rate constant for the association reaction $CH + N_2$ at 53 K and its relevance to Triton's atmosphere. Le Picard, S.D. and Canosa, A. *Geophys. Res. Letts.* **25,** 485 (1998).
Nitrogen airglow sources—comparison of Triton, Titan, and Earth. Strobel, D.F. et al. *Geophys. Res. Letts.* **19,** 669 (1991).
Triton's streaks as windblown dust. Sagan, C. and Chyba, C. *Nature, Lond.* **346,** 546 (1990).
Triton: do we see to the surface? Cruikshank, D.I., Brown, R.H., Giver, L.P., and Tokumaja, A.T. *Science* **245,** 283 (1989).
Triton stratospheric molecules and organic sediments. Thompson, W.R., Singh, S.K., Khare, B.N., and Sagan, C. *Geophys. Res. Letts.* **16,** 981 (1989).
Volatiles on Triton: the infrared spectroscopic evidence, 2.0 to 2.5 micrometers. Cruikshank, D.P., Brown, R.H., Tokunaga, A.T., Smith, R.G., and Piscitelli, J.R. *Icarus* **74,** 413 (1988).
Nitrogen on Triton. Cruikshank, D.P., Brown, R.H., and Clark, R.N. *Icarus* **58,** 293 (1984).

Pluto

Pluto and Charon. Stern, A. and Mitton, J. (Wiley, Chichester, 1998).
Pluto. Binzel, R. *Scient. Am.* **262,** 26 (June 1990).
Pluto's atmosphere. Elliott, J.L., Dunham, E.W., Bosh, A.S., Slivan, S.M., Young, L.A., Wasserman, L.H., and Millis, R.L. *Icarus* **77,** 148 (1989).
Photochemical models of Pluto's atmosphere. Lara, L.M., Ip, W.H., and Rodrigo, R. *Icarus* **130,** 16 (1997).
Detection of gaseous methane on Pluto. Young, L.A., Elliot, J.L., Tokunaga, A., De Bergh, C., and Owen, T. *Icarus* **127,** 258 (1997).
Synoptic CCD spectrophotometry of Pluto over the past 15 years. Grundy, W.M. and Fink, U. *Icarus* **124,** 329 (1996).
Rotationally resolved spectral studies of Pluto from 2500 to 4800 angstrom. Trafton, L.M. and Stern, S.A. *Astron. J.* **112,** 1212 (1996).
Hydrodynamic flow of N_2 from Pluto. Krasnopolsky, V.A. *J. geophys. Res.* **104,** 5955 (1999).
Seasonal nitrogen cycles on Pluto. Hansen, C.J. and Paige, D.A. *Icarus* **120,** 247 (1996).
Solar wind–Pluto interaction revised. Sauer, K., Lipatov, A., Baumgartel, K., and Dubinin, E. *Advan. Space Res.* **20,** 295 (1997).
On the vertical thermal structure of Pluto's atmosphere. Strobel, D.F., Zhu, X., Summers, M.E., and Stevens, M.H. *Icarus* **120,** 266 (1996).
The Pluto–Charon system. Stern, S.A. *Annu. Rev. Astron.* **30,** 185 (1992).

Pluto's atmosphere near perihelion. Trafton, L.M. *Geophys. Res.* Letts. **16**, 1213 (1989).

Upper limits on possible photochemical hazes on Pluto. Stansberry, J.A., Lunine, J.I., and Tomasko, M.G. *Geophys. Res. Letts.* **16**, 1221 (1989).

Methane absorption variations in the spectrum of Pluto. Buie, M.W., and Fink, U. *Icarus* **70**, 483 (1987).

Observations of Pluto and Charon by IRAS. Sykes, M.V., Cutri, R.M., Lebofsky, L.A., and Binzel, R.P. *Science* **237**, 1336 (1987).

Stability of Pluto's atmosphere. Hunten, D.M. and Watson, A.J. *Icarus* **51**, 665 (1982).

See also references under Sections 8.4 and 8.5

Section 8.7

See also references under Section 9.1.2.

Fire in the sky. Olson, R.J.M. and Pasachoff, J.M. (Cambridge University Press, Cambridge, 1998).

The origin of comets. Bailey, M.E., Clube, S.V.M., and Napier, W.M. (Pergamon Press, Oxford, 1990).

Comets and the origin of the solar system: reading the Rosetta stone. Mumma, M.J., Weissman, P.R., and Stern, S.A. in *Protostars and planets III*, Levy, E.H. and Lunine, J.I. (eds.) (University of Arizona Press, Tucson, AZ, 1993).

Implications of small comets for the noble gas inventories of Earth and Mars. Swindle, T.D. and Kring, D.A. *Geophys. Res. Letts.* **24,** 3113 (1997).

Chemistry in comets. Lust, R. *Topics curr. Chem.* **99**, 73 (1981).

Our cometary environment. Napier, W.M. and Clube, S.V.M. *Rep. Prog. Phys.* **60,** 293 (1997).

Protosolar nebula and composition of comets. Vanysek, V. and Moravec, Z. *Earth Moon and Planets* **68,** 95 (1995).

Very small bodies in the solar system: The Earth's atmosphere as detector. Elford, W.G. *Earth Moon and Planets* **71,** 245 (1995).

Comets, meteorites and atmospheres. Owen, T. and Bar-Nun, A. *Earth Moon and Planets* **72,** 425 (1996).

Physics and chemistry of comets: recent results from comets Hyakutake and Hale–Bopp — Answers to old questions and new enigmas. Crovisier, J. *Faraday Discs.* **134,** 249 (1998).

Cometary volatiles — The status after comet C/1996 B2 Hyakutake. Bockelee-Morvan, D. *IAU Symposia* **822,** 96 (1997).

Physical chemistry of a heterogeneous medium: Transport processes in comet nuclei. Bouziani, N. and Fanale, F.P. *Astrophys. J.* **499,** 463 (1998).

Time-dependent gas kinetics in tenuous planetary atmospheres: The cometary coma. Combi, M.R. *Icarus* **123,** 207 (1996).

The composition of comets. Jessberger, E.K., Kissel, J., and Rahe, J. in *Origin and evolution of planetary and satellite atmospheres*. Atreya, S.K., Pollack, J.B., and Matthews, M.S. (eds.) (University of Arizona Press, Tucson, 1989).

Did comets form from unaltered interstellar dust and ices? The evidence from infra-red spectroscopy. Tokunaga, A.T. and Brooks, T.Y., *Icarus* **86**, 208 (1990).

Comets and their composition. Spinrad, H. *Annu. Rev. Astron. Astrophys.* **25**, 231 (1987).

Composition measurements and the history of cometary matter. Geiis, J. *Astron. Astrophys.* **187**, 859 (1987).
On the temperature and gas composition in the region of comet formation. Bar-Nun, A. and Kleinfeld, I. *Icarus* **80**, 243 (1989).
Observations of cometary nuclei. A'Hearn, M.F. *Annu. Rev. Earth & Planet. Sci.* **16**, 273 (1988).
A postencounter view of comets. Mendis, D.A. *Annu. Rev. Astron. Astrophys.* **26**, 11 (1988).
Sources of cometary radicals and their jets: gases or grains. Combi, M.R. *Icarus* **71**, 178 (1987).
Infrared study of ion-irradiated water-ice mixtures with hydrocarbons relevant to comets. Moore, M.H. and Hudson, R.L. *Icarus* **135,** 518 (1998).
Visible spectroscopy of possible cometary candidates. Lazzarin, M., Barucci, M.A., and Doressoundiram, A. *Icarus* **122,** 122 (1996).
A survey of 39 comets using CCD spectroscopy. Fink, U. and Hicks, M.D. *Astrophys. J.* **459,** 729 (1996).
Comets as a source of low eccentricity and low inclination interplanetary dust particles. Liou, J.C. and Zook, H.A. *Icarus* **123,** 491 (1996).
Spectral analysis of two Perseid meteors. Borovicka, J. and Betlem, H. *Planet. Space Sci.* **45,** 563 (1997).

Individual chemical species in comets

Water vapour production in comets by Lyman alpha emission: SWAN experiment on board SOHO. Goukenleuque, C. *Comptes Rendus Ser. II, Fasc. B.* **327,** 635 (1999).
Hubble Space Telescope ultraviolet imaging and high-resolution spectroscopy of water photodissociation products in comet Hyakutake (C/1996 B2). Combi, M.R., Brown, M.E., Feldman, P.D., Keller, H.U., Meier, R.R., and Smyth, W.H. *Astrophys. J.* **494,** 816 (1998).
Spectroscopic evidence for interstellar ices in comet Hyakutake. Irvine, W.M., Bockelee-Morvan, D., Lis, D.C., Matthews, H.E., Biver, N., Crovisier, J., Davies, J.K., Dent, W.R.F., Gautier, D., Godfrey, P.D., Keene, J., Lovell, A.J., Owen, T.C., Phillips, T.G., Rauer, H., Schloerb, F.P., Senay, M., and Young, K. *Nature, Lond.* **383,** 418 (1996).
Comet Halley $O(^1D)$ and H_2O production rates. Magee-Sauer, K., Scherby, F., Roesler, F.L., and Harlander, J. *Icarus* **84,** 154 (1990).
Water clusters in the coma of comet Halley and their effect on gas density, temperature, and velocity. Crifo, J.F. *Icarus* **84,** 414 (1990).
Observations of the OH radical in comet C/1996 B2 (Hyakutake) with the Nancay radio telescope. Gerard, E., Crovisier, J., Colom, P., Biver, N., Bockelee-Morvan, D., and Rauer, H. *Planet. Space Sci.* **46,** 569 (1998).
Trails of OH emissions from small comets near Earth. Frank, L.A. and Sigwarth, J.B. *Geophys. Res. Letts.* **24,** 2435 (1997).
Deuterated water in comet C 1996 B2 (Hyakutake) and its implications for the origin of comets. Bockelee-Morvan, D., Gautier, D., Lis, D.C., Young, K., Keene, J., Phillips, T., Owen, T., Crovisier, J., Goldsmith, P.F., Bergin, E.A., Despois, D., and Wootten, A. *Icarus* **133,** 147 (1998).

Deuterium in comet C 1995 O1 (Hale–Bopp): Detection of DCN. Meier, R., Owen, T.C., Jewitt, D.C., Matthews, H.E., Senay, M., Biver, N., Bockelee-Morvan, D., Crovisier, J., and Gautier, D. *Science* **279,** 1707 (1998).

A determination of the HDO/H_2O ratio in comet C/1995 O1 (Hale–Bopp). Meier, R., Owen, T.C., Matthews, H.E., Jewitt, D.C., Bockelee-Morvan, D., Biver, N., Crovisier, J., and Gautier, D. *Science* **279,** 842 (1998).

Ion irradiation and extended CO emission in cometary comae. Brucato, J.R., Castorina, A.C., Palumbo, M.E., Satorre, M.A., and Strazzulla, G. *Planet. Space Sci.* **45,** 835 (1997).

Carbonyl sulfide in comets C/1996 B2 (Hyakutake) and C/1995 O1 (Hale–Bopp): Evidence for an extended source in Hale–Bopp. Dello-Russo, N., Di Santi, M.A., Mumma, M.J., Magee-Sauer, K., and Rettig, T.W. *Icarus* **135,** 377 (1998).

An attempt to detect the C_2 intercombination transition lines in comet Hale–Bopp. Rousselot, P., Laffont, C., Moreels, G., and Clairemidi, J. *Astron. Astrophys.* **335,** 765 (1998).

A critical study of molecular photodissociation and CHON grain sources for cometary C_2. Combi, M.R. and Fink, U. *Astrophys. J.* **484,** 879 (1997).

Gaseous CN, C_2 and C_3 jets in the inner coma of comet P/Halley observed from the Vega 2 spacecraft. Clairemidi, J., Moreels, G., and Krasnopolsky, V.A., *Icarus* **86**, 115 (1990).

On the role of solar EUV, photoelectrons, and auroral electrons in the chemistry of $C(^1D)$ and the production of C I 1931Å in the cometary coma: A case for comet 1P/Halley. Bhardwaj, A. *J. geophys. Res.* **104,** 1929 (1999).

Origin and production of $C(^1D)$ atoms in cometary comae. Tozzi, G.P., Feldman, P.D., and Festou, M.C. *Astron. Astrophys.* **330,** 753 (1998).

Production and emissions of atomic carbon and oxygen in the inner coma of Comet Halley: Role of electron impact. Bhardwaj, A., Haider, S.A., and Singhal, R.P. *Icarus* **120,** 412 (1996).

The photodissociation of interstellar and cometary CH_2. Van Dishoeck, E.F., Bearda, R.A., and van Hemert, M.C. *Astron. Astrophys.* **307,** 645 (1996).

Hydrocarbon radiation chemistry in ices of cometary relevance. Hudson, R.L. and Moore, M.H. *Icarus* **126,** 233 (1997).

The source of the high C_2H_6/CH_4 ratio in Comet Hyakutake. Notesco, G., Laufer, D., and BarNun, A. *Icarus* **125,** 471 (1997).

Detection of acetylene in the infrared spectrum of comet Hyakutake. Brooke, T.Y., Tokunaga, A.T., Weaver, H.A., Crovisier, J., Bockelee-Morvan, D., and Crisp, D. *Nature, Lond.* **383,** 606 (1996).

Polycyclic aromatic hydrocarbon lifetime in cometary environments. Joblin, C., Boissel, P., and de Parseval, P. *Planet. Space Sci.* **45,** 1539 (1997).

Production and chemical analysis of cometary ice tholins. McDonald, G.D., Whited, L.J., DeRuiter, C., Khare, B.N., Patnaik, A., and Sagan, C. *Icarus* **122,** 107 (1996).

NH_2 and its parent molecule in the inner coma of comet Hyakutake (C/1996 B2). Kawakita, H., and Watanabe, J. *Astrophys. J.* **495,** 946 (1998).

The NH and CH bands of Comet C/1996 B2 (Hyakutake). Meier, R., Wellnitz, D., Kim, S.J., and AHearn, M.F. *Icarus* **136,** 268 (1998).

HNC and HCN in comets. Rodgers, S.D. and Charnley, S.B. *Astrophys. J.* **501,** L227 (1998).

Giotto IMS measurements of the production rate of hydrogen cyanide in the coma of comet Halley. Ip, W.-H., Balsiger, H., Geiss, J., Goldstein, B.E., Kettmann, G., Lazarus, A.J., Meier, A., Rosenbauer, H., Schwenn, R., and Shelley, E. *Ann. Geophys.* **8**, 319 (1990).

Chemical processing in the coma as the source of cometary HNC. Irvine, W.M., Bergin, D.A., Dickens, J.E., Jewitt, D., Lovell, A.J., Matthews, H.E., Schloerb, F.P., and Senay, M. *Nature, Lond.* **393,** 547 (1998).

Sodium tails of comets: NaO and NaSi abundances in interplanetary dust particles. Rietmeijer, F.J.M. *Astrophys. J.* **514**, L125 (1999).

The spatial distribution of gaseous atomic sodium in the comae of comets: Evidence for direct nucleus and extended plasma sources. Combi, M.R., DiSanti, M.A., and Fink, U. *Icarus* **130,** 336 (1997).

H_2O^+, CO^+, and dust in Comet P/Swift–Tuttle. Jockers, K., and Bonev, T. *Astron. Astrophys.* **319,** 617 (1997).

Quantitative analysis of H_2O^+ coma images using a multiscale MHD model with detailed ion chemistry. Haberli, R.M., Combi, M.R., Gombosi, T.I., DeZeeuw, D.L., and Powell, K.G. *Icarus* **130,** 373 (1997).

HCO^+ imaging of comet Hale–Bopp (C/1995 O1). Lovell, A.J., Schloerb, F.P., Dickens, J.E., DeVries, C.H., Senay, M.C., and Irvine, W.M. *Astrophys. J.* **497,** L117 (1998).

Chemistry of the ions $\leq$40 amu in the inner coma of comet Halley. Haider, S.A. and Bhardwaj, A. *Advan. Space Res.* **20,** 291 (1997).

Near-infrared spectroscopy of low-albedo surfaces of the solar system: Search for the spectral signature of dark material. Dumas, C., Owen, T., and Barucci, M.A. *Icarus* **133,** 221 (1998).

Natural solid bitumens as possible analogs for cometary and asteroid organics: 1. Reflectance spectroscopy of pure bitumens. Moroz, L.V., Arnold, G., Korochantsev, A.V., and Wasch, R. *Icarus* **134,** 253 (1998).

Investigations of individual comets

Observation and analysis of high-resolution optical line profiles in comet Hyakutake (C/1996 B2). Combi, M.R., Cochran, A.L., Cochran, W.D., Lambert, D.L., and Johns-Krull, C.M. *Astrophys. J.* **512,** 961 (1999).

Evidence for interacting gas flows and an extended volatile source distribution in the coma of comet C/1996 B2 (Hyakutake). Harris, W.M., Combi, M.R., Honeycutt, R.K., Mueller, B.E.A., and Scherb, F. *Science* **277,** 676 (1997).

Spectroscopic observations of Comet C 1996 B2 (Hyakutake) with the Caltech submillimeter observatory. Lis, D.C., Keene, J., Young, K., Phillips, T.G., Bockelee-Morvan, D., Crovisier, J., Schilke, P., Goldsmith, P.F., and Bergin, E.A. *Icarus* **130,** 355 (1997).

Comet Hyakutake reveals new chemistry. Wilson, E. *Chem. Eng. News* **74,** 8 (1996).

Comet 46P/Wirtanen, the target of the Rosetta mission. Rickman, H. and Jorda, L. *Advan. Space Res.* **21,** 1491 (1998).

Temperature and gas production distributions on the surface of a spherical model comet nucleus in the orbit of 46P/Wirtanen. Enzian, A., Klinger, J., Schwehm, G., and Weissman, P.R. *Icarus* **138,** 74 (1999).

The dependence of the circumnuclear coma structure on the properties of the nucleus — III. First modeling of a CO-dominated coma, with application to comets 46 P Wirtanen and 29 P Schwassmann–Wachmann I. Crifo, J.F., Rodionov, A.V., and Bockelee-Morvan, D. *Icarus* **138,** 85 (1999).

On the outgassing profile of Comet Hale–Bopp. Flammer, K.R., Mendis, D.A., and Houpis, H.L.F. *Astrophys. J.* **494,** 822 (1998).

The heliocentric evolution of key species in the distantly-active comet C/1995 O1 (Hale–Bopp). Womack, M., Festou, M.C., and Stern, S.A. *Astron. J.* **114,** 2789 (1997).

Encounters with comet Halley (special reports). *Nature, Lond.* **321**, supplement to no. 6067 (1986).

Giotto encounter with comet Halley (special reports). *Astron. Astrophys.* **187**, nos. 1–2 (1987).

Spectroscopic observation of Comet P/de Vico: Comparison with P/Halley and P/Brorsen–Metcalf. Kawakita, H., Ayani, K., and Matsubara, K. *Pubs. Astron. Soc. Japan* **50,** 343 (1998).

Spectral survey of the Comet Halley atmosphere during an occultation of a star. Nazarchuk, H.K. and Shulman, L.M. *Astronomische Nachrichten* **318,** 45 (1997).

Solar nebula origin for volatile gases in Halley's comet. Engel, S., Lunine, J.I., and Lewis, J.S. *Icarus* **85**, 380 (1990).

The organic component in dust from comet Halley as measured by the PUMA mass spectrometer on board Vega 1. Kissel, J. and Krueger, F.R. *Nature, Lond.* **326**, 755 (1987).

Abundances in comet Halley at the time of the spacecraft encounters. Wyckoff, S., Tegler, S., Wehinger, P.A., Spinrad, H., and Belton, M.J.S. *Astrophys. J.* **325**, 927 (1988).

9 Evolution and change in atmospheres and climates

9.1 Sources of atmospheric constituents

9.1.1 Origin and development of atmospheres

Atmospheric compositions can offer valuable clues about the earliest processes in the formation of the solar system and its planets. Geological records are available for Earth, and in a very superficial way for Mars, and they inform us about the development of the planetary crust and of interactions in the crust–hydrosphere–atmosphere system. Planetary probes have returned data that complement chemical and isotopic analyses of meteorites and lunar samples to provide models of the primitive solar nebula and of planetary accretion. Some planets (e.g. Jupiter) appear to have retained the *primordial* atmosphere with which they were created. Others (e.g. Mars or the satellite Titan) seem to have atmospheres that have undergone much change. Evolution of our own atmosphere is of more than parochial interest, because the development of an environment suitable for the creation and support of life seems unique in the solar system. With life came the growth of our oxygen-rich atmosphere as it is today (Chapter 1). We have recently become increasingly aware that our atmosphere is susceptible to change both because of extraterrestrial influences, such as changing solar intensity or collision with asteroids or comets, and because of man's activities. Climates, depending as they do on atmospheric composition and temperature, can alter as atmospheres evolve. In the case of our own planet, modifications in climate may be partially controlled by man's deliberate or accidental intervention.

Radioactive dating of meteorites and lunar samples gives ample evidence that the solar system is 4.6 Gyr (4.6×10^9 yr) old. The 'Big Bang' created the Universe 10–20 Gyr ago (current best estimates, based on Hubble space-telescope data, are 12 Gyr). Ever since, the Universe has been expanding and evolving. During this evolution, nuclear fusion has transformed hydrogen and helium, present from the beginning, into heavier elements such as C, N, O, Mg, Si, and Fe which make up a planet like Earth. Supernova explosions—the death throes of the massive stars—produce elements heavier than iron, and scatter and disperse the heavier elements through the galaxies as tiny dust grains 10–1000 nm in diameter. Spectroscopic investigation suggests that the particles are mainly graphite (C), H_2O ice, and iron and magnesium silicates. Most of the mass of any galaxy still remains as hydrogen and helium, the heavier elements constituting only a very small fraction (about one per cent).

Dust and gas in the Universe are concentrated in the arms of spiral galaxies in which new stars are formed when local regions become so dense that attractive gravitational forces dominate internal gas pressures. Collapse of the material starts, and a flattened spinning disk—a *solar nebula*—is produced. Compression of the hydrogen gas at the centre of the nebula is ultimately sufficient for fusion processes to be sustained, and a star, in our case the Sun, is formed.

While the concept of a rotating disc-like nebula is common to most current models of the origin of the solar system, there is considerable divergence of opinion about the next stage in the evolution of the solid bodies. One picture maintains that the dust grains accreted into metre-sized *planetesimals* that were the building blocks of which the inner planets were made. Such planetesimals were largely silicate materials in the inner solar system, since ices would have evaporated at the high local temperatures. Collisional processes leading to planet formation were favoured for large objects, so that planets grew at the expense of the multitude of planetesimals by sweeping them up. Many planetesimals are still located in the asteroid belt where Jupiter's gravitational field just cancels the forces of accretion. Material equilibrating and accreting at high temperatures will contain proportionately more iron than material equilibrating at lower temperatures. Taking into account the effects of compression, the densities of Mercury, Venus, Earth, and Mars follow the decrease in density expected from their distances from the Sun. Grains of H_2O, NH_3, and CH_4 ices were available as well as silicates in the outer solar system. Jupiter and Saturn may have grown sufficiently rapidly that they attracted large quantities of hydrogen and helium from the nebula. Further out, Neptune and Uranus were less efficient at capturing the nebular gases. Present-day compositions of the outer planets in terms of H_2 and He are thought to be 80, 70, 15, and 10 per cent for Jupiter, Saturn, Uranus, and Neptune. Astronomical observations of stars that are now young suggest that our early Sun passed through a phase of vigorous solar activity in which a violent solar wind carried off a sizeable amount of the Sun, as well as sweeping away remaining gases that had not yet condensed, or accumulated on a solid body.

Events following the accumulation of dust and gas to form larger bodies are entirely different for the inner and the outer planets. The gaseous planets must have undergone an initial period of hot contraction, after which there has probably been little further change. Impact of infalling asteroidal, meteoric, and cometary bodies on the unprotected surface, together with decay of short-lived radioactive elements, heated the inner planets so much that they must have been molten to a considerable depth. Release of gases from the interior ('*outgassing*') may have been significant since the intense heating could have led to the dissociation of minerals containing bound H_2O and CO_2, as well as to degassing of physically trapped gases. Atmospheric outgassing is thus tied to the thermal evolution of a planet, and the present composition of the planetary atmosphere provides an indicator of past history. Venus, Earth, Mars, and Titan have atmospheres whose compositions are very different from those of the Sun or of the primitive solar nebula, as a result of loss of the primary atmosphere and the evolution of a secondary one.

9.1.2 Interstellar clouds and their chemistry

Stars and their planets account for about 90 per cent of the mass of our own galaxy, the remainder being scattered rather unevenly throughout space. Much of the interstellar medium is very tenuous, possessing between one and a hundred atoms or molecules for every 1000 cm^3. In some places, much more matter has accumulated to form *interstellar clouds*, of which *diffuse* and *dense* categories are recognized. Particle densities are typically 10^2 cm^{-3} in the diffuse clouds, and may be as high as 10^6 cm^{-3} in the dense ones, and the temperature may be from 100 K to perhaps as low as 10 K. While the fascinating subject of interstellar cloud chemistry lies largely outside the scope of our present enquiry, there are analogies with processes occurring in planetary atmospheres that should not go unnoticed. Several references are given in the Bibliography that provide a more extensive introduction to the subject.

For nearly 70 years, it has been realized that molecular as well as atomic species are present in the interstellar clouds, visible and ultraviolet absorption spectroscopy identifying several diatomic species. Chemical change is thus occurring in the clouds. Less than 40 years ago, however, it was thought out of the question that complex polyatomic molecules, especially highly evolved organic ones, could ever exist in such cold and relatively thin regions of space. Radio- and millimetre-wave astronomy has now completely altered our picture of interstellar chemistry. More than 118 molecular species, including free radicals and at least 12 positive ions, had been identified by 1999 as a result of increasingly sophisticated observational techniques and improved spectral data derived from laboratory experiments. The list includes many large polyatomic species as well as diatomics, and new molecules are being added constantly as observations proceed. Polyatomic species detected include HCN, HCHO, C_2H_2, C_3O, HNCO, HCOOH, NH_2CN, CH_3CN, CH_3NH_2, CH_3CHO, C_2H_5OH, SiC_3, and the cyanopolyacetylenes, $H(C{\equiv}C)_nCN$, with *n* up to 10. The 'molecular' dense clouds are the most massive objects in the galaxy, possessing up to 10^5 times the mass of our Sun, and they are very active regions in which new stars are being formed continuously. Chemical compounds evolving in the clouds could therefore be present in the gaseous envelopes surrounding any planetary objects formed together with the stars. The compounds may also become incorporated into the frozen volatile materials that partially make up the comets. The presence in space of the building blocks of conventional organic synthesis has obvious and interesting implications. Our concern here is primarily with how the molecular species come to be formed in the interstellar clouds, and how they survive there.

Survival of complex molecules is, in fact, intimately bound up with the density of the clouds in which they are found. Starlight provides a galactic ultraviolet flux that would photodissociate and photoionize all polyatomic molecules were they not protected in some way. For cloud densities greater than a few times 10^2 cm^{-3}, most of the hydrogen in the interstellar medium is in molecular form (possibly as a result of heterogeneous recombination of H atoms on interstellar grain surfaces). Accumulation of H_2, and perhaps dust grains, ensures that the dense clouds are opaque to the galactic ultraviolet radiation, so that the clouds and the molecules within them become self-shielded from photolysis. It is, indeed, the opacity of the clouds to visible and shorter wavelength radiation that prevented the detection of complex interstellar molecules by conventional optical techniques. Radio-frequency and microwave radiation is not strongly attenuated by H_2 or dust grains, and so can be used to probe the composition of the dense clouds.

Ion–molecule reactions appear to be responsible for the synthesis of a large number of the polyatomic molecules found in dense interstellar clouds, so that ionic processes are seen to play a very significant part in the chemistry of interstellar clouds just as they do in planetary atmospheres (Chapter 6). As explained in Section 3.4.1, collisions between ions and molecules are several times more rapid than collisions between neutral partners for identical concentrations of partners. Dipolar fields induced by the ion in the neutral collision partner lead to much longer range attractive forces than are provided by the van der Waals attraction between two neutral species. Not only is the collision frequency enhanced, but spiralling and multiple impact encounters facilitate formation of long-lived intermediate complexes that may be involved in bimolecular and associative ion–molecule reactions. The energy gained in bringing ion and neutral together is often enough to overcome any kinetic activation barrier to reaction, so that most ion–molecule reactions have little if any activation energy. At the temperatures of the interstellar clouds, most reactions between neutral partners would be at a standstill. A few radical–radical reactions of low activation energy form the only exceptions.

We have reviewed important types of ion chemistry in Section 6.2 and Table 6.1, and the principles expounded apply to interstellar as well as to ionospheric chemistry. One difference is that *radiative* association, recombination, and attachment processes were assumed to be of negligible importance in planetary atmospheres, because their rates are so small in comparison with other available reactive pathways. In the interstellar clouds, the temperatures, concentrations, and time-scales may actually make the radiative processes *relatively* more important than non-radiative ones.

Ionization itself is provided in the dense clouds by interaction of galactic cosmic rays with the most abundant species, hydrogen and helium. It is not our purpose here to give a detailed account of the route followed in the formation of each complex molecule from the primary ions, but it is worthwhile to try to identify some of the more important steps. For every H_2 molecule ionized, an H_3^+ ion is produced, whose function in more complex ion chemistry is to produce OH, H_2O, protonated species, and NH_2^+ and NH_3^+. Carbon in the clouds is mainly in the form of atoms and CO; C^+ ions are, however, one of the most abundant ionic species present, because collision of He^+ with CO generates the atomic ions. The C^+ ion is a cornerstone in the formation of organic species.

For example, reactions with NH_2 or NH_3 yield HCN^+ and H_2CN^+ ions that lead to HCN in a manner already discussed in connection with Titan's atmosphere (Section 8.5.1). Synthesis of C–C bonds is achieved by the attack of C^+ on carbon-containing neutral radicals to yield (directly or indirectly) first C_2H^+ then $C_2H_2^+$ and finally the ethynyl radical, C_2H, invoked in much of the neutral chemistry of Jupiter, Saturn, and Titan (Sections 8.4, 8.5.1). From that point on, it is obviously possible to postulate routes to most of the polyatomic species observed in the dense clouds. Some of the longest carbon chains are to be found in the cyanopolyacetylenes that can be synthesized from CN, C_2H, and C_2H_2, species all actually detected in the interstellar clouds. Tricarbon monoxide, C_3O, was the first interstellar carbon chain molecule to be detected that contains oxygen.

Abundances of the rare stable isotopes of several elements (e.g. D, ^{13}C, ^{15}N) relative to the more common isotope appear in certain interstellar molecules to be enriched over the abundances typical of the solar system. It is important to know if the isotope ratios in those molecules reflect the ratios in the entire cloud, or if some isotope-selective chemistry has favoured formation of the enriched molecules. Laboratory studies of isotope exchange in ion–neutral reactions have shown that chemical fractionation can indeed explain the apparent isotopic enrichment in some interstellar molecules. We shall meet the subject of isotopic enrichment again in Section 9.3, but for the time being we should heed the lesson learned from the interstellar studies that isotope ratios measured from a particular molecule may not be representative of the overall abundance of an element's isotope.

9.2 Noble gases and nitrogen in planetary atmospheres

9.2.1 Inner planets

Noble gases in planetary atmospheres provide valuable pointers concerning atmospheric origins and evolution. Except for the lightest noble gas, helium, they cannot readily escape to space from any of the inner planets, Venus, Earth, or Mars. Chemical inertness prevents loss of the noble gases (with the possible exception of xenon) to surface rocks. Abundances of the gases thus correspond to the cumulative quantity of gas present in and released to the atmosphere over the entire history of the planet. Two types of noble gas can be distinguished: primordial and radiogenic. Primordial isotopes such as ^{20}Ne, ^{36}Ar, ^{38}Ar, ^{84}Kr, and ^{132}Xe were present in the solar system from the time of its creation. Radiogenic isotopes, however, have built up from the decay of radioactive nucleides: ^{40}Ar from the decay of ^{40}K, and ^{4}He from the decay of ^{232}Th, ^{235}U, and ^{238}U.

Before the Viking and Pioneer–Venus missions (see Chapter 8), the only information we had about the isotopic composition of planetary atmospheres was for Earth. Table 9.1 shows that radiogenic argon is nearly 300 times more

Table 9.1 Abundance of noble gases

Gas	^{20}Ne	^{36}Ar	$^{36}Ar/^{12}C$	$^{36}Ar/^{38}Ar$	$^{40}Ar/^{36}Ar$	$^{20}Ne/^{22}Ne$
	kg per kg of object		number ratios			
Object						
Sun	2.2×10^{-3}	9.0×10^{-5}	2.3×10^{-2}	5.6	< 1.0	13.7
CI[a]	2.9×10^{-10}	1.3×10^{-9}	3.4×10^{-8}	5.3	–	8.9
Venus	2.9×10^{-10}	2.5×10^{-9}	9.7×10^{-5}	5.6	1.0	11.8
Earth	1.0×10^{-11}	3.5×10^{-11}	2.3×10^{-6}	5.3	296	9.8
Mars	4.4×10^{-14}	2.2×10^{-13}	1.9×10^{-5}	4.1	2840	10.1

Compiled from data given by Pepin, R.O. in Atreya, S.K., Pollack, J.B., and Matthews, M.S. (eds.) *Origin and evolution of planetary and satellite atmospheres*, University of Arizona Press, Tucson, 1989. Pepin quotes errors and indicates the uncertainties in measurements and interpretations. Argon isotope ratios are taken from Pollack, J.B. and Black, D.C. *Icarus* **51**, 169 (1982).

[a] CI: carbonaceous chondrites, a class of meteorite.

abundant than the primordial isotope. In comparison with solar abundances, on a mass per unit mass basis, the Earth's atmosphere is depleted of ^{36}Ar by a factor of more than two million. For ^{20}Ne, the depletion is 220 million, but for carbon, mainly bound up in involatile compounds, the depletion is only 260. Evidence of this kind is taken as clear proof that Earth has lost almost all its primordial atmosphere, if such an atmosphere existed at all, and that the present atmosphere has been acquired later. In 1975, it appeared that the depletion of noble gases would be even more marked for the hotter planet Venus, but that Mars, having formed in a cooler part of the solar system, might have retained more of its primordial components. Viking (1976) dispelled that idea for Mars, as can be seen in Table 9.1. Natural isotopes of argon and neon are even more deficient on Mars than on Earth, and the radiogenic isotopes relatively more important. Pioneer–Venus (1978) showed that the Mars results were not a freak, but that there was a real tendency for there to be greater abundances of the noble gases in the atmospheres of the planets closer to the Sun. Table 9.1 shows that the abundance of ^{20}Ne relative to ^{36}Ar is comparable on the three planets (in the range 0.1 to 0.3), but much less than the solar ratio (about 25). The patterns of abundances of the noble gases are seen in Fig. 9.1, which also illustrates the decreasing primordial gas residue on the planets at greater distances from the Sun The depletions relative to solar abundances are greatest for the lightest elements. Carbon is much less depleted on the planets than are the noble gases. The planetary patterns for the noble gases clearly resemble each other much more closely than they do the Sun's pattern.

Having established that the atmospheres of Venus, Earth, and Mars are not primordial remnants, it is necessary to examine ways in which the planets could have obtained secondary atmospheres. Several hypotheses exist, which

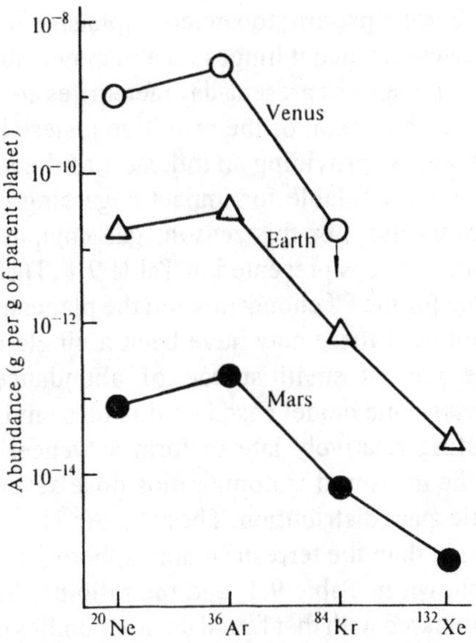

Fig. 9.1. Abundances of primordial gases observed in the atmospheres of Venus, Earth, and Mars, and solar abundances. Data from same sources as Table 9.1.

can be broadly classified as *solar nebula*, *solar wind*, *comet–asteroid*, and *accretion*. The first two of these hypotheses argue for gravitational capture and retention by the planets after their formation either of gases of the primordial solar nebula or of the solar wind that has flowed over them during their lifetimes. The differences in ratios of noble gases between the Sun and the planets that we have discussed in the last paragraph are evidence against these mechanisms. Substantial numbers of small bodies have impacted with the planets of the inner solar system over the planetary lifetimes, and the comet–asteroid hypothesis proposes that atmospheres were brought to the planets as a result of such impacts. However, Venus and Earth have a roughly equal chance of encountering comets and asteroids, and yet Earth has nearly two orders of magnitude less ^{36}Ar on a mass for mass basis than Venus, thus suggesting that the comet–asteroid hypothesis cannot account for a substantial proportion of the present-day atmospheres. The remaining hypothesis, which is that volatile materials were incorporated into the planet as it accreted, thus seems the most probable. If the planetesimals that formed the planets contained small amounts (perhaps a fraction 10^{-4} by mass) of volatile materials, then gases could be released from within the planet as it heated up. One explanation for the large excess of non-radiogenic noble gases on Venus compared to Earth could be that the planetesimals that formed Venus were exposed to an intense solar wind that was absorbed before it reached the part of the solar system where Earth (or Mars) formed. In this context, it is interesting that the pattern of noble-gas abundances for Venus (Fig. 9.1) shows a hint of a solar modification of the Earth's pattern; too much emphasis should not be given to this observation, however, since it hinges on a disputed abundance for ^{84}Kr.

The gaseous components of present-day meteorites are of interest because they may reflect the composition of the primitive materials out of which the planets accreted, as well as providing an indicator of the materials present in the solar system that are available for impact degassing (as required by the comet–asteroid hypothesis). For this reason, gas composition data for one important class of meteorite is presented in Table 9.1. The striking similarity in the $^{20}Ne/^{36}Ar$ ratios for the CI chondrites and the planetary atmospheres has prompted speculation that there may have been a single type of parent gas reservoir, with the present small spread of abundances determined by evolutionary processes. One model based on this idea envisages volatile-rich planetesimals accreting relatively late to form a 'veneer' on the planetary surface. However, the measured isotopic ratios pose severe constraints on a single type of volatile mass distribution. The ratio of $^{20}Ne/^{22}Ne$ is significantly higher in the Venusian than the terrestrial atmosphere (and than the ratio for CI chondrites), as shown in Table 9.1, and the ratio of $^{36}Ar/^{38}Ar$ on Mars is anomalously low compared with that found on other bodies of the solar system. It seems, then, that the similar elemental but disparate isotopic compositions cannot be a result of accretion of planetesimals with constant inventories of volatile species, at least if the compositions were to resemble those of present-day meteorites. The explanation of the similarities, as well as the differences, of atmospheric composition must therefore be a coincidental result of the fractionation and mixing processes that operated, both before and after accretion, on source gases with different origins. Escape of early solar-composition atmospheres from planetesimals and planets is currently thought to be the most likely mechanism that could have achieved the requisite fractionation. Whatever the detailed processes turn out to be, it is clear that the new atmospheric measurements obtained by the planetary missions have provided the basis for reasonable speculation about the origins of the planets and their atmospheres.

9.2.2 Titan

Considerable excitement followed the discovery of N_2 as the major component of Titan's atmosphere (Section 8.5.1) because of the possibility that Titan might resemble Earth in an early, or at least frozen, state of development. Insight into our own evolution might therefore be achieved by an understanding of the development of Titan's atmosphere to its present form. The first question is how did Titan, 50 times less massive than Earth, acquire so much gas from the preplanetary nebula that it now has an atmosphere one and a half times as dense as Earth's? Capture from the solar nebula (or an unfractionated proto-Saturn nebula) is excluded by the mean molecular weight (i.e. RMM) of the atmosphere, which is almost exactly 28. This value is not consistent with large amounts of neon in the atmosphere, although cosmic abundances would require an atmospheric composition of 67 per cent Ne and 33 per cent N_2, with a mean molecular weight of 22.6. Evidently the atmosphere represents a devolatilization of the ices and rocky material that accreted to form the satellite. Two possibilities can now be considered: either NH_3 may be the parent molecule that has undergone photolysis to yield the observed N_2, or N_2 could have been trapped in molecular form. We shall examine these alternatives.

Current models of the Saturnian and solar nebulae do not offer much guidance, some predicting predominantly NH_3 and others predicting predominantly N_2 in the Saturnian nebula during Titan's formation. Ammonia photolysis can form hydrazine (Section 8.4)

$$NH_3 + h\nu \rightarrow NH_2 + H \tag{9.1}$$

$$NH_2 + NH_2 \xrightarrow{(M)} N_2H_4. \tag{9.2}$$

Hydrazine itself yields N_2 as a photolysis product

$$N_2H_4 + h\nu \rightarrow N_2H_3 + H \tag{9.3}$$

$$N_2H_3 + N_2H_3 \rightarrow N_2 + H_2 + N_2H_4, \tag{9.4}$$

but these processes have to compete with physical removal of N_2H_4 by condensation. As we saw in Section 8.5.1, the present thermal structure of Titan would suggest that NH_3 photolysis generates only N_2H_4. An obvious requirement, if N_2 is to be formed from NH_3, is that during some period of Titan's evolution it was substantially warmer than it is now. Any warm period must have lasted long enough to yield Titan's N_2, but must obviously have been shorter than the age of the satellite. A surface temperature of 150 K or greater for 1.6×10^8 years suffices, and the restriction on time-scale is not contravened. If, as is thought likely, solar-ultraviolet fluxes were once much higher than they are now, the increased photolysis rate could reduce the time needed for NH_3 conversion to below 100 000 years.

The alternative hypothesis, that nitrogen was always present as a satellite material in the form of N_2, also turns out to be viable, and in some ways is more attractive since it does not demand an unproven thermal history. Molecular nitrogen cannot be condensed as pure solid N_2, but if Titan were formed at a nebular temperature of ~ 60 K, N_2 would condense out preferentially as a solid clathrate hydrate, $N_2.7H_2O$. Ammonia, methane, and other gases were also probably trapped as water clathrate ices. At $T \sim 60$ K, the vapour pressure of neon is > 40 bar, so that the formation of neon hydrate is out of the question, and the absence of neon in Titan's atmosphere is easily explained. Argon, on the other hand, is a conceivable constituent of the atmosphere. Although radio occultation experiments put the mean RMM of the atmosphere close to 28, near to the surface a high vapour pressure of CH_4 (RMM = 16) could be compensated by relatively large amounts of Ar to maintain the mean RMM. However, atomic spectral lines of Ar are absent from the ultraviolet emission spectrum from the upper atmosphere, showing that Ar is at most a minor constituent at the higher altitudes.

Two sources of argon can be envisaged: primordial, nebular, Ar trapped as clathrate hydrates, and radiogenic decay of ^{40}K. The actual amount of argon present provides a test of the origin of the atmospheric nitrogen. Nitrogen from nebular gases captured as clathrate must be accompanied by argon. Clathrate thermodynamics suggest that $[Ar]/[N_2]$ should be 1–10 per cent if all present-day N_2 is derived from clathrate. On the other hand, if most of the N_2 is a product of photochemical or chemical processing of NH_3, then the argon released from the clathrate is relatively much less important, perhaps making up as little as 10^{-3} per cent. Direct measurements of argon in Titan's atmosphere may thus provide a diagnostic test of the two models of the origin and early evolution of Titan's atmosphere.

9.3 Isotopic enrichment

Studies of noble gases and their isotopes in planetary atmospheres are particularly rewarding because of the chemical inertness of the elements over geologic time. Abundances of reactive elements also provide useful information, especially if comparisons are made between natural isotopes of the same element, since chemical losses will have been the same. We have seen an example of this kind of study in discussing the loss of nitrogen from Mars (Section 8.3.3). Escape of the ^{14}N isotope is slightly faster than that of ^{15}N because of the lower mass. With the atmosphere as the reservoir of nitrogen, the N_2 remaining will have become slowly richer in ^{15}N over the life of the planet. In comparison with nitrogen on Earth, where neither isotope escapes, the Martian ^{15}N is enriched by a factor of 1.6. It follows that Mars once had ten or more times as much nitrogen in its atmosphere as it has now. Continuous degassing of nitrogen from the planet's interior would tend to sustain the original isotopic ratio, so that the observed enrichment favours an evolutionary model in which Mars acquired its nitrogen atmosphere early in its history, with relatively little degassing in later epochs. By way of contrast to the nitrogen isotopes, the Martian $^{16}O/^{18}O$ ratio is almost exactly the same as that for Earth, and ^{18}O has been enriched by less than 5 per cent. Yet Mars is losing O atoms at present at the rate of 6×10^7 atoms s^{-1} for every square centimetre of surface (cf. Sections 8.3.2, 8.3.3). Lack of ^{18}O enrichment implies a source of 'new' oxygen, in a reservoir holding at least 4.5×10^{25} atoms cm^{-2}, presumably in the form of H_2O, since the escape of hydrogen and oxygen from Mars is constrained to have a 2:1 stoicheiometry. There is, however, apparently an enrichment of D over H by a factor of about 5 over the terrestrial value. If D/H for juvenile water on Mars is the same as that for Earth, the observed enhancement must be explained by a divergent history of atmospheric evolution on the two planets. Several steps in the escape of hydrogen favour loss of H over loss of D, but, even so, the observed enrichments can only be attained if some of the (D-enriched) atmospheric water can exchange back with the condensed phase. Model calculations, based on the assumption that D/H in primordial Martian H_2O is the same as the terrestrial value, imply an initial reservoir of hydrogen equivalent to a layer of water about 3.6 m thick, most of which has escaped, to leave a present-day residue that is 0.2 m thick. The calculations assume, amongst other things, that the escape rate has remained constant over geological time. Nevertheless, the required exchangeable surface layer thickness is almost two orders of magnitude less than geological inventories of subsurface water, so that it is not unreasonable. Presumably, the much weaker fractionation between ^{18}O and ^{16}O compared with that between D and H has prevented a measurable enhancement of the heavier oxygen isotope even in the presence of exchange with a modest surface reservoir.

Measurement of the extent of deuterium enrichment in atmospheres has, in fact, been regarded as an important tool in determining how much hydrogen has escaped, and thus how much water the planet or satellite originally possessed. To draw the proper inferences, it is obviously necessary to know (at least) the D/H ratio in juvenile water and the relative efficiency of loss of D and H by all plausible mechanisms. Some problems arise even with the 'starting' D/H ratio. Interstellar measurements give D/H = 1×10^{-5} at the present day; nuclear burning in stars reduces the relative abundance of deuterium, perhaps by a factor of two over the history of the solar system (4.6 Gyr). A likely value of D/H at the birth of the solar system is thus 2×10^{-5}. Measurements of CH_3D (but not of HD, which may be perturbed) give almost exactly this ratio for both Jupiter and Saturn, in accordance with the expectation that there would have been no fractionation on these gas giants. In Titan's atmosphere, in ocean water on Earth, and in hydrated minerals in meteorites, D/H is much higher, about 1.6×10^{-4}. For Halley's comet, values up to 5.4×10^{-4} have been reported (although other estimates put the ratio closer to the terrestrial value), while in organic molecules in carbonaceous chondrites the ratio can be as large as 2×10^{-3}. One interpretation of the observations is that ices in the solar system condensed from the solar nebula so as to favour partitioning of deuterium into heavier molecules such as water, methane, and ammonia. Solar system bodies that had a relatively large ice content in their make-up (Earth, Titan, meteorites, comets) will carry the signature of an elevated deuterium abundance. Uranus and Neptune, with a much higher ice/gas ratio than Jupiter or Saturn, show some enhancement in deuterium, with D/H about 9.0×10^{-5} in the atmosphere of Uranus and 1.2×10^{-4} in that of Neptune (Table 8.6). Mars and, as we shall see shortly, Venus show further enhancements in D because of preferential escape of H over geological time. The highest of all values of D/H are found in the molecules of the interstellar clouds (Section 9.1.2): ratios of several per cent, three orders of magnitude greater than the cosmic value, have been reported. It has been suggested that comets might become particularly deuterium rich by picking up large quantities of dust grains from interstellar clouds.

The question now arises how the ices became enriched in deuterium. Equilibrium at low enough temperatures is thermodynamically capable of giving the required partitioning into water, methane, and ammonia. However, the time taken to reach this equilibrium is too long compared with the lifetime of the nebula, even if the processes are catalysed by dust grains. An alternative explanation is based on the kinetic isotope effect. It is known from laboratory experiments that different isotopes participate in reactions at slightly different rates. The effects are particularly marked for H and D. Mechanisms that can produce the effect include differences in zero-point energy (which make activation energies smaller for molecular reactants containing hydrogen rather than deuterium), and greater quantum-mechanical tunnelling for hydrogen. Thus reactions involving breaking of H-bonds are faster than those involving the D-substituted analogues, especially at low temperatures. For example, an O–H bond is involved in the forward reaction

$$D + OH \rightarrow OD + H, \qquad (9.5)$$

which is expected to be more than seven times faster at temperatures below 300 K than its reverse, where the O–D bond is broken. Ultraviolet radiation present throughout the galaxy (the *interstellar radiation field*) can photolyse water, so that we can thus envisage a sequence of steps, initiated photochemically, that leads to preferential formation of a deuterated product

$$H_2O + h\nu \rightarrow OH + H \qquad (9.6)$$

$$OH + HD \rightarrow H_2O + D \qquad (9.7)$$

$$D + OH \rightarrow OD + H \qquad (9.5)$$

$$OD + H_2 \rightarrow HDO + H. \qquad (9.8)$$

A similar photochemically initiated series of steps can convert CH_4 to CH_3D preferentially, the key processes being

$$CH_3 + D \rightarrow CH_2D + H \qquad (9.9)$$

$$CH_2D + H + M \rightarrow CH_3D + M. \qquad (9.10)$$

Once again, the forward process in reaction (9.9) has a greater rate constant than its reverse because the zero-point energy of the CH_2–H bond contributes more to overcoming the activation barrier than does that of the CH_2–D bond. As a consequence, CH_3D is favoured kinetically over CH_4.

The explanation for fractionation into the D-rich species is not, ultimately, very different from that for equilibrium fractionation (since both phenomena go back to the vibrational zero-point energies of D- and H- substituted molecules). However, the formation of radicals photochemically allows the occurrence of radical–radical chemistry, which is relatively fast at low temperatures. Ion–molecule reactions are also fast at low temperatures (Section 3.4.1), and it is kinetic isotope effects acting on such reactions that are believed to be responsible for isotopic fractionation in interstellar molecules.

Chemical enrichment of the kind just outlined may be responsible for the much higher D/H ratio on Titan compared with that on its parent, Saturn. Preferential escape of H cannot be responsible because the low gravity allows both H and D to escape easily. However, the reactions

$$C_2H + CH_4 \rightarrow C_2H_2 + CH_3 \qquad (9.11)$$

$$C_2H + CH_3D \rightarrow C_2H_2 + CH_2D \qquad (9.12)$$

are the main atmospheric sinks for CH_4 and CH_3D, so that the expected larger rate of reaction (9.11) compared with (9.12) would lead to an enrichment of the deuterated compound over geological time.

Discussion of the loss of water from Mars leads us to the question of whether Venus once possessed large quantities of water. Venus contains quantities of carbon and nitrogen similar to Earth, but hydrogen is deficient. Water abundance on Venus is about 42 $kg\,m^{-2}$ compared with 2.7×10^6 $kg\,m^{-2}$ on Earth. There is certainly no liquid water on the surface of Venus today, and the mixing ratio for water in the atmosphere is probably not more than 2×10^{-4}. Either Venus had a low H_2O content from the outset because it was formed in a warmer region of the solar nebula, or whatever water was originally present has since disappeared. In the second case, the 'runaway greenhouse effect' (Section 2.2.5) would have demanded that all water was present as vapour, and the escape of hydrogen to space (and the escape and chemical reactions of oxygen) would have played a major role in the evolution of Venus.

Most mechanisms identified as potentially important for escape of hydrogen from Venus discriminate strongly against loss of deuterium, because of the large escape velocity (10.3 $km\,s^{-1}$; Table 2.1). Enrichment of deuterium might therefore be expected if Venus had originally possessed a water-rich atmosphere. Several pieces of evidence support deuterium enrichment, although they are not unequivocal. The ion mass spectrometer on Pioneer–Venus detected a signal at $m/e = 2$ from the upper atmosphere that can be attributed to D^+ (although the ion could be H_2^+). Interpretation of the intensity data would require D/H in the bulk atmosphere of $\sim 10^{-2}$. Mass peaks at $m/e = 18.01$ and 19.01 obtained in the lower atmosphere (below 63 km) with the large-probe neutral mass spectrometer may be caused by H_2O and HDO (although the $m/e = 19$ ion could be H_3O^+). If HDO is the source of the heavier ion, then D/H on Venus is $(1.6 \pm 0.2) \times 10^{-2}$, in agreement with the upper atmospheric ion data. On Earth, D/H $\sim 1.6 \times 10^{-4}$ overall (and perhaps twice that value in the upper atmosphere, according to Spacelab 1 observations), so that the deuterium enrichment on Venus is 50–100, implying large quantities of water in the early history of the planet. We shall see below that this enrichment factor is the *maximum* that could arise.

Molecular hydrogen would have been the dominant gas in the early upper atmosphere of Venus. Degassing might be expected to release materials with oxidation states similar to those for terrestrial volcanic gases ($[CO]/[CO_2] \sim 10^{-2}$; $[H_2]/[H_2O] \sim 10^{-2}$). However, at high Venusian temperatures, the gas-phase equilibrium

$$CO + H_2O \rightleftharpoons CO_2 + H_2 \tag{9.13}$$

and reactions such as

$$2FeO + H_2O \rightarrow Fe_2O_3 + H_2 \tag{9.14}$$

at the planetary surface would have increased the H_2 content relative to H_2O. Supersonic hydrodynamic outflow, powered by solar ultraviolet heating, would have resulted in the loss of H_2 to space. Interestingly, this flow would have entrained HD, thus sweeping deuterium away, *until* the mixing ratio of H_2 dropped below $\sim 2 \times 10^{-2}$. Only after this limit was passed would deuterium enrichment begin, regardless of how much water was originally present. Hydrogen, in the form of water, is now present at a mixing ratio of $\sim 2 \times 10^{-4}$, according to the Venera spectrophotometer data for 54 km altitude. Deuterium enrichment is thus limited to a factor of ~ 100, in accordance with the apparent measured value. The escape rate calculated for loss of H_2 would have exhausted the equivalent of Earth's oceans in about 280 million years.

As the Venusian atmosphere progressed towards its contemporary water vapour content, additional hydrogen loss processes probably began to operate. Translationally 'hot' hydrogen atoms can escape if their velocities exceed 10.3 km s^{-1}. Ion reactions are an obvious source of excess translational energy (cf. Sections 2.3.2 and 8.3.3). Recombination reactions such as

$$OH^+ + e \rightarrow O + H^* \tag{9.15}$$

can liberate kinetic energy directly, but a more important process on Venus may be elastic collision between 'hot' O* and ambient H

$$O^* + H \rightarrow O + H^*. \tag{9.16}$$

The atomic oxygen itself can be generated by dissociative recombination

$$O_2^+ + e \rightarrow O(^1D) + O(^3P), \tag{9.17}$$

which can provide 239 kJ mol^{-1} excess translational energy (Table 8.3), corresponding to a velocity of ~ 5.5 km s^{-1}. Approximately 15 per cent of collisions between H possessing thermal velocities (at 300 K) and O* will produce H* in reaction (9.16)with speeds in excess of the 10.3 km s^{-1} escape velocity. Mariner 5 Lyman-α (H resonance) airglow observations showed that there is an H atom component with an effective temperature of 1000 K in addition to the atoms that are thermally equilibrated at 300 K. Escape *via* this collisional mechanism could have reduced the hydrogen content from 2 per cent to the contemporary 0.02 per cent in about 4.2 Gyr, so that reactions

(9.16) and (9.17) alone could account for the observed deuterium enrichment. Probably both hydrodynamic and ionic–collisional mechanisms operated at the higher hydrogen abundances, with the hydrodynamic loss becoming less important as the water vapour content approached its present level. Whatever the detailed mechanism, the deuterium enhancement suggests that Venus was once much moister than it is now. The contemporary D/H ratio does *not* provide evidence for loss of several oceans' worth of water, although detailed models of escape of hydrogen suggest that large quantities of water might once have been present. Massive loss of hydrogen from water brings with it the problem of disposal of the oxygen. It may be that the oxygen escaped to space along with the hydrogen; alternatively, oxidation of surface material would provide a plausible sink if the surface were molten. Another problem concerns the present-day escape of hydrogen from the atmosphere of Venus. Calculations put the time taken to exhaust hydrogen from the atmosphere at the contemporary escape rate at between 500 to 1500 Myr.

The longer time is perhaps compatible with a gradual depletion of water over the life of the planet, but if the shorter time is correct, then the implication is that water is being replenished as fast as it escapes. Such replenishment could be provided by outgassing from the planetary interior (and possibly by cometary impacts). Mixing ratios for water vapour drop by a factor of about 5 between 10 km altitude and the surface, suggesting that there is a large flux of water from the atmosphere into the surface, which could nearly balance a relatively large flux of juvenile water from the interior. Substantial oxidation of the surface would be expected with large water fluxes through it, and some results (e.g. from Venera 13) indicate the presence of Fe(III) minerals that are consistent with a relatively highly oxidized surface.

9.4 Evolution of Earth's atmosphere

Access to the detailed geological record for Earth constrains speculation about our ancient atmosphere (*palaeoatmosphere*) and its development. Isotopic abundances of, for example, the noble gases (cf. Section 9.2.1) argue strongly for our atmosphere being of secondary origin, rather than being a primordial remnant of the solar nebula. Gases trapped in the interior of the Earth outgassed to form the atmosphere, but the composition of the pre-biological palaeoatmosphere did not reflect the composition of the outgassed volatiles. Almost all the H_2O condensed to form the oceans, and the bulk of the CO_2 formed carbonates in sedimentary rocks, leaving the outgassed N_2 to accumulate and become the most abundant species in the atmosphere. The pre-biological atmosphere thus seems to have been mildly reducing in nature. Strongly reducing atmospheres of CH_4 and NH_3 must have been very short-lived, even if they existed at all. Photolysis of NH_3 (ultimate product N_2: cf. Section 9.2.2) reduces the lifetime to a few years at most for NH_3 mixing ratios between 1 and 100 p.p.m., while the lifetime against rainout is ~ 10 days. Methane is photochemically stable near the surface (because short-wavelength ultraviolet radiation is filtered out by H_2O vapour). However, the lifetime against oxidation by OH, initiated by the reaction

$$OH + CH_4 \rightarrow CH_3 + H_2O \tag{9.18}$$

is only about 50 years. No known chemical processes generate either CH_4 or NH_3 in the atmosphere, and hydrocarbons (including CH_4) in geothermal emanations are ultimately of biological origin, according to carbon isotope measurements. Recent geological and geochemical data support the idea of a mildly reducing pre-biological atmosphere. The rock record begins 3.8 Gyr ago with highly metamorphosed sediments at Isua in West Greenland. These rocks were formed in the presence of H_2O and CO_2, but without abundant CH_4. Thus the starting point for the evolution of life and of the present atmosphere is likely to have been an atmosphere containing predominantly N_2, with some H_2O and CO_2. Minor constituents can include NO (formed by lightning acting on N_2, CO_2, and H_2O), HCl (from sea-salt spray), trace amounts of volcanic H_2 and CO; and formaldehyde produced by photochemical processes.

Accumulation of free oxygen in the atmosphere led to a transition from reducing to oxidizing conditions. While the bulk of oxygen in the contemporary atmosphere is photosynthetic in origin (Section 1.5.1), some O_2 can be formed by inorganic photochemistry. Photolysis of water vapour

$$H_2O + h\nu \rightarrow OH + H \tag{9.6}$$

$$OH + OH \rightarrow O + H_2O \tag{9.19}$$

$$O + OH \rightarrow O_2 + H \tag{9.20}$$

$$O + O \xrightarrow{(M)} O_2 \tag{9.21}$$

and of carbon dioxide

$$CO_2 + h\nu \rightarrow CO + O \tag{9.22}$$

$$O + O \xrightarrow{(M)} O_2 \qquad (9.21)$$

are the favoured routes, but two important limitations are placed on the amount of oxygen that can be produced. It has long been recognized that H_2O and CO_2 are photolysed by ultraviolet radiation in a spectral region that is absorbed by O_2 (say at $\lambda \leq 240$ nm for H_2O, and $\lambda \leq 230$ nm for CO_2). 'Shadowing' by the O_2 thus self-regulates photolysis at some concentration. Secondly, photolyses of H_2O vapour or CO_2 do not, on their own, constitute net sources of O_2. Reactions (9.6) and (9.19)–(9.21) have the effect of converting two H_2O molecules to one O_2 molecule and four H atoms. Only if atomic hydrogen is lost by exospheric escape is there a gain in O_2, because otherwise H_2O is re-formed. Addition of CO_2 photolysis [reaction (9.22)] to the scheme does not alter this conclusion, since the CO product interacts with OH

$$CO + OH \rightarrow CO_2 + H. \tag{9.23}$$

Water-vapour photolysis now follows a new route

$$3H_2O + 3h\nu \rightarrow 3OH + 3H \qquad 3 \times (9.6)$$

$$CO_2 + h\nu \rightarrow CO + O \qquad (9.22)$$

$$CO + OH \rightarrow CO_2 + H \qquad (9.23)$$

$$OH + OH \rightarrow H_2O + O \qquad (9.20)$$

$$\text{Net} \qquad 3\,(H_2O + h\nu_{H_2O}) + (CO_2 + h\nu_{CO_2}) \rightarrow 2O + 4H$$

but the outcome is still that for every O_2 molecule formed, four H atoms must be lost. Escape is thus the crucial event, and the rate is determined by the transport of all hydrogen species through lower levels of the atmosphere to the exosphere (Section 2.3.2). Loss of O_2, for example by reaction with crustal or oceanic Fe^{2+}, or with volcanic H_2, competes with production, and so further limits the amount of free O_2 that can build up without the help of photosynthesis.

Considerable difficulties arise in giving quantitative expression to the pre-biological formation of oxygen because of uncertainties in concentrations of precursor molecules (H_2O and CO_2), temperatures, and solar ultraviolet intensities. Concentrations of CO_2 might have been much greater before the gas was converted to carbonate deposits, and water-vapour levels would have been elevated had surface and atmospheric temperatures been higher than they are now. Young stars (the 'T-Tauri' stars), which resemble the Sun at the age of a few million years, emit 10^3 to 10^4 times as much *ultraviolet* radiation as the present Sun. If enhanced solar-ultraviolet intensity was available during the pre-biological evolutionary period of our atmosphere, then the rates of photolysis of H_2O and CO_2 are greatly enhanced, and become a significant source of O_2, especially if $[CO_2]$ is high. Photochemical models developed for interpretation of the modern atmosphere (Sections 3.6, 4.4.6, and 5.9.1) can be adapted for the palaeoatmosphere by incorporating appropriate source terms, temperature profiles, and boundary conditions. Recent results suggest that pre-biological O_2 at the surface would have been limited to about 2.5×10^{-14} of the present atmospheric level (PAL) had both $[CO_2]$ and solar ultraviolet intensities been at their current values. With 100 times more CO_2, and 300 times more ultraviolet radiation from the young Sun, the surface $[O_2]$ calculated is $\sim 5 \times 10^{-9}$ PAL. The geological record provides some further information about oxygen concentrations. The simultaneous existence of oxidized iron and reduced uranium deposits in early rocks ($>$ 2.2 Gyr old) requires $[O_2]$ to be more than 5×10^{-12} PAL, but less than 10^{-3} PAL; the values accommodated both by the model and by geochemistry thus seem to be roughly in the range 5×10^{-12} to 5×10^{-9} PAL. Other mineralogical evidence suggests that $[O_2]$ increased from $\leq 2 \times 10^{-3}$ PAL at 2.2.GYr BP to $\geq 3 \times 10^{-2}$ PAL at 2.0 Gyr BP.

Pre-biological oxygen concentrations in the palaeoatmosphere are of importance in two ways connected with the emergence of life. Organic molecules are susceptible to thermal oxidation and photo-oxidation, and are unlikely to have accumulated in large quantities in an oxidizing atmosphere. Living organisms can develop mechanisms that protect against oxidative degradation, but they are still photochemically sensitive to radiation at $\lambda \leq 290$ nm. Life-forms known to us depend on an ultraviolet screen provided by atmospheric oxygen and its photochemical derivative, ozone, because DNA and nucleic acids are readily destroyed (Section 1.4). Biological evolution therefore seems to have proceeded in parallel with the changes in our atmosphere from an oxygen-deficient to an oxygen-rich one.

Naturally occurring abiotic processes could have synthesized the organic molecules that were the precursors of terrestrial life. In the 1920s, the Russian biochemist A.I. Oparin and the British geneticist J.B.S. Haldane independently proposed that such synthesis could not take place in an oxidizing atmosphere, and they suggested that it required an atmosphere rich in methane, ammonia, hydrogen, and water. Stanley Miller and Harold Urey put these ideas on an experimental basis with their now-famous experiments in which they passed electric discharges through strongly reducing mixtures of CH_4, NH_3, H_2, and H_2O. Amongst the collection of products were found amino acids, essential to living systems. More recent research has shown that discharges through mixtures of CO_2, CO, N_2, and H_2O also result in all the gaseous precursors needed for the production of complex organic molecules. Thus, it seems that organic compounds could have been synthesized in a palaeoatmosphere of the composition now thought most probable (CO_2, N_2, and H_2O), and it is not *necessary* to look for non-atmospheric sources, such as deposition of cometary material, that have been proposed from time to time. The alternative view finds its most extreme expression in the suggestions of Hoyle and Wickramasinghe, who propose that interstellar molecules accumulated within the heads of comets (cf. Sections 8.7 and 9.1.2). Chemical evolution occurring a few hundred metres below the cometary surface is seen as progressing as far as biopolymers and micro-organisms. Hoyle and Wickramasinghe even contend that some past and present epidemics were initiated by the viruses and bacteria falling to Earth in cometary dust, although these ideas have gained little acceptance so far.

The stages between organic molecules and 'life' cannot concern us here, however fascinating speculation about the steps may be. Although the 'dawn of fossil records' has often been regarded as 550 Myr (5.5×10^8 yr) BP, with the emergence of the trilobites at the beginning of the *Cambrian* period, the large organisms that left conspicuous fossils were preceded by micro-organisms. Microscopic examination of rocks reveals much evidence of ancient microbial life. Fossils are found in the oldest sedimentary rocks and, so long as these fossils were laid down with the rocks, they show the presence of abundant life from 3.5 Gyr BP (Before Present). Another line of evidence involves geologically stable *biomarker molecules* that can act as 'molecular fossils'. Biomarker lipids discovered in northwestern Australia have been shown to be almost certainly contemporaneous with the 2.7 Gyr-old shales in which they are found, and thus provide fairly secure evidence for the presence of life. Some of the compounds are formed only by *cyanobacteria* ('blue-green algae'). Photosynthetic bacterial cells of this kind initially changed the atmosphere from its oxygen content of $< 10^{-8}$ PAL towards much higher concentrations. The earliest cells lacked a nucleus, and are classified as *prokaryotic*: bacteria fall into this category, and can operate anaerobically. Larger cells, that are almost certainly *eukaryotic*, or possessing a nucleus, are found in the fossil record for 1.4 Gyr BP. However, the Australian biomarkers include compounds most plausibly formed from eukaryotic membranes, so that these more complex cells may already have been present by 2.7 Gyr BP. The importance of this finding for the interpretation of atmospheric evolution is that almost all eukaryotic cells require large quantities of oxygen to function. Cell division is preceded by a clustering and splitting of the chromosomes within the nucleus (*mitosis*), a process dependent on the protein actomyosin that cannot form in the absence of oxygen. As the atmospheric oxygen reached about 10^{-2} PAL, a revolution occurred, because eukaryotic cells, and then animal and plant life, emerged. Respiration and large-scale photosynthesis became of importance, enough free oxygen became available for the fibrous protein collagen to be formed, and the scene was set for the appearance of *metazoans*, or multi-celled species. About 550 million years ago, the Cambrian period opened. According to earlier ideas, this period heralded an 'evolutionary explosion'. In many ways, the real significance of the Cambrian period is that the first animals with clear external skeletons are preserved as fossils whose identity has been recognized for centuries, while remains of earlier life-forms had not yet been discovered or understood. Metazoan fossils from the preceding 120 million years, the 'Ediacarian' period, are now known. Many of these are from species resembling jellyfish. Such organisms can absorb their oxygen through the external surfaces at concentrations of about seven per cent of PAL. A reasonable estimate for when this level of oxygen was reached in the atmosphere can thus be set at about 670 Myr BP. The relatively impervious surface coverings of the Cambrian metazoans suggest that 120 million years later the oxygen concentration was approaching 10^{-1} PAL. Following the opening of the Cambrian, the complexity of life is known to have multiplied rapidly and the foundations for all modern phyla were laid. 'Advanced' life forms (i.e. non-microscopic) were found ashore by the Silurian age (420 Myr BP), and by the Early Devonian, only 30 Myr later, great forests had appeared. Soon afterwards, amphibian vertebrates ventured onto dry land.

According to one interpretation, due first to Berkner and Marshall, the evolution of O_2, and hence of 'protective' O_3, controlled the migration of life from the safety of stagnant pools and lakes, to the oceans, and then to dry land. In this scenario, a depth of say 10 m of *liquid* water will filter out much of the damaging ultraviolet radiation while allowing photosynthetic visible light to

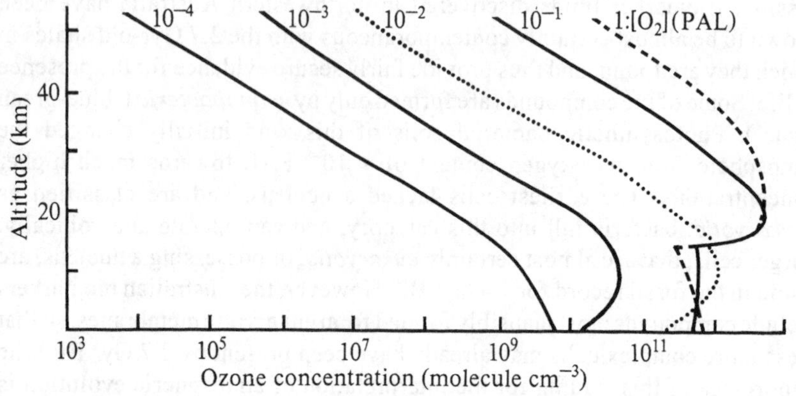

Fig 9.2. Vertical distribution of ozone for different total atmospheric oxygen contents ranging from the present atmospheric level (PAL) to 10^{-4} PAL. The model used to obtain these results includes nitrogen, hydrogen, carbon, and chlorine chemistry; allowance is made for ozone loss at the planetary surface. From Levine, J.S. *J. molec. Evol.* **18**, 161 (1982).

reach the organisms. Life in the oceans seems improbable at this stage, since organisms would be brought too near the surface by mechanical motions. When O_2 and O_3 had built up yet further, the ultraviolet zone of lethality would be restricted to a thin layer at the ocean surface, and life could spread to entire ocean areas, thus greatly enhancing photosynthetic activity. As the oxygen content of the atmosphere moved towards its present level, enough O_3 was available for no liquid-water filter to be needed for protection, and life could finally be supported on dry land.

A major question attached to the interpretation just presented is whether the biological evolutionary events were linked causally to the atmospheric changes that undoubtedly occurred. If they were, then some kind of feedback mechanism (Gaia; see Section 1.6) may have been in operation, since the atmospheric evolution was certainly mediated by the biota. Resolution of this question will require further information: in the first place, it is necessary to know the time history of growth of O_2 and O_3 in the atmosphere. What we *can* do is to use atmospheric photochemical models to calculate the ozone concentrations that accompanied smaller O_2 levels in the early atmosphere. Figure 9.2 shows the results of one such calculation, in which the full chemistry of catalytic cycles (HO_x, NO_x, and ClO_x; see Section 4.4) was incorporated. One interesting feature of the evolution of ozone concentrations in our atmosphere is immediately apparent. At low $[O_2]$, maximum ozone concentrations were found near the surface, but as oxygen concentrations increased, an ozone layer developed with its peak at successively higher altitudes.

From a biological standpoint, the characteristic of the ozone profiles that is important is the total column depth, since that is what determines how much

Table 9.2 Oxygen and ozone in the evolving atmosphere

$[O_2]$ PAL	$[O_3]$ column[a] molecule cm^{-2}	Fractional absorption by ozone present at:		Water depth (in m) that brings total attenuation to 'standard' value[b]	
		$\lambda = 250$ nm	$\lambda = 302$ nm	$\lambda = 250$ nm	$\lambda = 302$ nm
10^{-4}	5.2×10^{15}	0.06	0.00	6.0	5.4
10^{-3}	7.0×10^{16}	0.54	0.03	5.2	5.3
10^{-2}	1.6×10^{18}	1.00	0.45	–	4.2
10^{-1}	5.9×10^{18}	1.00	0.89	–	0.8
1	9.7×10^{18}	1.00	0.97	–	–

[a]Calculated for chemistry including chlorine species. From Levine J.S. *J. molec. Evol.* **18**, 161 (1982).

[b]'Standard' value is the screen provided by an ozone column density of 7×10^{18} molecule cm^{-2}. From Ratner, M.I. and Walker, J.C.G. *J. atmos. Sci.* **29**, 803 (1972).

ultraviolet radiation may leak through to the surface. Numerical values for the integrated areas under the curves of Fig. 9.2 are given in Table 9.2.

Figure 9.3 summarizes the material presented so far in this section. The growth in surface $[O_2]$ is identified on this plot by markers denoting the geochemical and fossil evidence. Column ozone abundances calculated using the model just described are also displayed in the figure. The problem is to decide what flux of ultraviolet radiation is tolerable to life, and hence what column density of ozone furnishes an adequate screen.

DNA is probably the most vulnerable and vital part of an evolving biological system. Absorption of radiation by DNA falls off at wavelengths longer than the broad maximum at $\lambda = 240$–280 nm, but most genetic damage may be caused by radiation at the longest absorbed wavelengths (say at $\lambda \sim 302$ nm). Damage would occur primarily at the few hours near mid-day, when the Sun is most nearly overhead, so that for all organisms in which the period between successive generations is less than a day, the maximum exposure is about four hours of high-intensity light. Based on experiments with genetic damage to corn pollen, an arbitrary damaging dose at $\lambda = 302$ nm of 1.6×10^3 $J m^{-2}$ may be specified. This translates to a flux of about 0.1 $J m^{-2} s^{-1}$ at or below a wavelength of 302 nm. An adequate ultraviolet screen would be provided by an ozone column density of 7×10^{18} molecule cm^{-2} for a mid-day solar zenith angle of 30°. Calculated ozone column densities are shown in Table 9.2 for several values of $[O_2]$: an effective ultraviolet shield is available from the atmosphere alone soon after $[O_2]$ exceeds 10^{-1} PAL. Also shown in the table are the fractional absorbances due to ozone at two wavelengths: $\lambda = 250$ nm corresponds to the maximum in DNA (and ozone) absorption, while $\lambda = 302$ nm is the estimate of the longest genetically active wavelength. The last two columns of the table show the thicknesses of liquid water needed

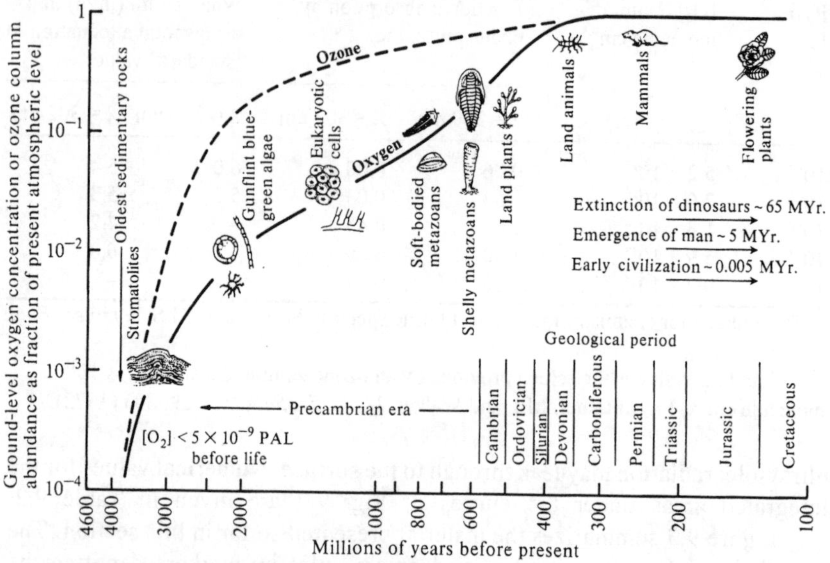

Fig. 9.3. Evolution of oxygen, ozone, and life on Earth. In the absence of life, surface oxygen concentrations are unlikely to have exceeded ~ 5×10^{-9} of the present value. The build-up of oxygen to its present level is largely a result of photosynthesis. Early organisms would have found high oxygen concentrations toxic, but eukaryotic (nucleated) cells require at least several per cent of the present level for their respiration. Soft-bodied metazoans could have survived at similar oxygen levels, but the reduced surface oxygen uptake area available once the species had developed shells must mean that the concentration was approaching one-tenth of its current value about 570 Myr ago. Considerations such as these are used in drawing up the oxygen growth curve. Ozone concentrations can be derived from a photochemical model. Life could not have become established on land until there was enough ozone to afford protection from solar ultraviolet radiation.

to make the total ozone + water absorption equivalent to that of the 'standard' column (7×10^{18} molecule cm^{-2}) of ozone. Protective layers of considerable depth do seem necessary for $[O_2] < 10^{-1}$ PAL. Does the transition to $[O_2] > 10^{-1}$ PAL then really explain the appearance of life on dry land? Shelled organisms require dissolved oxygen that would be in equilibrium with $> 10^{-1}$ PAL in the atmosphere, so that the critical level of O_3 for biological protection would have already been passed when the organisms appeared abundantly in the Cambrian period (550 Myr BP). But life was probably not firmly established on dry land for a further 170 Myr (late Silurian). Thus the possibility exists that the ozone screen may have been established before the

Silurian period, and was not directly linked with the spread of life onto land. The connection between the emergence of life out of water and the development of the ozone shield remains a tantalizing one. Whether or not life moved onto dry land in response to the presence of sufficient ozone, it should be clear by now that there is, and always has been, an intimate relation between the existence of the biota and the composition of our atmosphere. The evolution of the one, ever since life appeared, has been the story of the evolution of the other.

9.5 Climates in the past

Evolution on dry land brought with it an increased rate of photosynthesis, and it seems possible that oxygen concentrations may have exceeded 1 PAL during the lush growth of vegetation during the carboniferous period (340 Myr BP). A possible consequence of this rapid photosynthesis is that atmospheric carbon dioxide could have been depleted if oxidation of organic matter did not regenerate CO_2 at a sufficient rate. Reduced radiation trapping would cause a drop in the Earth's temperature, and the lower temperatures would then lead to a decreased rate of photosynthesis and a lowered atmospheric O_2 concentration. Cooling of the Earth from the depletion of CO_2 may well have been responsible for the Ice Ages of the Permian period (280 Myr BP). At the lower temperatures, CO_2 would be restored to a higher level, because of reduced rates of consumption in photosynthesis. Oxygen and CO_2 concentrations would thus swing in opposite senses, and may still be undergoing damped oscillations with a period of about 100 million years.

Changes of temperature, such as those described in the last paragraph, reflect one aspect of a changing *climate*. By 'climate', we understand the mean values and range of meteorological parameters such as temperature, pressure, humidity, precipitation, wind, and so on. A particular region, or even locality, can have its own climate, but the concept can also encompass wider areas, up to the global scale. Extrinsic factors that could alter the Earth's climate range from the well-established variations in solar luminosity and in our orbit round the Sun (probably responsible for the 10–50 000-year period 'Ice Ages' of the last two million years) to imaginative speculations such as alteration in atmospheric opacity resulting from dust clouds raised by impacting asteroids. Intrinsic factors include changes in surface albedo or in atmospheric reflectivity due to volcanic aerosol, variations in oceanic circulation, and alterations in atmospheric composition.

Although we called the glacial cold periods 'Ice Ages' in the last paragraphs, the term is often an exaggeration. In fact, one of the most interesting features of the Earth's climate is that it has been rather stable, at least since the emergence of life, some 3.5 Gyr ago. Ice Ages affect only those parts of the Earth at latitude > 45° north or south, corresponding to only 30 per cent of the surface, some of which is, in any case, often partially frozen between glacial periods (as now). Both the geochemical record and the persistence of life itself indicate that the oceans can never have either frozen or boiled in their entirety. Mean surface temperatures have probably never departed from the range of 5 to 50 °C, and may have been highest at very early periods. But this result leads to a riddle! Standard stellar evolutionary models predict that 4.5 Gyr BP the Sun's luminosity was lower than it is today by 25 to 30 per cent.[a] Such dimming has created the 'dim Sun paradox', because it translates into a decrease of Earth's effective temperature by 8 per cent, low enough to keep sea-water frozen for ~ 2 Gyr. Explanations proposed include changes in albedo or increasing the greenhouse efficiency. Alterations in clouds, for example, could exert a negative-feedback, stabilizing, effect, since lower temperatures would mean decreased cloud cover and reduced reflection away of solar radiation. Water vapour makes the largest contribution to the greenhouse effect in the contemporary atmosphere, but it is unlikely to be the agent of long-term temperature control. Its relatively high freezing and boiling points render its blanketing effect prone to unstable, *positive*, feedbacks by increasing ice and snow albedo at low temperatures (further reducing temperature), but increasing water vapour content at high temperatures (and yet further increasing the greenhouse effect). Whatever greenhouse gas or other mechanism kept the Earth warm, it must have been smoothly reduced to avoid exceeding the high-temperature limit for life. Carbon dioxide seems the most likely greenhouse gas to have exerted thermostatic control of our climate. Negative feedback mechanisms can be identified for this gas. Non-biological control might include acceleration of the weathering of silicate minerals to carbonate deposits in response to increased temperatures. However, as we discussed in Section 1.6 (p. 31), present-day weathering is biologically determined, and the biota both sense and amplify temperature changes. This feedback regulation of climate is seen by its proponents as evidence in support of the Gaia hypothesis (Section 1.6). Models that include not only the CO_2 greenhouse effect, but also the consequential changes in cloud albedo and water-vapour trapping, suggest that $[CO_2]$ might have been 10^3 PAL at ~ 4.2 Gyr BP, 10^2 PAL at ~ 3 Gyr BP, and 10 PAL at ~ 1.5 Gyr BP. Such concentrations seem reasonable if we accept that in Earth's pre-biological atmosphere CO_2 was a major component as it is on Venus or Mars today. Even 1000 times the present mixing ratio constitutes a partial pressure of CO_2 of only one-third of a (contemporary) atmosphere.

As we approach more closely (within a million years!) our own era, so the record of climate and atmospheric composition becomes richer and more detailed. Particularly fruitful sources of information have proved to be the

[a]Speculative new models of solar physics have led to suggestions that the Sun may have lost mass as it evolved. A larger initial mass and luminosity would then result in larger radiative fluxes at the Earth's surface than predicted by the 'standard' Sun model. The validity of such a model has not, however, been established.

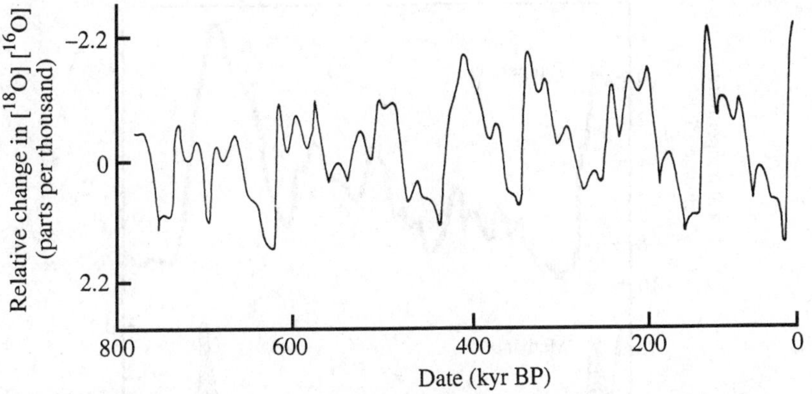

Fig. 9.4. Relative changes in the ratio of concentrations of ^{18}O to ^{16}O found in deep-sea sediment cores. The plot is arranged so that glacial periods correspond to the top of the figure and interglacial periods near the bottom. From Graedel, T.E. and Crutzen, P.J. *Atmosphere, climate, and change*, W.H. Freeman, New York, 1997

examination of cores of rock drilled deep into the ocean floors and cores obtained from the ice sheets covering Greenland and the Antarctic. The *Deep Sea Drilling Program* is an international cooperative venture that is recovering marine cores from around the world. Marine cores are relatively lacking in the folding and faulting, and so on, that complicate the records from exposed land rocks. They can, in fact, be used to examine climate in periods as long ago as 200 Myr. Figure 9.4, however, shows changes in ^{18}O content of the carbonate of fossil shells from marine sediments over the past 800 kyr (800 000 yr). These changes reflect changes in temperature, for reasons that will be explained shortly. There are clear fluctuations with different periods present in the data, and detailed mathematical analysis shows cycles between glacial and interglacial conditions with periods of 100, 41, 23, and 19 kyr. These periodicities can all be correlated with cyclic changes in the Earth's orbital eccentricity, precession, and tilt.

Ice cores provide another valuable source of information. For example, a large fraction of the world's fresh water is contained in the Antarctic ice sheet, which is up to 5 km thick in parts. Analysis has been conducted on samples from Greenland and from some mountain glaciers, as well as from Antarctica. Virtually no melting of snow occurs even in the summer, so that each year a new layer of snow is added to the ice caps and compressed into solid ice. As the snow falls, it scavenges aerosols from the atmosphere, and these aerosols are trapped together with bubbles of air in the ice. Chemical or biological alteration to the trapped material is not expected, so that a core taken from the ice provides a stratigraphic record of the atmosphere with a high degree of

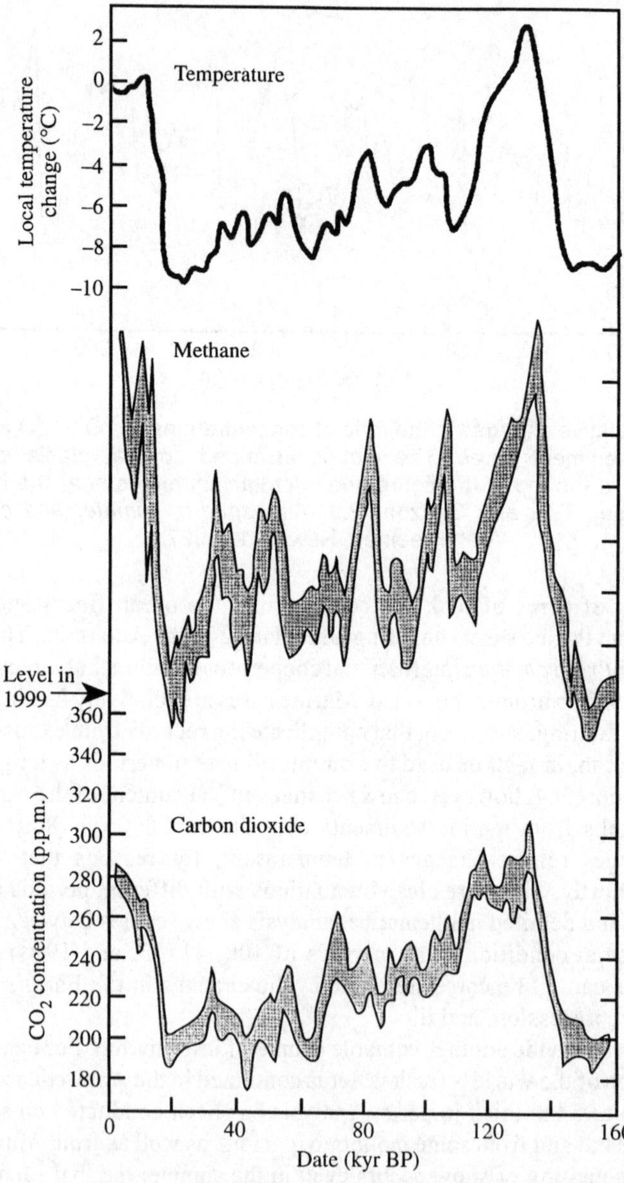

Fig. 9.5. Climatic record and trace-gas concentrations obtained from an ice core recovered at Vostock in Antarctica. Temperatures were calculated primarily from analysis of ^{18}O concentrations. The shaded bands for CH_4 and CO_2 concentrations represent the uncertainty range of the measurements. From Houghton, J.T. *Global warming: the complete briefing*, 2nd edn, Cambridge University Press, Cambridge, 1997.

resolution and stability. A limitation in the time-span of the record is ultimately imposed by the compression and resulting horizontal flow of the lowest layers of ice under the weight of newer ice deposited on top. Nevertheless, the oldest reliable samples date back to 250 kyr BP in cores from Greenland and 500 kyr BP in cores from Antarctica. The cores are up to 2500 m long.

Dating of ice cores may be achieved in several ways. The most ancient samples have been dated by an analysis of ice-sheet flow caused in response to horizontal stresses imposed by the overlying ice. Natural radioactive isotopes in the ice can be used for absolute dating of ice samples up to about five half-lives old. For ^{14}C (half-life 5600 years), the oldest datable sample thus approaches 30 000 years of age. Younger samples can be very accurately dated by stratigraphic methods, in which ice layers are counted in the same way as tree rings. Regular seasonal cycles, and 'catastrophic' events such as volcanic eruptions, can be recognized.

Ice cores provide information both on climate and on atmospheric composition. One method for unlocking the climatic record involves measurement of stable isotope ratios such as D/H or $^{18}O/^{16}O$ in the ice. The method depends on the slightly lower vapour pressure of the heavier isotope species (e.g. about 1 per cent less for $H_2{}^{18}O$ compared with $H_2{}^{16}O$). Water vapour is in equilibrium with the oceans in the subtropical source regions of atmospheric humidity, but as the air masses move towards the polar regions they lose water irreversibly by precipitation over the pack ice and ice sheets. The precipitation removes relatively more of the heavier isotopic H_2O, thus depleting the remaining vapour yet more. During a climatically cool period, or, indeed, during winter, there is more rapid cooling at high latitudes than there is nearer the source regions. As a result, the depletion of D and ^{18}O in the water of polar ice cores shows long-term trends in parallel with changing climate, as well as seasonal oscillations.

Figure 9.5 shows some results from a core taken at Russia's Vostock station in Antarctica. The record spans the most recent climatic cycle, with the temperatures reflecting the beginning of the last major ice age about 120 kyr ago and its end roughly 20 kyr ago. The fluctuations in temperature demonstrate that the climate over the past 150 kyr has not been at all stable, but rather has changed frequently. As in the example presented earlier, the climatic changes can be linked to variations in the Earth's orbit. An important feature that emerges, however, is that the *magnitude* of the climate changes is greater than could be produced by the altered radiation flux alone. Some amplification of the flux changes appears to have occurred. The figure also shows the very close correlation that exists between temperature and the atmospheric concentrations of carbon dioxide and of methane. It is thus tempting to explain the climate changes in terms of a positive-feedback mechanism involving the greenhouse heating brought about by increased concentrations of the gases. Successful modelling of climates of the past does, indeed, require the influence of changes of CO_2 concentrations on radiation trapping to be included (a complicating feature is that changes in global temperatures themselves influence the concentrations of CO_2 and CH_4 through biological feedbacks).

9.6 Climates of the future

Recognition of the influence that carbon dioxide is likely to have had on climates in the past has stimulated much thought into the possibility or even likelihood that future changes in CO_2 levels might also bring about climate change. Heightened significance is given to the topic by our becoming aware of the unprecedented rate at which CO_2 concentrations are increasing and that the effect is largely of Man's making (see Fig. 1.5 and the associated discussion). In response to the perceived impacts of global increases in temperature, a concerted international effort has been made to assess the current state of knowledge by the establishment of an *Intergovernmental Panel on Climate Change* (IPCC). The present section of the book provides a brief introduction to the problem, and largely reflects views put forward in the 1995 report of the IPCC. There are, however, dissenting views from several quarters, and where appropriate these opinions are also considered.

9.6.1 Radiatively active gases and particles in the atmosphere

The carbon dioxide concentrations shown for even the last 10 000 years of Fig. 9.5 fall a long way short of the present-day values shown in Fig. 1.5. Detailed comparisons of calibration techniques using ice-core samples from Antarctica and Greenland show that in the period 800–2500 years before present, the CO_2 mixing ratio in the atmosphere was 2.6×10^{-4} (260 p.p.m.). By 1999, the mixing ratio had exceeded 365 p.p.m. for the first time. The carbon dioxide increase has come about almost entirely since the Industrial Revolution, and mainly within the last 70 years, as a direct result of man's activities. Increased CO_2 emission is the primary cause, with burning of fossil fuels (98 per cent) and cement manufacture (2 per cent) as the major contributors. Deforestation in the tropics and changing agricultural practices reduce the efficiency of CO_2 recycling, thus providing a secondary aggravation of the CO_2 increase. Samples taken from individual tree rings show a decreased specific activity of ^{14}C over the last 100–150 years. The result suggests dilution of ordinary atmospheric CO_2 (in which the ^{14}C content is continuously renewed) by fossil-fuel releases, containing 'old' carbon whose ^{14}C has already decayed, rather than a reduced uptake of CO_2 by the biosphere.[a]

[a]Tree rings afford another important piece of evidence about past climates, since the temperature during any one season can be estimated from the growth. Some trees (the Bristle Cone Pines of California) reach ages of > 1000 years. However, sub-fossil pieces of dead wood can be used to piece together a continuous series of rings extending nearly 10 millennia.

Over the duration of the entire twentieth century, atmospheric CO_2 concentrations have increased on average at about 0.3 per cent per annum. There have, however, been very considerable fluctuations in the growth rate. For example, after the dramatic rise in fuel prices in 1973, at one stage the growth almost ceased for a short period. The early 1990s were also characterized by rather low growth rates, but were preceded and followed by periods of rapid growth (up to 2.5 times the 100-year average). At present, 2.0×10^{13} kg of CO_2 are produced annually as a result of fossil-fuel combustion and cement manufacture, while an effective emission of a further 0.6×10^{13} kg results from changes in tropical land-use. This 2.6×10^{13} kg of CO_2 released by man to the atmosphere each year corresponds to roughly 10 per cent of the rate of carbon assimilation by terrestrial plants, and adds to the 2.8×10^{15} kg of CO_2 already in the atmosphere. It is initially partitioned between atmosphere and ocean in the ratio 0.53:0.47, so that atmospheric concentrations are increasing at the rate of 0.5 per cent per year (~ 1.8 p.p.m. per year). Eventually, almost all of the added carbon dioxide will end up in the oceans, but removal from the atmosphere is limited by the slow mixing of the large volume of deep ocean water with the small volume surface layer, acting over a period of 500 years or more. Man is releasing carbon dioxide too fast for the oceans to cope, and it seems likely that in the next century the oceans will continue to take up less than half of the CO_2 added to the atmosphere each year.

Concern about increasing CO_2 levels is directed first towards possible consequences for our climate that would follow from increased radiation trapping, but CO_2 is not the only greenhouse gas whose concentrations have been increasing in recent years. Methane is another important contributor to the greenhouse effect whose concentrations have been rising. Greenland polar ice cores show that $[CH_4]$ is constant from depths of 250 m to 1950 m, corresponding to the period 1580 AD to about 20 000 BC! In 1580, methane levels began to rise from their mixing ratio of 0.7 p.p.m., at first slowly, but since 1918 much more rapidly, to reach their present mixing ratio of ~ 1.72 p.p.m. During the late 1970s, atmospheric methane levels increased at the rate of 1.2 per cent per year, and at about 0.8 per cent per year in the 1980s (although, as pointed out on p. 235, the rate of increase slowed to about one-third of this value over the period 1992–98). What has caused an accelerating increase in methane concentrations since the seventeenth century, after they had remained constant for 20 000 years? Man seems implicated because of the time-scale. Possibly some activity has reduced atmospheric [OH], which is the primary sink for CH_4 (Section 5.3.2) through reaction (9.18). Alternatively, increasing $[CH_4]$ could result from increasing production of the gas. Atmospheric methane has largely biogenic sources, arising from anaerobic bacterial fermentation in swamp lands, tropical rain forests, and the intestinal tracts of livestock and termites. The growth in rice-paddy cultivation and in cattle farming, which are a response to population growth, could account for increased methane production. If this explanation is correct, then it seems probable that methane levels will continue to rise in the decades ahead. The increase in [CH_4] already experienced is predicted to have increased global temperatures by about one-third of the effect estimated for the CO_2 increases since the industrial revolution to the present day (see later). Increases in the future can only further elevate temperatures. However, there are additional chemical effects about which we can only speculate. Methane plays a central role in tropospheric chemistry (see Sections 5.3 and 5.4.1). Increased concentrations of CH_4 may have caused consequential changes in the levels of other atmospheric constituents such as O_3, CO, and OH.

Input of nitrogen into cropland is an additional way in which average world temperatures might be raised in the future. Commercial fertilizers and nitrogen-fixing leguminous crops (Sections 4.4.3 and 4.6.8) both have the potentiality of increasing emissions of N_2O, a greenhouse gas, from soils. The nitrogen fixed annually by combustion and in manufacturing fertilizers has now reached almost half of what plants produce naturally. Atmospheric N_2O concentrations are now known to be increasing slowly (at ~ 0.2 per cent per annum for the period 1961–95). Once again, there is a coupling with chemistry, this time stratospheric, because N_2O is an important 'natural' source of the NO_x that controls ambient O_3 concentrations (Section 4.4.3).

Release by man to the atmosphere of halocarbons (CFCs, HCFCs, HFCs, and halons: see Sections 4.6.4–4.6.7) may represent another anthropogenic influence on climate. As described in Section 4.6.4, the concentration of these synthetic compounds is building up in the troposphere (cf. Fig. 4.23). The CFCs themselves possess strong absorption bands in the infrared region that, by chance, coincide with regions where CO_2 itself has relatively weak absorption. They thus have the potential of closing the atmospheric 'windows' through which radiation could escape to space, and the contributions of such compounds to greenhouse warming may consequently be much greater than the simple additive effect of the radiation trapped by the CFCs on their own. Chapter 4 concentrates on the impact that CFCs may have on depletion of ozone in the stratosphere, but it is evident that CFCs may have an additional adverse environmental impact by contributing to global warming. In general, the control strategies discussed in Section 4.6.6 are those needed to counter both threats, and reduced tropospheric lifetime is clearly one desirable quality in an 'alternative' halocarbon designed to replace the CFCs. However, it must be recognized that the compounds selected to reduce ozone depletion potentials (ODPs: Section 4.6.5) may not be ideal in terms of their infrared absorption intensities and wavelengths, and due consideration must be given to these factors in assessing overall environmental impact.

The complexity of the climate system is nicely exemplified by a further influence of the CFCs and other halocarbons. These compounds may certainly produce direct heating through the trapping of radiation, but those that deplete stratospheric ozone have a secondary effect. Ozone itself is an important greenhouse gas in both the stratosphere and the troposphere. Reductions of ozone concentrations in the lower stratosphere produce a net *cooling* of the atmosphere, which somewhat offsets the direct radiative heating. This process is an example of a *negative feedback* in the climate–chemistry system, and we shall return to the subject in Section 9.6.3.

Other factors are believed to have a cooling rather than a heating effect. For example, both sulphate aerosol and the aerosol from biomass burning (Sections 5.6, 5.8, 5.10.4, and 5.10.10) affect the energy balance. They absorb some radiation from the Sun, and reflect some back to space as well. Averaged over the planet, anthropogenic particles seem to reflect more than they absorb, and are thus direct contributors to some cooling. In a more complicated indirect mechanism, such particles also act as cloud-condensation nuclei and thus promote the formation of clouds and fogs. We discuss clouds further in Section 9.6.3, but, once again *on average*, the clouds nucleated by anthropogenic particles appear to exert a net cooling effect.

9.6.2 Radiative forcing

The term *radiative forcing* is used in discussions of increased or decreased greenhouse trapping of radiation to provide a simple quantitative measure of a potential mechanism for change of climate. It is defined as the perturbation to the energy balance of the Earth–atmosphere system, and is measured in units of power per unit area (watts per square metre: $W m^{-2}$). Figure 9.6 shows estimates presented by the IPCC in their 1995 assessment (see introduction to Section 9.6) for the radiative forcing due to different atmospheric changes that are thought to have occurred between the pre-industrial period and 1992. The various anthropogenic components are those introduced qualitatively in the previous section.

The values for radiative forcing can be compared with the total amount of radiation that arrives to heat the Earth, which is roughly 235 $W m^{-2}$ (cf. Fig. 2.4). Taking, as an example, the greenhouse gases alone, something like 3.5 $W m^{-2}$ less power would radiated back to space now than before the industrial revolution *if the surface temperatures had remained constant*. Since the radiation arriving and returning to space must be in balance, the conclusion has to be that temperatures must, in reality, rise if there is net positive radiative forcing. It is the correct calculation of this temperature rise that is the subject of Section 9.6.3.

There are substantial differences in the geographical distribution of the forcing due to the well-mixed greenhouse gases and that due to ozone and the aerosols. As a result, there could be differences in global and regional responses to the two groups of species, and the negative forcing from the aerosols cannot necessarily be regarded as an offset against the positive forcing of CO_2, CH_4, N_2O, and the halocarbons.

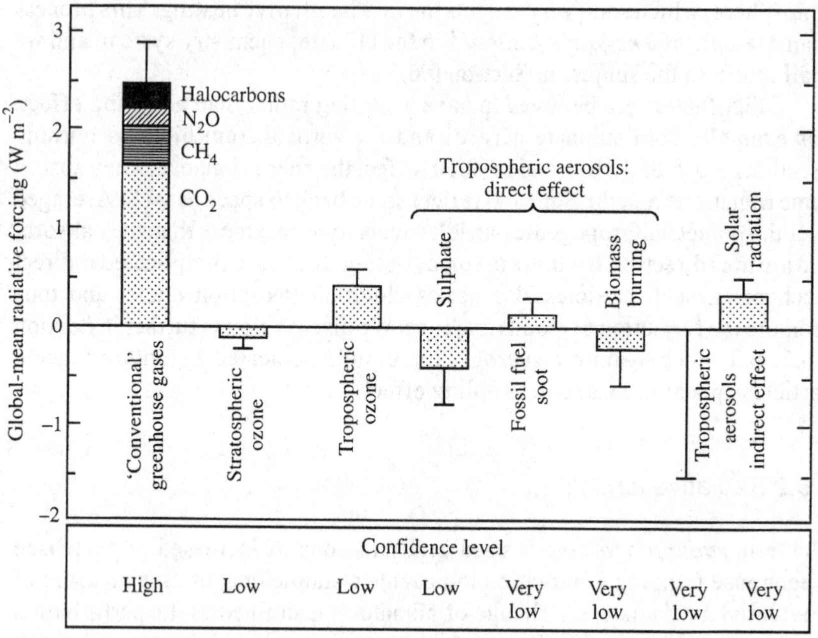

Fig. 9.6. Radiative forcing resulting from changes in concentrations of greenhouse gases and aerosols between the pre-industrial period and 1992. These estimates are global and annual averages. The shaded rectangles indicate the mid-range estimates of the forcing and the error bars are a suggested uncertainty range. Included on the diagram is the forcing resulting from natural changes in solar output since 1850. From Houghton, J.T. *et al.* (eds.), *Climate change 1995*, Cambridge University Press, Cambridge, 1996.

Figure 9.6 does not include any forcing associated with aerosol from volcanic eruptions. Over the period represented, volcanic activity has been highly variable. Frequent eruptions occurred between 1850 and 1920, and again since 1960. The volcanic eruption of Mount Pinatubo in 1991 initiated a global-scale negative radiative forcing that peaked at −3 to −4 $W m^{-2}$ in early 1992. That forcing is as large as anything shown in Fig. 9.6 for a century and a half. However, the volcanic forcing is transient, and that due to Mount Pinatubo had largely disappeared by 1995. Nevertheless, averaged over periods of a decade, the radiative forcings resulting from volcanic aerosol may have varied by as much as 1.5 $W m^{-2}$ since 1850, which is large compared to any other known forcing over the same time-span. It thus seems quite likely that volcanic activity plays a significant part in explaining fluctuations in surface temperature from decade to decade, and it may also have some relevance to the way in which gases such as CH_4 and N_2O have exhibited anomalous rates of growth since the eruption of Mount Pinatubo (see Section 9.6.1).

9.6.3 Feedbacks and models

The radiative forcing accompanying a doubling of atmospheric CO_2 concentration from its pre-industrial value is very roughly 4 Wm^{-2}. A simplified calculation shows that, in order to restore the energy balance, the surface and lower atmosphere would have to warm up by about 1.2 °C. One aspect of the simplifications made in reaching this conclusion is that the *only* change is temperature. Cloudiness, water vapour content, snow cover, and many other factors are assumed to be unaffected. In reality, of course, these assumptions will not be justified. According to current best estimates, the true average temperature change following a doubling of CO_2 concentrations would be nearer 2.5 °C. This change for doubled CO_2 is the *climate sensitivity*.

The difference between the results of the simple and the sophisticated calculations comes about because of the intervention of a variety of meteorological and biological feedbacks, both positive and negative in effect, that working together have a net positive amplification factor that more than doubles the temperature rise predicted without feedback.

The meteorological feedbacks include (i) *water-vapour feedback*; (ii) *cloud–radiation feedback*; (iii) *ocean–circulation feedback*; and (iv) *ice–albedo feedback*. Of these, the first is the most important, and in many ways the simplest. If the temperature rises, the vapour pressure of water is increased, and more water evaporates from the oceans and surface waters. Water is itself a major greenhouse gas, and the increased atmospheric concentrations produce a further increase in temperature. There is thus a strong positive feedback operating. Just this single influence increases the effect of doubling CO_2 concentrations by a factor of 1.6. Cloud–radiation feedback is more complex, because clouds affect atmospheric radiation in two opposing ways. They reflect some radiation back to space, thus reducing the solar energy available. On the other hand, they also absorb thermal energy emitted from the surface, and themselves emit thermal radiation, thus acting as a blanket to increase surface temperatures. The sign and magnitude of the feedback depends on the altitude of the cloud layer, how thick it is, the size of the particle droplets, and if the droplets are ice or water. In general, low clouds cool the system, while high clouds warm it. Climate turns out to be very sensitive to the amount and type of cloud cover. In turn, cloud formation is sensitive to water vapour content and the presence of particles that act as condensation nuclei, so yet further facets of the multiple feedbacks become apparent. The ocean feedbacks arise in three ways: oceans are the source of most of the atmospheric water vapour; they redistribute heat throughout the climate system as a result of their circulation; and they act as a buffer against rapid change because their heat capacity is much greater than that of the atmosphere. Finally, the albedo of ice and snow provides a positive-feedback mechanism. The surface of ice or snow is a strong reflector of radiation.

Melting of the cover as a result of warming permits the previously reflected radiation to be absorbed by the surface, and bring about yet further warming.

Biospheric feedbacks may also be of importance in the climate system. One of the most fascinating is the *plankton multiplier* effect. During cold periods (presumably associated with low atmospheric CO_2), convection in the upper layers of the oceans is enhanced, and the resulting thicker mixed layer near the surface promotes biological activity. Growth of marine plankton results in some long-term removal of CO_2 to the deep ocean. There is thus a positive feedback present that amplifies changes in atmospheric CO_2 levels. Other biological positive feedbacks that have been identified are the increased rates of respiration, leading to larger CO_2 emissions from soils, during periods of elevated temperature, and the reduced growth of forests in response to stress induced by climate change. Opposing these positive-feedback processes is an important biospheric negative-feedback process that is referred to as *carbon dioxide fertilization*. For certain plants (so-called 'C3 plants' such as wheat, rice, and soya bean), higher CO_2 concentrations stimulate photosynthesis, and carbon is then fixed at a higher rate, removing CO_2 from the atmosphere. CO_2 thus stimulates its own removal. For soya bean, up to 40 per cent increases in rates of CO_2 removal can accompany a doubling of CO_2 concentration, so that the magnitude of the negative feedback could be substantial.

Methane may be involved in a positive-feedback process that will amplify temperature changes. Enormous quantities of methane are locked up as the clathrate hydrates such as $CH_4.6H_2O$, which are inclusion compounds of methane trapped in the interstices of ice crystals. Gas hydrates below the permafrost of polar regions may account for 2×10^{15} kg of carbon (2.6 times the carbon content of the atmosphere), while estimates for the content of sea-floor sediments are even larger (up to 10^{17} kg). Phase diagrams for methane hydrate can be used to discover at what depths (≡pressures) and temperatures the compound is only marginally stable and might therefore be released in a warmer climate. Arctic Ocean sediments seem the most likely to be labile and to be exposed to higher temperatures. Decomposition over a 100 yr period from a layer 40 m thick at ~ 300 m depth and halfway round the Arctic Ocean could release ~ 8×10^{12} kg yr^{-1} of CH_4, enough to produce a significant positive-feedback effect on CO_2-induced warming. Because of the depth of the sediments, the feedback is unlikely to become important during the next century, but if global warming were to continue for longer, releases from hydrates might become the major contributor to atmospheric CH_4.

Numerical models are clearly the only available tool with which quantitative expression can be given to the climate system, involving as it does interacting, non-linear, positive- and negative-feedback components. The essence of the methods is that outlined in Section 3.6, and discussed several times elsewhere, especially in Chapters 4 and 5. Many of the feedback processes are already properly treated in weather-forecasting models. Others, such as the cloud–radiation and ocean–circulation feedbacks have had to be incorporated in new approaches. Increasing computational power is making such modelling feasible, and the results apparently reliable, as we shall see later. Our discussion has so far considered only radiation trapping by the greenhouse effect, together with direct chemical intervention. There follow a multitude of additional feedbacks and couplings, such as the effects of the greenhouse gases as stratospheric cooling agents (Section 4.6.9) which produce consequential changes in chemical kinetics. Models are now beginning to treat climate and chemistry together in an attempt to simulate the 'real life' situation (Section 4.6.9).

9.6.4 Detection of twentieth-century climate change

An obvious question to ask is if we have yet experienced a global warming attributable to the increased concentrations of trapping gases. Theoretically, warmings of 0.5–1.0 °C should have arisen over the course of this century. However, 'noise' in the form of natural influences on climate make it extremely difficult to establish the existence of an enhanced greenhouse effect. The climatic influences include fluctuations of solar radiation intensity, the release of aerosols from major volcanic eruptions, and oceanic warmings and coolings. Maverick weather patterns towards the end of the 1980s and in the 1990s, including very intense El Niño events (see Section 4.5.4), summer drought conditions in the USA, and hot summers and violent winter storms in Europe, have led to a popular conception that climate has already undergone a substantial change. Ground-based observations support this notion. Figure 9.7 shows the globally averaged temperature record for the period 1900–1997, plotted in the form of deviations ('temperature anomalies') from the mean of the years 1961–1990. Visual inspection alone would persuade the reader that temperatures in the final few years shown in the diagram were higher than in any that had gone before. Then in 1998, the year following the final one reported in the figure, temperatures were higher again. Indeed, it now seems that the 1990s were the warmest decade of the millennium, with 1998 the warmest year so far. The 1997–98 season also saw a massive El Niño event—one of the major climatic events of the century—that had major meteorological impacts, although the relation of El Niño to climate warming is still poorly understood.

Figure 9.7 certainly seems to show a signal of temperature increase emerging from the noise of year-to-year variability. Whether it can be attributed to net positive radiative forcing by greenhouse gases is another matter. The evidence has been hotly debated, of course. Even in 1990, the IPCC was disinclined to make an unequivocal statement linking climate change to anthropogenic releases of greenhouse gases. In their 1995 report, however, the IPCC cautiously concluded that 'the balance of evidence suggests a discernible influence on global climate'. Amongst the studies that support the

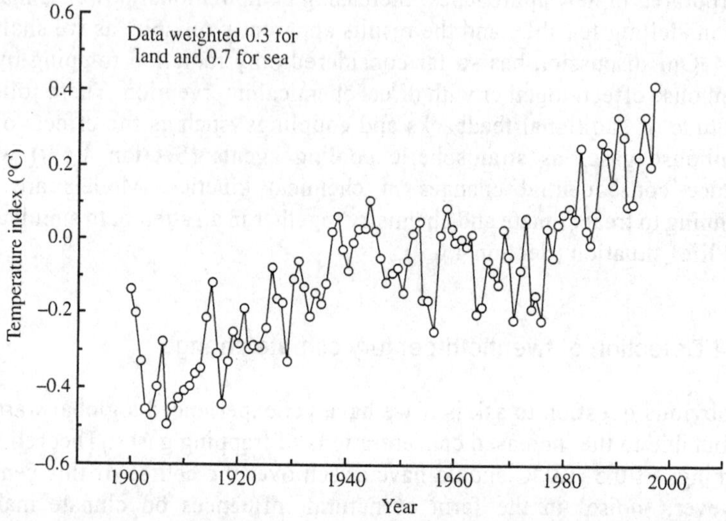

Fig. 9.7. Globally averaged land and sea-surface temperatures over the 20th century. The data are plotted as deviations in °C from the mean calculated for the period 1961–1990. Source: National Climatic Data Center, NOAA, Asheville, NC.

hypothesis, those comparing measured temperature changes with the predictions of models are especially significant. Modelled global temperature changes seem to be in accordance with observations, at least when aerosols are included in the treatment of forcing. Perhaps even more important, both geographical and vertical distributions of temperature seem to be well represented by the models.

The wording of the previous paragraph will have made it clear that there exists a minority view in the scientific community that the signature of anthropogenically induced warming has *not* been seen. Individuals on both sides of this argument have put forward their views with some vigour, and it has not gone unnoticed that the issues at stake could have very serious political and economic ramifications, since the whole question of controlling emissions from fossil fuels arises. The disagreement revolves in part around the model predictions that the IPCC and its supporters see as a key feature of their evidence. It is argued by the dissenters that the mainstream climatologists have manipulated their models unjustifiably to fit the observations, and that the models are not therefore either valid indicators of past climatic change nor good predictors of what will happen in future. Part of the problem arises because of satellite measurements of temperature that do not show the expected warming. A network of satellites, launched in 1979, carry microwave radiometers capable of measuring temperatures to 0.01 °C. According to the satellite results, there is apparently a slight *cooling* of the troposphere, corresponding to a temperature decrease of about 0.05 °C per decade. On the other hand, the models indicate that the warming seen at the surface should be even more pronounced and should increase with altitude in the lower troposphere. Balloon experiments also show apparent contradictions, with tropospheric warming up until the late 1970s, when there was surface cooling, and then a switch to tropospheric temperature stability, if not cooling, after the surface began to warm. One argument from the IPCC is that, if corrections are made for the known effects of El Niño events and the eruption of Mount Pinatubo, then surface air measurements, satellite, and balloon data all show an upward temperature trend, although not of the same magnitude. Another view is that the divergence in the surface and tropospheric temperature trends does not support the controversial view, but rather is indicative of an interesting complexity in atmospheric temperature structure that had not been expected.

Naturally enough, the dissenting view claims that the apparent failure of the models to incorporate atmospheric behaviour correctly precludes their use in interpreting the past or in predicting the future. A particularly significant aspect of the argument concerns the water-vapour positive feedback, presented as an amplifier of temperature changes in Section 9.6.3. The point made is that, if the free troposphere is largely cut off from the surface, as the satellite data suggest, water that evaporates from the oceans will not necessarily mean more water vapour in the troposphere. Other sceptics claim that the treatment of clouds is incorrect, and even that warm clouds will turn a greater proportion of their moisture into rain, thus causing the water-vapour process to exert a negative, rather than positive, feedback on the system! Yet other arguments run that the treatment of sulphate aerosol is incorrect. The lifetime of sulphate is relatively short; most sulphate is emitted in the Northern Hemisphere, so the cooling effect should be limited to that geographical region. Yet, since 1987, warming has virtually ceased in the Southern Hemisphere and has been very considerable in the North.

At the present time, it seems as though the dissenting opinion suffers from finding points of attack on the existing climate models, rather than providing a viable and quantitative alternative understanding of how the atmosphere behaves. At the same time, the 'establishment' modellers have undoubtedly been shaken in their belief in how well they can predict the future as a result of the continuing revisions that they have had to adopt in their models. It is to deal with this kind of situation that bodies such as the IPCC exist, in order that a consensus opinion can emerge that is based on the best scientific evidence available at the time. It is to the IPCC view that we return in the remaining parts of Section 9.6.

Table 9.3 Some of the assumptions regarding emissions and concentrations in emission scenario IS 92a

	1990	2025	2100
CO_2 emissions in 10^{12} kg C yr^{-1}			
fossil-fuel combustion	6.2	11.1	20.4
deforestation	1.3	1.1	−0.1
total	7.5	12.1	20.3
SO_2 emissions in 10^9 kg S yr^{-1}	98	141	169
CO_2 atmospheric mixing ratio (p.p.m.)	358	438	710
CH_4 atmospheric mixing ratio (p.p.b.)	1700	2242	3616
N_2O atmospheric mixing ratio (p.p.b.)	310	344	417

Data obtained from Houghton, J.T. *et al.* (eds.) *Climate change 1995*, Cambridge University Press, Cambridge, 1996.

9.6.5 Projected changes in concentrations, forcing, and climate

Regardless of the quantitative connection between concentrations of CO_2 in the atmosphere and climate, it seems certain that concentrations of greenhouse gases will continue to increase throughout the next century and beyond. As a consequence, radiative forcing will also increase. Various scenarios can be devised, of which one of the more reasonable is that referred to as IS 92a by the IPCC. For the purposes of explaining how climate change is predicted, this scenario is as convenient as any other and we shall adopt it for illustration here. In it, the world population is assumed to grow from 5.2 billion in 1990 to 11.3 billion in 2100, economic growth is maintained at a moderate level, and, perhaps most critically, no strong environmental action is taken to reduce CO_2 emissions. Table 9.3 shows some of the assumptions made about emissions and atmospheric concentrations for this particular scenario. Figure 9.8 converts the anticipated changes in atmospheric concentrations of the trace gases and loadings of aerosols into estimates of radiative forcing. The upper part of the figure shows the positive forcing from CO_2 itself, and from all other trace gases, together with the negative forcing produced from various contributions to the aerosol load. These aerosol contributions are broken down into the direct and indirect effects of sulphate aerosol (see p. 681) and the impact of aerosol derived from biomass burning. The lower part of the figure provides more detail about the forcing from non-CO_2 trace gases. Note here particularly the positive forcing in the early years brought about by the Cl- and Br-containing halocarbons, and the reduction of the forcing as these compounds begin to disappear from the atmosphere in response to the Montreal protocol and subsequent legislation (Section 4.6.6). At the same time, stratospheric ozone losses brought about by these same compounds exert a

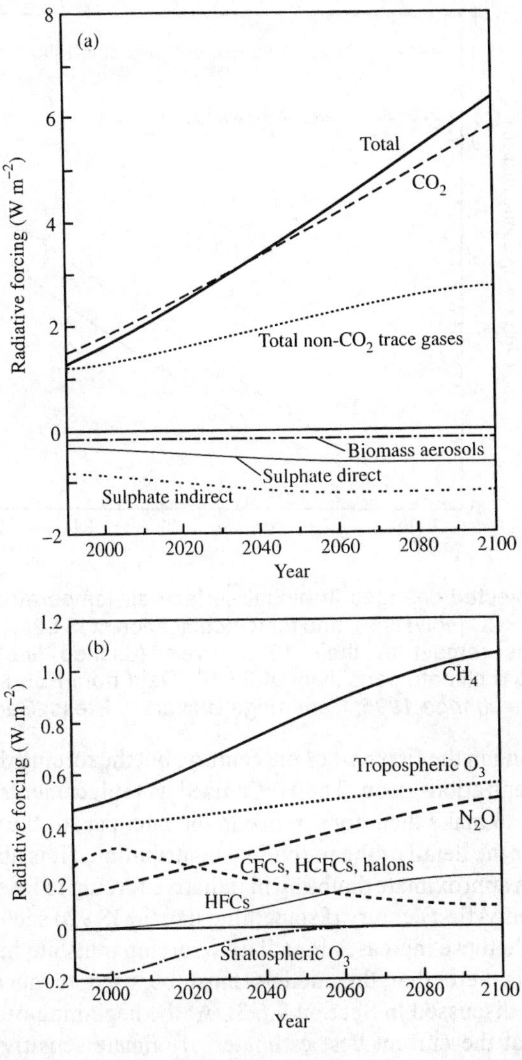

Fig. 9.8. Radiative forcing components resulting from the IS 92a emission scenario for the period 1990 to 2100: (a) effects of CO_2, summary of other trace greenhouse gases, and negative forcing from aerosol; (b) more detailed components of the non-CO_2 trace-gas forcing. The CH_4 curve includes methane-related increases in stratospheric water vapour. Emissions of halocarbons have been modified to take account of current legislation; effects of the ozone-depleting halocarbons are reflected in the curve for stratospheric ozone. Tropospheric ozone changes take account only of the indirect influence of changes in CH_4. From Houghton, J.T. *et al.* (eds.) *Climate change 1995*, Cambridge University Press, Cambridge, 1996.

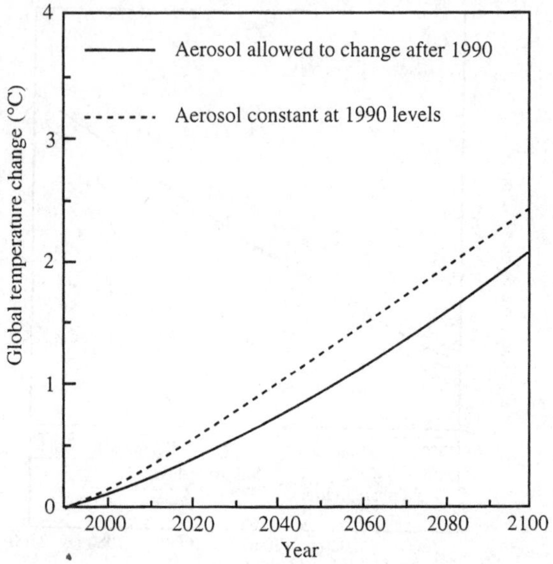

Fig. 9.9. Projected changes in global surface air temperature for scenario IS 92a of the IPCC (solid line), and for a modification of IS 92a in which aerosol concentrations remain at their 1990 levels (dashed line). The curves correspond to a climate sensitivity of 2.5°C. Data from Houghton, J.T. *et al.* (eds.) *Climate change 1995*, Cambridge University Press, Cambridge, 1996.

negative forcing in the first half of the century, but the forcing decreases as the catalyst concentrations drop. The HFCs used as replacements for CFCs and HCFCs begin to make their appearance in the later part of the period.

Whatever the detail of the individual contributions, it is abundantly clear that there is an approximate doubling in radiative forcing to be expected by the end of the twenty-first century if something like the IS 92a scenario unfolds in reality. How do these increases in radiative forcing translate into temperature changes? It is here that the models must be used to accommodate the feedbacks, as discussed in Section 9.6.3. At the beginning of that section, it was stated that the current best estimates of climate sensitivity (the global temperature change resulting from a doubling of CO_2 concentrations) was 2.5°C. Figure 9.9 shows in more detail projected changes in global average surface-air temperature for the IS 92a scenario using this value for the climate sensitivity. With the same climate sensitivity, the modelled temperature increase from pre-industrial times until the present is about 0.5°C if aerosol changes are included. The IPCC report of 1995 suggests that the extreme range of climate sensitivities lies between 1.5°C and 4.5°C; using the different values essentially scales the curves of Fig. 9.9 in the proportions 1.5/2.5 to 4.5/2.5.

As it happens, CO_2 concentrations nearly double between 1990 and 2100 in the IS 92a scenario. The ultimate temperature rise in Fig. 9.9 might thus be expected to approach 2.5 °C under steady-state conditions. In reality, the temperature rise by 2100 predicted in Fig. 9.9 is somewhat smaller, because the influence of the oceans reduces the rate at which temperature increases. However, what this conclusion means is that even if the CO_2 concentration in the atmosphere remained constant after 2100, global temperatures would continue to rise until the steady-state value was attained. This behaviour must therefore be taken into account when considering what controls are necessary and when they need to be applied.

The solid line in the figure is the true IS 92a prediction. A modification of the scenario is illustrated by the broken line, which allows atmospheric aerosol loadings to remain at their 1990 levels. The justification for this modification is that IS 92a projections of sulphur emissions may be much too high if action is taken, independent of the global-warming aspect, to alleviate the acid-rain problem (Section 5.10.6). For the IS 92a scenario, it turns out that the net effect of the non-CO_2 gases is approximately cancelled out at the global mean level by the negative forcing due to aerosols. At the regional level, however, there is no such cancellation, so there may be large spatial variations in forcing that will almost certainly have a major effect on the patterns of future climatic change. This variability is, in fact, a feature of all the other scenarios studied by the IPCC, and is further reason for showing the two curves in Fig. 9.9.

9.6.6 Aircraft

Section 4.6.2 discusses aircraft operation in relation to stratospheric ozone losses, but we documented there both the expansion of aviation over recent decades and the several ways in which aircraft emissions might affect the atmosphere. Passenger traffic has grown at nearly nine per cent per annum since 1960, which is more than twice the rate of increase of the average Gross Domestic Product. Freight traffic (80 per cent of which is carried by passenger aircraft) has grown in a similar way. It is expected that global passenger air travel will continue to grow by about five per cent per annum until at least the year 2015, although fuel use may increase by only three per cent per annum if anticipated improvements in aircraft efficiency materialize. Projections beyond this date are much more uncertain. Aircraft operations on this scale conceivably might make some contribution to climate change, and we consider very briefly suggestions about the extent of radiative forcing that can be estimated. As with the terrestrial sources of climatically active species, it is useful to consider a number of scenarios. In the case of aviation, the 'reference scenario' is labelled Fa1. It proposes an average traffic growth of 3.1 per cent per annum between 1992 and 2050, which translates into a factor of 6.4

increase in traffic over the period. With the hoped-for improvements in efficiency, the fuel usage might increase by a factor of 2.7. Other scenarios accommodate a range from 1.6 to 10 times the 1992 value. Emissions of CO_2 by aircraft were 0.14×10^{12} kg C yr^{-1} (0.51×10^{12} kg CO_2 yr^{-1}), so that the Fa1 scenario implies aircraft emissions of about 0.38×10^{12} kg C yr^{-1} in 2050. This quantity is four per cent of all the CO_2 deriving from human activities according to the mid-range IS 92a scenario. Aircraft could thus indeed make a small but significant contribution to global warming in terms just of the CO_2 emissions.

Emission of CO_2 alone is not, however, the end of the story. Aircraft are known to have several other potential impacts on radiative forcing. Subsonic aircraft emit NO_x, which generates ozone; ozone increases will, on average, tend to warm the Earth's surface. In the 1990s, aircraft operations are estimated to have increased O_3 by up to six per cent in northern mid-latitudes at cruise altitudes. The Fa1 scenario suggests the ozone increase will rise to 13 per cent by 2050. On the other hand, aircraft NO_x tends to decrease concentrations of CH_4, thus reducing the greenhouse heating from this gas. However, the changes in tropospheric O_3 occur mainly in the Northern Hemisphere, while the CH_4 changes are global, so that they do not offset one another.

Water vapour emitted in the troposphere is fairly rapidly released by precipitation; that released in the lower stratosphere can build up to higher concentrations. More important than the vapour-phase water may be the formation of contrails, which in turn can induce the development of extensive cirrus clouds. Contrail cover is expected to grow to a global average of 0.5 per cent by 2050, a factor of five increase; estimates of aircraft-induced cirrus cover range from 0 to 0.2 per cent globally for the late 1990s, and may increase by a factor of four by 2050. The radiative effect of contrails is currently uncertain, but increased cirrus-cloud cover tends to have a warming effect on the surface.

Aircraft engines emit soot and species that produce sulphate aerosol. Although aerosol accumulation is expected to grow with increased burning of aviation fuel, the mass concentrations are likely to remain small compared to surface sources, and direct radiative forcing by the aerosols can probably be neglected. However, aerosols from aircraft may play a role in enhanced cloud formation, which might have a more important effect on the radiation balance. At present, these factors are not understood in any quantitative form.

Leaving aside the possible changes in cirrus clouds, another area where quantitative knowledge is almost completely absent, estimates can be made for the radiative forcing produced by all the other influences discussed in this section. For 1992, the best estimate is 0.05 W m^{-2}, or about 3.5 per cent of the total anthropogenic forcing. In the Fa1 scenario, the radiative forcing by aircraft in 2050 becomes 0.19 W m^{-2}, or five per cent of the total forcing for that year according to the IS 92a scenario.

The discussion of aircraft presented so far concerns extensions to the subsonic fleet. As explained in Section 4.6.2, a new generation of high-speed civil transport (HSCT) supersonic aircraft is under consideration. These aircraft will consume twice the fuel per passenger-kilometre compared with subsonic aircraft, although their NO_x emissions might be kept lower by the use of new technology. Replacing part of the subsonic fleet by HSCTs increases the radiative forcing somewhat, mainly because of the accumulation of stratospheric water vapour. For example, replacing 11 per cent of the subsonic fleet in terms of the emissions of scenario Fa1 is projected to add a further 0.08 $W m^{-2}$ to the 0.19 $W m^{-2}$ forcing of the purely subsonic fleet.

An interesting feature of the calculations is that, for all the scenarios considered, the overall radiative forcing by aircraft is a factor of two to four larger than the contribution of the aircraft CO_2 alone. For all human activities taken together, the overall radiative forcing is *at most* a factor of 1.5 larger than that produced by the CO_2 alone. Thus aircraft apparently have a special propensity for effects that lead to positive forcing—and that is before any effects on cirrus clouds have been included.

9.6.7 Impacts of climate change

As presented in Figure 9.9, the current best estimate of the IPCC for global warming is a temperature increase of about 2 °C between now and the end of the 21st century. At first sight, this does not seem a very large increase. Normal temperature variations in one location, between day and night, between one day and the next, and between one season and another can be greatly in excess of one or two degrees Celsius. But the warming of 2 °C is a *global* average for all seasons. This temperature change within 100 years probably represents a greater rate of increase than any seen since the end of the last ice age (Fig. 9.5). Figure 9.5 shows that global average temperatures change 5–6 °C between the middle of an ice age and the warm periods in between, so that 2 °C is the climatic equivalent of one-third of an ice age, a very significant change. From an ecological point of view, one of the major problems concerns the *rate* of change. Systems can adapt to a few degrees Celsius of temperature change, but over tens of thousands of years, not hundreds.

Temperatures will rise more over continental areas than over the oceans, and the Northern Hemisphere will warm more quickly than the Southern. As the CO_2 levels progress to a doubling of pre-industrial values, the southern parts of Europe and the United States would become truly tropical. The climate would be getting close to that which existed at the beginning of recorded history, some 4500 to 8000 years ago (the *Altithermal period*), when the world was definitely warmer.

A warmer Earth will have a smaller equator-to-pole temperature contrast, because the excess heating is concentrated in polar regions. Seasonal contrasts will become less pronounced. With less potential energy available in the system to convert to kinetic energy, the atmospheric heat engine will run more slowly. Large-scale circulation patterns will be influenced, and some regional climatic changes may be larger than the average. A warmer atmosphere and ocean will result in more evaporation and precipitation, but the important question is where the rain and snow will fall. Some indication of the expected precipitation patterns can be gleaned from reconstructions of the Altithermal period (see previous page), or from examination of the meteorology of particularly warm years in the immediate past. Mid- to low-latitude land masses seem likely to experience a wetter climate than at present, with the deserts of North Africa, north-west India, and the south-west USA becoming prairie-like. Conversely, northern Europe and most of the central parts of North America and the Soviet Union would become drier.

Warming in the polar regions could approach as much as 10 °C by the middle of the next century. Formation of sea-ice will be reduced in both polar regions, and climatologists are tantalized by the possibility that Arctic Ocean ice might disappear and not return. Modification of the entire climate of the Arctic Basin would ensue, with profound ecological consequences on land and in the sea, and with possible release of much CH_4 from methane hydrate (cf. p. 684).

One of the most significant impacts that a warmer climate would have would be to increase the level of the oceans. Large changes in sea level have indubitably occurred in the past. Melting or growth of large ice sheets that cover the polar regions could be responsible for the long-term changes: for example, in bringing about the 100 m increase between 18 000 years ago and today that separated Britain from the rest of Europe. On a shorter time-scale, the most important contributor to a rise in sea level is thermal expansion of the water. Next in order of importance is the melting of glaciers and ice caps. According to the IS 92a scenario of the IPCC, sea level will rise about 12 cm by the year 2030, and 50 cm by 2100. Of this rise, thermal expansion will contribute 56 per cent, and melting of glaciers and ice caps outside Greenland or Antarctica a further 32 per cent. The end of the twentieth century has already seen evidence for some unprecedented melting of glaciers. Glaciers in the Himalayas make up the largest body of ice outside the polar caps, and they seem to be retreating faster than those anywhere else on earth; at the present rate of loss, they could disappear from the central and eastern Himalayas by 2035. The Alps have lost about 50 per cent of their ice in the past century, and 14 of the 27 glaciers that existed in Spain in 1980 have now disappeared. Glaciers on Mount Kilimanjaro, in Kenya, are only one-quarter as big as they were a century ago. Global warming or not, something is causing the ice to melt!

The ice sheets of the Antarctic and Greenland (see Section 9.5) are special cases. Over the relatively short term, they provide two competing influences. A warmer climate is damper, and there is more snowfall that builds up the ice sheets; but there is also more melting into the oceans near the boundaries of the sheets. For Greenland, the losses are probably greater than the ice accumulation, but the reverse is true in Antarctica. The ice sheets are more like geological formations than components of the hydrologic system. Since they rest on solid bedrock, their melting, or sliding into the ocean, would cause a rise in sea-level. Straightforward melting of the enormous masses of ice would probably take many thousands of years, but another mechanism could accelerate the process. Much of the West Antarctic ice sheet (and some of the East Antarctic ice sheet) rests on bedrock that is below sea-level. Warmer ocean water could work its way under the ice sheet, separating it from the bedrock and causing it to slide towards the ocean. Disintegration and melting of the ice would then be relatively rapid because of the more intimate contact with the water and because of the diminished thermal insulation. There is no sign that anything like this is happening now, and the time-scale involved for disintegration is probably more than 200 years. However, if the entire West Antarctic ice sheet were to disappear, a rise in sea-level of 5 to 7 m would result, with a serious impact on all the shorelines of the world.

The socio-economic ramifications of climate change lie outside the scope of this book. Many practical issues will have to be faced, with some people being better off, and some faring worse, on a warmer Earth. As stated earlier, it is the *rate* of change that presents the greatest challenge. Agriculture will be much affected, with mid-latitude farmers expecting about ten days longer growing season for every degree Celsius rise in temperature. Large areas of Africa, the Middle East, India, and central China might cease to be water-deficient. On the other hand, the 'food basket' areas of North America and the Soviet Union would become much drier, and it would be harder to grow grain and other major food crops. A higher level of CO_2 in the atmosphere and more rainfall in the subtropics would both enhance forest growth and allow it to spread to places that are now too arid. Present dangerous trends in deforestation might therefore be arrested. Costs of space heating in winter would decrease, but air conditioning, where used, would become more expensive. Heat stress is expected to have an adverse effect on human health, and a warmer climate might encourage the spread of certain tropical diseases, such as malaria, to regions that were previously free of these problems.

Two of the most important impacts of a rise in temperature follow from the changes in the water cycle. Cycling of water between the oceans and the land masses provides the fresh water that is essential for many forms of life, and most especially to fulfil the needs and expectations of humans. An increase in temperature means that more water will evaporate from the surface. Although most scenarios show more rainfall in some parts of the world, other parts would have less, especially in summer. Coupled with greater rates of evaporation in these places, the potential clearly exists for substantial exacerbation of drought conditions. A rise in sea-level of 50 cm might not bother people living inland, except indirectly. Unfortunately, half of the human population inhabits coastal zones, and some of the lowest lying areas are the most fertile and densely populated. Bangladesh, the Netherlands, and certain small oceanic islands are obvious examples. In Bangladesh alone, roughly six million people live within one metre of sea-level. A fraction of a metre rise in sea-level would obviously jeopardize their livelihoods and even existence.

9.6.8 Legislation and policy

In view of the conflict of views about whether greenhouse-related global warming has *already* occurred, it is evident that forecasts of the extent of global warming in the next century do not provide absolute certainties for policy-making. Furthermore, projections into the future about fossil-fuel use, and hence of carbon dioxide release, also depend on many assumptions. Yet it is now that action needs to be taken if the untoward consequences of global warming are to be averted or alleviated. We have already pointed out that there is an inevitable delay between cutting emissions of greenhouse gases and the point at which anthropogenic climate change will cease. For example, by 2030 there would already have been a temperature rise of about 0.5 °C since 1990, and sea-levels would have risen by roughly 12 cm. If CO_2 emissions were stabilized then, temperatures and sea-levels would both continue to rise, sea-level changes reaching 30 cm by 2100 (as against the 50 cm anticipated with no controls).

International action is beginning to be taken to control the emission of greenhouse gases, rather in the same manner as legislation has been enacted to reduce emissions of gases that deplete the ozone layer (Section 4.6.6). In December 1997, the *Convention on Climate Change* was held in Kyoto, Japan, and in the ensuing *Kyoto Protocol*, some consensus on reductions was agreed. The European Union pledged to cut greenhouse gas emissions by eight per cent from 1990 levels, the US by seven per cent, and Japan by six per cent. However, a number of battles remain to won at the time of writing, and it is by no means clear that all of the 166 governments represented will ratify even those measures approved in the Protocol. By April 1999, some 84 countries had signed. One of the issues of contention is whether or not individual countries should be allowed to compensate for increasing or unchanged emission levels by 'buying credit' from other nations. The US is in favour of trading emissions in this way, but other countries, especially China, believe the policy to be unfair. Underlying all the discussions are clear divisions between those who believe the threat of global warming to be real and those who do not, between environmentalists and business, and, in the last analysis, between the developed and the developing nations.

Inadvertent chemical modification by man of the stratosphere (Section 4.6) or troposphere (Section 5.10) seems likely to be relatively unimportant compared to climate change resulting from the burning of fossil fuels and the consequent release of CO_2. The cause is an international activity, the effect is global in character, and the remedy will require the concerted action of the countries of the world. A major alteration of the global environment seems possible, or even probable, unless we decide to change the course of events. Man's greed for energy will either have to be satisfied by non-fossil sources, or we shall have to prepare for the climatic change that is in store for us.

9.7 A doomed biosphere?

Man's interference with nature has been the theme of the last section. We should, however, recognize that nature herself may well make the environment hostile to life, even without our intervention. We saw in Section 9.5 that the Sun's luminosity is likely to have increased by 25 to 30 per cent over the past 4.5 Gyr, and there is no reason to suppose that the trend has ceased. Whether by active or passive control, changes in carbon dioxide concentration appear to have compensated for the increasing solar intensity in such a way as to keep the planetary temperature very nearly constant. But the capacity for control is now nearly exhausted, because [CO_2] is approaching the lower limit tolerable for photosynthesis. If that limit is taken to be 150 p.p.m., then the CO_2 control of a climate favourable for life can continue for another 30–300 Myr. Some adaptation to lower CO_2 concentrations and to higher temperature is possible, but it would not buy much time. In human terms, the crisis is rather distant. Over the next centuries, our problem is to cope with, and then prevent, loss of temperature control because of *increases* in [CO_2] (Section 9.6). There is some prospect that controlled nuclear fusion may be developed within the next century as a clean and almost inexhaustible supply of power. Our book therefore concludes by pointing out that the best hope for the protection of Earth's atmosphere may lie in harnessing, as a source of energy, the process out of which the Sun, planets, and atmospheres were born.

Bibliography

Section 9.1.1

Sources of atmospheric species and the evolution of atmospheres.

Physics and chemistry of the solar system. Lewis, J.S., 2nd edn. (Academic Press, San Diego, 1997).

Solar system evolution. Taylor, S.R. (Cambridge University Press, Cambridge, 1992).

The chemical evolution of protostellar and protoplanetary matter. Van Dishoeck, E.F., Black, G.A., and Lunine, J.I. in *Protostars and planets III*, Levy, E.H. and Lunine, J.I. (eds.) (University of Arizona Press, Tucson, AZ, 1993).

Planets and their atmospheres: origin and evolution. Lewis, J.S. and Prinn, R.G. (Academic Press, Orlando, 1984).
Physics and chemistry of the solar nebula. Lunine, J.I. *Origins Life Evol. Biosphere* **27,** 205 (1997).
The elements on Earth. Cox, P.A. (Oxford University Press, Oxford, 1995).
The elements: their origin abundance and distribution. Cox P.A. (Oxford University Press, Oxford, 1989).
Origin of the solar system. Cameron, A.G.W. *Annu. Rev. Ast. Astrophys.* **26**, 441 (1988).

The origins of stars, galaxies, and planets, and the dusts within them

Interplanetary dust particles. Rietmeijer, F.J.M. *Rev. Mineral.* **36**, B1 (1998).
In search of planets and life around other stars. Lunine, J.I. *Proc. Nat. Acad. Sci.* **96,** 5353 (1999).
The origin of carbon, investigated by spectral analysis of solar-type stars in the Galactic Disk. Gustafsson, B., Karlsson, T., Olsson, E., Edvardsson, B., and Ryde, N. *Astron. Astrophys.* **342,** 426 (1999).
Structure and evolution of nearby stars with planets. I. Short-period systems. Ford, E.B., Rasio, F.A., and Sills, A. *Astrophys. J.* **514,** 411 (1999).
Chemical evolution of star-forming regions. Van Dishoeck, E.F. and Blake, G.A. *Annu. Rev. Astron. Astrophys.* **36,** 317 (1998).
Temperatures in protoplanetary disks. Boss, A.P. *Annu. Rev. Earth planet. Sci.* **26,** 53 (1998).
Abundance ratios and galactic chemical evolution. McWilliam, A. *Annu. Rev. Astron. Astrophys.* **35,** 503 (1997).
The chemical evolution in a model of contracting cloud. Amin, M.Y. and El Nawawy, M.S. *Earth Moon and Planets* **75,** 25 (1997).
The galactic oxygen abundance gradient. Smart, S.J. and Rolleston, W.R.J. *Astrophys. J.* **481,** L47 (1997).
Chemical reactions in protoplanetary accretion disks. 3. The role of ionisation processes. Finocchi, F. and Gail, H.P. *Astron. Astrophys.* **327,** 825 (1997).
Chemical evolution in collapsing cores. El Nawawy, M.S., Howe, D.A., and Millar, T.J. *Monthly. Not. Royal Ast. Soc.* **292,** 481 (1997).
Evolution of the abundance of CO, O_2, and dust in the early universe. Frayer, D.T. and Brown, R.L. *Astrophys. J. Supp. Ser.* **113,** 221 (1997).
The chemical evolution of planetary nebulae. Bachiller, R., Forveille, T., Huggins, P.J., and Cox, P. *Astron. Astrophys.* **324,** 1123 (1997).
The stellar metallicity giant planet connection. Gonzalez, G. *Monthly. Not. Royal Ast. Soc.* **285,** 403 (1997).
Chemical evolution in preprotostellar and protostellar cores. Bergin, E.A. and Langer, W.D. *Astrophys. J.* **486,** 316 (1997).
Chemical and dynamical evolution of protostellar clouds: Initial stages of collapse. Wiebe, D.Z., Shematovich, V.I., and Shustov, B.M. *Astron. Zh.* **73,** 702 (1996).
The origin of galaxies and clusters of galaxies. Peebles, P.J.E. *Science* **224**, 1385 (1984).
Dust in galaxies. Stein, W.A. and Soifer, B.T. *Annu. Rev. Astron. & Astrophys.* **21,** 177 (1983).

General principles concerning the origin and evolution of atmospheres

Formation of atmospheres during the accretion of the terrestrial planets. Ahrens, T.J., O'Keefe, J.D., and Lange, M.A. in *Origin and evolution of planetary and satellite atmospheres.* Atreya, S.K., Pollack, J.B., and Matthews, M.S., (eds.) (University of Arizona Press, Tucson, 1989).

Escape of atmospheres and loss of water. Hunten, D.M., Donahue, T.M., Walker, J.C.G., and Kasting, J.F. in *Origin and evolution of planetary and satellite atmospheres.* Atreya, S.K., Pollack, J.B., and Matthews, M.S. (eds.) (University of Arizona Press, Tucson, 1989).

Impact erosion of planetary atmospheres: Some surprising results. Newman, W.I., Symbalisty, E.M.D., Ahrens, T.J., and Jones, E.M. *Icarus* **138,** 224 (1999).

Atmospheric losses under dust bombardment in the ancient atmospheres. Pavlov, A.K. and Pavlov, A.A. *Earth Moon and Planets* **76,** 157 (1997).

Planetary atmosphere evolution: Do other habitable planets exist and can we detect them? Kasting, J.F. *Astrophys. Space Sci.* **241,** 3 (1996).

Origin of the major planet atmospheres — clues from trace species. Atreya, S.K., Edgington, S.G., Gautier, D., and Owen, T.C. *Earth Moon and Planets* **67,** 71 (1995).

Origins of the atmospheres of the terrestrial planets. Cameron, A.G.W. *Icarus* **56,** 195 (1983).

Accumulation of planetesimals in the solar nebula. Nakagawa, Y., Hayashi, C., and Nakazawa, K. *Icarus* **54**, 361 (1983).

Origin and evolution of planetary atmospheres: An introduction to the problem. Prinn, R.G. *Planet. Space Sci.* **30,** 741 (1982).

Origin and evolution of planetary atmospheres. Pollack, J.B. and Tung, Y.L. *Annu. Rev. Earth & planet. Sci.* **8,** 425 (1980).

Formation of the terrestrial planets. Wetherill, G.W. *Annu. Rev. Astron. & Astrophys.* **18,** 77 (1980).

The involvement of meteors and comets. See also Section 9.1.2.

Meteor phenomena and bodies. Ceplecha, Z., Borovicka, J., Elford, W.G., Revelle, D.O., Hawkes, R.L., Porubcan, V., and Simek, M. *Space Sci. Revs.* **84,** 327 (1998).

Stellar nucleosynthesis and the isotopic composition of presolar grains from primitive meteorites. Zinner, E. *Annu. Rev. Earth planet. Sci.* **26,** 147 (1998).

Expected characteristics of cometary meteorites. Campins, H. and Swindle, T.D. *Meteoritics & planet. Sci.* **33,** 1201 (1998).

From the interstellar medium to planetary atmospheres via comets. Owen, T. and Bar Nun, A. *Faraday Disc.* **62,** 2555 (1998).

Comets, meteorites and atmospheres. Owen, T. and Bar-Nun, A. *Earth Moon and Planets* **72,** 425 (1996).

Atmospheres of individual planets and satellites. See also entries in Chapter 8.
Earth

Accretion of interplanetary dust particles by the Earth. Kortenkamp, S.J. and Dermott, S.F. *Icarus* **135,** 469 (1998).

On the possible influence of extraterrestrial volatiles on Earth's climate and the origin of the oceans. Deming, D. *Palaeogeography Palaeoclimatology Palaeoecology* **146,** 33 (1999).

Mantle degassing of major and minor volatile elements during the Earth's history. Tajika, E. *Geophys. Res. Letts.* **25,** 3991 (1998).

The birth of the Earth's atmosphere: the behaviour and fate of its major elements. Javoy, M. *Chem. Geol.* **147,** 11 (1998).

Nonthermal escape of the atmospheres of Venus, Earth, and Mars. Shizgal, B.D. and Arkos, G.G. *Rev. Geophys.* **34,** 483 (1996).

Formation of the Earth. Wetherill, G.W. *Annu. Rev. Earth & Planet. Sci.* **18,** 205 (1990).

Venus

Venus: a contrast in evolution to Earth. Kaula, W.M. *Science* **247,** 1191 (1990).

Origin and evolution of the atmosphere of Venus. Donahue, T.M. and Pollack, J.B. in *Venus*. Hunten, D.M., Colin, L., Donahue, T.M., and Moroz, V.I. (University of Arizona Press, Tucson, 1983).

Loss of water on the young Venus: The effect of a strong primitive solar wind. Chassefiere, E. *Icarus* **126,** 229 (1997).

Hydrodynamic escape of oxygen from primitive atmospheres: Applications to the cases of Venus and Mars. Chassefiere, E. *Icarus* **124,** 537 (1996).

Mars

The history of Martian volatile. Jakosky, B.M. and Jones, J.H. *Rev. Geophys.* **35,** 1 (1997).

Atmospheric loss since the onset of the Martian geologic record: Combined role of impact erosion and sputtering. Brain, D.A. and Jakosky, B.M. *J. geophys. Res.* **103,** 22689 (1998).

The ancient oxygen exosphere of Mars: Implications for atmosphere evolution. Luhmann, J.G. *J. geophys. Res.* **102,** 1637 (1997).

The loss of atmosphere from Mars. Johnson, R.E. and Liu, M. *Science* **274,** 1932 (1996).

Water on early Mars. Carr, M.H. *Ciba foundation symposia* **202,** 249 (1996).

Sputtering of the atmosphere of Mars — 1. Collisional dissociation of CO_2. Johnson, R.E. and Liu, M. *J. geophys. Res.* **103,** 3639 (1998).

Oxidation of organic macromolecules by hydrogen peroxide: Implications for stability of biomarkers on Mars. McDonald, G.D., de Vanssay, E., and Buckley, J.R. *Icarus* **132,** 170 (1998).

Early Mars climate models. Haberle, R.M. *J. geophys. Res.* **103,** 28467 (1998).

CO_2 greenhouse in the early Martian atmosphere: SO_2 inhibits condensation. Yung, Y.L., Nair, H., and Gerstell, M.F. *Icarus* **130,** 222 (1997).

Warming early Mars with carbon dioxide clouds that scatter infrared radiation. Forget, F., and Pierrehumbert, R.T. *Science* **278,** 1273 (1997).

Carbonates in Martian meteorite ALH84001: A planetary perspective on formation temperature. Hutchins, K.S. and Jakosky, B.M. *Geophys. Res. Letts.* **24,** 819 (1997).

The early environment and its evolution on Mars: implications for life. McKay, C.P. and Stoker, C.R. *Rev. Geophys.* **27**, 2 (1989).
Photochemistry and evolution of Mars' atmosphere: A Viking perspective. McElroy M.B., Kong, T.Y., and Yung, Y.L. *J. geophys. Res.* **82**, 4379 (1977).

Outer planets and satellites

Some speculations on Titan's past, present and future. Lunine, J.I., Lorenz, R.D., and Hartmann, W.K. *Planet. Space Sci.* **46,** 1099 (1998).
Origin and bulk chemical composition of the Galilean satellites and the primitive atmosphere of Jupiter: A pre-Galileo analysis. Prentice, A.J.R. *Earth Moon and Planets* **73,** 237 (1996).

Section 9.1.2

The physics and chemistry of interstellar molecular clouds. Winnewisser, G. and Pelz, G.C. (eds.) (Springer-Verlag, Berlin, 1995).
The physics and chemistry of interstellar molecular clouds. Winnewisser, G. and Armstrong, J.T. (eds.) (Springer-Verlag, Berlin, 1989).
Interstellar chemistry. Duley, W.W. and Williams, D.A. (Academic Press, London, 1984).
Chemical physics: Molecular clouds, clusters, and corrals. Herschbach, D. *Rev. modern Phys.* **71,** S411 (1999).
Frontiers of astrochemistry. Williams, D.A. *Faraday Disc.* **36,** 317 (1998).
Stochastic astrochemical kinetics. Charnley, S.B. *Astrophys. J.* **509,** L121 (1998).
Comets and interstellar ices: a cosmic connection. Brooke, T.Y. *Endeavour* **21,** 101 (1997).
Chemical inhomogeneities in interstellar clouds: The high latitude cloud MCLD 123.5+24.9. Gerin, M., Falgarone, E., Joulain, K., Kopp, M., Le Bourlot, J., and des Forets, G.P. *Astron. Astrophys.* **318,** 579 (1997).
Circumstellar photochemistry. Glassgold, A.E. *Annu. Rev. Astron. Astrophys.* **34**, 241 (1996).
A chemical survey of molecules in 'spiral arm' clouds. Greaves, J.S., and Nyman, L.A. *Astron. Astrophys.* **305,** 950 (1996).
Discovery of interstellar dust entering the Earth's atmosphere. Taylor, A.D., Baggaley, W.J., and Steel, D.I. *Nature, Lond.* **380,** 323 (1996).
Molecular clusters in interstellar clouds. Duley, W.W. *Astrophys. J.* **471,** L57 (1996).
The composition of the diffuse interstellar medium. Fitzpatrick, E.L. *Astrophys. J.* **473,** L55 (1996).
Sublimation from icy jets as a probe of the interstellar volatile content of comets. Blake, G.A., Qi, C., Hogerheijde, M.R., Gurwell, M.A., and Muhleman, D.O. *Nature, Lond.* **398,** 213 (1999).
Molecular abundance variations in the Magellanic Clouds. Heikkila, A., Johansson, L.E.B., and Olofsson, H. *Astron. Astrophys.* **344,** 817 (1999).
Molecular clouds, star formation and galactic structure. Scoville, N. and Young, J.S., *Scient. Am.* **250**, 30 (April 1984).

Interstellar clouds and the chemical species in them. See also Section 8.7.

Chemical abundances in molecular clouds. Irvine, W.M., Goldsmith, P.F., and Hjalmarson, A. in *Interstellar Processes*. Hollenbach, D.J. and Thronson, H.A. (eds.) (Reidel, Dordrecht, 1987).

Formation of interstellar ices behind shock waves. Bergin, E.A., Neufeld, D.A., and Melnick, G.J. *Astrophys. J.* **510,** L145 (1999).

The distribution of molecules in star-forming regions. Taylor, S.D., Morata, O., and Williams, D.A. *Astron. Astrophys.* **336,** 309 (1998).

Interstellar abundances in the Magellanic Clouds .1. GHRS observations of the Small Magellanic Cloud star Sk 108. Welty, D.E., Lauroesch, J.T., Blades, J.C., Hobbs, L.M., and York, D.G. *Astrophys. J.* **489,** 672 (1997).

Abundance of ^{3}He in the local interstellar cloud. Gloeckler, G. and Geiss, J. *Nature, Lond.* **381,** 210 (1996).

Detection of far-infrared water vapor, hydroxyl, and carbon monoxide emissions from the supernova remnant 3C 391. Reach, W.T., and Rho, J. *Astrophys. J.* **507,** L93 (1998).

ISO observations of interstellar ices and implications for comets. Ehrenfreund, P., d'Hendecourt, L., Dartois, E., de Muizon, M.J., Breitfellner, M., Puget, J.L., and Habing, H.J. *Icarus* **130,** 1 (1997).

Galactic OH absorption and emission toward a sample of compact extragalactic mm-wave continuum sources. Liszt, H. and Lucas, R. *Astron. Astrophys.* **314,** 917 (1996).

Atomic oxygen in molecular clouds? High-resolution spectroscopy of the [O I] 63 micron line toward DR 21. Poglitsch, A., Herrmann, F., Genzel, R., Madden, S.C., Nikola, T., Timmermann, R., Geis, N., and Stacey, G.J. *Astrophys. J.* **462,** L43 (1996).

Molecular gas in Markarian 109: Constraining the O_2/CO ratio in chemically young galaxies. Frayer, D.T., Seaquist, E.R., Thuan, T.X., and Sievers, A. *Astrophys. J.* **503,** 231 (1998).

Astronomical detection of the cyanobutadiynyl radical C_5N. Guelin, M., Neininger, N., and Cernicharo, J. *Astron. Astrophys.* **335,** L1 (1998).

Nitrogen sulfide in giant molecular clouds. McGonagle, D. and Irvine, W.M. *Astrophys. J.* **477,** 711 (1997).

Radio observations in NH_3 and C_2S towards small molecular clouds and around pre-main-sequence stars. Scappini, F., and Codella, C. *Monthly. Not. Royal Ast. Soc.* **282,** 587 (1996).

Chlorine-bearing molecules in cold and warm interstellar clouds. Amin, M.Y. *Earth Moon and Planets* **73,** 133 (1996).

The physics and chemistry of small translucent molecular clouds. X. SiO. Turner, B.E. *Astrophys. J.* **495,** 804 (1998).

The physics and chemistry of small translucent molecular clouds. 6. Organo-sulfur species. Turner, B.E. *Astrophys. J.* **461,** 246 (1996).

The SO-to-CS abundance ratio in molecular cirrus clouds. Heithausen, A., Corneliussen, U., and Grossmann, V. *Astron. Astrophys.* **330,** 311 (1998).

The physics and chemistry of small translucent molecular clouds. IX. Acetylenic chemistry. Turner, B.E., Lee, H.H., and Herbst, E. *Astrophys. J. Supp. Ser.* **115,** 91 (1998).

The physics and chemistry of small translucent molecular clouds. XI. Methanol. Turner, B.E. *Astrophys. J.* **501,** 731 (1998).

Deuterated methanol in the Orion compact ridge. Charnley, S.B., Tielens, A.G.G.M., and Rodgers, S.D. *Astrophys. J.* **482,** L203 (1997).

First astronomical detection of the cumulene carbon chain molecule H_2C_6 in TMC-1. Langer, W.D., Velusamy, T., Kuiper, T.B.H., Peng, R., McCarthy, M.C., Travers, M.J., Kovacs, A., Gottlieb, C.A., and Thaddeus, P. *Astrophys. J.* **480,** L63 (1997).

Interstellar PAH emission in the 11–14 micron region: New insights from laboratory data and a tracer of ionized PAHs. Hudgins, D.M. and Allamandola, L.J. *Astrophys. J.* **516,** L41 (1999).

The excitation temperatures of HC_9N and other long cyanopolyynes in TMC-1. Bell, M.B., Watson, J.K.G., Feldman, P.A., and Travers, M.J. *Astrophys. J.* **508,** 286 (1998).

Detection of HCCNC from IRC+10216. Gensheimer, P.D. *Astrophys. J.* **479,** L75 (1997).

Ions in the terrestrial atmosphere and in interstellar clouds. Smith, D. and Spanel, P. *Mass Spec. Revs.* **14,** 255 (1995).

Association reactions of Fe^+ with hydrocarbons: Implications for dense interstellar cloud chemistry. Petrie, S., Becker, H., Baranov, V., and Bohme, D.K. *Astrophys. J.* **476,** 191 (1997).

Observation of DCO^+ in the Large Magellanic Cloud. Heikkila, A., Johansson, L.E.B., and Olofsson, H. *Astron. Astrophys.* **319,** L21 (1997).

New observations of the $[HCO^+]/[HOC^+]$ ratio in dense molecular clouds. Apponi, A.J. and Ziurys, L.M. *Astrophys. J.* **481,** 800 (1997).

The photodissociation of SiH^+ in interstellar clouds and stellar atmospheres. Stancil, P.C., Kirby, K., Sannigrahi, A.B., Buenker, R.J., Hirsch, G., and Gu, J.P. *Astrophys. J.* **486,** 574 (1997).

Novel pathways to CN^- within interstellar clouds and circumstellar envelopes: Implications for IS and CS chemistry. Petrie, S. *Monthly. Not. Royal Ast. Soc.* **281,** 137 (1996).

Chemical conversions within interstellar clouds

The chemistry of interstellar gas and grains. Irvine, W.M. and Knacke, R.F. in *Origin and evolution of planetary and satellite atmospheres*. Atreya, S.K., Pollack, J.B., and Matthews, M.S. (eds.) (University of Arizona Press, Tucson, 1989).

Chemistry in the interstellar medium. Special issue of *J. Chem. Soc. Faraday Trans. 2*, Oct 1989.

Constraints on the abundances of various molecules in interstellar ice: laboratory studies and astrophysical implications. Boudin, N., Schutte, W.A., and Greenberg, J.M. *Astron. Astrophys.* **331,** 749 (1998).

The sensitivity of gas-phase chemical models of interstellar clouds to C and O elemental abundances and to a new formation mechanism for ammonia. Terzieva, R. and Herbst, E. *Astrophys. J.* **501,** 207 (1998).

The chemical composition and evolution of giant molecular cloud cores: A comparison of observation and theory. Bergin, E.A., Goldsmith, P.F., Snell, R.L., and Langer, W.D. *Astrophys. J.* **482,** 285 (1997).

Self-consistent model of chemical and dynamical evolution of protostellar clouds. Shematovich, V.I., Shustov, B.M., and Wiebe, D.S. *Monthly. Not. Royal Ast. Soc.* **292,** 601 (1997).

Fractional abundances of molecules in dense interstellar clouds: A compendium of recent model results. Lee, H.H., Bettens, R.P.A., and Herbst, E. *Astron. Astrophys. Supp. Ser.* **119,** 111 (1996).

Photodissociation regions in the interstellar medium of galaxies. Hollenbach, D.J. and Tielens, A.G.G.M. *Rev. modern Phys..* **71,** 173 (1999).

A study of the physics and chemistry of TMC-1. Pratap, P., Dickens, J.E., Snell, R.L., Miralles, M.P., Bergin, E.A., Irvine, W.M., and Schloerb, F.P. *Astrophys. J.* **486,** 862 (1997).

Photodissociation of H_2 and CO and time dependent chemistry in inhomogeneous interstellar clouds. Lee, H.H., Herbst, E., des Forets, G.P., Roueff, E., and Le Bourlot, J. *Astron. Astrophys.* **311,** 690 (1996).

Chemistry in the interstellar medium. Herbst, E. *Annu. Rev. phys. Chem.* **46,** 27 (1995).

Do H atoms stick to PAH cations in the interstellar medium? Herbst, E. and LePage, V. *Astron. Astrophys.* **344,** 310 (1999).

New H and H_2 reactions with small hydrocarbon ions and their roles in benzene synthesis in dense interstellar clouds. McEwan, M.J., Scott, G.B.I., Adams, N.G., Babcock, L.M., Terzieva, R., and Herbst, E. *Astrophys. J.* **513,** 287 (1999).

Neutral–neutral reactions in the interstellar medium.1. Formation of carbon hydride radicals *via* reaction of carbon atoms with unsaturated hydrocarbons. Kaiser, R.I., Stranges, D., Lee, Y.T., and Suits, A.G. *Astrophys. J.* **477,** 982 (1997).

Thermal chemistry of ice mixtures of astrophysical relevance. Schutte, W.A. *Advan. Space Res.* **20,** 1629 (1997).

Complex organics in laboratory simulations of interstellar/cometary ices. Bernstein, M.P., Allamandola, L.J., and Sandford, S.A. *Advan. Space Res.* **19,** 991 (1997).

Cyanopolyyne chemistry in TMC-1. Winstanley, N. and Nejad, L.A.M. *Astrophys. Space Sci.* **240,** 13 (1996).

On the stability of interstellar carbon clusters: The rate of the reaction between C_3 and O. Woon, D.E. and Herbst, E. *Astrophys. J.* **465,** 795 (1996).

Time dependent chemical study of contracting interstellar clouds. 1. Abundances of nitrogen- and carbon-bearing molecules. Amin, M.Y., El Nawawy, M.S., Ateya, B.G., and Aiad, A. *Earth Moon and Planets* **69,** 95 (1995).

Time dependent chemical study of contracting interstellar clouds.2. Abundances of oxygen- and sulphur-bearing molecules. Amin, M.Y., El Nawawy, M.S., Ateya, B.G., and Aiad, A. *Earth Moon and Planets* **69,** 113 (1995).

Some interstellar reactions involving electrons and neutral species: Attachment and isomerization. Petrie, S. and Herbst, E. *Astrophys. J.* **491,** 210 (1997).

The variation of the CO to H_2 conversion factor in two translucent clouds. Magnani, L., Onello, J.S., Adams, N.G., Hartmann, D., and Thaddeus, P. *Astrophys. J.* **504,** 290 (1998).

An extensive *ab initio* study of the $C^+ + NH_3$ reaction and its relation to the HNC/HCN abundance ratio in interstellar clouds. Talbi, D. and Herbst, E. *Astron. Astrophys.* **333,** 1007 (1998).

Selected ion flow tube studies of the atomic oxygen radical cation reactions with ethylene and other alkenes. Fishman, V.N., Graul, S.T., and Grabowski, J.J. *Int J. mass Spect.* **187,** 477 (1999).

Ion–molecule reactions in interstellar clouds: successes and problems. Herbst, E. *Advan. gas-phase ion Chem.* **3**, 1 (1998).
The amount of CH produced during CH^+ synthesis in interstellar clouds. Federman, S.R., Welty, D.E., and Cardelli, J.A. *Astrophys. J.* **481,** 795 (1997).
X-ray-induced chemistry of interstellar clouds. Lepp, S., and Dalgarno, A. *Astron. Astrophys.* **306,** L21 (1996).
Laboratory studies of isotope exchange in ion–neutral reactions: interstellar implications. Smith, D. *Phil. Trans. R. Soc.* **A303**, 535 (1981).
Some H/D exchange reactions involved in the deuteration of interstellar molecules. Smith, D., Adams, N.G., and Alge, E. *Astrophys. J.* **263**, 123 (1981).

Quite complex compounds can be built up in interstellar chemistry and the processes may at least mimic those responsible for the origin of life on Earth. The tholins are complex organic solids produced from cosmically abundant molecules.

Galactochemistry and the origin of life. Brown, R.D. *Chem. Br.* **15**, 570 (1979).
The diffuse interstellar bands: A tracer for organics in the diffuse interstellar medium? Salama, F. *Origins Life Evol. Biosphere* **28,** 349 (1998).
The formation of large hydrocarbons and carbon clusters in dense interstellar clouds. Bettens, R.P.A. and Herbst, E. *Astrophys. J.* **478,** 585 (1997).
The abundance of very large hydrocarbons and carbon clusters in the diffuse interstellar medium. Bettens, R.P.A. and Herbst, E. *Astrophys. J.* **468,** 686 (1996).
Interstellar matter and chemical evolution. Peimbert, M., Serrano, A., and Torres-Peimbert, S. *Science* **224**, 35 (1984).
Tholins: organic chemistry of interstellar grains and gas. Sagan, C. and Khare, B.N. *Nature, Lond.* **277**, 102 (1979).

Extraterrestrial matter is certainly brought in to the Earth's atmosphere. An extreme view is that life and even disease is borne in from space.

Comets and the origin of life. Ponnamperuma, C. (ed.) (D. Reidel, Dordrecht, 1981).
Does epidemic disease come from space? Hoyle, F. and Wickramasinghe, N.C. *New Scient.* **76,** 402 (1977).
Cometary delivery of organic molecules to the early Earth. Chyba, C.F., Thomas, P.J., Brookshaw, L., and Sagan, C. *Science* **249**, 366 (1990).
Interstellar dust, chirality, comets and the origins of life — life from dead stars. Greenberg, J.M., Kouchi, A., Niessen, W., Irth, H., Vanparadijs, J., Degroot, M., and Hermsen, W. *J. biol. Phys.* **20,** 61 (1994).
Comets as a possible source of prebiotic molecules. Huebner, W.E. and Boice, D.C. *Origins Life Evol. Biosphere* **21,** 299 (1992).
Comets and the formation of biochemical compounds on the primitive Earth — a review. Oro, J., Mills, T., and Lazcano, A. *Origins Life Evol. Biosphere* **21,** 267 (1992).
Comet impacts and chemical evolution on the bombarded Earth. Oberbeck, V.R. and Aggarwal, H. *Origins Life Evol. Biosphere* **21,** 317 (1992).
Dust particles in the atmospheres of terrestrial planets and their roles for prebiotic chemistry: An overview. Basiuk, V.A. and Navarro-Gonzalez, R. *Astrophys. Space. Sci.* **236,** 61 (1996).

Nontraditional pathways of extraterrestrial formation of prebiotic matter. Goldanskii, V.I. *J. phys. Chem.* A **101,** 3424 (1997).

Origins of life: A comparison of theories and application to Mars. Davis, W.L. and McKay, C.P. *Origins Life Evol. Biosphere* **26,** 61 (1996).

Exobiology, the study of the origin, evolution and distribution of life within the context of cosmic evolution—a review. Horneck, G. *Planet. Space Sci.* **43,** 189 (1995).

Life from space—a history of panspermia. Kamminga, H. *Vistas in Astronomy* **26,** 67 (1982).

The fusion crusts of stony meteorites: Implications for the atmospheric reprocessing of extraterrestrial materials. Genge, M.J. and Grady, M.M. *Meteoritics & planet. Sci.* **34,** 341 (1999).

Raman spectroscopic characterization of a Martian SNC meteorite: Zagami. Wang, A., Jolliff, B.L., and Haskin, L.A. *J. geophys. Res.* **104,** 8509 (1999).

The meteorites of Mars: Sample return this millenium. Sears, D. *Meteoritics & planet. Sci.* **34,** 313 (1999).

Near-infrared reflectance spectra from bulk samples of the two Martian meteorites Zagami and Nakhla. Schade, U. and Wasch, R. *Meteoritics & planet. Sci.* **34,** 417 (1999).

The composition and thickness of the crust of Mars estimated from rare-earth elements and neodymium-isotopic compositions of Martian meteorites. Norman, M.D. *Meteoritics & planet. Sci.* **34,** 439 (1999).

Argon-39 – argon-40 "ages" and trapped argon in Martian shergottites, Chassigny, and Allan Hills 84001. Bogard, D.D. and Garrison, D.L.H. *Meteoritics & planet. Sci.* **34,** 451 (1999).

Isotopic evidence for extraterrestrial non-racemic amino acids in the Murchison meteorite. Engel, M.H and Macko, S.A. *Nature, Lond.* **389,** 265 (1997).

The origin of organic matter in the Martian meteorite ALH84001. Becker, L., Popp, B., Rust, T., and Bada, J.L. *Earth planet. Sci. Letts.* **167,** 71 (1999).

Stable isotopic measurements of the low-temperature nitrogen components in ALH 84001. Wright, I.P., Grady, M.M., and Pillinger, C.T. *J. geophys. Res.* **104,** 1877 (1999).

The Shergottite age paradox and the relative probabilities for Martian meteorites of differing ages. Nyquist, L.E., Borg, L.E. and Shih, C.Y. *J. geophys. Res.* **103,** 31445 (1998).

A "unique" distribution of polycyclic aromatic hydrocarbons in Allan Hills 84001, or a selective attack in meteorites from Mars? Sephton, M.A. and Gilmour, I. *Meteoritics & planet. Sci.* **33,** A142 (1998).

Martian gases in an Antarctic meteorite. Bogard, D.D. and Johnson, P. *Science* **221,** 651 (1983).

Gas-rich meteorites: probe for particle environment and dynamical processes in the inner solar system. Goswami, J.N., Lal, D., and Wilkening, L.L. *Space Sci. Rev.* **37,** 111 (1984).

Interstellar matter in meteorites. Lewis, R.S. and Anders, E. *Scient. Am.* **249**, 54 (Aug 1983).

Evolution by bombardment. Prather, M.J. *Nature, Lond.* **308**, 604 (1984).

Extraterrestrial platinum group nuggets in deep-sea sediments. Brownlee, D.E., Bates, B.A., and Wheelock, M.M. *Nature, Lond.* **309**, 693 (1984).

Impacts with cometary objects may also have led to periodic mass extinctions for which there is geological evidence.

The galactic theory of mass extinctions: An update. Rampino, M.R. *Celest. Mech. dynamic. Astron.* **69,** 49 (1997).

Comet showers as a cause of mass extinctions. Hut, P., Alvarez, W., Elder, W.P., Hansen, T., Kauffman, E.G., Keller, G., Shoemaker, E.M., and Weissman, P.R. *Nature, Lond.* **329,** 118 (1987).

The causes of mass extinction. Hallam, A., *Nature, Lond.* **308,** 686 (1984).

Section 9.2.1

Atmospheric compositions: key similarities and differences. Pepin, R.O. in *Origin and evolution of planetary and satellite atmospheres.* Atreya, S.K., Pollack, J.B., and Matthews, M.S. (eds.) (University of Arizona Press, Tucson, 1989).

Terrestrial noble gases: constraints and implications on atmospheric evolution. Ozima, M. and Igarashi, G. in *Origin and evolution of planetary and satellite atmospheres.* Atreya, S.K., Pollack, J.B., and Matthews, M.S. (eds.) (University of Arizona Press, Tucson, 1989).

Noble gas components in planetary atmospheres and interiors in relation to solar wind and meteorites. Marti, K. and Mathew, K.J. *Proc. Ind. Acad. Sci.* **107,** 425 (1998).

Atmospheric evolution under dust bombardment and noble gas fractionation in the terrestrial atmospheres. Pavlov, A.K. and Pavlov, A.A. *Meteoritics & planet. Sci.* **32,** A104 (1997).

Escape of atmospheres, ancient and modern. Hunten, D.M., *Icarus* **85,** 1 (1990).

Mass fractionation of noble gases in diffusion-limited hydrodynamic hydrogen escape. Zahnle, K., Kasting, J.F., and Pollack, J.B. *Icarus* **84,** 502 (1990).

Circumstellar material in meteorites: noble gases, carbon, and nitrogen. Anders, E. in *Meteorites and the early solar system.* Kerridge, J.F., and Matthews, M.S. (eds.) (University of Arizona Press, Tucson, 1988).

Local and exotic components of meteorites, and their origin. Anders, E. *Phil. Trans Roy. Soc.* **A323,** 287 (1987).

Fractionation of noble gases by thermal escape from accreting planetesimals. Donahue, T.M. *Icarus* **66,** 195 (1986).

Noble gas geochemistry. Ozima, M. and Podosek, F.A. (Cambridge University Press, 1983).

Noble gas isotopic ratios from historical lavas and fumaroles at Mount Vesuvius (southern Italy): constraints for current and future volcanic activity. Tedesco, D., Nagao, K., and Scarsi, P. *Earth planet. Sci. Letts.* **164,** 61 (1998).

Intensive gas sampling of noble gases and carbon at Vulcano Island (southern Italy). Tedesco, D. and Scarsi, P. *J. geophys. Res.* **104,** 10499 (1999).

K–Ar and Ar-40/Ar-39 geochronology of weathering processes. Vasconcelos, P.M. *Annu. Rev. Earth planet. Sci.* **27,** 183 (1999).

A model to explain the various paradoxes associated with mantle noble gas geochemistry. Anderson, D.L. *Proc. Nat. Acad. Sci. USA* **95,** 9087 (1998).

Rare-gas solids in the Earth's deep interior. Jephcoat, A.P. *Nature, Lond.* **393,** 355 (1998).

Mantle degassing of noble gases—Implications of recent experiments. Sankaran, A.V. *Current Science* **77**, 13 (1999).

Helium isotope composition of the early Iceland mantle plume inferred from the tertiary picrites of West Greenland. Graham, D.W., Larsen, L.M., Hanan, B.B., Storey, M., Pedersen, A.K., and Lupton, J.E. *Earth planet. Sci. Letts.* **160,** 241 (1998).

Noble gases in the Earth's mantle. Farley, K.A. and Neroda, E. *Ann Rev. Earth planet. Sci.* **26,** 189 (1998).

Primordial helium and neon in the Earth — a speculation on early degassing. Honda, M. and McDougall, I. *Geophys. Res. Letts.* **25,** 1951 (1998).

Noble gas constraints on the evolution of the atmosphere – mantle system. Kamijo, K., Hashizume, K., and Matsuda, J. *Geochim. Cosmochim. Acta* **62,** 2311 (1998).

Helium isotopes in lithospheric mantle: Evidence from Tertiary basalts of the western USA. Dodson, A., DePaolo, D.J., and Kennedy, B.M. *Geochim. Cosmochim. Acta* **62,** 3775 (1998).

Plume-derived rare gases in 380 Ma carbonatites from the Kola region (Russia) and the argon isotopic composition in the deep mantle. Marty, B., Tolstikhin, I., Kamensky, I.L., Nivin, V., Balaganskaya, E., and Zimmermann, J.L. *Earth planet. Sci. Letts.* **164,** 179 (1998).

The evolution of terrestrial volatiles: a view from helium, neon, argon and nitrogen isotope modelling. Tolstikhin, I.N., and Marty, B. *Chem. Geol.* **147,** 27 (1998).

Mantle-derived noble gases in carbonatites. Sasada, T., Hiyagon, H., Bell, K., and Ebihara, M. *Geochim. Cosmochim. Acta* **61,** 4219 (1997).

Implications of small comets for the noble gas inventories of Earth and Mars. Swindle, T.D., and Kring, D.A. *Geophys. Res. Letts.* **24,** 3113 (1997).

Evolution of Earth's noble gases: Consequences of assuming hydrodynamic loss driven by giant impact. Pepin, R.O. *Icarus* **126,** 148 (1997).

The argon constraints on mantle structure. Allegre, C.J., Hofmann, A., and O'Nions, K. *Geophys. Res. Letts.* **23,** 3555 (1996).

Constraints on Venus evolution from radiogenic argon. Kaula, W.M. *Icarus* **139,** 32 (1999).

Martian stable isotopes: Volatile evolution, climate change and exobiological implications. Jakosky, B.M. *Origins Life Evol. Biosphere* **29,** 47 (1999).

Martian atmospheric xenon contents of Nakhla mineral separates: implications for the origin of elemental mass fractionation. Gilmour, J.D., Whitby, J.A., and Turner, G. *Earth planet. Sci. Letts.* **166,** 139 (1999).

Argon-39 – argon-40 "ages" and trapped argon in Martian shergottites, Chassigny, and Allan Hills 84001. Bogard, D.D. and Garrison, D.L.H. *Meteoritics & planet. Sci.* **34,** 451 (1999).

Noble gases and chemical composition of Shergotty mineral fractions, Chassigny, and Yamato 793605: The trapped argon-40/argon-36 ratio and ejection times of Martian meteorites. Terribilini, D., Eugster, O., Burger, M., Jakob, A., and Krahenbuhl, U. *Meteoritics & planet. Sci.* **33,** 677 (1998).

A nitrogen and argon stable isotope study of Allan Hills 84001: Implications for the evolution of the Martian atmosphere. Grady, M.M., Wright, I.P., and Pillinger, C.T. *Meteoritics & planet. Sci.* **33,** 795 (1998).

Signatures of noble gases and nitrogen in the atmosphere and the interior of Mars. Mathew, K.J. and Marti, K. *Meteoritics & planet. Sci.* **33,** A99 (1998).

Martian atmospheric and indigenous components of xenon and nitrogen in the Shergotty, Nakhla, and Chassigny group meteorites. Mathew, K.J., Kim, J.S., and Marti, K. *Meteoritics & planet. Sci.* **33,** 655 (1998).

Noble gas and carbon isotopes in Mariana Trough basalt glasses. Sano, Y., Nishio, Y., Gamo, T., Jambon, A., and Marty, B. *Appl. Geochem.* **13,** 441 (1998).

Mars, Modulus and MAGIC. The measurement of stable isotopic compositions at a planetary surface. Wright, I.P. and Pillinger, C.T. *Planet. Space Sci.* **46,** 813 (1998).

A nitrogen and argon stable isotope study of Allan Hills 84001: Implications for the evolution of the Martian atmosphere. Grady, M.M., Wright, I.P., and Pillinger, C.T. *Meteoritics & planet. Sci.* **33,** 795 (1998).

Life on Mars: chemical arguments and clues from Martian meteorites. Brack, A. and Pillinger, C.T. *Extremophiles* **2,** 313 (1998).

A reappraisal of the Martian Ar-36/Ar-38 ratio. Bogard, D.D. *J. geophys. Res.* **102,** 1653 (1997).

Impact of a paleomagnetic field on sputtering loss of Martian atmospheric argon and neon. Hutchins, K.S., Jakosky, B.M., and Luhmann, J.G. *J. geophys. Res.* **102,** 9183 (1997).

Ar–Ar chronology of the Martian meteorite ALH84001: Evidence for the timing of the early bombardment of Mars. Turner, G., Knott, S.F., Ash, R.D., and Gilmour, J.D. *Geochim. Cosmochim. Acta* **61,** 3835 (1997).

The xenon isotopic composition of the primordial Martian atmosphere: Contributions from solar and fission components. Swindle, T.D. and Jones, J.H. *J. geophys. Res.* **102,** 1671 (1997).

The contribution by interplanetary dust to noble gases in the atmosphere of Mars. Flynn, G.J. *J. geophys. Res.* **102,** 9175 (1997).

Isotopes, ice ages, and terminal Proterozoic Earth history. Kaufman, A.J., Knoll, A.H., and Narbonne, G.M. *Proc. Nat. Acad. Sci. USA* **94,** 6600 (1997).

Nitrogen and heavy noble gases in ALH 84001: Signatures of ancient Martian atmosphere. Murty, S.V.S. and Mohapatra, R.K. *Geochim. Cosmochim. Acta* **61,** 5417 (1997).

Helium on Mars: EUVE and PHOBOS data and implications for Mars' evolution. Krasnopolsky, V.A. and Gladstone, G.R. *J. geophys. Res.* **101,** 15765 (1996).

Evolution of Martian atmospheric argon: Implications for sources of volatile. Hutchins, K.S. and Jakosky, B.M. *J. geophys. Res.* **101,** 14933 (1996).

Section 9.2.2

Many more references to Titan's atmosphere and its evolution are given in connection with Section 8.5.1.

Present state and chemical evolution of the atmospheres of Titan, Triton and Pluto. Lunine, J.I., Atreya, S.K., and Pollack, J.B. in *Origin and evolution of planetary and satellite atmospheres*. Atreya, S.K., Pollack, J.B., and Matthews, M.S. (eds.) (University of Arizona Press, Tucson, 1989).

Some speculations on Titan's past, present and future. Lunine, J.I., Lorenz, R.D., and Hartmann, W.K. *Planet. Space Sci.* **46,** 1099 (1998).

Chemical evolution on Titan: Comparisons to the prebiotic Earth. Clarke, D.W. and Ferris, J.P. *Origins Life Evol. Biosphere* **27,** 225 (1997).
How primitive are the gases in Titan's atmosphere? Owen, T. *Adv. Space Res..* **7**, 51 (1987).
Clathrate and ammonia hydrate at high pressure: application to the origin of methane on Titan. Lunine, J.I. and Stevenson, D.J. *Icarus* **70**, 61 (1987).
The composition and origin of Titan's atmosphere. Owen T. *Planet. Space Sci.* **30**, 833 (1982).

Shock waves produced by impacting bodies have been suggested as a further means of initiating chemical change in Titan's atmosphere.

High temperature shock formation of N_2 and organics on primordial Titan. McKay, C.P., Scattergood, T.W., Pollack, J.B., Borucki, W.J., and van Ghysegham, H.T. *Nature, Lond.* **332,** 520 (1988).
Estimated impact shock production of N_2 and organic compounds on early Titan. Jones, T.D. and Lewis, J.S. *Icarus* **72,** 381 (1987).
Chemistry and evolution of Titan's atmosphere. Strobel, D.F. *Planet. Space Sci.* 30. 839 (1982).

Section 9.3

Mechanisms and observations for isotope fractionation in planetary atmospheres. Kaye, J.A. *Rev. Geophys.* **25**, 1609 (1987).
Isotopic reconstruction of past continental environments. Koch, P.L. *Annu. Rev. Earth planet. Sci.* **26,** 573 (1998).
Atmosphere science — Mass-independent isotope effects in planetary atmospheres and the early solar system. Thiemens, M.H. *Science* **283,** 341 (1999).
Characteristic modes of isotopic variations in atmospheric chemistry. Manning, M.R. *Geophys. Res. Letts.* 26, 1263 (1999).
The role of stable isotopes in geochemistries of all kinds. Epstein, S. *Annu. Rev. Earth planet. Sci.* **25,** 1 (1997).
Implications for the interpretation of ice-core isotope data from analysis of modelled Antarctic precipitation. Noone, D. and Simmonds, I. *Ann. Glaciol.* **27,** 398 (1998).
Martian stable isotopes: Volatile evolution, climate change and exobiological implications. Jakosky, B.M. *Origins Life Evol. Biosphere* **29,** 47 (1999).
Atmospheric loss since the onset of the Martian geologic record: Combined role of impact erosion and sputtering. Brain, D.A. and Jakosky, B.M. *J. geophys. Res.* **103,** 22689 (1998).
Nonprimordial deuterium in the interstellar medium. Mullan, D.J. and Linsky, J.L. *Astrophys. J.* **511,** 502 (1999).
Deuterium in the solar system. Yung, Y.L. and Dissly, R.W. in *Isotope effects in gas-phase chemistry*, Kaye, J.A. (ed.) (American Chemical Society, Washington, DC, 1992).
Deuterium in the outer solar system: evidence for two distinct reservoirs. Owen, T., Lutz, B.L., and de Bergh, C. *Nature, Lond.* **320**, 244 (1986).
Deuterium in the solar system. Geiss, J. and Reeves, H. *Astron. Astrophys.* **93**, 189 (1981).

Limits on HDS/H_2S abundance ratios in hot molecular cores. Hatchell, J., Roberts, H., and Millar, T.J. *Astron. Astrophys.* **346,** 227 (1999).
Origin of water in the solar system: Constraints from deuterium/hydrogen ratios. Drouart, A.R.M., Robert, F., Gautier, D., and Dubrulle, B. *Meteorics planet. Sci.* **33,** A42 (1998).
Desorption processes and the deuterium fractionation in molecular clouds. Willacy, K.and Millar, T.J. *Monthly. Not. Royal Ast. Soc.* **298,** 562 (1998).
A reestimate of the protosolar ($^2H/^1H$)(*p*) ratio from ($^3He/^4He$)(*sw*) solar wind measurements. Gautier, D. and Morel, P. *Astron. Astrophys.* **323,** L9 (1997).
Hydrogen and deuterium in the thermosphere of Venus: Solar cycle variations and escape. Hartle, R.E., Donahue, T.M., Grebowsky, J.M., and Mayr, H.G. *J. geophys. Res.* **101,** 4525 (1996).
Evolution of deuterium on Venus. Gurwell, M.A. *Nature, Lond.* **378,** 22 (1995).
Escape of hydrogen from Venus. McElroy, M.B., Prather, M.J., and Rodriguez, J.M. *Science* **215,** 1614 (1982).
Identification of deuterium ions in the ionosphere of Venus. Hartle, R. and Taylor, H.A., Jr. *Geophys. Res. Letts.* **10,** 965 (1983).
Venus was wet: A measurement of the ratio of deuterium to hydrogen. Donahue, T.M., Hoffman, J.H ., Hodges, R .R., Jr., and Watson, A.J. *Science* **216,** 630 (1982).
Deuteronomy? A puzzle of deuterium and oxygen on Mars. Yung, Y.L. and Kass, D.M. *Science* **280,** 1545 (1998).
Detection of atomic deuterium in the upper atmosphere of Mars. Krasnopolsky, V.A., Mumma, M.J., and Gladstone, G.R. *Science* **280,** 1576 (1998).
HDO in the Martian atmosphere: implications for the abundance of crustal water. Yung, Y.L., Wen, J.-S., Pinto, J.P., Allen, M.. Pierce, K.K., and Paulson, S. *Icarus* **76,** 146 (1988).
Deuterium on Mars: the abundance of HDO and the value of D/H. Owen, T., Maillard, J.P., de Bergh, C., and Lutz, B. *Science* **240,** 1767 (1988).
Water on Mars. Carr, M.H. *Nature, Lond.* **326,** 30 (1987).
Deuterium enrichment in giant planets. Lecluse, C., Robert, F., Gautier, D., and Guiraud, M. *Planet. Space Sci.* **44,** 1579 (1996).
Dynamical influence on the isotopic enrichment of CH_3D in the outer planets. Smith, M.D., Conrath, B.J., and Gautier, D. *Icarus* **124,** 598 (1996).
Monodeuterated methane in the outer solar system. IV Its detection and abundance on Neptune. De Bergh, C., Lutz, B.L., Owen, T. and Maillard, J.P. *Astrophys. J.* **35S,** 661 (1990).
Monodeuterated methane in the outer solar system. III Its abundance on Titan. De Bergh, C., Lutz, B.L., Owen, T., and Chauville, J. *Astrophys. J.* **329,** 951 (1988).
Kinetic isotopic fractionation and the origin of HDO and CH_3D in the solar system. Yung, Y.L., Friedl, R.R., Pinto, J.P., Bayes, K.D., and Wen, J.-S. *Icarus* **74,**121 (1988).
D to H ratio and the origin and evolution of Titan's atmosphere. Pinto, J.P., Lunine, J.I., Kim, S.-J., and Yung, Y.L. *Nature, Lond.* **319,** 388 (1986).
Cosmogonical implications of elemental and isotopic abundances in atmospheres of the giant planets. Gautier, D. and Owen, T. *Nature, Lond.* **304,** 691 (1983).
Deuterium in comet C 1995 O1 (Hale-Bopp): Detection of DCN. Meier, R., Owen, T.C., Jewitt, D.C., Matthews, H.E., Senay, M., Biver, N., Bockelee-Morvan, D., Crovisier, J., and Gautier, D. *Science* **279,** 1707 (1998).

Hydrogen isotope geochemistry of SNC meteorites. Leshin, L.A., Epstein, S., and Stolper, E.M. *Geochim. Cosmochim. Acta* **60,** 2635 (1996).
The D/H ratio in water from Halley. Eberhardt, P., Dolder, U., Schulte, W., Krankowsky, D., Lammerzahl, P., Hoffman, J.H., Hodges, R.R., Berthelier J.J., and Illiano, J.M. *Astron. Astrophys.* **187**, 435 (1987).
The ratio of ortho- to para-H_2 in photodissociation regions. Sternberg, A. and Neufeld, D.A. *Astrophys. J.* **516,** 371 (1999).
Carbon isotope evidence for early life. Eiler, J.M., Mojzsis, S.J., and Arrhenius, G. *Nature, Lond.* **386**, 665 (1997).
Carbon-isotope composition of Lower Cretaceous fossil wood: Ocean–atmosphere chemistry and relation to sea-level change. Grocke, D.R., Hesselbo, S.P., and Jenkyns, H.C. *Geol.* **27,** 155 (1999).
Isotopic constraints on the Cenozoic evolution of the carbon cycle. Francois, L.M. and Godderis, Y. *Chem. Geol.* **145,** 177 (1998).
Carbon isotope ratios of Phanerozoic marine cements: Re-evaluating the global carbon and sulfur systems. Carpenter, S.J. and Lohmann, K.C. *Geochim. Cosmochim. Acta* **61,** 4831 (1997).
Isotopic evolution of the biogeochemical carbon cycle during the Proterozoic Eon. Des Marais, D.J. *Org. Geochem.* **27,** 185 (1997).
Carbon 14 measurements of the Martian atmosphere as an indicator of atmosphere-regolith exchange of CO_2. Jakosky, B.M., Reedy, R.C., and Masarik, J. *J. geophys. Res.* **101,** 2247 (1996).
Ozone isotope enrichment: isotopomer-specific rate coefficients. Mauersberger, K., Erbacher, B., Krankowsky, D., Günther, J., and Nickel, R. *Science* **283**, 370 (1999).
Loss of oxygen from Venus. McElroy, M.B., Prather, M.J., and Rodriguez, J.M. *Geophys. Res. Letts.* **9**, 649 (1982).
Oxygen and carbon isotope ratios in Martian carbon dioxide: Measurements and implications for atmospheric evolution. Krasnopolsky, V.A., Mumma, M.J., Bjoraker, G.L., and Jennings, D.E. *Icarus* **124,** 553 (1996).
Oxygen isotopes in meteorites. Clayton, R.N. *Annu. Rev. Earth planet. Sci.* **21**, 112 (1993).
Hydrogen chloride detection in Orion A and Monoceros R2 and derivation of the $(HCl)^{35}Cl/(HCl)^{37}Cl$ isotopic ratio. Salez, M., Frerking, M.A., and Langer, W.D. *Astrophys. J.* **467,** 708 (1996).
Sulphur isotope fractionation in modern microbial mats and the evolution of the sulphur cycle. Habicht, K.S. and Canfield, D.E. *Nature, Lond.* **382,** 342 (1996).
Isotope specific kinetics of hydroxyl radical (OH) with water (H_2O): Testing models of reactivity and atmospheric fractionation. Dubey, M.K., Mohrschladt, R., Donahue, N.M., Anderson, J.G. *J. phys. Chem.* A **101,** 1494 (1997).

Section 9.4

Introductory reviews on the evolution of the Earth's atmosphere

Evolution of the atmosphere. Nunn, J.F. *Proc. Geol. Assoc.* **109,** 1 (1998).
The birth of the Earth's atmosphere: the behaviour and fate of its major elements. Javoy, M. *Chem. Geol.* **147,** 11 (1998).

Oceanic minerals: Their origin, nature of their environment and significance. Kastner, M. *Proc. Nat. Acad. Sci.* **96,** 3380 (1999).

Contemplation of things past. Wetherill, G.W. *Annu. Rev. Earth planet. Sci.* **26,** 1 (1998).

The young age of Earth. Zhang, Y.X. *Geochim. Cosmochim. Acta* **62,** 3185 (1998).

Application of the Pitzer model to the evolution of early planetary (Earth and Mars) oceans and atmospheres. Morse, J.W. and Marion, G.M. *Amer. Chem. Soc. Abstracts* **217,** 82 (1999).

Redox state of the Archean atmosphere: Evidence from detrital heavy minerals in ca. 3250–2750 Ma sandstones from the Pilbara Craton, Australia. Rasmussen, B. and Buick, R. *Geol.* **27,** 115 (1999).

Long-term fluxes and budget of ferric iron: implication for the redox states of the Earth's mantle and atmosphere. Lecuyer, C. and Ricard, Y. *Earth planet. Sci. Letts.* **165,** 197 (1999).

On the possible influence of extraterrestrial volatiles on Earth's climate and the origin of the oceans. Deming, D. *Palaeogeography Palaeoclimatology Palaeoecology* **146,** 33 (1999).

Extraterrestrial impact events: the record in the rocks and the stratigraphic column. Grieve, R.A.F. *Palaeogeography Palaeoclimatology Palaeoecology* **132,** 5 (1997).

Environmental perturbations caused by the impacts of asteroids and comets. Toon, O.B., Zahnle, K., Morrison, D., Turco, R.P., and Covey, C. *Rev. Geophys.* **35,** 41 (1997).

Primary sources of phosphorus and phosphates in chemical evolution. Macia, E., Hernandez, M.V., and Oro, J. *Origins Life Evol. Biosphere* **27,** 459 (1997).

Recognition of ≥3850 Ma water-lain sediments in West Greenland: their significance for the early Archaean Earth. Nutman, A.P., Mojzsis, S.J., and Friend, C.R.L. *Geochim. Cosmochim. Acta* **61,** 2475 (1997).

Atmospheric losses under dust bombardment in the ancient atmospheres. Pavlov, A.K., and Pavlov, A.A. *Earth Moon and Planets* **76,** 157 (1997).

Origin and evolution of the atmosphere. Wayne, R.P. *Chem. Br.* **24,** 225 (1988).

The photochemistry of the early atmosphere. Levine, J.S. in *The photochemistry of atmospheres*. Levine, J.S. (ed.) (Academic Press, Orlando, 1985).

The chemical evolution of the atmosphere and oceans. Holland, H.D. (Princeton University Press, Princeton, New Jersey, 1982).

Evolution of the atmosphere. Walker, J.C.G. (Macmillan, London, 1978).

The earliest atmosphere of the Earth. Walker, J.C.G. *Precambr. Res.* **17,** 147 (1982).

The photochemistry of the palaeoatmosphere. Levine, J.S. *J. molec. Evol.* **18,** 161 (1982).

Origin of organic compounds

Abiotic synthesis of bioorganic compounds in simulated primitive planetary environments. Kobayashi, K., Kaneko, T., Ponnamperuma, C., Oshima, T., Yanagawa, H., and Saito, T. *Nippon Kagaku Kaishi* **292,** 601 (1997).

Extraterrestrial organic matter: A review. Irvine, W.M. *Origins Life Evol. Biosphere* **28,** 365 (1998).

Complex organics in meteorites. Shimoyama, A. *Advan. Space Res.* **19,** 1045 (1997).

Circumstellar and interstellar synthesis of organic molecules. Tielens, A.G.G.M. and Charnley, S.B. *Origins Life Evol. Biosphere* **27,** 23 (1997).

The origin of life

The origin of life — a review of facts and speculations. Orgel, L.E. *Trends in Biochemical Sciences* **23,** 491 (1998).

Thermophiles, early biosphere evolution, and the origin of life on Earth: Implications for the exobiological exploration of Mars. Farmer, J. *J. geophys. Res.* **103,** 28457 (1998).

Is extraterrestrial organic matter relevant to the origin of life on Earth? Whittet, D.C.B. *Origins Life Evol. Biosphere* **27,** 249 (1997).

In search of planets and life around other stars. Lunine, J.I. *Proc. nat. Acad. Sci. USA* **96,** 5353 (1999).

Astronomical searches for Earth-like planets and signs of life. Woolf, N. and Angel, J.R. *Annu. Rev. Astron. Astrophys.* **36,** 507 (1998).

Small comets in the high atmosphere. Hoyle, F. and Wickramasinghe, C. *Astrophys. Space Sci.* **253,** 13 (1997).

Evidence for life on Earth more than 3850 million years ago. Holland, H.D. *Science* **275**, 38 (1997).

Evidence for life on Earth before 3800 million years ago. Mojsziz, S.J., Arrhenius, G., McKeegan, K.D., Harrison, T.M., Nutman, A.P., and Friend, C.R.L. *Nature, Lond.* **384,** 55 (1996).

Titan and the origin of life on Earth. Owen, T., Raulin, F., McKay, C.P., Lunine, J.I., Lebreton, J.P., and Matson, D.L. *ESA Bulletin* **24,** 2905 (1997).

The earliest atmosphere

Atmospheric chemistry: the evolution of our atmosphere. Wayne, R.P. *J. Photochem. Photobiol.* **A62,** 379 (1992).

Abiotic nitrogen reduction on the early Earth. Brandes, J.A., Boctor, N.Z., Cody, G.D., Cooper, B.A., Hazen, R.M., and Yoder, H.S. *Nature, Lond.* **395,** 365 (1998).

Evolution of the continents and the atmosphere inferred from Th–U–Nb systematics of the depleted mantle. Collerson, K.D. and Kamber, B.Z. *Science* **283,** 1519 (1999).

Chemical evolution on Titan: Comparisons to the prebiotic Earth. Clarke, D.W. and Ferris, J.P. *Origins Life Evol. Biosphere* **27,** 225 (1997).

Prebiotic phosphorus chemistry reconsidered. Schwartz, A.W. *Origins Life Evol. Biosphere* **27,** 505 (1997).

Possible role of volcanic ash–gas clouds in the Earth's prebiotic chemistry. Basiuk, V.A. and Navarro-Gonzalez, R. *Origins Life Evol. Biosphere* **26,** 173 (1996).

Dust particles in the atmospheres of terrestrial planets and their roles for prebiotic chemistry: An overview. Basiuk, V.A., and Navarro-Gonzalez, R. *Astrophys. Space Sci.* **236,** 61 (1996).

Earth's early atmosphere. Kasting, J.F. *Science* **259,** 920 (1993).

On the origin and early evolution of terrestrial planet atmospheres and meteoritic volatile. Pepin, R.O. *Icarus* **92,** 2 (1991).

The evolution of the prebiotic atmosphere. Kasting, J.F. *Origins Life Evol. Biosphere* **14,** 75 (1984).

Effects of high CO_2 levels on surface-temperature and atmospheric oxidation-state of the early Earth. Kasting, J.F., Pollack, J.B., and Crisp, D. *J. atmos. Chem.* **1,** 403 (1984).

Possible limits on the composition of the Archean ocean. Walker, J.C.G. *Nature, Lond.* **302,** 518 (1983).

Solar intensities may have varied considerably over geological time scales. The latest evidence suggests that the radiation in the ultraviolet may have been more intense than it is now.

The Sun as Variable Star: Solar and Stellar Variations. Pap, J.M. (ed.) (Cambridge University Press, Cambridge, 1994).

Solar ultraviolet and the evolutionary history of cyanobacteria. Garcia-Pichel, F. *Origins Life & Evol. Biosphere* **28**, 321 (1998).

Biological effects of high ultraviolet radiation on early Earth–a theoretical evaluation. Cockell, C.S. *J. theor. Biol.* **193,** 717 (1998).

The early faint sun paradox: Organic shielding of ultraviolet-labile greenhouse gases. Sagan, C. and Chyba, C. *Science* **276,** 1217 (1997).

The evolution of solar ultraviolet luminosity. Zahnle, K.J. and Walker, J.C.G. *Rev. Geophys. & Space Phys.* **20**, 280 (1982).

UV radiation from the young Sun and oxygen and ozone levels in the pre-biological palaeoatmosphere. Canuto, V.M., Levine, J.S., Augustsson, T.R., and Imhoff, C.L. *Nature, Lond.* **296**, 816 (1982).

The young Sun and the early Earth, its atmosphere and photochemistry. Canuto, V.M., Levine, J.S., Augustsson, T.R., Imhoff, C.L., and Giampapa, M.S. *Nature, Lond.* **305**, 281 (1983).

Formation of oxygen and ozone and fixation of nitrogen in the absence of life

Photochemistry of CO and H_2O: analysis of laboratory experiments and applications to the prebiotic Earth's atmosphere. Wen, J.-S., Pinto, J.P., and Yung, Y.L. *J. geophys. Res.* **94**, 14957 (1989).

Theoretical constraints on oxygen and carbon dioxide concentrations in the Precambrian atmosphere. Kasting, J.F. *Precambrian Res.* **34**, 205 (1987).

Photochemical reactions of water and CO in Earth's primitive atmosphere. Bar-Nun, A. and Chang, S. *J. geophys. Res.* **88**, 6662 (1983).

Oxygen and ozone in the early Earth's atmosphere. Canuto, V.M., Levine, J.S., Augustsson, T.R., and Imhoff, C.L., *Precambr. Res.* **20**, 109 (1983).

Prebiotic atmospheric oxygen levels. Carver, J.H. *Nature, Lond.* **292**, 136 (1981).

Limits on oxygen concentration in the prebiological atmosphere and the rate of abiotic fixation of nitrogen. Kasting, J.F. and Walker J.C.G. *J. geophys. Res.* **86**, 1147 (1981).

Rates of fixation by lightning of carbon and nitrogen in possible primitive atmospheres. Chameides, W.L. and Walker, J.C.G. *Orig. Life* **11**, 291 (1981).

Fixation of nitrogen in the prebiotic atmosphere. Yung, Y.L. and McElroy, M.B. *Science* **203**, 1002 (1979).

Evolution of the atmosphere after life was established

Chemical changes of the atmosphere on geological and recent time scales. Warneck, P. in *Global atmospheric chemical change.* Hewitt, C.N. and Sturges, W.T. (eds.) (Chapman and Hall, London, 1994).

Trends in global distributions of trace gases inferred from polar ice cores. Legrand, M., Raynaud, D., Barnola, J.-M., and Chappellaz, J. in *Low temperature chemistry of the atmosphere.* Moortgat, G.K., Barnes, A.J., Le Bras, G., and Sodeau, J.R. (eds.) (Springer-Verlag, Berlin, 1994).

Ice core chemistry: Implications for our past atmosphere. Legrand, M. in *Low temperature chemistry of the atmosphere.* Moortgat, G.K., Barnes, A.J., Le Bras, G., and Sodeau, J.R. (eds.) (Springer-Verlag, Berlin, 1994).

Miocene evolution of atmospheric carbon dioxide. Pagani, M., Arthur, M.A., and Freeman, K.H. *Paleoceanogr.* **14,** 273 (1999).

Modelling the concentration of atmospheric CO_2 during the Younger Dryas climate event. Marchal, O., Stocker, T.F., Joos, F., Indermuhle, A., Blunier, T., and Tschumi, J. *Clim. Dyn.* **15,** 341 (1999).

The carbon cycle and CO_2 over Phanerozoic time: the role of land plants. Berner, R.A. *Phil. Trans. Roy. Soc.* **B353,** 75 (1998).

Phototrophy, diazotrophy and paleoatmospheres — biological catalysis and the H, C, N and O cycles. Raven, J.A. and Sprent, J.I. *J. geol. Soc.* **146,** 161 (1989).

Sources and sinks for ammonia and nitrite on the early Earth and the reaction of nitrite with ammonia. Summers, D.P. *Origins Life Evol. Biosphere* **29,** 33 (1999).

Evolution of the nitrogen cycle and its influence on the biological sequestration of CO_2 in the ocean. Falkowski, P.G. *Nature, Lond.* **387,** 272 (1997).

Three-dimensional study of the relative contributions of the different nitrogen sources in the troposphere. Lamarque, J.F., Brasseur, G.P., and Hess, P.G. *J. geophys. Res.* **101,** 22955 (1996).

Long-term evolution of the biogeochemical carbon cycle. Des Marais, D.J. *Rev. Mineral.* **35,** 429 (1997).

Atmospheric and ionospheric response to trace gas perturbations through the ice age to the next century in the middle atmosphere. 1. Chemical composition and thermal structure. Beig, G. and Mitra, A.P. *J. atmos. solar-terrest. Phys.* **59,** 1245 (1997).

Precambrian glaciations and the evolution of the atmosphere. Carver, J.H. and Vardavas, I.M. *Ann. Geophys.* **12,** 674 (1994).

The influence of the biosphere on the development of the atmosphere. The first article discusses the evolution of life on Earth and its relationship to atmospheric oxygen concentrations. This together with the other papers also gives a guide to current thinking about the origins of life and the earliest living organisms.

The evolution of the biosphere. Budyko, M.I. (D. Reidel, Dordrecht, 1986).

The biosphere. Cloud, P., *Scient. Am.* **249**, 132 (Sept 1983).

The origin and early evolution of life on Earth. Oro, J., Miller, S.L., and Lazcano, A. *Annu. Rev. Earth & Planet. Sci.* **18**, 317 (1990).

Stromatolites in Precambrian carbonates: Evolutionary mileposts or environmental dipsticks? Grotzinger, J.P. and Knoll, A.H. *Annu. Rev. Earth planet. Sci.* **27,** 313 (1999).

Early evolution of land plants: Phylogeny, physiology, and ecology of the primary terrestrial radiation. Bateman, R.M., Crane, P.R., Di Michele, W.A., Kenrick, P.R., Rowe, N.P., Speck, T., and Stein, W.E. *Annu. Rev. Ecol. & Systematics* **29,** 263 (1998).

Quantification of the effect of plants on weathering: Studies in Iceland. Moulton, K.L. and Berner, R.. *Geol.* **26,** 895 (1998).

BIOME 6000: reconstructing global mid-Holocene vegetation patterns from palaeoecological records. Prentice, I.C. and Webb, T. *J. Biogeography* **25,** 997 (1998).

The past, present and future of nitrogenous compounds in the atmosphere, and their interactions with plants. Raven, J.A. and Yin, Z.H. *New Phytologist* **139,** 205 (1998).
Is there a link between biodiversity and greenhouse effect? Rozanov, S.I. *Biofizika* **43,** 1101 (1998).
Weathering, plants, and the long-term carbon-cycle. Berner, R.A. *Geochim. Cosmochim. Acta* **56,** 3225 (1992).
Hydrogen consumptions by methanogens on the early Earth. Kral, T.A., Brink, K.M., Miller, S.L., and McKay, C.P. *Origins Life Evol. Biosphere* **28,** 311 (1998).
The first organisms. Cairns-Smith, A.G. *Scient. Am.* **252** (3), 74 (June 1985).
How life affects the atmosphere. Walker, J.C.G. *Bioscience* **34,** 486 (1984).
Primordial organic chemistry. Ponnamperuma, C. *Chem. Br.* **15**, 560 (1979).
The early fossil record. Brasier, M.D. *Chem. Br.* **15**, 588 (1979).
The oldest eucaryotic cells. Vidal, G. *Scient. Am.* **250**, 32 (February 1984).
Designing the first organism. Andrew, S.P.S. *Chem. Br.* **15**, 580 (1979).
Organisms of the first kind. Cairns-Smith, G. *Chem. Br.* **15**, 576 (1979).

The evolution of Earth's atmosphere after life was established. Several of these papers seek a connection between the protection afforded by the ozone shield and the stages of evolutionary development of living organisms. The first paper is the prototype of such studies.

On the origin and rise of oxygen concentration in the Earth's atmosphere. Berkner, L.V. and Marshall, L.C. *J. atmos. Sci.* **22**, 225 (1965).
Paleosols and the rise of atmospheric oxygen: a critical review. Rye, R. and Holland, H.D. *Amer. J. Sci.* **298**, 621 (1998)
Late Proterozoic rise in atmospheric oxygen concentration inferred from phylogenetic and sulphur-isotope studies. Canfield, D.E. and Teske, A. *Nature, Lond.* **382,** 127 (1996).
Carbon isotopes and the rise of atmospheric oxygen. Karhu, J.A. and Holland, H.D. *Geology* **24**, 867 (1996).
Evolution of oxygen and ozone in Earth's atmosphere. Kasting, J.F. and Donahue, T.M. in *Life in the universe*. Billingham, J. (ed.) (MIT Press, Cambridge, Mass., 1981).
The evolutionary role of atmospheric ozone. Blake, A.J. and Carver, J.H. *J. atmos. Sci.* **34**, 720 (1977).
Possible variation of ozone in the troposphere during the course of geologic time. Chameides, W.L. and Walker, J.C.G. *Am. J. Sci.* **275**, 737 (1975).
Atmospheric constraints on the evolution of metabolism. Walker, J.C.G. *Orig. Life* **10**, 93 (1980).
Atmospheric ozone and the history of life. Ratner, M.I. and Walker, J.C.G. *J. atmos. Sci.* **29**, 803 (1972).
Atmospheres and evolution. Margulis, L. and Lovelock, J.E., in *Life in the universe* Billingham, J. (ed.) pp. 79–100. (MIT Press, Cambridge, Mass, 1981).

Section 9.5

An introduction to palaeoclimatology. Changes in the solar luminosity or in the Earth's orbit may have had an influence on climate. Much further information can be found in the books and papers cited in connection with Section 9.7.

The first book is a popular account of evolution and the climate.

Isotopes and climate. Bowen, R. (Elsevier Applied Science, London, 1990).
Understanding climate change. Berger, A., Dickinson, R.E., Kidson, J.W. (eds.) *Geophys. Monographs* Vol. 52 (1989).
Climate variability: past, present, and future. Mitchell, J.M. *Climate Change* **16**, 231 (1990).
Fossils and climate. Brenchley, P.J. (ed.) (Wiley, Chichester, 1984).
Climate and history. Wigley, T.M.L., Ingram, M.J., and Farmer, G. (eds.) (Cambridge University Press, Cambridge, 1981).
Climate variability on the scale of decades to centuries. Stockton, C.W. *Climate Change* **16**, 173 (1990).
Aspects of climate variability in the Pacific and the Western Americas. Peterson, D.H. (ed.) *Geophys. Monographs* Vol. 55 (1990).
Climates throughout geologic time. Frakes, L.A. (Elsevier, Amsterdam, 1979).
Climates: present past and future. Lamb, H.H. (2 vols) (Methuen, London, 1977).

The first article describes the transformation of our climate from that of a greenhouse to that of an icehouse and provides a summary of the kind of evidence used to infer climates of the past.

Fifty million years ago. McGowran, B. *Amer. Sci.* **78**, 30 (1990).
The Sun in time. Sonnett, C.P., Giampapa, M.S., and Mattews, M.S. (eds.) (University of Arizona Press, Tucson, 1991).
The ancient Sun. Pepin, R.O., Eddy, J.A., and Merrill, R.B. (eds.) (Pergamon Press, Oxford, 1981).
Simulated time-dependent climate response to solar radiative forcing since 1600. Rind, D., Lean, J., and Healy, R. *J. geophys. Res.* **104,** 1973 (1999).
Volcanic and solar impacts on climate since 1700. Bertrand, C., vanYpersele, J.P., and Berger, A. *Climate Dynamics* **15**, 355 (1999).
The Sun's variable radiation and its relevance for Earth. Lean, J. *Annu. Rev. Astron. Astrophys.* **35**, 33 (1996).
Solar–terrestrial influences on weather and climate. Gregory, J. *Nature, Lond.* **299**, 401 (1982).
The Earth's orbit and the ice ages. Covey, C. *Scient. Am.* **250**, 42 (Feb. 1984).
The variable sun. Foukal, P.V. *Scient. Am.* **262** (2), 26 (Feb. 1990).
What drives glacial cycles? Broecker, W.S. and Denton, G.H. *Scient. Am.* **262** (1), 42 (Jan. 1990).
Climate evolution on the terrestrial planets. Kasting, J.F., and Toon, O.B. in *Origin and evolution of planetary and satellite atmospheres*. Atreya, S.K., Pollack, J.B., and Matthews, M.S. (eds.) (University of Arizona Press, Tucson, 1989).
Spatial variability of climate and past atmospheric circulation patterns from central West Antarctic glaciochemistry. Reusch, D.B., Mayewski, P.A., Whitlow, S.I., Pittalwala, I.I., and Twickler, M.S. *J. geophys. Res.* **104,** 5985 (1999).

Glacial–interglacial changes in ocean surface conditions in the southern hemisphere. Vimeux, F., Masson, V., Jouzel, J., Stievenard, M., and Petit, J.R. *Nature, Lond.* **398,** 410 (1999).

Northern hemisphere temperatures during the past millennium: Inferences, uncertainties, and limitations. Mann, M.E., Bradley, R.S., and Hughes, M.K. *Geophys. Res. Letts.* **26,** 759 (1999).

Northern hemisphere storm tracks in present day and last glacial maximum climate simulations: A comparison of the European PMIP models. Kageyama, M., Valdes, P.J., Ramstein, G., Hewitt, C., and Wyputta, U. *J. Climate* **12,** 742 (1999).

Permian paleoclimate data from fluid inclusions in halite. Benison, K.C. and Goldstein, R.. *Chem. Geol.* **154,** 113 (1999).

On the possible influence of extraterrestrial volatiles on Earth's climate and the origin of the oceans. Deming, D. *Palaeogeography Palaeoclimatology Palaeoecology* **146,** 33 (1999).

Global-scale temperature patterns and climate forcing over the past six centuries. Mann, M.E., Bradley, R.S., and Hughes, M.K. *Nature, Lond.* **392,** 779 (1998).

High-resolution palaeoclimatic records for the last millennium: interpretation, integration and comparison with General Circulation Model control-run temperatures. Jones, P.D., Briffa, K.R., Barnett, T.P., and Tett, S.F.B. *Holocene* **8,** 455 (1998).

Climate change during the last 150 million years: reconstruction from a carbon cycle model. Tajika, E. *Earth planet. Sci. Letts.* **160,** 695 (1998).

Paleoenvironmental implications of the insoluble microparticle record in the GISP2 (Greenland) ice core during the rapidly changing climate of the Pleistocene–Holocene transition. Zielinski, G.A. and Mershon, G.R. *Geol. Soc. Amer. Bull.* **109,** 547 (1997).

Planetary atmospheres—Warming early Earth and Mars. Kasting, J.F. *Science* **276,** 1213 (1997).

Contrasting atmospheric and climate dynamics of the last-glacial and Holocene periods. Ditlevsen, P.D., Svensmark, H., and Johnsen, S. *Nature, Lond.* **379,** 810 (1996).

Oxygen-isotope record of sea level and climate variations in the Sulu Sea over the past 150,000 years. Linsley, B.K. *Nature, Lond.* **380,** 234 (1996).

Paleoclimatology and climate system dynamics. Overpeck, J.T. *Rev. Geophys.* **33,** 863 (1995).

Paleoecology, past climate systems, and C3/C4 photosynthesis. Spicer, R.A. *Chemosphere* **27,** 947 (1993).

Tectonic forcing of late Cenozoic climate. Raymo, M.E. and Ruddiman, W.F. *Nature, Lond.* **359,** 117 (1992).

Precambrian evolution of the climate system. Walker, J.C.G. *Global planet. Change* **82,** 261 (1990).

Modelling the concentration of atmospheric CO_2 during the Younger Dryas climate event. Marchal, O., Stocker, T.F., Joos, F., Indermuhle, A., Blunier, T., and Tschumi, J. *Clim. Dyn.* **15,** 341 (1999).

The variability in the carbon sinks as reconstructed for the last 1000 years. Joos, F., Meyer, R., Bruno, M., and Leuenberger, M. *Geophys. Res. Letts.* **26,** 1437 (1999).

Fossil plants record an atmospheric (CO_2)-^{12}C and temperature spike across the Palaeocene–Eocene transition in NW Europe. Beerling, D.J., and Jolley, D.W. *J. Geol. Soc.* **155,** 591 (1998).

The carbon cycle and CO_2 over Phanerozoic time: the role of land plants. Berner, R.A. *Phil. Trans. Royal Soc., Ser. B* **353,** 75 (1998).

Simulated influence of carbon dioxide, orbital forcing and ice sheets on the climate of the Last Glacial Maximum. Weaver, A.J., Eby, M., Augustus, F.F., and Wiebe, E.C. *Nature, Lond.* **394,** 847 (1998).

Terrestrial carbon storage during the past 200 years: A Monte Carlo analysis of CO_2 data from ice core and atmospheric measurements. Bruno, M. and Joos, F. *Global biogeochem. Cycles* **11,** 111 (1997).

C-4 photosynthesis, atmospheric CO_2 and climate. Ehleringer, J.R., Cerling, T.E., and Helliker, B.R. *Oecologia* **112,** 285 (1997).

Late Jurassic climate and its impact on carbon cycling. Weissert, H. and Mohr, H. *Palaeogeography Palaeoclimatology Palaeoecology* **122,** 27 (1996).

Natural and anthropogenic changes in atmospheric CO_2 over the last 1000 years from air in Antarctic ice and firn. Etheridge, D.M., Steel, L.P., Langenfelds, R.L., Francey, R.J., and Barnola, J.-M. *J.* geophys. *Res.* **101,** 4115 (1996).

Atmospheric carbon dioxide and the long-term control of the Earth's climate. Carver, J.H. and Vardavas, I.M. *Ann. Geophys.* **13,** 782 (1995).

Atmospheric carbon dioxide levels over Phanerozoic time. Berner, R.A. *Science* **249,** 1382 (1990).

Carbon redox and climate control through Earth history — a speculative reconstruction. Worsley, T.R. and Nance, R.D. *Global planet. Change* **75,** 259 (1989).

Long-term climate change and the geochemical cycle of carbon. Marshall, H.G., Walker, J.C.G., and Kuhn, W.R. *J. geophys. Res.* **93,** 791 (1988).

Feedbacks between weathering and atmospheric CO_2 over the last 100 million years. Volk, T. *Amer. J. Sci.* **287,** 763 (1987).

Atmospheric methane between 1000 AD and present: Evidence of anthropogenic emissions and climatic variability. Etheridge, D.M., Steele, L.P., Francey, R.J., and Langenfelds, R.L. *J. geophys. Res.* **103,** 15979(1998).

Analytic solutions for the antigreenhouse effect: Titan and the early Earth. McKay, C.P., Lorenz, R.D., and Lunine, J.I. *Icarus* **137,** 56 (1999).

Carbon-isotope composition of Lower Cretaceous fossil wood: Ocean – atmosphere chemistry and relation to sea-level change. Grocke, D.R., Hesselbo, S.P., and Jenkyns, H. *Geol.* **27,** 155 (1999).

Ocean influence on the atmosphere composition and climates of the Earth. Sorokhtin, O.G. and Ushakov, S.A. *Okeanologiya* **38,** 928 (1998).

A new model for Proterozoic ocean chemistry. Canfield, D.E. *Nature, Lond.* **396,** 450 (1998).

Isotopic evidence for palaeowaters in the British isles. Darling, W.G., Edmunds, W.M., and Smedley, P.L. *Appl. Geochem.* **12,** 813 (1997).

Vegetation–atmosphere interactions and their role in global warming during the latest Cretaceous. Upchurch, G.R., Otto-Bliesner, B.L., and Scotese, C. *Phil. Trans. Royal Soc., Ser. B* **353,** 97 (1998).

The influence of Carboniferous palaeoatmospheres on plant function: an experimental and modelling assessment. Beerling, D.J., Woodward, F.I., Lomas, M.R., Wills, M.A., Quick, W.P., and Valdes, P.J. *Phil. Trans. Royal Soc., Ser. B* **353,** 131 (1998).

The recent climate record: what it can and cannot tell us. Karl, T.R., Tarpley, J.D., Quayle, R.G., Diaz, H.F., Robinson, D.A., and Bradley, R.S. *Rev. Geophys.* **27,** 405 (1989).

Recent climatic change: a regional approach. Gregory, S. (ed.) (Bellhaven Press, London, 1988).

How climate evolved on the terrestrial planets. Kasting, J.F., Toon, O.B.. and Pollack, J.B. *Scient. Am.* **252** (2), 46 (Feb. 1988).

Volcanoes can certainly have an effect on climate. The last reference shows how an historical record of a stratospheric dust cloud can be associated with evidence of volcanic eruptions.

The Mount Pinatubo eruption: effects on the atmosphere and climate. Fiocco, G., Fu, D., and Visconti, G. (Springer-Verlag, Berlin 1996).

Atmospheric aerosols and climate. Toon, O.B. and Pollack, J.B. *Am. Scient.* **68,** 268 (1980).

Volcanoes and the climate. Toon, O.B. and Pollack, J.B. *Natural History* **86,** 8 (1977).

The faint young sun–climate paradox: volcanic influences. Schatten K.H. and Endal, A.S. *Geophys. Res.* Letts. **9**, 1309 (1982).

Volcanic CO_2 and solar forcing of Northern and Southern Hemisphere surface air temperatures. Gilliland, R.L. and Schneider, S.H. *Nature, Lond.* **310**, 38 (1984).

Mystery cloud of AD 536. Stothers, R.B. *Nature, Lond.* **307**, 344 (1984).

The use of ice cores to provide data on ancient atmospheric gas concentrations.

Ice time: climate science and life on Earth. Levenson, T. (Harper and Row, New York, 1989).

The environmental record in glaciers and ice sheets. Oeschger, H. and Langway, C.C., Jr. (eds.) (Wiley, New York, 1989).

Climate and atmospheric history of the past 420,000 years from the Vostok ice core, Antarctica. Petit, J.R., Jouzel, J., Raynaud, D., Barkov, N.I., Barnola, J.M., Basile, I., Bender, M., Chappellaz, J., Davis, M., Delaygue, G., Delmotte, M., Kotlyakov, V.M., Legrand, M., Lipenkov, V.Y., Lorius, C., Pepin, L., Ritz, C., Saltzman, E., and Stievenard, M. *Nature, Lond.* **399,** 429 (1999).

Implications for the interpretation of ice-core isotope data from analysis of modelled Antarctic precipitation. Noone, D. and Simmonds, I. *Ann. Glaciol.* **27,** 398 (1998).

Ice-core records of global climate and environment changes. Delmas, R.J. *Proc. Ind. Acad. Sci* **107,** 307 (1998).

Preliminary investigation of palaeoclimate signals recorded in the ice core from Dome Fuji station, east Dronning Maud Land, Antarctica. Aoki, S., Azuma, N., Fujii, Y., Fujita, S., Furukawa, T., Hondoh, T., Kamiyama, K., Motoyama, H., Nakawo, M., Nakazawa, T., Narita, N., Satow, K., Shoji, H., and Watanabe, O. *Ann. Glaciol.* **27,** 338 (1998).

Complexity of holocene climate as reconstructed from a Greenland ice core. O'Brien, S.R., Mayewski, P.A., Meeker, L.D., Meese, D.A., Twickler, M.S., and Whitlow, S.I. *Science* **270**, 1962 (1995).

The ice record of Greenland gases. Raynaud, D. et al. *Science* **259**, 926 (1993).

Antarctic ice: the frozen time capsule. Peel, D.A. *New Scient.* **98,** 477 (1983).

Ice core data of atmospheric carbon monoxide over Antarctica and Greenland during the last 200 years. Haan, D., Martinerie, P., and Raynaud, D. *Geophys. Res. Letts.* **23**, 2235 (1996).

Methane: the record in polar ice cores. Craig, H. and Chou, C.C. *Geophys. Res. Letts.* **9**, 1221 (1982).

Tree ring dating and the extraction of a climatic record.

Climate from tree rings. Hughes, M.K., Kelly, P.M., Pilcher, J.R., and LaMarche, V.C., Jr. (eds.) (Cambridge University Press, Cambridge, 1982).

Frost rings in trees as records of major volcanic eruptions. LaMarche, V.C., Jr. and Hirschboeck, K.M. *Nature, Lond.* **307**, 121 (1984).

Reconstruction of precipitation history in North American corn belt using tree rings. Blasing, T.J. and Duvick, D. *Nature, Lond.* **307**, 143 (1984).

July–August temperatures at Edinburgh between 1721 and 1975 from tree-ring density and width data. Hughes, M.K., Schweingruber, F.H., Cartwright, D., and Kelly, P.M. *Nature, Lond.* **308**, 341 (1984).

Trace elements in tree rings: evidence of recent and historical air pollution. Baes, C.F., III, and McLaughlin, S.B. *Science* **224**, 494 (1984).

This next reference is given to summarize the use of radiocarbon dating since the method is often employed in cross-calibrating tree-ring and ice-core records.

Radiocarbon dating. Seuss, H.E. *Endeavour (new series)* **4**, 113 (1980).

Climates in the past and their possible regulation by passive or active feedback mechanisms

Feedbacks between weathering and atmospheric CO_2 over the last 100 million years. Volks, T. *Am. J. Sci.* **287**, 763 (1987).

The carbonate–silicate geochemical cycle and its effect on atmospheric carbon dioxide over the past 100 million years. Berner, R.A., Lasaga, A.C., and Garrels, R.M. *Am. J. Sci.* **283**, 641 (1983).

Effects of increased CO_2 concentrations on surface temperature of the early Earth. Kuhn, W.R. and Kasting, J.F. *Nature, Lond.* **301**, 53 (1983).

Cloud Feedback: a stabilizing effect for the early Earth. Rossow, W.G., Henderson-Sellers, A., and Weinreich, S.K. *Science* **217**, 1245 (1982).

A negative feedback mechanism for the long-term stabilization of Earth's surface temperatures. Walker, J.C.G., Hays, P.B., and Kasting, J.F. *J. geophys. Res.* **86**, 9776 (1981).

The regulation of carbon dioxide and climate: Gaia or geochemistry? Lovelock, J.E. and Watson, A.J. *Planet. Space Sci.* **30**, 785 (1982).

Polar glaciation and the genesis of the ice ages. Hunt, B.G., *Nature, Lond.* **308**, 48 (1984).

Snow cover and atmospheric variability. Walsh, J.E. *Am. Scient.* **72**, 50 (1984).

Section 9.6

Introductory reviews. The first three books provide overviews of the subject of climatology, while the fourth discusses in more detail carbon dioxide, other greenhouse gases, and the supposed threat to world climate in the future.

Contemporary climatology. Robinson, P.J. and Henderson-Sellers, A. (Longmans, London, 1999).

Global physical climatology. Hartmann, D.L. (Academic Press, San Diego, 1994).
Physics of climate. Peixoto, J.P. and Oort, A.H. (American Institute of Physics, New York, 1992).
Composition, chemistry, and climate of the atmosphere. Singh, H.B. (ed.) (Van Nostrand, New York, 1995).

The reports of the Intergovernmental Panel on Climate Change (IPCC) represent a major source of information on the subjects of climates and climate change.

The first book listed here is written by the co-chairman of the Scientific Assessment Working Group of the IPCC, and is an exceptionally lucid and comprehensive introduction to the subject.

At the time of writing the most authoritative and strongly supported statement on climate change that has ever been made by the international scientific community is the1995 final report of Working Group I of the IPCC. The IPCC is sponsored jointly by the World Meteorological Organization and the United Nations Environment Programme and nearly one thousand scientists from more than 40 countries participated in the preparation and review of the scientific data. Issues confronted with full rigour include global warming, greenhouse gases, the greenhouse effect, sea level changes, forcing of climate, and the history of the Earth's changing climate.

Global Warming: the complete briefing. 2nd edn. Houghton, J.T. (Cambridge University Press, Cambridge, 1997).
Climate change 1995: the science of climate change. Houghton, J.T., Meira Filho, L.G., Callander, B.A., Harris, N., Kattenberg, A., and Maskell, K. (eds.) (Cambridge University Press, Cambridge, 1996).
Climate change 1995—impacts, adaptations and mitigations of climate change. Watson, R.T., Zinyowera, M.C., and Moss, R.H. (eds.) (Cambridge University Press, Cambridge, 1995).
Climate change 1994: radiative forcing of climate change and an evaluation of the IPCC IS92 emission scenarios. Houghton, J.T., Meira Filho, L.G., Bruce, J., Hoesung Lee, Callander, B.A., Haites, E., Harris, N., and Maskell, K. (eds.) (Cambridge University Press, Cambridge, 1994).
Climate change 1992: the supplementary report to the IPCC scientific assessment. Houghton, J.T., Callander, B.A., and Varney, S.K. (eds.) (Cambridge University Press, Cambridge, 1992).
Climate change: the IPCC scientific assessment. Houghton, J.T., Jenkins, G.J. and Ephraums, J.J. (eds.) (Cambridge University Press, Cambridge, 1990).

The following books and articles are amongst many that provide popular accounts of the greenhouse effect.

Anthropogenic climate change. Storch, H. and Flöser, G. (Springer, Berlin, 1999).
Historical perspectives on climate change. Fleming, J.R. (Oxford University Press, Oxford, 1998).
Too hot to handle? : the greenhouse effect. Gribbin, M., Gribbin, J.R., and Riley, J. (Corgi Books, London, 1992).

Primer on greenhouse gases. Wuebbles, D.J. and Edmonds, J. (Lewis Publishers, Chelsea, Mich., 1991).

Hothouse Earth: the greenhouse effect and Gaia. Gribbin, J. (Bantam Press, London, 1990).

Winds of change: living in the global greenhouse. Gribbin, J. and Kelly, M. (Hodder and Stoughton, London, 1989).

The greenhouse effect. Boyle, S. and Ardill, J. (Hodder and Stoughton, Sevenoaks, 1989).

Turning up the heat; our perilous future in the global greenhouse. Pearce, F. (The Bodley Head, London, 1989).

The greenhouse effect, climatic change, and ecosystems. Bolin, B. [SCOPE 29] (Wiley, Chichester, 1986).

The great climate debate. White, R.M. *Scient. Am.* **263**, 18 (July 1990).

Observational constraints on the global atmospheric CO_2 budget. Tans, P.P., Fung, I.Y., and Takahasi, T. *Science* **247**, 1431 (1990).

The greenhouse effect and climate change. Mitchell, J.F.B. *Rev. Geophys.* **27**, 115 (1989).

The changing climate. Schneider, S.H. *Scient. Am.* **261** (3), 38 (Sept. 1989).

Global climatic change. Houghton, R.A. and Woodwell, G.M. *Scient. Am.* **260** (4), 18 (April 1989).

Sun and dust versus greenhouse gases: an assessment of their relative roles in global climate change. Hansen J.E. and Lacis, A.A. *Nature, Lond.* **346**, 713 (1990).

Climate modelling. Schneider, S.H. *Scient. Am.* **256** (5), 72 (May 1987).

Carbon dioxide increase in the atmosphere and oceans and possible effects on climate. Chen, C.-T. and Drake, E.T. *Annu. Rev. Earth Planet. Sci.* **14**, 201 (1986).

Although deforestation may reduce the uptake of CO_2 from the atmosphere, increased emission rates seem largely responsible for elevated atmospheric concentrations.

Global deforestation: contribution to atmospheric carbon dioxide. Woodwell, G.M., Hobbie, J.E., Houghton, R.A., Melillo, J.M., Moore, B., Peterson, B.J., and Shaver, G.R. *Science* **222**, 1081 (1983).

Response of the global climate to changes in atmospheric chemical composition due to fossil fuel burning. Hameed, S., Cess, R.D., and Hogan, J.S. *J. geophys. Res.* **85**, 7537 (1980).

Sections 9.6.1 and 9.6.2

Introductory material

A search for human influences on the thermal structure of the atmosphere. Santer, B. et al. *Nature* **382**, 39 (1996).

Climate change report. Wigley, T.M.L., Santer, B.D., Mitchell, J.F.B., and Charlson, R.J. *Science* **271,** 1481 (1996).

Causes of twentieth-century temperature change near the Earth's surface. Tett, S.F.B., Stott, P.A., Allen, M.R., Ingram, W.J., and Mitchell, J.F.B. *Nature, Lond.* **399,** 569 (1999).

Relative detectability of greenhouse-gas and aerosol climate change signals. Wigley, T.M.L., Jaumann, P.J., Santer, B.D., and Taylor, K.E. *Climate Dynamics* **14,** 781 (1998).

There is more to climate than carbon dioxide. Walker, J.C.G. *Rev. Geophys.* **33**, 295 (1995).

The influence of changes in solar radiation

The role of the sun in climate change. Hoyt, D.V. and Schatten, K.H. (Oxford University Press, Oxford, 1997).

Solar cycle variability, ozone, and climate. Shindell, D., Rind, D., Balachandran, N., Lean, J., and Lonergan, P. *Science* **284**, 305 (1999).

Solar variability and its implications for the human environment. Reid, G.C. *Nature, Lond.* **61**, 3 (1999).

Modelling the impact of solar variability on climate. Haigh, J.D. *Nature, Lond.* **61**, 63 (1999).

Solar cycle variability, ozone, and climate. Shindell, D., Rind, D., Balachandran, N., Lean, J., and Lonergan, P. *Science* **284**, 305 (1999).

Climate forcing by changing solar radiation. Lean, J. and Rind, D. *J. Climate* **11**, 3069 (1998).

Greenhouse effect, atmospheric solar absorption and the Earth's radiation budget: From the Arrhenius-Langley era to the 1990s. Ramanathan, V. and Vogelmann, A.M. *Ambio* **26,** 38 (1997).

Radiatively active gases and aerosols

Global climate change due to radiatively active gases. Wuebbles, D.J. in *Global atmospheric chemical change.* Hewitt, C.N and Sturges, W.T. (eds.) (Chapman and Hall, London, 1994).

Climate response to increasing levels of greenhouse gases and sulphate aerosols. Mitchell, J.F.B., Johns, T.C., Gregory, J.M., and Tett, S.F.B. *Nature* **376**, 501 (1995).

The importance of atmospheric chemistry in the calculation of radiative forcing on the climate system. Hauglustaine, D.A., Granier, C., Brasseur, G.P., and Megie, G. *J. geophys. Res.* **99**, 1173 (1994).

Carbon dioxide

The CO_2 greenhouse-effect on Mars, Earth, and Venus. Idso, S.B. *Science of the total environment* **77**, 291 (1988).

Interannual extremes in the rate of rise of atmospheric carbon dioxide since 1980. Keeling, C.D., Whorf, T.P., Wahlen, M., and van der Plicht, J. *Nature, Lond.* **375,** 666 (1995).

Carbon dioxide is not the only greenhouse gas known to be increasing in concentration. Water vapour, CH_4, N_2O, and other trace gases can trap radiation so that chemical processes in the atmosphere have a direct influence on climate. The Bibliographies for Sections 1.5 and 5.4 provide further references to papers describing the changing atmospheric burden of trace gases.

Photochemistry, composition, and climate. Kuhn, W.R. in *The photochemistry of atmospheres.* Levine, J.S., (ed.) (Academic Press, Orlando, 1985).

The role of atmospheric chemistry in climate change. Wuebbles, D.J., Grant, K.E., Connell, P.S., and Penner, J.E. *APCA J.* **39**, 22 (1989).

Atmospheric trace gases and global climate: a seasonal model study. Wang, W.-C., Molnar, G., Ko, M.K.W., Goldenberg, S., and Sze, N.D. *Tellus* **42B**, 149 (1990).

A comparison of the contribution of various gases to the greenhouse effect. Rohde, H. *Science* **248**, 1217 (1990).

Relative contributions of greenhouse gas emissions to global warming. Lashof, D.A. and Ahuja, D.R. *Nature, Lond.* **344**, 529 (1990).

Carbon monoxide

On the climate forcing of carbon monoxide. Daniel, J.S., and Solomon, S. *J. geophys. Res.* **103**, 13249 (1998).

A comparison of climate forcings due to chlorofluorocarbons and carbon monoxide. Sinha, A. and Toumi, R. *Geophys. Res. Letts.* **23**, 65 (1996).

Relative radiative forcing consequences of global emissions of hydrocarbons, carbon monoxide and NO_x from human activities estimated with a zonally-averaged two-dimensional model. Johnson, C.E. and Derwent, R.G. *Climatic Change* **34,** 439 (1996).

An observation of the greenhouse radiation associated with carbon monoxide. Evans, W.F.J. and Puckrin, E. *geophys. Res. Let.* **22**, 925 (1995).

Ozone. There are effects from both tropospheric increases and stratospheric loss.

Atmospheric ozone as a climate gas. Wang, W.C. and Isaksen, I.S.A. (eds.) (Springer-Verlag, Berlin, 1995).

Cooling of the Arctic and Antarctic polar stratospheres due to ozone depletion. Randel, W.J. and Wu, F. *J. Climate* **12**, 1467 (1999).

Climate change and the middle atmosphere. Part III: The doubled CO_2 climate revisited. Rind, D., Shindell, D., Lonergan, P., and Balachandran, N.K. *J. Climate*, **11**, 876 (1998).

Radiative forcing due to tropospheric ozone and sulfate aerosols. Van Dorland, R., Dentener, F.J. , and Lelieveld, J. *J. geophys. Res.* **102**, 28079 (1997).

Radiative forcing and temperature trends from stratospheric ozone changes. Forster, P.M. de F., and Shine, K.P. *J. geophys. Res.* **102**, 10841 (1997).

An estimation of the climatic effects of the stratospheric ozone losses during the 1980s. MacKay, R.M., Ko, M.K.W., Shia, R-L., Yang, Y., Zhou, S., and Molnar, G. *J. Climate* **10**, 774 (1997).

Effects of anthropogenic emissions on tropospheric ozone and radiative forcing. Berntsen, T., Isaksen, I.S.A., Fuglestvedt, J.S., Myhre, G., Stordal, F., Freckleton, R.S., and Shine, K.P. *J. geophys. Res.* **102**, 28101 (1997).

Radiative forcing due to increased tropospheric ozone concentrations. Chalita, S., Hauglustaine, D.A., Le Treut, H., and Muller, J.-F. *Atmos. Environ.* **30**, 1641 (1996).

Radiative forcing of climate by changes of the vertical distribution of ozone. Lacis, A.A., Wuebbles, D.J., and Logan, J.A. *J. geophys. Res.* **95**, 9971 (1990).

Methane and other hydrocarbons

Climate effects of atmospheric methane. Lelieveld, J., Crutzen, P.J., and Briihl, C. *Chemosphere* **26**, 739 (1993).
Estimation of direct radiative forcing due to non-methane hydrocarbons. Highwood, E.J., Shine, K.P., Hurley, M.D., and Wallington, T.J. *Atmos. Environ.* **33**, 759 (1999).
Changing concentration, lifetime and climate forcing of atmospheric methane. Lelieveld, J., Crutzen, P.J., and Dentener, F.J. *Tellus* **B50**, 128 (1998).
Estimates of indirect global warming potentials for CH_4, CO and NO_x. Fuglestvedt, J.S., Isaksen, I.S.A., and Wang, W.C. *Climatic Change* **34,** 405 (1996).
Changes in CH_4 and CO growth rates after the eruption of Mt. Pinatubo and their link with changes in tropical tropospheric UV flux. Dlugokencky, E.J., Dutton, E.G., Novelli, P.C., Tans, P.P., Masarie, K.A., Lantz, K.O., and Madronich, S. *Geophys. Res. Letts.* **23,** 2761 (1996).
The growth rate and distribution of atmospheric CH_4. Dlugokencky, E.J., Steele, L.P., Lang, P.M. and Mesarie, K.A. *J. geophys. Res.* **99**, 17021 (1994).

Nitrous oxide

Increase in the atmospheric nitrous oxide concentration during the last 250 years. *Geophys. Res. Letts.* **22**, 2921 (1995).

Chlorofluorocarbons and the HFCs and HCFCs used as substitutes (see Section 4.6.6).

Infrared radiative forcing of CFC substitutes and their atmospheric reaction procucts. Pappasavva, S., Tai, S., Illinger, K.H., and Kenny, J.E. *J. geophys. Res.* **102**, 13643 (1997).
Radiative forcing of climate change by CFC-11 and possible CFC replacements. Christidis, N., Hurley, M.D., Pinnock, S., Shine, K.P., and Wallington, T.J. *J. geophys. Res.* **102**, 19597 (1997).
A measurement of the greenhouse radiation associated with carbon tetrachloride (CCl_4). Evans, W.F.J. and Puckrin, E. *Geophys. Res. Letts.* **23**, 1769 (1996).
Radiative forcing by hydrochlorofluorocarbons and hydrofluorocarbons. Pinnock, S., Hurley, M.D., Shine, K.P., Wallington, T.J., and Smyth, T.J. *J. geophys. Res.* **100**, 23227 (1995).
Radiative effects and halocarbon global warming potentials of replacement compounds for chlorofluorocarbons. Imasu, R., Suga, A., and Mastuno, T. *J. Meteor. Soc. Japan* **73**, 1123 (1995).
On the evaluation of halocarbon radiative forcing and global warming potentials. Daniel, J.S., Solomon, S., and Albritton, D.L. *J. geophys. Res.* **100**, 1271 (1995).
Infrared cross sections and global warming potentials of 10 alternative hydrohalocarbons. Clerbaux, C., Colin, R., Simon, P.C., and Granier, C. *J. geophys. Res.* **98**, 10491 (1993).

Aerosols

Aerosol Forcing of Climate. Charlson, R.J. and Heintzenberg, J. (eds.) (Wiley, New York, 1995).

Global climate change due to aerosols. Preining, O. in *Global atmospheric chemical change.* Hewitt, C.N and Sturges, W.T. (eds.) (Chapman and Hall, London, 1994).

On modification of global warming by sulfate aerosols. Mitchell, J.F.B. and Johns, T.C. *J. Climate* **10,** 245 (1997).

Climate forcing by stratospheric aerosols. Lacis, A., Hansen, J., and Sato, M. *Geophys. Res. Letts.* **19,** 1607 (1992).

Direct climate forcing by biomass-burning aerosols: Impact of correlations between controlling variables. Iacobellis, S.F., Frouin, R., and Somerville, R.C.J. *J. geophys. Res.* **104,** 12031 (1999).

Sea salt aerosols, tropospheric sulphur cycling, and climate forcing. Pszenny, A., Keene, W., O'Dowd, C., Smith, M., and Quinn, P. *IGBP Newsletter* No.33, 13 (1998).

Influence of sea-salt on aerosol radiative properties in the Southern Ocean marine boundary layer. Murphy, D.M., Anderson, J.R., Quinn, P.K., McInnes, L.M., Brechtel, F.J., Kreidenweis, S.M., Middlebrook, A.M., Posfai, M., Thomson, D.S., and Buseck, P.R. *Nature, Lond.* **392,** 62 (1998).

An assessment of the radiative effects of anthropogenic sulfate. Chuang, C.C., Penner, J.E., Taylor, K.E., Grossman, A.S., and Walton, J.J. *J. geophys. Res.* **102,** 3761 (1997).

Radiative forcing

The missing climate forcing. Hansen, J., Sato, M., Lacis, A., and Ruedy, R. *Phil. Trans Royal Soc.* **B352,** 231 (1997).

Radiative forcing from the 1991 Mount Pinatubo volcanic eruption. Stenchikov, G.L., Kirchner, I., Robock, A., Graf, H.F., Antuna, J.C., Grainger, R.G., Lambert, A., and Thomason, L. *J. geophys. Res.* **103,** 13837(1998).

Estimation of direct radiative forcing due to non-methane hydrocarbons. Highwood, E.J., Shine, K.P., Hurley, M.D., and Wallington, T.J. *Atmos. Environ.* **33,** 759 (1999).

Radiative effects of CH_4, N_2O, halocarbons and the foreign-broadened H_2O continuum: A GCM experiment. Schwarzkopf, M.D. and Ramaswamy, V. *J. geophys. Res.* **104**, 9467 (1999).

Influence of anthropogenic and oceanic forcing on recent climate change. Folland, C.K., Sexton, D.M.H., Karoly, D.J., Johnson, C.E., Rowell, D.P., and Parker, D.E. *Geophys. Res. Letts.* **25**, 353 (1998).

Estimates of radiative forcing due to modeled increases in tropospheric ozone. Haywood, J.M., Schwarzkopf, M.D., and Ramaswamy, V. *J. geophys. Res.* **103**, 16999 (1998).

Infrared radiative forcing and atmospheric lifetimes of trace species based on observations from UARS. Minschwaner, K., Carver, R.W., Briegleb, B.P., and Roche, A.E. *J. geophys. Res.* **103**, 23243 (1998).

Climate forcings in the Industrial era. Hansen, J.E., Sato, M., Lacis, A., Ruedy, R., Tegen, I., and Matthews, E. *Proc. Nat. Acad. Sci. USA* **95,** 12753 (1998).

Radiative forcing and climate response. Hansen, J., Sato, M., and Ruedy, R. *J. geophys. Res.* **102,** 6831 (1997).
Climate forcing–response relationships for greenhouse and shortwave radiative perturbations. Ramaswamy, V. and Chen, C.T. *Geophys. Res. Letts.* **24**, 667 (1997).
On aspects of the concept of radiative forcing. Forster, P.M. de F., Freckleton, R.S., and Shine, K.P. *Climate Dynamics* **13,** 547 (1997).
Indirect influence of ozone depletion on climate forcing by clouds. Toumi, R., Bekki, S., and Law, K.S. *Nature* **372**, 348 (1994).

Section 9.6.3

Two citations that illustrate chemical and physical feedbacks, respectively, in the climate system, followed by a more general reference to the types of feedback that are observed.

Indirect chemical effects of methane on climate warming. Lelieveld, J. and Crutzen, P.J. *Nature* **355**, 399 (1992).
Variability of Antarctic sea ice and changes in carbon dioxide. Zwally, H.J., Parkinson, C.L., and Comiso, J.C. *Science* **220**, 1005 (1983).
Feedback mechanisms in the climate system affecting future levels of carbon dioxide. Kellogg, W.W. *J. geophys. Res.* **88**, 1263 (1983).

Models and their validation

A climate modelling primer. Henderson-Sellers, A. and McGoffin, K. (John Wiley, Chichester, 1987).
An overview of the results of the Atmospheric Model Intercomparison Project (AMIP I). Gates, W.L., Boyle, J.S., Covey, C., Dease, C.G., Doutriaux, C.M., Drach, R.S., Fiorino, M., Gleckler, P.J., Hnilo, J.J., Marlais, S.M., Phillips, T.J., Potter, G.L., Santer, B.D., Sperber, K.R., Taylor, K.E., and Williams, D.N. *Bull. Amer. Meteorol. Soc.* **80,** 29 (1999).
Uses of satellite observations to validate climate/middle atmosphere models. Pawson, S. and Langematz, U. *Advan. Space. Res.* **22**, 1483 (1999).
A GCM study of climate change in response to the 11-year solar cycle. Haigh, J.D. *Q.J. Royal Meteor. Soc.* **125**, 871 (1999).
The National Center for Atmospheric Research Community Climate Model: CCM3. Kiehl, J.T., Hack, J.J., Bonan, G.B., Boville, B.A., Williamson, D.L., and Rasch, P.J. *J. Climate* **11**, 1131 (1998).
The second Hadley Centre coupled ocean–atmosphere GCM: Model description, spinup and validation. Johns, T.C., Carnell, R.E., Crossley, J.F., Gregory, J.M., Mitchell, J.F.B., Senior, C.A., Tett, S.F.B., and Wood, R.A. *Climate Dynamics* **13,** 103 (1997).
Time scales in atmospheric chemistry: Theory, GWPs for CH_4 and CO, and runaway growth. Prather, M.J. *Geophys. Res. Letts.* **23**, 2597 (1996).
Greenhouse warming and changes in the seasonal cycle of temperature: Model versus observations. Mann, M.E. and Park, J. *Geophys. Res. Letts.* **23,** 1111 (1996).

A model of biogeochemical cycling of phosphorus, nitrogen, oxygen, and sulfur in the ocean — one-step toward a global climate. Shaffer, G. *J. geophys. Res.* **94,** 1979 (1989).

Clouds

An interactive cirrus cloud radiative parameterization for global climate models. Joseph, E., and Wang, W.C. *J. geophys. Res.* **104,** 9501 (1999).
Cloud microphysics and climate. Baker, M.B. *Science* **276,** 1072 (1997).

Oceans

Aquatic sources and sinks of CO_2 and CH_4 in the polar regions. Semiletov, I.P., *J. atmos. Sci.* **56,** 286 (1999).
Ocean influence on the atmosphere composition and climates of the Earth. Sorokhtin, O.G., and Ushakov, S.A. *Okeanologiya* **38,** 928 (1998).
Marine biological controls on climate via the carbon and sulphur geochemical cycles. Watson, A.J. and Liss, P.S. *Phil. Trans. Royal Soc. Ser. B* **353,** 41 (1998).
A numerical study on the coupling between sea surface temperature and surface evaporation. Tsintikidis, D., and Zhang, G.J. *J. geophys. Res.* **103,** 31763 (1998).

The feedbacks between El Niño and climate are not clearly understood

Decadal El Niño. Navarra, A. (ed.) (Springer-Verlag, Berlin, 1998).
Currents of change: El Niño's impact on climate and society. Glantz, M.H. (Cambridge University Press, Cambridge, 1996).
Climate oscillations — Genesis and evolution of the 1997–98 El Niño. McPhaden, M.J. *Science* **283,** 950 (1999).
Recent changes in El Nino Southern Oscillation events and their implications for Southern African climate. Mason, S.J. *Trans. Royal Soc. South Africa* **52,** 377 (1997).

The rôle of coupling with the stratosphere

Stratospheric processes and their role in climate. Joint planning staff for World Climate Research Programme, WRCP–105 (World Meteorological Organization, Geneva, 1998).
The stratosphere and its role in the climate system. Brasseur, G.P. (ed.) (Springer-Verlag, Berlin, 1997).

Section 9.6.4

On the detection of temperature changes that may have occurred already as a result of enhanced greenhouse heating

Global temperature patterns. Mann, M.E., Bradley, R.S., Hughes, M.K., and Jones, P.D. *Science* **280,** 2029 (1998).
Decadal changes in the atmospheric circulation and associated surface climate variations in the Northern Hemisphere winter. Watanabe, M. and Nitta, T. *J. Climate* **12,** 494 (1999).

Optimal detection of global warming using temperature profiles: A methodology. Leroy, S.S. *J. Climate* **12,** 1185 (1999).
Baroclinic instability: An oceanic wavemaker for interdecadal variability. De Verdiere, A.C. and Huck, T. *J. phys. Oceanogr.* **29,** 893 (1999).
Short-term climatic fluctuations and the interpretation of recent observations in terms of greenhouse effect. Andre, J.C. and Royer, L.F. *Compt. Rend Ser. II Fasc. A* **328,** 261 (1999).
Optimal climate signal detection in four dimensions. Stevens, M.J. *J. geophys. Res.* **104**, 4089 (1999).
Surface air temperature and its changes over the past 150 years. Jones, P.D., New, M., Parker, D.E., Martin, S., and Rigor, I.G. *Rev. Geophys.* **37,** 173 (1999).
Arctic and Antarctic lakes as optical indicators of global change. Vincent, W.F., Laurion, I., and Pienitz, R. *Ann. Glaciol.* **27**, 691 (1998).
The potential effect of GCM uncertainties and internal atmospheric variability on anthropogenic signal detection. Barnett, T.P., Hegerl, G.C., Santer, B., and Taylor, K. *J. Climate* **11,** 659 (1998).
Scale-dependent detection of climate change. Stott, P.A., and Tett, S.F.B. *J. Climate* **11,** 3282 (1998).
Anthropogenic influence on the autocorrelation structure of hemispheric-mean temperatures. Wigley, T.M.L., Smith, R.L., and Santer, B.D. *Science* **282,** 1676 (1998).
Trends in daily wintertime temperatures in the northern stratosphere. Pawson, S. and Naujokat, B. *Geophys. Res. Letts.* **24**, 575 (1997).
Human influence on the atmospheric vertical temperature structure: detection and observations. Tett, S.F.B., Mitchell, J.F.B., Parker, D.E., and Allen, M.R. *Science* **274**, 1170 (1996).
Fingerprint of ozone depletion in the spatial and temporal pattern of recent lower-stratospheric cooling. Ramaswamy, V., Schwarzkopf, M.D., and Randel, W.J. *Nature, Lond.* **382**, 616 (1996).
A search for human influences on the thermal structure of the atmosphere. Santer, B.D., Taylor, K.E., Wigley, T.M.L., Johns, T.C., Jones, P.D., Karoly, D.J., Mitchell, J.F.B., Oort, A.H., Penner, J.E., Ramaswamy, V., Schwarzkopf, M.D., Stouffer, R.J., and Tett, S. *Nature, Lond.* **382,** 39 (1996).
Detecting greenhouse-gas-induced climate change with an optimal fingerprint method. Hegerl, G.C., von Storch, H., Hasselmann, K., Santer, B.D., Cubasch, U., and Jones, P.D. *J. Climate* **9,** 2281 (1996).
Towards the detection and attribution of an anthropogenic effect on climate. Santer, B.D., Taylor, K.E., Wigley, T.M.L., Penner, J.E., Jones, P.D., and Cubasch, U. *Climate Dynamics* **12,** 77 (1995).
Precise monitoring of global temperature trends from satellites. Spencer, R.W. and Christy, J.R. *Science* **247**, 1558 (1990).
Global warming trends. Jones, P.D. and Wigley, T.M.L. *Scient. Am.* **263**, 66 (Aug. 1990).

Representative publications that illustrate the argument about whether or not warming has really been observed.

Why is the global warming proceeding much slower than expected? Bengtsson, L., Roeckner, E., and Stendel, M. *J. geophys. Res.* **104**, 3865 (1999).

Uncertainties in observationally based estimates of temperature change in the free atmosphere. Santer, B.D., Hnilo, J.J., Wigley, T.M.L., Boyle, J.S., Doutriaux, C., Fiorino, M., Parker, D.E., and Taylor, K.E. *J. geophys. Res.* **104,** 6305 (1999).
Global warming – global climate data and models: A reconciliation. Hansen, J.E., Sato, M., Ruedy, R., Lacis, A., and Glascoe, J. *Science* **281,** 930 (1998).
CO_2-induced global warming: a skeptic's view of potential climate change. Idso, S.B. *Climate Research* **10,** 69 (1998).
Major greenhouse cooling (yes, cooling)—the upper-atmosphere response to increased CO_2. Roble, R.G. *Rev. Geophys.* **33**, 539 (1995).
Satellite and surface temperature data at odds? Hansen, J.E. et al. *Clim. Change* **30**, 103 (1995).

Section 9.6.5

Climate process and change. Bryant, E.A. (Cambridge University Press, Cambridge, 1997).
World survey of climatology. Vol. 16: Future climates of the world. Henderson-Sellers, A. (ed.) (Elsevier, Amsterdam, 1995).
Climate change scenarios for the United Kingdom : scientific report. Jenkins, G.J. and Hulme, M. (Climatic Research Unit, Norwich, 1998).
A nonlinear dynamical perspective on climate prediction. Palmer, T.N. *J. Climate* **12,** 575 (1999).
Climate evolution: from the recent past to the future. Jouzel, J. and Lorius, C. *Comptes Rendus Ser. II, Fasc. A* **328,** 229 (1999).
Climate assessment for 1996. Halpert, M.S. and Bell, G.D. *Bull. Amer. Meteor. Soc.* **78,** S1 (1997).
Uncertainties in projections of human-caused climate warming. Mahlman, J.D. *Science* **278**, 1416 (1997).
Predictions of climate changes caused by man-made emissions of greenhouse gases: a critical assessment. Mason, B.J. *Contemporary Physics* **36,** 299 (1995).
The impact of the future scenarios for methane and other chemically active gases on the GWP of methane. Bruhl, C. *Chemosphere* **26,** 731 (1993).
Uncertainties of estimates of climatic change: a review. Dickinson, R.E. *Climatic Change* **15,** 5 (1989).
Influences of mankind on climate. Kellogg, W.W. *Annu. Rev. Earth & Planet. Sci.* **7,** 63 (1979).

Section 9.6.6

The special IPCC report includes climatic effects in its assessment; the report provides an excellent survey of the impact of aircraft on the atmosphere. See also Section 4.6.2.

IPCC special report on aviation and the global atmosphere. IPCC (World Meteorological Organization, Geneva, 1999).
Ice particle habits in Arctic clouds. Korolev, A.V., Isaac, G.A., and Hallett, J. *Geophys. Res. Letts.* **26,** 1299 (1999).
An interactive cirrus cloud radiative parameterization for global climate models. Joseph, E., and Wang, W.C. *J. geophys. Res.* **104,** 9501 (1999).

Impact of future subsonic aircraft NO_x emissions on the atmospheric composition. Grewe, V., Dameris, M., Hein, R., Kohler, I., and Sausen, R. *Geophys. Res. Letts.* **26,** 47 (1999).

A diagnostic study of the global distribution of contrails. Part I: Present day climate. Sausen, R., Gierens, K., Ponater, M., and Schumann, U. *Theor. Appl. Climatol.* **61,** 127 (1998).

Atmospheric distributions of soot particles by current and future aircraft fleets and resulting radiative forcing on climate. Rahmes, T.F., Omar, A.H., and Wuebbles, D.J. *J. geophys. Res.* **103,** 31657 (1998).

Cloud condensation nuclei measurements in the high troposphere and in jet aircraft exhaust. Hudson, J.G., and Xie, Y.H. *Geophys. Res. Letts.* **25,** 1395 (1998).

A numerical study of aircraft wake induced ice cloud formation. Gierens, K.M., and Strom, J. *J. atmos. Sci.* **55,** 3253 (1998).

Shape and size of contrails ice particles. Goodman, J., Pueschel, R.F., Jensen, E.J., Verma, S., Ferry, G.V., Howard, S.D., Kinne, S.A., and Baumgardner, D. *Geophys. Res. Letts.* **25,** 1327 (1998).

Effect of aircraft on ultraviolet radiation reaching the ground. Plumb, I.C. and Ryan, K.R. *J. geophys. Res.* **103,** 31231 (1998).

Contrail-cirrus and their potential for regional climate change. Sassen, K. *Bull. Amer. Meteorol. Soc.* **78,** 1885 (1997).

Climatic impact of aircraft induced ozone changes. Sausen, R., Feneberg, B., and Ponater, M. *Geophys. Res. Letts.* **24,** 1203 (1997).

Climatic effect of water vapor release in the upper troposphere. Rind, D., Lonergan, P., and Shah, K. *J. geophys. Res.* **101,** 29395 (1996).

In situ observations of particles in jet aircraft exhausts and contrails for different sulfur-containing fuels. Schumann, U., Strom, J., Busen, R., Baumann, R., Gierens, K., Krautstrunk, M., Schroder, F.P., and Stingl, J. *J. geophys. Res.* **101,** 6853 (1996).

Could high-speed civil transport aircraft impact stratospheric and tropospheric temperatures measured by microwave sounding unit?. Shah, K.P., Rind, D., and Lonergan, P. *J. geophys. Res.* **101,** 28711 (1996).

Simulating the global atmospheric response to aircraft water vapour emissions and contrails: A first approach using a GCM. Ponater, M., Brinkop, S., Sausen, R., and Schumann, U. *Ann. Geophys.* **14,** 941 (1996).

The impact of aircraft NO_x emissions on tropospheric ozone and global warming. Johnson, C., Henshaw, J., and McInnes, G. *Nature, Lond.* **355,** 69 (1992).

Section 9.6.7

The articles to which references are given here explore not only the influences of man on climate but also some of the social and economic consequences of climate change.

Climate and human change: disaster or opportunity? Cowie, J. (Parthenon, New York, 1998).

Climate, change and risk. Downing, T.E., Olsthoorn, A.A., and Tol, R.S.J. (Routledge, London, 1998).

Human choice and climate change. Rayner, S. and Malone, E.L. (Batelle Press, Columbus, 1998).
Climate change and the global harvest : potential impacts of the greenhouse effect on agriculture. Rosenzweig, C. and Hillel, D. (Oxford University Press, Oxford, 1998).
Does the weather really matter?: the social implications of climate change. Burroughs, W.J. (Cambridge University Press, Cambridge, 1997) .
The regional impacts of climate change: an assessment of vulnerability. Watson, R.T., Zinyowera, M.C., and Moss, R.H. (Cambridge University Press, Cambridge, 1997).
Climate change 1995—impacts, adaptations and mitigations of climate change. Watson, R.T., Zinyowera, M.C., and Moss, R.H. (eds.) (Cambridge University Press, Cambridge, 1995).
Scientific perspectives on the greenhouse problem. Nierenberg, W.A., Jastrow, R., and Seitz, F. (Marshall Institute, Washington, DC, 1989).
Assessing the social implications of climate fluctuations: guide to climate impact studies. Riesbaume, W.E. (UNEP, Nairobi, 1989).
Greenhouse gas emissions: environmental consequences and policy responses. Oppenheimer, M. (ed.) *Climatic Change* Special Issue (Vol 15, Nos 1–2, 1989).
Carbon dioxide and other greenhouse gases: climatic and associated impacts. Fantechi, R. and Ghazi, A. (Kluwer Academic Publishers, Dordrecht, 1989).
Ozone depletion greenhouse gases and climate change. (National Academy Press, Washington 1989).
Quantifying the impact of global climate change on potential natural vegetation. Sykes, M.T., Prentice, I.C., and Laarif, F. *Climatic Change* **41,** 37 (1999).
Increased plant growth in the northern high latitudes from 1981 to 1991. Myneni, R.B., Keeling, C.D., Tucker, C.J., Asrar, G., and Nemani, R.R. *Nature, Lond.* **386,** 698 (1997).
Rapid sea-level rise soon from West Antarctic Ice Sheet collapse? Bentley, C.R. *Science* **275**, 1077 (1997)
Increased activity of northern vegetation inferred from atmospheric CO_2 measurements. Keeling, C.D., Chin, J.F.S., and Whorf, T.P. *Nature, Lond.* **382,** 146 (1996).
Effects of low and elevated CO_2 on C-3 and C-4 annuals.1. Growth and biomass allocation. Dippery, J.K., Tissue, D.T., Thomas, R.B., and Strain, B.R. *Oecologia* **101,** 13 (1995).
Climate change — does it matter? Firor, J. *Rev. Geophys*. **33**, 883 (1995).
Effects of low and elevated CO_2 on C-3 and C-4 annuals.2. Photosynthesis and leaf biochemistry. Tissue, D.T., Griffin, K.L., Thomas, R.B., and Strain, B.R. *Oecologia* **101,** 21 (1995).

Section 9.6.8

Climate change policy : facts, issues, and analyses. Jepma, C.J. and Munasinghe, M. (Cambridge University Press, Cambridge, 1998).
Economics and policy issues in climate change. (Resources for the future, Washington, DC, 1998).
Buying greenhouse insurance : the economic costs of carbon dioxide emission limits. Manne, A.S. and Richels, R.G. (MIT Press, Cambridge, MA, 1991).

Section 9.7

Carbon dioxide may not be able to act as a thermostatic regulator for more than a limited period in the future.

Life span of the biosphere. Lovelock, J.E. and Whitfield, M. *Nature, Lond.* **296**, 561 (1982).

The Goldilocks problem — climatic evolution and long-term habitability of terrestrial planets. Rampino, M.R and Caldeira, K. *Annu. Rev. Astron. Astrophys.* **32,** 83 (1994).

Failure of climate regulation in a geophysiological model. Lovelock, J.E. and Kump, L.R. *Nature, Lond.* **369,** 732 (1994).

Predicting Earth's life-span. Lapenis, A. and Rampino, M.R. *Nature, Lond.* **363,** 218 (1993).

The life-span of the biosphere revisited. Caldeira, K. and Kasting, J.F. *Nature, Lond.* **360,** 721 (1992).

When climate and life finally devolve. Volk, T. *Nature, Lond.* **360,** 707 (1992).

Chemical stability of the atmosphere and ocean. Kump, L.R. *Global planet. Change* **75,** 123 (1989).

Index